T0140484

Advances in Intelligent Systems and Computing

Volume 742

Series editor

The series "Advances in Intelligent Systems and Computing" contains publications on theory, applications, and design methods of Intelligent Systems and Intelligent Computing. Virtually all disciplines such as engineering, natural sciences, computer and information science, ICT, economics, business, e-commerce, environment, healthcare, life science are covered. The list of topics spans all the areas of modern intelligent systems and computing such as: computational intelligence, soft computing including neural networks, fuzzy systems, evolutionary computing and the fusion of these paradigms, social intelligence, ambient intelligence, computational neuroscience, artificial life, virtual worlds and society, cognitive science and systems, Perception and Vision, DNA and immune based systems, self-organizing and adaptive systems, e-Learning and teaching, human-centered and human-centric computing, recommender systems, intelligent control, robotics and mechatronics including human-machine teaming, knowledge-based paradigms, learning paradigms, machine ethics, intelligent data analysis, knowledge management, intelligent agents, intelligent decision making and support, intelligent network security, trust management, interactive entertainment, Web intelligence and multimedia.

The publications within "Advances in Intelligent Systems and Computing" are primarily proceedings of important conferences, symposia and congresses. They cover significant recent developments in the field, both of a foundational and applicable character. An important characteristic feature of the series is the short publication time and world-wide distribution. This permits a rapid and broad dissemination of research results.

More information about this series at http://www.springer.com/series/11156

Kanad Ray · Tarun K. Sharma
Sanyog Rawat · R. K. Saini
Anirban Bandyopadhyay
Editors

Soft Computing: Theories and Applications

Proceedings of SoCTA 2017

Editors
Kanad Ray
Department of Physics, Amity School of Applied Sciences
Amity University Rajasthan
Jaipur, Rajasthan
India

Tarun K. Sharma
Department of Computer Science and Engineering, Amity School of Engineering and Technology
Amity University Rajasthan
Jaipur, Rajasthan
India

Sanyog Rawat
Department of Electronics and Communication Engineering, SEEC
Manipal University Jaipur
Jaipur, Rajasthan
India

R. K. Saini
Institute of Basic Science
Bundelkhand University
Jhansi, Uttar Pradesh
India

Anirban Bandyopadhyay
Advanced Key Technologies Division, Nano Characterization Unit, Surface Characterization Group
National Institute for Materials Science
Tsukuba, Ibaraki
Japan

ISSN 2194-5357 ISSN 2194-5365 (electronic)
Advances in Intelligent Systems and Computing
ISBN 978-981-13-0588-7 ISBN 978-981-13-0589-4 (eBook)
https://doi.org/10.1007/978-981-13-0589-4

Library of Congress Control Number: 2018941538

This Springer imprint is published by the registered company Springer Nature Singapore Pte Ltd.
The registered company address is: 152 Beach Road, #21-01/04 Gateway East, Singapore 189721, Singapore

Preface

SoCTA 2017 is the second international conference of the series Soft Computing: Theories and Applications (SoCTA). SoCTA 2017 was organized at Bundelkhand University, Jhansi, India, during December 22–24, 2017. The conference is technically associated with MIR Labs, USA.

The objective of SoCTA is to provide a common platform to researchers, academicians, scientists, and industrialists working in the area of soft computing to share and exchange their views and ideas on the theory and application of soft computing techniques in multi-disciplinary areas. The conference stimulated discussions on various emerging trends, innovation, practices, and applications in the field of soft computing.

This book that we wish to bring forth with great pleasure is an encapsulation of research papers, presented during the three-day international conference. We hope that the effort will be found informative and interesting to those who are keen to learn on technologies that address to the challenges of the exponentially growing information in the core and allied fields of soft computing.

We are thankful to the authors of the research papers for their valuable contribution to the conference and for bringing forth significant research and literature across the field of soft computing. The editors also express their sincere gratitude to SoCTA 2017 patron, plenary speakers, keynote speakers, reviewers, program committee members, international advisory committee and local organizing committee, sponsors, and student volunteers without whose support the quality of the conference could not be maintained.

We would like to express our sincere gratitude to Prof. Ajith Abraham, MIR Labs, USA, for being the Conference Chair and finding time to come to Jhansi amid his very busy schedule and delivering a plenary talk.

We express our special thanks to Prof. Jocelyn Faubert, University of Montreal, Canada, and Prof. Badrul Hisam Ahmad, Universiti Teknikal Malaysia Melaka, for their gracious presence during the conference and delivering their plenary talks.

We express special thanks to Springer and its team for the valuable support in the publication of the proceedings. With great fervor, we wish to bring together researchers and practitioners in the field of soft computing year after year to explore new avenues in the field.

Jaipur, India	Kanad Ray
Jaipur, India	Tarun K. Sharma
Jaipur, India	Sanyog Rawat
Jhansi, India	R. K. Saini
Tsukuba, Japan	Anirban Bandyopadhyay

Organizing Committee

Chief Patron
Prof. Surendra Dubey, Vice Chancellor, Bundelkhand University, Jhansi

Patron
Shri C. P. Tiwari, Registrar, Bundelkhand University, Jhansi

Co-Patron
Prof. V. K. Sehgal, Bundelkhand University, Jhansi
Shri Dharm Pal, FO, Bundelkhand University, Jhansi

Chairman and Convener
Prof. R. K. Saini, Director IET, Bundelkhand University, Jhansi

Conference Chair
Prof. Ajith Abraham, Director, MIR Labs, USA
Prof. Kanad Ray, Amity University, Rajasthan

General Chair
Dr. Tarun K. Sharma, Amity University, Rajasthan
Dr. Sanyog Rawat, Manipal University Jaipur

Organizing Secretary
Dr. Lalit Kumar Gupta, Bundelkhand University, Jhansi

Joint Conveners
Dr. Saurabh Srivastava, Bundelkhand University, Jhansi
Dr. Ankit Srivastava, Bundelkhand University, Jhansi
Dr. Musrrat Ali, Glocal University, Saharanpur

Joint Organizing Secretary
Dr. Sachin Upadhyay, Bundelkhand University, Jhansi
Er. Brajendra Shukla, Bundelkhand University, Jhansi
Er. Lakhan Singh, Bundelkhand University, Jhansi

Program Chairs
Er. Rajesh Verma, Bundelkhand University, Jhansi
Er. Ekroop Verma, Bundelkhand University, Jhansi
Dr. Subhangi Nigam, Bundelkhand University, Jhansi
Dr. Sushil Kumar, Amity University, Noida, Uttar Pradesh
Dr. Pravesh Kumar, Rajkiya Engineering College, Bijnor
Mr. Jitendra Rajpurohit, Amity University, Rajasthan

Technical Committee
Dr. Deepak Tomar, Bundelkhand University, Jhansi
Dr. Sebastian Basterrech, Czech Technical University, Prague, Czech Republic
Prof. Steven Lawrence Fernandes, Sahyadri College of Engineering & Management, Mangalore

Web Administrators
Prof. Chitreshh Banerjee, Amity University, Rajasthan
Prof. Anil Saroliya, Amity University, Rajasthan
Er. Sadik Khan, Bundelkhand University, Jhansi
Mr. Dhram Das Prajapati, Bundelkhand University, Jhansi
Er. Satya Prakash Sahyogi, Bundelkhand University, Jhansi

Tutorials Committee
Prof. S. K. Srivastava, DDU University, Gorakhpur
Prof. Dharminder Kumar, GJU, Hisar

Special Sessions Committee
Prof. M. L. Mittal, Indian Institute of Technology Roorkee
Prof. Satyabir Singh, Punjabi University, Patiala

Competitive Committee
Prof. M. K. Gupta, CCS University, Meerut
Prof. Avnish Kumar, Bundelkhand University, Jhansi
Dr. Arvind Kumar Lal, Thapar University, Patiala

Poster Session Committee
Prof. Vishal Gupta, MMU, Ambala
Prof. V. K. Saraswat, Agra University

Student Committee
Er. A. P. S. Gaur, IET, Bundelkhand University, Jhansi
Er. B. B. Niranjan, IET, Bundelkhand University, Jhansi
Er. Priyanka Pandey, IET, Bundelkhand University, Jhansi

Women@SoCTA
Dr. Mamta Singh, Bundelkhand University, Jhansi
Dr. Gazala Rizvi, Bundelkhand University, Jhansi
Dr. Anu Singla, Bundelkhand University, Jhansi
Dr. Laxmi Upadhyay, Bundelkhand University, Jhansi
Dr. Anjita Srivastava, Bundelkhand University, Jhansi

Er. Sakshi Dubey, Bundelkhand University, Jhansi
Er. Toolika Srivastava, Bundelkhand University, Jhansi
Dr. Divya Prakash, Amity University, Rajasthan
Ms. Arpita Banerjee, St. Xavier's College, Jaipur

Publicity and Media Committee
Dr. Munna Tiwari, Bundelkhand University, Jhansi
Dr. Dheerandra Yadav, Bundelkhand University, Jhansi
Er. Rahul Shukla, Bundelkhand University, Jhansi
Mr. Vivek Agarwal, Bundelkhand University, Jhansi
Mr. Anil Saroliya, Amity University, Rajasthan
Mr. Jitendra Rajpurohit, Amity University, Rajasthan
Ms. Preeti Gupta, Amity University, Rajasthan
Mr. Varun Sharma, Amity University, Rajasthan

Registration Committee
Dr. Anjali Saxena, Bundelkhand University, Jhansi
Er. Vijay Verma, Bundelkhand University, Jhansi
Mr. Anil Kewat, Bundelkhand University, Jhansi
Er. Sashikant Verma, Bundelkhand University, Jhansi
Ms. Sonam Seth, Bundelkhand University, Jhansi
Mr. Puneet Matapurkar, Bundelkhand University, Jhansi
Mr. Amit Hirawat, Amity University, Rajasthan

Local Hospitality Committee
Dr. Prem Prakash Rajput, Bundelkhand University, Jhansi
Dr. Sunil Trivedi, Bundelkhand University, Jhansi
Dr. Sunil Niranjan, Bundelkhand University, Jhansi
Er. B. P. Gupta, Bundelkhand University, Jhansi
Mr. Kamal Gupta, Bundelkhand University, Jhansi
Mr. Anil Bohere, Bundelkhand University, Jhansi
Dr. Rajeev Sanger, Bundelkhand University, Jhansi

Cultural Activities
Dr. Vinamra Sen Singh, Bundelkhand University, Jhansi
Dr. Naushad Siddiqui, Bundelkhand University, Jhansi
Er. Anupam Vyas, Bundelkhand University, Jhansi
Dr. Atul Khare, Bundelkhand University, Jhansi

Sponsorship and Printing Committee
Dr. Rishi Saxena, Bundelkhand University, Jhansi
Er. Mukul Saxena, Bundelkhand University, Jhansi
Er. Dinesh Prajapati, Bundelkhand University, Jhansi

Coordinating Committee
Mr. Bilal, Indian Institute of Technology Roorkee
Mr. Nathan Singh, Indian Institute of Technology Roorkee
Mr. Sunil K. Jauhar, Indian Institute of Technology Roorkee

Ms. Hira Zahir, Indian Institute of Technology Roorkee
Prof. Bruno Apolloni, University of Milano, Italy
Dr. Mukul Prastore, Bundelkhand University, Jhansi
Dr. Zakir Ali, Bundelkhand University, Jhansi

Conference Proceedings and Printing and Publication Committee
Prof. Kanad Ray, Amity University, Rajasthan
Dr. Tarun K. Sharma, Amity University, Rajasthan
Dr. Sanyog Rawat, Manipal University Jaipur
Prof. R. K. Saini, Bundelkhand University, Jhansi
Dr. Anirban Bandyopadhyay, NIMS, Japan

Transportation Committee
Dr. O. P. Singh, Bundelkhand University, Jhansi
Mr. Vijay Yadav, Bundelkhand University, Jhansi
Er. Ram Singh Kushwaha, Bundelkhand University, Jhansi
Mr. Uday Pratap Singh, Bundelkhand University, Jhansi

Medical Committee
Dr. Kamad Dikshit, Bundelkhand University, Jhansi
Dr. Kumar Yashwant, Bundelkhand University, Jhansi

Finance Committee
Er. Ram Singh Kushwaha, Bundelkhand University, Jhansi
Mr. Vinay Varshney, Bundelkhand University, Jhansi

Steering Committee
Prof. S. K. Kabia, Bundelkhand University, Jhansi
Prof. M. L. Maurya, Bundelkhand University, Jhansi
Prof. Pankaj Atri, Bundelkhand University, Jhansi
Prof. Rochana Srivastava, Bundelkhand University, Jhansi
Prof. Poonam Puri, Bundelkhand University, Jhansi
Prof. Aparna Raj, Bundelkhand University, Jhansi
Prof. M. M. Singh, Bundelkhand University, Jhansi
Prof. Archana Verma, Bundelkhand University, Jhansi
Prof. S. P. Singh, Bundelkhand University, Jhansi
Prof. S. K. Katiyar, Bundelkhand University, Jhansi
Dr. C. B. Singh, Bundelkhand University, Jhansi
Dr. Pratik Agarwal, Bundelkhand University, Jhansi
Dr. D. K. Bhatt, Bundelkhand University, Jhansi
Dr. Puneet Bishariya, Bundelkhand University, Jhansi
Dr. V. K. Singh, Bundelkhand University, Jhansi
Dr. B. S. Bhadauria, Bundelkhand University, Jhansi
Dr. Vineet Kumar, Bundelkhand University, Jhansi
Mrs. Preeti Mittal, Bundelkhand University, Jhansi
Mr. Manoj Kumar, Bundelkhand University, Jhansi

Dr. Alok Verma, Bundelkhand University, Jhansi
Mr. Atul Sangal, Sharda University, G Noida

International Advisory Board
Prof. Aboul Ella Hassanien, University of Cairo, Egypt
Prof. Adel Alimi, University of Sfax, Tunisia
Prof. Aditya Ghose, University of Wollongong, Australia
Prof. Ashley Paupiah, Amity, Mauritius
Prof. Francesco Marcelloni, University of Pisa, Italy
Prof. Sang-Yong Han, Chung-Ang University, Korea
Prof. André Ponce de Leon F de Carvalho, University of São Paulo, Brazil
Prof. Bruno Apolloni, University of Milano, Italy
Prof. Francisco Herrera, University of Granada, Spain
Prof. Imre J. Rudas, Obuda University, Hungary
Prof. Javier Montero, Complutense University of Madrid, Spain
Prof. Jun Wang, Chinese University of Hong Kong, Hong Kong
Prof. Naren Sukurdeep, Amity, Mauritius
Prof. Mo Jamshidi, University of Texas, San Antonio, USA
Prof. Sebastián Ventura, University of Cordoba, Spain
Prof. Witold Pedrycz, University of Alberta, Canada
Prof. Tomas Buarianek, VSB-Technical University of Ostrava
Prof. Varun Ojha, Swiss Federal Institute of Technology Zurich, Switzerland
Prof. Francois Despaux, Researcher, Universite de Lorraine
Prof. Jan Drhal, Faculty of Electrical Engineering, Czech Technical University, Prague, Czech Republic
Prof. Pavel Kromer, Associate Professor, VSB-Technical University of Ostrava, Ostrava, Czech Republic
Prof. Doan Nhat Quang, Associate Professor, University of Science and Technology, Hanoi, Vietnam

National Advisory Board
Prof. A. K. Solnki, BIET Jhansi
Prof. Aruna Tiwari, IIT Indore
Prof. Ashish Verma, IIT Guwahati
Prof. Ashok Deshpande, Pune
Prof. D. Nagesh Kumar, IISc., Bangalore
Prof. Debasish Ghose, IISc., Bangalore
Prof. Deepti, AHEC, IIT Roorkee
Prof. Himani Gupta, IIFT, Delhi
Prof. Kanti S. Swarup, IIT Madras
Prof. M. P. Biswal, IIT Kharagpur
Prof. N. R. Pal, ISI, Kolkata
Prof. Rama Mehta, NIH IIT Roorkee
Prof. Ashok Kumar Singh, DST, New Delhi
Prof. S. K. Sharma, MITRC, Alwar
Dr. Santosh Kumar Pandey, Ministry of IT, New Delhi

Prof. B. K. Das, Delhi University
Prof. Anjana Solnki, BIET Jhansi
Prof. C. Thangaraj, IIT Roorkee
Prof. D. Gnanaraj Thomas, Christian College, Chennai
Prof. Vinod Kumar, Gurukul Kangri University, Haridwar
Prof. Dharmdutt, Saharanpur Campus, IIT Roorkee
Prof. Manoj Kumar Tiwari, IIT Kharagpur
Prof. Nirupam Chakraborti, IIT Kharagpur
Prof. Pankaj Gupta, Delhi University
Prof. Punam Bedi, University of Delhi
Prof. Raju George, IIST, Trivandrum

Contents

Meta-Heuristic Techniques Study for Fault Tolerance in Cloud Computing Environment: A Survey Work 1
Virendra Singh Kushwah, Sandip Kumar Goyal and Avinash Sharma

Complexity Metrics for Component-Based Software System 13
Rajni Sehgal, Deepti Mehrotra and Renuka Nagpal

Identification of Relevant Stochastic Input Variables for Prediction of Daily PM_{10} Using Artificial Neural Networks 23
Vibha Yadav and Satyendra Nath

An Efficient Technique for Facial Expression Recognition Using Multistage Hidden Markov Model 33
Mayur Rahul, Pushpa Mamoria, Narendra Kohli and Rashi Agrawal

Study and Analysis of Back-Propagation Approach in Artificial Neural Network Using HOG Descriptor for Real-Time Object Classification 45
Vaibhav Gupta, Sunita and J. P. Singh

Study and Analysis of Different Kinds of Data Samples of Vehicles for Classification by Bag of Feature Technique 53
Anita Singh, Rajesh Kumar and R. P. Tripathi

Hybrid Text Summarization: A Survey 63
Mahira Kirmani, Nida Manzoor Hakak, Mudasir Mohd and Mohsin Mohd

Field Based Weighting Information Retrieval on Document Field of Ad Hoc Dataset 75
Parul Kalra, Deepti Mehrotra and Abdul Wahid

Movie Rating System Using Sentiment Analysis 85
Abhishek Singh Rathore, Siddharth Arjaria, Shraddha Khandelwal, Surbhi Thorat and Vratika Kulkarni

Energy Consumption of University Data Centre in Step Networks Under Distributed Environment Using Floyd–Warshall Algorithm 99
Kamlesh Kumar Verma and Vipin Saxena

Modified Integral Type Weak Contraction and Common Fixed Point Theorem with an Auxiliary Function 113
R. K. Saini, Naveen Mani and Vishal Gupta

Crop Monitoring Using IoT: A Neural Network Approach 123
Rajneesh Kumar Pandey, Santosh Kumar and Ram Krishna Jha

Secure Image Steganography Through Pre-processing 133
Indu Maurya and S. K. Gupta

Understandable Huffman Coding: A Case Study 147
Indu Maurya and S. K. Gupta

Performance Investigations of Multi-resonance Microstrip Patch Antenna for Wearable Applications 159
Raghvendra Singh, Dambarudhar Seth, Sanyog Rawat and Kanad Ray

Fuzzy Counter Propagation Network for Freehand Sketches-Based Image Retrieval 171
Suchitra Agrawal, Rajeev Kumar Singh and Uday Pratap Singh

An Efficient Contrast Enhancement Technique Based on Firefly Optimization 181
Jamvant Singh Kumare, Priyanka Gupta, Uday Pratap Singh and Rajeev Kumar Singh

A Review of Computational Swarm Intelligence Techniques for Solving Crypto Problems 193
Maiya Din, Saibal K. Pal and S. K. Muttoo

Defense-in-Depth Approach for Early Detection of High-Potential Advanced Persistent Attacks 205
Ramchandra Yadav, Raghu Nath Verma and Anil Kumar Solanki

An Efficient Algorithm to Minimize EVM Over ZigBee 217
Anurag Bhardwaj and Devesh Pratap Singh

Performance Measure of the Proposed Cost Estimation Model: Advance Use Case Point Method 223
Archana Srivastava, S. K. Singh and Syed Qamar Abbas

Fractal Image Compression and Its Techniques: A Review 235
Manish Joshi, Ambuj Kumar Agarwal and Bhumika Gupta

Web Services Classification Across Cloud-Based Applications 245
M. Swami Das, A. Govardhan and D. Vijaya Lakshmi

Performance Evaluation and Analysis of Advanced Symmetric Key Cryptographic Algorithms for Cloud Computing Security 261
Abdul Raoof Wani, Q. P. Rana and Nitin Pandey

A Survey of CAD Methods for Tuberculosis Detection in Chest Radiographs 273
Rahul Hooda, Ajay Mittal and Sanjeev Sofat

Static Image Shadow Detection Texture Analysis by Entropy-Based Method 283
Kavita, Manoj K. Sabnis and Manoj Kumar Shukla

A PSO Algorithm-Based Task Scheduling in Cloud Computing 295
Mohit Agarwal and Gur Mauj Saran Srivastava

A Novel Approach for Target Coverage in Wireless Sensor Networks Based on Network Coding 303
Pooja Chaturvedi and A. K. Daniel

Natural Vibration of Square Plate with Circular Variation in Thickness 311
Amit Sharma and Pravesh Kumar

Majority-Based Classification in Distributed Environment 321
Girish K. Singh, Prabhati Dubey and Rahul K. Jain

Detection of Advanced Malware by Machine Learning Techniques 333
Sanjay Sharma, C. Rama Krishna and Sanjay K. Sahay

Inter- and Intra-scale Dependencies-Based CT Image Denoising in Curvelet Domain 343
Manoj Diwakar, Arjun Verma, Sumita Lamba and Himanshu Gupta

A Heuristic for the Degree-Constrained Minimum Spanning Tree Problem 351
Kavita Singh and Shyam Sundar

Resource Management to Virtual Machine Using Branch and Bound Technique in Cloud Computing Environment 365
Narander Kumar and Surendra Kumar

Anomaly Detection Using K-means Approach and Outliers Detection Technique 375
A. Sarvani, B. Venugopal and Nagaraju Devarakonda

A Steady-State Genetic Algorithm for the Tree *t*-Spanner Problem 387
Shyam Sundar

A Triple Band $ Shape Slotted PIFA for 2.4 GHz and 5 GHz WLAN Applications 399
Toolika Srivastava, Shankul Saurabh, Anupam Vyas and Rajan Mishra

A Min-transitive Fuzzy Left-Relationship on Finite Sets Based on Distance to Left 407
Sukhamay Kundu

Effective Data Clustering Algorithms 419
Kamalpreet Bindra, Anuranjan Mishra and Suryakant

Predictive Data Analytics Technique for Optimization of Medical Databases 433
Ritu Chauhan, Neeraj Kumar and Ruchita Rekapally

Performance Analysis of a Truncated Top U-Slot Triangular Shape MSA for Broadband Applications 443
Anupam Vyas, P. K. Singhal, Satyendra Swarnkar and Toolika Srivastava

Automated Indian Vehicle Number Plate Detection 453
Saurabh Shah, Nikita Rathod, Paramjeet Kaur Saini, Vivek Patel, Heet Rajput and Prerak Sheth

Named Data Network Using Trust Function for Securing Vehicular Ad Hoc Network 463
Vaishali Jain, Rajendra Singh Kushwah and Ranjeet Singh Tomar

An Intelligent Video Surveillance System for Anomaly Detection in Home Environment Using a Depth Camera 473
Kishanprasad Gunale and Prachi Mukherji

Vehicles Connectivity-Based Communication Systems for Road Transportation Safety 483
Ranjeet Singh Tomar, Mayank Satya Prakash Sharma, Sudhanshu Jha and Bharati Sharma

Some New Fixed Point Results for Cyclic Contraction for Coupled Maps on Generalized Fuzzy Metric Space 493
Vishal Gupta, R. K. Saini, Ashima Kanwar and Naveen Mani

DWSA: A Secure Data Warehouse Architecture for Encrypting Data Using AES and OTP Encryption Technique 505
Shikha Gupta, Satbir Jain and Mohit Agarwal

A New Framework to Categorize Text Documents Using SMTP Measure 515
M. B. Revanasiddappa and B. S. Harish

A Task Scheduling Technique Based on Particle Swarm Optimization Algorithm in Cloud Environment 525
Bappaditya Jana, Moumita Chakraborty and Tamoghna Mandal

Image Enhancement Using Fuzzy Logic Techniques 537
Preeti Mittal, R. K. Saini and Neeraj Kumar Jain

A Jitter-Minimized Stochastic Real-Time Packet Scheduler for Intelligent Routers 547
Suman Paul and Malay Kumar Pandit

Differential Evolution with Local Search Algorithms for Data Clustering: A Comparative Study 557
Irita Mishra, Ishani Mishra and Jay Prakash

A Review on Traffic Monitoring System Techniques 569
Neeraj Kumar Jain, R. K. Saini and Preeti Mittal

Large-Scale Compute-Intensive Constrained Optimization Problems: GPGPU-Based Approach 579
Sandeep U. Mane and M. R. Narsinga Rao

A Comparative Study of Job Satisfaction Level of Software Professionals: A Case Study of Private Sector in India 591
Geeta Kumari, Gaurav Joshi and Ashfaue Alam

Simultaneous Placement and Sizing of DG and Capacitor to Minimize the Power Losses in Radial Distribution Network 605
G. Manikanta, Ashish Mani, H. P. Singh and D. K. Chaturvedi

Integration of Dispatch Rules for JSSP: A Learning Approach 619
Rajan and Vineet Kumar

White Noise Removal to Enhance Clarity of Sports Commentary 629
Shubhankar Sinha, Ankit Ranjan, Anuranjana and Deepti Mehrotra

Analysis of Factors Affecting Infant Mortality Rate Using Decision Tree in R Language 639
Namit Jain, Parul Kalra and Deepti Mehrotra

Content-Based Retrieval of Bio-images Using Pivot-Based Indexing . . . 647
Meenakshi Srivastava, S. K. Singh and S. Q. Abbas

Road Crash Prediction Model for Medium Size Indian Cities 655
Siddhartha Rokade and Rakesh Kumar

Label Powerset Based Multi-label Classification for Mobile Applications 671
Preeti Gupta, Tarun K. Sharma and Deepti Mehrotra

Artificial Bee Colony Application in Cost Optimization of Project Schedules in Construction 679
Tarun K. Sharma, Jitendra Rajpurohit, Varun Sharma and Divya Prakash

Aligning Misuse Case Oriented Quality Requirements Metrics with Machine Learning Approach 687
Ajeet Singh Poonia, C. Banerjee, Arpita Banerjee and S. K. Sharma

Improvising the Results of Misuse Case-Oriented Quality Requirements (MCOQR) Framework Metrics: Secondary Objective Perspective 693
C. Banerjee, Arpita Banerjee, Ajeet Singh Poonia and S. K. Sharma

A Review: Importance of Various Modeling Techniques in Agriculture/Crop Production 699
Jyoti Sihag and Divya Prakash

Software-Defined Networking—Imposed Security Measures Over Vulnerable Threats and Attacks 709
Umesh Kumar, Swapnesh Taterh and Nithesh Murugan Kaliyamurthy

Working of an Ontology Established Search Method in Social Networks 717
Umesh Kumar, Uma Sharma and Swapnesh Taterh

Author Index 727

About the Editors

Dr. Kanad Ray is Professor and Head of Physics at the Amity School of Applied Sciences Physics, Amity University Rajasthan (AUR), Jaipur, India. He has obtained his M.Sc. and Ph.D. degrees in Physics from Calcutta University and Jadavpur University, West Bengal, India. In an academic career spanning over 22 years, he has published and presented research papers in several national and international journals and conferences in India and abroad. He has authored a book on the *Electromagnetic Field Theory*. His current research areas of interest include cognition, communication, electromagnetic field theory, antenna and wave propagation, microwave, computational biology, and applied physics. He has served as Editor of Springer Book Series. Presently, he is Associate Editor of the *Journal of Integrative Neuroscience* published by the IOS Press, the Netherlands. He has established a Memorandum of Understanding (MOU) between his University and University of Montreal, Canada, for various joint research activities. He has also established collaboration with the National Institute for Materials Science (NIMS), Japan, for joint research activities and visits NIMS as Visiting Scientist. He organizes international conference series such as SoCTA and ICoEVCI as General Chair. He is Executive Committee Member of IEEE Rajasthan Chapter.

Dr. Tarun K. Sharma has obtained his Ph.D. degree in Artificial Intelligence as well as M.C.A. and M.B.A. degrees and is currently associated with the Amity University Rajasthan (AUR), Jaipur. His research interests encompass swarm intelligence, nature-inspired algorithms and their applications in software engineering, inventory systems, and image processing. He has published more than 60 research papers in international journals and conferences. He has over 13 years of teaching experience and has also been involved in organizing international conferences. He is Certified Internal Auditor and Member of the Machine Intelligence Research (MIR) Labs, WA, USA, and Soft Computing Research Society, India.

Dr. Sanyog Rawat is currently associated with the Department of Electronics and Communication Engineering, SEEC, Manipal University Jaipur, India. He holds a B.E. degree in Electronics and Communication, an M.Tech. degree in Microwave

Engineering, and a Ph.D. degree in Planar Antennas. He has been involved in organizing various workshops on "LabVIEW" and antenna designs and simulations using FEKO. He has taught various subjects, including electrical science, circuits and system, communication system, microprocessor systems, microwave devices, antenna theory and design, advanced microwave engineering, and digital circuits.

Dr. R. K. Saini is Director of Institute of Engineering and Technology, Bundelkhand University, Jhansi, India. He specializes in the fields of real analysis, fuzzy set theory, operator theory, fuzzy optimization, and summability theory. He is currently working on fuzzy transportation and assignment problems, fuzzy logics, and soft computing. He has published 70 research papers.

Dr. Anirban Bandyopadhyay is Senior Scientist at the National Institute for Materials Science (NIMS), Tsukuba, Japan. He completed his Ph.D. degree in Supramolecular Electronics at the Indian Association for the Cultivation of Science (IACS), Kolkata, 2005. From 2005 to 2008, he was an independent researcher, as Research Fellow at the International Center for Young Scientists (ICYS), NIMS, Japan, where he worked on the brain-like bio-processor building. In 2008, he joined as Permanent Scientist at NIMS, working on the cavity resonator model of human brain and design synthesis of brain-like organic jelly. From 2013 to 2014, he was Visiting Scientist at the Massachusetts Institute of Technology (MIT), USA. He has received several honors, such as the Hitachi Science and Technology Award 2010, Inamori Foundation Award 2011–2012, Kurata Foundation Award, Inamori Foundation Fellow (2011–), and Sewa Society international member, Japan. He has patented ten inventions such as (i) a time crystal model for building an artificial human brain, (ii) geometric musical language to operate a fractal tape to replace the Turing tape, (iii) fourth circuit element that is not memristor, (iv) cancer and Alzheimer's drug, (v) nano-submarine as a working factory and nano-surgeon, (vi) fractal condensation-based synthesis, (vii) a thermal noise harvesting chip, (viii) a new generation of molecular rotors, (ix) spontaneous self-programmable synthesis (programmable matter), and (x) fractal grid scanner for dielectric imaging. He has also designed and built multiple machines and technologies: (i) terahertz (THz) magnetic nano-sensor, (ii) a new class of fusion resonator antennas, etc. Currently, he is building time crystal-based artificial brain using three ways: (i) knots of darkness made of fourth circuit element, (ii) integrated circuit design, and (iii) organic supramolecular structure.

Meta-Heuristic Techniques Study for Fault Tolerance in Cloud Computing Environment: A Survey Work

Virendra Singh Kushwah, Sandip Kumar Goyal and Avinash Sharma

Abstract This paper focuses on a study on meta-heuristic techniques for fault tolerance under cloud computing environment. Cloud computing introduces the support to increasing complex applications, the need to check and endorse handling models under blame imperatives which turns out to be more vital, meaning to guarantee applications execution. Meta-heuristics are problem-independent techniques and workload planning is known to be a NP-Complete problem, therefore meta-heuristics have been used to solve such problems. The idea behind using meta-heuristics is to increase the performance and decrease the computational time to get the job done and in our case, meta-heuristics are to be considered the robust solution of finding the right combinations of resources and tasks to minimize the computational expenses, cut costs and provide better services for users. Fault tolerance plays an important key role in ensuring high serviceability and unwavering quality in cloud. In these days, the requests for high adaptation to noncritical failure, high serviceability, and high unwavering quality are turning out to be exceptionally solid, assembling a high adaptation to internal failure, high serviceability and high dependability cloud is a basic, challenging, and urgently required task.

Keywords Meta-heuristic · Cloud computing · Fault tolerance · GA · PSO · ACO

V. S. Kushwah (✉) · S. K. Goyal · A. Sharma
Maharishi Markandeshwar University, Mullana, Ambala, India
e-mail: kushwah.virendra248@gmail.com

S. K. Goyal
e-mail: skgmmec@gmail.com

A. Sharma
e-mail: sh_avinash@yahoo.com

K. Ray et al. (eds.), *Soft Computing: Theories and Applications*, Advances in Intelligent Systems and Computing 742,
https://doi.org/10.1007/978-981-13-0589-4_1

1 Introduction

Cloud Computing is a computing environment which is consolidated by the conventional PC innovation, for Figuring, Disseminated Figuring, Parallel Computing, Utility Computing, Network Storage Technology, Virtualization, Load Balance along these lines on and the framework development. It coordinates various minimal effort PCs into a capable processing capacity framework through the system, and conveys these effective figuring abilities among those terminal customers by plans of action, for example, SaaS, PaaS, and IaaS. Distributed computing is the innovative computational worldview. It is a developing registering innovation that is quickly solidifying itself as the eventual fate of scattered on ask for figuring [1, 2]. Distributed computing is rising as an imperative spine for the assortments of web organizations utilizing the guideline of virtualization. Many registering structures are proposed for the colossal information stockpiling and exceptionally parallel figuring needs of distributed computing [2]. Then again, Internet empowered business (e-Business) is getting to be distinctly one of the best plans of action in the present period.

2 Meta-heuristic Techniques

Meta-heuristics are issue free strategies that can reply to an extensive extent of issues. We could state that a heuristic uses issue-related data to locate an "adequate" answer for a particular issue, while meta-heuristics are utilized to create general calculations that can be executed to a wide scope of issues. Mostly, meta-heuristics systems include Genetic Algorithm (GA), swarm insight calculations, for example, Ant Colony Optimization (ACO) and Particle Swarm Optimization (PSO), and other nature-roused calculations. The nonspecific structures of the GA, ACO, and PSO taking care of the cloud asset planning issues are shown in the accompanying figures. Their transformative procedures all experience the cycle of

1. Fitness evaluation,
2. Candidate selection, and
3. Trial variation

by a populace of hopeful hunting down potential arrangements in parallel and trading seek data among the applicants. These calculations are a posteriori and nondeterministic in nature. Thus, they require no earlier direction and are reasonable for complex spaces, and can offer various and multi-target arrangements.

2.1 Genetic Algorithm

A Genetic Algorithm (GA) is an exception for discovering answers for complex pursue issues. They are as often utilized as a bit of field, for example, attempting to make staggeringly prominent things because of their capacity to look a through a colossal blend of parameters to locate the best match. For instance, they can search through various blends of materials and graphs to search the ideal mix of both which could accomplish a more grounded, lighter, and better last thing. They can be like the way to be used to organize PC counts, to schedule endeavors, and to comprehend other advancement issues. Here, calculations depend on the procedure of advancement by normal determination. They copy the way in which life uses progression to find answers for bona fide issues. Shockingly though genetic computations can be used to find answers for unfathomably convoluted issues, they are themselves easy to utilize and use it.

The essential procedure for a hereditary calculation is:

(a) **Initialization Create**: An underlying populace. This populace is typically haphazardly produced and can be in any fancied size, from just a couple of people to thousands.
(b) **Evaluation**: Each individual from the populace is then assessed and we figure a "wellness" for that person. The wellness esteem is ascertained by how well it fits with our sought necessities. These necessities could be basic, "quicker calculations are better", or more intricate, "more grounded materials are better however they shouldn't be too overwhelming".
(c) **Selection**: We always need to enhance our populace's general wellness. Choice helps us to do this by disposing of the awful outlines and just keeping the best people in the populace. There are many techniques however the fundamental thought is the same, make it more probable that fitter people will be chosen for our people to come.
(d) **Crossover**: While using crossing over, we make new people by joining parts of our chose people. We need to decide now about that this as imitating how sex functions in nature. The trust is that by consolidating certain characteristics from at least two people, we will make a significantly "fitter" posterity, which will acquire the best attributes from each of its folks.
(e) **Mutation**: we need to incorporate a tiny bit haphazardness into our populaces' hereditary qualities generally, every mix of arrangements we can make would be in our underlying populace. Transformation commonly works by rolling out little improvements at irregular to an individual's genome.
(f) **Repeat**: Now we have our cutting edge from which we can begin again from step two until we achieve an end condition.

2.2 Partial Swarm Optimization

PSO depends on the aggregate conduct of a province of creepy crawlies, for example, ants, honeybees, and so forth; a run of feathered creatures; or a school of fish. The word molecule speaks to, for instance, a honeybee in a province or a fish in a school. Every person in a swarm carries on utilizing its own insight and the gathering knowledge of the swarm. In that capacity, in the event that one molecule finds a decent way whatever is left of the swarm will likewise have the capacity to take after the great way quickly regardless of the possibility that their area is far away in the swarm. The PSO first generates a random initial population of particles, each particle represents a potential solution of system, and each particle is represented by three indexes: position, velocity, fitness. Initially, each particle is assigned a random velocity, in flight, it progressively alters the speed and position of particles through their own flight involvement (individual best position), and their neighbors (worldwide best position). In this manner, the entire gathering will travel to the pursuit district with higher wellness through nonstop learning and upgrading. This procedure is rehashed until achieving the most extreme emphases or the foreordained least wellness. The PSO is hence a wellness based and amasses based streamlining calculation, whose favorable circumstances are effortlessness, simple actualizing, quick joining, and less parameter. The whole molecule swarm plays out a solid "joining", and it might be effectively got caught in neighborhood least focuses, which makes the swarm lose differing qualities (Figs. 1, 2 and 3).

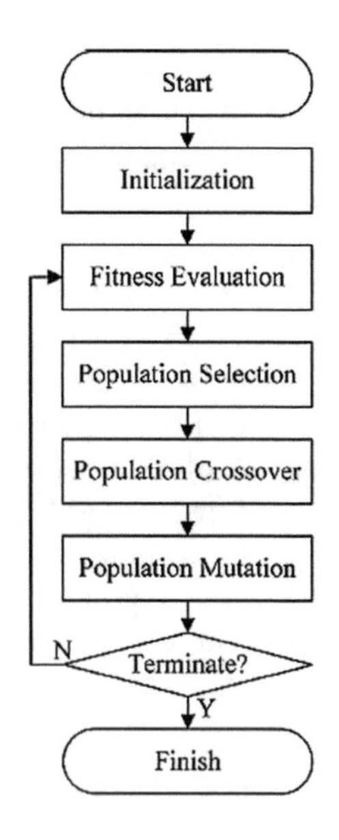

Fig. 1 Genetic Algorithm

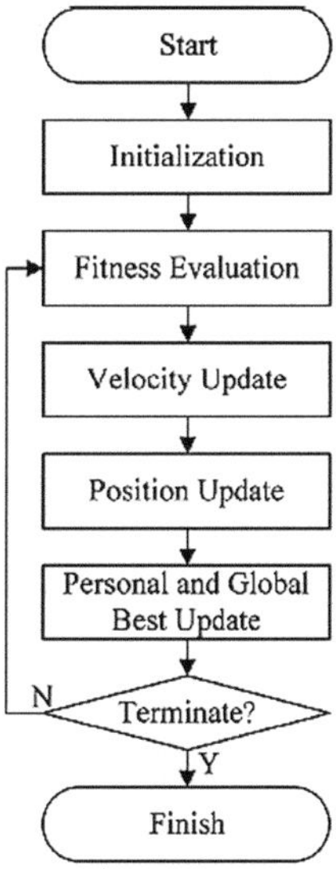

Fig. 2 Particle Swarm Optimization

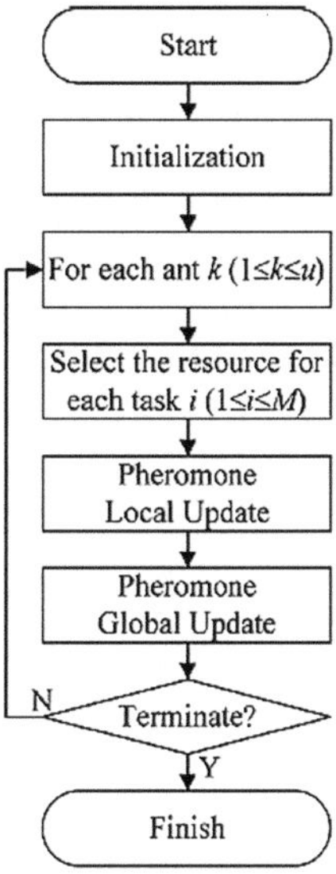

Fig. 3 Ant Colony Optimization

2.3 *Ant Colony Optimization*

ACO is a meta-heuristic technique for tackling improvement issues. A probabilistic system can be used for advancement issues by which their answer can be best depicted utilizing charts. It is enlivened by the genuine ants' conduct of how they rummage for sustenance. Genuine ants are fit for finding the briefest from a sustenance source to their home by misusing pheromone information, where pheromone is a concoction substance kept by ants. While strolling ants store pheromone on the ground and take after, probabilistically, ways that have more prominent measure of pheromone. In ACO, a limited size province of simulated ants is made. Every insect then forms an answer for the issue. While building its every insect gathers data in light of the issue attributes and all alone execution and store this data on the pheromone trails connected with the association of all edges. These pheromone trails assume the part of an appropriated long haul memory about the entire subterranean insect looks prepared. Edges can likewise have an extra related esteem speaking to from the earlier data about the issue. When all ants have registered their entire visit ACO calculation upgraded the pheromone trail utilizing all arrangements delivered by the subterranean insect province.

3 Fault Tolerance

Because of the application multifaceted nature, the majority of the circumstances blame in the framework is unavoidable and must be viewed as a typical part of the earth. Keeping in mind the end goal to legitimately manage flaws, we should appropriately comprehend them and the chain of occasions that lead from deformity to blame, mistake and disappointments [3]. Blame speaks to an inconsistency in the framework that causes it to act in a flighty and startling matter. Deficiencies can be characterized by the improvement stage they show up in, as advancement flaws or operational shortcomings. They can be outside or inward in light of the limits secured by the framework, equipment or programming, as indicated by what causes them and they can be deliberate or accidental, regardless of whether they show up in the testing stage or underway [3, 4].

In the event that the blunder cannot be dealt with, it will bring about a disappointment, which is a circumstance in which the outcomes delivered by the framework are changed somehow. Just particle disappointments may happen, yet it will at present cause either a misfortune in execution or even total shutdown of the framework. Keeping in mind the end goal to counteract disappointments in the application, three strategies can be utilized at various levels of the blame mistake disappointment chain: blame shirking, blame expulsion, and adaptation to noncritical failure. Blame shirking and blame evacuation both allude to the way that shortcomings ought to be managed at the earliest opportunity and not permit them to form into mistakes, as blunders are harder to deal with and can without much of a stretch prompt to disappointments [5, 6].

4 Literature Survey

(a) To tackle the issue utilizing GA, it starts with the introduction procedure to create a populace of chromosomes and after that starts the vectors haphazardly. The determination of parent chromosomes with a roulette wheel strategy is done by a likelihood appropriation of the wellness values [7]. To imitate another populace, there are two administrators to be utilized here, the cross-transform and scramble change, which are initially used to enhance the chromosome expansion for taking care of different building issues and here it is used for taking care of the site format issue. The cross-change comprises of three operations, i.e., flip, swap, and slide. These three operations are not utilized in grouping, rather, the program will haphazardly pick one operation to execute.

(b) In the basic Genetic Algorithm [8], the underlying populace is created haphazardly, so the distinctive calendars are less fit, so when these timetables are further changed with each other, there are especially less shots that they will deliver preferable youngster over themselves. In an enhanced hereditary calculation, the thought for creating beginning populace utilizes the Min-Min and Max-Min procedures for Genetic Algorithms.

(c) Authors for the most part pointed [9] at taking care of the issue of asset improvement inside a cloud. So they proposed a Hybrid Genetic Algorithm (HGA for short in the accompanying), which goes about as a free module in the cloud supervisor. This planning arrangement module is executed routinely and the outcome contains an arrangement of sensible virtual machine movement plans. The outcome can be utilized as the director's reference, or it can be straightforwardly used to relocate the virtual machines consequently. The base of this calculation is that the three load measurements: CPU stack, arrange throughput, and circle I/O heap of all the virtual machines carried on one particular physical machine, can be coordinated and computed to get the ideal relocation exhortation. The calculation depends on hereditary calculation with multiple wellnesses. In the plan of this calculation, three sub-wellness capacities are embraced. Sub-wellness work 1 speaks to the virtual machine stack complementation. Sub-wellness work 2 speaks to that the measure of physical machines is least after relocation. Sub-wellness work 3 speaks to that the measure of virtual machines that should be moved.

(d) A hereditary calculation-based load adjusting procedure for cloud computing has been produced [10] to give a proficient usage of asset in cloud environment. Investigation of the outcomes shows that the proposed methodology for load adjusting beats a couple existing procedures as well as assures the QoS prerequisite of client occupation. In this, the work is a novel loads adjusting system utilizing Genetic Algorithm (GA) [11]. The calculation develops to adjust the heap of the cloud framework while taking a stab at limiting the make traverse of a given undertakings set. The proposed stack adjusting technique has been mimicked utilizing the cloud analyst test system. And reproduction would come about for an ordinary specimen application demonstrates that the proposed calculation beat the current methodologies like First Come First Serve (FCFS), Round Robin (RR) and a neighborhood look calculation Stochastic Hill Climbing (SHC).

(e) The PSO is a populace based stochastic streamlining method roused by social conduct of fowl running or fish tutoring to a promising position for specific goals [12]. The position of a molecule can be utilized to speak to an applicant to answer for the current issue. A swarm of particles with haphazardly introduced positions would fly toward the ideal position along a way that is iteratively redesigned in view of the present best position of every molecule, i.e., neighborhood best and the best position of the entire swarm, i.e., worldwide best [13]. PSO has known the possibility to take care of the office design issue, especially in the development field.

(f) A particle swarm optimization proposed by author [14] and it is based on calculation. In the planning of utilizations, considering execution and information exchange cost both. Work contrasted the cost investment funds and existing 'Best Resource Selection' (BRS) calculation. PSO accomplished better circulation of workload on assets with three circumstances cost funds.

(g) The ACO is an organic enlivened meta-heuristic that mirrors the conduct of ants hunting down nourishment. This calculation has been established on the perception of genuine subterranean insect settlements by [15]. In the normal world, ants begin to search haphazardly for sustenance. Every insect would choose a way haphazardly, and the subterranean insect on the most limited way will tend to store pheromone with higher focus than whatever remains of the settlement. In that capacity, the ants adjacent will notice the exceptionally thought pheromone and join the most limited way, which will build the pheromone fixation. More ants will continue joining until most of the settlement meets to the briefest way. The possibility of ACO is to copy this conduct with reenacted ants strolling around the diagram speaking to the issue to comprehend. To apply ACO, the enhancement issue is changed into the issue of finding the best way on a weighted graph. By utilizing an assortment of viable abuse instruments and tip top techniques, scientists proposed many refined ACO calculations, and gets better outcomes in tests. By utilizing the self-versatile system and the pheromone upgrading guideline of problematic arrangements, which is dictated by the enrollment work uncertainly, the cloud-based structure can make ACO calculation traveler look space even more adequately. Hypothetical examination on the cloud-based system for ACO demonstrates that the structure is merged, and the reenactment demonstrates that the system can enhance the ACO calculations apparently.

(h) The designation of distributed computing asset is critical in the distributed computing. Presently, creators attempted to do some change in the insect state calculation to advance the allotment of distributed computing asset. The simulation shows that the throughput is expanded and the reaction time is lessened contrasted and the OSPF [16].

(i) Ideas of numerous meta-heuristic calculations have been given and focused on laying out the center wellspring of motivation as it were. The glossary may

Table 1 Pros and cons of meta-heuristics techniques

Algorithm name	Pros	Cons
Genetic Algorithm (GA)	1. Exchange data (hybrid or change) 2. Efficient to tackle consistent issues	1. No Memory 2. Premature meeting 3. Weak nearby pursuit capacity 4. High computational exertion 5. Difficult to encode an issue as a chromosome
Particle Swarm Optimization (PSO)	1. Has a memory 2. Easy execution because of the use of straightforward administrator 3. Efficient to take care of ceaseless issues	1. Premature merging 2. Weak neighborhood look capacity
Ant Colony Optimization (ACO)	1. Has a memory 2. Rapid disclosure of good arrangements 3. Efficient in tackling TSP issue and other discrete issues	1. Premature merging 2. Weak neighborhood seek capacity 3. Probability conveyance changes by emphases 4. Not compelling in taking care of the ceaseless issues

flabbergast youthful scientists and spur them to investigate their environment, discover their wellspring of motivation and utilize it to grow new and more productive meta-heuristics [11].

5 Pros and Cons of Meta-Heuristics Techniques

Meta-heuristic techniques are vital optimization techniques. The following table shows the pros and cons of the algorithms (Table 1).

6 Performance Evaluation

We have compared GA, PSO, and ACO under CloudSim tool for fault tolerance. The algorithms are simulated under various parameters like RAM size, number of VM, number of cloudlets, etc. The details are given in Table 2.

Now, we have a graphical representation of the performance evaluation and data taken from Table 2.

From Figs. 4 and 5, it is easy to state that ACO is better than GA and PSO and ACO has a lesser average number of failed cloudlets and average total cost from both GA and PSO.

Table 2 Tabular data for comparison between GA, PSO and ACO

Algorithm name	Average no. of failed cloudlets	Average total cost	Average total completion time (s)
GA	15.40	3948.33	5541.68
PSO	10.5	3766.38	5073.29
ACO	5.95	3622.16	4739.97

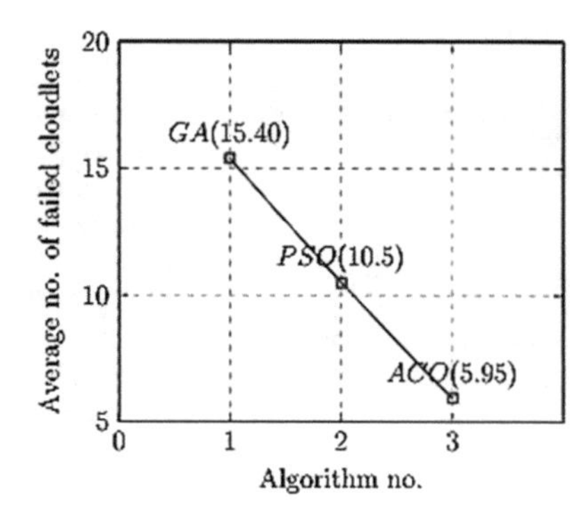

Fig. 4 Average no. of failed cloudlets

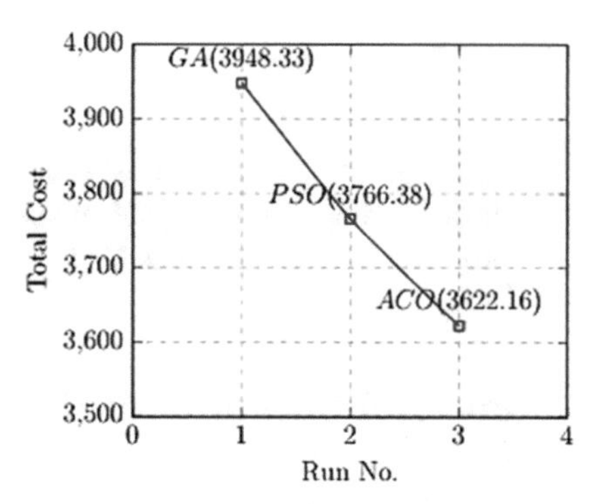

Fig. 5 Average total cost

7 Conclusion and Future Work

In this paper, we have given the basic details on only GA, PSO, and ACO techniques, which are supported by meta-heuristic concepts. Cloud computing itself is an emerging model that is elaborated on services. These services may have been failed due to some faults. So fault tolerance is necessary in cloud computing. We have also shown the results of average number of failed cloudlets and average total cost.

As far as future scope is concerned, we would like to introduce new fault-tolerant algorithm and will also compare with existing meta-heuristic based fault-tolerant algorithms.

Acknowledgements Authors are thankful to Department of Computer Science & Engineering at Maharishi Markandeshwar University, Ambala for giving high motivational supports.

References

1. Dean, J., Ghemawat, S.: MapReduce: simplified data processing on large clusters. Commun. ACM **51**(1), 107–113 (2008)
2. Buyya, R., et al.: Cloud computing and emerging IT platforms: vision, hype, and reality for delivering computing as the 5th utility. Future Gener. Comput. Syst. **25**(6), 599–616 (2009)
3. Lavinia, A., et al.: A failure detection system for large scale distributed systems. In: 2010 International Conference on Complex, Intelligent and Software Intensive Systems (CISIS). IEEE (2010)
4. Nastase, M., et al.: Fault tolerance using a front-end service for large scale distributed systems. In: 2009 11th International Symposium on Symbolic and Numeric Algorithms for Scientific Computing (SYNASC). IEEE (2009)
5. Palmieri, F., Pardi, S., Veronesi, P.: A fault avoidance strategy improving the reliability of the EGI production grid infrastructure. In: International Conference on Principles of Distributed Systems. Springer, Berlin, Heidelberg (2010)
6. Saha, G.K.: Software fault avoidance issues. In: Ubiquity 2006, Nov 2006, 5
7. Zhu, K., et al.: Hybrid genetic algorithm for cloud computing applications. In: 2011 IEEE Asia-Pacific on Services Computing Conference (APSCC). IEEE (2011)
8. Kumar, P., Verma, A.: Independent task scheduling in cloud computing by improved genetic algorithm. Int. J. Adv. Res. Comput. Sci. Softw. Eng. **2**(5) (2012)
9. Chen, S., Wu, J., Lu, Z.: A cloud computing resource scheduling policy based on genetic algorithm with multiple fitness. In: 2012 IEEE 12th International Conference on Computer and Information Technology (CIT). IEEE (2012)
10. Dasgupta, K., et al.: A genetic algorithm (GA) based load balancing strategy for cloud computing. Procedia Technol. **10**, 340–347 (2013)
11. Rajpurohit, J., et al.: Glossary of metaheuristic algorithms. Int. J. Comput. Inf. Syst. Ind. Manag. Appl. **9**, 181–205 (2017). ISSN 2150-7988
12. Li, H.-H., et al.: Renumber strategy enhanced particle swarm optimization for cloud computing resource scheduling. In: 2015 IEEE Congress on Evolutionary Computation (CEC). IEEE (2015)
13. Rodriguez, M.A., Buyya, R.: Deadline based resource provisioning and scheduling algorithm for scientific workflows on clouds. IEEE Trans. Cloud Comput. **2**(2), 222–235 (2014)
14. Pandey, S., et al.: A particle swarm optimization-based heuristic for scheduling workflow applications in cloud computing environments. In: 2010 24th IEEE International Conference on Advanced Information Networking and Applications (AINA). IEEE (2010)
15. Dorigo, M., Blum, C.: Ant colony optimization theory: a survey. Theoret. Comput. Sci. **344**(2), 243–278 (2005)
16. Gao, Z.: The allocation of cloud computing resources based on the improved Ant Colony Algorithm. In: 2014 Sixth International Conference on Intelligent Human-Machine Systems and Cybernetics (IHMSC), vol. 2. IEEE (2014)

Complexity Metrics for Component-Based Software System

Rajni Sehgal, Deepti Mehrotra and Renuka Nagpal

Abstract Today, softwares are influencing almost every process involved in our day-to-day life. The dependence of our routine processes of software system makes the reliability of these softwares a major concern of industry. Various metrics and benchmark are designed to ensure the smooth design and implementation of software, among which complexity is one. It is always a desire software architect to design software with lesser complexity. In this paper, component-based software is considered and metrics to measure the complexity of the software is proposed. Complexity needs to be measured at component level and its relationship with other components. UML diagram is drawn to propose a new metrics, and dependency of one component to another component is measured. Various complexity metrics namely Method Complexity MCOM, Number of Calls to Other Methods (NCOM), Component Complexity (CCOM) is evaluated, and Emergent System Complexity (ESCOM) and overall complexity of the system are evaluated incorporating the contribution of each.

Keywords Complexity · UML · CPDG · MCOM · NCOM · CCOM

1 Introduction

Software complexity is one of the biggest challenges for the software industry to maintain the quality of the software. As the software grows in size, software complexity also increases and becomes difficult to measure [1]. The complexity of a

R. Sehgal (✉) · D. Mehrotra · R. Nagpal
Amity School of Engineering and Technology, Amity University,
Noida, Uttar Pradesh, India
e-mail: rsehgal@amity.edu

D. Mehrotra
e-mail: dmehrotra@amity.edu

R. Nagpal
e-mail: rnagpal1@amity.edu

K. Ray et al. (eds.), *Soft Computing: Theories and Applications*,
Advances in Intelligent Systems and Computing 742,
https://doi.org/10.1007/978-981-13-0589-4_2

software system developed based on the component-based software engineering is dependent on the various parameters: (i) Number of components (ii) Number of interfaces (iii) Usage of a component (iv) and Number of dependency of one component on another. Software metrics plays a major role in measuring the attributes of software quality. Complexity of a software system is a measurable parameter of software engineering and can be measured well with the help of complexity metrics.

This paper considers component-based software which is having 15,000 LOC and 12 components. Various complexity metrics namely Method Complexity MCOM, Number of Calls to Other Methods (NCOM), Component Complexity (CCOM), Emergent System Complexity (ESCOM) are evaluated. A new metrics to measure the complexity of the software is proposed based on the metrics that can be calculated at the time of design of the software using the UML diagram [2]. A Component Dependency Graph (CPDG) is constructed based on the sequence diagram showing the relative dependencies between several components and how many times these components interact with one another [3]. It gives an overall description of the entire system that includes all the components involved in the system. This paper is divided into six sections. Section 2 presents the work done by the various authors in the field of software complexity. Section 3 describes the framework to propose the new metric. In Sect. 4, the methodology is discussed. Section 5 presents the results obtained during the study. Finally, the conclusion is discussed in Sect. 6.

2 Related Work

Shin and Williams [1] perform the statistical analysis on nine software complexities and analyze that the software complexity plays a major role in software quality. Nagappan et al. [4] apply the principal component analysis to find out the cyclomatic complexity which strongly affects the complexity of the code. D'Ambrogio et al. [2] discussed a method of prediction of software reliability which predicts the reliability at earlier stages in the development process using the fuzzy method. This paper contains a more generalized approach, and the reliability is calculated using the fault tree method. This approach principally makes use of UML diagrams developed in the earlier stages of software development lifecycle making it easier to predict the software reliability at the more initial stages saving cost, time, and money. Ali et al. [5] discussed a technique for early reliability prediction of software components using behavior models. The behavior models are the diagrams which predict the reliability by showing the interaction between these components, and how these components work with each other to perform the desired functionalities. Bernardi et al. [6] predict the software reliability by using fuzzy methods. The paper contains a more generalized approach, and the reliability is calculated using the fault tree method with the use of UML diagrams. System dependability is defined as the degree to which the system can be considered dependable. System dependability is measured using the following factors, or it considers the following factors: fault tolerance, fault prevention, fault removal, and fault forecasting. Sahoo and Mohanty [7] predict the

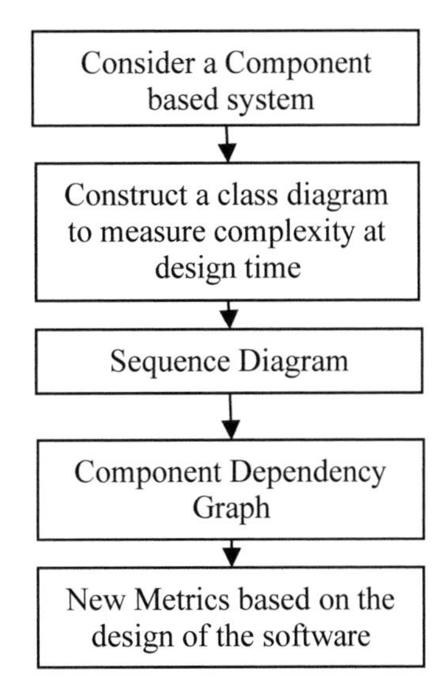

Fig. 1 Frame work to propose a new complexity metric

testing effort of software tester using UML Diagrams. Software testing is essential to maintain the quality of the software. Accurate test cases defined for the system are significant for checking as to what degree the software is reliable. Estimation of test cases done at earlier stages of the SDLC [8] is very challenging and has its issues because very less knowledge is available about the system in these initial phases of SDLC [9], [10]. The functional requirements are ready during the requirement phase. Hence, there is a need of the hour for early test prediction at the requirements specification stage and are accomplished using UML diagrams. UML gives enough information about the system which is necessary for the prediction of test cases [11].

3 Framework for Proposed Model

In this study, various complexities are evaluated at various levels which include

(i) Level-1 Complexity called Method Complexity (MCOM).
(ii) Level-2 Complexity Component Complexity (CCOM).
(iii) Level-3 Complexity Emergent System Complexity (ESCOM) of a component-based system is measured by using various design metrics which is given in the literature.
(iv) Level-4 Complexity metrics is proposed by using the parameter coupling and interaction probabilities between the components [12]. To evaluate this metric, framework is given in Fig. 1.

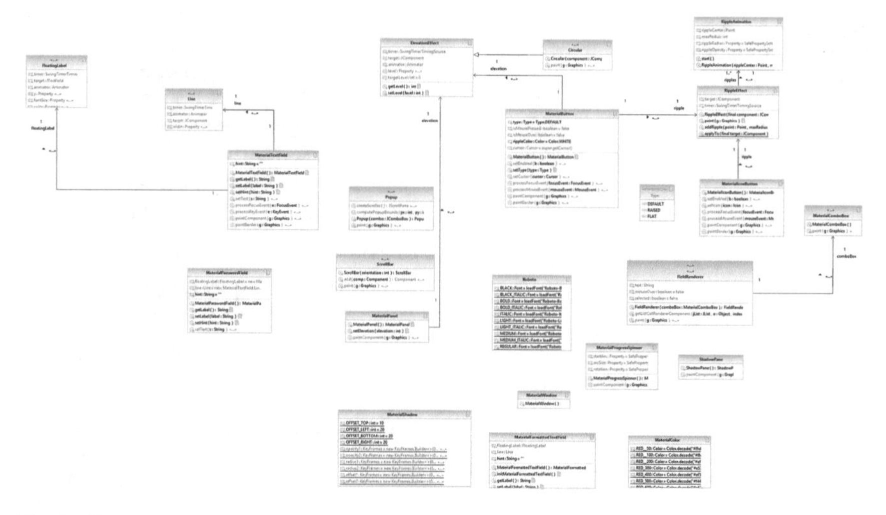

Fig. 2 Class diagram

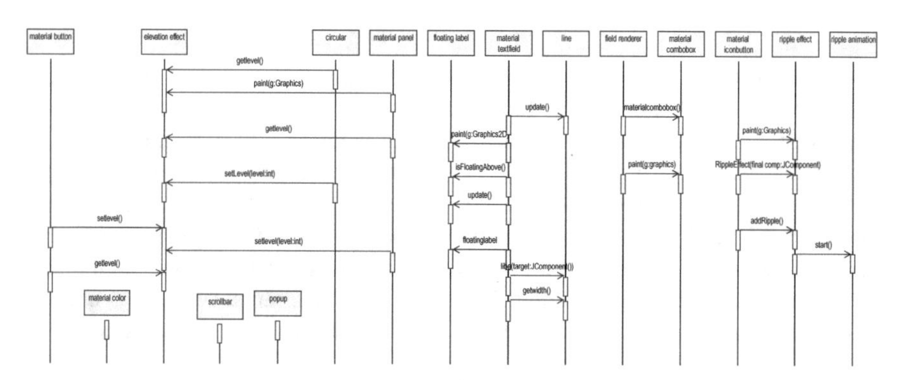

Fig. 3 Sequence diagram

4 Methodology

In order to calculate metric, the following steps are followed.

Step1: Component-based software is considered which is having 15,000 LOC is considered and various UML diagrams: (i) Class diagram and (ii) Sequence diagram is drawn in order to achieve the design level complexity by calculating the dependency of one component to another which is shown in Figs. 2 and 3.

Step2: A component dependency graph (CPDG) is developed which gives us the estimation of dependency of components with each other. This is given in Fig. 4.

Step3: Various complexities are measured at level-1, level-2 and Level-3 for component-based software, which is given in Eqs. 1–3 and new metrics which is called Level-4 complexity metrics is proposed in Eq. 6.

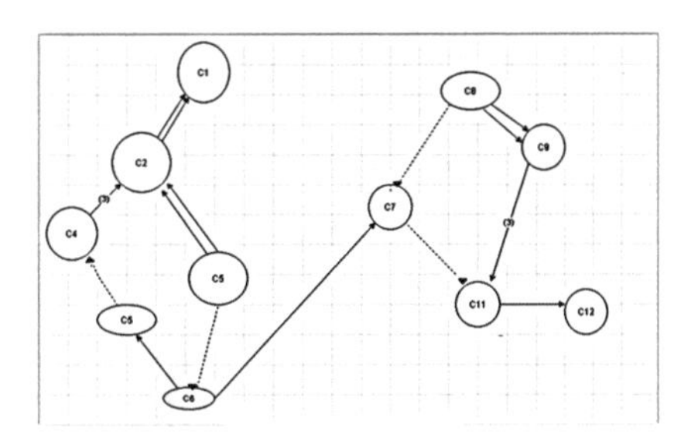

Fig. 4 CPDG graph

Table 1 Method complexity

Class name	Cyclomatic Complexity CC(mi)	Components	NCOM(mi)
Roboto	14	C1	2
Material Color	2	C2	5
Material Shadow	14	C3	0
Material Button	27	C4	1
Material Formatted Text Field	27	C5	1
Material progress Spinner	2	C6	1
Material Password Field	15	C7	2
Material Text Field	15	C8	0
Line	1	C9	2
Elevation Effect	9	C10	0
Ripple Effect	9	C11	1
Material Combobox	3	C12	1
Icon button	10		
Icon button Window	2		
Icon button Panel	4		
Ripple Animation	1		

(i) **Level-1 Method Complexity (MCOM)** is calculated as given in Eq. (1) and shown in Table 1.

$$\mathrm{MCOM}(\mathrm{m_i}) = \mathrm{CC}(\mathrm{m_i}) + \mathrm{NCOM}(\mathrm{m_i}) \tag{1}$$

where CC is the Cyclomatic Complexity, NCOM is the Number of Calls to other methods (Table 1).

The total method complexity of a component Cj (TMCOM) is the sum of all complexities of individual methods and is estimated as:

Table 2 Component complexity

Component name	Components	NIMC	NOIC	NCOC V
MaterialFormattedTextField	C1	2	1	0
ElevationEffect	C2	5	2	1
MaterialButton	C3	0	0	2
MaterialCombobox	C4	1	1	1
MaterialColor	C5	1	2	2
Circular	C6	1	1	0
FieldRenderer	C7	2	3	2
RippleEffect	C8	0	2	2
RippleAnimation	C9	2	1	0
MaterialPanel	C10	0	0	2
FloatingLabel	C11	1	1	2
Line	C12	1	0	0

$$\mathbf{TMCOM(Cj)} = \sum_{\mathbf{i=1}}^{\mathbf{n}} \mathbf{MCOM(m_i)} = \mathbf{155 + 16 = 171}$$

(ii) **Level-2 Component Complexity (CCOM)** which can be calculated in Eq. 2 and given in Table 2.

$$\mathrm{CCOM(C_i) = NIMC(C) + NOIC(C) + NCOC(C)} \quad (2)$$

where

NIMC Number of Internal Method Calls
NOIC Number of Interfaces of a Component
NCOC Number of Couplings to Other Components

The Total Component Complexity (TCCOM) based on all components Ci in a component-oriented software system S is estimated as

$\mathrm{TCCOM(S)} = \sum_{i=1}^{n} \mathrm{CCOM}\ (c_i) = 16 + 14 + 14 = 44$ for all components Ci of a software system.

(iii) **Level-3** Emergent System Complexity (ESCOM) of a component-oriented system S could be represented by a function f for the two given values (i) Number of Links Between Components (NLBC) and (ii) Depth Of the Composition Tree (DOCT) can be calculated in Eq. 3 and shown in Table 3.

$$\mathrm{ESCOM(S) = f(NLBC(S), DOCT(S))} \quad (3)$$

where Depth Of the Composition Tree (DOCT): It is the number of levels in the entire system. Thus, emergent complexity is the summation of NLBC and DOCT values, ESCOM = 22 + 5 = 27.

Table 3 NLBC values

Interacting components	NLBC value
C1,2	2
C2,4	3
C2,3	2
C5,4	1
C6,5	4
C6,3	1
C6,7	1
C7,8	1
C7,11	1
C8,9	2
C10,11	3
C11,12	1

Thus, the overall complexity of the system is TMCOM $(C_j)+$TCCOM(S)$+$ ESCOM$=177+44+27=248$.

(iv) **Level-4 New Proposed metrics to measure the complexity of component-based system**

The new proposed metrics is based on the two parameters (i) coupling between the components (ii) Interactions probabilities between the components which are described below

(a) Coupling: coupling is defined as the degree to which the components present in the system are related to each other i.e., how much dependency is present among the components Eq. (4) [13]. A highly coupled system is one in which the directed graph is dense and it requires more precise reliability estimation since the components are dependent on each other vice versa id for low coupling systems

$$NU_j = \frac{\left|\cup_{1 \leq i \leq m \cap i \neq j} NU_{j,i}\right|}{|N_j|+|U_j|} \tag{4}$$

where N_J and U_J are set of methods and instance variable of component.

(b) Interaction probabilities: Interaction probability that is p_{ij} is the probability of transition tij being executed, where i and j represent the source component and destination component, respectively, and it is estimated from the percentage of the number of messages that the source component C_i sends to the target component C_j to the total number of messages that the source component C_i sends to all other components C_k in the architecture as given in Eq. 5 [14]

Table 4 Probability distribution table

1		C1	C2	C3	C4	C5	C6	C7	C8	C9	C10	C11	C12
2	C1	0	0	0	0	0	0	0	0	0	0	0	0
3	C2	1	0	0	0	0	0	0	0	0	0	0	0
4	C3	0	1	0	0	0	1	0		0	0	0	0
5	C4	1	0	0	0	0	0	0	0	0	0	0	0
6	C5	0	0	0	0	0	0	0	0	0	0	0	0
7	C6	0	0	0	1	1	0	1	0		0	0	0
8	C7	0	0	0	0	0	0	0	0	0	0	0	0
9	C8	0	0	0	0	0	0	1	0	1	0	0	0
10	C9	0	0	0	0	0	0	0	0	0	0	0	0
11	C10	0	0	0	0	0	0	0	0	0	0	1	0
12	C11	0	0	0	0	0	0	1	0	0	0	0	1
13	C12	0	0	0	0	0	0	0	0	0	0	0	0

$$p_{ij} = \frac{\text{Interact}(c_i, c_j)}{\sum \text{Interact}(c_i, c_j)} \tag{5}$$

Hence, final New Proposed Level-4 complexity is calculated as Eq. 6

$$\text{Level} - 4\,\text{Complexity} = \sum_{i=1}^{n} \prod C_i * P_{ij} \tag{6}$$

where

C_i Coupling between the components
p_{ij} Interaction Probablities.

5 Results

In order to calculate the probabilities of interaction between the components, pairwise comparison is done using Fig. 3 which is given in Table 4.

After the pairwise comparison of interaction between the component, probability and coupling are calculated using Eqs. 4 and 5. Level 4 complexity is calculated using Eq. 6 shown in Table 5.

The overall complexity of the given software using the new proposed complexity metrics is 10.6.

Table 5 Probability of interaction between the components

Cij (Pairwise components)	Pij	Coupling between objects (CBO)	Level 4 complexity
C2,1	1	2	1
C3,2	0.66	4	2.64
C4,2	1	1	0.66
C5,4	1	2	1.32
C6,5	0.4	2	0
C6,7	0.5	1	0.66
C8,9	0.66	2	1.32
C8,7	0.33	2	1
C10,11	1	2	1
C11,7	0.5	1	1
C11,12	0.5	0	0
		Overall complexity	10.6

6 Conclusion

A new level of complexity component-based software is proposed. We calculated the level 4 complexity for each component in the system. The components whose complexity came out to be zero can be removed as they do not contribute to system's performance. By eliminating such component, we make our system's performance fast and reliable, also we need less of testing time, manpower, and resources when such components are removed. The calculations and results we obtained are of great importance and significance in the field of future research. This area of study needs more of practical implementation and validation from large and complicated data that is used for industry purposes. We will be able to find a good relationship between metrics values and development and maintenance efforts, which are the most cost-determining factors in software projects when used on implemented and large projects. There is another significant point that must be taken into consideration is that it is difficult to determine metric values from system designs and usually developers do complain that collection of metrics is a very boring and time-consuming task.

References

1. Shin, Y., Williams, L.: Is complexity really the enemy of software security?. In: Proceedings of the 4th ACM Workshop on Quality of Protection. ACM (2008)
2. D'Ambrogio, A., Iazeolla, G., Mirandola, R.: A method for the prediction of software reliability, Department of Computer Science, S&P University of Roma, Italy Conference Paper, Nov 2012

3. Balsamo, S., et al.: Model-based performance prediction in software development: a survey. IEEE Trans. Softw. Eng. **30**(5), 295–310 (2004)
4. Nagappan, N., Ball, T., Zeller, A.: Mining metrics to predict component failures. In: Proceedings of the 28th International Conference on Software Engineering (ICSE'06), pp. 452–461, Shanghai, China, 20–28 May 2006
5. Ali, A., Jawawi, D.N.A., Isa, M.A., Babar, M.I.: Technique for Early Reliability Prediction of Software Components Using Behaviour Models, Published: 26 Sept 2016
6. Bernardi, S., Merseguer, J., Petriu, D.C.: 1Centro Universitario de la Defensa, Dependability Modeling and Assessment in UML-Based Software Development, Department of Systems and Computer Engineering, Carleton University, Ottawa, ON, Canada, 25 May 2012
7. Sahoo, P., Mohanty, J.R.: Early Test Effort Prediction using UML Diagrams, School of Computer Engineering, KIIT University, Campus-15, Bhubaneswar (2017)
8. Software Safety, Nasa Technical Standard. NASA-STD- 8719.13A, 15 Sept 2000
9. Yacoub, S.M., Ammar, H.H.: A methodology for architecture level reliability risk analysis. IEEE Trans. Softw. Eng. **28**(6), 529–547 (2002)
10. Ammar, H.H., Nikzadeh, T., Dugan, J.B.: Risk assessment of software system specifications. IEEE Trans. Reliab. **50**(2) (2001)
11. Jatain, A., Sharma, G.: A systematic review of techniques for test case prioritization. Int. J. Comput. Appl. (0975 – 8887) **68**(2) (2013)
12. Chen, J., Wang, H., Zhou, Y., Bruda, S.D.: Complexity metrics for component-based software systems. Int. J. Digit. Content Technol. Appl. **5**(3), 235–244 (2011)
13. Talib, M.A.: Towards early software reliability prediction for computer forensic tools (case study), June 2016
14. Vinod, G.: An Approach for Early Prediction of Software Reliability, Nov 2011

Identification of Relevant Stochastic Input Variables for Prediction of Daily PM_{10} Using Artificial Neural Networks

Vibha Yadav and Satyendra Nath

Abstract Air pollution has a great impact on environment and humans. It is necessary to analyze air pollutant (AP) data but these data are not available in most of the site. Therefore, prediction of AP becomes an important research to solve the problem of time series data. For this, seven ANN models (ANN-1, ANN-2, ANN-3, ANN-4, ANN-5, ANN-6 and ANN-7) with different input variables are developed using Levenberg–Marquadt (LM) algorithm to predict daily A.P. The ANN models are developed using daily measured value of PM_{10}, solar radiation, vertical wind speed, and atmospheric pressure for Ardhali Bazar, Varanasi India. ANN models incorporate 473 data points for training and 31 data points for testing. The results show that vertical wind speed is found to be most influencing variable for PM_{10} prediction. Apart from this first input combination of solar radiation, wind speed and second input combination of solar radiation, and atmospheric pressure can be used for PM_{10} prediction.

Keywords PM_{10} · Prediction · ANN

1 Introduction

Air pollution is a worldwide growing threat to the life and natural environment. The occurrence of rapid urbanization and flaming of industrialization contributing the release of many air pollutants in the atmosphere. Different pollutants of the atmosphere are particulate matter, oxides of carbon, oxides of sulfur, oxides of nitrogen, ozone, and hydrocarbons. Particulate matter (PM) is one of the major air pollutants

V. Yadav (✉) · S. Nath
Department of Environmental Sciences and NRM, College of Forestry,
Sam Higginbottom University of Agriculture, Technology and Sciences,
Allahabad 211007, Uttar Pradesh, India
e-mail: yvibha3@gmail.com

S. Nath
e-mail: satyendranath2@gmail.com

K. Ray et al. (eds.), *Soft Computing: Theories and Applications*,
Advances in Intelligent Systems and Computing 742,
https://doi.org/10.1007/978-981-13-0589-4_3

contributed by dust particle arises from different sources, fibrous material of the plants and animal origin, combustion of coal and oil or from mining operations. To determine the behavior of pollutant in atmosphere, the size of particle is important. PM is categorized into PM_{10} and $PM_{2.5.}$ Particle larger than 10 um tends to settle down rapidly whereas particle smaller than 10 um in size usually remain suspended in air. These PMs are highly associated with cardiovascular and respiratory disorders, therefore, it is essential to predict an accurate pollutant concentration prediction by means of time and location identification. Several tools adopted by different authors are given.

Baawain et al. [1] developed ANN model for prediction of daily value of CO, PM_{10}, NO, NO_2, NOx, SO_2, H_2S, and O_3 using different combinations of input variable (AP and meteorological variables). The regression (R) value is more than 70% for training and testing data. Elminir and Galil [2] developed ANN model to predict daily PM_{10}, NO_2, and CO for Egypt. The inputs for ANN are WS, RH, T, and WD and output data are PM_{10}, NO_2, CO. The model predicts PM_{10}, NO_2, and CO with accuracy of 96%, 97%, and 99.6%, respectively. Filho and Fernandes [3] performed daily prediction of CO, SO_2, $NO_{2,}$ and PM_{10} using proposed particle swarm optimization (PSO) and ANN model. The pollutant data are taken from São Paulo, Brazil. The results show that PSO-ANN method has accurate and fair forecast. Skrzypski and Szakiel [4] used MLP and RBF for prediction of daily PM_{10} concentration. The input variables are daily value of rainfall, WS, SR, vertical wind velocity, T, and atmospheric pressure. The prediction accuracy of MLP and RBF models are 3.7% and 1.9%, respectively. Sahinet al. [5] used cellular neural network (CNN) for prediction of daily value SO_2 and PM_{10} for Istanbul Turkey. The CNN model predicts SO_2 better than PM_{10} and is also used to find out missing air pollutant data. The R value for CNN model varies from 54 to 87%.

From preceding discussion, daily prediction accuracy of ANN models changes with the influence of geographical meteorological variables. Therefore, it is necessary to select relevant input variables for which prediction error is minimum. In this study, seven ANN models are developed. The input variables for ANN-1 are solar radiation (SR), wind speed (WS) and atmospheric pressure (AP). SR and WS are inputs for ANN-2 model. The inputs for ANN-3 are SR, AP and for ANN-4 inputs are WS, AP. The ANN-5, ANN-6 and ANN-7 models utilize SR, WS, and AP as inputs.

This paper is organized as follows: Methodology is shown in Sect. 2. Research and discussions are presented in Sect. 3 followed by conclusions in Sect. 4.

2 Methodology

2.1 Data Collection

A real-world dataset from 1 January 2016 to 6 June 2017 is used in this study. The daily value of solar radiation (SR), wind speed (WS), atmospheric pressure (AP), and PM_{10} for Ardhali Bazar Varanasi, India are collected from Central Pollution Control Board (CPCB), Delhi [6]. The daily value of SR, WS, AP, and PM_{10} are shown in Figs. 1, 2, 3 and 4. The PM_{10} varies from 12.06 to 991.37 and average value is 276.12.

2.2 Artificial Neural Network

The ANN models for prediction can be developed with neural network fitting tool (nftool). It consists of two layers feed forward neural network and training algorithm is Levenberg–Marquardt (LM) which is suitable for static fitting problems. The target

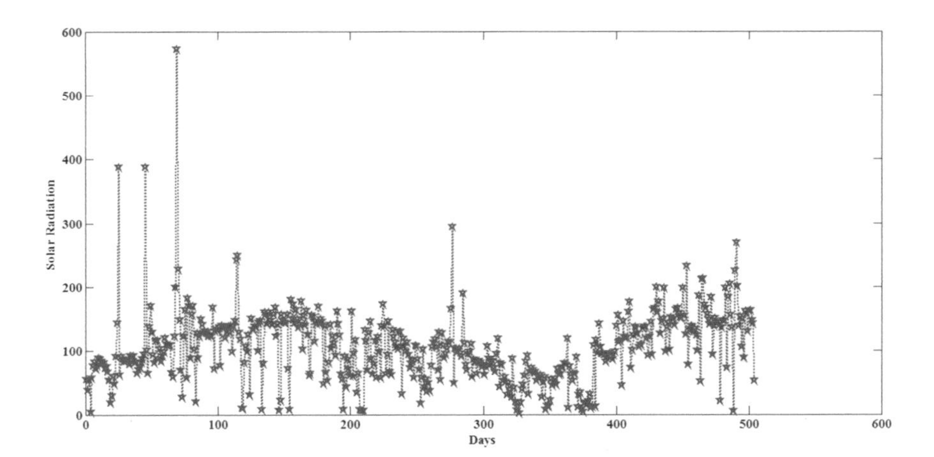

Fig. 1 Daily solar radiation

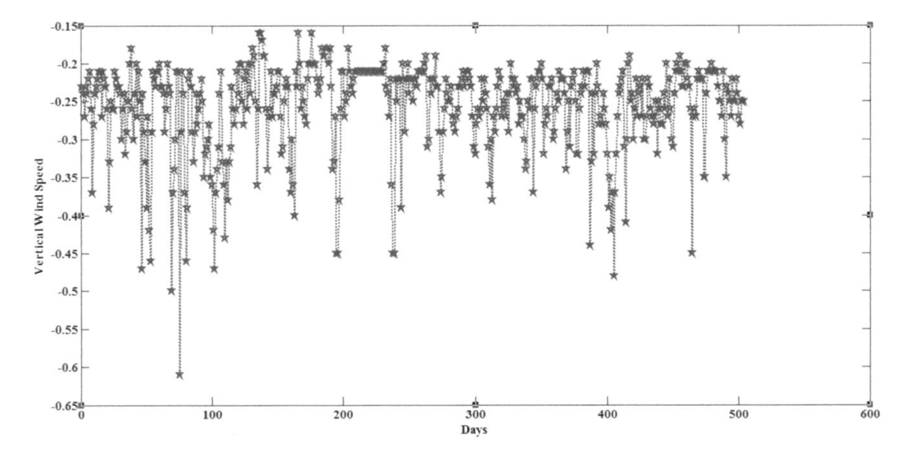

Fig. 2 Daily vertical wind speed

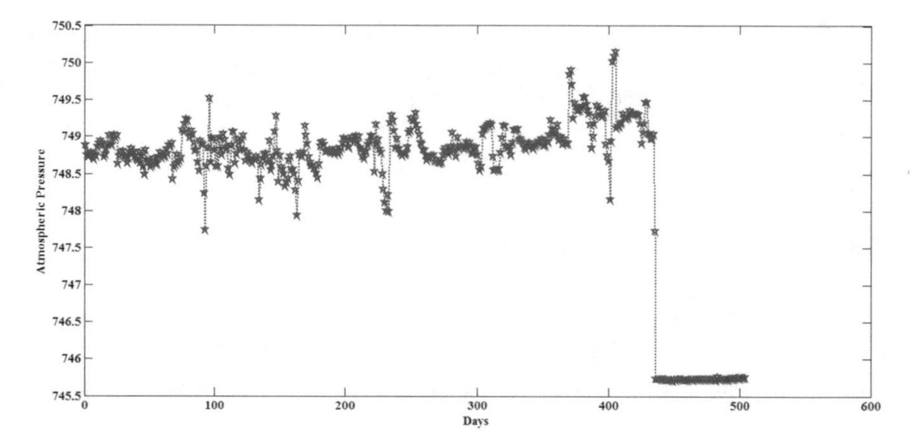

Fig. 3 Daily atmospheric pressure

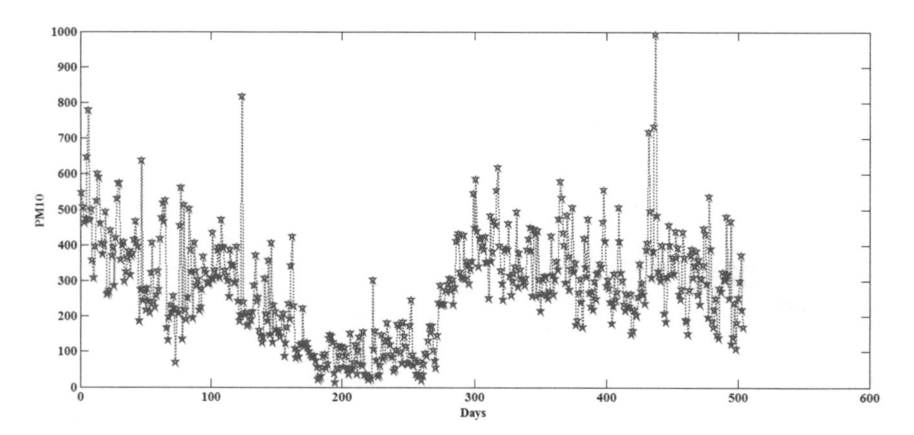

Fig. 4 Daily PM_{10}

and input data are scaled from −1 to 1 and 60%, 20%, 20% of randomly divided data are used in training, testing and validation, respectively. The stepwise methods to implement nftool are shown by Yadav et al. [7]. The number of neurons in the hidden layer is evaluated in Eq. 1 Chow et al. [8] where H_n and S_n are a number of hidden layer neurons and number of data samples used in ANN model, and I_n and O_n denote number of input and output parameters.

$$H_n = \frac{I_n = O_n}{2} + \sqrt{S_n} \quad (1)$$

The calculated value of H_n is changed from ±5. The measures of ANN model performance are the correlation coefficient (R), index of agreement (IOA), normalized

Table 1 Statistical performance measures for ANN models

Models	Input variables	IOA	NMSE	FB	R
ANN-1	SR, WS, AP	0.61	0.13	-2.69×10^4	1.88
ANN-2	SR, WS	0.44	0.20	1.83×10^4	0.89
ANN-3	SR, AP	0.40	0.18	1.17×10^4	0.84
ANN-4	WS, AP	0.31	0.15	-1.94×10^4	1.53
ANN-5	SR	0.23	0.22	1.37×10^4	−0.76
ANN-6	WS	0.30	0.17	8.98×10^4	0.91
ANN-7	AP	0.22	0.25	5.37×10^4	1.39

mean square error (NMSE), and fractional bias (FB) which are as follows, Mishra and Goyal [9].

$$R = \frac{\sum (AP_{i(ms)} - \overline{AP_{i(ms)}})(AP_{i(pred)} - \overline{AP_{i(pred)}})}{\sqrt{\sum (AP_{i(ms)} - \overline{AP_{i(ms)}})^2 (AP_{i(pred)} - \overline{AP_{i(pred)}})^2}} \tag{2}$$

$$IOA = 1 - \frac{\sum_{i=1}^{n} (AP_{i(pred)} - AP_{i(ms)})^2}{\sum_{i=1}^{n} (|AP_{i(pred)} - \overline{AP_{i(ms)}}|) + (|AP_{i(pred)} - \overline{AP_{i(ms)}}|)^2} \tag{3}$$

$$NMSE = \frac{\overline{(AP_{i(ms)} - AP_{i(pred)})^2}}{\overline{AP_{i(ms)}} - \overline{AP_{i(pred)}}} \tag{4}$$

$$FB = \frac{(\overline{AP_{i(pred)}} - \overline{AP_{i(ms)}})}{0.5(\overline{AP_{i(pred)}} + \overline{AP_{i(ms)}})} \tag{5}$$

where $AP_{i(pred)}$ is the predicted air pollutant by ANN, $AP_{i(ms)}$ is the measured value of air pollutant, n is the number of data samples.

3 Results and Discussions

The seven ANN models (ANN-1, ANN-2, ANN-3, ANN-4, ANN-5, ANN-6 and ANN-7) are developed using MATLAB software (R 2011a). For training ANN models, 473 data points are used and for testing, 31 data points are utilized from Figs. 1, 2, 3 and 4. The statistical analysis of ANN models is shown in Table 1. The R values for ANN-2, ANN-3, and ANN-6 are found to be 0.89, 0.84, and 0.91 showing predicted value that are close to measured value. The comparison between predicted and measured air pollutant is shown in Figs. 5, 6 and 7.

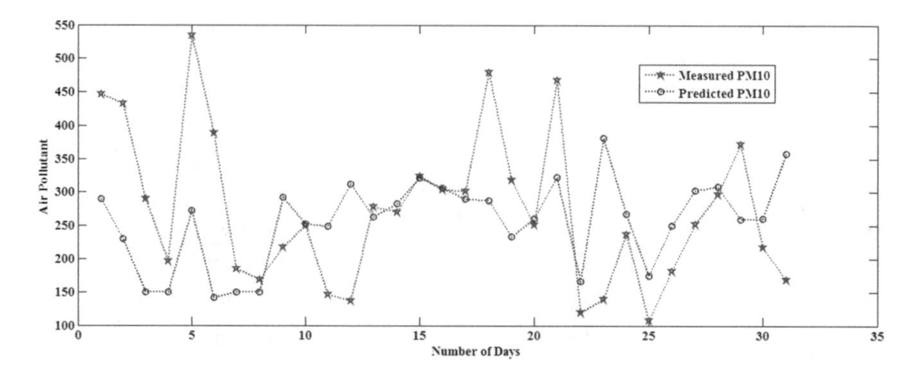

Fig. 5 Comparison between predicted and measured air pollutant for ANN-2

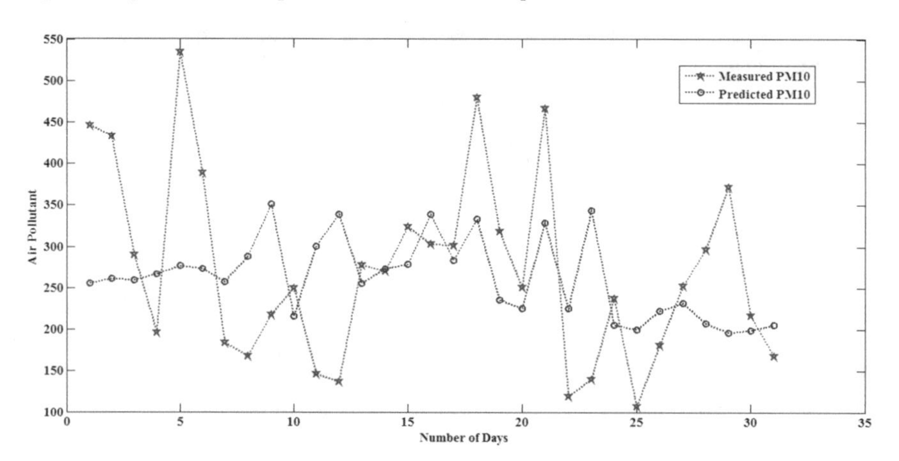

Fig. 6 Comparison between predicted and measured air pollutant for ANN-3

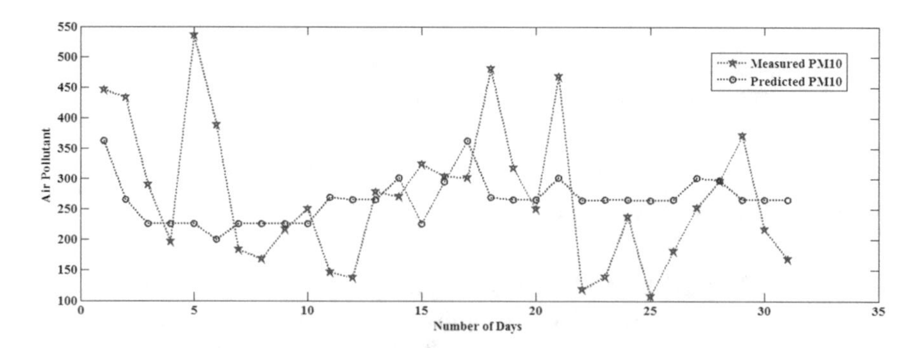

Fig. 7 Comparison between predicted and measured air pollutant for ANN-6

The number of hidden layer neurons of ANN models is calculated using Eq. 1. The performance plot of three ANN models (ANN-2, ANN-3, and ANN-6) shows

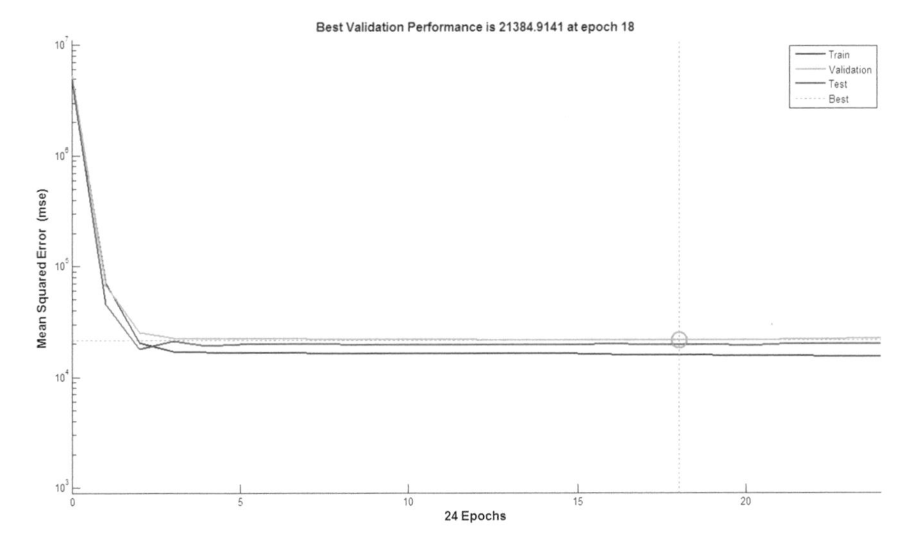

Fig. 8 Performance plot of ANN-2

that mean square error becomes minimum as the number of epochs is increasing (Figs. 8, 9 and 10). The epochs are one complete sweep of training, validation and testing data set. The test set error and validation set error has comparable characteristics and no major overfitting has happened near epoch 4, 5 and 6 (where best validation performance has taken place). The performance curve is a plot of mean square error (MSE) with number of epochs. The MSE plot in training data has lower curve and has an upper curve in validation data set. The network with least MSE in validation curve is called as trained ANN model. The training of ANN models stops automatically when validation error stops improving as indicated by an increase in MSE of validation data samples.

4 Conclusion

In this study, seven ANN models are proposed with different input combinations to find out relevant input variable for PM_{10} prediction. The ANN models are developed with nftool. The input variables for ANN-1 are solar radiation (SR), wind speed (WS), and atmospheric pressure (AP). SR and WS are inputs for ANN-2 model. The inputs for ANN-3 are SR, AP and for ANN-4 inputs are WS, AP. The ANN-5, ANN-6 and ANN-7 models utilize SR, WS, and AP as inputs. A good agreement was observed between the predicted and measured air pollutant concentrations at Ardhali Bazar Varanasi, India. The PM_{10} is found to vary from 12.06 to 991.37 and average value is 276.12, showing air quality index is poor. The R values for ANN-1, ANN-2, ANN-3, ANN-4, ANN-5, ANN-6, and ANN-7 are 1.88, 0.89, 0.84, 1.53,

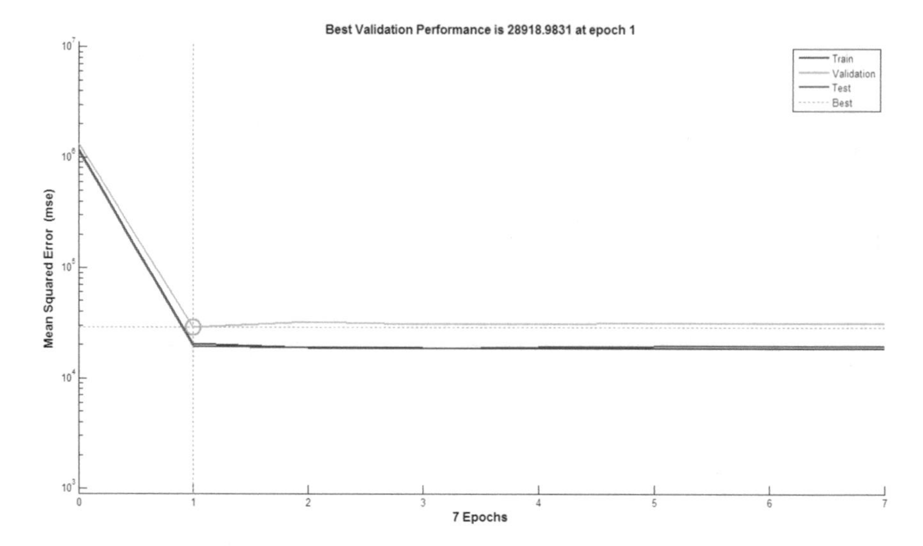

Fig. 9 Performance plot of ANN-6

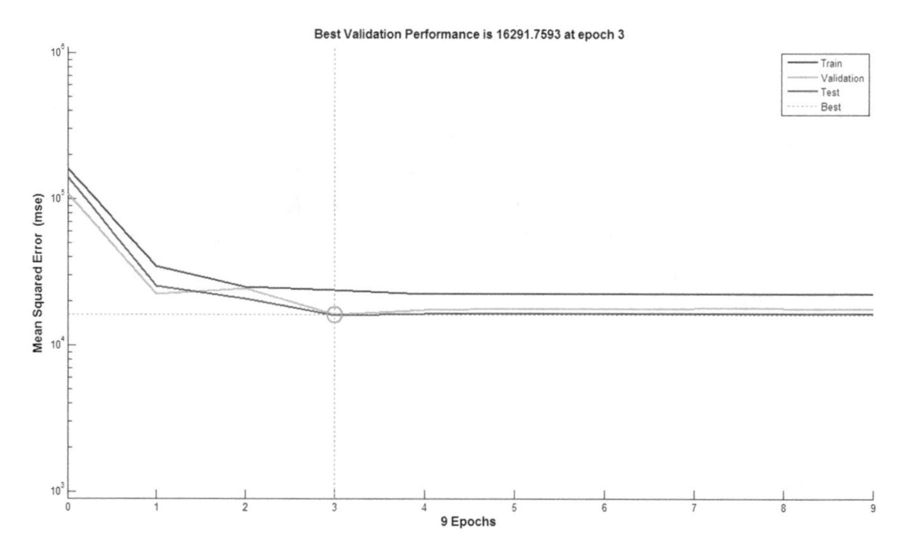

Fig. 10 Performance plot of ANN-7

−0.76, 0.91 and 1.39. The R value of ANN-6 is found to be more close to 1, showing vertical wind speed are the most relevant input variable for PM_{10} prediction. The R value for ANN-2 and ANN-3 are 0.89 and 0.84 showing input combinations of solar radiation, wind speed, and solar radiation, and the atmospheric pressure can be used to predict daily PM_{10} with acceptable accuracy. It is suggested that PM_{10} should be control as soon as it is possible in Varanasi by introducing CNG and electric vehicle.

Acknowledgements The authors would like to acknowledge Department of Science and Technology, New Delhi-110016 India for providing inspire fellowship with Ref. No. DST/INSPIRE Fellowship/2016/IF160676. We would also like to thanks Central Pollution Control Board, New Delhi India for providing online time series data for this study.

References

1. Baawain, S.M., Serihi, A.S.A.: Systematic approach for the prediction of ground-level air pollution (around an industrial port) using an artificial neural network. Aerosol Air Qual. Res. **14**, 124–134 (2014)
2. Elminir, K.H., Galil, A.H.: Estimation of air pollutant concentration from meteorological parameters using Artificial Neural Network. J. Electr. Eng. **57**(2), 105–110 (2006)
3. Filho, A.S.F., Fernandes, M.F.: Time-series forecasting of pollutant concentration levels using particle swarm optimization and artificial neural network. J. Quim Nova **36**(6), 783–789 (2013)
4. Skyrzypski, J., Szakiel, J.E.: Neural network prediction models as a tool for air quality management in cities. J. Environ. Protect. Eng. **34**, 130–137 (2008)
5. Sahin, A.U., Bayat, C., Ucan, N.O.: Application of cellular neural network (CNN) to the prediction of missing air pollutant data. J. Atmos. Res. **101**, 314–326 (2011)
6. http://www.cpcb.gov.in/CAAQM/frmUserAvgReportCriteria.aspx
7. Yadav, A.K., Malik, H., Chandel, S.S.: Selection of most relevant input parameters using WEKA for artificial neural network based solar radiation prediction models. Renew. Sustain. Energy Rev. **31**, 509–519 (2014)
8. Chow, S.K.H., Lee, E.W.M., Li, D.H.W.: Short-term prediction of photovoltaic energy generation by intelligent approach. Energy Build. **55**, 660–667 (2012)
9. Mishra, D., Goyal, P.: Development of artificial intelligence based NO2 forecasting models at Taj Mahal, Agra. Atmos. Pollut. Res. **6**, 99–106 (2015)

An Efficient Technique for Facial Expression Recognition Using Multistage Hidden Markov Model

Mayur Rahul, Pushpa Mamoria, Narendra Kohli and Rashi Agrawal

Abstract Partition-based feature extraction is widely used in the pattern recognition and computer vision. This method is robust to some changes like occlusion, background, etc. In this paper, partition-based technique is used for feature extraction and extension of HMM is used as a classifier. The new introduced multistage HMM consists of two layers. In which bottom layer represents the atomic expression made by eyes, nose, and lips. Further upper layer represents the combination of these atomic expressions such as smile, fear, etc. Six basic facial expressions are recognized, i.e., anger, disgust, fear, joy, sadness, and surprise. Experimental result shows that proposed system performs better than normal HMM and has the overall accuracy of 85% using JAFFE database.

Keywords JAFFE · Gabor wavelets transform · PCA · Local binary patterns SVM · FAPs · Cohn-Kanade database · Markov process · Static modeling

---

M. Rahul (✉)
AKTU, Lucknow, India
e-mail: mayurrahul209@gmail.com

P. Mamoria
Department of Computer Applications, UIET, CSJMU, Kanpur, India
e-mail: p.mat76@gmail.com

N. Kohli
Department of Computer Science & Engineering, HBTU, Kanpur, India
e-mail: kohli.hbti@gmail.com

R. Agrawal
Department of IT, UIET, CSJMU, Kanpur, India
e-mail: dr.rashiagarwal@gmail.com

K. Ray et al. (eds.), *Soft Computing: Theories and Applications*,
Advances in Intelligent Systems and Computing 742,
https://doi.org/10.1007/978-981-13-0589-4_4

1 Introduction

The facial expression is the most useful medium for nonverbal communication in human daily life. Facial expression recognition can be applied in many interesting areas such as in human–computer interface, telecommunications, behavioural science, video games, animations, psychiatry, automobile safety, affect sensitive music jukeboxes and televisions, educational software, etc. Better classifier and better feature extraction method is very important because it significantly improves the accuracy and speed for identification and classification.

The hidden Markov modelling is useful in spectral and dynamic nature of speech pattern using acoustic state model [1]. The HMM becomes the most useful in speech recognition. Due to the handling of time series data modelling and learning capability, HMM is very useful in classifying unknown feature vector sequence. Only small number of training samples required to train full HMM [2, 3]. The discriminative strength of normal HMM is less suitable to handle more difficult task. Due to dynamic statistical modelling and time sequence pattern matching principle, HMM matches the most similar signal state as recognition results [4, 5]. Hidden Markov model is capable for processing continuous dynamic signal, and can effectively use the timing signal moments before the state transition and after the state transition.

In this paper, partition-based feature extraction method and two-layer extension of HMM is introduced. New introduced HMM consists of two layers. Bottom layer represents the atomic facial expressions such as expressions involving eyes, nose, and lips separately. Upper layer represents the expressions which are basically the combination of atomic expressions such as smile, fear, etc. Our assumption is that every facial expression is the combination of expressions of eyes, nose, and lips. Our newly introduced system is capable of handling almost every situation like occlusion, brightness, background etc.

The rest of the paper are as follows: Basics of HMM is explained in Sect. 2, Related works are discussed in Sect. 3, the proposed system in Sect. 4, and experiments and results are discussed in Sect. 5 and finally concluded in Sect. 6.

2 Basics of HMM

Markov process is used for time series data modelling. This model is applicable for situations where there is a dependability of present state to the previous state (Fig. 1). For example, bowler bowls the ball in cricket game; ball first hits the ground then hits the bat or pad. This situation can be easily modelled by time series data modelling. In other examples, such as in a spoken sentence, the present pronounced word is depending on the previous pronounced word. Markov process effectively handles such situations.

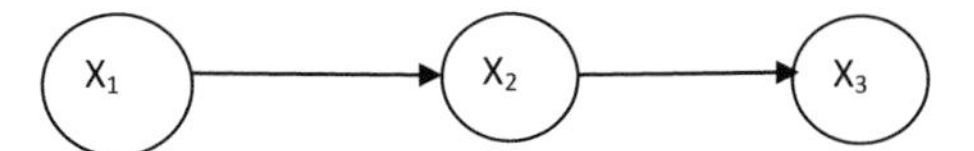

Fig. 1 A Markov model for 3 observations variables

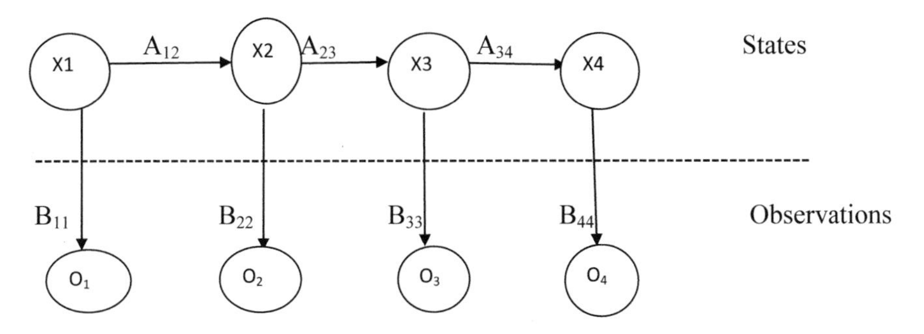

Fig. 2 A hidden Markov model for four variables

When an observation variable is dependent only on previous observation variable in a Markov model is called first-order Markov chain. The joint probability distribution is given by

$$p(X_1, X_2, \ldots, X_N) = p(X_1) \prod p(X_N|X_{N-1}) \tag{1}$$

where $p(X_1, X_2, \ldots, X_N)$ is the joint probability distribution of states $X_1, X_2, \ldots, X_N$ and p(X1) is the probability of state X_1 and $p(X_N|X_{N-1})$ is the probability of state X_N given state X_{N-1}.

A hidden Markov model is a statistical model which is used to model a Markov process with some hidden states. This model is widely used in speech and gesture recognition. Hidden Markov model is a set of observable states are measured and these variables are assumed to depend on the states of a Markov process which are hidden to the observer (Fig. 2). Hidden states are states above the dashed lines.

$$p(X_{t+1} = j|X_t = i) = a_{ij} \tag{2}$$

where $p(X_{t+1} = j|X_t = i) = aij$ is the conditional probability of the state X_{t+1} given the state X_t and a_{ij} is the ith row, jth column entry of the transition matrix $A = (a_{ij})$.

$$p(O_t = k|X_t = l) = b_{lk} \tag{3}$$

where $p(O_t = k|X_t = l) = b_{lk}$ is the conditional probability of the observation variable O_t given the state X_t and b_{lk} is the lth row, kth column entry of the emission matrix $B = (b_{lk})$.

An HMM is defined by A, B, and $\prod$. The HMM is denoted by $\partial = (A, B, \prod)$.

3 Related Works

The facial expression and physiognomy was first started in the early fourth century. The physiognomy is the study of the person's character from their outer appearance [6]. The interest on physiognomy is minimized and the studies on facial expressions become active. The studies of facial expressions have been found in seventeenth century back. John Bulwer in 1649 was given the detail theory about the facial expressions and head movement in his book 'Pathomyotomia'. Le Brun in 1667 gave a detailed note on facial expression, then it was later reproduced as a book in 1734 [7].

In the beginning of nineteenth century, Charles Darwin gives a theory which produced the greatest impact in today's automatic facial expression recognition field. Darwin created a benchmark for various expressions and means of expressions in both animals and humans [8]. In 1970, Ekman and his colleagues done a great job in this field and further their work has great influence for today's new era of facial expression recognition. Today, this field is used in almost all area for example in clinical and social psychologists, medical practitioners, actors, and artists. However in the end of the twentieth century, with the advances in the fields of robotics, computer graphics and computer vision, animators and computer scientists started showing interest in the field of facial expressions. Suwa et al. in 1978 create a framework in automatic facial expression recognition by tracking 20 points to analyse facial expressions from the sequence of images.

Ekman and Freisan introduced a framework in Psychology to recognize facial expression [9]. In 1990, researcher used Ekman's framework to recognize facial expression in video and image [10]. Hidden Markov model was also used to recognized facial expressions [11]. Further, Cohen et al. proposed a multilevel HMM to recognize emotions and further optimize it for better accuracy as compared to emotion-specific HMM [10]. Pardas et al. (2002) used the automatic extraction of MPEG-4 Facial Animation Parameters(FAP) and also proved that the FAPs provide the necessary information required to extract the emotions [12]. FAPs were extracted using an improved active contour algorithm and motion estimation. They used the HMM classifier. They also used the Cohn-Kanade database for recognition. Their recognition rate was 98% for joy, surprise and anger and 95% for joy surprise and sad. The average recognition rate was 84%.

Aleksic et al. (2006) showed their performance improvement using MS-HMM [13]. They also used PCA to reduce dimensionality before giving it to HMM. They used MPEG-4 FAPs, outer lip (group 8) and eyebrow (group 4) followed by PCA to reduce dimensionality. They used HMM and MS-HMM as classifiers. They used Cohn-Kanade databases with 284 recordings of 90 subjects. Their recognition rate using HMM was 88.73% and using MS-HMM was 93.66%. Shang et al. (2009) proposed an effective nonparametric output probability estimation method to increase the accuracy using HMM [14]. They worked on CMI database and get the accuracy of 95.83% as compared to nonparametric HMMs. Jiang (2011) proposed a method based on code-HMM and KNN further applied some discrimination rules [15]. Their proposed method achieves better accuracy to some extent. Suk et al. proposed a real-time temporal video segmenting approach for automatic facial expression recognition [16]. They used SVM as a classifier and get the accuracy of 70.6%.

Wu et al. (2015) incorporated a multi-instance learning problem using CK+ and UNBC-McMaster shoulder pain database [17]. They combine both multi-instance learning and HMM to recognize facial expression and outperforms state of the arts. Sikka et al. (2015) proposed a model based similarity framework and combine SVM and HMM [18]. They worked on CK+ and OULU-CASIA databases and get the accuracy of 93.89%. Ramkumar et al. incorporated the Active Appearance Model(AAM) to identify the face and extracted its features [19]. They used both KNN and HMM to recognize facial expression and get the better results. Senthil et al. proposed a method for recognizing facial expression using HMM [20]. Xufen et al. incorporated some modification in HMM to get better recognition rate [21].

Xiaorong et al. proposed a framework for partially occluded face recognition using HMM and get the better result to some extent [22]. Punitha et al. proposed a real-time facial expression recognition framework and get the better results as compared to existing framework [23]. Pagariya et al. also proposed a system using Multilevel HMM for recognizing facial expression from the video [24]. Islam et al. incorporated both PCA and HMM for appearance and shape based facial expression recognition framework [25]. Singh et al. introduced a 3 state HMM for recognizing faces under various poses [26].

4 Proposed System

Different databases require different feature extraction method and different classifiers. It is not easy to conclude which feature extraction methods and classifiers best suited the situation. HMMs have been used in speech recognition so far. Due to handling of time series data modelling and learning capability, HMM is very useful in classifying unknown feature vector sequence. Only small number of training samples required to train full HMM [2, 3]. The discriminative strength of normal HMM is less suitable to handle more difficult task. In this paper, we introduce a new framework which uses the most powerful partition based feature extraction method and modified HMM as a classifier. New Multistage HMM is the two-layer extension of normal HMM. Bottom layer represents the atomic expressions made by eyes, lips, and nose and upper layer represents the combination of these atomic expressions such as smile, fear, etc. The proposed framework uses the partition based feature extraction method, and then these extracted features are used to train the classifier. Some datasets are used to test the classifier. This framework is robust to occlusion, background, and orientation. In this proposed system, Baum–Welch method is used for parameter estimation. Viterbi method and forward procedure are used for calculating the optimal state sequence and probability of the observed sequence respectively. The proposed framework is shown in Fig. 3.

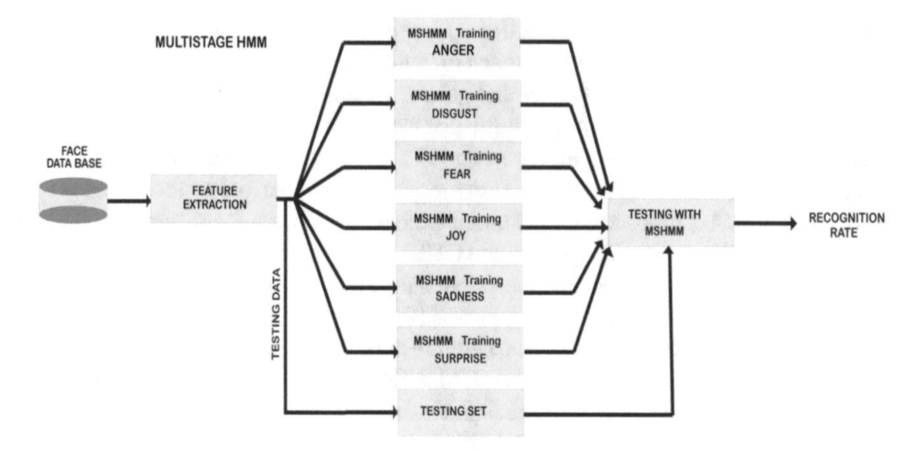

Fig. 3 Block diagram for facial expression recognition using proposed system

5 Experiments and Results

Following conditions have been applied during the experiments:

(1) Faces are analysed from frontal view only.
(2) No movement in the head.
(3) No conversation during image taken.
(4) No glasses during image capturing.

JAFFE database [27–29] has 152 images from 13 subjects. They are 16 of anger, 27 of disgust, 19 of fear, 33 of joy, 30 of sadness, and 27 of surprise. To train the HMM, we follow N fold cross validation rule. For example if there is k fold, then k – 1 fold are used to train the HMM and remaining 1 fold is used to test HMM. For example in the set of nine images of anger. These nine images are divided into ninefolds. We are using eight images to train and 1 for test. Further final result is the mean of all the 9 results.

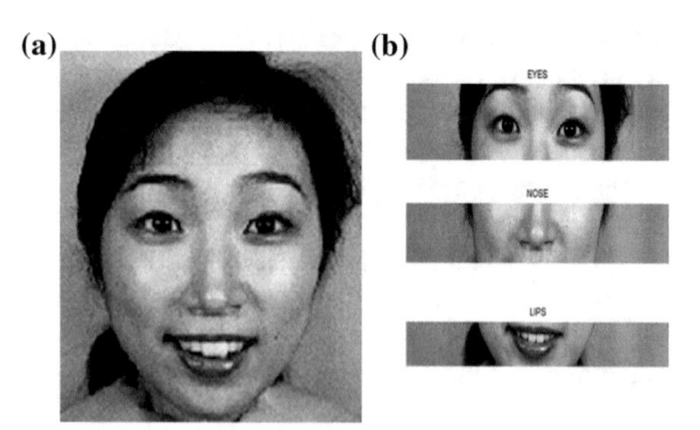

Fig. 4 **a** Original image [27–29]. **b** Partitioned image

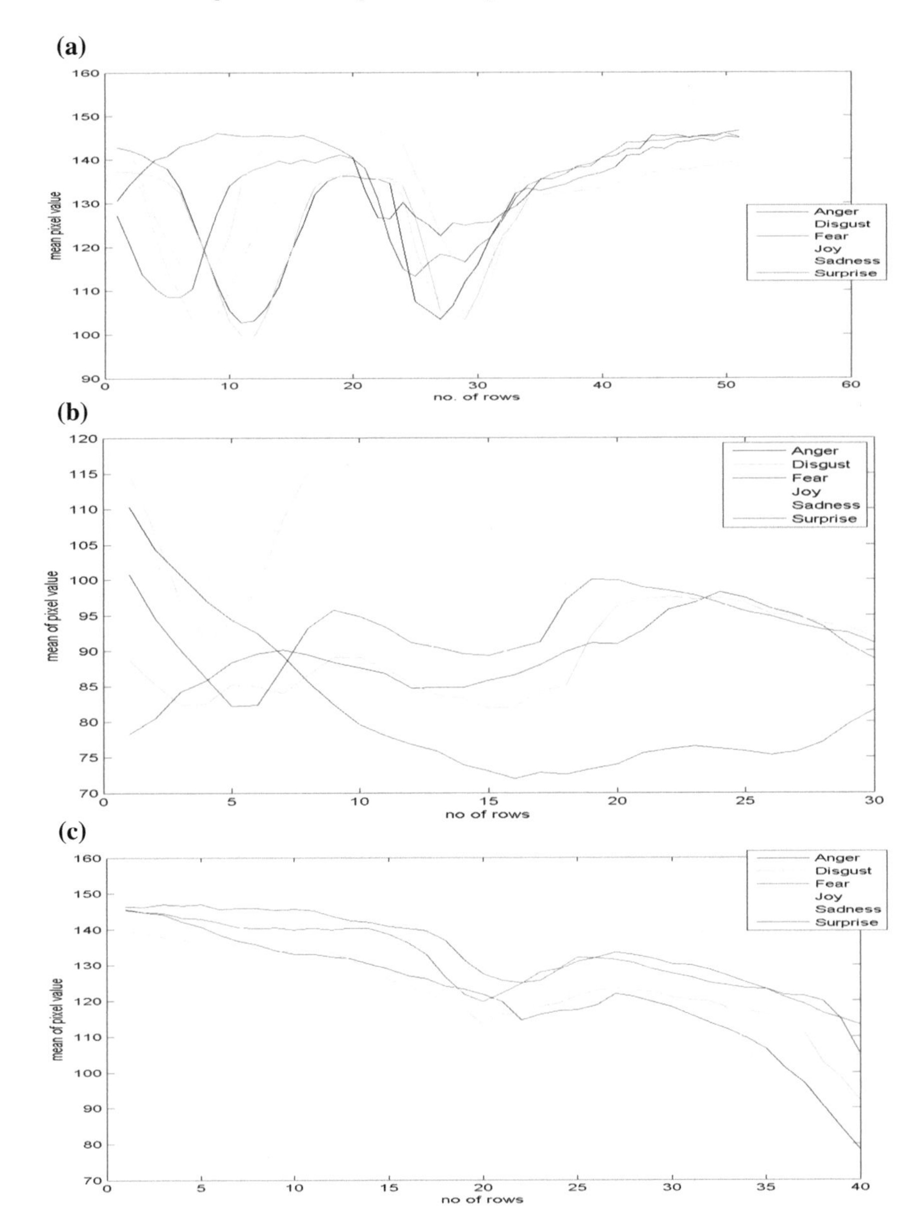

Fig. 5 **a** Plot of eyes partition. **b** Plot of lips partition. **c** Plot of nose partition

Images are taken from the JAFFE database [27–29] of facial expression. Each image is partitioned into three parts (see Fig. 4a, b). Partitioning of image is very effective method as compared to other state-of-the-art feature extraction methods.

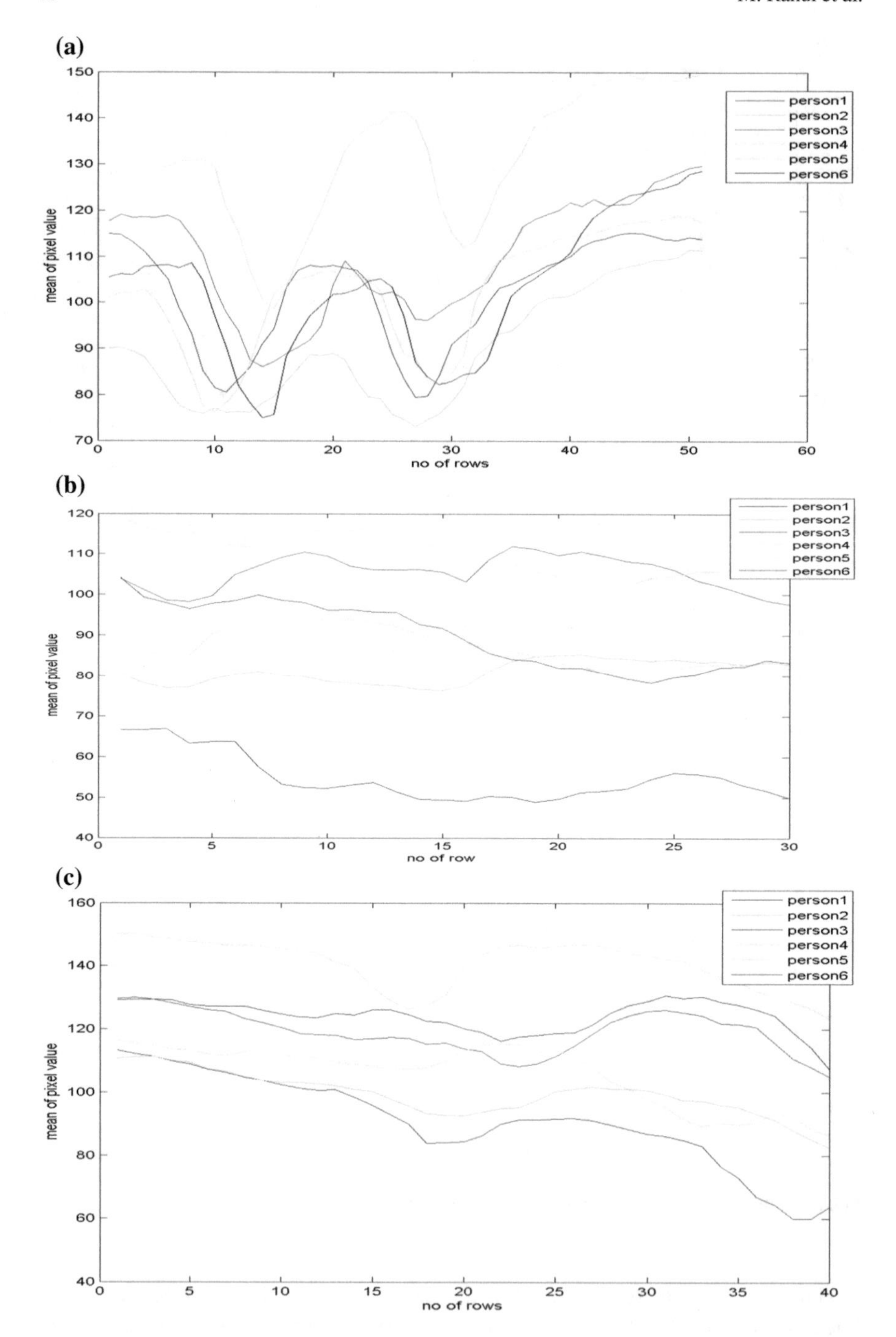

Fig. 6 **a** Plot of eyes partition for "Disgust" expression of different persons. **b** Plot of lips partition for "Disgust" expression of different persons. **c** Plot of nose partition for "Disgust" expression of different persons

Table 1 Confusion matrix for FER using proposed system

	Anger	Disgust	Fear	Joy	Sadness	Surprise	Recognition rate (%age)
Anger	12	0	2	0	0	2	75
Disgust	0	22	0	0	2	3	81
Fear	1	0	12	4	2	0	64
Joy	0	0	0	33	0	0	100
Sadness	6	0	0	0	24	0	80
Surprise	0	0	0	5	0	22	81

These partitions are easily distinguishable to achieve the higher accuracy (see Fig. 5). Plot shows the eyes, nose, and lips of all six expressions of the same person.

Again next plot shows the eyes, nose, and lips partition of same 'disgust' expression for different persons (see Fig. 6).

These plots show the difference between different expressions of the same person and same expressions of the different persons. This partition based technique when incorporated with new introduced HMM gives much better results as compared to other state-of-the-art methods.

The recognition results using partition based feature extraction technique and modified HMM as a classifier is shown in Table 1. From the results, we can see that using the proposed system will improve the recognition rate.

The receiver operating characteristics curve or ROC curve is a graph used to represent the performance of the classifiers (see Fig. 7). It is the plot between True Positive Rate(TPR) versus False Positive Rate(FPR). TPR is the proportion of positive samples identified as positive samples and FPR is the proportion of negative samples

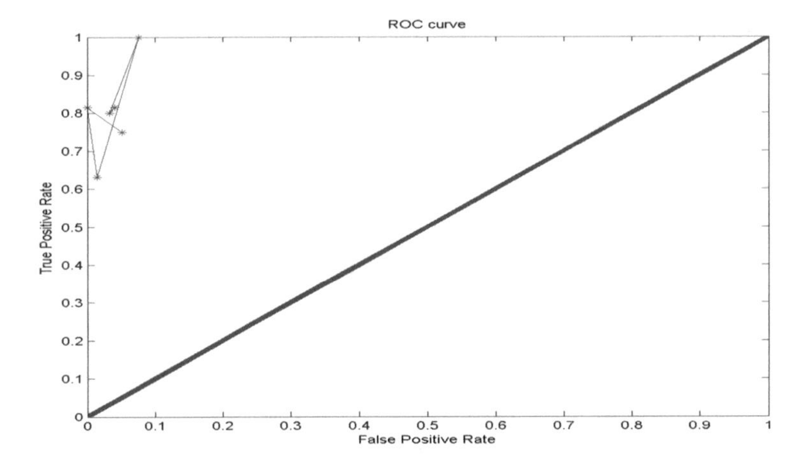

Fig. 7 ROC curve for the given confusion matrix

identified as positive. The 45-degree line is called 'line of no-discrimination'. The ROC curve above this line represents the goodness of the classifier and also indicates the classifying rate of identifying positive samples as positive is greater than its rate of misclassifying negative samples. This given ROC curve indicates the goodness of our classifier.

6 Conclusions

It is not easy to conclude which feature extraction method and classifier is best for this situation, but still modification of HMM significantly improves accuracy for the recognition of six basic facial expressions. This paper introduced a first framework for partition-based method as a feature extraction and modified HMM as a classifier for facial expression recognition. HMM provides some modelling advantages over other feature-based methods in facial expression recognition. Due to generative nature of HMM, it is weak classifier as compared to other discriminative classifiers such as SVM. Our proposed method makes HMM more powerful as a classifier with compared to other discriminative classifiers. We have also shown its strength using confusion matrix and Receiver Operating Characteristics (ROC) curve and found the overall accuracy of 82%.

References

1. Katagiri, S., Lee, C.-H.: A new hybrid algorithm for speech recognition based on HMM segmentation and learning vector quantization. IEEE Trans. Speech Audio Process. **1**(4), 421–430 (1993)
2. Tsapatsoulis, N., Leonidou, M., Kollias, S.: Facial expression recognition using HMM with observation dependent transition matrix. In: 1998 IEEE Second Workshop on Multimedia Signal Processing, pp. 89–95, 7–9 Dec 1998
3. Rabiner, L.R.: A tutorial on hidden Markov models and selected applications in speech recognition. Proc. IEEE **77**(2), 257–286 (1989)
4. Atlas, L., Ostendorf, M., Bernard, G.D.: Hidden Markov models for monitoring, machining tool-wear. In: IEEE International Conference on Acoustics, Speech, and Signal Processing **6**, 3887–3890 (2000)
5. Hatzipantelis, E., Murray, A., Penman, J.: Comparing hidden Markov models with artificial neural network architectures for condition monitoring applications. In: Fourth International Conference on Artificial Neural Networks, vol. 6, pp. 369–374 (1995)
6. Highfield, R., Wiseman, R., Jenkins, R.: How your looks betray your personality, New Scientist, Feb 2009
7. http://www.library.northwestern.edu/spec/hogarth/physiognomics1.html
8. Darwin, C.: In: Darwin, F. (ed.) The Expression of the Emotions in Man and Animals, 2nd edn. J. Murray, London (1904)
9. Ekman, P., Friesen, W.: Facial Action Coding System: A Technique for the Measurement of Facial Movement. Consulting Psychologists Press, Palo Alto (1978)

10. Cohen, I., Sebe, N., Chen, L., Garg, A., Huang, T.S.: Facial expression recognition from video sequences: temporal and static modelling. In: Computer Vision and Image Understanding, pp. 160–187 (2003)
11. Lien, J.J.-J., Kanade, T., Cohn, J.F., Li, C.-C.: Detection, tracking, and classification of action units in facial expression. Robot. Auton. Syst. **31**(3), 131–146 (2000)
12. Pardas, M., Bonafonte, A.: Facial animation parameters extraction and expression recognition using Hidden Markov Models. Signal Process.: Image Commun. **17**, 675–688 (2002)
13. Aleksic, P.S., Katsaggelos, A.K.: Automatic facial expression recognition using facial animation parameters and multistream HMMs. IEEE Trans. Inf. Forensics Secur. **1**(1), 3–11 (2006)
14. Shang, L., Chan, K.P.: Nonparametric discriminant HMM and application to facial expression recognition. In: IEEE Conference on CVPR (2009)
15. Jiang, X.: A facial expression recognition model based on HMM. In: International Conference on Electronic and Mechanical Engineering and Information Technology (2011)
16. Suk, M., Prabhakran, B.: Real-time facial expression recognition on smartphones. In: IEEE Winter Conference on Applications of Computer Vision (2015)
17. Wu, C., Wang, S., Ji, Q.: Multi-instance Hidden markov models for facial expression recognition, FG2015
18. Sikka, K., Dhall, A., Bartlett, M.: Exemplar HMM for classification of facial expressions in video. In: IEEE Conference on CVPRW (2015)
19. Ramkumar, G., Lagashanmugam, E.: An effectual facial expression recognition using HMM. In: ICACCCT (2016)
20. Vijailakshmi, M., Senthil, T.: Automatic Human Facial Expression Recognition Using HMM. IEEE (2014)
21. Jiang, X.: A Facial Expression Recognition Model Based on HMM. IEEE (2011)
22. Xiorong, P., Zhiku, Z., Heng, T.: Partially Occluded Face Recognition Using Subspace HMM. IEEE (2014)
23. Punitha, A., Geetha, K.: HMM based peak time facial expression recognition. In: IJETAE, vol. 3, Jan 2013
24. Pagaria, R.R., Bartere, M.M.: Facial emotion recognition in video using HMM. In: IJCER, vol. 3, issue 4
25. Islam, M.R., Toufiq, R., Rahman, M.F.: Appearance and Shape Based Facial System Using PCA and HMM. IEEE (2014)
26. Singh, K.R., Zaveri, M.A., Raghuwanshi, M.M.: Recognizing Face Under Varying Poses with Three States HMM. IEEE (2014)
27. Lyons, M.J., Akamatsu, S., Kamachi, M., Gyoba, J.: Coding facial expressions with gabor wavelets. In: Proceedings, Third IEEE International Conference on Automatic Face and Gesture Recognition, 14–16 Apr 1998, Nara Japan. IEEE Computer Society, pp. 200–205
28. Lyons, M.J., Budynek, J., Akamatsu, S.: Automatic classification of single facial images. IEEE Trans. Pattern Anal. Mach. Intell. **21**(12), 1357–1362 (1999)
29. Dailey, M.N., Joyce, C., Lyons, M.J., Kamachi, M., Ishi, H., Gyoba, J., Cottrell, G.W.: Evidence and a computational explanation of cultural differences in facial expression recognition. Emotion **10**(6), 874–893 (2010)
30. Rajpurohit, J., Sharma, T.K., Abraham, A., Vaishali: Glossary of metaheuristic algorithms. Int. J. Comput. Inf. Syst. Ind. Manag. Appl. **9**, 181–205 (2017)

Study and Analysis of Back-Propagation Approach in Artificial Neural Network Using HOG Descriptor for Real-Time Object Classification

Vaibhav Gupta, Sunita and J. P. Singh

Abstract The proposed work summarizes approach of using histogram of gradients as descriptor which are taken as training features for the neural network. This paper describes object classification using artificial neural network with back-propagation as a Feed-Forward network. HOG features were extracted from the images to pass on to this feed-forward network and this neural network has been used to classify different categories of objects based on the features extracted and trained. The converging condition is determined and analyzed for the designed approach. The experimental neural network comprises of 64 neurons in the input layer and 16 neurons in the hidden layer and the output layer has 4 neurons, which is the number of classes. The accuracy for training as well as testing will be discussed and provided in a tabular form up to 3500 epochs. All experimental results are shown in form of graphs and tables.

Keywords Neural networks · Artificial intelligence · Object recognition · HOG (histogram of gradients) · Stochastic gradient descent · Backpropagation

1 Introduction

Human brain is believed to be the most powerful processor, which perceives stimulus from the sensing environment and perform trillions of calculations in just few seconds. Think of a system that can mimic human brain. One cannot imagine how powerful will be the developed system. Here, Neural Network is the answer which can truly automate the task as human beings performs. Neural Network is a biological inspired network, which works on the principle, the way human brain take decisions. The beauty of the neural network is that it has adaptiveness to learn, same as we learn from our experiences. Object detection is one of the most active research and application areas of neural networks. A part of detection is classification of segmented parts

V. Gupta (✉) · Sunita · J. P. Singh
Instruments Research & Development Establishment, DRDO, Ministry of Defence, Dehradun 248008, Uttarakhand, India
e-mail: vaibhav.drdo@gmail.com

K. Ray et al. (eds.), *Soft Computing: Theories and Applications*, Advances in Intelligent Systems and Computing 742,
https://doi.org/10.1007/978-981-13-0589-4_5

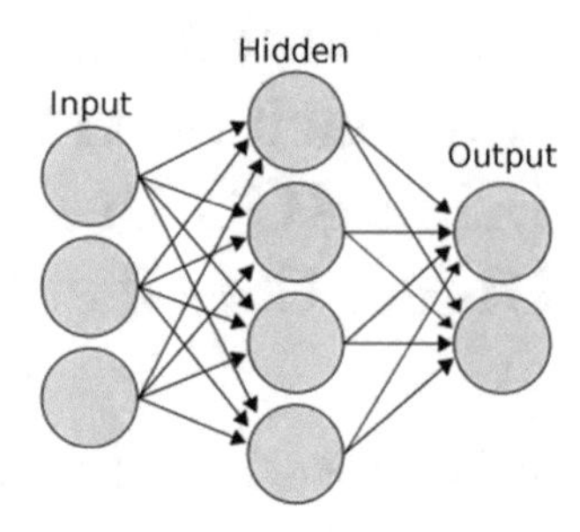

Fig. 1 A typical ANN with one hidden layer

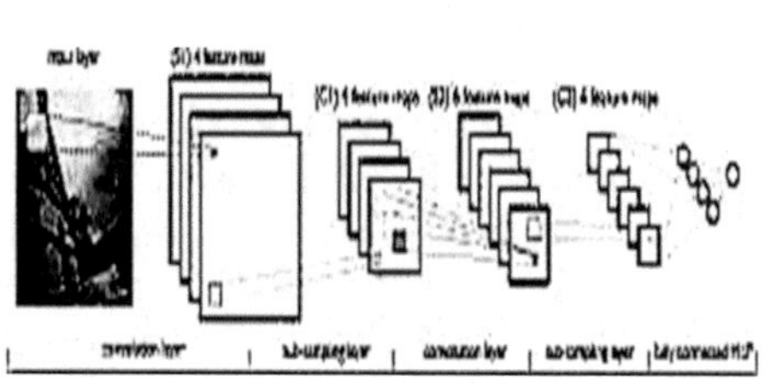

Fig. 2 LeCun's convolution network architecture

of the image that are likely to contain objects. A classifier such as an artificial neural network can help to recognize objects once the candidates have been localized using crowd sourcing, segmentation, etc., just to name a few. In this paper training features are extracted using histogram of gradients from the training samples of one or more categories to build up the neural classifier [10]. Once, the classifier is ready the test data is passed to the feed-forward network which uses back-propagation approach and classification is obtained [2] (Fig. 1).

1.1 Neural Networks

Neural Network has come up as a vital tool for classification. Past work has covered various ways to detect objects. From Marvin Minsky creating neural networks using vacuum tubes and pioneering the beginning of a new way to look at artificial intelligence. Recent work includes recursive neural networks which are control systems with active feedback from the output layers (Fig. 2).

Other architectures such as deep neural networks and convolution neural networks, also known as CNNs perform multi-level convolution with learned image processing in multi-dimensional data. This results in high level abstracted features and translational invariance. These high-level features also ensure that high dimensional data is compressed and prevent bad performance in the networks due to the curse of dimensionality.

The research activities in the recent past in neural classification have confirmed that neural networks [2] are trustworthy alternative to various classification methods which are conventionally used. In theoretical aspects the main advantage of the neural networks is that these networks are data driven with self-adaptive methods i.e. the data can be adjusted without specifying any functional or distributional form

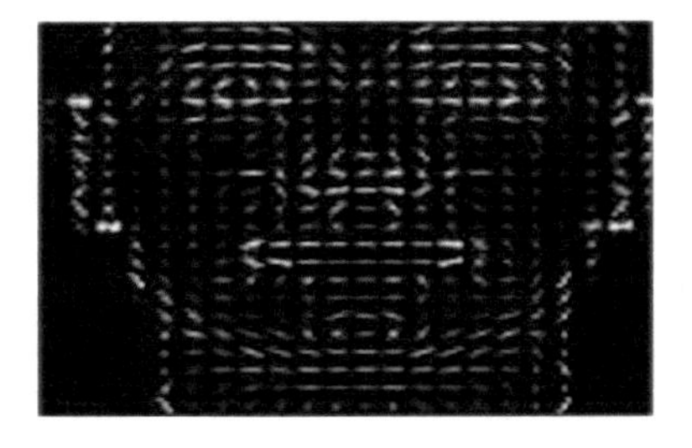

Fig. 3 Gradients of a human face

explicitly. The effectiveness of neural network is that the system learns by observation and experience. The application of neural network is in variety of real world task such as in industry, science and business. To name a view neural network can be applied in the field of speech recognition, fault detection, product inspection, handwriting recognition, medical diagnosis and etc. There are many comparisons given in the studies between neural classifier and conventional classifier on the basis of performance and accuracy. Several experiments have also been done to evaluate neural networks on basis of test conditions. A self-comparative study of result has also been put forward for the support of presented algorithm in this paper (Table 2).

Artificial neural networks (ANN) are studied under machine learning and cognitive science which are inspired by biological neural networks. The biological neural network comprises brain which functions on the basis of large number of inputs for estimation and approximation, hence the artificial neural network behaves the same. In ANN there are numerous neurons interconnected with each other through which the messages are exchanged between each other. These connections are associated with numeric weights, which can be adjusted on basis of experience, thus making it adaptive for learning.

The pixel of an input image in case of handwriting recognition behaves as a neuron in neural network. This neuron is associated with a weight which is activated and transformed by a function. Similarly many neurons are activated and passed on to other neurons and the process continues till the final output neuron is activated and the character is read [7, 9].

2 Histogram of Gradient

The neural network requires a set of inputs to apply weights and non-linear functions on to produce the expected classification. Instead of providing the raw pixel intensity values of the image as a row vector, a HOG [3] transform is performed to get a better set of features. This also reduces the number of input neurons in the initial layer [4] (Fig. 3).

The resized image is divided into 4×4 pixel grids. The various orientations of the pixels are obtained by calculating dx and dy values among the adjacent pixels.

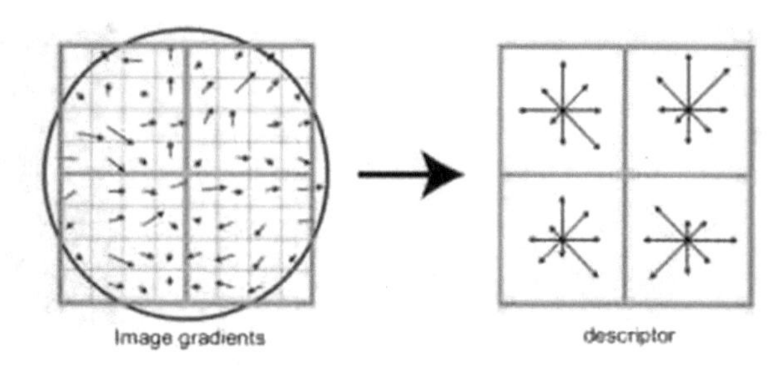

Fig. 4 HOG on 4 × 4 grids and select the most frequent orientation

These are first order derivatives through which the angle is obtained. The angles are binned into 8 directions [4].

The magnitude m(x, y) and angle are governed by the following equations:

$$m(x, y) = \sqrt{[L(x+1, y) - L(x-1, y)]^2 + [L(x, y+1) - L(x, y-1)]^2}$$

$$\theta(x, y) = \tan^{-1}\{[L(x+1, y) - L(x-1, y)]/[L(x+1, y) - L(x-1, y)]\}$$

We choose the most frequent orientation in a 4 × 4 grid and make it a feature of that particular grid. This is done for all 4 × 4 sub grids in the image (Fig. 4).

3 Network Architecture

A neural network's important design component is the architecture. This primarily includes the number of hidden layers and the number of neurons per layer. Larger, more complicated architectures performs better when there is a large dataset to train with the availability of powerful GPU and parallelizable code.

$$J(\theta) = \frac{1}{m}\sum_{i=1}^{m}\left(h_\theta\left(x^{(i)}\right) - y^{(i)}\right)^2$$

Error function [1].

After trying out various architectures, quick training was obtained with better results and convergence on a single hidden layer network. The reason behind the performance was less complicated function ensuring less chance of getting stuck in local minima, which is highly likely in more complicated neural networks [12].

The network had 64 input neurons each of which took the orientation calculated for a 4 × 4 grid on the image. The next layer which is the hidden layer had 16 neurons. The final layer had 4 neurons which is the number of classes that the network has to classify [3, 7].

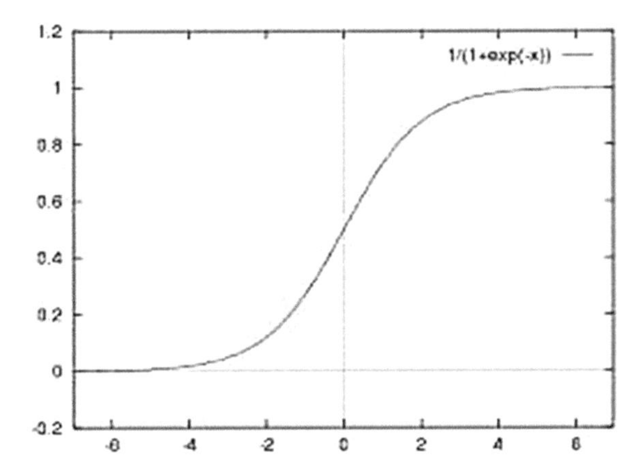

Fig. 5 Sigmoid activation function

4 Training

The implementation was done using NumPy and Python. The code is in vectorized form which makes it modular and more efficient than the normal loop structure.

Back propagation with stochastic gradient descent algorithm was used to train the neural network. The following weight update equations were utilized to reach convergence [8].

$$\frac{\delta E_p}{\delta w_{ij_x}} = \frac{\delta E_p}{\delta a_{jz}} \frac{\delta a_{jz}}{\delta w_{ij_x}}$$

$$\Delta w_{ij_x} = -\varepsilon \frac{\delta E_p}{\delta w_{ij}} = \varepsilon \delta a_{ix}$$

where $\delta j_p = a_{j_p}(1 - a_{j_p})(t_{j_p} - a_{j_p})$, if the output node.

The activation function used for firing the receptors was the sigmoid function (Fig. 5)

$$S(t) = 1/1 + e^{-t}$$

5 Results

The training was done with nearly 3000 images on Intel i7 3.60 GHz 64bit Processor with 8 GB RAM and the classifier obtained was stored. There were 4 classes with cars, airplanes, motorbikes and face images in the dataset respectively. The images were obtained from a subset of Caltech 101 dataset which has 100 categories of objects and UIUC car dataset.

The data was shuffled randomly and divided into a test and training set in 1:9 ratio. The test data was simulated on raspberry pi3 board (Quad Core 1.2 GHz Broadcom BCM2837 64 bit CPU having 1 GB RAM) with the generated classifier. After each epoch the test error was calculated. The learning rate is taken as 0.5. The number of receptors in the hidden layer were changed to see the performance of the network.

The following graph plots the error versus iteration (Fig. 6).

The blue lines indicate the error in classification of the training data, whereas the green curve represents the intermediate error of the neural net on the test data (which was not the part of training set). The test data is converged to 97% accuracy in less than 3500 epochs [14] (Tables 1 and 2).

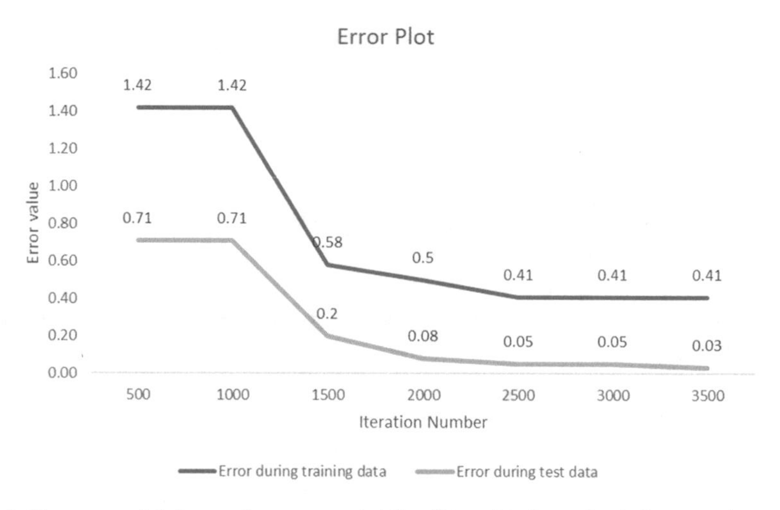

Fig. 6 The average label error changes over training & test data loops for 4-class experiment

Table 1 Error/Iteration

Iterations	Error during training data	Error during test data
500	1.42	0.71
1000	1.42	0.71
1500	0.58	0.20
2000	0.50	0.08
2500	0.41	0.05
3000	0.41	0.05
3500	0.41	0.03

Table 2 Comparison with state of art techniques

Classification technique	No. of input layer neurons	No. of hidden layer neurons	Recognition rate	Epoch	Classifying pattern	Samples training images
By Zheng et al. [13]	768 (32 × 24)	384	92.11%	1000	3	150
By Jeyanthi Suresh et al. [1]	750	6	89.8%	–	6	600
By Rajvir Kaur et al. [11]	81	20	95%	–	9	360
by Vardhan et al. [6]	900	10	90% (Probabilistic Neural Network)	–	–	900
			93.82% (Radial Basis Probabilistic Neural Networks)			
Classification technique by Badami et al. [5]	–	–	53.1% (MIT Flickr Material Database)	–	–	4752
Proposed technique	64	16	97%	3500	4	3000

6 Conclusion

This proposed approach achieved more accuracy on testing data while training set data is relatively less accurate. The approach achieved good results with just 64 input neurons and 16 neurons in the hidden layer. The complexity of the system is highly reduced in real-time environment, since the system is converging to 97% accuracy within 3500 epochs which is highly acceptable [9]. The future scope is to further optimize the number of epochs with the accuracy.

References

1. Jeyanthi Suresh, A., Asha, P.: Human action recognition in video using histogram of oriented gradient (HOG) features and probabilistic neural network (PNN). Int. J. Innov. Res. Comput. Commun. Eng. **4**(7) (2016)
2. Bishop, C.: Neural Networks and Pattern Recognition. Wiley
3. Dalal, N., Briggs, B.: Histogram of Gradients
4. Fard, H.M., Khanmohammadi, S., Ghaemi, S., Samadi, F.: Human age-group estimation based on ANFIS using the HOG and LBP features. Electr. Electron. Eng.: Int. J. (ELELIJ) **2**(1) (2013)
5. Badami, I.: Material recognition: Bayesian inference or SVMs?. In: Proceedings of CESCG 2012: The 16th Central European Seminar on Computer Graphics, Apr 2012

6. Vardhan, J.V., Kumar, U.: Plant recognition using hog and artificial neural network. Int. J. Recent Innov. Trends Comput. Commun. **5**(5), 746–750 (2017). ISSN: 2321-8169
7. Neural Networks for Classification: A Survey, Peter Zhang
8. Devijver, P.A., Kittler, J.: Pattern Recognition: A Statistical Approach. Prentice-Hall, Englewood Cliffs, NJ (1982)
9. Duda, P.O., Hart, P.E.: Pattern Classification and Scene Analysis. Wiley, New York (1973)
10. Pradeebha, R., Karpagavalli, S.: Prediction of lung disease using HOG features and machine learning algorithms. Int. J. Innov. Res. Comput. **4**(1) (2016)
11. Kaur, R., Singh, S., Sohanpal, H.: Human activity recognition based on Ann using Hog features. In: Proceedings of the International Conference on Advances in Engineering and Technology—ICAET-2014 Copyright© Institute of Research Engineers and Doctors. All rights reserved. ISBN: 978-1-63248-028-6. https://doi.org/10.15224/978-1-63248-028-6-02-35
12. Tsai, W.-Y., Choi, J., Parija, T., Gomatam, P., Das, C., Sampson, J., Narayanan, V.: Co-training of feature extraction and classification using partitioned convolutional neural networks. In: DAC' 17 Proceedings of the 54th Annual Design Automation Conference 2017, Article No. 58, Austin, TX, USA, 18–22 June 2017. ISBN: 978-1-4503-4927-7
13. Zheng, Y., Meng, Y., Jin, Y.: Object recognition using neural networks with bottom-up and top-down pathways. Neurocomputing **74**, 3158–3169 (2011)
14. Zheng, Y., Meng, Y., Jin, Y.: Fusing bottom-up and top-down pathways in neural networks for visual object recognition. In: Proceedings of the International Joint Conference on Neural Networks, pp. 1–8 (2010). https://doi.org/10.1109/ijcnn.2010.5596497

Study and Analysis of Different Kinds of Data Samples of Vehicles for Classification by Bag of Feature Technique

Anita Singh, Rajesh Kumar and R. P. Tripathi

Abstract This paper presents a new method and data analysis for reducing false alarm of vehicle targets in real-time classification [1] by introducing two new parameters, i.e., average confidence and decision parameter. The main challenge is to do tracking and classification by using infrared sensor. The greater the number of data set provided, greater is the accuracy of classification. The confidence factor in this algorithm is defined as the percentage of target occurrences in the past 25 frames and varies linearly with number of detections. An optimization of the confidence factor could be an area of further work in the proposed algorithm.

Keywords Vehicle detection · Classification · Background subtraction
Surveillance · Feature extraction

1 Introduction

The Bag of Features (BOF) approach is similar to that of object characterization, where it uses features of images to extract the information. The bag of features uses image features as the key parameter to detect an image. It analyzes the image and concludes in the form of collection of features [2]. BOF is very popular in robotic vision tasks such as image classification, image retrieval, video searches, and texture recognition. Due to the formation of visual vocabulary the simplicity, efficiency and performance of BOF technique are enhanced. Defining the visual vocabulary tends to improve the computational efficiency as well as the accuracy for classification in object recognition.

A. Singh (✉)
Instruments Research & Development Establishment, DRDO, Ministry of Defence,
Dehradun 248008, Uttarakhand, India
e-mail: anitashakarwal@yahoo.com; anitashakarwal@gmail.com

A. Singh · R. Kumar · R. P. Tripathi
Graphic Era University, Dehradun, Uttarakhand, India

K. Ray et al. (eds.), *Soft Computing: Theories and Applications*,
Advances in Intelligent Systems and Computing 742,
https://doi.org/10.1007/978-981-13-0589-4_6

Most of the current research is focused on classification of the right target based on parametric values. The vehicle target classification systems based on different kind of techniques include cascade classifier, SVM classifier, SHIFT [3], Convolution Neural Networks (CNN), etc. The efficiency of target surveillance systems is totally depending on efficiency of image processing techniques. The surveillance system acquires real-time videos to track the moving objects, extracts trajectories, and also finds target intensity or estimates target velocity, etc. The developed system has been used for the real-time surveillance capability enhancement. This kind of developed surveillance system is not only applicable in military applications but also in civilian applications. To achieve big applications through surveillance, the radar-based system is also admired. The proposed algorithm cost is low as compared to radar-based system. However, it requires periodical maintenance and used to acquire surface target.

The proposed algorithm is based on deep learning. The efficiency will increase by introducing two new parameters to BOF technique, i.e., average confidence and decision parameter. The BOF uses the key point detection, descriptor extraction and matching to detect objects. This kind of classifier uses multiple algorithms to do each step like SIFT, SURF, SVM etc.

There are many other techniques developed for moving target detection. Initially, this technique extracts the background from video and then processes it by applying filters. After that, the video is again processed to get the features of the target. In foreground finally, it is able to detect a number of parametric features like area, major axis, minor axis and aspect ratio, etc. These parameters classify the target according to the training set.

2 Problem Description

The proposed method is used for reducing the false alarm rate of detection of the target tracked by infrared sensor. The developed algorithm determines a confidence factor, defined as the percentage of target occurrences in the past selected frames. The optimum value of confidence factor is to analyze the false detection rate. Another innovative factor, decision parameter is designed and analyzed to determine its limits for true target category. Some other features like mean and variance are also analyzed to conclude the accuracy.

3 Algorithm Design and Implementation

The real-time videos processed are taken as an input in BOF on MATLAB2015 platform. The BOF classifier in MATLAB-2015 platform is used to get various parameters of target for classification. The real-time videos captured by tracking surveillance system are continuously processed through the developed algorithm. The simulator is developed on the basis of some pre-assumptions:

- Place surveillance system simulator at a fixed position.
- Place the desired target object for the simulator to find in the controlled area and ensure that the object is in the field of view.
- Initiate autonomous exploration of position and environment on the simulator.
- Set a background upgradation timer for 5 s.
- To classify objects trained for a particular range and particular vehicle.

The development of parameters by BOF [4] algorithm covers all the theoretical aspects of image processing required for understanding the steps followed in video processing. Starting with background subtraction, it covers the basics of morphological operations and blob detection. The input video is processed by the following algorithms step wise step to achieve classification:

(a) ***Preprocessing***: This algorithm works on grayscale images. Hence the first step is to convert RGB images into grayscale called preprocessing.
(b) ***Background modeling***: The background extraction is called modeling of background. The background extraction is done by means of few numbers of starting frames of the video.
(c) ***Object Segmentation***: The object segmentation is carried out by implementing background subtraction. The estimated background obtained from above step is subtracted from each frame of the video and the resultant image is converted into a binary image in which the object appears.

4 Detailed Description of Model and Implementation

Bag of Features described has two main steps:

(a) ***Bag of Features***: Bag of features is an important concept to classify [5] the content of an image object. Bag of Features (BOF) uses the privilege of defined visual descriptors for visual data classification. Bag of features is mainly inspired by Bag of words model that is used for character classification in text for retrieval and text analysis. A bag of words is a vector of counts of occurrences of words in histogram over the text. Similarly, in computer vision a bag of features is a vector of occurrence counts of image features.
The video is recorded by surveillance camera and preprocessing of the video is converted into the frames by image processing techniques. The various parameters such as number of frames, frame rate, color format and frame size are determined with the BOF technique.
(b) ***Bag of Features descriptor***:
It describes the set of bags of features. This actually describes some particular features as a bag, required for creating the BOF descriptor. After that, a set of given features converted into set of bags that were created in first step and

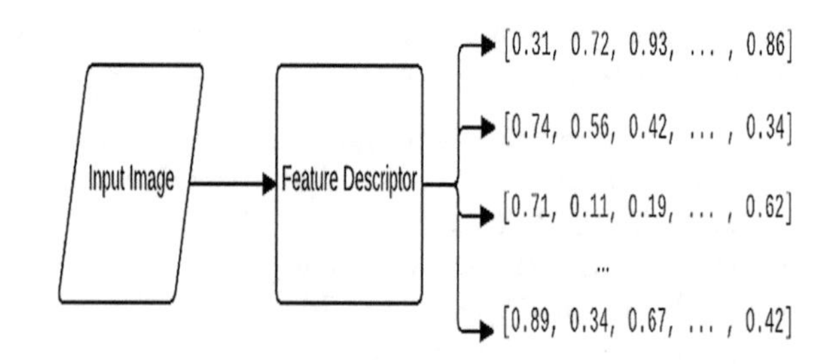

Fig. 1 Multiple feature vectors extracted per image

then plot histogram taking the bags as the bins (consecutive, nonoverlapping intervals of variables). Now, this histogram can be used to classify the image or video frame.

The above two steps can be broken down into further four steps process:

Step1: ***Feature extraction***: The detection of key points is done by extracting SIFT features from salient regions of the images, applying a grid at regularly spaced intervals (called dense key point detector) and further extracting another form of local invariant descriptor like extract mean RGB values from random locations in the image (Fig. 1).

Step2: ***Code book construction***: The extracted feature vectors out of the images produce training data set that we need to construct our vocabulary of possible features.

Step3: ***Vocabulary construction***: It is mainly accomplished via the k-means clustering algorithm, where we cluster the feature vectors obtained from the above step. The resulting cluster centers (i.e., centroids) are treated as our dictionary of features.

Step4: ***Vector quantization***: A random image can quantify and categorize as a class of image using our bag of features model by applying the following process: Extract feature vectors from the image and compute its nearest neighbor defined created dictionary. This is accomplished by using the Euclidean Distance. The set of nearest neighbor labels build a histogram of length k (the number of clusters generated from k-means).

5 Best Features of Algorithm

- The best features of the algorithms are
- The algorithm works effectively on both digital videos and thermal videos.
- The algorithm has given satisfactory results during both day and night.
- Due to regular updating of background, the algorithm is adaptive to slowly varying lightening conditions.
- The confidence factor used in the above algorithm reduces the probability of false detections, as it ideally takes 9 frames (approx. 0.3 s) to confirm its detection.

6 Limitations of Algorithm

- The algorithm needs a static background to work properly. It fails to work on a moving background video.
- The object to be classified must be at a specified distance range of the camera for which it has been trained. The area to perimeter ratio is not the best criteria for separation of trucks from tanks.
- The database of the algorithm is limited to the moment with only about 50 images of tanks. A database of about 300–1000 images would magnify the efficiency of the detector in a great amount.
- The confidence factor used here is a simple linear variable. A better measure of confidence would decrease the errors made by the algorithm.

7 Results Evaluation

The generated experimental data has been used for training as well as testing of the algorithm. The developed algorithm creates a visual vocabulary of 500 words after extracting the Speeded Up Robust Features (SURF) from the training images. The developed visual vocabulary helps in defining feature vector of each image. This vector can be represented as histogram [6] as shown in Fig. 2.

Number of features: 607,130
Number of clusters (K): 500

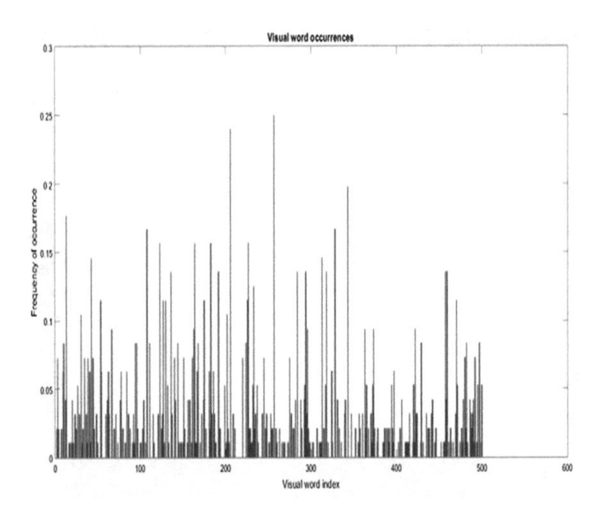

Fig. 2 Feature histogram of a tank image prepared using the 500-word visual vocabulary

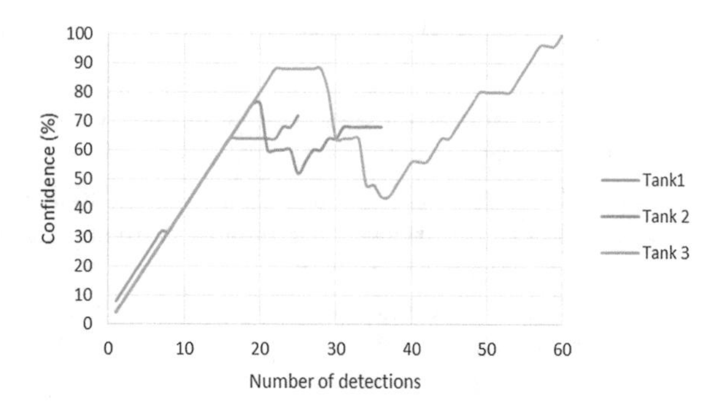

Fig. 3 Confidence variation with number of detections for videos of Targeted vehicle

To get an estimate of the efficiency of the classifier, confusion matrix is created. It is calculated by evaluating the classifier for the training/test images. Another parameter defined as trace of the matrix gives the statistical average accuracy of the category classifier. The confusion matrix for the test set provided is defined as (Tables 2, 3 and 4).

Average Accuracy = Trace (Confusion matrix) = 90%

From the above table, it is observed that the decision parameter ratio of targeted vehicle will mostly lie between 7 and 10. It is a deciding factor for removal of false alarms.

Thus, the area to perimeter ratio of a target varies from 10 to 21, so there is some overlap in detection. To avoid confusion due to false detections, a confidence factor is used. The confidence in observation is defined here as the percentage of times a target has been detected in the past 25 frames (<5 s). The detection is displayed only if the prediction of target is continuous and a confidence of more than 35% is achieved, i.e., target is detected in more than 8 out of the previous 25 frames. Table 5 shows the average confidence of detection of vehicles predecided as 'Target' and 'Nontarget'.

Thus, the average confidence of a correct detection is higher than that of an incorrect detection, and the code predicts with much higher confidence if the object in the video input is actually a tank. The following curves show the variation of confidence level with the number of detections for the above cases (Figs. 3 and 4).

Thus, the detections of truck are weak and the confidence level is low. Hence the confidence factor is important in removing false detections. Using the data in the above videos, a confidence factor of 35% is optimum for removing false detections (Figs. 5, 6 and 7).

8 Conclusion

The algorithm is based on deep learning [7], efficiency will increase, if more data is provided to the Bag of Features. The greater the number of images of the target in the data set, greater will be the efficiency of the image category classifier. The

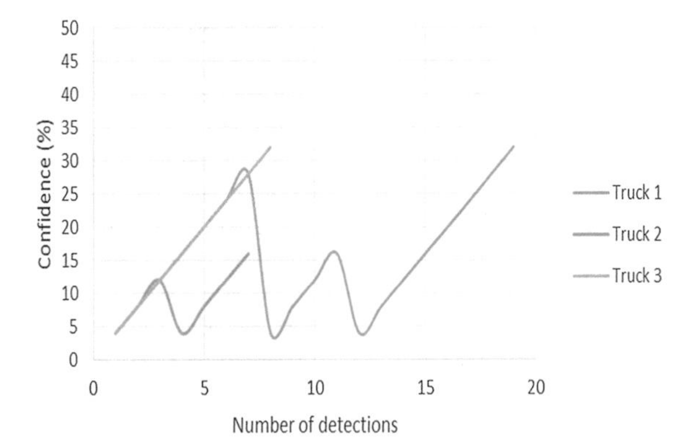

Fig. 4 Confidence variation with number of detections for videos of Nontarget vehicles

Fig. 5 Confidence variation with number of detections for videos of Target

Fig. 6 Confidence variation with number of detections for videos of Target

accuracy of true target detection is 98%, while the accuracy of nontarget detection is 82% as shown in Table 1. The calculated confidence factor in this algorithm is defined as the percentage of target occurrences in the past 25 frames. This confidence factor up to 35% is optimum for removing false detections that varies linearly with number of detections as shown in Table 5. Apart from this, the decision parameter

Fig. 7 Confidence variation with number of detections for videos of Target

Table 1 Calculated average confidence of detection for different Targeted vehicle

Predicted		
Known	Target	Not Target
Target (%)	98	2
Not target (%)	18	82

Table 2 Calculated Mean and Variance for different Targeted vehicle videos

Targeted vehicle	Mean	Variance
Video 1	8.71	1.33
Video 2	8.85	0.84
Video 3	7.49	0.47

Table 3 Calculated mean and variance for different nontargeted vehicle videos

Nontargeted (TRUCK)	Mean	Variance
Video 1	16.11	5.61
Video 2	12.97	3.62

lies between 7 and 10 for true target. This confidence helps in enhancing the accuracy of true target detection, while the calculated confidence factor for nontarget class is less approximate 15%.

Hence, it is concluded that the BOF accuracy for positive class is higher than accuracy for negative class. The inferences that proposed algorithm implemented, performed, and supervised well were for predecided target. For, nontarget category, the accuracy is enhanced more by training data. The optimization of the confidence factor and decision parameter can be an area of further research work in this algorithm.

Table 4 Ratio of area to perimeter for different targeted/nontargeted vehicle videos

Object	Decision parameter = Area/Perimeter Ratio
Target	7–10
Non Target	11–21

Table 5 Average confidence of detection of predecided as 'target' and 'nontarget'

Number of observations	Target confidence (%)	Non target confidence (%)
1	37.4	15.6
2	42.5	9.14
3	63.1	14.7

References

1. Bhanu, B., Jones, T.L.: Image understanding research for automatic target recognition. IEEE Trans. Aerosp. Electron. Syst. **8**(10), 15–23 (1993)
2. Csurka, G., Bray, C., Dance, C., Fan, L.: Visual categorization with bags of keypoints. In: ECCV Statistical Learning in Computer Vision Çalıştayı, pp. 59–74 (2004)
3. Lowe, D.G.: Distinctive image features from scale-invariant keypoints. Int. J. Comp. Vis. **60**, 91–110 (2004)
4. Nowak, E., Jurie, F., Triggs, B.: Sampling strategies for bag-of-features image classification. In: ECCV Statistical Learning in Computer Vision Çalıştayı, pp. 696–709 (2006)
5. Weber, M., Welling, M., Perona, P.: Unsupervised learning of models for recognition. In: ECCV Statistical Learning in Computer Vision Çalıştayı, pp. 18–32 (2000)
6. Dalal, N., Triggs, B.: Histograms of oriented gradients for human detection. In: IEEE Computer Society Conference on Computer Vision and Pattern Recognition, vol. 1, pp. 886–893 (2005)
7. Burden, M.J.J., Bell, M.G.H.: Vehicle classification using stereo vision. In: Proceedings of Sixth International Conference on Image Processing and Its Applications, vol. 2, pp 881–885 (1997)

Books

8. Gonzales, R.C., Woods, R.E.: Digital Image Processing, 2nd edn. Prentice Hall (2002)
9. Tcheslavski, G.V.: Morphological Image Processing: Basic Concepts. Springers (2009)

Hybrid Text Summarization: A Survey

Mahira Kirmani, Nida Manzoor Hakak, Mudasir Mohd and Mohsin Mohd

Abstract Text summarization is the technique of shirking the original text document in such a way that its meaning is not altered. Summarization techniques have become important for information retrieval as large volumes of data are available on Internet and it is impossible for a human to extract relevant information from enormous amount of data in a time-bound situation. Thus, automatic text summarizer is a tool for reducing the information available on Internet by providing nonredundant and salient sentence extracted from a single or multiple text documents. Text summarization has two approaches: extractive and abstractive. Extractive approach generates the summary by selecting subsets of words, sentences, and phrases of text documents whereas abstractive approach understands the main idea of the document and then represents that idea in a natural language using natural language generation technique to create summaries. This paper represents a Survey of Automatic Hybrid Text Summarization.

Keywords Text summarization · Extractive summary · Abstractive · Features
Machine learning

1 Introduction

As an enormous amount of data is available over the Internet, the summarization technique is becoming most popular tool for compressing the volume of data in a timely and efficient manner as it is very difficult for humans to summarize such large text documents manually.

M. Kirmani (✉) · N. Manzoor Hakak
Department of CSE, Maharishi Dayanand University, Rohtak, Haryana, India
e-mail: mahira.kirmani@yahoo.com

M. Mohd
Department of Computer Science, Kashmir University, Srinagar, India

M. Mohd
Department of CSE, Kurukshetra University, Kurukshetra, India

K. Ray et al. (eds.), *Soft Computing: Theories and Applications*,
Advances in Intelligent Systems and Computing 742,
https://doi.org/10.1007/978-981-13-0589-4_7

The aim of text summarizer is to compress large text documents into shorter text which includes importance sentences and retains its original meaning [1]. Researchers have put much effort in automatic summarization techniques and have tried different combinations of both statistical and semantic features along with machine learning approaches for generating summaries [2]. Summarization can be done on a single document or multiple documents. This survey tries to find out some of the most relevant techniques used for both single-document and multiple document summarizations.

Text summarization methods can be classified [3–6] into extractive and abstractive summarization. An abstractive summarization [7] attempts to develop an understanding of the main concepts in a document and then express those concepts in clear natural language. It uses linguistic methods to examine and interpret the text and then finds the new concepts and expressions to best describe it by generating a new shorter text that conveys the most important information [8] of the original text document.

Extractive summaries [7] are formed by extracting stand out sentences from the text documents and then a generic method is used for scoring these sentences [6]. These sentences are scored based on statistical as well semantic features [9]. Statistically, each sentence is analyzed based on features such as sentence length, sentence to sentence cohesion, cue phrases, verb phrases, etc.

A summary can be employed in an indicative way or informative way [10]. Indicative waypoints to some important parts of the original document while informative way covers all the information of text that is relevant to a summary. Summary generated automatically by certain methods and procedures has following advantages

(1) Summary size can be controlled.
(2) Summary contents are deterministic.
(3) The link between a text element in the summary and its position in the original text can be established earlier.

2 Related Work

Automatic text summarization came into existence in 1950s with the work of Luhn [11]. In his work, he suggested that salient sentences [7] can be identified by the frequency of a particular word in that text document. Sentences containing words whose frequency is more in a particular document are considered as important and must be included in a summary. Other features for extracting salient sentences were highlighted by Edmundson and Wyllys [12]. These were the word frequency, count of title or heading words in a sentence, sentence position, and cue phrases in a document. Yong et al. [13] contributed to the work by developing an automatic text summarization system which integrates the learning ability by combining a statistical approach with a neural network and keywords extraction with unsupervised learning. Further, Alguliev et al. [14] have defined automatic summarization as an interdisciplinary research area of computer science that includes artificial Intelligence, data mining, statistics as well as psychology. Goldstein et al. [15] defined MMR Model which is popularly used to reduce redundancy of sentences in a summary. The MMR (Maximal Marginal Relevance) criteria "strive to reduce redundancy while maintaining query relevance in rearranging retrieved documents and in selecting relevant passages for text summarization". This technique gives better result for Multi-Document Summarization.

Mohd et al. [16] proposed another method for automatic text summarization. In this method, the NLP features and machine learning techniques are used. The process of summarization is divided into three stages, preprocessing, assigning ranks, and postprocessing. Preprocessing involves segmentation, synonym removal, initial ambiguity removal, stop words removal, POS tagging, and word stemming [17]. The second step is divided into two stages, first words are ranked according to different features and then the sentences are ranked by using the features like sentence length, summation of TF-DF, existence of noun and verb phrases, and summation of power factor. The last step of this method is postprocessing which includes four steps for the generation of the final summary

(a) *Sentence extraction*
Every sentence is ranked in this step. The sentence with the highest rank is selected to be included in the summary.
(b) *Dealing with connecting words*
Sentences that contain words like however, although, but, etc., signifies that the meaning of these sentences is incomplete without previous sentences.
(c) *Removing additional information*
When sentences contain words like typically, moreover, additionally, etc. it means the author is giving some extra information which is not important for the summary.
(d) *Using WordNet*
WordNet is used for getting the Synsets of a particular word and Lesk algorithm [18] is employed for word sense disambiguation to get synonym of the word to be replaced. By using above two novel structure generators, a final summary is obtained which is partially abstract.

Neto et al. [10] proposed a method for automatic text summarization using a machine learning approach. The method is divided into three steps. Preprocessing, processing, and generation. In preprocessing step, the original documents are represented in a structured way that helps in generating quality summaries. This also reduces the dimensionality of the representation space; it includes stop word elimination, case folding and stemming. Each sentence is represented in terms of vector and the similarity between sentences is determined by cosine similarity measure. The method uses the concept of reference summary that evaluates the performance of automatic summary objectively using classical and recall measures. Second step that is processing step involves two key points of the method, the first being a set of features used for extracting important sentence and the second is to define a framework for the trainable summarizer and classifier. This method also employs two classical algorithms "Naive Bayes" and "C4.5". In the last step, final summary is derived from the summary structure.

Cheng [19] proposed a data-driven approach of text summarization based on neural network [13] and continuous sentence features. It is composed of a hierarchical document encoder and an attention-based extractor modeled by recurrent neural networks [20]. Document Encoder peruses the sequence of sentence in continuous space representation. The decoder generates the target sequence from the continuous space representation of original document. A general frame is built for extracting words and sentences. The proposed method employs a Document Reader, Convolutional Sentence Encoder, Recurrent Document Encoder, Sentence Extractor, and Word Extractor. The purpose of document reader is to provide meaningful representation of document from its sentences. Convolutional Sentence Encoder uses a convolutional neural network that is used to obtain representation vectors at sentence level. Several Features are used to compute a list of features that match the dimensionality of a sentence under each kernel width. All these sentences vectors are summed to get the final sentence representation. Recurrent document encoder employs a recurrent neural network that composes of sequence of sentence vectors into document vectors. The RNN used has an LSTM memory activation unit for overcoming vanish gradient problem when training long sequences. Sentence extractor is another recurrent neural network that performs a sequence labeling task for extraction of meaningful sentences. It is used not only to label the sentence that are important to summary but also extracts mutually redundant sentences. Word extractor, instead of labeling the sentence, directly provides the next word that is relevant for the summary. Two datasets were used for training sentence and word-based summarization models. These datasets were taken from DailyMail News. Models were then evaluated on DUC-2002 single-document summarization task.

Sutskever [21] describes a method that uses machine learning model which is extremely significant as it attempts to model high-level abstractions in data. This method uses a multilayered Long Short-Term Memory (LSTM). One LSTM is used to read the input sequence, one time step, at a time, to get a vector representation and then to use another LSTM to extract the output sequence from that vector. It was found that deep LSTM performance was better than shallow LSTM, so the method uses LSTM with four layers. The second LSTM is essentially a recurrent neural network language model except that it is conditioned on the input sequence. In this method, WMT'14 English to French dataset is used. It was discovered that LSTM learned better when the source sentences were reversed.

Yadav et al. [22] proposed an approach for single text document summarization. The approach is extraction based that depends on combination of the statistical and semantic features. Text summarization procedure in this method consists of three steps. In first step, important sentences are extracted by using sentence score, according to linear combination of different features. Some of the features used are aggregate cosine similarity, sentence position, sentiment of sentence, centroid score [23]; [24], and TF-IDF. For finding sentiment score of sentences, the first sentiment of each entity is determined and then the sum of all these entities give the sentiment score of each sentence. Sentence sentiment in this method is either positive or negative. The second step uses an algorithm for redundancy removal. The sentence with the highest rank is added to summary and then the next sentence, whose similarity is less than predefined threshold and length of summary<L, is included in the summary. Evaluation of the summary in this method is done using MEAD, MICROSOFT, and OPINOSIS (Table 1).

3 Extractive Text Summarization Features

(1) *Sentence Length*: Length of a sentence is considered to be important for making a decision regarding which sentence is to be included in the summary and which not.
(2) *Sentence Position*: Importance of a sentence is also determined by its position as suggested by the Edmundson.
(3) *Cue Phrases*: It is one of the important features that help us in extracting useful information from the text document. Edmundson used this feature in 1968. The sentences that begin with phrases such as "in particular", "significantly", "surely", "the best", "impossible", etc. can provide useful information to us.
(4) *Noun and Verb phrases*: Sentences containing noun and verb phrases are considered important. These sentences are included in summary as they contain valuable information.
(5) *Similarity to Title*: Similarity of document title is computed by querying the title of document against all sentences of a document. Similarity between title and document is measured by cosine similarity. A sentence that contains title words is considered important for the summary.
(6) *Upper Case*: The sentences of the input text containing words starting with the capital letter are considered important for the summary. This feature is not applicable for Hindi text summarization.
(7) *Centroid feature*: Similarity between set of words in a cluster is determined and based on its value importance of sentence is determined.
(8) *Sentence to Sentence cohesion*: This feature allows us to identify sentences that are important for a summary producer. First, the similarity between each sentence "S" and every other sentence "s" of a document is computed and then these similarity values are added up to generate the final value of this feature for the sentences.
(9) *Similarity to Keywords*: Sentences containing keywords are considered important. Similarity between keywords and sentences is obtained by using a query against all sentences of a document [25].
(10) *Sentence to Centroid Cohesion*: This feature is obtained for a sentence as follows: first, we compute the vector representing the centroid of the document, which is the arithmetic average over the corresponding coordinate values of all the sentences of the document; then we compute the similarity between the centroid and each sentence, obtaining the raw value of this feature for each sentence.

Table 1 Comparison of work done by researchers

Researchers	Input	Methods	Datasets	Results
Padma Priya G. et al.	Multiple documents	RBM and Deep Learning Algorithm	Multiple documents from each of the different domains are collected and processed	The maximum recall values marked for the existing approach is 0.72, while for the proposed approach it Comes around 0.62
Jianpeng Cheng et al.	Single document	Data-driven approach based on neural networks and continuous sentence features	Two Datasets created from DailyMail News, Models Extracted on DOC 2002	The NN-SE outperforms the LEAD and LREG baselines with a significant margin, while performing slightly better than the ILP model
Chendra Shekhar Yadav et al.	Single document	Combination of Statistical and Semantic technique	Self Designed (Taken from various newspapers)	Summary length is nearly 27% but high precision, and F-score w.r.t. MEAD, reference summary and high recall w.r.t Microsoft-generated summary
Arlay Barrera et al.	Single Document	POS, Name entity recognition, Text Rank Word Extraction, Sense Learner	Two datasets used DUC 2002 and Scientific Magazine article set	Their system outperforms MEAD and Text Rank sentence extraction in all experiments and is consistently higher than the baseline. Its ROUGE scores are also statistically higher than the baseline for the scientific magazine article set

(continued)

Table 1 (continued)

Researchers	Input	Methods	Datasets	Results
Joel Larocca Neto et al.	Multiple documents	Two machine learning algorithms used: Naïve Bayes and C4.5	TIPSTER Document base Data used	The values of precision and recall for all the methods are significantly higher with the rate of 20% with the compression rate of 10%

(a) Mean TF-ISF
It is a feature that computes the importance of a word in whole document. The importance increases with the number of times a word appears in a sentence (TF) but is balanced by the frequency of the term in the document.

(b) *Power Factor*
Sentences which contain important information are written in Capital, Bold or Italics. This feature helps in extraction of such important sentences.

(c) Biased words
Biased words are those words which are previously defined and may contain substantial information. If a word appearing in a sentence is from biased word list, then the sentence is important and should be included in the summary

(11) *Sentiment features*
Sentiments play a vital role in extracting important sentences from a document. Thus, sentences with emotional content are important to an author and should be included in the summary. Several emotion classes are used such as (positive, negative, fear, joy, surprise, hate, disgust) as seed words which are used to identify emotions in text.

4 Comparison of Different Types of Summarization

See Table 2.

5 Techniques Used in Automatic Summarization

(1) Term Frequency-Inverse Document Frequency (TF-IDF) Method
Term frequency (TF) measures how frequently a term occurs in a document

Table 2 Types of summarization

Types of summarization	Description
Single document	Summary extracted from a single document
Multiple document	Multiple documents are used for extracting a summary
Extractive	Important sentences are determined from the document
Abstractive	Develops main idea of document and expresses that idea in natural language
Indicative	Only some important parts of the text are covered that gives a main idea to the user
Informative	All relevant information of text is covered
Generic	Summary are generated containing main topics of original document
Query based	Generates summaries that contain sentences that are relevant to the given queries
Domain dependent	Knowledge of domain and text structure is needed
Domain independent	Does not require prior knowledge of text structure and can accept any type of text

$$\{[\mathrm{Tf(t)} = (\text{No of times term t appears in a document}) / (\text{total no of terms in a document})]\}$$

Inverse document frequency (IDF) measures how important a term is, while computing
TF, all terms are considered equally important. However, it is necessary to weigh down the frequency of those terms that are not so important for the document.

$$\{\mathrm{IDF(t)} = \mathrm{loge}(\text{total no of documents}) / (\text{no of documents with term t in it})\}$$

The method is used to identify the importance of sentences. Based on the score of TF-IDF, sentences that get highest score are included in summary.

(2) Cluster-Based Method
Normally, if documents are written for different topics, they are divided into sections either implicitly or explicitly to generate a significant summary. This aspect is known as clustering. The overall score of a sentence is calculated as the weighted sum of three factors: similarity of the sentence to the theme of a particular cluster, location of the sentence in the document, and similarity of the sentence to the first sentence in the document to which it belongs.

(3) Machine Learning Method
Machine learning approach takes training dataset as input and uses their extractive summaries for generating the summary of the original text document. The process of summarization is modeled as classification problem. The classification is obtained by application of a trainable machine learning algorithm on the document. Therefore, the summarization task in machine learning method can be envisaged as a "two" class classification problem. If sentence belongs to the extractive reference summary then these sentences are marked as correct and

if not then they are marked as incorrect. Training data is used to learn the pattern (classification) of sentences which is based on classification problem that results in the generation of summaries. Summarization based on the application of trainable machine learning algorithm [2, 23] uses set of features directly extracted by the original text.

(4) Text Summarization with Neural Networks
Using neural networks [13] for text summarization has become a popular tool for generating automatic summaries. Neural networks represent the human brain structure and these artificial structures are trained to learn types of sentences which are important for summary. Neural network consists of multiple layers and single path traversal, from front to back. The goal of the neural network is to solve problems in the same way as the human brain would. Neural networks are based on real numbers, with the value of the core and of the axon typically being a representation between 0.0 and 1. Neural networks are fed with some training data which includes information about which sentence should be included and which should not be included, human experts are employed for this work. The neural networks learn from the pattern of the sentences in this training data.

(5) *Query-based extractive text summarization*
Here, sentences are extracted by using a query based approach. Features like similarity to sentence, similarity to keywords, and similarity to title use the query based extraction are used for identifying important sentences in the text document. In this method, the title or keywords of the document is used as query against all sentences. Sentences which contain query words are ranked and based on this ranking sentences are included in the summary. Cosine similarity measure is used to compute similarity between document title or keyword and sentences.

6 Conclusion

Text summarization has become a main source of interest for researchers from past few years due to gigantic amount of information available on Internet. It is impossible for humans to manually summarize such a large amount of information and get precise and meaningful summaries in less time. Hence, automatic text summarizer is needed to get the job done in less time. This survey paper has presented all the features and methods used by researchers over years for extracting a proper and meaningful summary. Text summarization is classified into three categories Extractive, Abstractive, and Hybrid summarization. All the three categories have been used for generating summaries. Some of the machine learning approaches combined with certain features have also been covered in this paper. Features can be statistical and semantic. In future, we can combine deep learning, artificial neural network along with semantic feature which can provide us quality summaries that will be less redundant, processing time will be less and such summaries will be more human oriented.

References

1. Dalal, V., Malik, L.G.: Proceedings of Emerging Trends in Engineering and Technology (ICETET), 2013 6th International Conference on Emerging Trends in Engineering and Technology
2. Yadav, C.S., Sharan, A.: Hybrid approach for single text document summarization using statistical and sentiment features, Jawaharlal Nehru University, Delhi, India. Int. J. Inf. Retr. Res. (IJIRR) **5**(4), 46–70 (2015)
3. Das, D., Martins, A.F.: A survey on automatic text summarization. Literature Survey for the Language and Statistics II course at CMU, vol. 4, pp. 192–195 (2007)
4. Amini, M.R., Usunier, N., Gallinari, P.: Automatic text summarization based on word-clusters and ranking algorithms. In: Proceedings of the 25th ACM SIGIR, pp. 105–112 (2002)
5. Patil, V., Krishnamoorthy, M., Oke, P., Kiruthika, M.: A statistical approach for document summarization. Department of Computer Engineering Fr. C. Rodrigues Institute of Technology, Vashi, Navi Mumbai, Maharashtra, India. Int. J. Adv. Comput. Technol. (IJACT). ISSN 2319-7900
6. Gupta, V.: A survey of text summarization extractive techniques. J. Emerg. Technol. Web Intell. **2**(3), 258–268 (2010)
7. Ren, F.: Automatic abstracting important sentences. Int. J. Inf. Technol. Decis. Making **4**(1), 141–152 (2005)
8. Alguliev, R.M., Aliguliyev, R.M.: Effective summarization method of text documents. In: Proceedings of IEEE/WIC/ACM International Conference on Web Intelligence (WI'05), pp. 1–8 (2005)
9. Gupta, V., Lehal, G.S.: A survey of text summarization extractive techniques. J. Emerg. Technol. Web Intell. **2**(3) (2010)
10. Neto, J.L., Freitas, A.A., Kaestner, C.A.A.: Automatic text summarization using a machine learning approach. In: Advances in Artificial Intelligence: Lecture Notes in Computer Science, vol. 2507, pp. 205–215. Springer, Berlin, Heidelberg (2002)
11. Luhn, H.P.: The automatic creation of literature abstracts. IBM J. Res. Dev. **2**, 159–165 (1958); Baxendale, P.B.: Machine-made index for technical literature: an experiment. IBM J. Res. Dev. **2**, 354–361 (1958)
12. Edmundson, H., Wyllys, R.: Automatic abstracting and indexing—survey and recommendations. Commun. ACM **4**(5), 226–234 (1961)
13. Yong, S.P., Abidin, A.I.Z., Chen, Y.Y.: A neural based text summarization system. Int. J. Eng. Trends Technol. (IJETT) (2005). ISSN 2231-5381
14. Alguliev, R.M., Aliguliyev, R.M., Hajirahimova, M.S., Mehdiyev, C.A.: MCMR: maximum coverage and minimum redundant text summarization model. Expert Syst. Appl. **38**(12), 14514–14522 (2011). https://doi.org/10.1016/j.eswa.2011.05.033
15. Goldstein, J., Mittal, V., Carbonell, J., Callan, J.: Creating and evaluating multidocument sentence extract summaries. In: Proceedings of the Ninth International Conference on Information and Knowledge Management, pp. 165–172. ACM (2000). https://doi.org/10.1145/354756.354815
16. Mohd, M., Shah, M.B., Bhat, S.A., Kawa, U.B., Khanday, H.A., Wani, A.H., Wani, M.A., Hashmy, R.: Sumdoc a unified approach for automatic text summarization. In: Fifth International Conference on Soft Computing for Problem Solving, SocProS 2015, At Indian Institute of Technology Roorkee, vol. 1. Springer
17. Porter, M.F.: An algorithm for suffix stripping. Program **14**, 130–137 (1980). Reprinted in: Sparck-Jones, K.
18. Banerjee, S.: Adapting the Lesk algorithm for word sense disambiguation to WordNet. In: Proceeding of the Third International Conference on Linguistic and Intelligent Text Processing, pp. 136–145
19. Cheng, J.: Neural summarization by extracting sentences and words. In: ACL2016 Conference Paper. arXiv:1603.07252v3 [cs.CL]. Accessed 1 July 2016

20. Bengio, S., Vinyals, O., Jaitly, N., Shazeer, N.: Scheduled sampling for sequence prediction with recurrent neural networks. In: Proceedings NIPS'15 of the 28th International Conference of Neural Information Processing System, pp. 1171–1179 (2015)
21. Sutskever, I.: Sequence to sequence learning with neural networks. In: Proceeding NIPS; 14 of the 27th International Conference on Neural Information Proceeding Systems, pp. 3104–3112
22. Yadav, C.S., Sharan, A., Kumar, R., Biswas, P.: A new approach for single text document summarization. In: Second International Conference on Computer and Communication Technologies, pp. 401–411
23. Fattah, M.A., Ren, F.: Automatic text summarization. Proc. World Acad. Sci. Eng. Technol. **27**, 192–195 (2008). ISSN 13076884
24. Radev, D.R., Jing, H., Stys, M., Tam, D.: Centroid-based summarization of multiple documents. Inf. Process. Manage. **40**(6), 919–938 (2004)
25. Tonelli, S., Planta, M.: Matching documents and summaries uses key-concepts. In: Proceedings of the French Text Mining and Evaluation Workshop, pp. 1–6 (2011)

Field Based Weighting Information Retrieval on Document Field of Ad Hoc Dataset

Parul Kalra, Deepti Mehrotra and Abdul Wahid

Abstract Information retrieval is a process of representing, retrieving and normalising data items. The retrieval system is a method that verifies how a system responds against users' needs. The accessing of useful information is directly related by the user's job and the conceptual view of the information possessed by the retrieval system. In order to increase the efficiency of the retrieval system, the authors have considered new fields (TITLE and DESC) for evaluating the recall and precision parameters. This paper demonstrates the comparison of baseline probabilistic models with document fields in the retrieval process and for experimental analysis. The authors' have used the standard TREC Ad hoc test collections based on the weighting and field models.

Keywords Information retrieval · Divergence from randomness
Probabilistic models · Multinomial from randomness · TREC

1 Introduction

Retrieving actual and accurate information from the large corpus is a tedious task. Information Retrieval (IR) system takes in a query, based on which it returns some corpus that are assumed to hold high significance to the user. The documents that are returned are ranked on the basis of their relevance using a weighting model. Information Retrieval models have many ways by which the information can be accessed from the system.

P. Kalra (✉) · D. Mehrotra
Amity University, Uttar Pradesh, India
e-mail: parulkalra18@gmail.com

D. Mehrotra
e-mail: mehdeepti@gmail.com

A. Wahid
Maulana Azad National Urdu University, Hyderabad, India
e-mail: wahidabdul76@gmail.com

K. Ray et al. (eds.), *Soft Computing: Theories and Applications*,
Advances in Intelligent Systems and Computing 742,
https://doi.org/10.1007/978-981-13-0589-4_8

The first method is weight based information retrieval (i.e. DFR, TF_IDF, PL2, LGD, DPH, BM25 etc.) and the second method is field based retrieval model [1]. The field model provide us with the term in a field and also tell us the accuracy of the frequency of the term's occurrence in that field. For instance, in the corpus, the query term exists in the description and narration of the document and then the authors have negligible probability that the text is actually related to that query term. However, if the query term appears in the title of the body, then the prospect of the same increases. The motive of this paper is to compare and analyse effective and efficient retrieval of field and baseline models in Terrier. Information retrieval models offer a way to integrate the document-fields into the structure of the document. In XML corpus, the corpus records are related to the contents of the specified XML tags, such as TOP, NUM, TEXT and TITLE of the tags. The retrieval effectiveness can be improved by using document fields for XML collection. The baseline model processes TrecQueryTags as TOP, NUM and TITLE. Further, the field model utilises TITLE and TEXT. To improve the precision the authors have customised the models by processing DESC (description) instead of TEXT in the FieldQueryTags process. In the coming sections the authors would see how the recall and the precision, that is the efficiency of the system is being affected by taking into account the improved query tags that are TITLE and DESC [2].

Section 2 gives a detailed explanation of the DFR Models, and the related field weighting models. Section 3 explains the multinomial randomness distribution with multinomial models and approximation. Section 4 presents the experiment evaluation, experimental setting with results of the baseline and field models on standard Ad Hoc collection. Section 5 explain the conclusion remarks with future scope of the field models in cognitive area respectively.

2 The DFR Models

In information retrieval, weighting based models have high significance. One of the crucial frameworks in weighing models is Divergence from Randomness or DFR [1, 3, 4]. DFR framework belongs to the probabilistic weighing models with its constituents as follows:

1. Arbitrariness Model (ARM)
2. Accusation Model (ACM)
3. Normalised TF Model (NTFM)

A set of corpus C, has number of terms t. The Arbitrariness Model (ARM) calculates the likelihood. $P_{ARM}(t \in c|C)$ with term frequency t, f occurs in the corpus record c, it is related to the informative notion $-\log_2(P_{ARM}(t \in c|C))$. The sequence of the independent term t with Bernoulli's experiment, Arbitrariness Model (ARM) with Binomial Probability Distribution:

$$P_B(t \in c|C) = \binom{TF}{tf} p^{tf}(1-p)^{TF-tf} \tag{1}$$

Here, the TF refers to the term frequency of t in the corpus collection C, where $p = \frac{1}{N}$ is a consistent possibility in the term t which appears in the corpus record c and N is the frequency of records in the corpus collection C.

The Poisson distribution P with binomial distribution:

$$P_B(t \in c|C) \approx P_P(t \in c|C) = \frac{G^{tf}}{tf!}e^{-G} \quad \text{where,} \quad G = \frac{TF}{N} \tag{2}$$

Accusation Model (ACM) approximates the explanatory notion $1 - P_{(R)}$ of the probability $p_{(R)}$. If term t has high frequency in the record, then it has less risk and vis-a-vis.

$$P_{(R)} = \frac{tf}{tf + 1} \tag{3}$$

$p_{(R)}$ approximates the likelihood of frequency of the term in the record, when *tf* is looked upon previously. DFR framework's last model is Normalised TF Model (NTFM). This model calibrates the occurrence of term t in c, it also manages the given length lg of c and the average length $\overline{\text{lg}}$ in C. Now, *NTFM* calculates the density of the NTF of the corpus record length lg. The NTF is:

$$NTF = tf \cdot \log_2(1 + h \cdot \frac{\overline{\text{lg}}}{\text{lg}}) \tag{4}$$

where h is a hyper-parameter. In NTFM, *tf* is plugged in Eqs. (2) and (3) with *NTF*. The applicable score of a corpus record c for a query q is $w_{c,q}$:

$$w_{c,q} = \sum qtw \cdot w_{c,t} \quad \text{Here,} \quad w_{c,t} = (1 - P_{(R)}) \cdot (-\log_2 P_{ARM}) \tag{5}$$

where $w_{c,t}$ is the heft of the term t in corpus record c,$qtw = \frac{qtf}{qtf\text{max}}$, qtf is the term recurrence and qtf max is maximal frequency in the query q. the P_{ARM} is calculated using the Poisson's Indexing Model on Arbitrariness Model (ARM), and the $P_{(R)}$ is calculated using the Laplace and NTF are evaluated to NTFM, for resulting weighting model PL2 [2]. In the next section NTFM is extended to manage the term frequencies of record field.

2.1 DFR Record Fields Using BM25F and PL2F

DFR Record Fields manages numerous record fields to link per-field NTF. For this concept, NTF is extended to NTFF [5]. The record has m fields. Only single field is associated with the frequency of the each term. The term frequency tf_i where n is the n-th field of the term is normalised and weighted individually and merged with single spurious occurrence i.e., on NTFF.

$$NTFF = \sum_{n=1}^{m} w_n \cdot tf_n \log_2(1 + h_n \cdot \frac{\overline{\lg_n}}{\lg_n}) \tag{6}$$

where w_n is weight and tf_n is the term frequency and lg_n is the length of the n-th field of corpus record c. lg_n is the mean length of the corpus collection C and h_n is a hyper parameter for the n-th field. BM25F and PL2F are based on per-field normalisation models which are actually based on PL2 and BM25 [3]. It calculates the term frequency with length normalisation from the each field of the document [6, 7].

3 Multinomial Randomness Models: Multinomial Distribution

This section explains that the term frequency in document fields use multinomial and approximate multinomial distribution. For the calculation of probability of the frequency of term in the record field, the authors use Multinomial distribution (ML2) [5]. For instance, the corpus record c has m fields. Then the probability of a term tf_i occurs in the n-th field f_i.

$$P_M(t \in c|C) = \binom{TF}{tf_1 tf_2 \dots tf_m tf'} p_1^{tf_1} p_2^{tf_2} \dots p_m^{tf_m} p'^{tf'} \tag{7}$$

Here, TF is the term frequency of t in corpus, $P_n = \frac{1}{m \cdot N}$ is the probability of term occurs in corpus record c and N is the frequency of the corpus collection C. The frequency of $tf' = TF - \sum_{i-1}^{k} tf_i$ related with the occurrence of t in the corpus record c and $p' = 1 - m\frac{1}{m \cdot N} = \frac{N-1}{N}$ is related to the likelihood of the term t's frequency in the field c. Multinomial distribution is inherited from DFR weighting model of randomness Eq. (7) and Laplace Equation and then the term frequency tf_i is replaced with *NTF* and NTFM from Eq. (4) is applied. Now, the relevance score of the corpus record c is calculated as:

$$\begin{aligned} w_{c,q} &= \sum_{t \in q} qtw \cdot w_{c,t} = \sum_{t \in q} qtw \cdot (1 - P_{(R)}) \cdot (-\log_2(P_M(t \in c|C)) \\ &= \sum_{t \in q} \frac{qtw}{\sum_{n-1}^{m} NTF_n + 1} \cdot \Bigg(-\log_2(TF!) + \sum_{n-1}^{m} (\log_2(NTF_n!) \\ &\quad - NTF_n \log_2(p_n)) + \log_2(NTF_n'!) - NTF_n' \log_2(p') \Bigg) \end{aligned} \tag{8}$$

In Eq. (8), weight of the query q, term t and equation is derived from the weighting model for ML2. ML2 calculate the weight of the terms in the field with two ways. One of the method is to proliferate NTFs with consistent weight like in NTFM in Eq. (6) and the other method is to find the terms having less frequency and thereafter changing the scores of those terms as seen in Eq. (8). ML2 also computes the approximate the factorial using Lanczos to the Γ function in the Eq. (8) [5, 8]. The DFR PL2 weighting models use Stirling's formula for approximation: $\left(tf! = \sqrt{2\pi} tf^{tf+0.5} e^{-tf}\right)$.

3.1 Multinomial Distribution with Approximation: MDL2

The multinomial randomness model ML2 is replaced by new weighting model with theoretic approximation of the probability p_i [9].

$$\frac{TF}{tf_1!tf_2!\cdots tf_m!tf'!} p_1 tf_1 p_2 tf_2 \ldots p_m^{tf_m} p'^{tf'} \approx \frac{1}{\sqrt{2\pi TF^m}} \frac{2^{-TF \cdot C(\frac{tf_n}{TF}, p_n)}}{\sqrt{p_{t1} p_{t2}} \cdots p_{tm} p'_t} \tag{9}$$

The above Eq. (9) explains the concept of information theoretic divergence of the probability and terms occurs in the field of the document with prior probability function is given below:

$$C\left(\frac{tf_n}{TF}, p_i\right) = \sum_{n=1}^{m} \left(\frac{tf_n}{TF} \log_2 \frac{tf_n}{TF \cdot p_n}\right) + \frac{tf'}{TF} \log_2 \frac{tf_n}{TF \cdot p'} \tag{10}$$

The ML2 can take the place of approximation randomness model M from Eq. (9):

$$w_{c,q} = \sum_{t \in q} qtw \cdot \frac{m/2 \log_2(2\pi TF)}{\sum_{n=1}^{m} NTF_n + 1} \tag{11}$$

$$\cdot \left(\sum_{n=1}^{m} \left(NTF_n \log_2 \frac{NTF_n/TF}{p_n} + \frac{1}{2} \log_2 \frac{NTF_n}{TF}\right) + NTF' \log_2 \frac{NTF_n/TF}{p'} + \frac{1}{2} \log_2 \frac{NTF'}{TF}\right)$$

The above model defined MDL2. In this model the authors compute the query term t in the field in which the term t turns up. In the entire ad hoc dataset the query terms may appear in its description or the title. The system calculates the heft of the field by the proliferation of the recurrence of a term in a field by a static value or by calibrating the field [5]. The effectiveness of the retrieval is comparatively improved by using DFR and multinomial randomness models. Further, these results are confirmed in the upcoming observations [1, 3].

4 Experiment Observations

The module analyses the performances of the DFR field models (BM25F, PL2F) with multinomial randomness models (ML2, MDL2). First, compare the retrieval effectiveness of baseline weighting models (PL2, BM25) with DFR field models (BM25F, PL2F) [9, 10]. Second, retrieval effectiveness comparison between DFR document fields with Multinomial randomness document field. The evaluation of these models was conducted on the dataset provided by TREC [6]. The data used is from Associated Press News Wire Disk 1. Various parameters like, TITLE, HEAD, TEXT and DESCRIPTION were combined in order to track the best recall and precision scores. These scores help in predicting the finest combination which may give the best result. The data availability has exceeded the power to analyse linearly. It is thus necessary to figure out the results that matches the needs. Various field models were tested and compared with optimising parameter i.e. Mean Average Precision (MAP) for the set of fields given. For BM25 and PL2, HEAD, TITLE, TEXT and DESCRIPTION was used. Also, for field models, combination of two fields was used i.e. TEXT, TITLE and TITLE, DESCRIPTION. The indexing of the corpus record was done after eliminating 'bag of words' and implementing the stemming algorithm given by Porter. The platform used by the authors for performing the analysis is an IR tool 'TERRIER' [11].

Baseline DFR Model. The normalised term frequency (NTF) is the complex concern that is related to the parameters of information retrieval. In the baseline model, the tuning parameters are difficult to normalize. DFR is a probabilistic model which has been inherited from Hater's 2 Poisson model. While using the models, the authors are processing the query tags in TREC dataset that are TEXT, TITLE, HEAD and NUM. These four parameters are used to evaluate the results. After evaluation the system provides MAP (Mean Average Precision) as an observation results given in Table 1.

4.1 Comparison Between Baseline and Field Models

This part demonstrates the execution of the above two models chosen, along with their comparison. As per the analysis, if the query tag fields are changed then the result gives best match MAP in the context of baseline model.

Table 1 Comparison between Baseline and Field Models

#	Baseline model with query tags—TEXT, TITLE, NUM, HEAD	Field models with field tags—[TEXT, TITLE]	Field models with field tags—[TITLE, DESC]
Number of documents	84,678	84,678	84,678
Number of unique terms	138,809	138,809	164,197
Number of pointers	13,384,542	13,384,542	15,561,373
Number to tokens	22,290,723	22,290,723	24,250,948

Table 2 Baseline models with query tags

S. No	Model name	Retrieved	Relevant Retrieved	MAP	R Precision
1	BM25	149,458	4737	0.0382	0.0848
2	PL2	149,458	4753	0.0399	0.0859

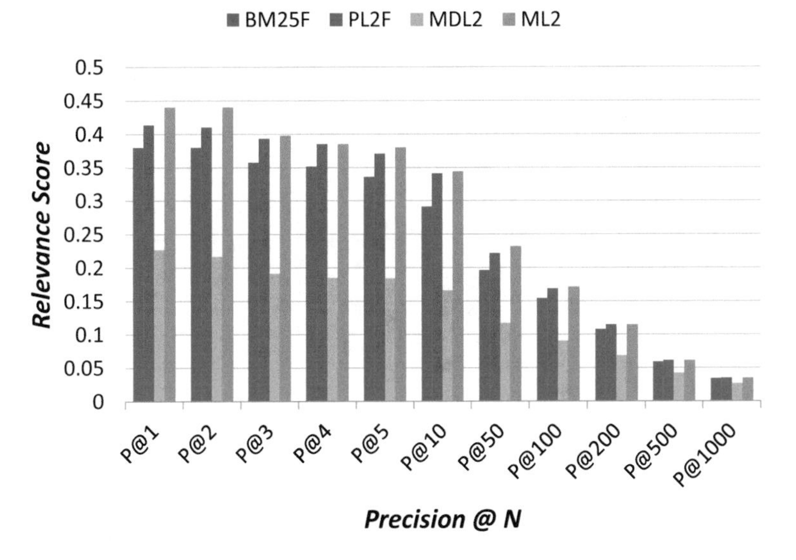

Fig. 1 Graph showing Relevance Score for P@n

The tables below clearly shows the difference in the values obtained in the baseline model and the field models which further explains that how the retrieval precision is affected when the tags are changed.

Considering, MAP for BM25 and PL2 which is 0.0382 and 0.0399 respectively. Then inspecting the MAP obtained for the Field Query Tags TEXT and TITLE, it is conspicuous that the MAP is decreasing in Table 2.

Now looking at the MAP evaluated for the Field Query Tags TITLE and DESC, it is seen that the MAP as well as the Recall Precision has a noticeable increase. This is clearly visible in the below Table 3 (Fig. 1).

Table 3 Field models with field query tags

Field models		Field query tags TEXT and TITLE				Field query tags TITLE and DESC			
S. No	Model name	Retrieved	Relevant retrieved	MAP	R Precision	Retrieved	Relevant retrieved	MAP	R Precision
1	BM25F	149,459	4689	0.0366	0.0828	150,000	5106	**0.0416**	**0.0914**
2	MDL2	149,459	3691	0.0158	0.0497	150,000	5199	**0.0466**	**0.0965**
3	PL2F	149,459	4712	0.0387	0.0835	150,000	3926	**0.0189**	**0.0569**
4	ML2	149,459	4734	0.0389	0.0843	150,000	5230	**0.0477**	**0.0974**

5 Conclusion

In the paper, the authors have discussed the importance of the Field models along with their field tags for improving the search results of the IR system. For the experiment, the authors have compared the baseline DFR models, DFR models with document field, multinomial randomness distribution and approximation with the field of IR models. The experiment was conducted on the TREC ad hoc dataset of Association Press news wire for finding out the efficiency of the system. In addition, the study lead us concluding the fact out that by changing the query tags with the field tags the results so obtained were improvised. The normalisation parameters rely profoundly on the datasets. If these parameters are recognised then the effectiveness of IR system can be improved incredibly which can be studied in future work.

References

1. Amati, G., van Rijsbergen, C.J.: Probabilistic models of information retrieval based on measuring divergence from randomness. ACM TOIS **20**, 357–389 (2002)
2. Chowdhury, G.: TREC: experiment and evaluation in information retrieval. Online Inf. Rev. http://trec.nist.gov/ (2013)
3. Macdonald, C., Plachouras, V., He, B., Lioma, C., Ounis, I.: University of Glasgow at WebCLEF 2005: experiments in per-field normalisation and language specific stemming. In: CLEF, vol. 4022, pp. 898–907 (2005)
4. Clinchant, S., Gaussier, E.: Information-based models for ad hoc IR. In: Proceedings of the 33rd International ACM SIGIR Conference on Research and Development in Information Retrieval, pp. 234–241. ACM (2010)
5. Plachouras, V., Ounis, I.: Multinomial randomness models for retrieval with document fields. Adv. Inf. Retr. 28–39 (2007)
6. Robertson, S., Zaragoza, H., & Taylor, M.: Simple BM25 extension to multiple weighted fields. In: Proceedings of the Thirteenth ACM International Conference on Information and Knowledge Management, pp. 42–49. ACM (2004)
7. Mishra, A., Vishwakarma, S.: Analysis of tf-idf model and its variant for document retrieval. In: 2015 International Conference on Computational Intelligence and Communication Networks (CICN), pp. 772–776. IEEE (2015)
8. Lin, Y.S., Jiang, J.Y., Lee, S.J.: A similarity measure for text classification and clustering. IEEE Trans. Knowl. Data Eng. **26**(7), 1575–1590 (2014)
9. Lioma, C.: Dependencies: Formalising Semantic Catenae for Information Retrieval. arXiv:17 09.03742 (2017)
10. Petersen, C., Simonsen, J. G., Järvelin, K., Lioma, C.: Adaptive distributional extensions to DFR ranking. In: Proceedings of the 25th ACM International on Conference on Information and Knowledge Management, pp. 2005–2008. ACM (2016)
11. Ounis, I., Amati, G., Plachouras, V., He, B., Macdonald, C., Lioma, C.: Terrier: A high performance and scalable information retrieval platform. In: Proceedings of the OSIR Workshop, pp. 18–25. http://terrier.org (2006)

Movie Rating System Using Sentiment Analysis

Abhishek Singh Rathore, Siddharth Arjaria, Shraddha Khandelwal, Surbhi Thorat and Vratika Kulkarni

Abstract The proposed work aims to collect correct reviews on movies using sentiment analysis. It classifies movie review into polarity classes based on the sentences and considers the biasness of user with respect to star cast of the film. Additionally, personality vector is also added to improve results. These two factors are used to minimize the biasness in the feature set to create hypothesis. The results with and without attribute of biasness are evaluated with different classification algorithms on twitter dataset. Results of the proposed work suggest that when removing biasing from movie review can give honest reviews.

Keywords Sentiment analysis · Movie rating · Biasing · Classification

1 Introduction

Nowadays, customers take purchase decisions on the basis of reviews and opinions available on social media. They usually ask people around them for recommendations and what do they think about a particular entity, whether to buy it or not. Market research has become an important step before buying a stuff so as to obtain its

A. S. Rathore (✉)
SVIIT, SVVV, Indore, India
e-mail: abhishekatujjain@gmail.com

S. Arjaria
Department of CSE, TIT, Bhopal, India
e-mail: arjarias@gmail.com

S. Khandelwal · S. Thorat · V. Kulkarni
Department of CSE, ATC, Indore, India
e-mail: kshraddha7890@gmail.com

S. Thorat
e-mail: surbhithorat1997@gmail.com

V. Kulkarni
e-mail: vratika8@gmail.com

K. Ray et al. (eds.), *Soft Computing: Theories and Applications*,
Advances in Intelligent Systems and Computing 742,
https://doi.org/10.1007/978-981-13-0589-4_9

comparative features and also to seek better deals. These opinions and reviews are required to be monitored and analyzed, which requires a lot of work. Hence, there is a need for an automated system which would help the user in avoiding catastrophic results and in doing reliable deals.

Sentiment analysis is the process of computational identification and categorization of opinions regarding certain facts or attributes so as to determine the writer's tone or orientation including further initialization of its polarity.

The opinions of people about a movie are considered to determine its quality on different parameters like music, story, direction, acting, editing, and background. The proposed work will analyze and evaluate user reviews of movies and predicts the opinion of people with the help of sentiment rating associated with each movie. It garners all the comments for a particular movie and then calculates an average rating to score it along with genre of the movie and personality of the user.

Sentiment analysis of movies review aims to automatically infer the opinion of the movie reviewer and often generates a rating on a predefined scale. Automated analysis of movie reviews is quite challenging in text classification due to the various nuances associated with a critic' review.

1.1 Solution Strategy Applied

The first task is sentiment or opinion detection, which may be viewed as a classification of text as objective or subjective. Usually, opinion detection is based on the examination of the adjective and sentences. The second task is that of polarity classification. The goal is to classify the opinion as falling under one of two opposing sentiment polarities, or located position between two polarities.

The above two tasks can be done at several levels: term, phrase, and sentence or document level. Different techniques are suitable for different levels. Techniques using n-gram classifiers or lexicon, usually work on term level, whereas part of speech tagging is used for phrase and sentiment analysis. The third task is to use supervised machine learning methods for classification and for unlabeled data clustering algorithm is used.

The remaining section of the paper is structured as follows. Section 2 discusses related work done in the field and provides a comparative study on their work. Section 3 explains our proposed work, followed by Sect. 4 to discuss results.

2 Related Published Work

The proposed work aims to classify the sentiments initially, and then identify a place where a new contribution could be made. This section discusses the different works done in this field so as to identify the appropriate approaches taken for the proposed work and the further approaches which can be taken.

Pang et al. [1] used three machine learning algorithms Naïve Bayes, support vector machines, and maximum entropy with a Bag of Words assumption and support vector machine gives more promising results as compared to Naïve Bayes.

Baroni and Vegnaduzzo [2] proposed a large list to rank adjectives according to a subjective score by employing a small set of manually selected subjective adjectives and computing the mutual information of pairs of adjectives using frequency and co-occurrence frequency counts on the Web.

Turney [3] considered the algebraic sum of the orientation of terms as respective of the orientation of the documents for classification.

Salvetti et al. [4] applied an Overall Opinion Polarity (OvOp) concept with Naïve Bayes and Markov model with the hypernym from wordnet and Part of Speech (POS) as lexical filter. The result obtained by wordnet filter is less accurate in comparison with that of POS filter.

Beineke et al. [5] applied Naïve Bayes model for sentiment classification with Turney [3] approach and anchor words. This system acted as a probabilistic model with improved accuracy.

Text classification is sometimes difficult to create these labeled training documents, but it is easy to collect the unlabelled documents, unsupervised learning methods overcome it presented by Youngjoong and Jungyun [6]. They divide the documents into sentences, and categorize each sentence using keyword lists of each category and sentence similarity measure.

Habernal et al. [7] employed different feature selection techniques namely mutual information, information gain, chi-square, odds ratio, and relevancy score. Mutual information was found to perform the best, whereas chi-square was the worst performer.

Blascovich and Mendes [8] concluded that computer-based sentiment analysis techniques may not provide sufficient results alone and therefore some other human-based technique should be used in assistance. Read [9] used the emoticons that often appears in the movie rating positive and negative opinions.

A lexical based method for sentiment analysis with a semantic orientation calculator (SO-CAL) was developed by Taboda el al. [10] for the detection of sentiment orientation. Separate dictionaries were created and opinions were ranked manually. It lacks detailed information about how the manual ranking was carried out.

Montejo-Raez et al. [11] explored WordNet graph using random walk algorithm to compare the words with the synsets. Preprocessing techniques, like spell correction, lemmatization, slang/abbreviation expansion, etc. were not considered in this approach.

Ortega et al. [12] employed an unsupervised technique based on Senti-WordNet, a sentiment lexicon with prior polarities. The proposed work analyzes sentiments in three phases; data preprocessing, SentiWordNet based sentiment identification and rule-based labeling. The final result was the detection of sentiment orientation of the input data.

SentiWord-Net was also explored to estimate sentence level polarity with Stanford POS tagger for adjectives only. However, the intensity of the sentiment was not considered by the authors and there was no justification given for assigning class labels based on an even/odd count [13].

Lin et al. [14] applied supervised sentiment analysis; term sentiment scores were weighed with respect to the mutual information to attain a high accuracy level.

Another supervised approach was proposed by Xu et al. [15] S-HAL algorithm uses pointwise mutual information. Algorithm assumes terms to subjective and term's sentiment orientation is estimated on its statistical relationship with positive and negative sets.

Chikersal et al. [16] proposed another supervised learning approach with linguistic rules to classify tweets. SentiWordNet, Sen-ticNet and Bing Liu list of positive and negative words were used to determine the polarity of word/phrase.

2.1 Comparative Study

A comparative study is performed on different techniques is shown in Table 1.

Various methods have been used for classification and feature selection purpose. These methods have their respective advantages and limitations. The comparative study shows these approaches with their corresponding year and on what basis they are evaluated. If accuracy is considered to be the result evaluation parameter than the SentiWordNet Adjectives approach proves to be the best suited approach for clustering of unlabeled dataset. Achieving accuracy level up to 85%, Pointwise MI—S-HAL approach gives more promising result for classification of labeled dataset.

In this study, we find that users' reviews are considered to be correct. None of the studies identifies the biased nature of user towards particular. Our aim is to develop a new model that gives semantically meaningful opinions as per the end user expectations; so that accuracy of the topic model will be enhanced by including graph structure, "Bag of words" representation thus providing a more accurate feature selection.

3 Proposed Work

The ratings and opinion are usually biased. Person from Tamil Nadu might have some good emotions for South Indian war films than person from northern side. Even the person likes war films will surprisingly give good rating to poor low budget films too. Another problem is that rate only those films which they like most or hated most. Another problem is with children and/or cartoon movies, gets only average score.

Another issue is person's choice, precisely saying "genres." Some people likes comedy, other likes romantic and so on. So, one that stuck to one genre will give average or poor ratings to other genres.

First thing is to get personality of the user. Find out personality by big five model [17] of personality traits. The big five model of personality dimensions is one of the

Table 1 Comparative study of different techniques

Method/technique	Year	Feature selection	Clustering/classification	Dataset used	Result evaluation techniques
Classify the dataset using different machine learning algorithms and n-gram model	2002	Naive Bayes (NB), maximum entropy (ME), support vector machine (SVM)	Classification	Internet movie database (IMDb)	Unigram: SVM (82.9), Bigram: ME (77.4), Unigram + Bigram: SVM (82.7)
Accessed overall opinion polarity(OvOp) concept using machine learning algorithms	2004	Naive Bayes (NB) and Markov model (MM)	Lexical filter classification	Internet movie database (IMDb)	NB: 79.5, MM: 80.5
Syntactic relationship among words used as a basis of document-level sentiment analysis	2005	Support vector machine	Unigram approach classifier	Internet Movie Database (IMDb), Polarity dataset	Unigram: 83.7, Bigram: 80.4, Unigram + Bigram: 84.6
Syntactic relationship among words used as a basis of document-level sentiment analysis	2005	Support vector machine	Unigram approach classifier	InternetMovie database (IMDb), polarity Dataset	Unigram: 83.7, Bigram: 80.4, Unigram + Bigram: 84.6
Lexicon based semantics	2010		Supervised (classification)	Web-based semantic labelling and rule-based	
Weakly and supervised classification	2010		Supervised (classification)	IMDB, Amazon.com	Accuracy 67%
SentiWordNet random walk	2012		Clustering (semi-supervised)	Self-collected twitter corpus	Accuracy 63%
Pointwise MI—S-HAL	2012		Supervised (classification)	Chinese language SogouCS corpus	Accuracy 85%

(continued)

Table 1 (continued)

Method/technique	Year	Feature selection	Clustering/classification	Dataset used	Result evaluation techniques
SentiWordNet rule-based labeling	2013		Clustering (semi-supervised)	SemEval 2013 Twitter and SMS	Accuracy 50%
SentiWordNet adjectives Only	2013		Clustering (semi-supervised)	Self-collected customer reviews	Accuracy 69%
N-gram, lexicon, POS, tweet-based features and SentiWordnet	2013	CRF, SVM and heuristic method	Supervised (classification)	Sentiment analysis in twitter	P-measure 49%
Emoticons, SentiWordNet SenticNet Bing Liu words list	2015		Supervised (classification)	Stanford twitter dataset	F-measure 79.8%
Lexicon based analysis to transform data into the required format	2016	BOW feature with TF and TF-IDF approach	statistical learning methods (Classification)	Manually annotated twitter data	Text: 71.9, Visual Feature: 68.7, Multi-view:75.2

most emerging area for researcher to personality identification in recent years. The Big Five model is taxonomy of personality traits, maps which traits go together in people's descriptions of ratings of one another. Factor analysis, a statistical method is used to analyze different personality traits, is employed to discover the big five factors [17]. Today, many researchers are agreed on the five core models of personality, openness, conscientiousness, extroversion, agreeableness, and neuroticism. Let x_i be the vector associated with personality of the user by $x_i = \{x_1, x_2, \ldots, x_5\}$. This can help to understand person's specific choice to understand genres he likes.

The methodology for proposed work consists of two phases: In the first phase, the reviews are gathered from various social networking platforms. These reviews are then analyzed to search keywords and classify them according to the attributes. Selected features are then rated according to the scale and the dataset is formed. In the second phase, the features are classified with different classification algorithms.

3.1 Feature Selection

For Feature Selection, the following algorithm has been opted:

1. Let F be a feature set of with attributes.
   $F = \{music, story, direction, acting, editing, background\}$

2. Let the dataset T consists
   $T=\{T_1, T_2, ..., i\ T_n\}$
3. for each T_i
   3.1 Identify each F_j and extract 3-gram features with respect to F_j.
   3.2 If F_k and F_j are co-occurred then mark for both features for T_i.
   3.3 Let r be the aggregate rating of the tweet with respect to AFINN [18] scale.
   3.4 Identify biasing of user **Bias()**.
   3.5 Calculate the cost function C.

For identification of biasing following algorithm is used **Bias()**

1. Identify Star cast of the film.
2. Let S be the set that contains all the star cast of a movie w.r.t F such that $S=\{S_1 i\ S_2 i, ..., S_n\}$.
3. Check if the user u is a follower of any member of S.
   3.1 For each S_i
      3.1.1 Calculate aggregate polarity of past tweets using AFINN scale
      3.1.2 Let s_{ui} be the normalized the rating.
4. Let x_{uj} be the Personality traits of a user u from 5 parameters of Big Five Model.
5. Return total biasness r'

where r' for user u is calculated as:

$$r'_u = \sum_{j=1}^{5} \sum_{i=1}^{n} x_{uj} s_{ui} \tag{1}$$

Our idea is to find x_{uj} and s_{uj} which minimizes the cost function C to minimum biasness. C is calculated as:

$$C = \sum_{u,i,j} r_u - x_{uj} s_{ui} + \lambda \left(\sum_{u,j} \left\| x_{uj}^2 \right\| + \sum_{u,i} \left\| s_{ui}^2 \right\| \right) \tag{2}$$

where λ is regularization parameter for x and s as they get very large magnitude. It will reduce the overfitting. For the randomly taken small sample of data, the value of regularization parameter λ is estimated. For very smaller value, let $\lambda = 0$, it will overfit the cost function while higher value increases variance. We used $\lambda = 0.5$.

3.2 *Classification*

After getting different degrees of biasness, classification algorithms are applied on ratings obtained biased opinion and proposed unbiased (rating of minimal cost function) opinion for each feature. Naive Bayes classifier, multiclass classifier, SGD classifier, and J48 classifier are used to classify the features to prove the assumption.

4 Results and Discussion

Twitter Dataset on movie is used for evaluation of proposed work with 2 classes, 2000 instances (1000 positive and 1000 negative). The dataset is transformed into plaintext by converting into lowercase, removal of stop words, emoticons, accents and URLS and then porter stemmer is applied to get tokens.

Each tweet is labeled with the rating for each attribute of the movie on a scale of -5 to +5. The ratings are given on the basis of a manual scale which has been suggested from AFINN, which is the official list of English words or valence with an integer between minus five (negative) and plus five (positive).

Weka tool is used to apply different classification algorithms. The performance of various classifiers is evaluated on the basis of these parameters such as F-measure, precision, recall, and FP-Rate.

The results are compared with applying biasing attribute in classification versus without biasing attribute. The results are shown in Table 2, 3, 4, 5, 6, 7, 8, 9, 10, 11, 12, 13, 14, 15, 16, 17, 18, 19, 20, 21, 22, 23, 24, 25, 26, 27, 28, 29, 30, 31, 32, and 33. Applying biasing attribute performs better in classification algorithms.

4.1 Naïve Bayes (with 10 Fold Cross Validation)

See Tables 2, 3, 4 and 5.

Table 2 F-measure

Class	With biasing attribute	Without biasing attribute
Positive	0.892	0.886
Negative	0.887	0.886

Table 3 Recall

Class	With biasing attribute	Without biasing attribute
Positive	0.841	0.841
Negative	0.833	0.833

Table 4 Precision

Class	With biasing attribute	Without biasing attribute
Positive	0.950	0.938
Negative	0.943	0.938

Table 5 FP rate

Class	With biasing attribute	Without biasing attribute
Positive	0.167	0.174
Negative	0.041	0.041

Table 6 F-measure

Class	With biasing attribute	Without biasing attribute
Positive	0.934	0.887
Negative	0.934	0.880

Table 7 Recall

Class	With biasing attribute	Without biasing attribute
Positive	0.891	0.873
Negative	0.837	0.830

Table 8 Precision

Class	With biasing attribute	Without biasing attribute
Positive	0.982	0.945
Negative	0.982	0.937

Table 9 FP rate

Class	With biasing attribute	Without biasing attribute
Positive	0.133	0.192
Negative	0.012	0.014

Table 10 F-measure

Class	With biasing attribute	Without biasing attribute
Positive	0.923	0.895
Negative	0.727	0.667

Table 11 Recall

Class	With biasing attribute	Without biasing attribute
Positive	0.947	0.895
Negative	0.667	0.667

Table 12 Precision

Class	With biasing attribute	Without biasing attribute
Positive	0.900	0.895
Negative	0.800	0.667

Table 13 FP rate

Class	With biasing attribute	Without biasing attribute
Positive	0.333	0.333
Negative	0.053	0.105

Table 14 F-measure

Class	With biasing attribute	Without biasing attribute
Positive	0.90	0.862
Negative	0.882	0.723

Table 15 Recall

Class	With biasing attribute	Without biasing attribute
Positive	0.833	0.813
Negative	0.812	0.791

Table 16 Precision

Class	With biasing attribute	Without biasing attribute
Positive	0.98	0.92
Negative	0.967	0.667

Table 17 FP rate

Class	With biasing attribute	Without biasing attribute
Positive	0.029	0.037
Negative	0.167	0.310

Table 18 F-measure

Class	With biasing attribute	Without biasing attribute
Positive	0.948	0.903
Negative	0.774	0.727

Table 19 Recall

Class	With biasing attribute	Without biasing attribute
Positive	0.947	0.907
Negative	0.723	0.667

Table 20 Precision

Class	With biasing attribute	Without biasing attribute
Positive	0.951	0.900
Negative	0.833	0.800

Table 21 FP rate

Class	With biasing attribute	Without biasing attribute
Positive	0.167	0.333
Negative	0.053	0.129

Table 22 F-measure

Class	With biasing attribute	Without biasing attribute
Positive	0.833	0.772
Negative	0.745	0.715

Table 23 Recall

Class	With biasing attribute	Without biasing attribute
Positive	0.833	0.833
Negative	0.845	0.841

Table 24 Precision

Class	With biasing attribute	Without biasing attribute
Positive	0.835	0.721
Negative	0.667	0.623

Table 25 FP rate

Class	With biasing attribute	Without biasing attribute
Positive	0.059	0.046
Negative	0.167	0.167

Table 26 F-measure

Class	With biasing attribute	Without biasing attribute
Positive	0.878	0.824
Negative	0.895	0.770

Table 27 Recall

Class	With biasing attribute	Without biasing attribute
Positive	0.947	0.911
Negative	0.933	0.913

Table 28 Precision

Class	With biasing attribute	Without biasing attribute
Positive	0.818	0.753
Negative	0.861	0.667

Table 29 FP rate

Class	With biasing attribute	Without biasing attribute
Positive	0.062	0.172
Negative	0.103	0.193

Table 30 F-measure

Class	With biasing attribute	Without biasing attribute
Positive	0.779	0.753
Negative	0.835	0.684

Table 31 Recall

Class	With biasing attribute	Without biasing attribute
Positive	0.733	0.712
Negative	0.819	0.667

4.2 Naïve Bayes Classifier (with Percentage Split)

See Tables 6, 7, 8 and 9.

4.3 Multiclass Classifier (with 10-Fold Cross Validation)

See Tables 10, 11, 12 and 13.

Table 32 Precision

Class	With biasing attribute	Without biasing attribute
Positive	0.833	0.801
Negative	0.853	0.702

Table 33 FP rate

Class	With biasing attribute	Without biasing attribute
Positive	0.040	0.059
Negative	0.167	0.249

4.4 Multiclass Classifier (with Percentage Split)

See Tables 14, 15, 16 and 17.

4.5 SGD (Stochastic Gradient Descent) Classifier (with 10 Fold Cross Validation)

See Tables 18, 19, 20 and 21.

4.6 SGD (Stochastic Gradient Descent) Classifier (with Percentage Split)

See Tables 22, 23, 24 and 25.

4.7 J48 Classifier (with 10 Fold Cross Validation)

See Tables 26, 27, 28 and 29.

4.8 J48 Classifier (with Percentage Split)

See Tables 30, 31, 32 and 33.

5 Conclusion

The proposed work aims to uncover the attitude of the user on a particular dimension of the movie. It takes plaintext as input with twitter graph structure to identify biasness of user with a particular cast of the movie. It applies simple learning techniques to find statistical and linguistic patterns in the text that reveal attitudes with Big Five model. The results of different standard classification algorithms on feature set proposed with unbiased attributes confirm our assumption and thus we are able to provide more accurate results.

References

1. Pang, B., Lee. L., Vaithyanathan. S.: Thumbs up?: sentiment classification using machine learning techniques. In: Proceedings of the ACL-02 Conference on Empirical Methods in Natural Language Processing (2002)
2. Baroni, M., Vegnaduzzo, S.: Identifying subjective adjectives through web-based mutual information. In: Proceedings of the 7th Konferenz zur Verarbeitung Natrlicher Sprache (2004)
3. Turney, P.: Thumbs up or thumbs down: semantic orientation applied to unsupervised. In: Proceedings of the 40th Annual Meeting on Association for Computational Linguistics (2002)
4. Salvetti, F., Reichenbach, C., Lewis, S.: Automatic opinion polarity classification of movie review. Colorado Res. Linguist. **17**, 420–428 (2004)
5. Beineke, P., Hastie, T., Vaithyanathan, S.: The sentimental factor: Improving review classification via human-provided information. In: Proceedings of the 42nd Annual Meeting on Association for Computational Linguistics (2004)
6. Youngjoong, K., Jungyun. S.: Automatic text categorization by unsupervised learning. In: Proceeding of the 18th International Conference on Computational Linguistics (2000)
7. Habernal, I., Ptacek, T., Steinberger, J.: Sentiment analysis in czech social media using supervised machine learning. Inf. Process. Manage. **51**, 532–546 (2014)
8. Blascovich, J., Mendes, W.: Challenge and threat appraisals: The role of Affective cues. In: Forgas, J. (ed.) Feeling and Thinking: The Role of Affect in Social Cognition, pp. 59–82. Cambridge University Press, Paris (2000)
9. Read, J.: Using emoticons to reduce dependency in machine learning techniques for sentiment classification. In: Proceedings of the ACL Student Research Workshop (2005)
10. Taboada, M., Brooke, J., Tofiloski, M., Voll, K., Stede, M.: Lexicon-based methods for sentiment analysis. Comput. Linguist. **37**, 267–307 (2011)
11. Montejo-Ráez, A., Martinez-Camara, E., Martin-Valdivia, M., Urena-Lopez, L.: Random walk weighting over SentiWordNet for sentiment polarity detection on twitter. In: Proceedings of the 3rd Workshop in Computational Approaches to Subjectivity and Sentiment Analysis (2012)
12. Ortega, R., Fonseca, A., Gutierrez, Y., Montoya, A.: SSA-UO: Unsupervised Twitter Sentiment Analysis. In: Second Joint Conference on Lexical and Computational Semantics (2013)
13. Jain, A.K., Pandey, Y.: Analysis and implementation of sentiment classification using lexical POS markers. Int. J. Comput. Commun. Netw. **2**, 36–40 (2013)
14. Lin, Y., Zhang, J., Wang, X., Zhou, A.: An information theoretic approach to sentiment polarity classification. In: Proceedings of the 2nd Joint WICOW/AIRWebWorkshop on Web Quality (2012)
15. Xu, T., Peng, Q., Cheng, Y.: Identifying the semantic orientation of terms using S-HAL for sentiment analysis. Knowl.-Based Syst. **35**, 279–289 (2012)
16. Chikersal, P., Poria, S., Cambria, E., Gelbukh A.F., Siong, C.E.: Modelling public sentiment in Twitter: using linguistic patterns to enhance supervised learning. In: 16th International Conference CICLing (2015)
17. Barrick, M.R., Mount, M.K.: The Big Five personality dimensions and job performance: a meta-analysis. Pers. Psychol. **44**, 1–26 (1991)
18. Nielsen, F.A.: AFINN. IMM Publication. http://www2.imm.dtu.dk/pubdb/views/publication_details.php?id=6010

Energy Consumption of University Data Centre in Step Networks Under Distributed Environment Using Floyd–Warshall Algorithm

Kamlesh Kumar Verma and Vipin Saxena

Abstract Many of the computer data centres across world are interconnected of network systems. In the network connection, the distributed systems of multiprocessors are arranged for time-dependent run of tasks through task scheduling algorithms of effective networks topology. Time to time, the energy consumption of distributed computing is a big problem of few years back and onwards. The energy is a concept of any network is very precious and the quality of services (QoS) of any computer networks. In the present work, a section of computer centre is considered as a data centre which contains many electrical devices which emit static and dynamic energies. There is a big challenge to optimize the power consumption in the computer centre. In this paper, an energy consumption of the multiple frequencies is considered in which devices are arranged under distributed environment for providing better facilities to the performance of the computer. Each processor has distributed frequency and the energy model is proposed for optimization of power consumption. The results are represented in the form of tables and graphs.

Keywords Distributed computing · Multiple frequencies · Step network topology · Power consumption

1 Introduction

Distributed computing is an emerging area of research. The application of distributing computing devices power consumption is increasing day by day in all areas of network application throughout the world. The broadband wired communication network including the home personal computers and number of popular devices are based on electrical application that gathered information from the surrounding. These devices are highly energy efficient due to uses of motherboard, processor,

K. K. Verma (✉) · V. Saxena
Department of Computer Science, Babasaheb Bhimrao Ambedkar University, Raebareli Road, Lucknow, India
e-mail: kamalca11@gmail.com

V. Saxena
e-mail: vsax1@rediffmail.com

K. Ray et al. (eds.), *Soft Computing: Theories and Applications*, Advances in Intelligent Systems and Computing 742,
https://doi.org/10.1007/978-981-13-0589-4_10

electromechanical devices and hard disks throughput the information processing. The proposed constraint makes energy consumption one of the most predicted areas among researcher and scientists. In these computing networks, there have been various techniques for reducing energy consumption in previous and depend basically to traffic load. The networks consume the largest amount of energy consumption which are observed to fix and variable in wired networks with a huge number of distributed networks. The highest energy consumption has seen in data ware centre and IP backbone networks. The method for reducing energy consumption is various types and matrices found for energy related of networks. It is divided into main categories (1) Network workload. (2) The sleep mode approach. These two solutions can find the energy to save extra energy. Therefore, a technique to find for energy minimization or reducing is the research region in networks. The current work is focused on computational techniques, which minimization the extra energy due to traffic load. In the computers, the CPU works as major role of the power effective in consumption mode. So the energy consumption of CPU and networks load for energy is given below

$$Energy = Packets\ Load \times \frac{Packets\ Throughput}{Data\ Packets\ Input} \tag{1}$$

The energy has high impact on computer networks. In the present paper, different methods for energy consumption are described. Let us describe some of the important references related to the work.

2 Related Work

Moharir et al. [1] have given a dense distributed scheduling algorithm for large-scale networks. They have described the scheduling algorithms are two types (1) High efficiency in mobile association with timescale of seconds (2) Multinode connectivity where multiple mobile nodes are by time. Xiao et al. [2] described the energy consumption for heterogeneous distributed networks by the use of scheduling algorithms. In the algorithms, problems have been divided two subproblems (1) Task-based energy consumption (2) Task scheduling of low time complexity and application of Fourier transform parallel and distributed computing. Devi et al. [3] have developed a cost-efficient dynamic batch mode scheduling for assigning task. They have used scheduling as many types (1) Round Robin (2) min-min (3) max-min to use the energy performance cost-effective scheduling. Shi et al. [4] have developed echo state networks (ESNs). Huang et al. [5] have suggested the balancing of the energy consumption in network transportation. The energy consumption of the mobile sinks during the data transmits across the network. They have proposed an energy-aware clustering algorithms and energy aware routing algorithm. Kaswan et al. [6] have developed an algorithm for designing efficient of mobile sinks. The algorithm use with k-means clustering for network efficiency. They have proposed also scheduling algorithms for effective data transportation. Jiang et al. [7] have formulated a mathematical methodology of Integer Linear Programming (ILP). The computational complexity of the network utilization is maximum if network is peak form but the low network utilization the unnecessary energy is waste in data centre. Imran et al. [8] have proposed a Hybrid Optical Switch Architecture. The hybrid system design works as slow and fast optical switches of the performance and reliability. They have also investigated the scalability cost and power consumption of design of the Hybrid Optical Switch Architecture for the data centre by the network level simulation at the time of workload parameter latency, high throughput and communication load. Harbin et al. [9] have proposed the Transaction-Level Model algorithms for the networks. They have given two application cases (1) lightweight Transaction-Level Model simulation models can produce latency of low energy. (2) Dynamic power consumption model. In the network performance of latency dynamic power modelling algorithms. Alonso et al. [10] have described a methodology of managing the power consumption of fat tree networks interconnections. In the methodology, the network traffic depends on bandwidth and routing algorithm for power saving. This methodology uses in fat tree networks then the performance of power saving up to 36% energy. Zhang et al. [11] have investigated a multi-ring based on optical circuit networks. They have developed Integer Linear Program (ILP) and heuristic algorithm to solve of the RSA problem. They evaluate the traffic latency by the integer linear program and two spectrum allocation for network traffic the inter Rack traffic solution to the data centre and flexible spectrum represents increase throughput, power consumption and computer complexity. Avci et al. [12] have described a distributed algorithm for detecting the motion of the network transportation. The transportation network is two types (1) continuously moving towards (2) persistently moving towards. Above these methods uses for the energy versus latency and other power saving. Khelladi et al. [13] have proposed an algorithm for on-demand multinode charging solution. These techniques are one of the methods for the reduction of energy consumed. The method also reduced the time complexity and efficiency. Heddeghem et al. [14] have described the mathematical power consumption model for large telecommunication network for the power consumption in the computer devices such as internet protocol, switching, ethernet and wavelength division multiplexing equipment. Martins et al. [15] have described a social network analysis method for of multi-scalar energy networks of global data centre network and subnetwork systems. Jiwei et al. [16] have given Markov Decision Model (MDM) for energy consumption in distributed computing. The Markov Decision Process uses in the computing devices as the task scheduling and algorithms for optimal solution for the process throughput. Lin et al. [17] have given a theory of routing for mobile-based agent in wireless sensor networks. They have depicted the unbalanced energy of a sensor network with uniformly distribution in data centre. The method of energy finding of cluster routing based on a mobile agent in data centre. Zhuo et al. [18] have given an algorithm a theory for energy consumption of system level CPU energy-efficient algorithm. Zomaya et al. [19] have evaluated the energy consumption of the data centre using by the multiple frequency for multiple processors to minimize the energy consumption.

3 Research Methodology

The cost of energy is a unique feature of the physical devices, which acts as inclusive and during the time of application. So the energy is a vast challenge in distributed computing world; the energy is an important feature of networks consumption by the devices or other peripheral devices. The peripheral device architecture is important role of the action of network application.

$$Total\ Systems\ Energy\ (E_T) = CPU\ Energy + (Power\ Consumption)_{Computer\ Components} \quad (2)$$

The CPU energy is calculated by dynamic power measurement $p\,\alpha\,c\,v^2\,f$ and system power consumption by different types of hardware component like memory, input/output, network, hard disks and fan, algorithms and scheduling algorithms for data packets.

3.1 Energy Model for University Data Centre

The distributed computing system is an autonomous collection of different devices having different frequencies which are given below

$$y = f(x) = f_1 < f_2 < f_3 < f_4 < f_5 < f_6 < f_7 < f_8 < f_9 \ldots f_{N-1} < f_N \quad (3)$$

Let us consider, take the range of the frequency is given as 1000 MHz to 2000 MHz which has dynamic part of the data centre architecture and static part of the data centre. In dynamic case, the processor is frequently changed by applying the different set of frequency and stability of power consumption varies from maximum and minimum. While in the static case, the processor of power consumption is negligible because leakage voltage of the power consumption is considered as negligible (Fig. 1).

The frequency of processor p_1 is f_1

$$f = \frac{p \times w}{c \times v^2} \quad (4)$$

where w is the weight of the processor considered in terms of distance from one node to next node.

Let us differentiate (4) with respect to the time t

$$\frac{df}{dt} = \frac{dp}{dt} \cdot \frac{w}{c\,v^2} \quad (5)$$

In the above, the energy consumption $E = \frac{dp}{dt}$, $\Delta f \approx f$ and $\Delta t \approx t$, then

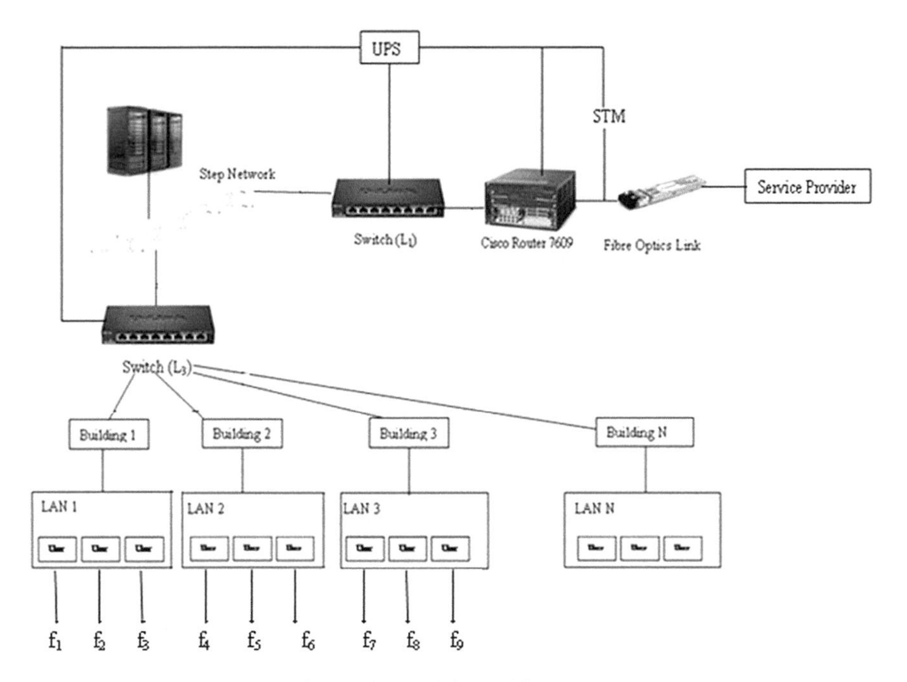

Fig. 1 University data centre with multiple weights and frequency

$$\frac{df}{dt} = \frac{E \times w}{c \times v^2} \tag{6}$$

$$\frac{df_1}{dt} = \frac{E \times t_1}{c\,v^2 w_1}, \frac{df_2}{dt} = \frac{E \times t_2}{c\,v^2 w_2}\, \frac{df_3}{dt} = \frac{E \times t_3}{c\,v^2 w_3} \tag{7}$$

$$\frac{df_4}{dt} = \frac{E \times t_4}{c\,v^2 \times w_4}, \frac{df_5}{dt} = \frac{E \times t_5}{c\,v^2 w_5}, \frac{df_6}{dt} = \frac{E \times t_6}{c\,v^2 w_6} \tag{8}$$

$$\frac{df_7}{dt} = \frac{E \times t_7}{c\,v^2 w_7}, \frac{df_8}{dt} = \frac{E \times t_8}{c\,v^2 w_8}, \frac{df_9}{dt} = \frac{E \times t_9}{c\,v^2 w_9} \tag{9}$$

The dynamic power consumption of the processor enabled by DVFS (Dynamic Voltage Frequency Scaling)

Now the total frequency of all processors,

$$\sum_{i=1}^{i=9} \frac{df_i}{dt} = \frac{E}{cv^2}\left[\frac{t_1}{w_1} + \frac{t_2}{w_2} + \frac{t_3}{w_3} + \frac{t_4}{w_4} + \frac{t_5}{w_5} + \frac{t_6}{w_6} + \frac{t_7}{w_7} + \frac{t_8}{w_8} + \frac{t_9}{w_9}\right] \tag{10}$$

where Energy = Energy during frequency selection; C = capacitance; V = Voltage;

$$Packet\ Delivery\ Ratio = \frac{Packet\ \text{Received}}{Packet\ Send} \tag{11}$$

$$Throughput = \frac{Amount\ of\ Data\ send}{Time} \quad (12)$$

$$Energy\ Consumption = Final\ Energy - Initial\ Energy \quad (13)$$

Now find the energy of all processors putting the values from table.

$$E = \frac{cv^2 \times f}{w \times t} \quad (14)$$

3.2 Floyd–Warshall Algorithm

1. The shortest path in a weighted graph with positive or negative edge weight with no negative cycles.
2. The single execution of the algorithm will find the length of the shortest path between all pairs of vertices.
3. The algorithm computes the distance between each pair of nodes in O (N3).
4. The weight is calculated by node distance from source to destination.

3.3 Dynamic Scheduling Algorithm

The application of multiple frequencies, weight and time for the Dynamic Scheduling Algorithm.

1. Receive a task.
2. Search and select with multiple frequency available servers from servers for requirement of the task.
3. Frequency matches from the servers.
4. If frequency gap for penalty group in servers.
5. Now return the frequency assign of the tasks.
6. If the requirement of the task is not satisfied then goes to step 4.
7. Receive the frequency lies servers or not lies server or task is failed, it goes to step 8.
8. It sends the assignment task of frequency lies distribution.

3.4 Static Scheduling Algorithm

1. Receive a task.
2. Search and select with fixed frequency available from the server connected processor.
3. If assignment frequency is failed then go to step 1.

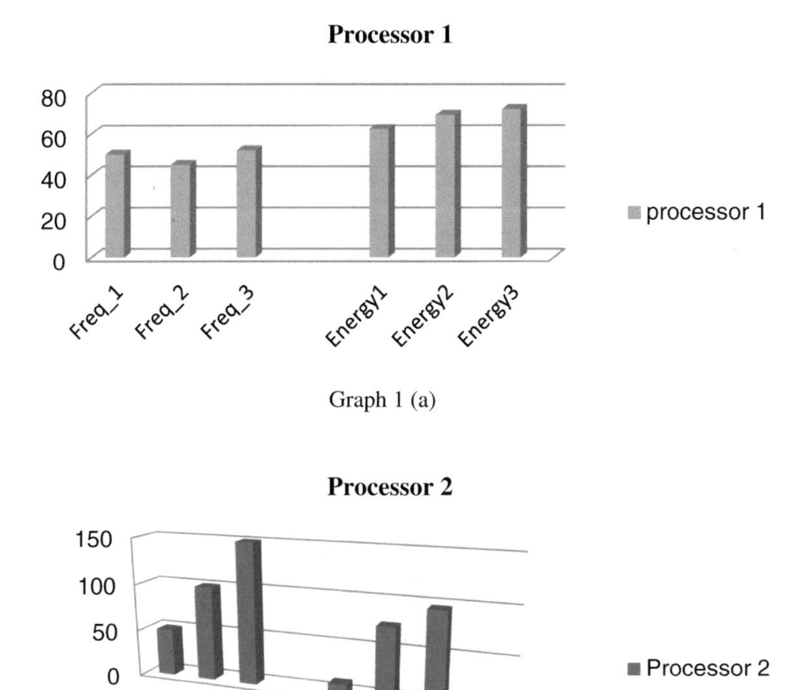

Graph 1 (a)

Graph 1(b)

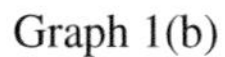

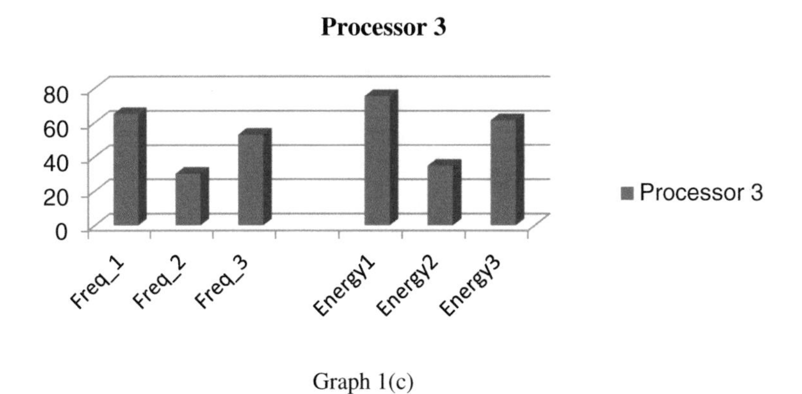

Graph 1(c)

The following graph represents the calculation by Table 1, based on multiple frequency and different energy results by the calculated.

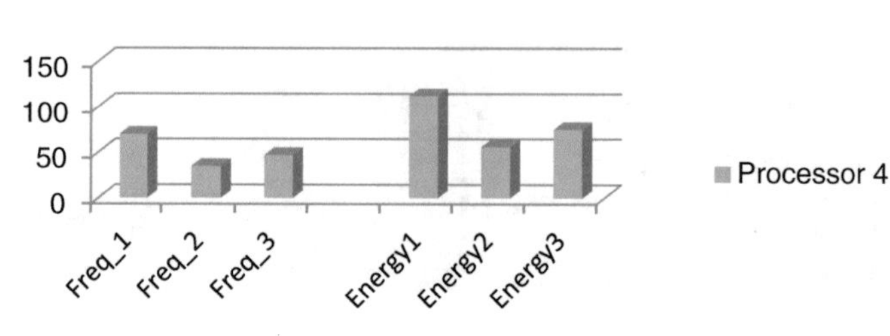

Graph 1(d)

Processor 5

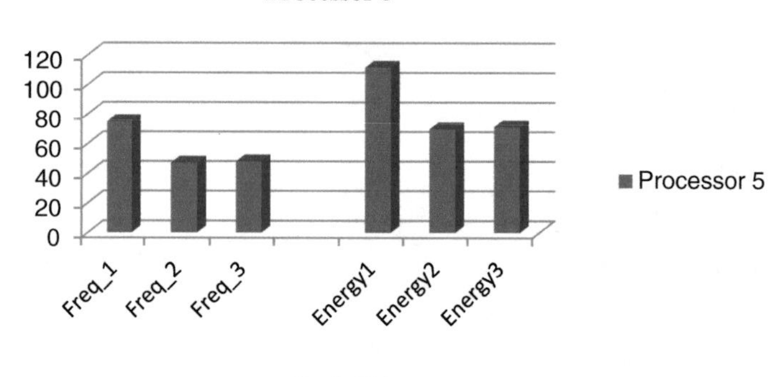

Graph 1(e)

Processor 6

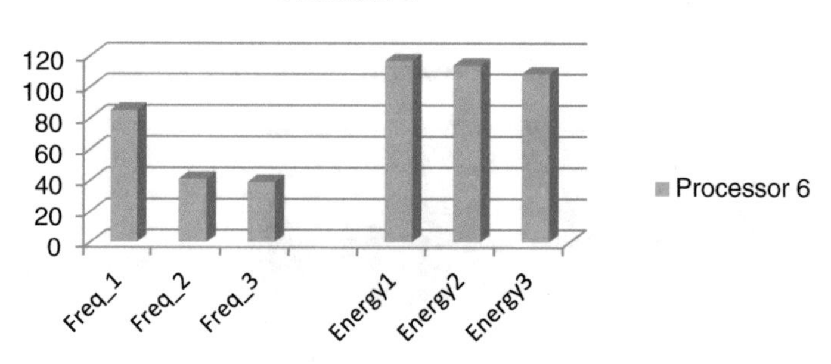

Graph 1(f)

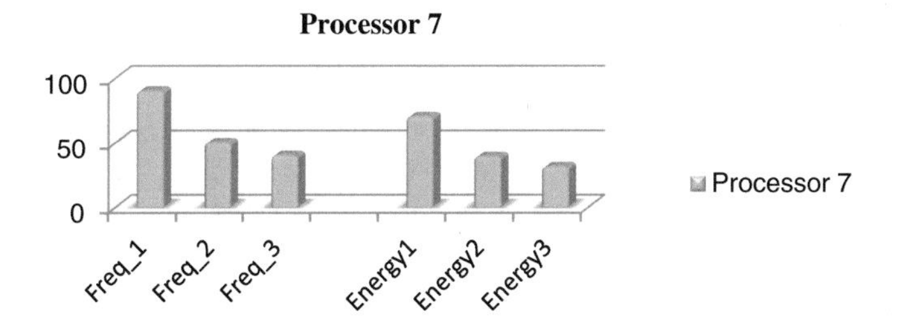

Graph 1 (g)

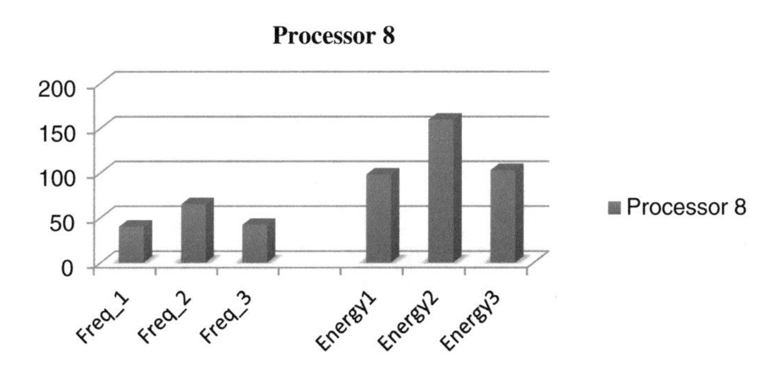

Graph 1(h)

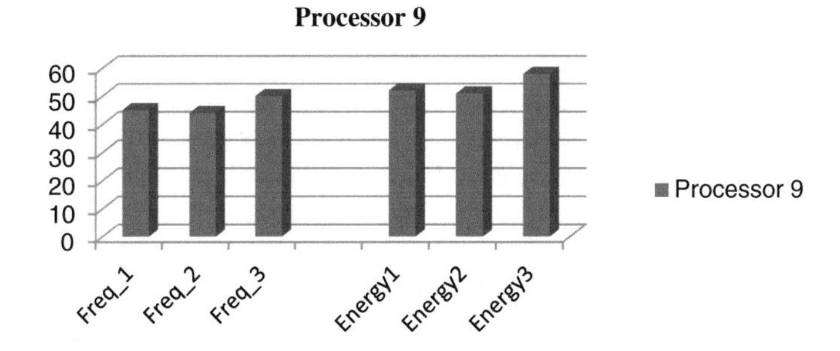

Graph 1(i)

4 Results and Remarks

In below graph, the results have shown the consolidated graph by the result of 9 processors and the measures values in Table 1. In given below, the graph shown as the multiple energy computed by the multiple frequencies with multiple time and different weight measures in the neighbour node. The different processor and distinct energy consume. By apply multiple frequency and different weighted for computed results in Table 1, the results shows the comparison of the processor for minimum multiple frequency and low energy in data transmission (Table 2).

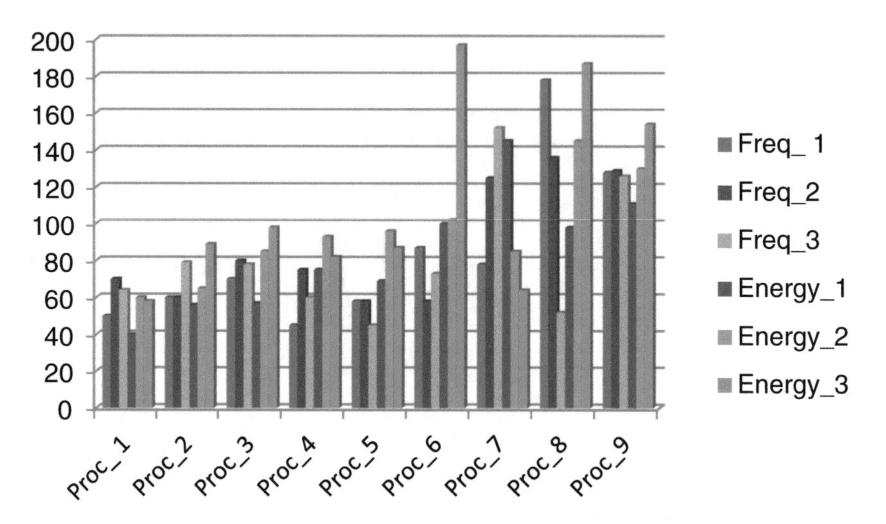

The energy consumption shown in above graph in the nine processors gives the minimum energy of the distributed frequency of the Processor 7 and distributed energy $E_3 = 31.25$ W and frequency is 40 Hz and weight = 10 units. The more energy results shown is the maximum weight of the processor connected nodes at the server distance the energy is more consumption. There are three multiple frequencies applied to the data centre, we have found that multiple energy in the distributed data centre. The graph represents the highest peak shows the maximum energy of the processor 6 and lowest point of the graph represents minimum energy consumption as per results of above graph. So as well as the maximum frequency point indicates the graph in processor 8 and minimum frequency represents the graph in processor 5. The consolidated results of all 9 processors and gives the energy to reduce using the Floyd–Warshall Algorithm.

Table 1 The parameter taken by the CPU and multiple frequencies, time and V

Proc_ (p)	Cap_(c)	Input Voltage (V)	Time (s)	Freq F_1 (Hz)	Freq F_2 (Hz)	Freq F_3 (Hz)	$Wt_{(i,j,k)}$	Energy (E_1) (W)	Energy (E_2) (W)	Energy (E_3) (W)
Proc_1	50	5	100	50	45	52	9	69.44	62.50	72.22
Proc_2	50	5	110	60	40	51	6	11.36	75.75	96.59
Proc_3	50	5	120	65	30	53	9	75.33	34.72	61.34
Proc_4	60	5	130	70	35	47	6	112.17	56.08	75.32
Proc_5	65	5	140	75	47	48	6	111.60	69.94	71.42
Proc_6	50	5	150	85	41	39	3	116.66	113.88	108.33
Proc_7	50	5	160	90	50	40	10	70.31	39.06	31.25
Proc_8	50	5	170	40	65	42	3	98.03	159.31	102.94
Proc_9	50	5	180	45	44	50	6	52.00	50.92	57.87

Table 2 Data centre parameters

Data centre parameter	
Topology	Step network
Numbers of nodes	9 or verified as required
Number of destination nodes	All nodes (9)
MAC protocol	802.3 IEEE (Wired), 802.11 IEEE (Wireless)
Maximum simulation time	250 s
Maximum packet size	65 bytes
Connection types	TCP/UDP
Graph	Excel graph

5 Conclusion

This paper is comparison [19] of the multiple processors energy consumptions of the different level frequency distribution. This paper used a new concept named weight (wt). The weight is dependent on the node by node distance to connect the topology graph. If the weight is maximum, then energy consumption will be maximum and now the weight measured by the neighbour node is low and the energy consumption will be low. In this paper a new concept implemented and comparison with research literature [19]. The futuristic research work of the resulting requirement is mandatory for power efficiency and consumption partially dependent on the weight. In futuristic work, we can implement Floyd–Warshall Algorithm to reduce the energy in any networks. Aghbari et al. [20] have proposed a grid-based clustering algorithm for optimizing energy consumption in wireless sensor network. Agnihotri and Venkatachalapathy [21] have described a distribution function for information sink with N correlated source to compute energy. They have also proposed the distribution function for two class (1) compressible function and (2) incompressible function in worst-case computation.

References

1. Moharir, S., Krishnasamy, S., Shakkottai, S.: Scheduling in densified networks: algorithms and performance. IEEE/ACM Trans. Netw. **25**(1), 164–178 (2017)
2. Xiao, X., Xie, G., Li, R., Li, K.: Minimizing schedule length of energy consumption constrained parallel applications on heterogeneous distributed systems. In: IEEE Trust-Com/BigDataSE/ISPA (2016)
3. Devi, R.K., Devi, K.V., Arumugam, S.: Dynamic batch mode cost-efficient independent task scheduling scheme in cloud computing. Int. J. Adv. Soft Comput. Appl, **8**(2) (2016). ISSN 2074-8523
4. Shi, G., Liu, D., Wei, Q.: Energy consumption prediction of office buildings based on echo state networks. Neurocomputing **216**, 478–488 (2016)

5. Huang, H., Savkin, A.V.: An Energy Efficient Approach for Data Collection in Wireless Sensor Networks Using Public
6. Kaswan, A., Nitesh, K., Jana, P.K.: Energy efficient path selection for mobile sink and data gathering. Wirel. Sens. Netw. Int. J. Electron. Commun. (AEÜ) **73**, 110–118 (2017)
7. Jiang, H.-P., Chuck, D., Chen, W.-M.: Energy-aware data center networks. J. Netw. Comput. Appl. **68**, 80–89 (2016)
8. Imran, M., Collier, M., Landais, P., Katrinis, K.: Performance evaluation of hybrid optical switch architecture for data center networks. Opt. Switch. Netw. **21**, 1–15 (2016)
9. Harbin, J., Indrusiak, L.S.: Comparative performance evaluation of latency and link dynamic power consumption modelling algorithms in wormhole switching networks on chip. J. Syst. Architect. **63**, 33–47 (2016)
10. Alonso, M., Coll, S., Martínez, J.M., Santonja, V., Lopez, P.: Power consumption management in fat-tree interconnection networks. Parallel Comput. **48**, 59–80 (2015)
11. Zhang, Z., Hu, W., Ye, T., Sun, W., Li, Z., Zhang, K.: Routing and spectrum allocation in multi-ring based data center networks. Opt. Commun. **360**, 25–43 (2017)
12. Avci, B., Trajcevski, G., Tamassia, R., Scheuermann, P., Zhou, F.: Efficient detection of motion-trend predicates in wireless sensor networks. Comput. Commun. **101**, 26–43 (2017)
13. Khelladi, L., Djenouri, D., Rossi, M., Badache, N.: Efficient on- demand multi-node charging techniques for wireless sensor networks. Comput. Commun. **101**, 44–56 (2017)
14. Van Heddeghem, W., et al.: Power consumption modelling in optical multilayer networks. Photon Netw. Commun. **24**, 86–102 (2012)
15. Martinus, K., Sigler, T.J., Searle, G., Tonts, M.: Strategic globalizing centers and sub-network geometries: a social network analysis of multi-scalar energy networks. Geoforum **64**, 78–89 (2015)
16. Huang, J., Lin, C., Bo, C.: Energy efficient speed scaling and task scheduling for distributed computing systems. Chin. J. Electron. **24**(3) (2015)
17. Lin, K., Chen, M., Zeadally, S., Rodrigues, J.J.P.C.: Balancing energy consumption with mobile agents in wireless sensor networks. Future Gener. Comput. Syst. **28**, 446–456 (2012)
18. Zhuo, J., Chakrabarti, C.: Energy-efficient dynamic task scheduling algorithms for DVS systems. ACM Trans. Embedded Comput. Syst. (TECS) **7**(2), 17 (2008)
19. Zomaya, A.Y., Choon Lee, Y.: Multiple frequency selection in dvfs enabled processors to minimize energy consumption. In: Energy- Efficient Distributed Computing Systems, 1st edn. Wiley (2012)
20. Al Aghbari, Z., Kamel, I., Elbaroni, W.: Energy-efficient distributed wireless sensor network scheme for cluster detection. Int. J. Parallel Emergent Distrib. Syst. **28**(1), 1–28 (2013)
21. Agnihotri, S., Venkatachalapathy, R.: Worst-case asymmetric distributed function computation. Int. J. Gener. Syst. **42**(3), 268–293 (2013)

Modified Integral Type Weak Contraction and Common Fixed Point Theorem with an Auxiliary Function

R. K. Saini, Naveen Mani and Vishal Gupta

Abstract Nowadays, the utility of metric space and fixed point iterative techniques in computer sciences increased rapidly. Iterative results offer the mathematical root for a novel method to take out intricacy analysis of algorithms via metrics. This manuscript develops a new framework to derive common fixed point results, by replacing modified integral type contraction in metric spaces with an auxiliary function, without assuming continuity, commutative and compatible property of maps. Additionally, we demonstrate and tried to authenticate a theory which validates our result to deduce some innovative and interesting results that are novel in nature. To give some assistance to our findings, an illustrative example is given.

1 Introduction

Banach [1] established an extraordinary fixed point theorem which is the most commonly applied fixed point result in different areas of mathematics and in engineering sciences. Matthews [2, 3] in 1994, led down the idea of Scott-like topology and extended Banachs fixed point theorem after motivated by the applications to program verification, which is one of the most significant finding of analysis, in framework of partial metric spaces. For some more results, we refer to see [4–9].

Branciari [10], in 2002, introduced an integral type contraction, and proved a unique fixed point result in complete metric spaces. Rhoades [11] and Loung and

R. K. Saini
Department of Mathematics, Bundelkhand University, Jhansi, Uttar Pradesh, India
e-mail: rksaini03@yahoo.com

N. Mani (✉)
Department of Mathematics, Sandip University, Nashik, Maharashtra, India
e-mail: naveenmani81@gmail.com

V. Gupta
Department of Mathematics, Maharishi Markandeshwar (Deemed to be University), Mullana, Haryana, India
e-mail: vishal.gmn@gmail.com

K. Ray et al. (eds.), *Soft Computing: Theories and Applications*,
Advances in Intelligent Systems and Computing 742,
https://doi.org/10.1007/978-981-13-0589-4_11

Thuan [12] extends the main result of [10] and prove a result with the help of control functions. Gupta and Mani [13, 14] generalized and extended the results of Branciari [10], Rhoades [11] and Loung and Thuan [12] for pair of two maps. For some more results on integral type contractions, we refer to see [11, 13–20].

Let us denote $\Psi_1 = \{\omega | \omega : R^+ \to R^+\}$, which is Lebesgue-integrable, summable on subsets of R^+ which is compact and also $\forall\, \epsilon > 0$, $\int_0^\epsilon \omega(l)dl > 0$.

Recently, Bhardwaj et al. [19] proved the following result in complete metric spaces satisfying a new rational contraction for pair of self maps.

Theorem 1 *[19] Suppose S and T are self-maps on $\mathscr{Y}$, and let δ is a distance defined on $\mathscr{Y}$ s.t. $(\mathscr{Y}, \delta)$ is a complete. If possible for each $r, l \in \mathscr{Y}$*

$$\int_0^{\delta(Sr,Ts)} \omega(l)dl \le \kappa \int_0^{\lambda(r,s)} \omega(l)dl + \eta \int_0^{\delta(r,s)} \omega(l)dl \tag{1}$$

and

$$\lambda\,(r, s) = \frac{\delta\,(s, Ts)\,[1 + \delta(r, Sr)]}{[1 + \delta(r, s)]}, \tag{2}$$

where, constants $\kappa, \eta > 0$ with $\kappa + \eta < 1$ and $\omega \in \Psi_1$. Then there exist a coincidence point μ, which is common and unique s.t. $S\mu = T\mu = \mu$.

Ansari [21] in 2014, defined a new notion known as C-class function, as one of the extraordinary generalization of Banach contraction principle. In general, we denote it by $\mathscr{C}$.

Definition 1 [21] We say $\phi : [0, +\infty) \to [0, +\infty)$ ultra distance function if it is continuous and $\phi(0) \ge 0$, and $\phi(l) > 0, l > 0$.

Remark 1 We let Φ_r denote the class of the ultra distance functions.

Definition 2 [21] A family of mappings $G : [0, \infty)^2 \to R$ are known as C-class functions if

1. G is continuous;
2. $G(u, l) \le u$;
3. $G(u, l) = u$ only if either $u = 0$ or $l = 0$; $\forall\ u, l \in [0, \infty)$.

Remark 2 Clearly, for some G we have $G(0, 0) = 0$.

Some beautiful examples of C-class functions are given in [21].

Remark 3 Consider a rational expression:

$$\begin{aligned} N(r, s) &= \frac{\delta(s, Ts) \cdot \delta(r, Sr)}{1 + \delta(r, s)} < \frac{\delta(s, Ts)}{1 + \delta(r, s)} + \frac{\delta(s, Ts) \cdot \delta(r, Sr)}{1 + \delta(r, s)} \\ &= \frac{\delta\,(s, Ts)\,[1 + \delta(r, Sr)]}{[1 + \delta(r, s)]} \\ &= \lambda\,(r, s) \end{aligned}$$

This concludes that expression given by $N(r, s)$ in Remark 3 is weaker than expression given by $\lambda(r, s)$ on the right-hand side. We also mention here an important lemma that are helpful in deducing our main result.

Lemma 1 *[22] Let $\left(a_p\right)_{p\in N}$ be a nonnegative sequence with $\lim_{p\to\infty} a_p = k$. If $\omega \in \Psi_1$, then*

$$\lim_{p\to\infty} \int_0^{a_p} \omega(l)dl = 0 \ \ iff \ \ \lim_{p\to\infty} a_p = 0.$$

The main objective of this paper is to prove a new common fixed point theorem for pair of self-maps satisfying a weaker rational expression (given by $N(r, s)$, see equation (4)) for integral type contraction with C- class function. We prove our result by omitting the completeness of spaces, continuity, commutativity, and compatibility of maps. Section 3 contains an example and some applications of our findings which demonstrated that our findings are new, weaker and different from other results (such as [19, 23, 24]).

2 Main Results

Our main result is the following theorem.

Theorem 2 *Suppose S and T are self-maps on $\mathscr{Y}$, and let δ is a distance defined on $\mathscr{Y}$ s.t. $(\mathscr{Y}, \delta)$ is a complete. If possible for each $r, l \in \mathscr{Y}$*

$$\int_0^{\delta(Sr,Ts)} \omega(l)dl \leq G\Bigg(\int_0^{N(r,s)} \omega(l)dl, \phi\Bigg(\int_0^{N(r,s)} \omega(l)dl \Bigg)\Bigg), \tag{3}$$

where,

$$N(r, s) = \frac{\delta(s, Ts) \cdot \delta(r, Sr)}{1 + \delta(r, s)}, \tag{4}$$

where $G \in \mathscr{C}$, $\phi \in \Phi_r$ and $\omega \in \Psi_1$.
Then there exists a coincidence point μ, which is common and unique s.t. $S\mu = T\mu = \mu$.

Proof Choose $r_0 \in \mathscr{Y}$ such that $Sr_0 = r_1$ and $Tr_1 = r_2$. Continuing like this, we can construct sequence $\{r_p\}$ in $\mathscr{Y}$ such as

$$r_{2p+1} = Sr_{2p} \quad and \quad r_{2p+2} = Tr_{2p+1}, \quad where \ \ p = 0, 1, 2, \cdots. \tag{5}$$

If possible, assume that for no $p \in N$

$$r_{2p+1} = r_{2p+2}. \tag{6}$$

Consider

$$\int_0^{\delta(r_{2p+1},r_{2p+2})} \omega(l)dl = \int_0^{\delta(Sr_{2p},Tr_{2p+1})} \omega(l)dl$$

$$\leq G\left(\int_0^{N(r_{2p},r_{2p+1})} \omega(l)dl, \phi\left(\int_0^{N(r_{2p},r_{2p+1})} \omega(l)dl\right)\right), \tag{7}$$

From (4)

$$N\left(r_{2p}, r_{2p+1}\right) = \frac{\delta(r_{2p+1}, Tr_{2p+1}) \cdot \delta(r_{2p}, Sr_{2p})}{1 + \delta(r_{2p}, r_{2p+1})} = \frac{\delta(r_{2p+1}, r_{2p+2}) \cdot \delta(r_{2p}, r_{2p+1})}{1 + \delta(r_{2p}, r_{2p+1})} \tag{8}$$

Since δ is a metric, therefore for all p $\frac{\delta(r_{2p},r_{2p+1})}{1+\delta(r_{2p},r_{2p+1})} < 1$, and hence

$$\frac{\delta(r_{2p+1}, r_{2p+2}) \cdot \delta(r_{2p}, r_{2p+1})}{1 + \delta(r_{2p}, r_{2p+1})} < \delta(r_{2p+1}, r_{2p+2})$$

Therefore, from (8)

$$N\left(r_{2p}, r_{2p+1}\right) \leq \delta(r_{2p+1}, r_{2p+2}).$$

Hence from (7),

$$\int_0^{\delta(r_{2p+1},r_{2p+2})} \omega(l)dl \leq G\left(\int_0^{\delta(r_{2p+1},r_{2p+2})} \omega(l)dl, \phi\left(\int_0^{\delta(r_{2p+1},r_{2p+2})} \omega(l)dl\right)\right).$$

This is only feasible if $\int_0^{\delta(r_{2p+1},r_{2p+2})} \omega(l)dl = 0$, and therefore $\delta(r_{2p+1}, r_{2p+2}) = 0$. Thus, assumption in (6) is wrong and therefore $r_{2p+1} = r_{2p+2}$ for some $p = i \in N$. Accordingly, $r_{2i+1} = r_{2i+2}$ and $\mu = r_{2i+1}$, we obtain $T\mu = \mu$ from (5).
Also from (3), if we consider

$$\int_0^{\delta(S\mu,\mu)} \omega(l)dl = \int_0^{\delta(S\mu,T\mu)} \omega(l)dl$$

$$\leq G\left(\int_0^{N(\mu,\mu)} \omega(l)dl, \phi\left(\int_0^{N(\mu,\mu)} \omega(l)dl \right) \right) \tag{9}$$

Using (4),

$$N(\mu, \mu) = \frac{\delta(\mu, T\mu) \cdot \delta(\mu, S\mu)}{[1 + \delta(\mu, \mu)]} = 0.$$

Thus from (9), we get

$$\int_0^{\delta(S\mu,\mu)} \omega(l)dl \leq G(0, 0) = 0.$$

This is possible only if $\int_0^{\delta(S\mu,\mu)} \omega(l)dl = 0$ implies that $\delta(S\mu, \mu) = 0$, implies $S\mu = \mu$.
Assume that, obtained fixed point is not unique. That is, there exists another point $\nu \neq \mu$ s. t. $S\nu = T\nu = \nu$.
Again from (3),

$$\int_0^{\delta(\nu,\mu)} \omega(l)dl = \int_0^{\delta(S\nu,T\mu)} \omega(l)dl$$

$$\leq G\left(\int_0^{N(\nu,\mu)} \omega(l)dl, \phi\left(\int_0^{N(\nu,\mu)} \omega(l)dl \right) \right), \tag{10}$$

where,

$$N(\nu, \mu) = \frac{\delta(\mu, T\mu) \cdot \delta(\nu, S\nu)}{[1 + \delta(\nu, \mu)]} = 0. \tag{11}$$

Hence from (10)

$$\int_0^{\delta(\nu,\mu)} \leq G(0, 0) = 0.$$

This is only feasible if $\delta(\nu, \mu) = 0$. This completes the proof of main result.

3 Applications

Here, we give several interesting corollaries, as an application of our main result, in the underlying spaces. Some of them are novel in literature. One example is given in support of our findings.

First we take $G(u, s) = \frac{u}{1+s}$ in Theorem 2, to get one of the innovative result as follows.

Corollary 1 *Suppose S and T are self-maps on $\mathscr{Y}$, and let δ is a distance defined on $\mathscr{Y}$ s.t. $(\mathscr{Y}, \delta)$ is a complete. If possible for each r, l $\in \mathscr{Y}$*

$$\int_0^{\delta(Sr,Ts)} \omega(l)dl \le \frac{\int_0^{N(r,s)} \omega(l)dl}{1+\phi\left(\int_0^{N(r,s)} \omega(l)dl\right)},$$

where $N(r, s)$ is given by Eq. (4), $\phi \in \Phi_r$ and $\omega \in \Psi_1$.
Then there exist a coincidence point μ, which is common and unique s.t. $S\mu = T\mu = \mu$.

Again, we take $G(u, l) = u - \frac{l}{k+l}$ in Theorem 2, to get one of the new and interesting result.

Corollary 2 *Suppose S and T are self-maps on $\mathscr{Y}$, and let δ be a distance defined on $\mathscr{Y}$ s.t. $(\mathscr{Y}, \delta)$ is a complete. If possible for each r, l $\in \mathscr{Y}$*

$$\int_0^{\delta(Sr,Ts)} \omega(l)dl \le \int_0^{N(r,s)} \omega(l)dl - \frac{\phi\left(\int_0^{N(r,s)} \omega(l)dl\right)}{k+\phi\left(\int_0^{N(r,s)} \omega(l)dl\right)},$$

where $N(r, s)$ is given by Eq. (4), $\phi \in \Phi_r$ and $\omega \in \Psi_1$.
Then there exists a coincidence point μ, which is common and unique s.t. $S\mu = T\mu = \mu$.

Also, we take $G(u, l) = \frac{u}{(1+u)^2}$ in Theorem 2 to get one of the innovative and interesting result.

Corollary 3 *Suppose S and T are self-maps on $\mathscr{Y}$, and let δ is a distance defined on $\mathscr{Y}$ s.t. $(\mathscr{Y}, \delta)$ is a complete. If possible for each r, l $\in \mathscr{Y}$*

$$\int_0^{\delta(Sr,Ts)} \omega(l)dl \le \frac{\int_0^{N(r,s)} \omega(l)dl}{\left(1+\int_0^{N(r,s)} \omega(l)dl\right)^2},$$

where $N(r,s)$ is given by Eq. (4) and $\omega \in \Psi_1$.
Then there exist a coincidence point μ, which is common and unique s.t. $S\mu = T\mu = \mu$.

On taking $G(u,l) = u - \theta(u)$ in Theorem 2, we have the following innovative and novel result.

Corollary 4 *Suppose S and T are self-maps on $\mathscr{Y}$, and let δ is a distance defined on $\mathscr{Y}$ s.t. $(\mathscr{Y},\delta)$ is a complete. If possible for each $r, l \in \mathscr{Y}$*

$$\int_0^{\delta(Sr,Ts)} \omega(l)dl \leq \int_0^{N(r,s)} \omega(l)dl - \theta\Big(\int_0^{N(r,s)} \omega(l)dl\Big),$$

where $N(r,s)$ is given in Eq. (4), $\omega \in \Psi_1$ and $\theta : R^+ \to R^+$ is a mapping which is continuous, and s.t. $\theta(l) = 0$ iff $l = 0$.
Then there exists a coincidence point μ, which is common and unique s.t. $S\mu = T\mu = \mu$.

On taking $G(u,l) = u - l$ in main result 2, we get the following generalized new result.

Corollary 5 *Suppose S and T are self-maps on $\mathscr{Y}$, and let δ be a distance defined on $\mathscr{Y}$ s.t. $(\mathscr{Y},\delta)$ is a complete. If possible for each $r, l \in \mathscr{Y}$*

$$\int_0^{\delta(Sr,Ts)} \omega(l)dl \leq \int_0^{N(r,s)} \omega(l)dl - \phi\Big(\int_0^{N(r,s)} \omega(l)dl\Big),$$

where $N(r,s)$ is given in Eq. (4), $\phi \in \Phi_r$ and $\omega \in \Psi_1$.
Then there exists a coincidence point μ, which is common and unique s.t. $S\mu = T\mu = \mu$.

4 Example

Example 1 Let $\mathscr{Y} = [1,\infty)$ be a space. Define a metric $\delta(r,s)$ as

$$\delta(r,s) = \left\{\begin{matrix} r+s, & if \;\; r \neq s; \\ 0, & if \;\; r = s; \end{matrix}\right\}$$

Then clearly δ is a metric on $\mathscr{Y}$ and $(\mathscr{Y},\delta)$ is a metric space.
Define maps $S, T : \mathscr{Y} \to \mathscr{Y}$ by

$$S(1) = 1; \;\; S(r) = \frac{r^{\frac{1}{4}}}{4}, \;\; if\ 1 < r < \infty;$$

$$T(1) = 1; \quad T(r) = \frac{r^{\frac{1}{5}}}{5}, \quad if\ 1 < r < \infty;$$

Let $\phi, \omega : [0, +\infty) \rightarrow [0, +\infty)$ be defined as $\omega(l) = 2$ and $\phi(l) = \frac{l}{10}$ for all $l \in R^{+}$, then $\phi \in \Phi_r$ and for each $\epsilon > 0$, $\omega \in \Psi_1$ such that $\int_0^{\epsilon} \omega(l)dl = 2\epsilon > 0$.

CASE: 1 If define function $G : [0, \infty)^2 \rightarrow R$ as $G(u, l) = u - l$, for all $u, l \in [0, \infty)$,

CASE: 2 If define function $G : [0, \infty)^2 \rightarrow R$ as $G(u, l) = u - \omega(u)$, for all $u, l \in [0, \infty)$, where, $\omega : [0, \infty) \rightarrow [0, \infty)$ is a continuous function defined by $\omega(u) = \sqrt{u}$.

Then clearly, above in all the two cases, G is a C-class function.
Also, all the conditions of Theorem 2 are satisfied for all the two cases given above, and clearly for $r = 1$, $Sr = Tr = r$. in $\mathscr{Y}$.

5 Conclusion

Main result of this article has been proved by omitting the completeness of spaces, continuity, commutativity, and compatibility of maps. Section 3 contains a number of applications of our findings which demonstrated that our findings are new, weaker, and different from other results (such as [19, 23, 24]). In Sect. 4, a beautiful example illustrating and justifying our main finding has been given.

References

1. Banach, S.: Sur les operations dans les ensembles abstraits et leur application aux equations integrals. Fundam. Math. **3**, 133–181 (1922)
2. Matthews, S.G.: Partial metric topology. Ann. N.Y. Acad. Sci. **728**, 183–197 (1994)
3. Matthews, S.G.: An extensional treatment of lazy data flow deadlock. Theor. Comput. Sci. **151**, 195–205 (1995)
4. Gupta, V., Shatanawi, W., Mani, N.: Fixed point theorems for (ψ, β)-Geraghty contraction type maps in ordered metric spaces and some applications to integral and ordinary differential equations. J. Fixed Point Theory Appl. **19**(2), 1251–1267 (2017)
5. Seda, A.K., Hitzler, P.: Generalized distance functions in the theory of computation. Comput. J. **53**, 443–464 (2010)
6. Abbas, M., Nazir, T., Romaguera, S.: Fixed point results for generalized cyclic contraction mappings in partial metric spaces. Rev. R. Acad. Cienc. Exactas Fs. Nat. Ser. A Mat. **106**, 287–297 (2012)
7. Romaguera, S., Tirado, P., Valero, O.: New results on mathematical foundations of asymptotic complexity analysis of algorithms via complexity spaces. Int. J. Comput. Math. **89**, 1728–1741 (2012)

8. Cerda-Uguet, M.A., Schellekens, M.P., Valero, O.: The Baire partial quasi-metric space: a mathematical tool for the asymptotic complexity analysis in computer science. Theory Comput. Syst. **20**, 387–399 (2012)
9. Gupta, V., Mani, N., Devi, S.: Common fixed point result by using weak commutativity. In: AIP Conference Proceedings **1860**(1) (2017)
10. Branciari, A.: A fixed point theorem for mappings satisfying a general contractive condition of integral type. Int. J. Math. Math. Sci. **29**, 531–536 (2002)
11. Rhoades, B.E.: Two fixed point theorems for mappings satisfying a general contractive condition of integral type. Int. J. Math. Math. Sci. **63**, 4007–4013 (2003)
12. Loung, N.V., Thuan, N.X.: A fixed point theorem for weakly contractive mapping in metric spaces. Int. J. Math Anal. **4**, 233–242 (2010)
13. Gupta, V., Mani, N.: Existence and uniqueness of fixed point for contractive mapping of integral type. Int. J. Comput. Sci. Math. **4**(1), 72–83 (2013)
14. Gupta, V., Mani, N.: A common fixed point theorem for two weakly compatible mappings satisfying a new contractive condition of integral type. Math. Theory Model. **1**(1), 1–6 (2011)
15. Gupta, V., Mani, N., Tripathi, A.K.: A fixed point theorem satisfying a generalized weak contractive condition of integral type. Int. J. Math. Anal. **6**, 1883–1889 (2012)
16. Altun, I., Turkoglu, D., Rhoades, B.E.: Fixed points of a weakly compatible maps satisfying a general contractive condition of integral type. Fixed Point Theory Appl. **2007**, 9 (2007). Article Id 17301
17. Liu, Z.Q., Zou, X., Kang, S.M., Ume, JS.: Fixed point theorems of contractive mappings of integral type. Fixed Point Theory Appl. **2014** (2014). Article Id: 394
18. Alsulami, H.H., Karapinar, E., O Regan, D., Shahi, P.: Fixed points of generalized contractive mappings of integral type. Fixed Point Theory Appl. **2014** (2014). Article Id 213
19. Bhardwaj, V.K., Gupta, V., Mani, N.: Common fixed point theorems without continuity and compatible property of maps. Bol. Soc. Paran. Mat. (3s.) **35(3)**, 67–77 (2017)
20. Gupta, V., Mani, N.: Common fixed point for two selfmaps satisfying a generalized $^{\psi}\int_{\phi}$ weakly contractive condition of integral type. Int. J. Nonlinear Sci. **16**(1), 64–71 (2012)
21. Ansari, A.H.: Note on φ -ψ- contractive type mappings and related fixed point. In: The 2nd Regional Conference on Mathematics And Applications. PNU, pp. 377–380 Sept 2014
22. Mocanu, M., Popa, V.: Some xed point theorems for mapping satisfying implicit relations in symmetric spaces. Libertas Math. **28**, 1–13 (2008)
23. Dass, B.K., Gupta, S.: An extension of Banach contraction principle through rational expression. Indian J. of Pure Appl. Math. **12**(6), 1455–1458 (1973)
24. Jaggi, D.S.: On common fixed points of contractive maps. Bull. Math. de la Soc. Sct. Math.de la R.S.R. **20**(1–2), 143 – 146 (1975)

Crop Monitoring Using IoT: A Neural Network Approach

Rajneesh Kumar Pandey, Santosh Kumar and Ram Krishna Jha

Abstract For a decent yield, the farmer needs to monitor the field continuously. Internet of Things (IoT) plays an important role in enhancing the yield by changing the conventional method of monitoring. This paper centers on crop monitoring using IoT which will help farmers to continuously monitor the crops and take various decisions according to the needs. The system will provide data of soil moisture, leaf area index, leaf water area index, plant water content, and vegetation water mass. The data are collected using Arduino microcontroller along with the sensors deployed in the field remotely. The data once received are analyzed by applying feed forward, cascade forward, and function fitting neural network. After analyzing the data, cascade forward neural network gave the best result.

Keywords Soil moisture · Leaf area index (LAI) · Leaf water area index (LWAI) Plant water content (PWC) · Neural network

1 Introduction

The Internet has been around for a while, but it has been mostly a product of people that was created by people for people and about people. Internet is one of the most important and transformative technology ever invented. The Internet is like digital fabric that woven into the life of all of us in one way or another. The Internet of people changed the world. The new Internet is emerging and it is poised to change the world again. This new Internet is not just about connecting people but it is about connecting things and so its name is the "Internet of Things" [1]. The benefit of

R. K. Pandey (✉) · S. Kumar · R. K. Jha
Graphic Era University, 566/6 Bell Road, Clement Town, Dehradun, India
e-mail: rajneesh.p22@gmail.com

S. Kumar
e-mail: amu.santosh@gmail.com

R. K. Jha
e-mail: vickeyjha22@gmail.com

K. Ray et al. (eds.), *Soft Computing: Theories and Applications*,
Advances in Intelligent Systems and Computing 742,
https://doi.org/10.1007/978-981-13-0589-4_12

connecting things by Internet is sharing of information between them by giving them the ability to sense and communicate. Agriculture provides us food grains and other raw material for survival of human being and thus acts as the most important sector for humans [2]. In a country like India which is known to be the land of farmers, around one-third of the population depends on the agriculture and its related field, it provides the employment and thus is the source of income to many. The available area for agricultural purpose is decreasing day by day as the population is increasing because the land is being compromised for housing and other industrial needs. So there is a need for increasing the yield from the available field to meet the needs of increasing population. Also changing environmental factors makes it harder for the farmers to increase the yield using traditional farming process. Researchers are trying to produce new verities to increase the yield using minimal resources but these crops need to be monitored properly to get the desired result otherwise the outcome may not be as per as expectations. The manual monitoring of crop is very tedious and time consuming. The present work is done to address these problems as they were the motivation behind it. Modern technology like IoT can help farmers in monitoring the crop remotely and also in taking timely action. This paper presents an IoT-based crop monitoring system which would help the farmers to monitor the crop remotely. The soil moisture sensor along with the Arduino microcontroller measures the soil moisture. The system then checks for the water content in the crop using various parameters like LAI, LWAI, PWC, and it alerts farmers in a needy situation. Also, the data gathered can be stored and used for future prediction. LAI is an important biophysical variable used for monitoring of crop growth, health and estimation of yield.

Rest of the paper is organized as follows: Sect. 2 provides background, Sect. 3 is proposed solution has been detailed, Sects. 4 and 5 describe implementation and result. Finally, conclusion forms the essence of Sect. 6.

2 Background

The leading challenge identified in the area of agriculture is to provide farmers essential data in time. It is hard to find knowledge to help practical farming as it may not lie or burdensome to find. Here, the web application comes up with a good solution to enhance distribution of intelligence about feasible farming procedures. Joaquín et al. developed an algorithm to control water quantity with the use of threshold values of soil moisture and temperature which can be programmed into a microcontroller-based gateway. Photovoltaic panels can be used to power the system which has duplex communication link based on a cellular–Internet interface which allows programming through a web page for data inspection and irrigation [3, 4]. Y. Kim developed an irrigation system with the help of distributed wireless sensor network with the intent of changing rate of irrigation, real time in field sensing to increase the yield with minimum application of water [5]. Zhiqiang Xiao et al.

developed a method to estimate LAI from time series remote sensing data using general regression neural networks (GRNNs) [6].

3 Proposed Solution

Water is the main resource required for the healthy growth of plant. The intake of all the nutrients by the crop from the soil is done with the help of water. Also, water is necessary for photosynthesis. It is the building block of cells and thus the most important resource for the production of crop. The main aim is to continuously monitor the level of water in the plant so as to gain maximum yield from the crops [3, 7]. The user can monitor the crop through electronic devices and take necessary action when needed. This system will save time and use the water more precisely and also keep the crop needs fulfilled.

The leaf area index (LAI) [8] is defined as the one-sided green leaf area per unit ground area.

$$\mathrm{LAI} = N_a * A_1 \tag{1}$$

where A_1 is the one-sided area of the leaf and N_a is the areal density which is calculated as

$$\mathrm{N_a} = \text{No. of leaves/plant} \times \text{No. of plants, per m}^2 \tag{2}$$

Plant water content (PWC) defined as the total amount of water (or moisture) is present in the samples of plant leaves and stem.

$$\mathrm{PWC} = \frac{W_f - W_d}{W_d} \tag{3}$$

where W_f and W_d are the oven-dried weights of freshly picked plant samples collected in the field.

As shown in Fig. 1, soil moisture sensor along with Arduino UNO R3 microcontroller was used for monitoring the moisture level of soil in the field so to predict the amount of water available in soil for crops. The data collected were stored on the cloud database, where data of other parameters were also stored. This data can be accessed by the system where the code is written and can be used for monitoring and analyzing the same. The sensors and board are not so costly so can be easily purchased by the farmers. The value of LAI, LWAI, VWM, and PWC is compared with the soil water content. Many times, though water is present in the soil, it is not absorbed by crop due to problem in roots or any obstacle in the body of crop. By comparing the values the farmers can take decision when required and keep the crop healthy and get optimum yield.

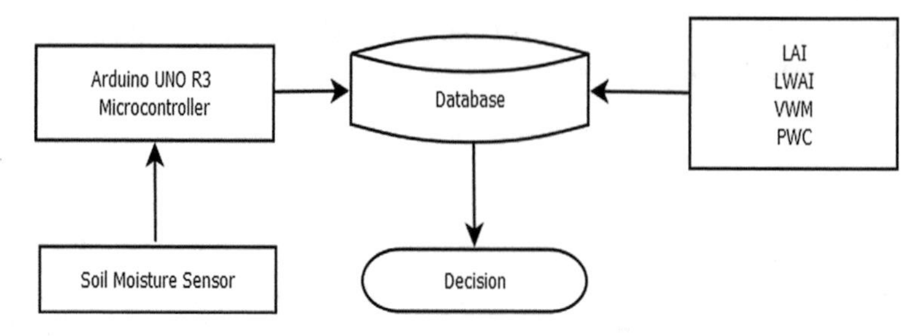

Fig. 1 Block diagram of the proposed system

The system works as shown in the Fig. 2 with the help of soil moisture sensor and Arduino microcontroller, the value of soil moisture content is acquired when switched on. The data gathered are transmitted to the system database and it compares the various factors. The data are analyzed and any deviation from the ideal condition is found. If deviation is found more than acceptable value, then appropriate action is being taken according to the situation that also includes warning to the farmers.

4 Implementation

A wheat crop was monitored from the system and the results were carried out in the form of soil moisture, LAI, LWAI, PWC, and VWM. The crop was monitored for 140 days and the data were stored in the database for analyzing.

Figure 3 shows a plot of soil moisture data received over 140 days. The plot shows the variations in the moisture. When the value goes below 10, farmers need to take the action.

Figure 4 shows a plot of LAI data received over 140 days. The plot shows the variations in the LAI. When the value goes below 1, it signifies unhealthy leaf.

Figure 5 shows a plot of LWAI data received over 140 days. The plot shows the variations in the LWAI. When the value goes below 5, it signifies that the water content in the leaf is less.

Figure 6 shows a plot of PWC data received over 140 days. The plot shows the variations in the PWC. When the value goes below 3, it shows that there is not enough water in the plant. It signifies that soil is dry.

Figure 7 shows a plot of VWM data received over 140 days. The plot shows the variations in the VWM. When the value goes below 1000, it may hinder the proper growth of the plant.

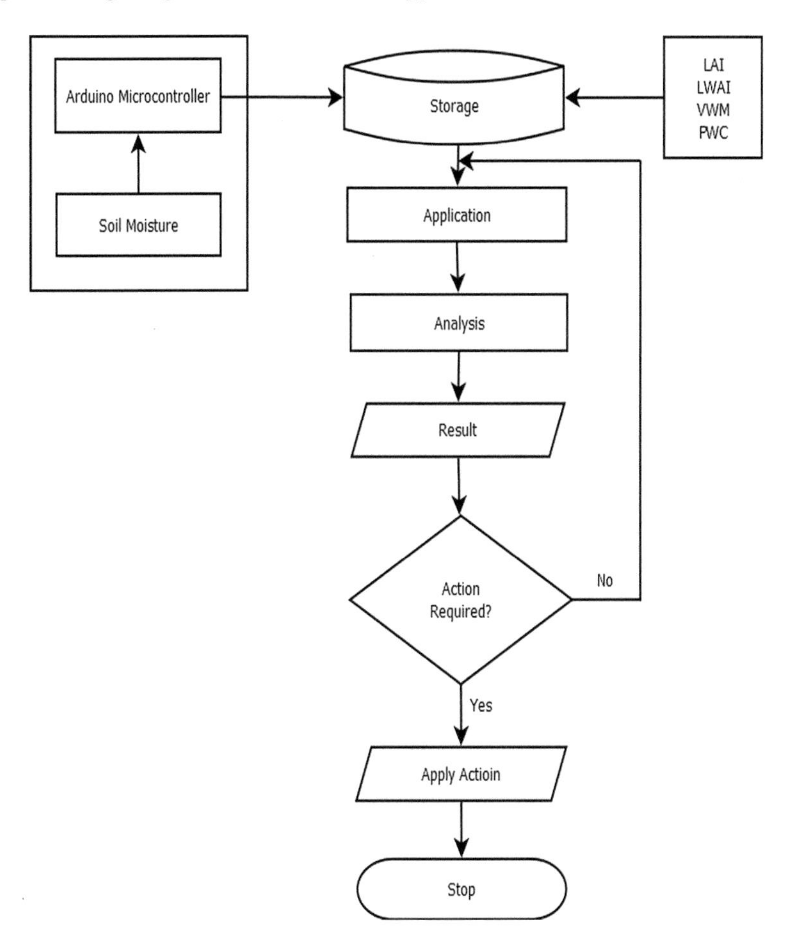

Fig. 2 Proposed system model

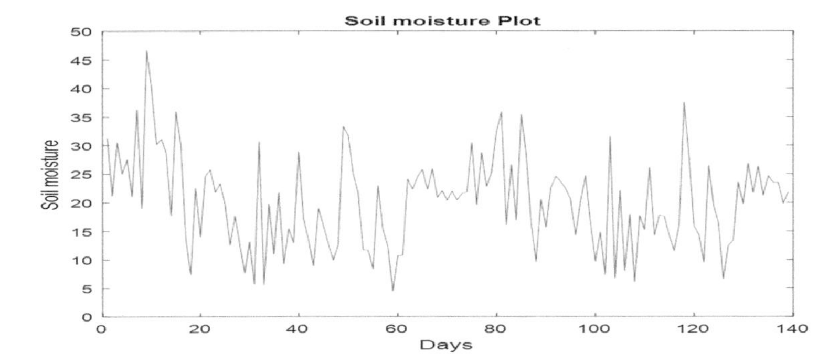

Fig. 3 Soil moisture monitoring for 140 days

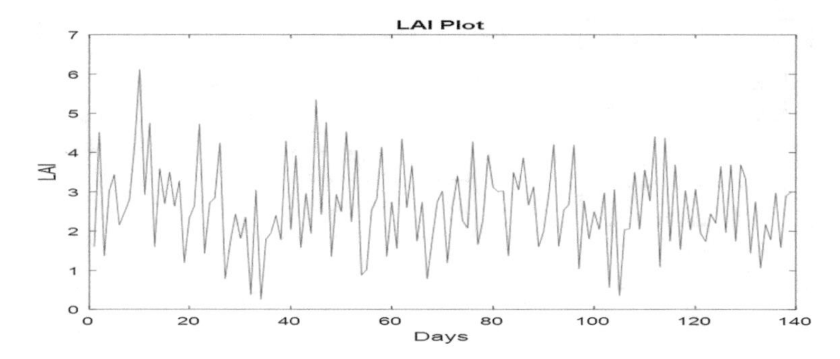

Fig. 4 Leaf area index monitoring for 140 days

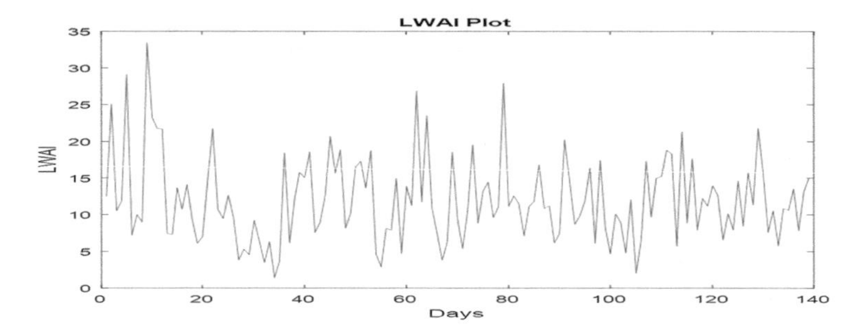

Fig. 5 Leaf water area index monitoring for 140 days

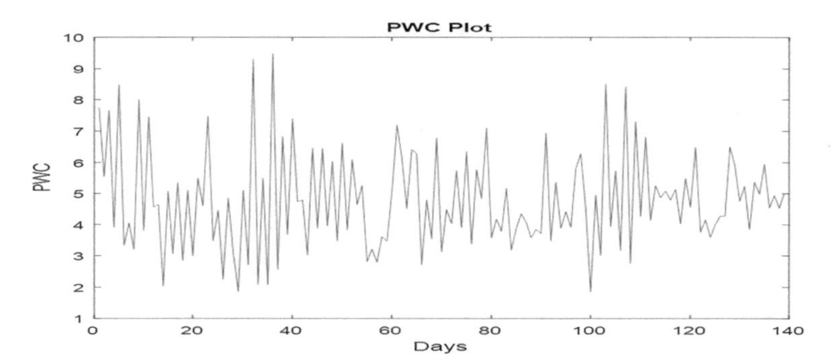

Fig. 6 Plant water content monitoring for 140 days

5 Result Analysis

After analyzing the crop data for 140 days, an artificial neural network was used for prediction. The workflow of the network has following steps: collect data, create the network, configure the network, initialize the weights and biases, train the network, and validate the network. Performance, training state, error histogram, and regression plot can be obtained from the training window.

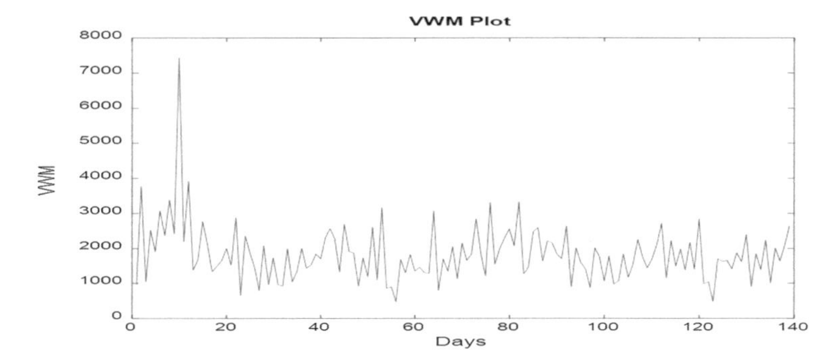

Fig. 7 Vegetation water mass monitoring for 140 days

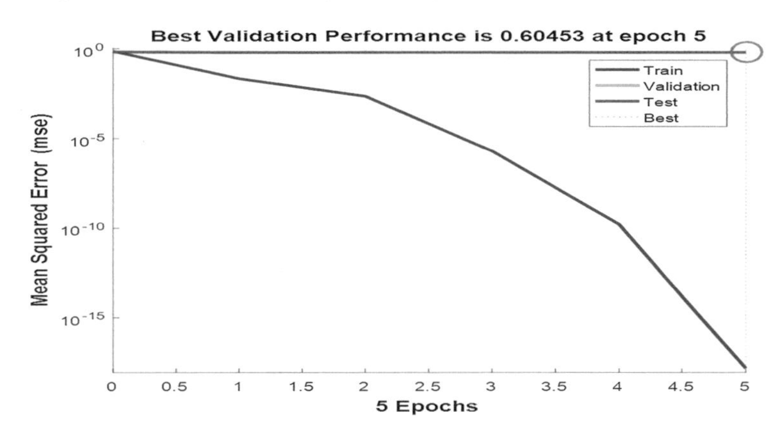

Fig. 8 Performance plot

Figure 8 shows variation in mean square error with respect to epochs. As the number of epochs for training and testing increases the error rate decreases.

Figure 9 shows variation in gradient coefficient with respect to number of epochs. Minimum the value of gradient better will be the training and testing. We can see from the figure the value of gradient goes decreasing as the number of epoch increases.

Figure 10 shows the error histogram formed after loading training set of data into Matlab workspace. The various blocks with different colors show the range of variation in the data. The training data is represented by blue bars, validation data by green bars and testing data by red bars. From the graph, it is evident that most errors fall between −1.023 and 0.4052. The training error is 1.502 and target was considered 10 but it came up as 8.

The subsequent stage in approving the network is to make a regression plot as shown in Fig. 11 which shows the association between the outputs of the network and the objective. The four plots represent training, validation, testing of data and the fourth one combines them. The dotted line represents the ideal result. Thus, data point falls along 45° it is called a perfect fit. If the training is best, the network outputs

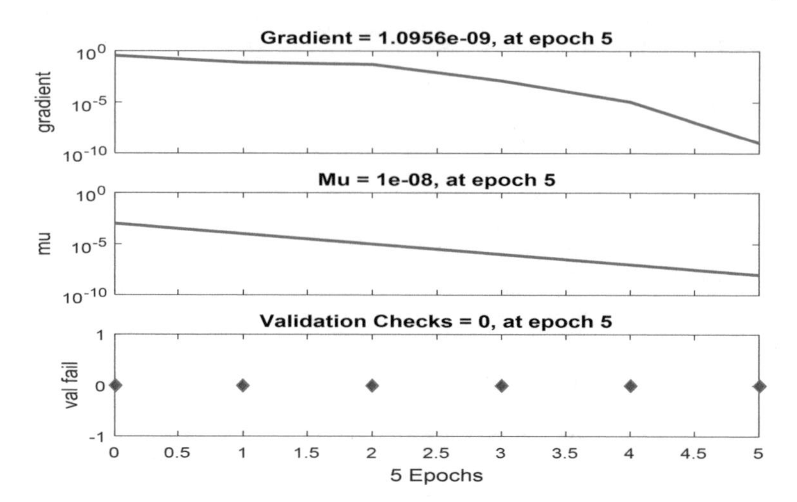

Fig. 9 Training state

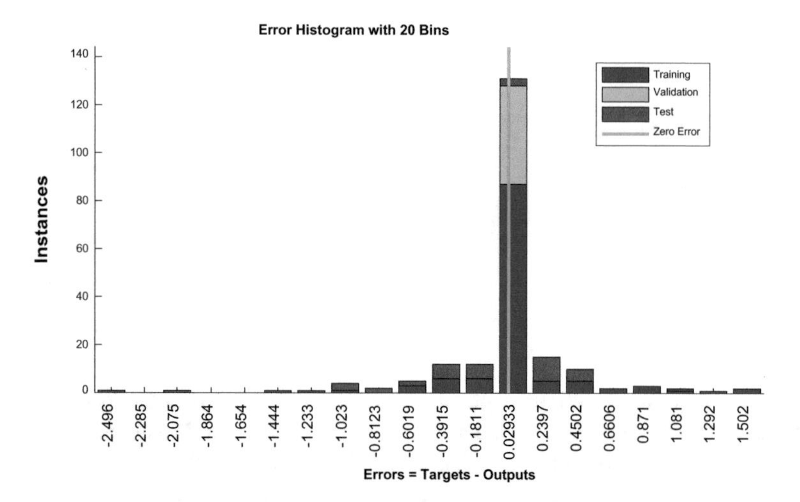

Fig. 10 Error histogram

and the objective would be precisely equivalent, however, the relationship is once in a while idealize in practice The performance of the neural network can be measured by the value of output in the regression plot. If this value is one then the performance is 100% accurate. In the Fig. 11, the value is 0.96 so its performance is 96%. As most of the data points falls around the ideal line, the farmer does not need to worry still an alarm can be triggered to keep an eye on the crop.

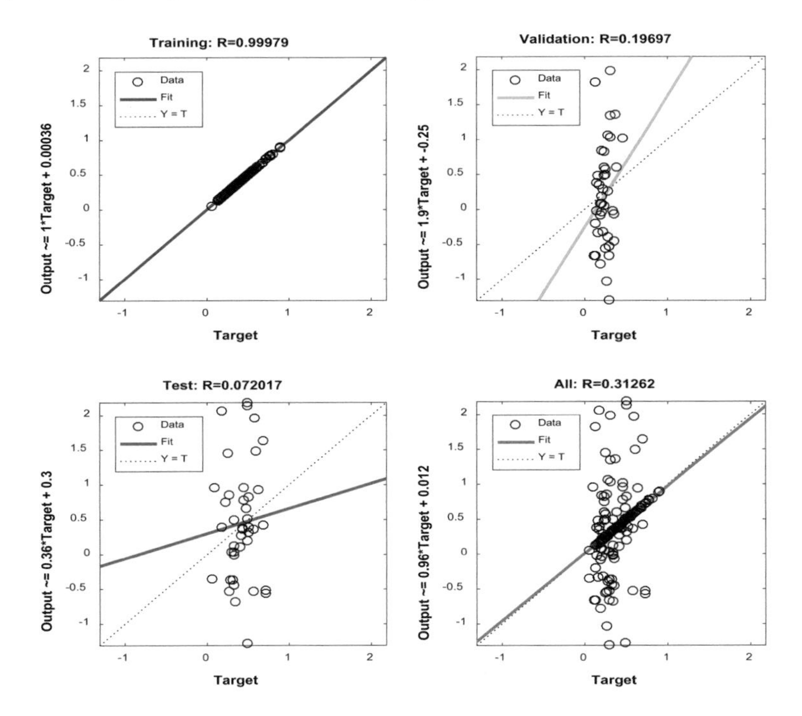

Fig. 11 Regression plot

6 Conclusion

Farmers need the support of technology for better decision-making and guidance for taking timely actions according to situation. Sensors collecting data remotely and constantly monitor the crop acts as a helping hand to farmers. The system is trained perfectly and thus shows 96% accuracy in regression plot. This system constantly monitors the various factors and triggers an alarm to the farmer when the crop condition deviates from a certain limit as compared to ideal condition, i.e., the dotted 45° line so as to take proper countermeasure within the time. This helps in keeping the crop healthy and thus increases the yield. Also, the system is more efficient than previous systems as it considers other parameters apart from soil moisture for decision-making. The gathered information could be utilized in future for better decision-making.

References

1. Patil, V.C., Al-Gaadi, K.A., Biradar, D.P., Rangaswamy, M.: Internet of things (IoT) and cloud computing for agriculture: an overview In: Agriculture (AIPA 2012) (2012)
2. Sarkar, P.J., Chanagala, S.: A survey on IOT based digital agriculture monitoring system and their impact on optimal utilization of resources. IOSR J. Electron. Commun. Eng. (IOSR-JECE) **11**(1), 01–04, Ver.II (Jan–Feb 2016), e-ISSN: 2278-2834,p- ISSN: 2278-8735.
3. Gondchawar, N., Kawitkar, R.S.: IoT based Smart Agriculture In: International Journal of Advanced Research in Computer and Communication Engineering Vol. 5, Issue 6, June 2016
4. Juan Francisco V.-M
5. Kim, Y., Evans, R., Iversen, W.: Remote sensing and control of an irrigation system using a distributed wireless sensor network. IEEE Trans. Instrum. Measur. 1379–1387 (2008)
6. Xiao, Z., Liang, S., Wang, J., Chen, P., Yin, X., Zhang, L., Song, J.: Use of general regression neural networks for generating the glass leaf area index product from time-series MODIS surface reflectance. IEEE Trans. Geosci. Remote Sens. **52**(1) (2014)
7. Na, A., Isaac, W., Varshney, S., Khan, E.: An IoT based system for remote monitoring of soil characteristics. In: 2016 International Conference on Information Technology (InCITe)—The Next Generation IT Submit
8. Chen, J., Black, T.A.: Defining leaf area index for non-flat leaves. Plant, Cell Environ. **15**(4), 421–429 (1992)

Secure Image Steganography Through Pre-processing

Indu Maurya and S. K. Gupta

Abstract The security of information is required by internet technologies and its application during transmission over the nonsecured communication channel. Steganography means is to hide secret data into another media file. In the spatial domain, Least Significant Bit (LSB) is used. LSB is the most common approach because of its easiness and hiding capacity. Preprocessing is required for better security but all current techniques of Image Steganography mainly concentrate on the hiding strategy and less concern about encryption of secret image. For digital Image Steganography security, a unique technique based on the Data Encryption Standard (DES) is introduced in this paper. Plaintext with a block size of 64 bit and 56 bits of the secret key is used by DES. Better security and a high level of secrecy are achieved by pre-processing as it is not possible to extract the image without knowledge of the S-Box mapping rules and a secret key of the function. The discussion shows that the Steganographic technique with pre-processing is better than LSB steganography algorithms directly.

Keywords Data encryption standard · Least significant bit · Steganography

1 Introduction

The exceptional means of security is required by the communications system. The security of communication system is securing priority because the information being exchanged over the web increases. Therefore, two things are needed as a safeguard over nonsecured communication channels: confidentiality and integrity. Confidentiality and Integrity have resulted in an unstable success for information hiding and covers many applications like Cryptography, Steganography, and Digital Watermark-

I. Maurya (✉) · S. K. Gupta
CSE Department, B.I.E.T, Jhansi, India
e-mail: indumaurya42@gmail.com

S. K. Gupta
e-mail: guptask_biet@rediffmail.com

K. Ray et al. (eds.), *Soft Computing: Theories and Applications*,
Advances in Intelligent Systems and Computing 742,
https://doi.org/10.1007/978-981-13-0589-4_13

ing, Fingerprinting and copyright protection for digital media. These information hiding applications are entirely varied. Cryptographic techniques conceal the content of the secret information. It modifies the information so that it is not easy to understand but by doing this, the interest level of an intruder increases. The concept of Steganography [1] is hiding the confidential information within any other media like audio, video, and text files known as the cover media. Steganography is a Greek word which is a combination of two words (Steganos + Grafia). Steganos refer to the "covered/secret" and Grafia refers to the "writing", simply Steganography means "covered writing" or "secret writing". For information hiding, two popular schemes are cryptography and steganography and have acquired important scrutiny from both academia and management in the recent past. Steganography with Cryptography provides a high level of security and secrecy of information. Steganography is used for secret communication and has been widely used to secretly convey data among the people.

2 Steganography (Definition and Types)

In Steganography the confidential information (secret) is hidden into a cover file in order to produce a secret cover media known as stego media or stego image [2]. So, the secret information is not recognized or recovered by the intruders. High level of security and secrecy is provided by the Steganography [3].

1. **Text**: Inside the text files, it comprises of masking data. Behind every nth letter of every word of the text message, the confidential information is concealed. There are some methods that are used to conceal information in the text file and defined as fellow: (i) Format-Based Method; (ii) Random and Statistical Method; (iii) Linguistics Method.
2. **Image**: Image steganography is to as concealing the information by taking the secret image and cover image and inside the cover image the secret image is embedded. To conceal the information inside the cover object the pixel intensities of the secret image is used. In digital steganography, the cover source is used by images because a number of bits are present in a digital representation of an image.
3. **Audio**: As the name implies, audio file stores the concealed information. For concealing information, it uses sound files like WAV, AU, MP3, etc. Audio steganography consists of several methods. These methods are (i) LSB (ii) Phase Coding (iii) Spread Spectrum (iv) Echo hiding and a summary of audio steganography methods shown in Table 1.
4. **Video**: In order to conceal information, this technique uses a digital video format. For concealing the information carrier is used which is a combination of pictures (case video). DCT alter the values which are used to conceal information in each of the images in the video and the human eye cannot recognize it. Video steganography formats are H.264, Mp4, MPEG, and AVI (Table 2).

Table 1 Summary of audio steganography methods

Methods	Data hiding techniques	Strength	Weakness
LSB	Using one-bit of concealed information each and every sample in the audio file is substituted	Straightforward and accessible	Simple to extract
Phase coding	Phase of the cover signal is restrain	Robust against signal processing operation	Less amplitude
Spread spectrum	Transmit data overall signal regularity	Implement sophisticated robustness and transparency	Weak time rate correction and involve higher transmission
Echo hiding	Conceal data by proposing echo in the cover signal	Adaptive noise problem is avoided	Less information security and capacity

Table 2 Summary of video steganography analysis features

Algorithm	Potent	Amplitude	Protected	Cover image	Secret image
Secured data transmission (SDT)	Potent	Bigger	Less protected	AVI	Image
Hybrid encryption and steganography (HES)	Potent	Large	Protected	AVI	Text
LSB polynomial equation algorithm (LSBPOLY)	Less Potent	Small	Less protected	AVI	Text
Hash-based LSB (HLSB)	Potent	Excellent	Less protected	AVI	Text
Multiple LSB (MLSB)	Potent	Excellent	Less protected	AVI	Text
LSB matching-revised algorithm (LSBMR)	Potent	Large	Protected	AVI	Text
Novel video steganography (NVS)	Potent	Large	Highly protected	AVI, MPEG, MOV, FLV	Text, image, audio and video

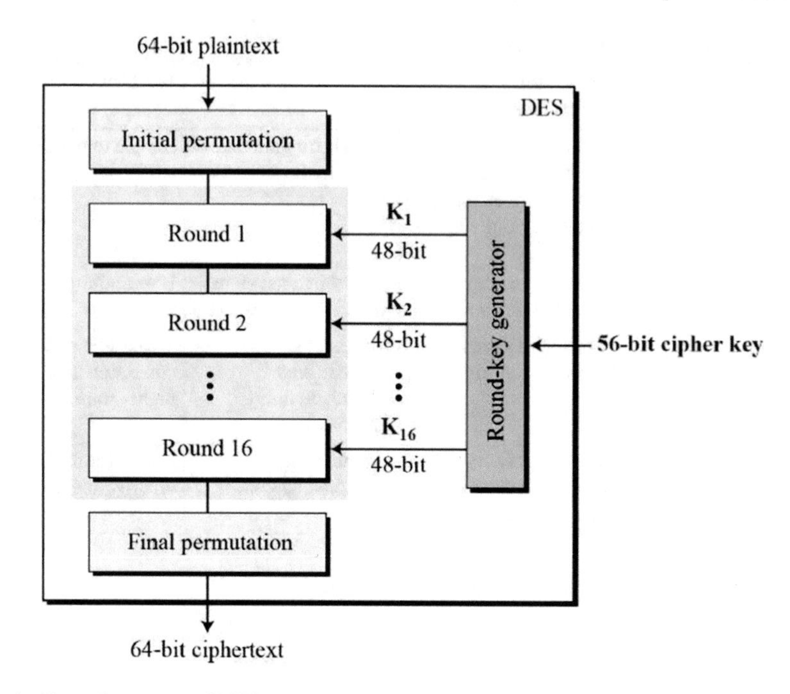

Fig. 1 General structure of DES

3 DES with Image Steganography

3.1 Data Encryption Standard (DES)

DES [4] is a block cipher and uses a 56-bit key and 8-bit parity bit, resulting in the largest 64-bit packet size. DES uses the Feistel cipher technology; using this the text block half is encrypted. The use of sub-key pair, half of which application circulatory function, and then the output with the other half to "XOR" operator followed by the exchange of the two and a half, this process is repeated again and again, at the end a cycle of nonexchange. A total number of cycles is 16 in DES and uses operations such as XOR, replacement, substitution, four basic arithmetic shift operations [5] (Fig. 1).

Feistel Cipher consists of following, to specify DES these are required

- Round function
- Key schedule
- Any additional processing—Initial and final permutation

Initial and Final Permutation

The initial and final permutations are inverse of each other. They are straight P-boxes (Permutation boxes). In DES, there is no cryptography significance of Initial and final

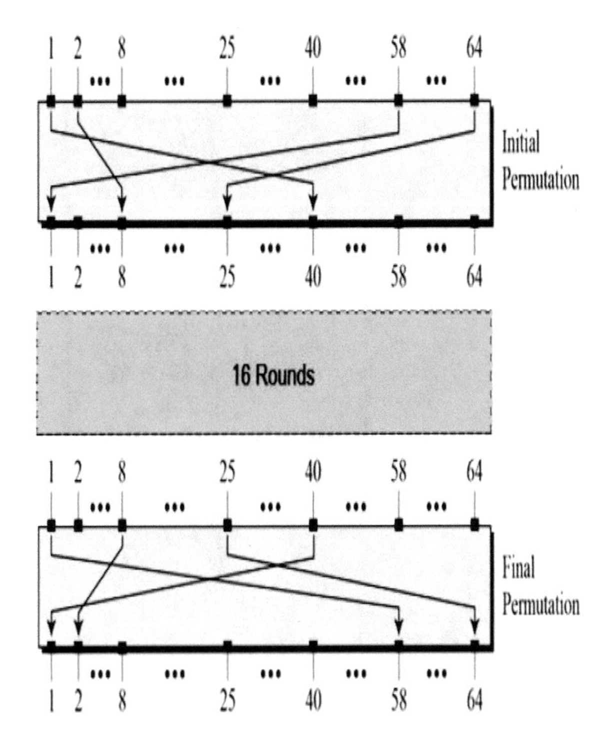

Fig. 2 Block diagram of initial and final permutation

permutation. The block diagram of initial and final permutation is as follows (Figs. 2, 3, 4, and 5):

Round Function

The DES function f is the heart of this cipher. 48-bit key is applied by the DES function to the rightmost 32 bits order to achieve a 32-bit output as shown in Fig. 3. **Expansion Permutation Box**—as 32-bit is the right input and 48-bit round key, so first expand 32-bit right input to 48-bit. Permutation logic is shown in the diagram graphically:

- **XOR—**after performing the expansion operation, XOR operation is applied on the expanded right section and the round key. In this operation, the round key is used only.
- **S-Boxes**—the S-boxes accomplish the confusion (real mixing). 8 S-boxes are used by DES, which accepts 6-bit as input and produces 4-bit as output. Shown in below (Fig. 6):

The S-box rule is illustrated below

- The total number of table for S-boxes is eight. All eight s-boxes output is then mixed into 32-bit segment.

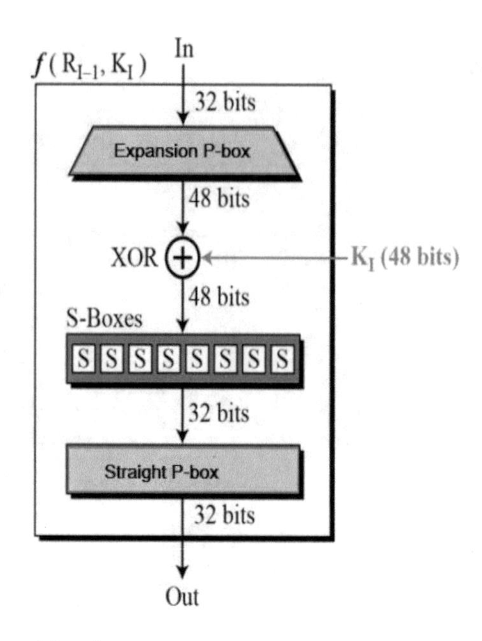

Fig. 3 Structure of round function

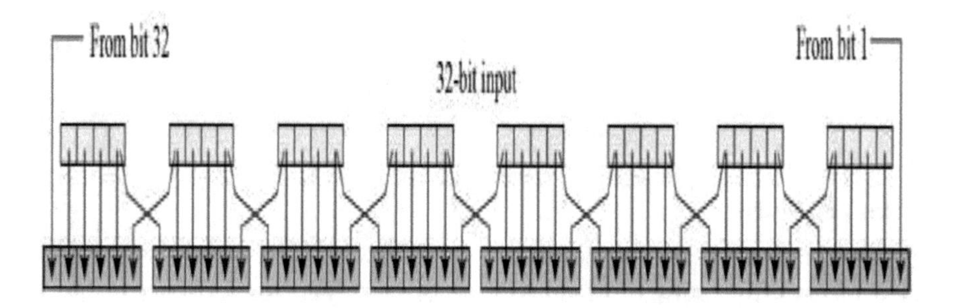

Fig. 4 Structure how to expand 32-bit input to 48-bit using expansion box

- **Straight Permutation**—The 32-bit output of eight S-boxes is managed to the straight permutation along the direction as shown below (Fig. 7):

Key Generation

The 48-bit keys in association with a 56-bit cipher key are created by round key generator. The key generation procedure is shown below (Fig. 8):

32	01	02	03	04	05
04	05	06	07	08	09
08	09	10	11	12	13
12	13	14	15	16	17
16	17	18	19	20	21
20	21	22	23	24	25
24	25	26	27	28	29
28	29	31	31	32	01

Fig. 5 Permutation logic table

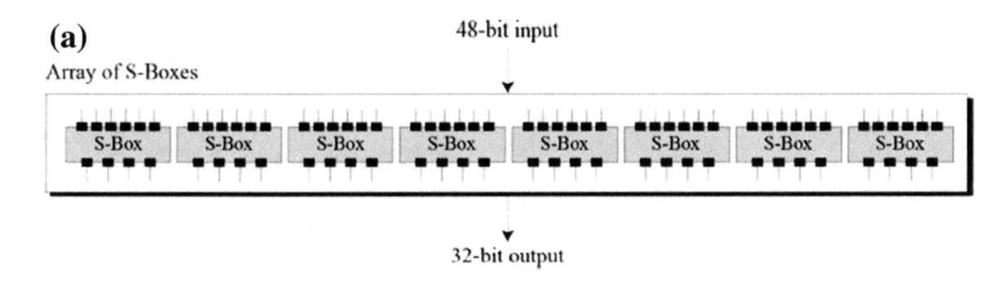

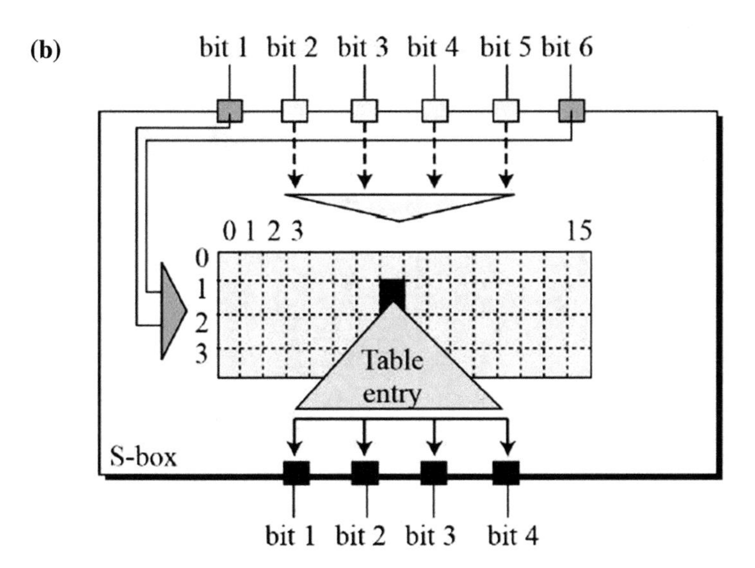

Fig. 6 **a** and **b** Structure of S-boxes

16	07	20	21	29	12	28	17
01	15	23	26	05	18	31	10
02	08	24	14	32	27	03	09
19	13	30	06	22	11	04	25

Fig. 7 32-bit S-box output

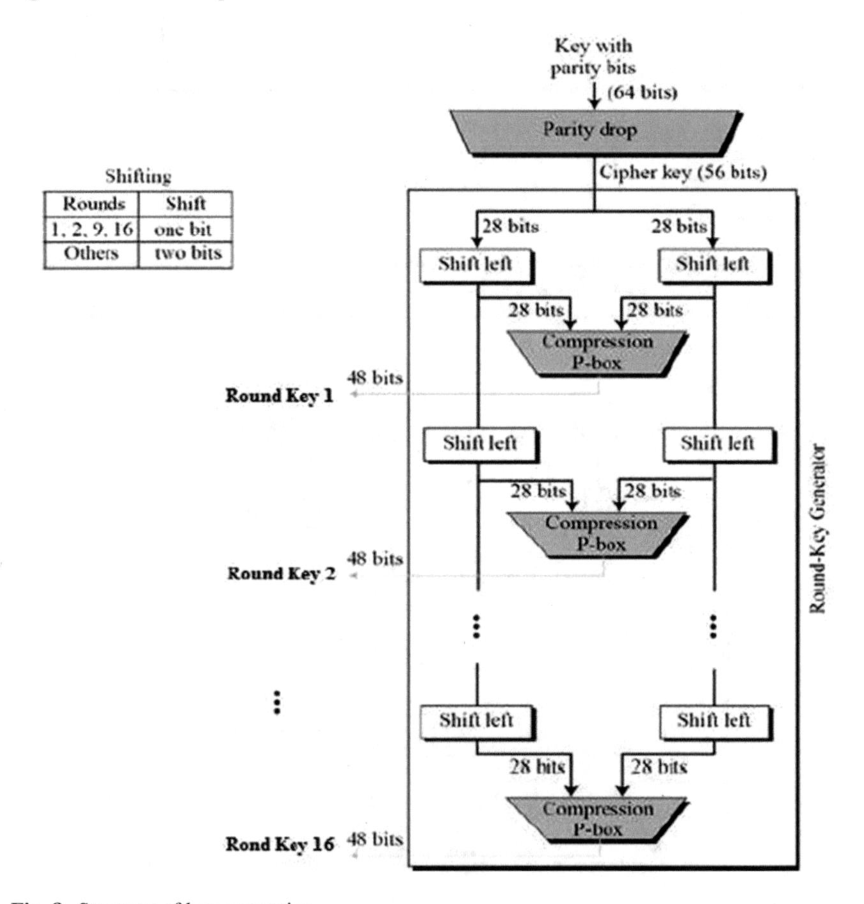

Fig. 8 Structure of key generation

3.2 *Digital Image with LSB*

The Least Significant Bit (LSB) is the most simplest and straightforward steganography algorithm in which the confidential information is concealed to the LSB bit of the carrier image [6]. It is still most widely used algorithm up to today. Rather than the last bit of the image pixels value, it uses the confidential information, which is identical to cover a weak signal consequent to the authentic carrier signal, and hence it is not easy to recognize visually. But when the embedding operation is performed, the lowest bit of the confidential information is checked from that of pixel value, and if it is different from the pixel value, then we have to change the lowest bit of the image pixel value, named as 2i to 2i + 1 or 2i + 1 to 2i. The parity value produces parity asymmetry and the parity value transform in the LSB steganography. RS attack [2] is the method of using this asymmetrical parity value transform of LSB proposed by Fridirich et al. and LSB steganography algorithm attack using Chi-square test [3] is proposed by Westfield et al. If the distortion of the histogram can be minimized by Steganographic method, the attack of analysis methods can be essentially prevented. So this paper presents a Steganographic technique combined with pre-processing. In this paper, a unique technique based on the Data Encryption Standard (DES) pre-processing. 64-bit of plaintext and a secret key of 56-bits are used by DES. Pre-processing is required for better security and secrecy of data as to regain the original image is not possible without knowledge of the S–Box mapping rules and also the secret key of the function.

3.3 Encryption with Steganography

For improving the imperceptibility of steganography algorithm this paper presents the steganography technique with encryption, which reaches the goal by changing the matching relationship between carrier image and secret information. At the encoding end, we first select the carrier image, and then use DES encryption [4] to encrypt the text information followed by LSB steganography algorithm to conceal the encrypted secret information in the carrier image. At the decoding end, the encrypted secret information from the carrier image is extracted by the algorithm which is opposed to steganography algorithm, and then apply DES decryption algorithm to regain the original hidden information.

4 Proposed Approach

Secrecy and accuracy with an excellent image quality of the resulted image are the main reason for choosing this model. The proposed work is a security approach because steganography and cryptography which are two different techniques are combined in order to achieve a high level of secrecy and security of information. For better security he pre-processing such as encryption of the secret image is used which provides a high level of security and secrecy of information. On the other hand, Steganography [7] provides secrecy of information in that for decrypting the information the cryptographic key is required that is kept secret and must not be

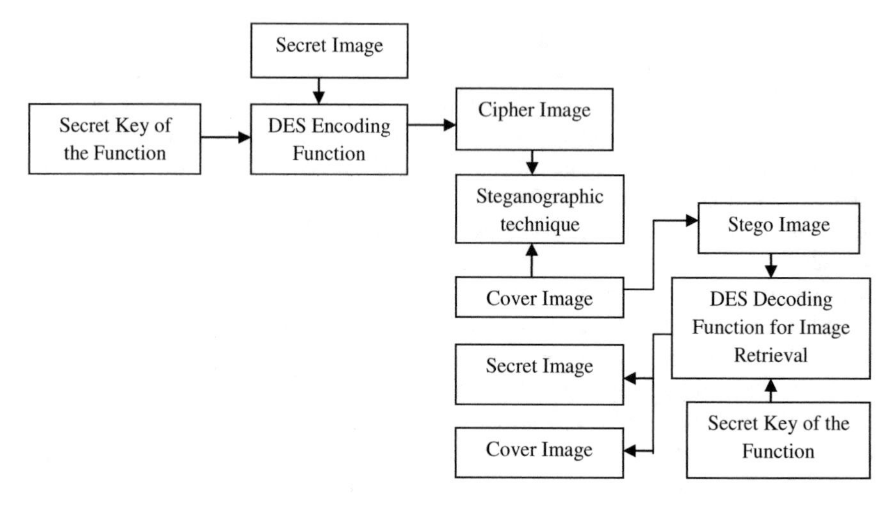

Fig. 9 Proposed system model

shared with the third party. In this proposed work two approaches are applied on a confidential (secret) image defined as follows:

(a) Encoding and Decoding using DES Algorithm
(b) Steganography.

The proposed system model is shown in Fig. 9, which describes the whole concept of this proposed work.

DES encryption algorithm is applied after selecting the secret image and passes through the encoding procedure which includes Inverse/Initial permutation, the function f, and S-boxes. This encoding procedure converts original selected secret image into the cipher image, after complete execution of DES encryption. This cipher image is the encrypted secret image. The Steganography method is called after getting an encrypted secret image (cipher image) where cover image is taken as an input and steganography technique read the least significant bits from the cover image and replace from cipher bit value. Continue this procedure until the last pixel value of cipher image is taken. The resultant image will be known as stego image.

4.1 *Encoding Algorithm (Image Encryption and Embedding)*

Input: Two images are selected (a) a grayscale image of m × n size (b) a grayscale cover image of 2m × 2n size.

Output: Resultant image or stego image of 2m × 2n size.

Procedure:

1. Taking a secret image of size 64 × 64 bits and applying DES encoding function on the secret image which will produce the encrypted secret image, also known as a ciphered image.
2. After getting encrypted secret image divide each pixel value of it into 4 parts. Each part contains 2 bits.
3. One by one all these pixel values are inserted into LSB of the cover image.
4. End.

4.2 Decoding Algorithm (Image Retrieval)

Input: Stego image of size 2m × 2n is selected first.

Output: A secret image of m × n size.

Procedure:

1. Taking the 2 bits LSB from the Stego image.
2. Concatenate the LSB in order to get the encrypted secret image or cipher image.
3. For obtaining a secret image, use DES function in reverse order.
4. End.

5 Experimental Outcomes

Experimental results show that Steganographic technique is stronger, because to extract the secret image from the stego image it is important to know the secret keys, function f and S-Box mapping which is very difficult to determine (Figs. 10 and 11).

6 Conclusion

The mapping rule of S-Boxes and a secret key of the function is the strength of this proposed DES-based Steganographic model and it increases security and secrecy of information and image quality compared to existing algorithms to predict the percentage of improvements. Steganography with Cryptography is considered as a powerful tool for secretly communicate information. Pre-processing is required for better security because all current techniques of Image Steganography mainly concentrate on the hiding strategy. For the security of image steganography, a unique technique based on the Data Encryption Standard (DES) is introduced this paper. DES is a block cipher which uses a 64-bit block size of plaintext and 56-bits of a secret key. Better security and a high level of secrecy are achieved by pre-processing. Results show the stronger Steganographic technique because to extract the secret

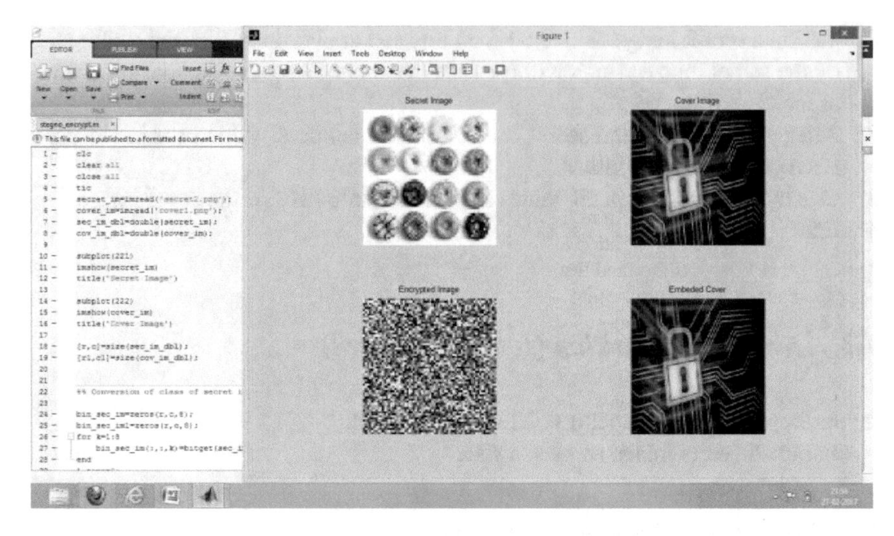

Fig. 10 Encoding function (secret image, cover image, encrypted image and embedded cover image)

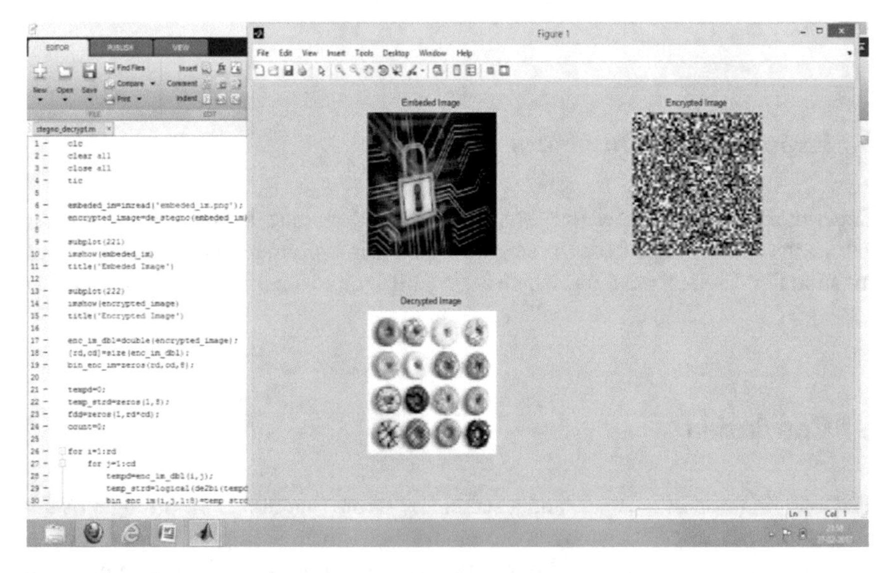

Fig. 11 Decoding function (embedded image, encrypted image and decrypted image which is a secret image)

image from the stego image it is important to know the secret keys, function f and S-Box mapping which is very difficult to determine.

References

1. Moerland, T.: Steganography and Steganalysis. Leiden Institute of Advanced Computing Science. www.liacs.nl/home/trnoerl/privtech.pdf
2. Provos, N., Honeyman, P.: Hide and seek: an introduction to steganography. IEEE Secur. Priv. 32–44 (2003)
3. Babu, K.S., Raja, K.B., Kiran, K.K., Devi T.H.M., Venugopal, K.R., Pathnaik, L.M.: Authentication of secrete information in image steganography. IEEE Trans. 13
4. Schaefer, E.F.: A Simplified Data Encryption Standard Algorithm. Cryptologia, January 1996
5. Mielikainen, J.: LSB matching revisited. IEEE Signal Process. Lett. **13**(5), 285–287 (2006)
6. Khosravirad, S.R., Eghlidos, T., Ghaemmaghami, S.: Higher order statistical of random LSB steganography. IEEE Trans. 629–632 (2009)
7. Ge, H., Huang, M., Wang, Q.: Steganography and steganalysis based on digital image. IEEE Transaction on International Congress on Image and Signal Processing, pp. 252–255 (2011)
8. Luo, W., Huang, F., Huang, J.: Edge adaptive image steganography based on LSB matching revisited. IEEE Trans. Inf. Forensics Secur. **5**(2), 201–214 (2010)
9. Zhu, G., Wang, W.: Digital image encryption algorithm based on pixel. In: ICIS—2010 IEEE International Conference, pp. 29–31 Oct 2010, pp. 769–772
10. Cosic, J., Bacai, M.: Steganography and steganalysis does local web site contain "Stego" contain. In: 52th IEEE Transaction International Symposium ELMAR-2010, Zadar, Croatia, pp. 85–88 (2009)
11. Yun-peng, Z., Wei, L.: Digital image encryption algorithm based on chaos and improved DES. In: IEEE International Conference on System, Man and Cybernatics (SMC 2009), pp. 474–479, 11–14 Oct 2009

Understandable Huffman Coding: A Case Study

Indu Maurya and S. K. Gupta

Abstract For the compression of images, the needs of some techniques for compressing the images is always increasing as when the images are raw, they do need additional space in the disk and as far as the transmission and their storage is concerned it is a big disadvantage. Though for the compression of images, there are a lot of latest techniques available nowadays, but the fast and efficient one is the technique which generally goes well with the client's requirements. The lossless method of compression and decompression of images is studied in this paper with the help of a simple and efficient technique for coding known as Huffman's coding. As far as implementation is concerned, the technique is simple and comparatively lesser memory is needed for its execution. There has been a special software algorithm which is generated and also implemented to either compress or decompress the images with the help of Huffman techniques of coding images in a platform known as MATLAB.

Keywords Encoding and decoding of Huffman · Symbol · Source reduction

1 Introduction

There is a need of enormous storage in obtaining a digital image via sampling and quantizing with a picture containing continuous tone. As, for example, an image of 24-bit colors containing 512 × 512 image pixels is supposed to capture 768 Kb of disk storage, the image which is double to the mentioned size cannot adjust well into a simple and single floppy disk. For the transmission of this type of images with a modem of 28.8 Kbps, as much as 4 min are needed. The aim behind compression of images is to lessen the data amount which is needed for representation of digital and sampled images and hence the cost of storage and transmission is also reduced. In

I. Maurya (✉) · S. K. Gupta
CSE Department, B.I.E.T, Jhansi, India
e-mail: indumaurya42@gmail.com

S. K. Gupta
e-mail: guptask_biet@rediffmail.com

K. Ray et al. (eds.), *Soft Computing: Theories and Applications*,
Advances in Intelligent Systems and Computing 742,
https://doi.org/10.1007/978-981-13-0589-4_14

some of the highly significant applications, this technique of image compression is very important; such applications include database images, image communication, and remote sensing (the utilization of satellite imagery for earth resource applications including weather). To be compressed images are grayscale with the values of pixels ranging in between 0 and 255. There is always a requirement of efficient compression technique for images as the more disk space and transfer time is needed when raw images are used. There are so many modern and efficient techniques for compression of data of an image but the best one is a technique which is faster in execution and is more efficient in terms of memory and suits what the user needs. For image compression and decompression, some techniques are used that are known as Lossless image compression techniques with a help of simple coding methodology known as Huffman coding. It is easy and simpler to be implemented as comparatively lesser memory is used in it. For this purpose, algorithmic software is created. There are so many techniques for the image compression. Lossless and lossy compression techniques are two very important classes of compression techniques. As it is clear in the name, the technique of compression which is lossless involves the compression where no information of image is lost at all. Or we can also say that, from the compression, the image which is reconstructed is similar to the initial and original image in all the respects. On the other hand, in the compression technique which is known as Lossy, the stored information is basically lost. It means that the images which are reconstructed images obtained after compression are somewhat similar to the initial images but they are not much identical to the original ones. In the task, we are going to make use of lossless compression and decompression with a help of Huffman's coding techniques (also known as encoding and decoding). It is also an established fact that algorithm of Huffman is producing least redundancy codes in contrast to other algorithms. The coding technique of Huffman is used efficiently and effectively in the images, texts as well as video compression, in the systems of conferencing like JPEG, H.263, MPEG-2, MPEG-4, etc. The unique symbols are collected via Huffman's coding techniques and these symbols are obtained via source images and estimate the values of probability for every single symbol and their sorts. Along with the symbol for lowest probability value to the symbol of highest probability value, these two symbols are aggregated at a time while forming a binary tree. Also to the left node, zero is allocated and one to the right side node originating from the tree root. For a particular symbol while obtaining the Huffman's code, all the zeros and the zero which is derived from the root to a specific node are placed in a similar order. This paper aims to highlight the importance and techniques of image compression by reduction in amount of per-pixel bits for its representation and also to lessen the time needed for the transmission of such images along with the reconstruction of images back with the help of decoding of the Huffman's codes. The paper is organized as follows. The needs of compression are mentioned in Sect. 1 and types of data redundancies are expressed in details in Sect. 2. The methodology which can be used for compression is explained in Sect. 3 while in Sect. 4, we will highlight the implantation of lossless compression and decompression method. (i.e., Coding and Decoding of Huffman) is performed in Sect. 5. The outcomes are represented with due explanation in Sect. 6. In the end of this paper, you will find the references.

2 Need for Compression

Some examples below represent the compression requirements for images which are digital [1].

a. For the storage of moderate-sized colored image, for example, a disk space of 0.75 MB is needed for 512 × 512 pixels.
b. 18 MB is needed for a digital slide of 35 mm containing 12 μm of resolution.
c. A digital pal of one second which is also known as Phase alternation line needs 27 Mb for a video.

For the storage of such images and to ensure their network availability we always need the special techniques for an image compression. In order to ensure the storage of digital images, one has to address the concerns which are involved in the data reduced. The removal of redundant information is the underlying basis of the data reduction. From the mathematical point of view, this amounts to the transfer of a pixel array which is two dimensional into a data set which is uncorrelated statistically. Before the storage or image transmission, this transformation is applied. The compressed image is then decompressed at the receiver to reconstruct the genuine image or if not original it must be very close to it. The importance of compression is clearly shown in the example given below. 1024 pixel × 1024 pixel × 24 bit containing image without any compression needs storage of approximately 3 MB and 7 min will be needed for the transmission of such images if one is using high-speed internet (i.e., 64 Kbits/s, ISDN line). If there is an image compression at 10:1 ratio of compression, the needs of its storage are considerably reduced to 300 KB and for the transmission. The importance of image compression is shown clearly in the example given below. We can consider a simple case here, an image of pixel 1024 × 1024 × 24 bit if not compressed will need 7 min of transfer and 3 MB disk space if one is making use of high-speed Internet ISDN line of 64 Kbits/s. when the image compression takes place at 10:1 ratio for compression, the amount the same image needs for storage is 300 KB disk space and it will need as little as 6 s to transfer it.

3 Redundancy Types

As far as a compression of a digital image is concerned, 3 (three) main information redundancies are distinguished and exploited:

- Redundancy coding
- Interpixel redundancy
- Psychovisual redundancy

While removing one or more redundancies or either by decreasing them, we can get data compression.

3.1 Coding Redundancy

An image of gray level with pixels ($\boldsymbol{n}$) is kept under consideration. $\boldsymbol{L}$ is the number of gray level inside an image ($\boldsymbol{0}$ $\boldsymbol{to}$ $\boldsymbol{L-1}$ is the range of gray level); also the pixels number with $\boldsymbol{r_k}$ gray level is $\boldsymbol{n_k}$. In this case the occurring probability of gray level is $\boldsymbol{r_k}$ or $\boldsymbol{P_r(r_k)}$. If the number of bits used to represent the gray level $\boldsymbol{r_k}$ is $\boldsymbol{l(r_k)}$, then the average number of bits required to represent each pixel.

$$\boldsymbol{L_{avg}} = \sum_{\boldsymbol{k=0}}^{\boldsymbol{L-1}} \boldsymbol{l(r_k)}\mathbf{P_r}\boldsymbol{(r_k)} \tag{1}$$

where,

$$P_r(r_k) = \frac{n_k}{n}$$
$$k = 0, 1, 2 \ldots L - 1 \tag{2}$$

In this way we can say that the amount of bits which are needed for the representation of entire image is $n\ X\ L_{avg}$. When L_{avg} is minimized, we can get maximum ration of compression (i.e., when the length of gray-level representation function $l(r_k)$, leading to minimal L_{avg}, is found). An image containing coding redundancy is obtained by coding the gray levels with Lavg not minimized. In general, there is an occurrence of coding redundancy as the codes (with lengths represented as $l(r_k)$ assigned to gray levels fail to take maximum benefit of probability of gray level ($P_r(r_k)$ function). Hence, it is seen around always especially when the gray level of images is represented with a natural binary code or either straight [2]. Gray levels natural binary coding ensures that each of the bits has been assigned with the same bits number for the probable values containing the most and least values. Hence, there is no minimizing of the (1) equation and the outcome is the redundancy coding. Coding redundancy examples: as shown in the table, an 8 level image has distribution of gray level. If kept neutral a binary code of 3 bits, (see code 1 and $l\,rk$ in Table 1) is used to represent 8 possible gray levels, L_{avg} is 3-bits, because $l\,rk = 3$ bits for all rk. If code 2 in Table 1 is used, however the average number of bits required to code the image is reduced to $L_{avg} = 2(0.19) + 2(0.25) + 2(0.21) + 3(0.16) + 4(0.08) + 5(0.06) + 6(0.03) + 6(0.02) = 2.7$ bits. From the equation of compression ratio (n2/n1) the resulting compression ratio C_R is 3/2.7 or 1.11. Thus, approximately 10% of the data resulting from the use of code 1 is redundant. The exact level of redundancy can be determined from Eq. (1).

In the initial data set with a data of around 9.9% the redundancy is to be removed in order to get the compression.

Table 1 Representation of variable length coding with examples

r_k $l_2(r_k)$	$P_r(r_k)$		Code1	$l_1(r_k)$	Code2
r0 = 0	0.19	000	3	11	2
r1 = 1/7	0.25	001	3	01	2
r2 = 2/7	0.21	010	3	10	2
r3 = 3/7	0.16	011	3	001	3
r4 = 4/7	0.08	100	3	0001	4
r5 = 5/7	0.06	101	3	00001	5
r6 = 6/7	0.03	110	3	000001	6
r7 = 1	0.02	111	3	000000	6

$R_D = 1 - 1/1.1 = 0.099 = 9.9\%$

3.1.1 Coding Redundancy Reduction

In order to minimize the redundancy from a given image, Huffman technique is applied and in this fewer bits are assigned to the gray level in contrast to the ones which are lesser probable and in this way the data is compressed. This process is generally known as variable length coding. Several techniques are there either optimal or near optimal for constructs like code (Huffman coding and arithmetic coding).

3.2 Interpixel Redundancy

Data redundancy of another type is known as pixel redundancy, it is related directly to the correlations which exist as inter-pixel in an image. From the value of its neighbors one can reasonably predict the value of a given pixel; the individual pixels carry relatively smaller data. Single pixel's visual contribution to an image is generally redundant; and we can have its estimation on the values obtained from the neighborhood. There are much name variations like spatial redundancy, interframe or geometric redundancy etc. the redundancy is coined to refer to the dependencies of these Interpixels. In an image, if this interpixel redundancy is to be removed, the pixel array for 2-D which is generally used for the viewing along with its interpretation is to be converted into nonvisual format which is generally more efficient. For example, one can represent an image on the basis of differences between the adjacent pixels. The special name assigned to such transformation is mapping. If the genuine image elements can be reconstructed from the data set which is transformed they are also said to be reversible [2, 3].

3.2.1 Inter-Pixel Redundancy Reduction

Various techniques are applied for the reduction of inter-pixel redundancy: these include [4],

Table 2 Reduction of Huffman source

Original source			Source reduction		
S	P	1	2	3	4
a_2	0.4	0.4	0.4	0.4	0.6
a_6	0.3	0.3	0.3	0.3	0.4
a_1	0.1	0.1	0.2	0.3	
a_4	0.1	0.1	0.1		
a_3	0.06	0.1			
a_5	0.04				

Where S represents source and P represents probability

- Coding via Run length.
- Coding via Delta compression.
- Coding via Constant area.
- Predictive coding.

3.3 *Psychovisual Redundancy*

As far as the data is concerned, the human perception regarding an image lacks the quantitative analysis of all the pixels or either the value of luminance in an image. From the perspective of an observer, few qualities are searched like textural regions or edges and combine them mentally into the groups which are recognizable. With prior knowledge, the brain correlates these groupings so as to complete the process of image interpretation. In normal visual processing, some information has lesser importance as compared to other information and this is known as psychovisual redundancy. Without significantly impairing the image perception quality one can eliminate it. Coding and psychovisual redundancies are different from each other and also from inter-pixel redundancy. Psychovisual redundancies are related to the quantifiable visual data unlike coding and inter-pixel redundancies. It is possible to eliminate it as the data is not needed for a normal visual processing. There is a loss of quantitative information in the elimination of psychovisual redundancy data, the process is said to be irreversible.

3.3.1 Psycho Visual Redundancy Reduction

A quantizer is generally used to lessen the psychovisual redundancy. As the removal of data which is psycho visually redundant, it results generally in the loss of quantitative data. This process is generally known as quantization. Since the operation is not reversible because of the loss of information, the lossy data compression is obtained by quantization (Table 2).

Table 3 Lossless and lossy techniques

Techniques for lossless coding	Techniques for lossy coding
4.1. Encoding via run-length	4.1. Predictive coding 4.2. FT/DCT/Wavelets (transform coding)
4.2. Encoding via Huffman	
4.3. Encoding via arithmetic	
4.4. Coding via entropy	
4.5. Encoding via area coding	

4 Classification of Compression

See Table 3.

5 Lossless Compression and Decompression Techniques Implementation

5.1 Huffman's Coding

There are two observations on which the Huffman's coding procedure is based on [1].

a. The symbols which are occurring more frequently do have codes which are much shorter, which occur less frequently.
b. Two frequently occurring symbols will show the similarity in their lengths.

The code of Huffman is designed simply by joining the symbol with least probability and the procedure is continued for a long till the only two compound symbol probabilities are left behind and therefore a coding tree is produced. From the labeling of the code trees, the Huffman codes are obtained. It is shown with an example in Table 2.

In Table 1 at a far left, the listed symbols along with the probabilities for corresponding symbols are arranged in descending order. The lowest probabilities are adjoined here are 0.06 and 0.04. As a result, with a probability of 0.1 a compound symbol is obtained and the probability of the compound symbol is allowed to settle at column 1 of the source reduction which ensures that once again the probabilities are arranged in descending order. The process is continued and repeated as long as only two probabilities are left in the far right as mentioned in table 0.6 and 0.4. The step II for carrying out the procedure of Huffman is coding all the reduced sources; this is one by starting from the lowest source and executing back to the genuine source [1].

The length which is minimal for the binary codes for a source containing two symbols is 0 and 1. It is also mentioned in the given Table 4 that these all symbols are allocated to two right sided symbols (arbitrary assignment; reversal of the order of the 0 is supposed to work well). As with the probabilities 0.6 the reduced source symbols to the left of it, 0 is utilized for its coding as it is allocated to the two symbols and 0 as well as 1 are arbitrarily appended to each for making them easy to be distinguishable. The execution is done yet again for all the reduced sources, until the time when the original course is obtained. At the far left Table 4, the ultimate code is represented. The code's net length is obtained by the average of symbol's probability product and the bits number which is used for encoding it. The calculations are mentioned as below:

L_{avg} = (0.4) (1) + (0.3) (2) + (0.1) (3) + (0.1) (4) + (0.06) (5) + (0.04) (5) = 2.2 bits/symbol and the entropy of the source is 2.14 bits/symbol, the resulting Huffman code efficiency is 2.14/2.2 = 0.973.

Entropy,

$$\mathbf{H} = \sum_{\mathbf{k}=1}^{\mathbf{L}} \mathbf{P}(\mathbf{a_j}) \log \mathbf{P}(\mathbf{c_j}) \tag{3}$$

Optimal codes for the Huffman's procedure are created for symbol sets along with the probabilities which are subjected to the constraint which can be coded for at least once.

5.2 *Huffman's Decoding*

Once after the creation of a code, in the simple look up Table manner [5] the coding and decoding is accomplished. The code obtained itself is immediate exclusively decodable black code. As each source symbol is mapped into a fixed sequence of code symbols it is known as a block code. Also it is generally immediate as in a string of code; each codeword can be decoded without referencing the subsequent symbols. We can also call it exclusively decodable as in only one way the string of code symbols can be decoded. Therefore, any Huffman encoding symbols and their string can be simply decoded by evaluation of the string's individual symbol in a manner from left to right. For a Table 4 as far as binary code is concerned, a scanning from left to right of the string encoded 010100111100 gives us a clue that the code word which is initially valid is 01010 which for a symbol a_3 is a code. The code which is valid next is 011 which are corresponding to the a_1 symbol. For a symbol a_2 the code which is valid is 1. For a symbol a_6 the code which is valid is 00, the code which is valid for symbol a_6 is in a similar manner continuing and also highlights the message of an entire decode a_5 a_2 a_6 a_4 a_3 a_1, hence in a manner similar to it, the genuine information or image can be decompressed with the help of decoding via Huffman as mentioned in details above. We at first have much of the compressor

Table 4 Procedure of Huffman code assignment

Original source reduction				Source reduction					
4	S	P	code		1		2		3
	a_2	0.4	1	0.4	1	0.4	1	0.4	1
0.6 0									
	a_6	0.3	00	0.3	00	0.3	00	0.3	00
0.4 1									
	a_1	0.1	011	0.1	011	0.2	010	0.3	01
	a_4	0.1	0100	0.1	0100	0.1	011		
	a_3	0.06	01010	0.1	0101				
	a_5	0.04	01011						

S source, *P* probability

which allows the probability distribution. A code table is formed via compressor. In this method, the decompressor is not suitable. Rather it keeps the entire binary tree for Huffman and also a root indicator for the continuation of the recursion procedure. As far as our implementation is concerned, we are making a tree as usual and then you can successfully store a pointer in the list to the last node which generally is considered root here. At this phase, the process can start. One can easily navigate the tree with the help of pointers to the children which are contained by every node. The recursive function is a key to this process which fits as a parameter for a pointer to the present node and moves back to the symbol.

5.3 Huffman Coding and Decoding Algorithm Implementation

Step1 An image reading at the MATLAB's workspace.

Step2 The Conversion of an image of color (RGB) to an image of gray level.

Step3 Calling some functions so as to find out the symbols, for example non-repeated pixel values.

Step4 Calling a function so as to evaluate the symbol's probability.

Step5 Symbol's probability is compiled in a descending order and the probabilities which are least are merged. The procedure is repeated unless only two probabilities are left behind and allocated codes are in accordance to the rule that the shorter code lengths are obtained for the symbols of highest probability.

Step6 Additional encoding for Huffman is done for code mapping in correspondence to the symbols, these results in the compression of data.

Step7 The image which is genuine is reformed again, the decompression of an image takes place with the help of Huffman's decoding.

Step8 The generation of an encoding equivalent tree.

Step9 Reading of the character-wise input and to Table 2 left unless one gets the last element in the table.

Step10 Character encoding in the leaf output and move back to the root, the step 9 is to be continued as long as the corresponding symbol codes are known.

6 Experimental Outcomes

In Fig. 1, the input image is shown and Huffman's algorithmic coding is applicable for the code generation and later on for the algorithmic decompression (also known as Huffman's decoding) which is generally applied so as to get the genuine and original image back from the codes which are generated and this is depicted in the Fig. 2. The difference between the numbers of bits needed for the image representation is the

Fig. 1 Input image [6]

Fig. 2 Decompressed image

number of bits saved; it is shown below in the Fig. 1, keeping in consideration every symbol which can take a code of maximum length containing 8 bits and the amount of bits it takes to represent the compressed image by the Huffman's code. For example, bits saved = (8 * (r * c) − (l1 * l2)) = 3212, the input matrix size is represented by r and c on the other hand, size of Huffman's code is represented by l1 and l2. On the other hand, the ratio of compression is the number of bits which are needed for the representation of image making use of coding technique of Huffman to the number of bits which are needed to indicate the input image. For example, compression ratio = (11 * 12)/(8 * r * c) = 0.8455. The decompressed one is the resultant image (from Fig. 2). It is also evident that the image which is decompressed is almost same in length to that of the input image.

7 Conclusion

In this study, it is shown clearly that when there is a redundancy of higher data, it helps to get much more compression. In the topic mentioned above, the techniques of compression and decompression which make use of coding and decoding via Huffman for testing scan so as to reduce the volume of test data and time of test application. It is shown in the outcomes of experiments that 0.8456 is the ratio of compression for the image obtained as above. It can, therefore, be concluded that Huffman's coding is an ideal approach for the compression as well as decompression of images to some extent. The work to be carried out in future regarding the image compression for storage and their transmission is done by lossless methods of image compression as we have seen in above conclusion that decompressed image is similar to the input image and no information is lost during the transmission of the image. The other image compression techniques can also be used which are commonly known as JPEG image and entropy coding.

References

1. Compression Using Fractional Fourier Transform A Thesis Submitted in the partial fulfillment of requirement for the award of the degree of Master of Engineering In Electronics and Communication. By Parvinder Kaur
2. Ternary Tree & FGK Huffman Coding Technique Dr. Pushpa R. Suri † and Madhu Goel Department of Computer Science & Applications, Kurukshetra University, Kurukshetra, India
3. Massachusetts Institute of Technology Department of Electrical Engineering and Computer Science
4. Huffman, D.A.: A Method for the construction of minimum-redundancy codes. Proc. IRE **40**(10), 1098–1101 (1952)
5. Efficient Huffman decoding by Manoj Aggrawal and Ajai Narayan
6. Kanzariya, N., Nimavat, A., Patel, H.: Security of digital images using steganography techniques based on LSB, DCT, and Huffman encoding. In: Proceedings of International conference on Advances in Signal Processing and Communication (2013)

Performance Investigations of Multi-resonance Microstrip Patch Antenna for Wearable Applications

Raghvendra Singh, Dambarudhar Seth, Sanyog Rawat and Kanad Ray

Abstract Wearable innovation has risen as a sharply debated issue in recent years around the world and the enthusiasm for this segment keeps on creating and will also overwhelm later on. This investigation work shows a multi-resonance monopole microstrip patch antenna for body wearable applications. The proposed antenna works at multi frequencies for GSM band 235 MHz (1754–1989 MHz), ISM band 561 MHz (2283–2844 MHz) and C band 545 MHz (5174–4629 MHz) covering all requirements of body-centric wireless communication using on-body wearable antennas. In particular, the designing and investigation are concentrated upon the performance of on-body and off-body wearable antenna in terms of multi-frequency operation. The performance of the proposed antenna is explored using equivalent four-layered phantom model of the human body as far as reflection coefficient, radiation pattern, electric field pattern, gain and current density, all outcomes are in great understanding.

Keywords Body-centric wireless communication · On-body wearable antenna ISM band · GSM band · Multi-resonance · Return loss

R. Singh · D. Seth
Electronics and Communication Engineering,
JK Lakshmipat University, Jaipur, India
e-mail: planetraghvendra@gmail.com

D. Seth
e-mail: dseth@jklu.edu.in

S. Rawat
Department of Electronics and Communication Engineering, Manipal University Jaipur, Jaipur, India
e-mail: sanyograwat@gmail.com

K. Ray (✉)
Amity School of Applied Sciences, Amity University, Jaipur, Rajasthan, India
e-mail: kanadray00@gmail.com

K. Ray et al. (eds.), *Soft Computing: Theories and Applications*,
Advances in Intelligent Systems and Computing 742,
https://doi.org/10.1007/978-981-13-0589-4_15

1 Introduction

Body-centric wireless communication is now showing high potential in the field of health monitoring, sports, entertainment multimedia and data transfer. There are some design constraints for these antennas: compact size, robustness, flexibility, less power consumption and comfortness [1, 2]. All these constraints are correlated to the requirements of the consumers in future. A detailed structure of body-centric wireless communication antennas is required to do analysis and design it perfectly.

Antennas integration over the wearable clothing is reported by many researchers including [3–5] which present linearly polarized wearable patch antennas. Fabric-based antennas are also combined with electromagnetic bandgap structures reported in [6]. As the body-centric wireless communication is divided into three parts: In-body communication, On-body communication and off-body communication according to the application of antennas in and around of user. MICS band antennas are designed and analysed by [7] which are working in 402–405 MHz.

The human body interaction with the antenna is discussed by Terence and Chen [8] and the performance of body-worn planar inverted F antenna (PIFA) is investigated. Several ISM band antennas are investigated in [9] like monopole, patch array and patch array for on-body communication. The position of these body-worn antennas is investigated and different favourable results are obtained for on-body communication.

In current era, worldwide interoperability for microwave access (WiMAX) operate at 2.5 and 5.5 GHz band is becoming very renounced due to its usage [10]. A 50×50 mm^2 slot ring antenna integrated with the capacitive patch is investigated for WLAN and WiMAX applications [11]. Coplanar Waveguide (CPW)-fed slotted patch antennas of 23×30 mm^2 to operate in 2.4–2.63, 3.23–3.8 and 5.15–5.98 GHz bands [12], and 25×25 mm^2 to cover 2.14–2.85, 3.29–4.08, and 5.02–6.09 GHz bands [5] have been developed. In above-discussed antenna proposals, the researchers are achieving multi-band antennas comprising radiation performance or gain or fabrication cost. There is always scope for researchers to do work designing and fabrication of low profile, better radiation profile good gain value antennas, so that the wireless communication can be more efficient.

Based on the above-depicted investigations this paper is proposing a multi-resonance microstrip patch antenna (MRMPA) which is designed on Rogers 5880 substrate. The performance criteria of MRMPA is critically analysed and compared with other variants of this antenna. This paper is proposing MRMPA as a multi-purpose antenna for on-body and off-body communication because it is covering RF band (GSM), ISM band and C band. It can be used by patients to send their physiological data to a physician who is connected via RF network (GSM) or WiFi Network (ISM) or C band devices. MRMPA is a compact, low profile and multi-band operational antenna for wearable applications.

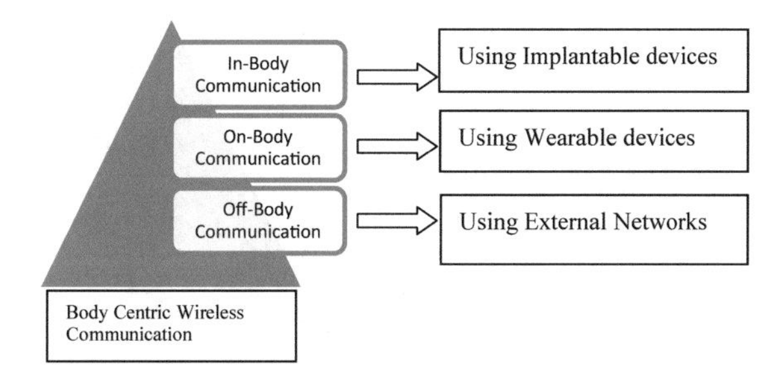

Fig. 1 Body-centric wireless communication

2 Body-Centric Wireless Communication

The body-centric wireless communication can be utilized for many causes as tracking, healthcare systems and identification. Due to growing fabrication technology, novel and appropriate antenna requirements in the biomedical field is also growing rapidly. For this particular purpose, antennas need to be lightweight, low cost and small size. The principle highlight of wearable Biomedical devices is to permit remote correspondence from or to the body by means of wearable receiving antennas [13, 14].

Body-driven remote correspondence networks are turning into the essential part for future correspondence. Body-centric wireless communication is divided into three types of correspondences: In-Body communication, On-Body communication and Off-Body communication which is shown in Fig. 1. In-body by the name itself recommends correspondence with the implantable devices or sensor. On-body essentially alludes to the correspondence between On-body/wearable devices which is the focal point of talk. Off-body alludes to the correspondence from off-body to On-body devices.

3 Antenna Design Strategy

Antenna is the most vital part in the field of remote correspondence for Biomedical applications. Design Strategy for multi-resonance microstrip patch antenna (MRMPA) is inspired from [14–18] in which a strategy is proposed for implanted antennas. Here in this paper, a rectangular microstrip patch antenna (MSPA) is designed initially for single band (Fr = 2.45 GHz, ISM) operation then converted into multi-resonance microstrip patch antenna (MRMPA) using horizontal slots, inverted hook slot in the patch and modified ground plane. This MRMPA is placed over the canonical model of human tissue which is the electrical equivalent of the human body [16, 19]. The antenna parameters are checked for MRMPA with and without the canonical layers. The parameters are compared and the favourable results are obtained using MRMPA. This complete strategy is shown in Fig. 2.

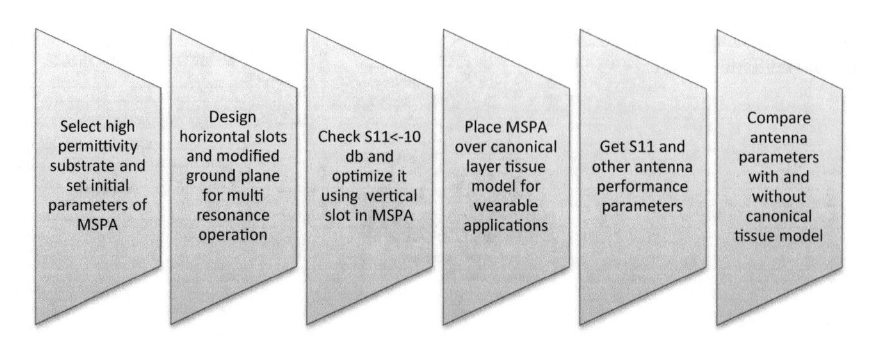

Fig. 2 Strategy of designing the proposed MRMPA

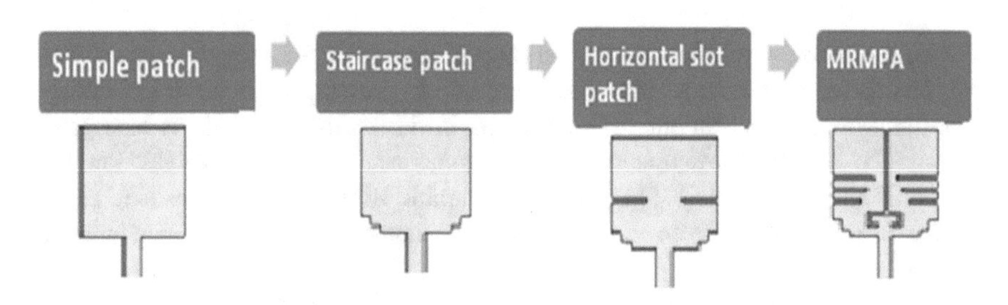

Fig. 3 Sequence of designing MRMPA

According to the flow sequence of Fig. 2. rectangular microstrip patch antenna is designed for the single band (ISM), the dimensions and other parameters of antenna are according to single-band and staircase patch is designed using 1 mm × 1 mm rectangular slots. A rectangular slot is cut at the middle of a patch of width 1 mm and named as horizontal slot patch antenna, this antenna is showing two bands (ISM and GSM). Favourable results are shown by MRMPA which is containing one hook slot and three horizontal slots at both the sides of hook slot. The sequence of designing MRMPA is shown in Fig. 3. In which MRMPA is the final design followed by horizontal slot patch, staircase patch and simple rectangular patch.

3.1 Equivalent Resonant Circuit

A rectangular microstrip patch antenna can be represented by a parallel combination of R_p, L_p and C_p, depicted in Eqs. 1–3 [20].

$$\mathrm{Rp} = \frac{Qr}{\omega Cp} \tag{1}$$

$$Lp = \frac{1}{\omega 2 \cdot \mathrm{Cp}} \tag{2}$$

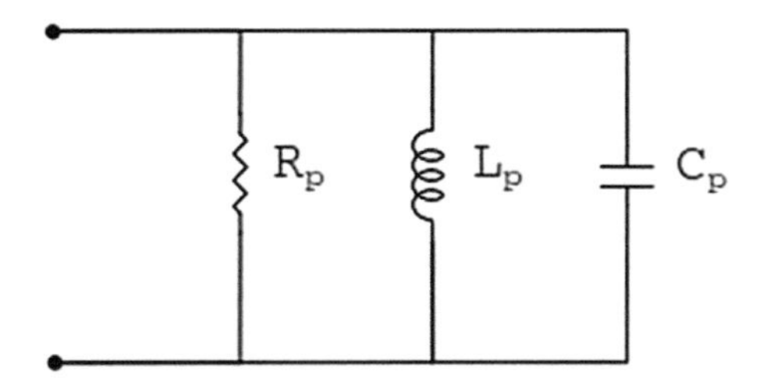

Fig. 4 Equivalent circuit of rectangular patch antenna [20]

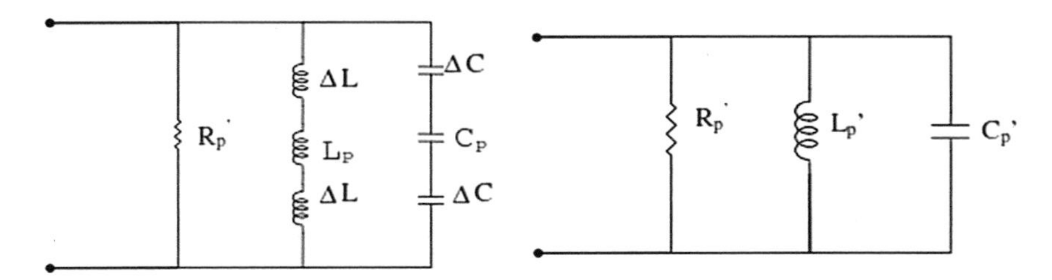

Fig. 5 Equivalent circuit of the proposed antenna using horizontal slots [20]

$$\mathrm{Cp} = \frac{\varepsilon 0 \varepsilon e L W}{2h} Cos^{-2}(\Pi x 0) \tag{3}$$

In which L = length of rectangular patch, W = width of rectangular patch X_0 = feed point location, h = thickness of substrate material

$$Qr = \frac{c\sqrt{\varepsilon e}}{fh} \tag{4}$$

where c = velocity of light, f = frequency of operation of antenna

The equivalent circuit of antenna is given in Fig. 4.

The horizontal slots generate extra value of capacitance and inductance (L and C) and the circuit is changed as in Fig. 5. R_p', L_p' and C_p' are the equivalent lumped parameters of the equivalent circuit of MRMPA [20]. The horizontal and vertical slots are generating the extra capacitance C and inductance L values and making the antenna multi resonator.

3.2 MRMPA Design

MRMPA is designed using a substrate Rogers 5880 of height 1.57 mm, permittivity 2.25 and tangent delta 0.002. The volume of the proposed antenna is $30 \times 32 \times 1.57$ mm^3, which is very compact antenna for using On-body or Off-body communication in body-centric wireless communication. The three horizontal slots at

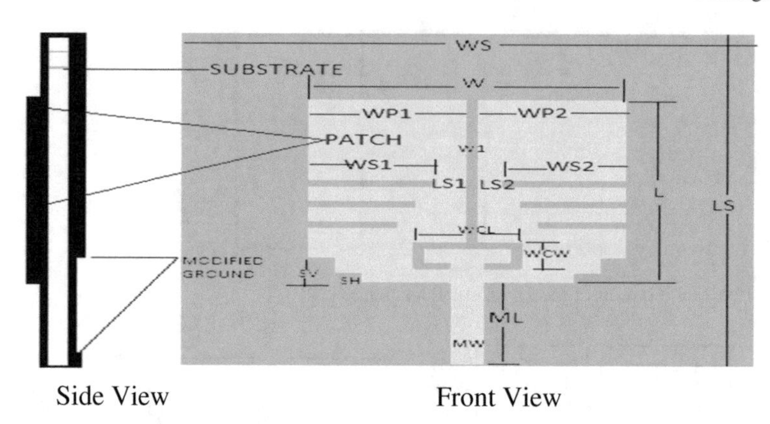

Fig. 6 MRMPA side view and front view

Table 1 Geometric design parameters for MRMPA

Parameter	Ws	Ls	W	L	Wp1	Wp2	Ws1	Ws2
Value (mm)	30	32	18	18	8.5	8.5	7	7
Parameter	Ls1	Ls2	Wcl	Wcw	Sv	S_H	M_L	M_W
Value (mm)	1	1	6	3.5	2	1	8.5	2

left and right side of the proposed antenna are introducing extra capacitance values hence multi-resonance operation is taking place. The vertical hook slot in MRMPA is providing better impedance matching. The modified ground plane is designed using three vertical boxes of width 4 mm and two slots of 1 mm width. The front and side view of MRMPA is shown in Fig. 6. Geometric design parameters of the proposed MRMPA are given in Table 1.

4 Results and Discussion

The proposed MRMPA is designed and simulated in CST microwave studio for wearable applications. Body-centric wireless communication requires compact, low profile antennas which can operate for On-body and Off-body communication. MRMPA is operating in radio frequency (GSM), ISM band and C band which is a multipurpose antenna.

4.1 Return Loss

Return loss characteristics are shown in Fig. 7, the −10 db bandwidth is obtained in RF band 235 MHz (1754–1989 MHz), ISM band 561 MHz (2283–2844 MHz) and C band 545 MHz (5174–4629 MHz). The multi-resonance is achieving −25.5 db, −23 db and −21.6 db return losses for RF band, ISM band and C band respectively.

This Antenna is made using various capacitive slots, hence there are simple patch antennas followed by horizontal slot, staircase and made in the sequence of making of MRMPA. The return loss comparison of these four variants is shown in Fig. 8. MRMPA is a refined antenna which is showing multi-resonance operation for aforementioned bands. Staircase antenna is showing resonance at 2 and 2.5 GHz, a horizontal slot antenna is also operating at 2.5 GHz, hence MRMPA is refined and following all agreements for multi-resonance antenna.

MRMPA is tested using a four-layered canonical model, called as Phantom. This Phantom is the electrical equivalent of human tissue as mentioned by Italian National Research Council [16] and all equivalent electrical parameters are given by them. The thickness of skin, fat, muscle and bone is 2 mm, 8 mm, 10 mm and 5 mm respectively

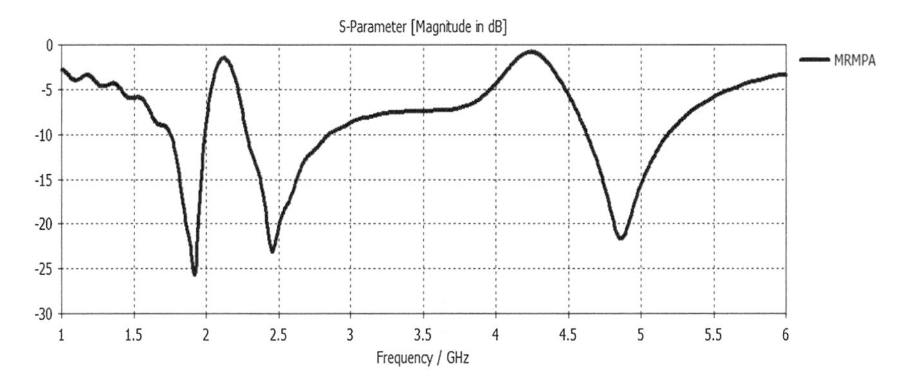

Fig. 7 Return loss versus frequency graph for MRMPA

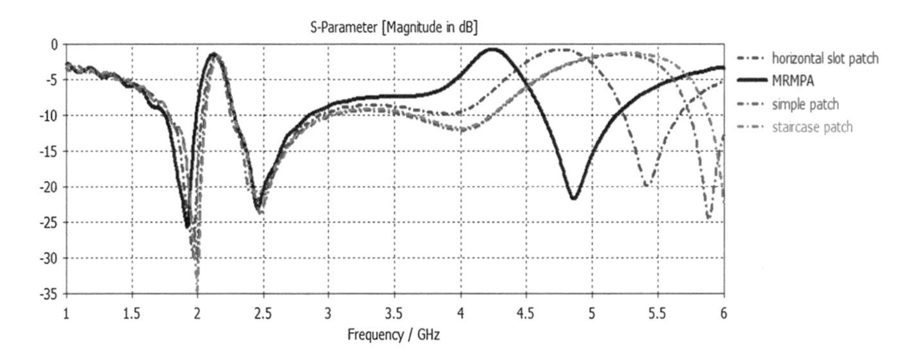

Fig. 8 Return loss comparison of variants of MRMPA

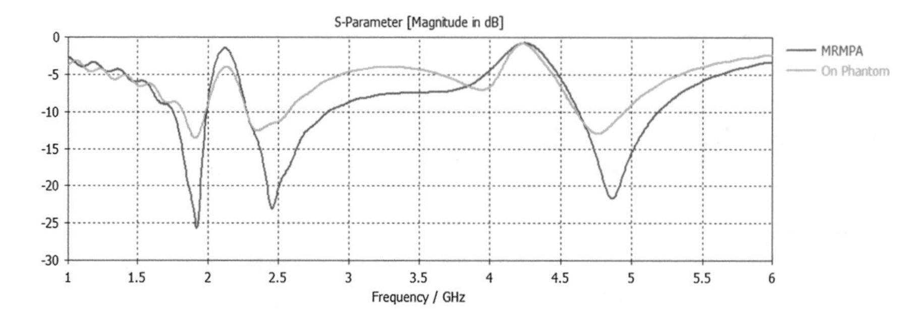

Fig. 9 Return loss comparison of MRMPA in air and on phantom

and MRMPA is simulated on this Phantom. The Return loss comparison of MRMPA in the air and on Phantom is shown in Fig. 9. Due to increased permittivity values of different layers, the return loss value is decreased using this Phantom.

4.2 Current Distribution and Electric Field

The current distribution on the surface of MRMPA is shown in Fig. 10. At 1.8 GHz the highest current is around microstrip line, vertical hook slot and three horizontal slot. At 2.45 GHz the horizontal slots are drawing maximum current, similarly at 5.1 GHz the vertical hook slot modified ground and horizontal slots are at maximum current.

Electric field distribution is shown in Fig. 11 at concerning frequencies. The field variation is in three colours; blue, pink or its mixture. The highest field around the patch is at 2.45 GHz and most of the area of patch is radiating.

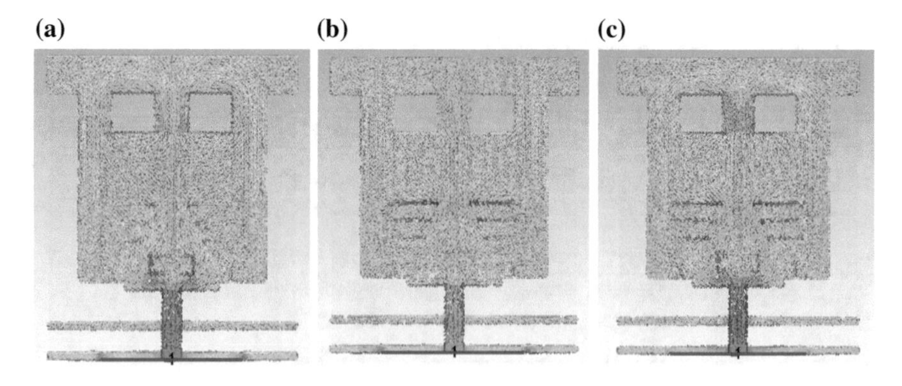

Fig. 10 Surface current distribution at **a** 1.8 GHz **b** 2.45 GHz **c** 5.1 GHz

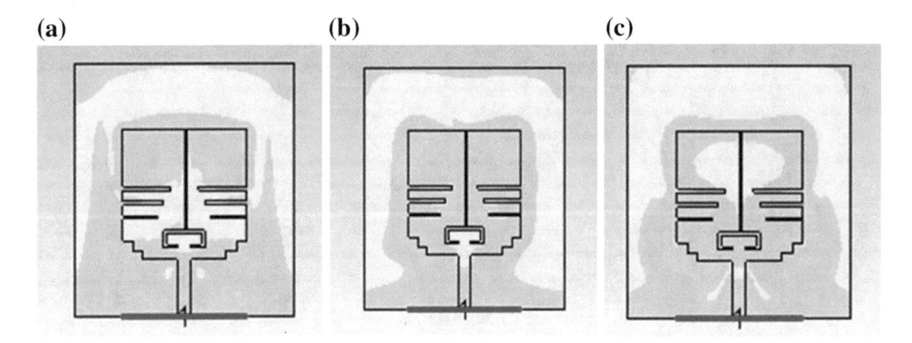

Fig. 11 Electric field distribution at **a** 1.8 GHz **b** 2.45 GHz **c** 5.1 GHz

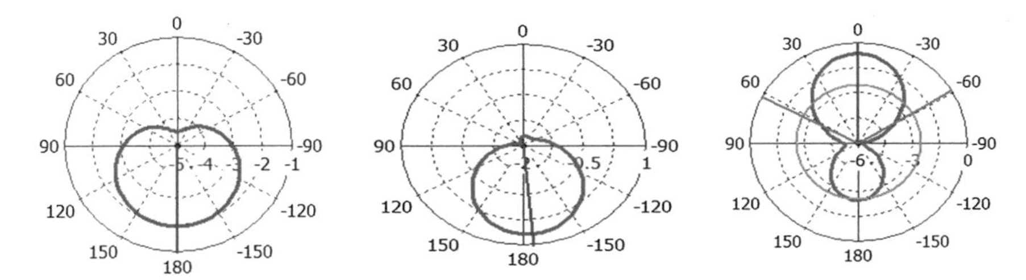

Fig. 12 Polar plot of MRMPA

4.3 Radiation Pattern

The two-dimensional polar plot of the MRMPA is shown in Fig. 12. It is unidirectional at 1.8 and 2.45 GHz but back lobe is observed at 5.1 GHz. The wearable antenna requirement is always unidirectional radiation pattern to decrease the field impact of radiation on the human body and this antenna is showing it successfully.

4.4 Gain of MRMPA

The gain of MRMPA is shown in Fig. 13 which is depicting the stable nature of antenna in resonant frequency ranges, 1.8 and 2.45 GHz. This antenna is behaving as a stable resonant system for application in wearable applications. The effect of body proximity on the electrical properties of antenna is adverse but MRMPA is stable throughout the simulated frequency ranges.

The maximum IEEE gain value 1.6 is achieved at frequency 2.5 GHz, this lower value is due to the body proximity of MRMPA.

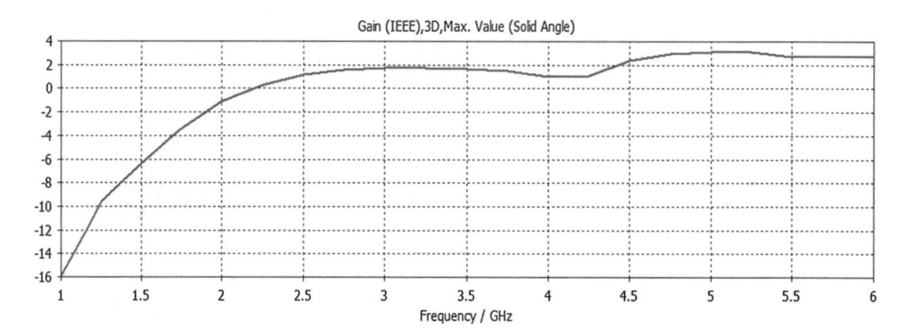

Fig. 13 Gain versus frequency plot of MRMPA

5 Conclusion

In this paper a $30 \times 32 \times 1.57$ mm^3 multi-resonance microstrip patch antenna operating in GSM band, ISM band and C band for wearable applications is proposed. The proposed MRMPA is designed for the body-centric wireless communication so that the vital data of patient or end user can be sent to the physician using any of the aforementioned bands. The multi-resonance characteristics are designed using a simple rectangular patch and introducing horizontal and vertical hook slot. The return loss (with phantom and without phantom), radiation pattern, electric field and surface current density of proposed MRMPA antenna are in good agreement. The simulated results for the proposed antenna reveal the operating bandwidths GSM band 235 MHz (1754–1989 MHz), ISM band 561 MHz (2283–2844 MHz) and C band 545 MHz (5174–4629 MHz). The performance comparison of this antenna with other antennas and its variants is showing unidirectional radiation pattern and consistent return loss for GSM band ISM band and C band operation in body-centric wireless communication.

References

1. Conway, G.A., Scanlon, W.G.: Antennas for over-body-surface communication at 2.45 GHz. IEEE **57**, 844–855 (2009)
2. Correia, L.M., Mackowiak, M.: A Statistical Model for the Influence of Body Dynamics on the Gain Pattern of Wearable Antennas in Off-Body Radio channels, pp. 381–399. Springer (2013)
3. Salonen, P., Kim, J., Rahmat-Samii, Y.: Dual-band E-shaped patch wearable textile antenna. In: IEEE Proceedings of APS, Washington (2005)
4. Ouyang, Y., Karayianni, E., Chappell, W.J.: Effect of fabric patterns on electrotextile patch antennas. In: IEEE Proceedings of APS, Washington (2005)
5. Tanaka, M., Jang, J.-H.: Wearable microstrip antenna. In: IEEE Antennas and Propagation International Symposium, vol. 2, pp. 704–707
6. Zhu, S., Langley, R.: Dual-band wearable antennas over EBG substrate. Electron. Lett. **43**(3), 141–142 (2007)

7. Guo, Y.-X., Xiao, S., Liu, C.: A Review of Implantable Antennas for Wireless Biomedical Devices. Forum for Electromagnetic Research Methods and Application Technologies
8. Terence, S.P., Chen, Z.N.: Effects of human body on performance of wearable PIFAs and RF transmission. In: IEEE Proceedings of APS, Washington (2005)
9. Kamarudin, M.R., Nechayev, Y.I., Hall, P.S.: Performance of antennas in the on-body environment. In: IEEE APS International Symposium, Washington (2005)
10. Liu, H.-W., Ku, C.-H., Yang, C.-F.: Novel CPW-fed planar monopole antenna for WiMAX/WLAN applications. IEEE Antennas Wirel. Propag. Lett. **9**, 240–243 (2010)
11. Sim, C.Y.D., Cai, F.R., Hsieh, Y.P.: Multiband slot-ring antenna with single-and dual-capacitive coupled patch for wireless local area network/worldwide interoperability for microwave access operation. IET Microw. Antennas Propag. **5**, 1830–1835 (2011)
12. Wang, P., Wen, G.-J., Huang, Y.-J., Sun, Y.-H.: Compact CPW-fed planar monopole antenna with distinct triple bands for WiFi/WiMAX applications. Electron. Lett. **48**, 357–359 (2012)
13. Jain, L., Singh, R., Rawat, S., Ray, K.: Miniaturized, meandered and stacked MSA using accelerated design strategy for biomedical applications. In: 5th International Conference of Soft Computing for Problem Solving, pp. 725–732 (2016)
14. Singh, R., Rawat, S., Ray, K., Jain, L.: Performance of wideband falcate implantable patch antenna for biomedical telemetry. In: 5th International Conference of Soft Computing for Problem Solving, pp. 757–765
15. Kiourti, A., Nikita, K.S.: Miniature implantable antennas for biomedical telemetry: from simulation to realization. IEEE Trans. Biomed. Eng. **59**(11) (2012)
16. Italian National Research Council, Institute for Applied Physics, homepage on Dielectric properties of body tissues (Online). http://niremf.ifac.cnr.it
17. Singh, R., Kumari, P., Rawat, S., Ray, K.: Design and performance analysis of low profile miniaturized MSPAs for body centric wireless communication in ISM band. Int. J. Comput. Inf. Syst. Ind. Manag. Appl. **9**, 153–161 (2017)
18. Jain, L., Singh, R., Rawat, S., Ray, K.: Stacked arrangement of meandered patches for biomedical applications. Int. J. Syst. Assur. Eng. Manag. (Springer). https://doi.org/10.1007/s13198-016-0491-6
19. Singh, R., Seth, D., Rawat, S., Ray, K.: In body communication: assessment of multiple homogeneous human tissue models on stacked meandered patch antenna. IJAEC **9**(1) (2018), Article 4 (Accepted)
20. Bahal, I.J.: Lumped Elements for RF and Microwave Circuits. Artech House, Boston (2003)

Fuzzy Counter Propagation Network for Freehand Sketches-Based Image Retrieval

Suchitra Agrawal, Rajeev Kumar Singh and Uday Pratap Singh

Abstract In this paper, we present Fuzzy Counter Propagation Network (FCPN) for Sketch-Based Image Retrieval (SBIR) with collection of freehand sketches; trademark and clip art, etc., using feature descriptors. FCPN is combination of Counter Propagation Network (CPN) and Fuzzy Learning (FL). We use features descriptor like Histogram of Gradient (HOG) for freehand sketches/images and these features are used to the training of FCPN. Flicker dataset containing 33 different shape categories, is used for training and testing. Different similarity measure functions are discussed and used similarity between query by nonexpert sketchers and database. We compare proposed FCPN method with other existing Feed-forward Networks (FFN) and Pattern Recognition Network (PRN). Experimental results show that FCPN methods outperform over networks.

Keywords FCPN · SBIR · HOG · Sketch image · Flickr dataset

1 Introduction

In the era of rapidly increasing technology and availability of a huge number of multimedia data search an image of certain categories is a very cumbersome job. The other important aspect is that most of the person is not an expert sketcher. These are two important points shows the importance of sketch-based image retrieval

S. Agrawal · R. K. Singh
Department of CSE & IT, Madhav Institute of Technology and Science, Gwalior 474005, India
e-mail: suchiagrawal0007@gmail.com

R. K. Singh
e-mail: rajeev.mits1@gmail.com

U. P. Singh (✉)
Department of Applied Mathematics, Madhav Institute of Technology and Science, Gwalior 474005, India
e-mail: usinghiitg@gmail.com

K. Ray et al. (eds.), *Soft Computing: Theories and Applications*, Advances in Intelligent Systems and Computing 742,
https://doi.org/10.1007/978-981-13-0589-4_16

system. The key challenge of SBIR is the ambiguous sketch of an image by different nonexpert sketchers. The key idea for sketch-based image retrieval is based on extract edges of the sketch images using image feature descriptor and then compare via some similarity/distance measure function between sketch feature and image features. The problem lies with the system is that the query image is a freehand-drawn sketch, the edge matching between the images is quite hard to provide the accurate results. In this paper, we propose the implementation of FCPN, FFN and PRN that allows learning a function which performs mapping between query image and database images. The database images serve as the target vectors for input vectors being the sketch image mapping each of them to an output vector evaluated by the network. The proposed system introduces the fuzzy counter propagation network applicability in the field of SBIR, which has been applied for the first time to the best of our knowledge. The key idea and motivation behind sketch-based image retrieval is based on extract edges of the sketch images using image feature descriptor and then compare via some similarity measure between sketch and image features. The problem lies with the system is that the query image is a freehand-drawn sketch, the edge matching between the images is quite hard to provide the accurate results.

The remaining part of this paper been organized as follows: Sect. 2 gives an overview of the existing work, Sect. 3 presents the sketch and image feature descriptors and FCPN algorithm. We discuss experimental results and comparison with other existing network and methods in Sect. 4, followed by conclusion and future work in Sect. 5.

2 Previous Work

In last few decades, lots of research has been done in CBIR systems, but in all these systems, the input query required is a digital image with low-level visual features such as shape, texture, and color. Sometimes, the user might not have a clear image of his query but may draw a sketch and search for similar images from the database. Few researchers have tried to bring the topic in light and have carried out remarkable research in the field. A brief and concise list of research based on content or sketch-based is described below.

A large number of scale feature vectors can be used for transforming image data [1] into scale-invariant co-ordinates. Corresponding features for each image/sketch pair are stored in a database and the query sketch/image features are compared to find the matching by a distance measure between them. The local self-similarity descriptor [2] which allows differentiating between internal geometric layout within images or videos has been implemented. The general benchmark [3] for assessing the execution of any SBIR framework has been characterized. For coordinating pictures SBIR framework utilizes the data contained in the outlined 3-D shape. The framework that can powerfully create shadows [4] generated or extracted from the huge database of images has been proposed. This framework retrieves related images continuously in view of inadequate sketches provided by the user. The problem of fast, large-scale database searching [5, 6] for over one million images have resulted in the development of SBIR to ease the user's understanding. The descriptors are constructed in such a way that the sketch as well as color image have to undergo the same preprocessing level. A bag-of-regions [7] has been presented to construct a SBIR framework. This system encodes the eminent shapes on different levels of points of interest by considering as part of locales.

Indexing structure [8] and raw shape-based coordinating algorithm [9] to compute similarities amongst natural and sketch images query, and make SBIR adaptable to different types of images were presented. To recognize similarities [10] between a hand-drawn sketch and the images in a database is an important process. A descriptor on tensor-based image [11, 12] for huge scale SBIR was built-up. Final effect of the trial result is little furthermore there is need for change in geometric sifting. An image retrieval framework for freehand sketches providing the interactive search [13] of image database depicting shape has been developed. Gradient Field HOG (GF-HOG); derived from HOG descriptor for sketch-based image retrieval along with Bag of Visual Words (BoVW) retrieval framework have controlled the system for providing a robust SBIR, and for identifying sketched objects within an image. Bag-of-features [14] model can also be utilized to develop feature set for human sketches, in a large database of sketches gathered to assess human identification. Hand-drawn sketch in light of stroke components is presented for keeping up basic data of visual words maintained in a hierarchical tree structure. A strategy to lessen the appearance gap [15] in SBIR was proposed for sketches and extracted edges of images are dealt with as a combination of lines, establishing the framework for providing a better view on edges. Using convolutional neural networks [16], from 2D human sketches, 3D models can be retrieved.

Generating a sketch by viewing the details from a real image can allow obtaining a better edge map [17] for each image. Edge grouping thus performed should minimize the overall energy function. After so much of research, SBIR is still a challenging task due to the ambiguity produced in sketches when compared with images. Instead of traditional image retrieval methods some soft computing techniques hybrid neural network, fuzzy inference system [18–20] are used for nonlinear approximation like SBIR was to pull out the output feature vectors with minimum distance from input feature vector. Sketch-image pairs that hold minimum distance on the basis of the feature vector are labeled as relevant otherwise irrelevant.

3 Feature Descriptors

In order to select the input/output vector for fuzzy counter propagation network for sketch-based image retrieval system, we consider some important feature descriptors to retrieve similar image to the query image (freehand sketch) from the database.

3.1 HOG Feature Descriptor

HOG is one of the simple yet powerful feature descriptors for facial images [21]. It forms a histogram on the basis of orientations of the magnitude of the gradient of each pixel.

$$\begin{bmatrix} g_x(x, y) \\ g_y(x, y) \end{bmatrix} = \begin{bmatrix} \frac{\partial I(x,y)}{\partial x} \\ \frac{\partial I(x,y)}{\partial y} \end{bmatrix} \tag{1}$$

$$\varphi = \tan^{-1}\left(\frac{g_y}{g_x}\right) \tag{2}$$

$$\rho = \left(g_x^2 + g_y^2\right)^{0.5} \tag{3}$$

In this paper, we use HOG feature descriptor due to simplicity and effectiveness. Klare et al. [22, 23], on matching forensic sketches to mugshot photos shows that HOG has larger the distance between features of sketch and photos, therefore more suitable for retrieval.

3.2 FCPN Learning Algorithm

The technology which resembles the human brain to the decision making and learning is known as artificial intelligence. One such important neural network is known as FCPN which is an effective method to recognize the SBIR. FCPN network [24] we used fuzzy logic relation IF-THEN rules i.e. input feature vector $X = (x_1 \cdot x_2, \ldots, x_n)$ of the query image is transformed to the output vector $= (y_1 \cdot y_2, \ldots, y_m)$. The FCPN network has n neurons in the input layer and $X = (x_1 \cdot x_2, \ldots, x_n)$ is HOG feature vector of sketch image and Grossberg layer has m neurons and $Y = (y_1 \cdot y_2, \ldots, y_m)$ is HOG feature vector of the target image. Input vector X is transformed to Y using FCPN through a logical rule known as IF-THEN rule. The connection weight matrix between input to competitive layer is denoted via W is called IF or antecedent and weight matrix between competitive to hidden layer is denoted via V is called THEN or consequent. Rule is defined as IF X is W THEN Y is V. First, evaluate distance between input vector X and weight matrix W, find the winning node with the smallest distance. The weights of the ***winning node*** (say jth node) in the competitive layer has been adjusted using IF-THEN rules. If the distance between HOG feature vector $X = (x_1 \cdot x_2, \ldots, x_n)$ of sketch image and weight vector w_j is less than the length of interval of triangular function δ i.e. $X - w_j < \delta$, THEN the weight

$$w_j = \frac{\sum_{i=1}^{n} \mu(x_i, w_j) x_i}{\sum_{i=1}^{n} \mu(x_i, w_j)}; j = 1, 2, \ldots, m \tag{4}$$

where $\mu(x_i, w_j)$ is the degree of membership defined as

$$\mu(x_i, w_j) = \frac{\|x_i - w_j\|}{\sum_{k=1}^{m} \|x_i - w_j\|} \tag{5}$$

where $\|X - w_j\|$ is distance between sketch feature X and w_j. IF $X - w_j < \delta$, THEN the weight is updated as follows:

$$w_j^{new} = w_j^{old} + \alpha(X - w_j^{old}) \tag{6}$$

where $\alpha \in (0, 1]$, X is sketch feature vector and w_j weight vector from the input layer to ***winning node***.

If $X - w_j > \delta$ then the weight is updated as follows:

$$w_j^{new} = w_j^{old} \tag{7}$$

The training of Grossberg layer is done, IF $\|X - w_j\| < \delta$, THEN the weight

$$v_j = \frac{\sum_{i=1}^{n} \mu(y_i, v_j) y_i}{\sum_{i=1}^{n} \mu(y_i, v_j)}; j = 1, 2, \ldots, m \tag{8}$$

where $\mu(y_i, v_j) = \frac{\|y_i - v_j\|}{\sum_{k=1}^{m} \|y_i - v_j\|}$ is updated as follows:

$$v_j^{new} = v_j^{old} + \beta(Y - v_j^{old}) \tag{9}$$

where $\beta \in (0, 1]$, Y is target feature vector and v_j is weight at jth neuron. IF $\|X - w_j\| > \delta$, THEN the weight is updated as:

$$v_j^{new} = v_j^{old} \tag{10}$$

Training and testing phase of FCPN can be described in following steps:

Training Phase:

1. First, divide the given dataset into two parts (i) training (ii) testing.
2. Initialize all the parameters of FCPN and start the training of FCPN with the training data from step 3–step 7.
3. Calculate ***winning node*** using smaller distance input feature vector and weights.
4. Competitive layer or divides the training data into similar type of clusters.
5. ***Winning node*** is used for inhibitory interconnections among hidden nodes and output nodes, when input data are presented at input layer.
6. Let jth neuron in the hidden layer is ***Winning node***, then weights between hidden and output layer is adjusted according to sketch and target feature vector using step 7.

7. Weights w_j and v_j (Eqs. 4 and 8) between layers are updated via selecting appropriate learning rate and IF-THEN rules as follows:

   7.1 **Input to competitive layer**

   IF $\left\| X - w_j \right\| < \delta$, THEN
   $w_j^{new} = w_j^{old} + \alpha\left(X - w_j^{old}\right)$
   where $\alpha \in (0, 1]$, X is sketch feature vector and w_j weight vector from the input layer to ***winning node***.
   IF $\left\| X - w_j \right\| > \delta$, THEN
   $w_j^{new} = w_j^{old}$

   7.2 **Competitive to output layer**

   IF $\left\| X - w_j \right\| < \delta$, THEN
   $v_j^{new} = v_j^{old} + \beta\left(Y - v_j^{old}\right)$
   where $\beta \in (0, 1]$, Y is target feature vector and v_j is weight at jth neuron.
   IF $\left\| X - w_j \right\| > \delta$, THEN
   $v_j^{new} = v_j^{old}$

8. Total input between kth hidden node is calculated as follows: $\Gamma_k = \sum_{i=1}^{N}\left(x_i - v_k^i\right)^2$, this represents the distance between input x_i and kth hidden neuron. BMN is denoted as Γ_k and defined as

$$\Gamma_k = \begin{cases} 1;\, if\ \Gamma_k is\ smallest\ among\ all\ k \\ 0;\, otherwise \end{cases} \tag{11}$$

9. Output at kth neuron is obtained as follows:

$$y_k = \sum_i z_i w_i^k \tag{12}$$

10. Test the stopping criteria. /*The goal of the training algorithm is to find the FCPN network with minimum MSE value based on training samples in the dataset.*/

Testing Phase:

1. Once the network training is complete, set weight and biases of network the initial weights, i.e., the weights obtain during training.
2. Apply FCPN to the input vector X.
3. Find unit J that is closet to vector X.
4. Set activations of output units.
5. Apply activation function at y_k, where $y_k = \sum_j z_j v_j^k$ (Fig. 1).

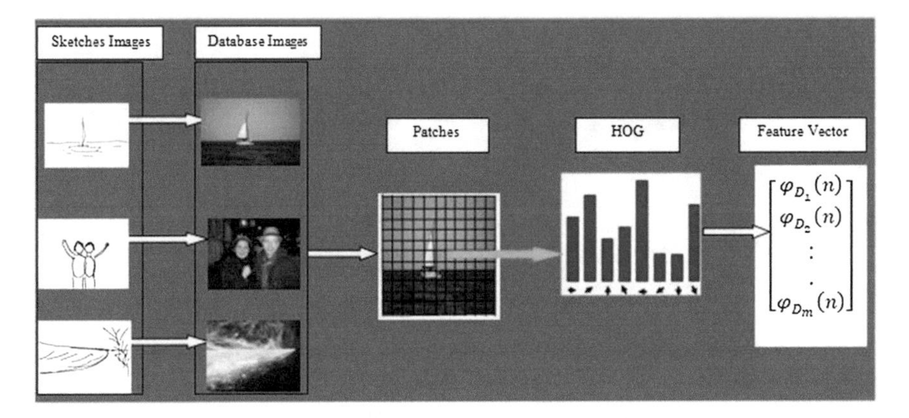

Fig. 1 Sketch and corresponding target image for FCPN

4 Experimental Results

We present, experimental results using FCPN approach on Flickr dataset [25] containing 33 different categories of images. HOG feature vector is computed for sketches and database images using open source library of MATLAB. The vector thus generated serves as the input/output vector for the FCPN, FFN, and PRN. Results thus computed have been formed by performing multiple iterations to give a more general output. The parameters computed for achieving the efficiency and precision of the system has been manually calculated based on different factors such as number of images to be retrieved, number of relevant images present in the database for a certain sketch. Then, the average has been computed over the entire query sketches present (Figs. 2, 3 and 4).

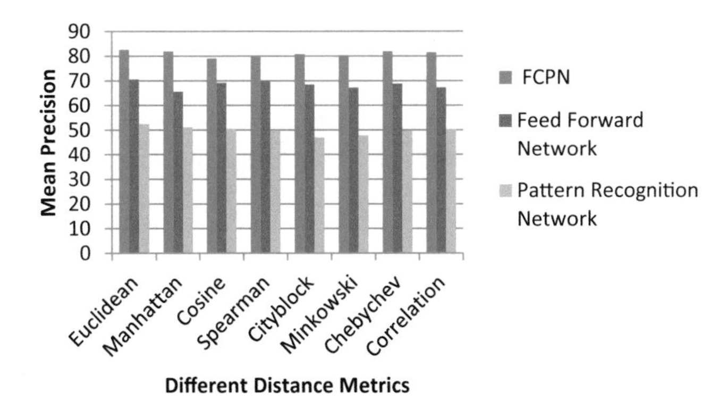

Fig. 2 Comparison between FCPN, FFN, and PRN

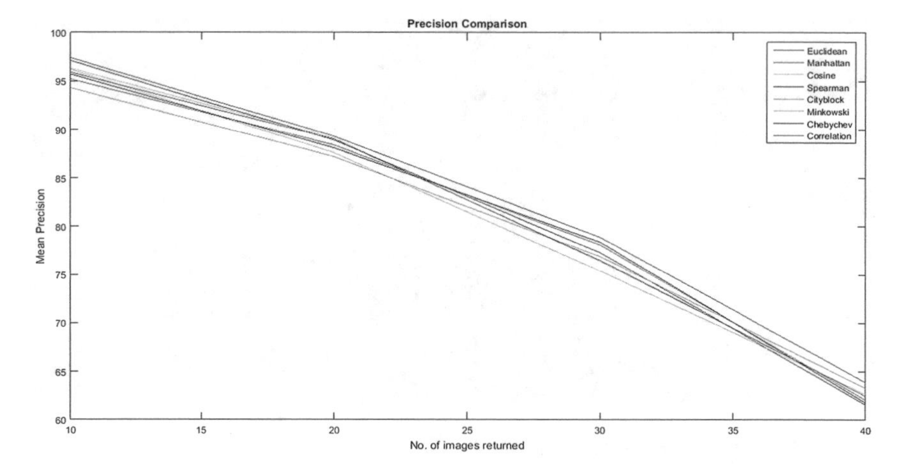

Fig. 3 Mean precision using different similarity measure of FCPN method

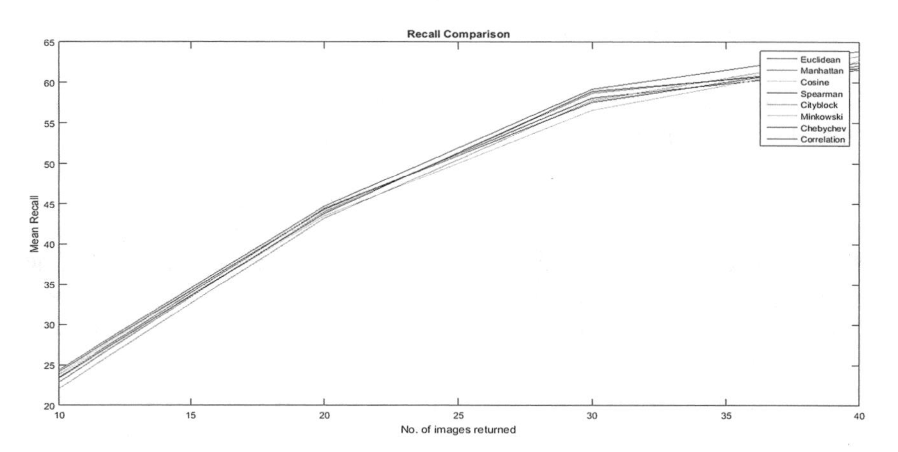

Fig. 4 Recall of using different similarity measure of FCPN method

$$Precision = \frac{Number\ of\ relevant\ images\ retrieved}{Total\ number of\ images retrieved}$$

On an average, system takes 1.511 s to retrieve the images containing similar object like sketch as shown in Fig. 5.

Mean precision and recall using different similarity measure function is shown in Figs. 3 and 4. Comparison of proposed FCPN method with FFN and PRN is shown in Fig. 2.

Fig. 5 Comparison of **a** freehand sketch drawn by nonexpert sketchers and **b** similar photos recognized by FCPN

5 Conclusion

In this paper, we present FCPN approach for sketch-based image retrieval system. Fuzzy IF-THEN rule is used for updating weights of CPN. HOG is used for input and output feature vector of sketch/query image on Flickr dataset comprises 33 shape categories. We have considered different distance measure to find the similarity between sketches and images. The experimental results demonstrate that the FCPN approach for SBIR gives improved mean precision over feed-forward and pattern recognition network. The proposed approach has been tested on Flickr dataset and compared with other existing networks. The experimental result shows that FCPN achieves high accuracy rate. In future various image descriptors as well networks can be considered for sketch-based image retrieval systems.

References

1. Lowe, D.G.: Distinctive image features from scale-invariant keypoints. Int. J. Comput. Vis. **60**(2), 91–110 (2004)
2. Shechtman, E., Irani, M.: Matching local self-similarities across images and videos. In: IEEE Conference on Computer Vision and Pattern Recognition, pp. 1–8 (2007)
3. Eitz, M., Hildebrand, K., Boubekeur, T., Alexa, M.: A descriptor for large scale image retrieval based on sketched feature lines. In: Proceedings of 6th Eurographics Symposium on Sketch-Based Interfaces Model, pp. 29–36 (2009)
4. Zitnick, C.L.: Binary coherent edge descriptors. In: Proceedings of 11th European Conference on Computer Vision, pp. 170–182 (2010)
5. Eitz, M., Hildebrand, K., Boubekeur, T., Alexa, M.: An evaluation of descriptors for large-scale image retrieval from sketched feature lines. Comput. Gr. **34**(5), 482–498 (2010)
6. Eitz, M., Hildebrand, K., Boubekeur, T., Alexa, M.: Sketch based image retrieval: benchmark and bag-of-features descriptors. IEEE Trans. Vis. Comput. Gr. **17**(11), 1624–1636 (2011)
7. Hu, R., Wang, T., Collomosse, J.: A bag-of-regions approach to sketch-based image retrieval. In: 18th IEEE International Conference on Image Process, pp. 3661–3664 (2011)

8. Cao, Y., Wang, C., Zhang, L., Zhang, L.: Edgel index for large-scale sketch-based image search. In: IEEE Conference on Computer Vision and Pattern Recognition, pp. 761–768 (2011)
9. Sun, X., Wang, C., Xu, C., Zhang, L.: Indexing billions of images for sketch-based retrieval. In: Proceedings of 21st ACM International Conference on Multimedia, pp. 233–242 (2013)
10. Bozas, K., Izquierdo, E.: Large scale sketch based image retrieval using patch hashing. Adv. Vis. Comput. 210–219 (2012)
11. Eitz, M., Hays, J., Alexa, M.: How do humans sketch objects. ACM Trans. Gr. **31**, 4 (2012)
12. Sousa, P., Fonseca, M.J.: Sketch-based retrieval of drawings using spatial proximity. J. Vis. Lang. Comput. **21**(2), 69–80 (2010)
13. Hu, R., Collomosse, J.: A performance evaluation of gradient field hog descriptor for sketch based image retrieval. J. Comput. Vis. Image Underst. **117**(7), 790–806 (2013)
14. Ma, C., Yang, X., Zhang, C., Ruan, X., Yang, M.H.: Sketch retrieval via local dense stroke features. Image Vis. Comput. **46**, 64–73 (2016)
15. Wang, S., Zhang, J., Han, T.X., Miao, Z.: Sketch-based image retrieval through hypothesis-driven object boundary selection with HLR descriptor. IEEE Trans. Multimed. **17**(7), 1045–1057 (2015)
16. Wang, F., Kang, L., Li, Y.: Sketch-based 3d shape retrieval using convolutional neural networks. In: IEEE Conference on Computer Vision and Pattern Recognition, pp. 1875–1883 (2015)
17. Qi, Y., Song, Y.Z., Xiang, T., Zhang, H., Hospedales, T., Li, Y., Guo, J.: Making better use of edges via perceptual grouping. In: IEEE Conference on Computer Vision and Pattern Recognition, pp. 1856–1865 (2015)
18. Singh, U.P., Jain, S.: Modified chaotic bat algorithm-based counter propagation neural network for uncertain nonlinear discrete time system. Int. J. Comput. Intell. Appl. **15**(3) (2016). https://doi.org/10.1142/s1469026816500164
19. Mori, G., Belongie, S., Malik, J.: Efficient shape matching using shape contexts. IEEE Trans. Pattern Anal. Mach. Intell. **27**(11), 1832–1837 (2005)
20. Singh, U.P., Jain, S.: Optimization of neural network for nonlinear discrete time system using modified quaternion firefly algorithm: case study of indian currency exchange rate prediction. Soft. Comput. (2017). https://doi.org/10.1007/s00500-017-2522-x
21. D´eniz, O., Bueno, G., Salido, J., Torre, F. de la: Face recognition using histograms of oriented gradients. Pattern Recognit. Lett. **32**(12), 1598–1603 (2011)
22. Klare, B., Li, Z., Jain, A.K.: Matching forensic sketches to mug shot photos. IEEE Trans. Pattern Anal. Mach. Intell. **33**(3), 639–646 (2011)
23. Rajpurohit, J., Sharma, T.K., Abraham, A., Vaishali, A.: Glossary of metaheuristic algorithms. Int. J. of Comput. Inf. Syst. Ind. Manag. Appl. **9**, 181–205 (2017)
24. Sakhre, V., Singh, U.P., Jain, S.: FCPN approach for uncertain nonlinear dynamical system with unknown disturbance. Int. J. Fuzzy Syst. **19**, 4 (2017). https://doi.org/10.1007/s40815-016-0145-5
25. Flicker dataset. http://personal.ee.surrey.ac.uk/Personal/R.Hu/SBIR.html

An Efficient Contrast Enhancement Technique Based on Firefly Optimization

Jamvant Singh Kumare, Priyanka Gupta, Uday Pratap Singh and Rajeev Kumar Singh

Abstract In the modern environment, digital image processing is a very vital area of research. It is a process in which an input image and output might be either any image or some characteristics. In image enhancement process, input image, therefore, results are better than given input image for any particular application or set of objectives. Traditional contrast enhancement technique results in lightning of image, so here Discrete Wavelet transform is applied on image and modify only Low–Low band. In this presented technique, for enhancement of given image having low contrast apply Brightness Preserving Dynamic Histogram Equalization (BPHDE), Discrete Wavelet Transform (DWT), Thresholding of sub-bands of DWT, Firefly Optimization and Singular Value Decomposition (SVD). DWT divides image into 4 bands of different frequency: High–high (HH), High–low (HL), Low–high (LH), and Low–low (LL). First apply a contrast enhancement technique named brightness preserving dynamic histogram equalization technique for enhancement of a given low-contrast image and boosts the illumination, then apply Firefly optimization on these 4 sub-bands and thresholding applied, this optimized LL band information and given input image's LL band values are passed through SVD and new LL band obtained. Through inverse discrete wavelet transform of obtained new LL band and three given image's HH, HL, and LH band obtained an image having high contrast. Quantitative metric and qualitative result of presented technique are evaluated and compared with other existing technique. A result reveals that presented technique

J. S. Kumare · P. Gupta (✉) · R. K. Singh
Department of CSE/IT, Madhav Institute of Technology & Science, Gwalior, India
e-mail: guptapriya071@gmail.com

J. S. Kumare
e-mail: jamvantsingh09@gmail.com

R. K. Singh
e-mail: rajeev.mits1@gmail.com

U. P. Singh
Department of Applied Mathematics, Madhav Institute of Technology & Science, Gwalior, India
e-mail: usinghiitg@gmail.com

K. Ray et al. (eds.), *Soft Computing: Theories and Applications*,
Advances in Intelligent Systems and Computing 742,
https://doi.org/10.1007/978-981-13-0589-4_17

is a more effective strategy for enhancement of image having low contrast. The technique presented by this study is simulated on Intel I3 64-bit processor using MATLAB R2013b.

Keywords Image enhancement · DWT · Firefly optimization · SVD · BPDHE IDWT

1 Introduction

In image-based application, image having better contrast and preserve the brightness of the image is primary necessities. It has several applications like medical science, identification of fingerprints, remote sensing and earth science. Producing well-contrast images is the strong necessity in various areas like medical images, remote sensing, and radar system. Common methods that are used in the enhancement of image contrast are linear contrast stretching as well as histogram equalization (HE), but the images produced by this procedure have over lightning and unnatural contrast. To deal with the drawback of these techniques, so many enhancement strategies of the image have been proposed [1–10]. Here, we attempted to develop an efficient image enhancement technique by using our proposed methodology and performed the experimental analysis by comparing it with other existing methods.

G. Anbarjafari, H. Demirel, presented a technique in 2010 for improving the resolution of images based on CWT. In this method, given image is decomposed into 4 sub-bands using DTCWT. Then, we can get high-resolution image by the interpolation of given image and sub-bands with high frequency and finally by performing inverse DT CWT. The qualitative metric and quantitative parameter, i.e., PSNR shows that proposed technique provides better enhancement over the traditional resolution enhancement techniques [11].

Demirel, Ozcinar, Anbarjafari, proposed a technique in 2010 which input image having low contrast decomposes into the four frequency sub-bands by using DWT and after the decomposition, in next step, the matrix of SVD can be evaluated for the LL sub-band image. This method is called (DWT–SVD) reforms the enhanced output image by performing the IDWT [12] for removing the drawback of this method. Demirel and Anbarjafari [13], proposed a new strategy in 2011 for improving resolution of satellite images based on DWT and through interpolated output of given image and sub-bands with high frequency. In this method, given image decomposes into sub-bands of different frequencies after that sub-bands having high frequency and given an image having low resolution interpolated and then perform IDWT for generating an image of high resolution [13].

Sunoriya et al. [14] performed the study of SWT, DWT, DWT, and SWT and DTCWT-based resolution enhancement of satellite image in 2013 and got better enhanced and sharper image using these wavelet-based enhancement techniques. As a result, enhancement by DWT-SWT has high resolution than enhancement using DWT.

In 2014, Shanna and Venna [15] proposed a modified algorithm proposed in [14]. This technique based on gamma correction for enhancement in contrast to satellite images using SVD and DWT. In this method, intensity transformation done by gamma correction improves illumination by using SVD. This presented technique confirmed the effeteness of its method by comparing with Sunoriya et al. [14] by calculating entropy, PSNR and EME (Measure of enhancement). Akila et al. [16] presented a method in 2014 based on DWT, SWT, and BPDHE. Image having low contrast and low resolution is decomposed by using SWT and DWT then the sub-bands are interpolated and some intermediate process is performed to produce an image with better resolution image but having less contrast. The contrast of the image is improved by using BPDHE and output image after applying BPDHE technique and split into 4 bands by DWT and LL sub-band is acquired by SVD. The inverse wavelet transform is performed to generate an output image. Given image contrast and resolution are enhanced by this technique. The quantitative and qualitative parameters are measured to depict that the presented technique gives a better result than traditional method.

Priyadarshini et al. [17] presented a technique in 2015 which is a variation of the SVD-based contrast enhancement methods. SVD is used for preservation of mean brightness.of given image. Now this technique, the weighted aggregation of input image's singular matrices of given image and histogram equalization (HE) of the image is considered to attain the equalized image's matrix having singular values. Through results, it depicts that the presented method deals with the brightness of image more specifically with relatively minor pictorial artifacts.

In 2016, Atta and Abdel-Kader [18] presented a resolution and contrast enhancement technique based on bi-cubic interpolation for resolution enhancement which is applied on sub-bands having high frequency of given image and difference image obtained by subtraction of LL sub-band of given image and given image itself parallel the SVD deal with brightness enhancement which is applied on LL sub-band of equalized image and LL sub-band of input image. LL sub-band and components having a high frequency of an image can be obtained by DWT transform. Whereas resolution enhancement works on HL, LH, HH sub-bands and contrast enhancement works on LL sub-band.

In 2016, Demirel and Anbarjafari [19] presented a technique for mammographic images. In this, RMSHE technique is used for contrast enhancement and the SWT-DWT filtering applied for edge preservation and resolution enhancement. EME results depict that proposed method gives a better result than HE and RMSHE technique.

There are various modules in this paper. Part II explains related work of various researchers in image enhancement. Section 3 describes DWT, BPDHE, and SVD with our suggested work. The experimental work is done in part IV. Part V mentions conclusion with the future scope of this work followed by references.

2 Proposed System

This technique is using DWT, SVD, and BPDHE. First, DWT applied on the captured image and contrast-enhanced image by BPDHE technique. Through DWT we get 4 frequency sub-bands and new LL component of an image obtained by performing SVD on a lower component of both images as depicted in the figure.

2.1 Discrete Wavelet Transform

DWT is used to transform an input signal into 4 sub-bands known as LL, LH, HL, and HH with help of some basic wavelets known as mother wavelets. Among different mother wavelets, a suitable wavelet is used to obtain wavelet coefficients. Filter bank DWT diagram for the first level is shown in Fig. 1. LL frequency band is protecting the edge information from degradation and reconstructed image is obtained using IDWT. Flowchart of the proposed method is depicted in Fig. 2.

2.2 Singular Value Decomposition

SVD can be used in various fields like for face recognition, feature extraction, and compression [21] and also for the enhancement of the images having less resolution and low contrast. SVD preserves the maximum signal energy of an input picture into a number of coefficients they are more optimal. Through scaling of singular value matrix, SVD-based methods expand the contrast of images by the decomposition of the image's matrix in a minimum square sense into an optimal matrix. The information of illumination represented by maximum signal energy. DWT decomposes the given image into 4 frequency bands and SVD will work only on the LL band component for the purpose of illumination enhancement. SVD deal with only the LL band because edge info contained by other frequency band and under any condition

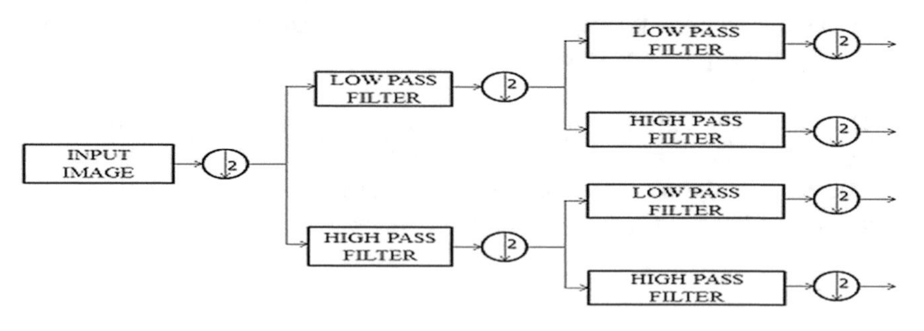

Fig. 1 The 2D-DWT decomposition [15]

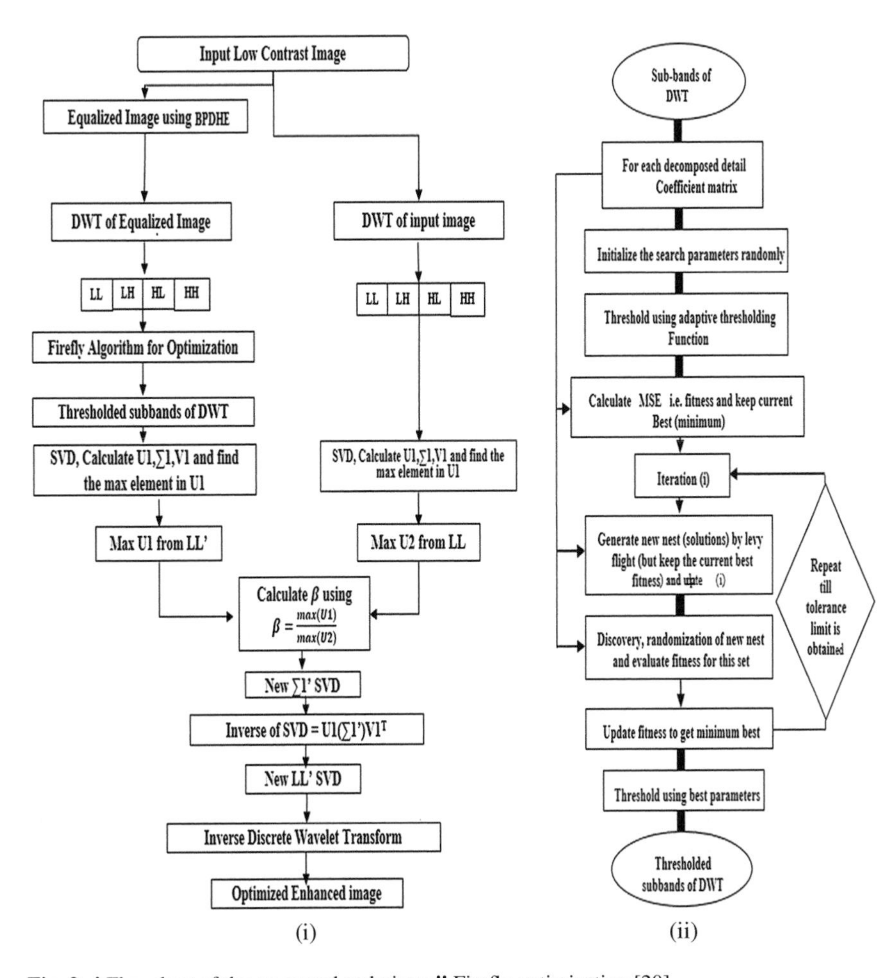

Fig. 2 **i** Flowchart of the presented technique **ii** Firefly optimization [20]

we don't want to distorted image's edge information. The SVD of an image can be depicted in the following form:

$$I = U_i \bar{\Sigma}_I V_{\mathrm{i}} \quad (1)$$

where U_i and V_{i} are matrices which deliberated as the Eigenvectors and $\bar{\Sigma}_I$ is a diagonal matrix that has singular value and its diagonal have nonzero value's square root [1].

In our suggested method, the given image contrast is enhanced by BPDHE and after this new LL sub-band is generated by taking the SVD of given image and enhanced image. The coefficient ratio can be evaluated by using Eq. 1.

$$\beta = \frac{max\left(\sum_{LL_{\hat{A}}}\right)}{max\left(\sum_{LL_A}\right)} \tag{2}$$

where the $LL_{\hat{A}}$ is high contrast image's SVM and LL is SVM of the given input image. After calculating the correction coefficient (β) composed a new LL sub-band by Eqs. 2 and 3

$$\bar{\Sigma}_{LL_A} = \beta \sum_{LL_A} \tag{3}$$

$$\overline{LL}_A = U_{LL_A} \bar{\Sigma}_{LL_I} V_{LL_A} \tag{4}$$

Then to generate output image recombine this generated LL sub-band and high-frequency sub-bands by IDWT.

$$\bar{A} = IDWT\left(\overline{LL}_A, LH_A, HL_A, HH_A\right) \tag{5}$$

Through the inverse DWT output image having better contrast obtained. The results by vision point of view through the suggested method are depicted in Figs. 3, 4, and 5. By all these results, we are able to perceive the variance between original given image and output image having better contrast. Images captured by Satellite as well as simple low contrast images improved by the suggested method.

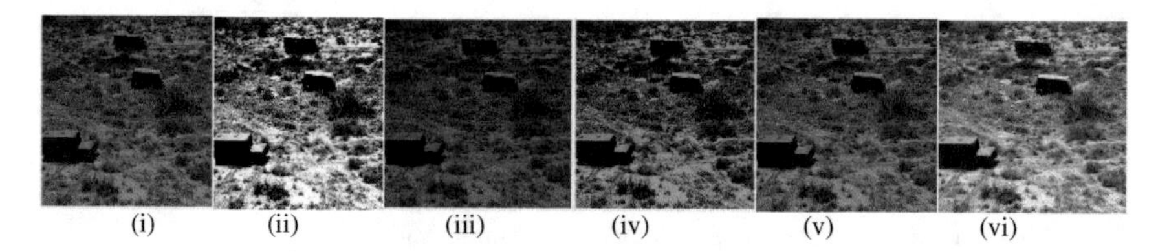

Fig. 3 **i** Original image **ii** HE [22] **iii** BPDHE [23] **iv** Demirel's [12] **v** Nitin's [15] **vi** the proposed technique [24]

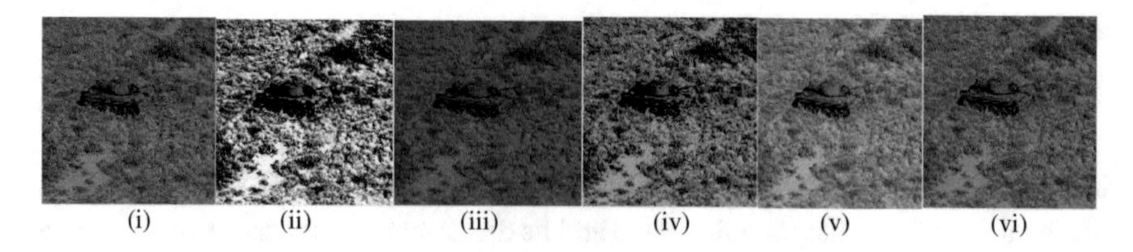

Fig. 4 **i** Original image **ii** HE [22] **iii** BPDHE [23] **iv** Demirel's [12] **v** Nitin's [15] **vi** the proposed technique [24]

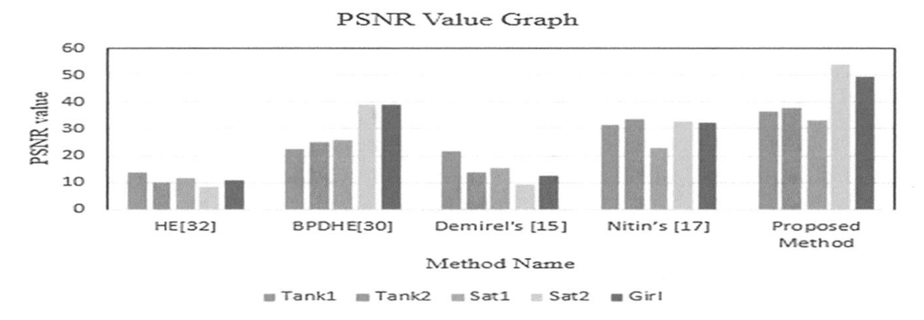

Fig. 5 PSNR value graph

2.3 Brightness Preserving Dynamic Histogram Equalization (BPDHE)

This technique is presented by Nicholas Kong et al. in 2008 an it is upgraded version of the DHE that can deliver the output image with nearly same mean intensity as that of the input image that means it completes the requirement of preserving the mean brightness of the image. It comprises of few steps. The initial step is the image smoothing. By using one-dimensional Gaussian filter histogram of the image is smoothed. This function removes the redundant and noisy high and low peaks from the histogram of the image. Through the removal of image histogram's jagged points which are generated by high-frequency components of the image histogram smoothies. The main cause of jagged shape of the histogram is caused mainly by noise. The Gaussian filter used for smoothness. The second step by tracing the histogram of the smoothed version of the image selects the local maximum points of the histogram. A point on the histogram is a local maximum if its amplitude is more than its neighbors. Next, the image histogram is divided according to the found maximum points. Each interval is the distance between two successive local maxima. Every segment will be assigned a new dynamic range by the highest intensity value contained in the sub histogram and by the lowest intensity value. Step three is HE. This step equalizes the histogram of each interval separately. The last step of this method involves the normalization of the output intensity by approximates the mean of the input image to the output one, by multiplying the intensity of each pixel to the ration of the mean intensity of the input and the output one. Now, result is the average intensity of the produced image will be same as the input image. With this measure BPDHE will yield better improvement compared with CLAHE, and better for preserving the mean brightness compared with DHE [25].

2.4 Firefly Optimization

Among several metaheuristic optimization techniques [26, 27], the Firefly Algorithm (FA) is a famous metaheuristic optimization algorithm [26, 28]. FA is gradient free and population-based method used for solving optimization problems. FA is based on the characteristics of flashing lights by fireflies. Few assumptions are considered for FA given below:

1. All fireflies are unisex, so one firefly is attracted to other firefly despite of their sex.
2. Attractiveness is proportional to a firefly's intensities. Thus, for any two flashing fireflies, the less intensity one will move towards the higher intensity one.
3. Intensities of every firefly indicate the quality of the solution.

Pseudocode of FA is given below:

1. Let the objective function for FA is MSE (Eq. 8)
2. Initialize population parameters of FA
3. Calculate light intensity.

$$I(r) = I_0\, e^{-\gamma r^2} \tag{6}$$

where I_0 is the source light intensity, γ is absorption coefficient and r is the distance between two fireflies.

1. While ($t \leq MaxGen$) // initially $t=0$
2. for $i=1$: n
3. for $j=1$: n
4. if ($I_i < I_j$) then
5. Move ith firefly toward jth firefly
6. end if
7. Evaluate new solution and light intensity
8. end for j
9. end for i
10. Update $t=t+1$
11. end while

2.5 Algorithm of the Proposed Technique

Step 1: Input an image having low contrast, I = Lowcontrast_image
Step 2: Perform the preprocessing step on the image. First, resize the given image to 256 × 256 and then transform it from RGB image to GRAY image.
Step 3: Apply BPDHE contrast enhancement technique.

Enhanced_I = Apply BPDHE technique on image I

Step 4: Apply Wavelet Transform on contrast-enhanced image and image having low contrast. Obtaining LL sub-band of equalized image and original image.

[LL_orig, HL_orig, LH_orig, HH_orig, Sub-bands_orig] = Sub-bands (I)
[LL_bpdhe, HL_bpdhe, LH_bpdhe, HH_bpdhe, Sub-bands_ghe] = Sub-bands (Enhanced _I)

Step 5: Apply Firefly Optimization on sub-bands of contrast-enhanced image

Objective function for FA is MSE (Eq. 8)
Initialize population parameters of FA
Calculate light intensity $I(r) = I_0 e^{-\gamma r^2}$ (Eq. 6)

Step 6: Apply SVD on LL component of images.

(i) Performing SVD on LL sub-band of the original image and equalized image.
[U_orig, S_orig, V_orig] = svd (LL_orig)
[U_bpdhe, S_bpdhe, V_bpdhe] = svd(LL_bpdhe)
(ii) Dividing the max singular value of the equalized image and max singular value of the original image.
β = max (S_bpdhe)/max (S_orig)
(iii) Calculating new Sigma
new_sigma = β * S_orig
(iv) Obtaining new LL sub-band
new_LL = U_orig * new_sigma * V_orig

Step 7: Reconstruct new enhanced image by IDWT
Contrast_enhanced_img = idwt2 (new_LL, HL_orig, LH_orig, HH_orig, 'db1')

3 Experimental Result

3.1 Implementation Details

Various tests were accomplished to evaluate the effect the enhancement techniques in these input images of 256 × 256 pixel resolutions and having low contrast. In this paper, 20 images have been tested but few results are shown to reduce the size of paper. The performance of the presented mechanism compared with HE, BPDHE, DWT-SVD and gamma correction-based method. It is very difficult to evaluate the degree of enhancement in images but, for comparing the result of several contrast enhancement methods; we required to have an unbiased parameter. Although we do not have any objective method to give consistent and meaningful results for all input images, but metrics which are based on mean and standard deviation are widely used. With these metrics, histogram equalization can attain the better result but it may generate artifacts in vision and produce an abnormal look.

Table 1 Comparison of PSNR values of images

Images	Method				
	HE [22]	BPDHE [23]	Demirel's [12]	Nitin's [15]	Proposed method
Tank1	13.83	22.60	21.66	31.64	36.43
Tank2	10.07	25.12	13.93	33.67	37.98
Sat1	11.83	25.81	15.49	22.93	33.50
Sat2	8.66	39.10	9.20	32.92	53.83
Girl	11.15	39.08	12.56	32.61	49.50

All these results represent that the output image, i.e., contrast improved by the presented method is brighter and sharper than the other traditional methods. Not only visually but also quantitatively the confirmation of the better effect of the proposed method through the result of PSNR and MSE. By using the following formula, we can calculate the PSNR value of an image:

$$PSNR = 10log_{10}\left(\frac{R^2}{MSE}\right) \tag{7}$$

Here, R is taken as 255 because images are signified by 8 bit and by using the following formula, the MSE of given image Ii and the produced image Io can be calculated as follows:

$$MSE = \frac{\sum_{i,j}(I_i(i,j) - I_o(i,j))^2}{M \times N} \tag{8}$$

3.2 *Subjective and Objective Assessments*

Table 1 is showing the comparative study of the presented method and HE, SVE, BPDHE, Demirel's method [12] and Nitin's method [15] by means of calculating PSNR for various images.

Figures 3 and 4 represent a low-contrast input image (i) and in (ii) image obtained after HE in (iii) processed image by BPDHE and Demirel's technique in (iv), (v) image after Nitin's method (vi) shows the output image by the proposed technique. The HE image, i.e., Tank1 image Tank1.jpg has a PSNR value 13.83 while proposed method having PSNR value 36.43.

Figure 5 depicts the graph of PSNR value of all method on 5 images and result represents that our proposed method performs better over the mentioned and traditional techniques.

Hence, by observing the output and corresponding values, we can say that brightness of image has been enhanced twice that of the given image as well as. This work finds applications in various fields mainly where the satellite images have been used. Images that are captured by satellite have numerous applications in geology, agriculture, landscape, meteorology regional, and conservation planning of biodiversity, education, intelligence, forestry, and warfare. Contrast-enhanced image by the presented technique leads to improvement in all of the above fields.

4 Conclusion and Future Scope

The proposed image enhancement method is based on Firefly Optimization of DWT components. Through the modification in singular value, the matrix obtained enhanced result. The SVD contains illumination-based information of picture, and LL band of DWT contains enlightenment information. BPHDE is a new technique for enhancement of given image having low contrast based on, SVD, DWT, and Firefly optimization. The proposed work is mostly applicable for medical images, satellite images, and radar images. Quantitative metric and qualitative result of presented technique are evaluated and compared with other existing techniques. The experimental results depict superiority of the proposed method in presence of PSNR and MSE.

References

1. Kim, Y.T.: Contrast enhancement using brightness preserving bi-histogram equalization. IEEE Trans. Consum. Electron. **43**, 1–8 (1997)
2. Chen, S.D., Ramli, A.R.: Minimum mean brightness error bi-histogram equalization in contrast enhancement. IEEE Trans. Consum. Electron. **49**, 1310–1319 (2003)
3. Chen, S., Ramli, A.: Preserving brightness in histogram equalizationbased contrast enhancement techniques. Digit. Signal Process. **14**, 413–428 (2004)
4. Isa, N.A.M., Ooi, C.H.: Adaptive contrast enhancement methods with brightness preserving. IEEE Trans. Consum. Electron. **56**, 2543–2551 (2010)
5. Kim, J.Y., Kim, L.S., Hwang, S.: An advanced contrast enhancement using partially overlapped sub-block histogram equalization. IEEE Trans. Circuits Syst. Video Technol. **11**, 475–484 (2001)
6. Ibrahim, H., Kong, N.S.P.: Brightness preserving dynamic histogram equalization for image contrast enhancement. IEEE Trans. Consum. Electron. **53**, 1752–1758 (2007)
7. Kim, T.K., Paik, J.K., Kang, B.S.: Contrast enhancement system using spatially adaptive histogram equalization with temporal filtering. IEEE Trans. Consum. Electron. **44**, 82–86 (1998)
8. Sun, C.C., Ruan, S.J., Shie, M.C., Pai, T.W.: Dynamic contrast enhancement based on histogram specification. IEEE Trans. Consum. Electron. **51**, 1300–1305 (2005)
9. Wan, Y., Chen, Q., Zhang, B.M.: Image enhancement based on equal area dualistic sub-image histogram equalization method. IEEE Trans. Consum. Electron. **45**, 68–75 (1999)
10. Wadud, M.A.A., Kabir, M.H., Dewan, M.A.A., Chae, O.: A dynamic histogram equalization for image contrast enhancement. IEEE Trans. Consum. Electron. **53**, 593–600 (2007)
11. Demirel, H., Anbarjafari, G., Jahromi, M.N.: Image equalization based on singular value decomposition. In: Proceedings of 23rd IEEE International Symposium on Computer Information Science, Istanbul, Turkey, pp. 1–5 (2008)

12. Demirel, H., Ozcinar, C., Anbarjafari, G.: Satellite image contrast enhancement using discrete wavelet transform and singular value decomposition. IEEE Geosci. Remote Sens. Lett. **7**, 333–337 (2010)
13. Demirel, H., Anbarjafari, G.: Discrete wavelet transform-based satellite image resolution enhancement. IEEE Trans. Geosci. Remote Sens. **49**(6), 1997–2004 (2011)
14. Sunoriya, D., Singh, U.P., Ricchariya, V.: Image compression technique based on discrete 2-D wavelet transforms with arithmetic coding. Int. J. Adv. Comput. Res. **2**(2), 92–99 (2012)
15. Shanna, N., Venna, O.P.: Gamma correction based satellite image enhancement using singular value decomposition and discrete wavelet transform. In: IEEE International Conference on Advanced Communication Control and Computing Technologies (ICACCCT) 2014. ISBN No. 978-1-4799-3914-5/14/$31.00 ©2014 IEEE
16. Akila, K., Jayashree, L.S., Vasuki, A.: A hybrid image enhancement scheme for mammographic images. Adv. Nat. Appl. Sci. **10**(6), 26–29 (2016)
17. Priyadarshini, M., Sasikala, M.R. Meenakumari, R.: Novel Approach for Satellite Image Resolution and Contrast Enhancement Using Wavelet Transform and Brightness Preserving Dynamic Histogram Equalization. IEEE (2016)
18. Atta, R., Abdel-Kader, R.F.: Brightness preserving based on singular value decomposition for image contrast enhancement. Optik **126**, 799–803 (2015)
19. Demirel, H., Anbarjafari, G.: Image resolution enhancement by using discrete and stationary wavelet decomposition. IEEE Trans. Image Process. **20**(5), 1458–1460 (2011)
20. Bhandari, A.K., Soni, V., Kumar, A., Singh, G.K.: Cuckoo search algorithm based satellite image contrast and brightness enhancement using DWT-SVD. ISA Trans. **53**, 1286–1296 (2014)
21. Agaian, S.S., Silver, B., Panetta, K.A.: Transform coefficient histogram-based image enhancement algorithms using contrast entropy. IEEE Trans. Image Process. **16**, 741–758 (2007)
22. Gupta, P., Kumare, J.S., Singh, U.P., Singh, R.K.: Histogram based image enhancement techniques: a survey. Int. J. Comput. Sci. Eng. **5**(6), 175–181 (2017)
23. Sheet, D., Garud, H., Suveer, A., Chatterjee, J., Mahadevappa, M.: Brightness preserving dynamic fuzzy histogram equalization. IEEE Trans. Consum. Electron. **56**(4), 2475–2480 (2010). http://dx.doi.org/10.1109/TCE.2010.5681130
24. Satellite Image got from—http://www.satimagingcrop.com//
25. Rajesh, K., Harish, S., Suman: Comparative study of CLAHE, DSIHE & DHE schemes. Int. J. Res. Manag. Sci. Technol. **1**(1)
26. Singh, U.P., Jain, S.: Modified chaotic bat algorithm-based counter propagation neural network for uncertain nonlinear discrete time system. Int. J. Comput. Intell. Appl. (World Scientific), SCI Index, IF: 0.62, **15** (3) (2016), 1650016. https://doi.org/10.1142/s1469026816500164
27. Singh, U.P., et. al.: Modified differential evolution algorithm based neural network for nonlinear discrete time system. In: Recent Developments in Intelligent Communication Applications. ISBN: 9781522517856
28. Atta, R., Ghanbari, M.: Low-contrast satellite images enhancement using discrete cosine transform pyramid and singular value decomposition. IET Image Proc. **7**, 472–483 (2013)

A Review of Computational Swarm Intelligence Techniques for Solving Crypto Problems

Maiya Din, Saibal K. Pal and S. K. Muttoo

Abstract The nature-inspired computational field is now being popularized as Computational Swarm Intelligence in research communities and gives a new insight in the amalgamation of nature and science. Computational swarm intelligence has also been used to solve many practical and difficult continuous and discrete optimization problems. The past decade has witnessed a lot of interest in applying computational swarm intelligence for solving crypto problems. This review paper introduces some of the theoretical aspects of Swarm Intelligence and gives a description about the various swarm based techniques and their applications for solving crypto problems.

Keywords Swarm intelligence · Stream ciphers · Block ciphers · Public key cryptosystem · Cryptanalysis

1 Introduction

In the field of computational sciences, there is always an increasing demand of systems to solve complex tasks with efficiency. This requirement had paved way for the field of Artificial intelligence. Artificial intelligence is defined as the study and design of 'intelligent' agents, where 'intelligent' refers to the ability of technically sound machines to learn from their environment and bring improvement in their working. This practice is almost similar to that of humans, and thus the term 'intelligence'. In the initial stages, this learning of machines was restricted to the already designed algorithms which were used as the basic working principle of the various intelligent devices like robots, washing machines, speech-to-text converters, etc. But with the advancement on studies, the intelligence is now being derived from biological aspects

M. Din (✉) · S. K. Pal
DRDO, New Delhi, Delhi, India
e-mail: anuragimd@gmail.com

S. K. Muttoo
Delhi University, New Delhi, Delhi, India
e-mail: skptech@yahoo.com

K. Ray et al. (eds.), *Soft Computing: Theories and Applications*,
Advances in Intelligent Systems and Computing 742,
https://doi.org/10.1007/978-981-13-0589-4_18

of nature. These biological aspects precisely involve using the collective behaviour projected by the various group of insects, birds and animals in nature. These groups are called swarm and the collective pattern shown by all the members of the group is termed as 'Swarm intelligence' [1–3]. Swarm Intelligence has found its application in many real-life problems and applications.

In cryptology [4], the techniques for ensuring the secrecy and/or authenticity of information are studied. The two main branches of cryptology are cryptography and cryptanalysis. In cryptography, the design of such techniques is studied but for cryptanalysis, the defeating of such techniques is studied to recover sensitive information.

Cryptanalysis [5] is the science of breaking cipher text to get its corresponding plain text without prior knowledge of the secret key. It is a technique used for finding out the drawbacks in the design of a cryptosystem. It is considered as one of the interesting research areas in information security. Cryptanalysis of any cryptosystem can be formulated as an optimization problem. Evolutionary computation algorithms [6] are employed in an attempt to find an optimal solution to the cryptanalysis problem. Evolutionary computation based optimization techniques are inspired by the paradigm of natural evolution/social behaviour of swarms/animals. A common characteristic of all such algorithms is that they do not employ information from the derivatives of the objective function. Therefore, they are applicable to hard optimization problems that involve discontinuous objective functions and/or disjoint search spaces.

This review paper describes the basic underlying concept behind the Swarm Intelligence in Sect. 2 and gives a brief description of various algorithms [3, 7–9] mentioned in Sect. 3. A review of SI techniques for solving cryptosystems is presented in Sect. 4 and cryptosystems are described in Sect. 5. In the last Section, the paper is concluded with current trends and future scope.

2 Swarm Intelligence

The term "Swarm Intelligence" was first coined by G. Beni and J. Wang in 1989. The same was referred as a set of algorithms to control robotic swarm in the global optimization problem.

2.1 Understanding Swarm Intelligence

A swarm is defined as a set of a large number of homogenous, simple agents which exchange information locally among themselves and their environment, without any leadership to achieve useful behaviour globally. Although these agents (insects or swarm individuals) have some limit on their capabilities, they communicate together with some behavioural patterns to achieve tasks needed for survival.

Some common examples include swarms of ants, honey bees or a flock of birds etc. Simulating and modelling this behavioural pattern of swarms to be applied in complex computational tasks to provide low cost, fast, efficient, optimized and robust solutions is termed as Swarm Intelligence (SI). Millonas (1994) gave an idea behind swarm intelligence as employing many simple agents present in nature and lead to an emergent global behaviour. The algorithms encompassed in this subfield of Artificial intelligence are purely nature-inspired and population based with simple yet unfailing underlying principles.

2.2 *Underlying Principle*

In 1986, Craig Reynolds first simulated swarm behaviour [7] on a computer with the 'boids' simulation program. This program simulated the movement of simple agents according to basic rules as follows:

1. All the members move in the same direction.
2. Members remain close to their neighbours.
3. Members should avoid collision with the neighbours.

Biologists and natural scientists have been analyzing the behaviour of swarms due to the effective efficiency of these natural swarm systems in the past decades. They have proposed that the social interactions can be of two types which are direct or indirect. Direct interaction is obtained through audio/visual contact. Indirect interaction is taking place when one swarm agent changes the environment and the other react to the new environment. This indirect type of interaction is known as 'stigmergy' [8], which means interaction among swarms through the environment.

Over the years of research using the different social psychology theories, we have some general principles over which Swarm Intelligence works:

1. **Proximity Principle**—The entities of the swarm should be able to perform 'computation', i.e. provide a response to their surroundings, based on the complexity of the task.
2. **Quality Principle**—Quality factors like food, safety etc. should be responded by swarms.
3. **Principle of diverse response**—Resources should be distributed properly to increase the probability of the agents not adapting to environmental fluctuations. A diverse response may also lead to improvement in their working abilities.
4. **Principle of stability and adaptability**—Swarms are expected to adapt to the environmental changes and remain stable in their modus operandi.
5. **Self-organization and Emergence**—Through self–organization and emergence, the otherwise disordered entities of a swarm locally interact with each other and bring out a coordinated system as a whole.

Using these principles and the observed swarm behaviour, analogies can be drawn between the computational sciences and Swarm Intelligence which in fact form the backbone for the application of these bio-inspired computing techniques.

3 Swarm Intelligence Based Algorithms

Many algorithms based on swarm intelligence have been devised and are being applied successfully in many complex computational problems. Some of the major algorithms [9–11] are briefly discussed in following subsections.

3.1 Ant Colony Optimization (ACO)

This was the first effective technique based on swarm intelligence presented by M. Dorigo and colleagues as a novel nature-inspired meta-heuristic to solve hard combinatorial optimization problems [11]. This technique is based on fact that ants, which 'seems to blind' can learn the shortest path between nest and food source without any vision.

Ants share information with each other with the help of 'Pheromone' (Volatile chemical substance), whose intensity and direction is perceived using antennas. This chemical is used as both for safety and food searching. A crushed ant releases an 'Alarm pheromone' which alerts the other ants to disperse. Ants search for the shortest route between food and nest and reinforce the path by leaving the pheromone on their way back to the nest and this path is then followed by other ants. The pheromone on other paths evaporates with time, thus leaving no space for confusion. This social behaviour, called 'Stigmergy' is exploited in solving Travelling Salesman Problem and other hard computational problems.

3.2 Particle Swarm Optimization (PSO)

This successful swarm intelligent technique was introduced in 1995 by James Kennedy and Russell Eberhart [11]. It was initially used to solve nonlinear continuous optimization problems, but now the application of this algorithm has been extended in many other real-life problems. PSO is inspired by the flocking of birds and schooling of fishes, penguins etc. and their sociological behaviour.

Bird flocking or schooling can be defined as the social and collective motion behaviour shown by the group members to achieve a common objective. Scientists with their research work have proposed that these social yet local interactions work on the 'Nearest Neighbour Principle' which works on three basic rules: Flock centering,

Collision avoidance and Velocity matching. The novel variants of PSO have been developed for solving combinatorial and crypto problems [12–14].

3.3 Cuckoo Search Algorithm (CSA)

Cuckoo Search algorithm was introduced by Yang and Deb in 2009. This is entirely based on the breeding behaviour of cuckoos [15–19]. These birds lay their eggs in foreign nests and the host bird incubates and feeds the chicks of the cuckoo bird, taking them as his own. The reproduction time of the cuckoo and the appearance and the initial behaviour showcased by the chicks of the cuckoo is similar to that of the host and thus the chances of the host bird getting deceived increases. The host bird dumps the nest in case the cuckoo's eggs are discovered. The algorithm, when applied to computational problems, uses Levy distribution [15] in order to search for random solutions.

3.4 Bee Colony Algorithm

The Bee colony algorithm [11] was first introduced by Pham and applied to mathematical functions. This class of algorithms is inspired from the 'foraging' behaviour of honeybees, in which some honeybees are sent out from hives to search for food. After the bees locate the food source, they come back to the hives and perform the 'waggle' dance to spread the information about the food source. These dances communicate three important aspects (Distance, Direction and Quality of food source) which help in locating the best food source. The polarization of bees and the objective of 'Nectar Maximization' also simulate the allocation of bees along flower patches which can be termed as the 'Search regions' in 'Search Spaces' in provided computational problems.

There have been significant improvements made over the original bee algorithm to give more efficient algorithms as Honeybee based algorithm, Virtual bee algorithm, and Artificial Bees colony (ABC) algorithm [10, 11] based on dividing the colony of bees in three groups, namely, Employed bees, Onlooker bees and Scouts. Several variants of ABC algorithm [20, 21] have been developed for solving optimization problems.

3.5 Firefly Algorithm

In 2008, Yang proposed this class of algorithms [9] which use flashing patterns and behaviours of the fireflies. The algorithms follow three idealized rules:

1. A firefly is unisexual, i.e. it can be attracted to any other fly.
2. The firefly will move randomly as less bright firefly moves to higher one. Attractiveness is directly proportional to the brightness and is indirectly proportional to the distance.
3. Brightness is to find out using the objective function.

Apart from the above-mentioned swarm intelligence techniques, there are other techniques which are very useful for solving optimization problems. These techniques [10, 11] are Bacterial algorithms (Bacterial foraging and Bacteria Evolutionary algorithms), Bat algorithms, Artificial Fish algorithms (AFSA), Glow-Warm Swarm optimization algorithm, Roach infestation optimization (RIO) and [22] Shuffled Frog-Leaping Algorithm (SFLA). Glossary of meta-heuristic algorithms [23] contains meta-heuristic algorithms in alphabetic order, which are useful for solving optimization problems.

4 Type of Cryptosystems

Cryptography [4] is concerned with designing cryptosystems for securing sensitive information. Two branches of cryptography are as Symmetric Key Cryptography (SKC) and Asymmetric Key Cryptography also known as Public Key Cryptography (PKC) [5, 24]. In SKC, the sender and receiver use the same key for encrypting and decrypting messages. In PKC, a pair of two keys is used consisting of encryption and decryption keys. During eighteenth–nineteenth century, different classical cryptosystems were used like Simple Substitution, Vigenere, Transposition, Playfair and Hill cipher. All of these were symmetric key cryptosystems. Symmetric Key Cryptosystems are broadly categorized into Stream ciphers and Block Ciphers. In a stream cipher, message is encrypted at bit/byte level with pseudorandom key sequence, while in Block cipher, a complete block is encrypted. Encryption rate of stream ciphers is faster than a block cipher. Examples of stream ciphers [25] are LFSR-based ciphers, Geffe Generator-based cipher, GRAIN-128, Trivium, etc. Block cipher examples are Data Encryption Standard (DES), International Data Encryption Algorithm (IDEA) and Advanced Encryption Standard (AES) etc. Basic block diagram of a cryptosystem given in Fig. 1 describes the process of encryption/decryption of plain message/ciphertext using a key. Figure 2, shows pseudorandom number (PN) key sequence generation using Linear Feedback Shift Register (LFSR). LFSRs are used extensively in the design of stream ciphers.

Various Public Key Cryptosystem (RSA, Diffie–Hellman Key Exchange and DSA) make use of modular exponentiation. Discrete Log and Elliptic Curve Discrete Log-based cryptosystems are also PKC based systems.

Cryptanalysis [5, 25–27] is concerned with the analysis of cryptosystems to get sensitive information by finding weaknesses of the used cryptosystem. The most common types of attack model are as follows:

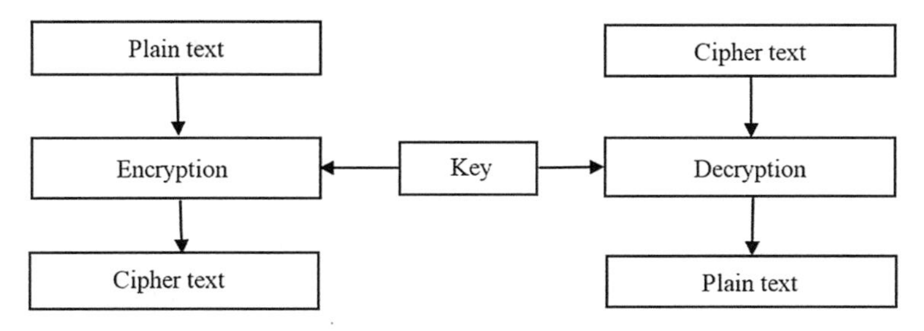

Fig. 1 Block diagram of a cryptosystem

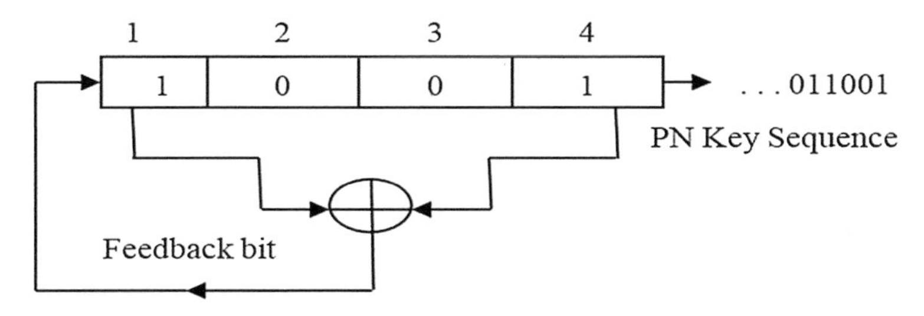

Fig. 2 Linear feedback shift register crypto primitive

1. **Ciphertext-only attack**: The opponent possesses only the ciphertext.
2. **Known-plaintext attack**: The opponent possesses several plaintext and the corresponding ciphertext pairs.
3. **Chosen plaintext**: The opponent has temporary possession to the encryption machine. So he can choose a plaintext and construct the corresponding ciphertext.
4. **Chosen ciphertext attack**: The opponent has temporary access to the decryption machine. So he can choose a ciphertext and construct the corresponding plaintext.

Identifying the correct Key or other parameters of the cryptosystem for the purpose of cryptanalysis require exhaustive search, which is impractical for modern cryptosystems. The other solution is to use discrete optimization technique for solving such problems. Swarm intelligence techniques offer directed and efficient search towards cryptanalysis of these systems.

In the following Section, various SI based techniques are described to analyze above-mentioned cryptosystems.

5 Swarm Intelligence Based Techniques for Crypto Problems

Swarm intelligence techniques have been successfully deployed in various scientific problems. These techniques are suitable to hard optimization problems. The use of automated techniques in solving crypto problems is required to minimize extra time taken in search process due to human interaction. Thus, the applications of efficient and effective techniques such as Swarm Intelligence (SI) based techniques are very suitable for solving crypto problems. Danziger and Henriques [28] give an insight that swarm intelligence computing algorithms can be applied in the field of cryptology.

The work of Laskari E. C. (2005) about Evolutionary Computing [6] based cryptanalysis is considered as an important initiative towards solving crypto problems. They applied evolutionary techniques in cryptanalysis of simplified DES, in this work PSO was applied as an optimization technique and promising results reported. Uddin M. F. and Youssef A. M. (2010) used PSO technique for breaking Simple Substitution, can be considered as a significant effort to the application of SI techniques in Cryptanalysis. Heyadri M. et al. (2014) applied Cuckoo Search in automated cryptanalysis of Transposition Ciphers [29]. Bhateja A. K. et al., applied "PSO technique with Markov Chain random walk" for Cryptanalysis of Vigenere Cipher. Vigenere cipher is a polyalphabetic cipher having very large key space. Authors proposed PSO with a random walk for enhancing performance of PSO algorithm. According to reported results, it is shown that proposed technique performed better compared to ordinary PSO technique in cryptanalysis of Vigenere cipher. In 2015, Bhateja A. K. et al. [30], also applied Cuckoo Search in cryptanalysis of Vigenere cipher.

The cryptanalytic work by Nalini N. et al. (2008) "Cryptanalysis of Block Ciphers via Improvised PSO", Shahzad W. et al. (2009) "Cryptanalysis for 4-Round DES using binary PSO" and Vimalathithan R. and Valarmathi M. L. (2011) PSO [12–14, 31] based cryptanalysis of Data Encryption Standard (DES) are considered as significant contribution in cryptanalysis of Block ciphers. Wafaa G. A. et al. (2011) presented Known-Plaintext Attack on DES-16 using PSO technique [31] and proposed technique is able to recover most root key bits of DES-16. Khan S., Ali A. and Durrani M. Y. (2013) proposed a novel Ant Colony Optimization based cryptanalytic attack on DES. The authors presented Ant-Crypto for cryptanalysis of DES and experimental results. Ant-Crypto is based on Known-Plaintext attack to recover the secret key of DES, required for breaking the confidential message. In 2014 Dadhich A. and Yadav S. K. also applied SI and evolutionary technique for cryptanalysis of DES (4-Round) cryptosystem. The authors reported PSO based cryptanalysis on the cryptosystem and demonstrated it as an effective technique which takes less convergence time. Apart from PSO, Ant colony optimization is also proposed to be used for cryptanalysis.

Swarm Intelligence-based techniques are also useful in analysis of Stream ciphers like Linear Feedback Shift Register (LFSR) based Crypto Systems. In this direction, Din M. et al. (2016) applied Cuckoo Search [29, 30, 32] based technique in the analysis of LFSR-based cryptoystem for finding initial states of used LFSRs by varying

Cuckoo search parameters (Levy Distribution parameter and Alien eggs discovering probability. The technique may also be applied in cryptanalysis of Nonlinear Feedback Shift Register-based cryptosystems. Din M. et al. [26] also applied Genetic Algorithm for analysis of Geffe Generator-based stream cipher.

Apart from Symmetric Key Cryptosystems, SI techniques are also applied for analysis of Public Key Cryptosystems. PKC systems are based on hard mathematical problems from the field of Number theory, Probability theory, Algebra and Algebraic Geometry. These mathematical problems are Integer factorization, Discrete Logarithm and Elliptic curves. Cryptosystems rely on the fact that these problems cannot be completed in polynomial time. In 2012, Sarkar and Mandal [33] developed SI based technique for faster Public Key Cryptography in wireless communication by minimizing a number of modular multiplication. Jhajharia et al. [34] proposed an algorithm for PKC using the hybrid concept of two Evolutionary Algorithms: PSO and Genetic Algorithm (GA) to find out keys with optimum fitness value. The keys obtained by the hybrid algorithm were tested on several randomness tests (Autocorrelation test, frequency test, gap test, serial test, run test, change point test and binary derivative test) and checked for their practical use.

Sai G. J. and George Amalarethinam D. I. (2015) proposed ABCRNG-Swarm intelligence technique [35] in PKC for random number generation which increases the key strength and security. It is based on ABC algorithm that provides quick and improved performance results and it is easy to implement. The random numbers generated by the technique are random and cryptographically strong used in the design of secure PKC System and many other applications.

The Firefly algorithm (FFA) is a novel nature-inspired algorithm, which is being applied to solve many optimization problems. Singh A. P et al. (2013) applied Firefly algorithm in cryptanalysis of mono-alphabetic Substitution cipher [36]. The authors reported that FFA converge faster compared to the random algorithm and FFA is better than random algorithm in terms of percentage of correctly decrypted text. The authors also suggested that variants of Firefly such as Gaussian Firefly, Chaotic Firefly and Hybrid GA-Firefly can be applied for improving percentage of corrected decrypted text. Ali et al. [37] described suitability of Firefly algorithm is suitable for analysis of cryptosystems. Pal et al. [38] compared FFA with PSO technique and reported that both algorithms seem to be not so different in finding optimum when there was no noise in the process. In case of multi-peaks function, FFA performs better. The authors have also used the chaotic firefly to solve the integer factorization problem [39].

6 Conclusion

The past decade has witnessed an increasing interest in applying Computational Swarm Intelligence to solve crypto problems. In this review paper, the concepts of SI techniques have been described with a review of work carried out in the field of Cryptology using SI techniques. SI techniques do not require mathematical properties like continuity and differentiability for the objective function of the optimization problem. So these techniques are suitable to computationally hard optimization problems like crypto problems where solution key is searched in very large and disjoint search space.

According to literature study, SI based techniques are applied in design and analysis of Crypto Systems. The techniques have been applied to Symmetric Key and Public Key Cryptosystems. Cryptanalysis has been formulated by several researchers as an interesting optimization problem. Several heuristic based SI techniques have been applied in cryptanalysis. The results reported in the referenced papers are very encouraging compared to statistical techniques. It is likely that the power and collective effort of swarms would be optimally used to solve some of the hard Cryptanalysis problems. Parallel implementation of these algorithms for Cryptanalysis is another direction for improving computational efforts involved in solving these problems.

References

1. Beni, G., Wang, J.: Swarm intelligence in cellular robotics systems. In: Proceedings of NATO Advanced Workshop on Robots and Biological System (1989)
2. Millonas, M.: Swarms, Phase Transitions, and Collective Intelligence. Addison-Wesley (1994)
3. Bela, M., Gaber, J., El-Sayed, H., Almojel, A.: Swarm Intelligence in Handbook of Bio-inspired Algorithms and Applications. Series: CRC Computer & Information Science, vol. 7. Chapman & Hall (2006). ISBN 1-58488-477-5
4. Menzes, A., Oorschot, V., Vanstone, S.: Handbook of Applied Cryptography. CRC Press (1996)
5. Stinson, D.R.: Cryptography: Theory and Practice. 3rd edn. Chapman & Hall/CRC Publication (2013)
6. Laskari, E.C., Meletiou, G.C., Stamatiou, Y.C., Vrahatis, M.N.: Evolutionary computing based cryptanalysis—a first study. Non Linear Anal. **63**(5–7), 823–830 (2005)
7. Blum, C., Merkle, D.: Swarm Intelligence—Introduction and Applications. Natural Computing. Springer, Berlin (2008)
8. Lim, C.P., Jain, L.C., Dehuri, S.: Innovations in Swarm Intelligence. Springer (2009)
9. Panigrahi, B.K., Shi, Y., Lim, M.H.: Handbook of Swarm Intelligence Series: Adaptation, Learning, and Optimization, vol. 7. Springer, Heidelberg (2011)
10. Yang, X.S., Deb, S.: Nature-Inspired Meta-Heuristic Algorithms, 2nd edn. Luniver Press, United Kingdom (2010)
11. Yang, X.S., Cui, Z., Xiao, R., Gandomi, A.H.: Swarm Intelligence and Bio-Inspired Computations: Theory and Applications. Elsevier (2013)
12. Kennedy, J., Eberhart, R.C.: A discrete binary version of the particle swarm optimization. IEEE Magazine, 4104–4108 (1997)
13. Khanesar, M.A.: A novel binary particle swarm optimization. In: Proceedings of 15th Mediterranean Conference on Control and Automation, Athens, Greece (2007)
14. Nalini, N., Rao, G.R.: Cryptanalysis of block ciphers via improvised particle swarm optimization and extended simulated annealing techniques. Int. J. Netw. Secur. **6**(3), 342–353 (2008)
15. Yang, X.S., Deb, S.: Cuckoo search via Levy flights. In: Proceedings of World Congress on Nature & Biologically Inspired Computing (NaBIC), India. IEEE Publications, USA, pp. 210–214 (2009)
16. Yang, X.S., Deb, S.: Engineering optimization by cuckoo search. Int. J. Math. Modeling Numer. Optim. **1**(4), 330–343 (2010)
17. Rajabioun, R.: Cuckoo optimization algorithm. Appl. Soft Comput. **11**, 5508–5518 (2011)

18. Milan, T.: Cuckoo search optimization meta-heuristic adjustment. Recent Advances in Knowledge Engineering and System Science (2013). ISBN 978-1-61804-162-3
19. Ouaarab, A., Ahiod, B., Yang, X.S.: Discrete cuckoo search algorithm for the travelling salesman problem. Neural Comput. Appl. **24**(7), 1659–1669 (2014). Springer
20. Sharma, T.K., Millie, Pant.: Differential operators embedded artificial bee colony algorithm. Int. J. Appl. Evol. Comput. (IJAEC), **2**(3), 1–14 (2011)
21. Bansal, J.C., Sharma, H., Nagar, A., Arya, K.V.: Balanced artificial bee colony algorithm. Int. J. AI Soft Comput. **3**(3), 222–243 (2013)
22. Sharma, T.K.: Performance optimization of the paper mill using opposition based shuffled frog-leaping algorithm. Int. J. Comput. Inf. Syst. Ind. Manag. Appl. **9**, 173–180 (2017)
23. Rajpurohit, J., Sharma, T.K., Abraham, A., Vaishali: Glossary of metaheuristic algorithms. Int. J. Comput. Inf. Syst. Ind. Manag. Appl. **9**, 181–205 (2017)
24. Stallings, W.: Cryptography and Network Security. Pearson Publications, London (2012)
25. Klein, A.: Stream Ciphers. Springer, London (2013)
26. Din, M., et al.: Cryptanalysis of geffe generator using genetic algorithm. In: Proceedings of International Conference SocProS-2013, AISC-258, pp. 509–516. Springer (2014). ISBN 978-81-322-1601-8
27. Ahmad, B.B., Aizaini Mohd, M.B.: Cryptanalysis using biological inspired computing approaches. Postgraduate Annual Research Seminar (2006)
28. Danziger, M., Henriques, M.A.: Computational Intelligence Applied on Cryptology—A Brief Review. CIBSI, Bucaramanga, Colombia (2011)
29. Heydari, M., Senejani, M.N.: Automated cryptanalysis of transposition ciphers using cuckoo search algorithm. Int. J. Comput. Sci. Mobile Comput. **3**(1), 140–149 (2014)
30. Bhateja, A.K., et al.: Cryptanalysis of Vigenere cipher using cuckoo search. Appl. Soft Comput. **26**, 315–324 (2015). Elsevier
31. Wafaa, G.A., Ghali, N.I., Hassanien, A.E., Abraham, A.: Known-plaintext attack of DES using particle swarm optimization. nature and biologically inspired computing (NaBIC). In: Third World Congress, pp. 12–16. IEEE (2011)
32. Din, M., et al.: Applying cuckoo search in analysis of LFSR based cryptosystem communicated to Elsevier journal. Perspect. Sci. **8**, 435–439 (2016)
33. Sarkar, A., Mandal, J.K.: Swarm intelligence based faster public key cryptography in wireless communication (SIFPKC). Int. J. Comput. Sci. Eng.Technol. **3**(7), 267–273 (2012)
34. Jhajharia, S., Mishra, S., Bali, S.: Public key cryptography using particle swarm optimization and genetic algorithms. Int. J. Adv. Res. Comput. Sci. Softw. Eng. **3**(6), 832–839 (2013)
35. Sai, G.J., George Amalarethinam, D.I.: ABCRNG—swarm intelligence in public key cryptography for random number generation. Int. J. Fuzzy Math. Arch. **6**(2), 177–186 (2015). ISSN 2320 –3242 (P)
36. Singh, A.P., Pal, S.K., Bhatia, M.P.S.: The firefly algorithm and application in cryptanalysis of mono-alphabetic substitution cipher. **1**, 33–52 (2013)
37. Ali, N., Othaman, M.A., Hussain, M.N., Misran, M.H.: A review of firefly algorithm. ARPN J. Eng. Appl. Sci. **9**(10), 1732–1736 (2014)
38. Pal, S.K., Rai, C.S., Singh, A.P.: Comparative study of firefly algorithm and PSO algorithm for noisy non-linear optimization problems. Int. J. Intell. Syst. Appl. **10**, 50–57 (2012)
39. Mishra, M., Chaturvedi, U., Pal, S.K.: A multithreaded bound varying chaotic firefly algorithm for prime factorization. In: Proceedings of International Conference on IACC-2014, pp. 1322–1325. IEEE Explore (2014)

Defense-in-Depth Approach for Early Detection of High-Potential Advanced Persistent Attacks

Ramchandra Yadav, Raghu Nath Verma and Anil Kumar Solanki

Abstract Cyber security has gained high level of attention due to its criticality and increased sophistication on organizations network. There is more number of targeted attacks happening in recent years. Advanced Persistent Threats (APTs) are the most complex and highly sophisticated attack in present scenario. Due to the sophistication of these attacks, it can be able to bypass the deployed security controls and more stealthily infiltrate the targeted internal network. Detection of these attacks are very challenging because they treated normal behaviors to hide itself from traditional detection mechanism. In this paper, we analyze the 26 APT campaigns reports and shows the different methods and techniques that are used by attacker to perform the sophisticated attacks. Our research is mainly focused on the three levels of investigation of APT campaigns that give some common characteristics of them such as APT attack usage zero-day vulnerability or not. Furthermore, according to their characteristics, we propose a novel approach that is capable to early detection of APTs and also suggest concrete prevention mechanism that make it possible to identify the intrusions as early as possible.

Keywords Advanced persistent attacks · Sophisticated attacks · Attack vector Attack phases · Vulnerability

R. Yadav (✉) · R. N. Verma · A. K. Solanki
Computer Science and Engineering Department, Bundelkhand Institute of Engineering and Technology, Jhansi, India
e-mail: ramchandra.iiita@gmail.com

R. N. Verma
e-mail: rnverma160@gmail.com

A. K. Solanki
e-mail: solankibiet13@gmail.com

K. Ray et al. (eds.), *Soft Computing: Theories and Applications*,
Advances in Intelligent Systems and Computing 742,
https://doi.org/10.1007/978-981-13-0589-4_19

1 Introduction

In recent years, there are more number of advanced attacks happening and various APT campaigns are revealed in digital worlds. These campaigns use the high level of techniques and tactics to compromise the target organization. Attacker's main objective is to exfiltration of crucial information. Due to their high sophistication, most of the security controls could not able to detect such type of attacks [1]. These APTs are able to compromise the particular company, government services and public properties. Generally, security investigation organization publish the APT campaigns detailed report and their analyzed results. This published report will help the other organization to analyze the scenario of their organization with corresponding to characteristics of APTs campaigns.

One important point is that these attacks are regularly changing and upgrading their techniques and procedures. Also very easily bypass the security system based on signature or anomaly-based detection. These targeted attacks generally use the social engineering method to gather important information about targets. Further, accessing long-term the target system, attackers may compromise the multiple internal systems. Sometime attacker uses the installed applications and tools for legitimate login to a system that hides the malicious activities. Due to this reason, new technology and controls need to deploy for identification and detection of malicious acts.

In this scenario, we analyzed the 26 APT campaigns that are recently active. Our analysis mainly focuses on APT campaign effects of only the windows based platform and excluded other affected platforms. We discussed the different phases of APTs and their important characteristics, their common techniques and methodology that are used in different phases. Further, we proposed the different approaches for early detection and prevention on the bases of analyzed characteristics of APT attacks.

This paper is structured as follows: Sect. 2 shows the state of the art, Sect. 3 describes the typical phases of an APT. Section 4 presents a general overview of various APT campaigns and used tactics. Section 5 proposed important approaches for early detection and prevention of potential attacks. Section 6 presents conclusions and possible direction for future research.

2 State of the Art

The first major APT attacks came into picture was publicized by Google in January 2011, although the attack has begun in the second half of 2009. Known as Operation Aurora [2], this attack was impacted widely and targeted more than 34 big organizations, including Morgan Stanley, Yahoo, Dow Chemical, Northrop Grumman and Google Also. After analysis of Operation Aurora, the report said that, attack is highly sophisticated and used new tactics, which defeated the McAfee defense system. The attack exploits a 0-day vulnerability, which was present in the Internet

Explorer browser. After some time, another such type of APT attacks is reported by industry [3–5].

In early 2011, many more persistent attacks are reported by Symantec. This was the in-depth analysis of APTs by authors in [6]. Symantec analyzed the advanced persistent threats that were forming a campaign with quite adverse impact on governments and organizations.

Kaspersky Lab's Global Research and Analysis Team analyze and published comprehensive report in October 2012, based on different attacks happens against computer network of several organizations [3]. Red October is the best example that is the biggest cyber espionage exposed by Kaspersky Lab's Team. No one confirms that how long they initiated but attackers have been active for several years. The main goal in this is to steal the secret credentials and make a compiled file which was later used against government organizations.

A USA-based information Security Company, Mandiant released an APT report on China's cyber espionage in February 2013 [4], which basically stole the tones of critical data from 20 global organizations. This attack has the capability to steal the data simultaneously from different organizations. They stole the company's intellectual property, including blueprints of technology, business files etc. This was the one of the large-scale thefts where almost hundreds of terabytes data dumping hapeneddd collectively.

In major cases, the main goal of an APT campaign is to get the critical information about targets although some attacks have other motives like Stuxnet virus that disturb the nuclear program of Iran [7].

3 Typical Stages of APTs

3.1 Reconnaissance

This is the initial phase to perform the targeted attack. In this phase attacker gather important information about their targeted organization. Selection of specific areas where an attack may perform, identification of weak points, identification of operating systems, publicly accessible services running on them, vulnerable software and security misconfiguration are crucial for preparing a successful attack. Findings of normal instruments used for security services (antivirus software, firewall, intrusion detection and intrusion prevention systems implemented) are other important matters the attacker must have, that helps them to save time to achieve their long-term goal. These basic information is collected by some open-source scanning tools and social engineering methods.

Additionally the gathering information about organization's employees, their position with the organization, their expertise and their relations with other employees. Using these information, attacker may create more highly realistic targeted spear-phishing campaigns. For example, in banking scenario, if an attacker identified an employee working as a bank manager, he can send a spoofed email from the email address of the related bank to their customer, asking him to read the attached file (e.g. account statement). The attachment can be a simple PDF or world file embedded with malicious codes, that moment when file opened attacker's payload executes. In fact, the source of generation of email from an organization known to customer significantly increases the likelihood of its legitimation.

3.2 Establishing Foothold

After getting the intelligence about a target, APT attacker prepares a plan and necessary tools to perform the attack. In this steps usually does not require any exploitation of services, instead of convincing the user to open an attachment that actually malicious or click on an instructed link they are not supposed to open. In case of failure, they can make a strategy using different attack vectors.

3.3 Delivery of Attack

In this phase, APT attacker delivers their malicious code to the targets. Delivery of attack is performed generally by two ways—if an attacker sends the exploits to their target via spear-phishing email called the direct delivery method and second indirect way of delivery is by watering-hole attack.

Spear Phishing—This is targeted form of phishing where an attacker sends the specific group of peoples and convince them real email comes from their related personals. In APT attack, the user is generally downloaded the harmful attachment (e.g., progress reports, business documents, and resumes) that contains exploitation code, or they click a link to a malicious site where ask the user to fill their sensitive information and act like drive-by-download exploits [8].

Watering-Hole Attack—The concept of this attack seems like the predator is waiting for the victim at a watering hole in a desert because predator knows that the victims definitely comes to the watering hole. Similarly, instead of frequently sending malicious emails, the attacker often infect the third-party websites using malware that are regularly visited by the targeted user. At the moment, when the infected web pages are visited by victims, the delivery accomplished. There are many APT campaigns [9–11] are happening in the recent year based on the watering-hole attacks.

3.4 Maintaining Access and Exploitation

After initial intrusion malicious software is installed on victim host that is referred to as RAT (remote access Trojan). RAT takes the responsibility to connect with the attacker and regularly performed the actions that instructed by an attacker. At this intruder take the full command and control (C2) over target host. The fact is that the initial connection is established by victim host, not by the attacker [12]. This will happens mainly for two reasons: (i) organizations firewall usually allows the connections initialized by internal hosts, and (ii) this will help the attacker to not to detected easily. Because intrusion detection systems [13] can easily detect the extremely suspicious activity such as downloads from outside hosts.

In a targeted attack, the intruder generally focuses on vulnerabilities presents in Microsoft Word, Internet Explorer, Adobe PDF Reader, and operating systems. But several APT attacks [14, 15] have taken the advantages of zero-day vulnerability to exploiting targets and many other go to older unpatched applications.

3.5 Lateral Movement

Once the intruder established the communication between victims host and command and control server, then more threat actors are placed inside the network to expand their control. In this phase, there are several activities: (i) try to gain access to another host with more privileges (ii) try to initialize new connection for another backdoor in case the first one is detected (e.g. via SSH) (iii) internal reconnaissance to figure out internal networks and collect important information (e.g. trade secrets, development plans, and valuable assets, etc.).

At this stage, the activities are designed in such a way that "run low and slow" to avoid early detection. The attacker tries to stay long period of time to stealth maximum valuable information. APT actors use more legitimate operating-system features and applications which are generally allowed or they may crack the legitimate software to gain authenticity. Both activities may lead to undetectable and untraceable. Hydraq (e.g. Operation Aurora or Google attacks) uses many confusing techniques to keep itself undetectable inside victim organization.

3.6 Data Exfiltration

Once the full C2 is accomplished by an attacker then final phase is to transfer stolen confidential data back to attacker centers. This phase can be done by either burst mode where all data transfer at once or more stealthy mechanism low-and-slow is used for data exfiltration. For example [16], in Adobe Leak in 2013 9 GB of encrypted password data were stolen and another case where 40 GB of database leaked known as Ashley Madison in 2015. Sometime these data may be sent in clear form (by web-based) or encrypted and zipped files with secure password protection. In case of Hydraq, there is different techniques used for sending the data back to attacker centers; one of them using port 443 act as a primary channel for upload of sensitive data, often uses a secure protocol like SSL/TLS, and also encrypted the content by private cipher which was left on victim organization.

Each APT stage has a distinct characteristic that may show some information that may be helpful for preparing defense system and early detection of APTs. In the coming section, we show how APTs campaigns happen in recent year.

4 Analysis of APT Campaigns

To create a general overview of various APT campaigns and used strategies, methodology and techniques, we analyzed the 26 different recently published reports. On the bases of revealed data, we focus only the APT campaigns on windows platform and excluded the other type of sophisticated attacks. For easy to understand, we segregate the APT phases into three main part: (1) initial compromise (2) lateral movement (3) command and control (C2) and also shows the active time of APT campaign in Table 1.

After analysis of whole APTs campaign report, we characterize the relevant attributes of APTs and form a prevention and detection mechanism effectively. Table 1 clearly shows that most of the APT campaigns used the spear-phishing techniques for initial compromise. There are 23 campaigns used the email attachments and 11 campaigns used the watering-hole attack for initial phase. There are only four campaigns used in the infected storage media and many of them used the multiple combinations of techniques to access the target network.

The analysis of various APTs campaign shows that to compromise the additional systems inside the internal network will help the attacker to persist for long time and collect more confidential data. These will be achieved by internal reconnaissance and try to exploit the known vulnerabilities or zero-day vulnerability. Most of the time the attacker usage the standard operating-system methods and tools for lateral movement because it is difficult to detect the genuine traffic and traffic generated by standard operating system tools. 12 campaign used the tools as Windows Management Instrumentation (WMI), Remote Desktop Protocol (RDP), PsExec, Powershell to gain the accessing remotely and the further move is to exploit the vulnerabilities. The analysis of report shows that only two campaign used the 0-day vulnerability exploit for lateral movement.

The analysis of report shows that 12 campaigns used the HTTPS or HTTP secure protocol to establish the communication with command and control servers. Because most of the companies only allow to access the Internet over specific web ports and rest of the port will be blocked. But some campaign attacker used the FTP to steal the data and two campaign used the custom protocols. Most protocols used encryption and decryption techniques to hide from detection.

Table 1 Techniques used in APT campaign

APT campaign (Active)	Initial compromise			Lateral movement			C2			Discovery	Report
	Spear-phishing	Watering-hole attacks	Storage media	Standard OS tolls	Hash and password dumping	Exploit vulnerabil-ities	HTTP/ HTTPS	Others	Custom protocols		
CosmicDuke	✓			✓	✓		✓			2012–2013	[17]
MiniDuke	✓			✓			✓	✓		2008–2013	[18]
Dark hotel	✓	✓				✓*	✓			2007–2014	[19]
Turla	✓	✓				✓		✓		2007–2014	[20]
Epic turla	✓	✓				✓	✓	✓		2012–2014	[21]
Crouching yeti	✓	✓				✓				2010–2014	[22]
Adwind	✓			✓		✓				2012–2013	[23]
Blue termite	✓	✓				✓		✓		2013–2014	[24]
Sofacy	✓		✓			✓		✓		2008–2014	[25]
Equation	✓		✓			✓				2002–2014	[26]
NetTraveler	✓	✓				✓		✓		2004–2013	[27]
Regin	✓			✓			✓	✓		2003–2012	[27]
Duke 2.0	✓			✓	✓	✓*	✓	✓	✓	2014–2015	[28]
Wild neutron	✓	✓				✓		✓		2011–2013	[29]
Winnti	✓			✓				✓		2009–2012	[30]
Desert falcons	✓			✓	✓		✓			2011–2014	[31]

(continued)

Table 1 (continued)

APT campaign (Active)	Initial compromise			Lateral movement			C2			Discovery	Report
	Spear-phishing	Watering-hole attacks	Storage media	Standard OS tolls	Hash and password dumping	Exploit vulnerabilities	HTTP/HTTPS	Others	Custom protocols		
Poseidon	✓			✓		✓		✓		2005–2015	[32]
FinSpy	✓				✓			✓		2007–2011	[33]
Black energy	✓		✓	✓	✓				✓	2010–2013	[34]
Hacking team RCS	✓		✓			✓	✓			2008–2011	[34]
Naikon	✓	✓		✓			✓			2009–2011	[35]
CozyDuke	✓	✓					✓			2014–2015	[36]
Cloud atlas	✓			✓	✓	✓		✓		2014–2014	[37]
Hellsing	✓	✓						✓		2012–2014	[38]
Kimsuky					✓		✓			2011–2014	[39]
Carbanak	✓	✓		✓	✓		✓	✓		2013–2014	[40]

✓* = Zero-day Vulnerability Exploitation

5 Approaches for Detection and Prevention

The detailed analysis of APT campaigns clearly shows that most of the time attacker try to hide their malicious activities inside internal networks and act like not to discriminate from the legitimate actions. More important, usage of standard operating systems and tools are rarely identified because they have the administrative right to access services. Mostly they exploit the unpatched vulnerability for lateral movement and very few campaigns use the 0-day vulnerability exploitation.

There are some proposed approaches to prevention and detection of APT campaigns such as the deployment of security systems as like firewalls, NIDS, HIDS, anomaly-based IDS and signature-based systems. Although these systems are automatically detected the malicious activities but required some level of expertise and concrete approach to full utilization of the systems. The following approaches will consider by the organization for early detection and prevention from highly potential advanced attacks.

5.1 To Prevent Vulnerability Exploitation

Our analysis shows that there are 14 campaigns used exploits of vulnerability in initial phases and lateral movement phase, but there is only two campaign take the advantage of 0-day exploits. Thus patches of already known vulnerability will prevent this problem.

There are number tools that can detect particular exploitation techniques like Malware bytes offers for their customers to detect and block the memory corruption exploits. Another Microsoft offers the Enhanced Mitigation Experience Toolkit (EMET) to protect the windows applications. In case of 0-day exploits, EMET could also helpful for detection, due to its multiple generic features.

5.2 Hashed-Password Dumping

Currently, there are various techniques available for password dumping in Windows environment, like mimikatz that have the capability to dump the clear text password and offers for further usage. Another, Windows Local security Authority Subsystem Service (LASASS) is most widely used for extracting the credentials. So, implementation of security monitors that restrict these tools or block these suspicious actions will prevent the attacker's movements.

5.3 Monitoring of Standard Operating Systems Tools and Techniques

Distinction between standard benign and malicious usage is very difficult. The monitoring of traffic generated by standard operating tools analyze generated log information will help to detect the potential lateral movements. Therefore, a correct logging policy should be implemented to ensure the genuine events and regular security configuration audit will help to achieve the objectives.

6 Conclusion

The analysis of 26 various recently active APT campaign published report we conclude that most of the campaigns take the advantage of spear-phishing attack for initial compromise, standard operating system tools for lateral movement and Secure HTTPS protocol for command and control. Some few lateral movements are dumping the password and credentials, such techniques can easily hide in between legitimate traffic and activities and help them to bypass strong security systems.

Another fact comes out from this analysis is that Zero-day vulnerabilities are not the main reason for such type of advanced attacks. Maximum time already known vulnerabilities which are not patched timely, exploited by attackers. Further, based on the characteristics, identification of the common type of malicious activities become easy and we will make strong defense mechanism for prevention and detection of sophisticated attacks.

In future work, more APT campaigns with relevant attack vector will be analyzed and their correlation with different advanced attacks will help to design the strong defense mechanism.

References

1. Mandiant: M-trends—a view from the front lines. Mandiant, Technical Report (2015)
2. Tankard, C.: Advanced persistent threats and how to monitor and deter them. **2011**(8), 16–19 (2011)
3. Kaspersky Lab: ZAO. Red October diplomatic cyber-attacks investigation (2014). http://www.securelist.com/en/analysis/204792262/Red_October_Diplomatic_Cyber_Attacks_Investigation
4. Mandiant Intelligence Center: Apt1: exposing one of China's cyber espionage units. Technical Report, Mandiant (2013)
5. Ronald, D., Rafal R.: Tracking ghost net: investigating a cyber-espionage network. Inf. Warf. Monitor, 6 (2009)
6. Thonnard, O., Bilge, L., O'Gorman, G., Kiernan, S., Lee, M.: Industrial espionage and targeted attacks: understanding the characteristics of an escalating threat. In: Research in Attacks, Intrusions, and Defenses, pp. 64–85. Springer, Berlin (2012)

7. Chien, E., OMurchu, L., Falliere, N.: W32.Duqu: the precursor to the next stuxnet. In: 5th USENIX Workshop on Large-Scale Exploits and Emergent Threats, Berkeley, CA, USENIX (2012). https://www.usenix.org/w32duqu-precursor-next-stuxnet
8. TrendLabs: Spear-Phishing Email: Most Favored APT Attack Bait (2012)
9. Will Gragido: Lions at the watering hole the VOHO affair (2012). http://blogs.rsa.com/lions-at-the-watering-hole-the-voho-affair
10. Haq, T., Khalid, Y.: Internet explorer 8 exploit found in watering hole campaign targeting Chinese dissidents (2013)
11. Kindlund, D., et al.: Operation Snowman: deputydog actor compromises US veterans of foreign wars website (2014)
12. Brewer, R.: Advanced persistent threats: minimising the damage. Netw. Secur. **4**, 5–9 (2014)
13. Denning, D.E.: An intrusion-detection model. IEEE Trans. Softw. Eng. **2**, 222–232 (1987)
14. McAfee Labs: Protecting your critical assets: lessons learned from operation aurora (2010)
15. Uri Rivner: Anatomy of an attack (2011). https://blogs.rsa.com/anatomy-of-an-attack
16. World most popular data breaches (2015). http://www.informationisbeautiful.net/visualizations/worlds-biggest-data-breaches-hacks
17. Baumgartner, K., Raiu, C.: The Cozy-Duke APT, Kaspersky Lab, April 2015
18. Kaspersky Labs: Global Research and Analysis Team. miniduke-is-back-nemesis-gemina-and-the-botgen-studio, July 2014
19. Kaspersky Labs: Global Research & Analysis Team. The Darkhotel APT—a story of unusual hospitality, Nov 2014
20. Kaspersky Labs: Global Research & Analysis Team. turla-apt-exploiting-satellites, Sept 2015
21. Kaspersky Labs: Global Research & Analysis Team. epic-turla-snake-malware-attacks (2015)
22. Kaspersky Labs: Global Research & Analysis Team. Energetic bear: more like a Crouching Yeti, July 2014
23. Kaspersky Labs: Global Research & Analysis Team. Adwind: malware-as-a-service platform (2014)
24. Kaspersky Labs: Global Research & Analysis Team. New activity of the blue termite APT, August 2015
25. Kaspersky Labs: Global Research & Analysis Team. Sofacy APT hits high profile targets with updated toolset, Dec 2015
26. Kaspersky Labs: Global Research & Analysis Team. Equation: the death star of malware galaxy, Feb 2015
27. Kaspersky Labs: Global Research & Analysis Team. NetTraveler is back: the 'red star' APT returns with new tricks, Sept 2013
28. Kaspersky Labs: Global Research & Analysis Team. The Duqu 2.0, June 2015
29. Kaspersky Labs: Global Research & Analysis Team. Wild neutron—economic espionage threat actor returns with new tricks, July 2015
30. Kaspersky Labs: Global Research & Analysis Team. Winnti FAQ. More than just a game, April 2013
31. Kaspersky Labs: Global Research & Analysis Team. The desert falcosn targeted attacks, Feb 2015
32. Kaspersky Labs: Global Research & Analysis Team. Poseidon Group: a targeted attack boutique specializing in global cyber-espionage, Feb 2016
33. Kaspersky Labs: Global Research & Analysis Team. Mobile malware evolution: part 6, Feb 2013
34. Kaspersky Labs: Global Research & Analysis Team. BE2 custom plugins, router abuse, and target profiles, Nov 2014
35. Baumgartner, K., Golovkin, M.: The MsnMM campaigns—the earliest naikon APT campaigns, Kaspersky Lab, May 2015
36. Kaspersky Labs: Global Research & Analysis Team. The CozyDuke APT, April 2015
37. Kaspersky Labs: Global Research & Analysis Team. Cloud atlas: RedOctober APT is back in style, Dec 2014

38. Raiu, C., Golovkin, M.: The chronicles of the hellsing APT: the empire strikes back (2015). https://securelist.com/analysis/publications/69567/the-chronicles-of-the-hellsing-apt-the-empire-strikes-back
39. Kaspersky Labs: Global Research & Analysis Team. The "Kimsuky" operation: a North Korean APTs, Sept 2013
40. Kaspersky Labs: Global Research & Analysis Team. Carbanak APT—the great bank robbery, Feb 2015

An Efficient Algorithm to Minimize EVM Over ZigBee

Anurag Bhardwaj and Devesh Pratap Singh

Abstract Advanced technology market desires more durability and reliability from wireless technologies, but during the transmission different factors, i.e., node failure, delay, data traffic, noise, distortion, etc., reduce the quality of communication. To overcome from such factors, some relatable parameters like signal-to-noise ratio (SNR), Error Vector Magnitude (EVM), Bit Error Rate (BER), etc., have been defined to describe the performance of a signal in the transmission. ZigBee is a wireless personal area network standard, which is designed for monitoring and control applications, i.e., home and industrial automation, grid monitoring, asset tracking, etc., because it is less expensive and more efficient than other wireless technologies. In this work, Error Vector Magnitude (EVM) is considered as a performance measuring metric for the quality of digitally modulated transmission in ZigBee wireless network and it is minimized for the improvement in reliability and efficiency of the transmission. However, by simulating the ZigBee sensor network in 2.4 GHz frequency by Offsetting the Quadrature Phase Shift Keying (OQPSK) modulation with half-sine pulse shaping filter minimizes the root mean square value of EVM is to 0.1% which is more efficient than Binary Phase Shift Keying (BPSK) with Root Raised Cosine (RRC) filter.

Keywords WPAN · ZigBee · EVM · BER · OQPSK · BPSK

1 Introduction

WSN play a vital role in the application domain of security, surveillance, monitoring of environmental conditions, health care and tracking of objects at a particular location. WSNs have various characteristics like autonomy, energy, and resource-

A. Bhardwaj (✉) · D. P. Singh
Graphic Era University, Dehradun, India
e-mail: er.anuragshr@live.com

D. P. Singh
e-mail: devesh.geu@gmail.com

K. Ray et al. (eds.), *Soft Computing: Theories and Applications*,
Advances in Intelligent Systems and Computing 742,
https://doi.org/10.1007/978-981-13-0589-4_20

constrained nodes with a dynamic topology. Sensor nodes are able to detect different phenomenal conditions like temperature, humidity, pressure, etc., Data which is sensed collaboratively sent to a base station or sink through single hop or multi-hop connectivity [1]. Nowadays, ZigBee which is the commercial term for IEEE 802.15.4 standard-based LR-WPAN is used for monitoring and controlling applications. It reduces the latency, battery consumption, and power consumption which lead to an efficient communication network. To measure performance, different matrices are used, i.e., BER, Acknowledgment wait duration, EVM, Frame retrievals, etc., In this work, EVM rate as a performance metric is used to improve the transmission in ZigBee. EVM is also known as receive constellation error (RCE) which quantifies the performance of digital radio transmission. Simulation of the ZigBee communication model with OQPSK half-sine pulse shaping filter in which I and Q are the In-Phase and Quadrature values of an ideal signal. In QPSK, I and Q transits to another signal occurs simultaneously. In OQPSK, it transmits for offset value by half the signal time. Since one OQPSK signal have two bits, then the offsetting time corresponds to each bit period. This bit period is referred to as chip period. Since I and Q are decoupled, only one can change at a given time. This limits phase changes to a maximum of 90°. In QPSK, maximum phase shifting is 180° and relates to a zero crossing. However, in OQPSK, it eliminates any amplitude variation and turns the OQPSK into a constant modulation which results EVM RMS value reduced and improvement in transmission.

2 Related Problem

In 2006, JoonHyung Lim et al. proposed that the compliant single-chip RF and Modem transceiver is completely implemented using 0.18 μ CMOS process and has the tiny chip size with low-power utilization for ZigBee. The minimum sensitivity of the receiver is less than 94.7 dBm for the 1% packet error rate (PER). This is 14.7 dB better than the required sensitivity. When the output power is 0 dBm, a maximum EVM of 10% is achieved [2].

In 2011 Dan Lei Yan et al. designed, implemented and measured the results of 2-point modulator for ZigBee transmitter. The chip consumes 5 mA from a single +1.8 V power supply achieving −109 dBc/HZ at 1 MHz offset over the entire band from 2405 MHz to 2480 MHz band. Measured result shows transmitter signal EVM is 6.5%, and Spectrum of modulation signal meet the PSD mask [3].

In 2014, Rohde et al. simulated the ZigBee sensor network in different radio frequencies, i.e., 868 and 908 MHz using BPSK modulation with RRC filter which results in 0.19% root mean square value of EVM but 0.29% peak value for 2.4 GHz frequency. Minimizing the EVM for higher frequency will improve the signal strength and results in better transmission and it can be done by modulating the ZigBee wireless sensor network with different phase shifting modulation techniques, i.e., QPSK, MSK, OQPSK, etc., in OFDM communication [4].

3 Proposed Algorithm

Setup

Initialize required variables

Setup WS nodes in 100m X 100m area

Map Chip values

Design Transmission and Receiving filter

Define Normalize Error Vector Magnitude (EVM)

Routing

Step 1. Generate random data bits to transmit data over the network

Step 2. Simulate number of symbols in a frame

Step 3. Initialize peak Root Mean Square EVM to –infinity

Step 4. Create Differential Encoder object to differentially encode data

Step 5. Create AWGN (Gaussian Noise) and set its noise method property to SNR

Step 6. Loop from 1 through number of frames

Generate random data

Differentially encode

Modulate using OQPSK

Convert symbols to chips (spread)

Reshape signal

Shape Pulse

Filter transmitted data

Calculate and set the 'SignalPower' property of the channel object

Add noise

Down sample received signal.

Account for the filter delay.

Measure using the EVM System object using eq(1)

Update peak RMS EVM calculation

if (peak RMS EVM < rms EVM) then

peakRMSEVM ⊠ rmsEVM

end if

end loop

4 Results

After Simulating ZigBee in OQPSK modulation, results show that the error rate reduced to 0.1% for the modulated signal. EVM RMS value for BPSK is 0.19% which is minimized to 0.1% (Figs. 1, 2 and 3).

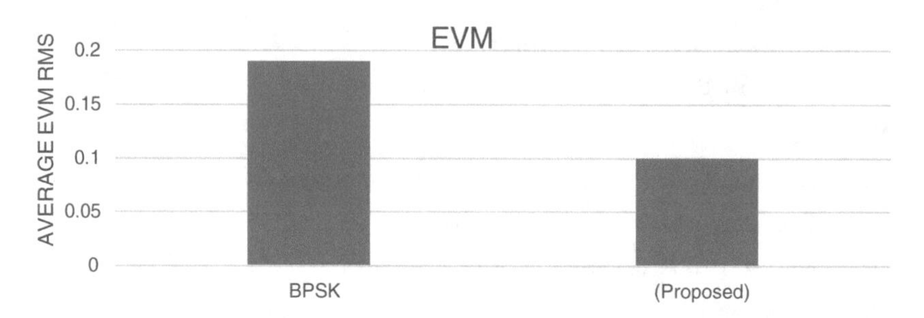

Fig. 1 Comparison of EVM by average RMS values of BPSK modulation and proposed OQPSK modulation

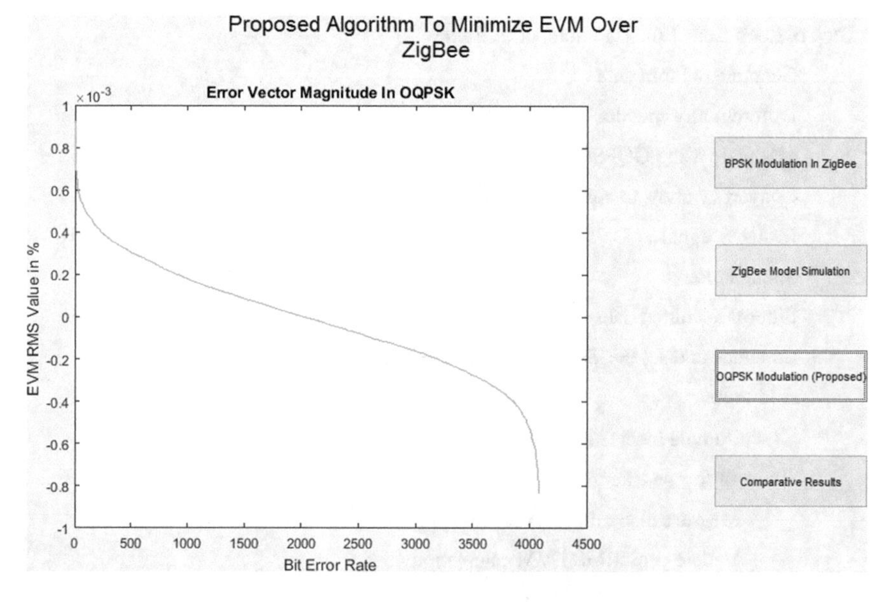

Fig. 2 Error rate for OQPSK modulated signal in ZigBee simulation

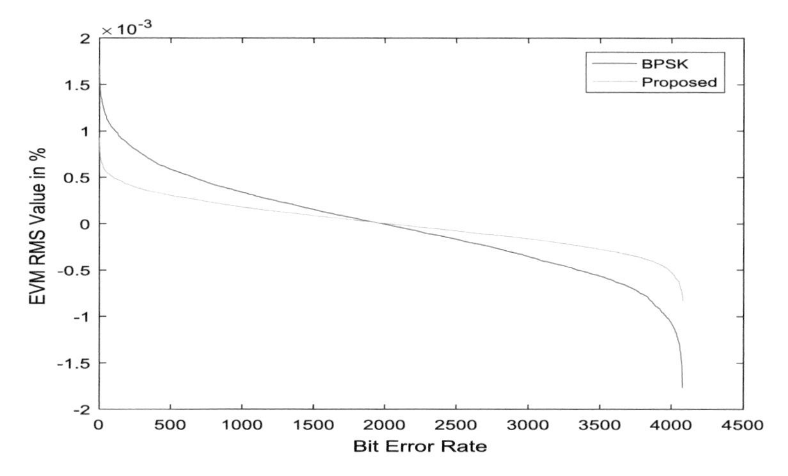

Fig. 3 Error rate after minimized EVM in OQPSK-modulated signal

5 Conclusion

802.15.4 Standardization-based ZigBee wireless sensor network is designed in 2.4 GHz radio frequency. To improve the signal strength, EVM is taken as a performance measure which has to be minimized. It is observed that using BPSK modulation with RRC filter in OFDM medium provides 0.19% root mean square value. However, for further improvement, Offset QPSK with half-sine wave pulse shaping filter is used instead of BPSK modulation, which minimizes the root mean square value to 0.1% and also reduced the BER value up to 2.5 dB which results in more efficient and reliable transmission. EVM can be minimized for higher frequencies by using other modulation techniques. It is also possible to take two or more different performance measures together to analyze and quantify the efficiency and reliability of the transmission which can help in real-life implementation in ZigBee applications.

References

1. Martincic, F., Schwiebert, L.: Introduction to Wireless Sensor Networking, vol. 1, pp. 1–34 (2005)
2. Lim, J.H., Cho, K.S.: A fully integrated 2.4 GHz IEEE 802.15.4 transceiver for zigbee applications. In: Proceedings of IEEE Asia-Pacific Conference on Microwave, pp. 1779–1782, Dec 2006
3. Yan, D.L., Zhao, B.: A 5 mA 2.4 GHz 2-point modulator with QVCO for zigbee transceiver in 0.18-μm CMOS. In: Proceedings of International Symposium on Integrated Circuits (ISIC), pp. 204–207 (2011)
4. Schmitt, H.: EVM Measurements for ZigBee Signals. Rohde and Schwarz (2014)

Performance Measure of the Proposed Cost Estimation Model: Advance Use Case Point Method

Archana Srivastava, S. K. Singh and Syed Qamar Abbas

Abstract Estimating size and cost of a software system is one of the primary challenges in software project management. An estimate is of critical significance to a project's success, hence the estimate should go through a rigorous assessment process. The estimate should be evaluated for its quality or accuracy, and also to ensure that it contains all of the required information and is presented in a way that is easily understandable to all project stakeholders. Software cost model research results depend on model accuracy measures such as MRE, MMRE and PRED. Advance Use Case Point Method (AUCP) is an enhancement of UCP. AUCP is our previously proposed and published model (Srivastava et al in Int. J. Control Theor. Appl. Eval. Softw. Project Estimation Methodol. AUCP 9(41):1373–1381, 2017) [1]. In this paper, performance evaluation of AUCP is carried out using the three widely accepted metrics including MRE, MMRE and percentage of the PRED.

Keywords Use case point method (UCP) · Advanced use case point method (AUCP) · End user development (EUD) · EUD_technical factors
EUD_environmental factors · MRE (Magnitude of relative error)
MMRE (Mean magnitude of relative error)

1 Introduction

The Use Case Points Method (UCP) is an effort estimation algorithm originally recommended by Gustav Karner (1993). The UCP method analyzes the project's use

A. Srivastava (✉) · S. K. Singh
Amity University, Lucknow, India
e-mail: srivastavaarchana891@gmail.com

S. K. Singh
e-mail: sksingh1@lko.amity.edu

S. Q. Abbas
Ambalika Institute of Management & Technology, Lucknow, India
e-mail: qrat_abbas@yahoo.com

K. Ray et al. (eds.), *Soft Computing: Theories and Applications*,
Advances in Intelligent Systems and Computing 742,
https://doi.org/10.1007/978-981-13-0589-4_21

case, actors, scenarios and various technical and environmental factors and abstracts them into an equation. As the end users are becoming more computer literate day by day the demand of end user development features is increasing [2–4]. End User Development features if incorporated in a software or website, enhances the quality and thus increases end user satisfaction. End users concerned here may not be professional developers but have may have sufficient knowledge of their respective domains and like to do coding or use various wizards to customize things as per their own requirements [1]. These additional EUD features need to be designed, coded and tested properly using verification and validations techniques, increases the development effort but ensures high quality software that is measured by the fulfillment of end user requirements.

Advanced Use Case Method enhances UCP by taking into consideration the extra effort required in adding End User Development features into the software along with the required quality parameters. End User Development features in software or website enhances the quality and increases end user satisfaction exponentially hence is an important additional cost driver.

2 Advanced Use Case Method

AUCP includes the additional effort required in incorporating end user development features in the software for overall project effort estimation. End user likes some flexibility of programming in software so as to satisfy their requirements. The additional technical and environmental cost drivers considered while providing end user development features in software are introduced below.
These EUD features can be classified into two categories as UCP model, i.e.
EUD TECHNICAL FACTORS (EUD_TF)
EUD ENVIRONMENTAL FACTORS (EUD_EF)
Total seventeen EUD_TF and eight EUD_EF are considered to be included as End User Development Cost Drivers [5]. AUCP method is as follows:

Use Case Point Method (UCP) is calculated as follows [6]:

(a) Calculating Unadjusted Use Case Weight (UUCW).
(b) Calculating Unadjusted Actor Weight (UAW)
(c) Calculating the Technical Complexity Factor (TCF)
(d) Calculating Environment Complexity Factor (ECF)
(e) Calculating Unadjusted Use Case Points (UUCP), where UUCP = UAW + UUCW.
(f) Calculating Complexity Factor, where:

   a. TCF = 0.6 + (0.01 * TF)
   b. ECF = 1.4 + (−0.03 * TF)

(g) Calculating the Use Case Point (UCP), where:

a. UCP = UUCP * TCF * ECF

(h) Identify the End User Development features required by the customers.

Advance Use Case Point Method (AUCP) is an extension of UCP and is further calculated as follows [7]:

T1 to T17 EUD_Technical factors (EUD_TF) are identified and weights are assigned to all the factors as in Table 1 after thorough analysis of its impact on the development:

(i) If EUD_Technical factors (EUD_TF) is applicable for the particular module it will be rated as 1 else 0 and multiply it with weights of EUD_TF. Take the summation of all factors.
F1 to F8 EUD_ENVIRONMENTAL FACTORS are identified and weights are assigned to them as in Table 2 after thorough analysis of its impact on the development.

Table 1 EUD_technical factors

EUD_Ti	EUD_technical factors	Weight	EUD_Ti	EUD_technical factors	Weight
T1	Creating throw away codes	1	T10	Error detection tools	1.5
T2	Creating reusable codes	1.2	T11	Availability of online help	1.5
T3	Sharing reusable code	1.4	T12	Self-efficacy	1.11
T4	Easily understandable codes	1.2	T13	Perceived ease of use	1.20
T5	Inbuilt security features in codes	1.5	T14	Perceived usefulness	1
T6	Authentication features	1.12	T15	Flexible codes	1.2
T7	Inbuilt feedback about the correctness	1.3	T16	Scalability features	1.25
T8	Testable codes	1.2	T17	Ease of maintenance	1.2
T9	Tools for analyzing and debugging	1.4			

Table 2 EUD_environmental factors

Fi	EUD_environmental factors	Weight	Fi	EUD_environmental factors	Weight
F1	Content level of EUP	1.4	F5	End user training and learning time constraint	1.12
F2	End user computing capability	0.75	F6	Reliability of end user code	1.2
F3	Ease of use and feedback	1.5	F7	End user storage constraint	1.02
F4	Inbuilt system assistance for EUP	1.25	F8	Risk factors	1.5

Table 3 Calculation of UCP

No	Project ID	UUCP	TCF	ECF	UCP
1	A	480	1.015	0.89	433.61
2	B	287	1.055	0.65	196.81
3	C	279	1.005	0.995	278.99
4	D	292	1.045	0.875	267.00
5	E	322	1.025	1.055	348.20
6	F	307	1.035	0.695	220.83
7	G	214	1.045	0.875	195.68
8	H	307	1.065	0.845	276.28

(j) If EUD_Technical factors (EUD_TF) is applicable for the particular module it will be rated as 1 else 0 and multiply it with weights of EUD_TF. Take the summation of all factors.

(k) Calculate EUD Technical Complexity Factor, EUD_TCF = 0.6 + (0.01 * EUD_TF)

(l) Calculate EUD Environmental Complexity Factor, EUD_ECF = 1.4 + (0.03 * EUD_EF)

(m) Calculation of AUCP is performed as

$$\text{AUCP} = \text{UCP} \times (\text{EUD\_TCF} \times \text{EUD\_ECF})$$

3 Result Analysis

This study was based on eight government website development projects without EUD features. The project was given ID from A to H. As the project developed by a small team with a number of personnel with 3 to 5 people [8]. UCP was calculated and is given in Table 3.

Now, if we have to incorporate EUD features in the existing websites having various EUD quality factors requirements using the AUCP method. We will mark the technical quality factors required in the software as 1 and if that particular quality factor is not required it will be marked as 0. Then, all the values will be multiplied by the weights assigned to each factor. Summation of the values will be taken for each project.

EUD_Technical Factors

Now, we will mark the environmental quality factors if required in the software as 1 and if that particular quality factor is not required it will be marked as 0. Then, all the values will be multiplied by the weights assigned to each factor. Summation of the values will be taken for each project A to H (Table 4).

Table 4 EUD_technical factors

EUD Ti	EUD technical factors	Weight	A	Val	B	Val	C	Val	D	Val	E	Val	F	Val	G	Val	H	Val
T1	Creating throw away codes	1	1	1	0	0	0	0	0	0	0	0	1	1	1	1	0	0
T2	Creating reusable codes	1.2	0	0	1	1.2	1	1.2	1	1.2	1	1.2	0	0	0	0	1	1.2
T3	Sharing reusable code	1.4	0	0	1	1.4	1	1.4	1	1.4	1	1.4	0	0	0	0	1	1.4
T4	Easily understandable codes	1.2	1	1.2	1	1.2	0	0	1	1.2	0	0		1.2		1.2	1	1.2
T5	Security features in codes for more control by end users	1.5	1	1.5	1	1.5	1	1.5	1	1.5	1	1.5		1.5		1.5	0	0
T6	Authentication features	1.12	1	1.12	1	1.12	0	0	1	1.12	1	1.12		1.12		1.12	0	0
T7	Inbuilt feedback about the correctness	1.3		0	1	1.3	1	1.3	1	1.3	1	1.3		1.3		1.3	1	1.3
T8	Testable codes	1.2	1	1.2	1	1.2	1	1.2	1	1.2	1	1.2		1.2		1.2	1	1.2
T9	Tools for analyzing by debugging	1.4	1	1.4	1	1.4	0	0	0	0	1	1.4		1.4		1.4	0	0
T10	Error detection tools	1.5	1	1.5	1	1.5	1	1.5	0	0	1	1.5		1.5	0	0	0	0
T11	Availability of online help	1.5	1	1.5	1	1.5	0	0	1	1.5	0	0		1.5	0	0	0	0
T12	Self - efficacy: high sense of control over the environment	1.11	1	1.11	1	1.11	1	1.11	1	1.11	0	0		1.11	0	0	0	0
T13	Perceived ease of use	1.20	1	1.2	1	1.2	0	0	1	1.2	0	0		1.2		1.2	1	1.2
T14	Perceived usefulness	1	1	1	1	1	1	1	1	1	1	1		1		1	1	1
T15	Flexible codes	1.2	1	1.2	1	1.2	0	0	1	1.2	1	1.2		1.2		1.2	0	0
T16	Scalability features	1.25	1	1.25	1	1.25	1	1.25	1	1.25	1	1.25		1.25		1.25	0	0
T17	Ease of maintenance	1.2	0	0	1	1.2	0	0	1	1.2	1	1.2		1.2		1.2	0	0
	Summation of EUD technical factors			16.18		20.28		11.46		17.38		15.27		18.68		14.57		8.5

EUD_Environmental Factors

Calculations done as per the given formulas (Tables 5 and 6):

$$\text{EUD\_TCF} = 0.6 + (0.01 * \text{EUD\_TF})$$

$$\text{EUD\_ECF} = 1.4 + (0.03 * \text{EUD\_EF})$$

$$\mathbf{AUCP} = \mathbf{UCP} \times (\mathbf{EUD\_TCF} \times \mathbf{EUD\_ECF})$$

4 Performance Measure

Performance Evaluation Metrics: The following evaluation metrics are adapted to assess and evaluate the performance of the effort estimation models.

Magnitude of Relative Error (MRE)

$$\text{MRE} = \left| \frac{\text{Actual effort} - \text{Estimated effort}}{\text{Actual Effort}} \right|$$

Mean Magnitude of Relative Error (MMRE)

$$\text{MMRE} = \sum_{\text{i}=0}^{\text{n}} \text{MREi}$$

The MMRE calculates the mean for the sum of the MRE of n projects. Specifically, it is used to evaluate the prediction performance of an estimation model. Conte et al. [9] consider MMRE $\leq$0.25 as an acceptable level for effort prediction models.

PRED (x) [10] considers the average fraction of the MRE's off by no more than x as defined by

$$PRED(x) = \frac{1}{N} \sum_{i=0}^{n} \begin{cases} 1\, if\; MRE \leq x \\ 0\, otherwise \end{cases}$$

Typically PRED (0.25) is used, but some studies also look at PRED (0.3) with little difference in results. Generally PRED (0.3) $\geq$ 0.75 is considered an acceptable model accuracy.

PRED(x) $= \frac{\text{k}}{\text{n}}$, where k is the number of projects in a set of n projects whose MRE$<$x. They suggested that an acceptable accuracy for a model is PRED (0.25)$>$0.75, which is seldom reached in reality [11].

Table 5 EUD_environmental factors

F1	Content level of EUP	1.4	0	0	1	1.4	1	1.4	1	1.4	1	1.4	0	0	1	1.4	1	1.4
F2	End User computing capability	0.75	1	0.75	1	0.75	0	0	1	0.75	1	0.75	0	0	1	0.75	0	0
F3	Ease of use and feedback	1.5	1	1.5	1	1.5	1	1.5	1	1.5	1	1.5	1	1.5	1	1.5	1	1.5
F4	Inbuilt system assistance for EUP	1.25	1	1.25	0	0	0	0	0	0	1	1.25	1	1.25	1	1.25	0	0
F5	End user training and learning time constraint	1.12	1	1.12	1	1.12	1	1.12	0	0	1	1.12	1	1.12	1	1.12	1	1.12
F6	Reliability of end user code	1.2	0	0	1	1.2	1	1.2	1	1.2	1	1.2	0	0	1	1.2	1	1.2
F7	End user storage constraint	1.02	1	1.02	1	1.02	0	0	0	0	0	0	0	0	1	1.02	0	0
F8	Risk factors	1.5	0	0	1	1.5	0	0	0	0	1	1.5	1	1.5	1	1.5	0	0
	Summation of EUD_environmental factors	EUD_EF		5.64		8.49		5.22		4.85		8.72		5.37		9.74		5.22

Table 6 Calculating AUCP

No	Project ID	UCP	EUD_TCF	EUD_ECF	AUCP
1	A	433.61	0.76	1.57	518.34
2	B	196.81	0.80	1.65	261.44
3	C	278.99	0.71	1.56	310.34
4	D	267.00	0.77	1.55	319.30
5	E	348.20	0.75	1.66	435.49
6	F	220.83	0.79	1.56	271.24
7	G	195.68	0.75	1.69	246.92
8	H	276.28	0.69	1.56	294.59

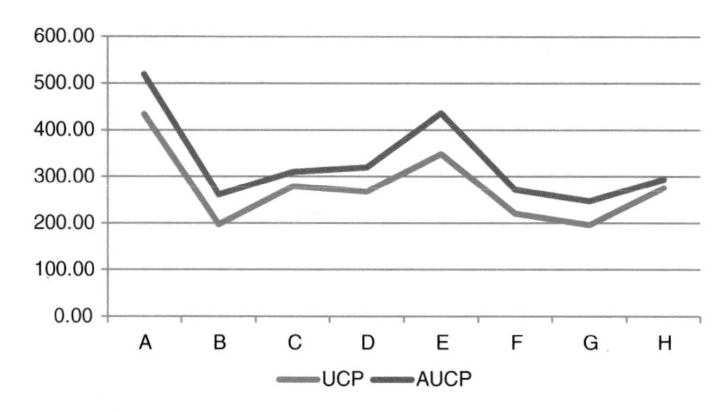

Fig. 1 The line graph of UCP and AUCP is shown below

Table 7 UCP versus AUCP

No	Project ID	UCP	AUCP	MRE	PRED(0.25)
1	A	433.61	518.34	0.195	1
2	B	196.81	261.44	0.328	0
3	C	278.99	310.34	0.112	1
4	D	267.00	319.30	0.196	1
5	E	348.20	435.49	0.251	0
6	F	220.83	271.24	0.228	1
7	G	195.68	246.92	0.262	0
8	H	276.28	294.59	0.066	1

5 Results

See Fig. 1 and Table 7.

Table 8 Observations

Variable	Observations	Minimum	Maximum	Mean	Std. deviation
UCP	8	195.676	433.608	277.175	80.975
AUCP	8	246.919	518.342	332.208	95.149
MRE	8	0.066	0.328	0.205	0.084
One-sample t-test/Two-tailed test (UCP):					
95% confidence interval on the difference between the means:					
] 209.478,			344.872 [		
Difference			277.175		
t (Observed value)			9.682		
|t| (Critical value)			2.365		
DF			7		
p-value (Two-tailed)			<0.0001		
Alpha			0.05		
Test interpretation:					
H0: the difference between the means is equal to 0.					
Ha: the difference between the means is different from 0.					
As the computed p-value is lower than the significance level alpha = 0.05, one should reject the null hypothesis H0, and accept the alternative hypothesis Ha.					
The risk to reject the null hypothesis H0 while it is true is lower than 0.01%.					

6 Result Analysis

See Table 8.

This result analysis shows minor increase in the overall effort required for development of EUD features. Average increase will be approximately 5–12%. In extreme cases, it may be around 20–33%. End user computing enriches end user satisfaction which is the final goal of Information system [7]. In end user development (EUD), the practice of users creating, modifying, or extending programs for their personal use [12, 13] is a very promising approach. This approach has two main benefits. One, it puts systems design in the hands of the domain experts who are most familiar with what requirements must be built. Two, it scales with both a rapid increase in users and the increasing rate of change of many business processes [14] (Fig. 2).

Mean magnitude of relative error is 0.048. Conte et al. [9] consider MMRE $\leq$0.25 as an acceptable level for effort prediction models. When EUD features are incorporated in the system or tools the development effort will increase slightly as EUD features make system more generalized for further customization. The cost incurred

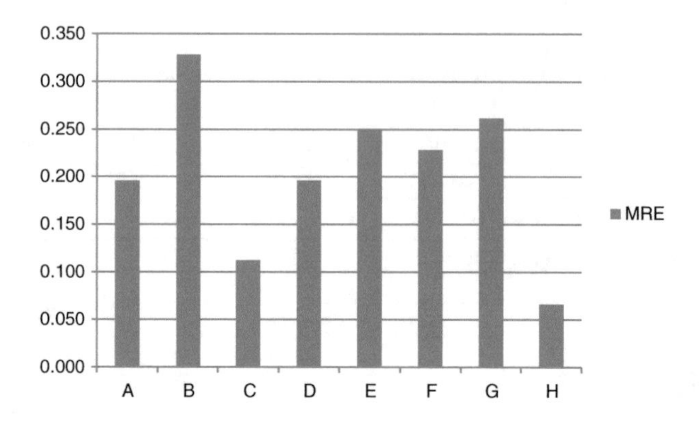

Fig. 2 MRE

in providing additional tools required for end user computing will be of less significance compared to the benefits that end user will get in return [15].

7 Conclusion

Software cost estimation is simple in concept, but difficult and complex in reality. The difficulty and complexity required for successful estimates exceed the capabilities of most software project managers. At the time of requirement engineering if end users demand for end user development facilities then they should be paying a little extra cost in incorporating those extra features. AUCP calculates the effort considering the additional End User Development features that are incorporated in the software considering the additional development requirements of the end user. Based on the above results, the proposed method for effort estimation incorporating EUD features is little higher to the result of UCP estimation models. Hence this type of Estimation may be recommended for the software development including EUD features. This MRE graph justifies that EUD features increase a little development effort but enhance End User Satisfaction exponentially, hence the little additional effort is justified.

References

Srivastava, A., Singh, S.K., Abbas, S.Q.: Int. J. Control Theor. Appl. Eval. Softw. Project Estimation Methodol. AUCP **9** (41),1373–1381 (2017)

Ko, A.J., Abraham, R., Beckwith, L., Blackwell, A., Burnett, M., Erwing, M., Scaffidi, C., Lawrance, J., Lieberman, H., Myers, B., Rosson, M.B., Rothermel, G., Shaw, M., Wiedenbeck, S.: The state of the art in end-user software engineering. ACM Comput. Surv. (2010)

Gelderman, M.: The relation between user satisfaction, usage of information systems and performance. Inf. Manag. **34**, 11–18 (1998)

Lieberman, H., Paterno, F., Klann, M., Wulf, V.: End-user development: an emerging paradigm. End User Development, pp. 1–8 (2006)

Srivastava, A., Singh, S.K., Abbas, Q.: airccse.org/journal/ijsea/papers/6215ijsea01.pdf

Karner, G.: Objective systems resource estimation for objectory projects SF AB (1993)

Srivastava, A., Singh, S.K., Abbas, S.Q.: Evaluation of software project estimation methodology. In: AUCP 2nd International Conference on Sustainable Computing Techniques in Engineering, Science and Management (SCESM-2017), 27–28 Jan 2017

Sholiq, Sutanto, T., Widodo, A.P., Kurniawan, W.: Effort rate on use case point method for effort estimation of website development. J. Theor. Appl. Inf. Technol. **63**(1) (2014)

Conte, S.D., Dunsmore, H.E., Shen, V.Y.: Software Engineering Metrics and Models. Benjamin-Cummings Publishing (1986)

Jørgensen, M.: Experience with the accuracy of software maintenance task effort prediction models. IEEE Trans. Softw. Eng. **21**(8) (1995)

Conte, S.D., Dunsmore, H.E., Shen, V.Y.: Software Engineering Metrics and Models. Benjamin/Cummings Publishing Company Inc, Menlo Park, Calif (1986)

Menzies, T., Port, D., Chen, Z., Hihn, J., Stukes, S.: Validation methods for calibrating software effort models. In: Proceedings of the 27th International Conference on Software Engineering (2005)

Wieczorek, I., Ruhe, M.: How valuable is company-specific data compared to multi-company data for software cost estimation? In: Proceeding for the Eights IEEE Symposium on Software Metrics (METRICS 02) (2002)

Jørgensen, M., Shepperd, M.: A systematic review of software development cost estimation studies. IEEE Trans. Softw. Eng. **33**(1) (2007)

Srivastava, A., Abbas, S.Q., Singh, S.K.: Enhancement in function point analysis. Proc. Int. J. Softw. Eng. Appl. **3**, 129–136

Fractal Image Compression and Its Techniques: A Review

Manish Joshi, Ambuj Kumar Agarwal and Bhumika Gupta

Abstract The main issue about an image is its size. Fractal Image Compression is a developing procedure which may represent an image by a contractive change on an image space for which the settled point is close to the primary image. This wide standard conceals a wide variety of coding designs, a robust segment of which have been explored in the rapidly creating collection of appropriated look into. While certain theoretical parts of this depiction are settled in, for the most part little thought has been given to the advancement of an understandable fundamental picture exhibit that would legitimize its use. Most basically fractal based plans are not forceful with the present best in class, yet these designs combining fractal compression and alternative techniques have gained widely more important ground. This review addresses an investigation of the most basic advances, both useful and theoretical in one of a kind fractal coding design. In this paper, we survey the fundamental models of the advancement of fractal objects with iterated function system (IFS) utilizing ICA and DBSCAN algorithms.

Keywords Fractal · Contractive · Robust · Iterated function system

1 Introduction

Information Compression has changed into a basic issue for information accumulating and transmission. This is particularly liberal for databases containing unending PC pictures. Beginning late, a liberal measure of strategies has shown up over the span of activity for completing high weight degree for compacted picture securing

M. Joshi
Uttrakhand Technical University, Dehradun, Uttrakhand, India

A. K. Agarwal (✉)
Teerthanker Mahaveer University, Moradabad, UP, India
e-mail: ambuj4u@gmail.com

B. Gupta
G. B. Pant Engineering College, Pauri, Garhwal, Uttrakhand, India

K. Ray et al. (eds.), *Soft Computing: Theories and Applications*,
Advances in Intelligent Systems and Computing 742,
https://doi.org/10.1007/978-981-13-0589-4_22

and among them, the fractal approach change into a possible and promising weight framework. The field of picture coding (or weight) guides possible systems for tending to pictures for transmission and farthest point purposes. The focal concentration of video coding is to pack the information rate by discharging wealth data. There are two essential classes of coding outlines (i.e., source coding and entropy coding). Mixed media information requires wide utmost most remote point and transmission data trade tie. The information include outlines, sound, video, and picture. These sorts of information must be compacted amidst the transmission system. Monstrous measure of information cannot be secured if there is low gathering limit show up. Weight requests that an approach oversees decrease the cost of motivation behind constrainment and overhaul the speed of transmission. Picture weight is utilized to limit the size in bytes of a depictions record without undermining the pixel thought of the photograph. There are two kinds of picture weight approaches exist. They are lossy and lossless. In lossless weight, the reiterated picture after weight is numerically foggy to the fundamental picture.

In lossy weight strategy, the copied picture contains degradation concerning the first. Lossy methodology causes picture quality contamination in each weight or decompression step. Everything considered, lossy structures suit more major weight degrees than lossless 5, i.e., Lossless weight gives breathtaking nature of stuffed pictures, yet yields basically less weight, however, the lossy weight frameworks provoke loss of informationwith higher weight degree. The systems for lossless picture weight blend variable length encoding, adaptive word reference figurings, for example, LZW, bit-plane coding, lossless sensible coding, and so on. The logic for lossy weight joins lossy sensible coding and change coding. Change coding, which applies a Fourier-related change, for example, DCT and Wavelet Transform, for example, DWT are the most routinely utilized approach. Over the explore starting late years, a system of gifted and complex fractal picture weight structure for picture weight have been passed on and comprehends it. The complement work structure gives a supernatural quality in the photographs. Source coding directs source material and yields works out not surprisingly which are lossy (i.e., picture quality is debased). Entropy coding completes weight by utilizing the quantifiable properties of the signs and is, on a key level, lossless. Unmistakable video weight methodology has been proposed over the most recent two decades and new ones are being made each day. For adequate picture quality, these systems can just aggregate energize diminishment in the source information not beating 25 and 200 times with still and chose pictures, solely (for instance by using an adaptable discrete cosine transform (ADCT) coding follows). Shockingly this is wound up being not elegant to change according to the developing sales in the use of transmission channels and most inaccessible point media. Along these lines, there is an unending fundamental for help diminishment in picture information reviewing the bona fide center to profit by the sharp progress in demonstrate day correspondence advance in the most ideal way. This supposition is moved around watching its idea into end client things, for example, Microsoft's Encarta or as a Netscape module by Iterated Systems Inc. Fractal picture weight abuses the normal relative wealth show up in like path pictures to satisfy high weight degrees in a lossy weight plan. The principal thought of the framework contains in finding an advance pick that passes on a fractal picture, approximating to the first. Fractal picture coding has its key foundations in the numerical theory of iterated work frameworks (IFS) made by Barnsley while the central absolutely robotized check was made by Jacquin. Fractal picture coding consolidates finding a strategy of changes that passes on a fractal picture which approximates the standard picture. Emphasis diminish is expert by plotting the overwhelming picture through humbler duplicates or parts of the photograph. Iterated limits structures (IFS) theory, decidedly identified with fractal geometry, has beginning late found a fascinating application in picture weight. Barnsley and Jacquin drove the field, trailed by various responsibilities. The approach combines giving a photograph as the attractor of a contractive cutoff centers structure, which can be recovered on an exceptionally essential level by complementing the diagram of breaking points beginning from any fundamental discretionary picture. The kind of emphasis mishandled is named piecewise self-transformability. This term suggests a property that each piece of a photograph can be genuinely permitted as a basic refinement in another piece of higher assertion. Vulnerabilities based still picture weight structures can claim to have wonderful execution at high weight degrees (around 70–80). The essential issue with fractal based coding procedures is that of multifaceted nature at the encoding stage. Notwithstanding, the multifaceted graph of the decoder stays sensible when showed up contrastingly in relationship with the encoding. Fractal develop structures run as for extraordinary outcomes like weight in pictures, holding an impossible to miss condition of self-closeness. Another spellbinding region of fractal based frameworks is their capacity to make a not very awful quality rendered picture for a subjective scaling factor. Fractal picture weight is dull in the encoding structure. The time is on an especially fundamental level spent on the compass for the best influence piece in a wide space to pool. In this paper, we audit the major standards of the improvement of fractal objects with iterated work systems (Uncertainties), by then we clear up how such a framework has been grasped by Jacquin for the coding (weight) of mechanized pictures.

1.1 Iterated Function Systems

The essential instrument utilized as a part of portraying pictures with iterated work frameworks is the relative change. This change is utilized to express relations between various parts of a picture. Relative changes can be depicted as blends of revolutions, scalings, and interpretations of facilitating tomahawks in n-dimensional space [1]. For instance, in two measurements a point (x, y) on the picture can be spoken to by (xn, yn) under relative change. This change can be portrayed as takes after: The parameters a, b, c, and d play out a revolution, and their sizes result in the scaling. For the entire framework to work appropriately, the scaling must dependably bring about shrinkage of the separations between focuses; generally rehashed cycles will bring about the capacity exploding to interminability. The parameters e and f cause a straight interpretation of the fact being worked upon. In the event that this change is connected to a geometric shape, the shape will mean another area and there turned and contracted to another, littler size. Keeping in mind the end goal to delineate source picture onto a coveted target picture utilizing iterated work frameworks, more than one change is frequently required and every change, i, must have a related likelihood, 9, deciding its relative significance regarding alternate changes. The irregular emphasis calculation given by Barnsley [1] can be utilized to decipher an IFS code with a specific end goal to reproduce the first picture. This calculation is given in the accompanying pseudocode.

1.2 Self-similarity Property

To encode a picture as indicated by self-likeness property. Each piece to be encoded must hunt in an extensive pool to locate the best match. For the standard full pursuit technique, the encoding procedure is tedious in light of the fact that a lot of calculations of comparability measure are required. Here, the picture will be shaped by duplicates of legitimately changed parts of the first. These changed parts do not fit together, when all is said in done, to frame a precise of the first picture, thus it must permit some blunder in our portrayal of a picture as an arrangement of changes.

2 Literature Review

The literature review is available to study fractal image compression.

Dan Liu, Peter K Jimack in his paper titled *A Survey of Parallel Algorithms for Fractal Image Compression* review the various techniques associated with parallel approach using fractal image compression. It is introduced due to the high encoding cost of fractal image compression. These techniques have been discussed from the viewpoints of granularity, load balancing, data partitioning and complexity reduction. It merits watching however that large portions of the properties of the present parallel hardware are not reliable with presumptions made in a portion of the past work. Specifically, it is incomprehensible today that a parallel processor would not have adequate of its own primary memory to store its own duplicate of the whole uncompressed picture. The advantages from parallel fractal image decoding are not prone to be so incredible as with the coding algorithm, since the decoding algorithm is proved to be fastest in any case. This is one of the primary attractions of fractal image compression after all considerations.

D. Sophin Seeli, Dr. M. K. Jeyakumar in his paper titled *A Study on Fractal Image Compression using Soft Computing Techniques* analyze the current FIC techniques that has been developed with the end goal of expanding compression ratio and shorten the computational time. Generally, these FIC techniques used the optimization methods to locate the ideal best matching blocks. Each of the FIC procedures and their execution are analyzed as far as their compression ratio, encoding time and PSNR

(Peak Signal-to-Noise Ratio) esteem. In view of these parameters, the execution of the FIC systems was contemplated and a comparative study of these procedures was given. The analysis demonstrates that the current techniques need to improve to accomplish the higher compression ratio. This lower execution in similar investigation process has roused to do another powerful heuristic FIC procedure for achieving the higher compression ratio. The new developed fractal image compression method used the most eminent technique to play out the image compression process. The execution of the most prestigious technique gave higher image compression ratio than the techniques discussed.

Gaurish Joshi in his paper titled *Fractal Image Compression and its Application in Image Processing* proposed two methods RDPS and ERB to make changes in the encoding time of fractal image compression. RDPS primarily focused on diminishing the encoding time and ERB concentrate on expanding compression ratio alongside slight change in encoding time. Thus, the two techniques are consolidated to frame new strategy RDPS-ERB to acquire the best outcomes. It has been demonstrated that compression ratio expanded to twofold of that of existing work with minimum loss in image quality. The paper suggested that affine parameter of FIC methods vary with image quality estimation like SSIM and MSE. There is a positive correlation between affine parameters and image contrast, i.e., by expanding parameter image upgrades and by diminishing parameter smoothing of image occurs. The new system becomes more faster and has better compression ratio with acceptable loss in image quality. Other feature extraction techniques like skewness; neighbor contrast, etc. may also be used to obtain the outcomes in near future.

John Kominek in his paper titled *Advances in Fractal Compression for Multimedia Applications* suggested that fast encoding is the need of fractal image compression in the modern age. Latest methods of fractal image compression are five times faster than the older ones. The paper begins with basic problems and then move it towards speed factor. The Fast Fractal Image Compression algorithm is a critical progress in this direction. Some aspects are deserving of further examinations. Alternative partitioning structures, especially HV partitioning, should be contrasted with the full quadtree deterioration utilized as a part of this work. Second, the domain pool filters might be refined, or in view of some other amount than block. Third, there might justify in planning a hybrid algorithm by combining FFIC with a local spiral search. Also, fourth, the augmentation to bilinear fractal transforms surely appears to be beneficial.

Jaseela C C and *Ajay James* in their research paper entitled "*A New Approach to Fractal Image Compression Using DBSCAN*" suggested using DBSCAN algorithm to pack a picture with fractal compression strategy. The change is connected to decline encoding time by decreasing the successive pursuits through the entire picture to its neighbors. This strategy packs and decompresses the shading pictures rapidly. The execution time of the compression algorithm is diminished essentially contrasted with the customary fractal picture compression. In web picture database is exceptionally large. So putting away pictures in less space is a test. Picture compression gives a potential cost investment funds connected with sending less information over exchanged phone system where the expense of call is truly generally based upon its duration. It diminishes capacity necessities as well as general execution time. The proposed strategy pack pictures in square by piece premise, rather than checking in pixel by pixel. This strategy packs the shading pictures and interprets them rapidly by considering their RGB values independently. This strategy is exceptionally helpful for putting away pictures in picture.

Ali Nodehi, Ghazali Sulong, Mznaah al rodhaan, Abdullah-Al-dheelan, Amjad Rehmaan and Tanjila Saba in their research paper entitled "*Intelligent fuzzy approach for fractal image compression*" suggested a two-phase algorithm to perform fractal picture compression which decreases the MSE computations. In the first phase, all picture pieces were partitioned into three classes as indicated by picture squares edge property utilizing DCT coefficients. From the spatial domain, a square of picture could be changed to the recurrence domain by method for DCT transformation. Within the recurrence domain, the DCT coefficients that are arranged in the upper left of the picture piece implies the picture piece's low recurrence data and its harsh shape whereas the DCT that is arranged in the lower right of the picture piece means the picture piece's high recurrence data and its fine texture. Therefore we can rexplore the class of picture square by thinking of its lower-higher recurrence DCT coefficients. In the second stage, the ICA algorithm found the reasonable domain pieces utilizing the outcome acquired as a part of the principal phase. It is discernible that the measure of MSE calculations in the boolean picture has been 502 times more than that of the fundamental FIC algorithm while the PSNR worth is only 0.41 not as much as that of essential FIC algorithm. ICA additionally works the idea of least optima in the entire locale.

Yuanyuan Sun, Rudan Xu, Lina Chen and Xiaopeng Hu in their research paper entitled "*Image compression and encryption scheme using fractal dictionary and Julia set*" recommended a novel compression—encryption plan utilizing a fractal word reference and Julia set. For the compression in this plan, fractal word reference encoding decreases time utilization, as well as gives great quality picture reproduction. For the encryption in the plan, the key has extensive key space and high affectability, even to little annoyance. In addition, the stream figure encryption and the dispersion procedure received in this study spread bother in the plaintext, accomplishing high plain affectability and giving a successful imperviousness to picked plaintext assaults. In the encryption procedure, the key has almost 2272 key space with a 10×10 Julia set. As examined in Sect. 4.3, the encryption framework has a high plaintext affectability and key affectability to a minor bother. Consequently, we completed a point by point investigation of the Julia set size, and the exploratory results demonstrate that when K is equivalent to 8 or 10, it has a high affectability for figure to both key and plaintext. Also, the ciphertext breezed through the sp 800–22 test suite, demonstrating that the ciphertext has a decent arbitrariness. At last, we tried the encryption/decoding time utilization of various pictures. The test results show that the encryption operation is <15% of compression procedure, implying that it does not postpone the compression procedure and this makes it simple to acknowledge continuous compression and encryption.

Sarabjeet Kaur and Er. Anand Kumar Mittal in his paper entitled "*Improved Fractal Based Image Compression for Grayscale Using Combined Shear and Skew*

Transformations" proposed that the force of fractal encoding is appeared by its capacity to outflank utilizing the DCT, which shapes the premise of picture compression. The shear and skew-based fractal picture compression is another algorithm yet is not without issues. Most fundamentally, quick encoding is required for it to discover wide use in mixed media applications. The outcomes procured demonstrate that the proposed approach uses between pixel redundancies to render fabulous de-relationship for characteristic pictures. The higher measure of the applicable pixels is related to the expansion of the shear and skews changes. These changes permit to frame lesser number of the reach obstructs in the montage thus the compression result is higher. In this way, all the uncorrelated change coefficients can be encoded autonomously without trading off coding proficiency. Additionally, a portion of the high recurrence substance can be disposed of without noteworthy quality debasement. Fractal picture compression gives speedier compression in dark scale when contrasted with RGB because of single plane multifaceted nature when contrasted with the three-plane many-sided quality in the shading picture.

All these considerations include uncertain trade-offs between quality and speed with complexity factor associated with the used algorithm.

3 Results

S no	Paper name	Authors	Findings	Research gap
1.	*A survey of parallel algorithms for fractal image compression*	*Dan Liu, Peter K Jimack*	Analyze the performance of various algorithms working on parallel approach of fractal image compression	A procedure to reduce the encoding cost of fractal compression is required
2.	*A study on fractal image compression using soft computing techniques*	*D. Sophin Seeli, Dr. M. K. Jeyakumar*	Analyze the different fractal compression techniques based on their compression ratio, encoding time and PSNR (Peak Signal-to-Noise Ratio)	An algorithm or mechanism is required to gain high compression ratio
3.	*Fractal image compression and its application in image processing*	*Gaurish Joshi*	Proposed two methods RDPS and ERB to make changes in the encoding time of fractal image compression	A procedure associated with other features like skewness, feature contrast should be suggested
4.	*Advances in fractal compression for multimedia applications*	*John Kominek*	Fast encoding method of fractal image compression are suggested	Concepts related to HV partitioning, augmentation to bilinear fractal transform should be implemented

S no	Paper name	Authors	Findings	Research gap
5.	*A new approach to fractal image compression using DBSCAN*	*Jaseela C C and Ajay James*	DBSCAN can pack the shading pictures and interpret them rapidly by considering their RGB values independently	It requires an efficient algorithm to separate R, G, B values
6.	*Intelligent fuzzy approach for fractal image compression*	*Ali Nodehi, Ghazali Sulong, Mznaah al rodhaan, Abdullah-Al-dheelan, Amjad Rehmaan and Tanjila Saba*	A two-phase algorithm to perform fractal image compression which decreases the MSE computations using DCT and ICA algorithm	A procedure for optimality of ICA is required
7.	*Image compression and encryption scheme using fractal dictionary and Julia set*	*Yuanyuan Sun, Rudan Xu, Lina Chen and Xiaopeng Hu*	It suggested a novel compression-encryption plan utilizing a fractal word reference and Julia set	A deterministic procedure for Julia set is required
8.	*Improved fractal based image compression for grayscale using combined shear and skew transformations*	*Sarabjeet Kaur and Er. Anand Kumar Mittal*	The force of fractal encoding is appeared by its capacity to outflank utilizing the DCT, which shapes the premise of picture compression having skewness	A mechanism for Identification of skewness level is required

4 Conclusion

After analyzing the outcomes of different methods suggested in various research papers, we conclude that the primary thought of DBSCAN is that, for each protest of a bunch the area of a given sweep; Eps must contain no less than a base number of focuses, min pts to pack the fractal of a picture while ICA works in two stages. In the first stage, it segments the picture in view of DCT coefficients and in the second stage, it performs imperialistic operations on DCT squares to diminish the fractal estimate. So the proposed work is to make the half-breed structure utilizing these two calculations, applying alteration on the same and to make another approach that will build the pressure proportion, diminish the mistake rate, and in the long run enhance the fractal image compression ratio comes about on an image.

References

1. Hitashi, kaur G., Sungandha S.: Fractal image compression—a review. Int. J. Adv. Res. Comput. Sci. Softw. Eng. (2012)
2. Sun, Y., Xu, R., Chen, L., Hu, X.: Image compression and encryption scheme using fractal dictionary and Julia set. IET Image Process. 9(3), 173–183 (2015)
3. Sarabjeet, K., Anand Kumar, M.: Improved fractal based image compression for grayscale using combined shear and skew transformations. GJRA Global J. Res. Anal. 156–160 (2015)
4. Nadira Banu Kamal, A.R.: Iteration free fractal image compression for color images using vector quantization, genetic algorithm and simulated annealing. TOJSAT Online J. Sci. Technol. 5(1), 39–48 (2015)
5. Nodehi, A., Sulong, G., Al rodhaan, M., Abdullah-Al-dheelan, Rehmaan, A., Saba, T.: Intelligent fuzzy approach for fractal image compression; EURASIP-J. Adv. Sig. Process. 1–9 (Springer) (2014)
6. Jaseela, C.C., James, A.: A new approach to fractal image compression using DBSCAN. Int. J. Electr. Energy 2(1), 18–22 (2014)
7. Nadira Banu Kamal, A.R., Priyanga, P.: ICTACT J. Image Video Process. 4(3), 785–790 (2014)
8. Michael Vanitha, S., Kuppusamy, K.: Survey on fractal image compression. Int. J. Comput. Trends Technol. (IJCTT), 4(5), 1462–1464 (2013)
9. Sophin Seeli, D., Jeyakumar, M.K.: A study on fractal image compression using soft computing techniques. IJCSI Int. J. Comput. Sci. Iss. 9(6), 420–430, no 2 (2012). ISSN 1694-0814
10. Negi. A., Chauhan. Y.S, Rana, R.: Complex dynamics of ishikawa iterates for non integer values. Int. J. Comput. Appl. 9(2), 0975–8887 (2010)
11. Negi, A., Rani, M., Mahanti, P.K.: Computer simulation of the behavior of Julia sets using switching processes. Chaos Solitons Fractals **37**, 1187–1192 (2008)
12. Negi, A.: Fractal generations and applications. Ph.D. thesis, Department of Mathematics, Gurukula Kangri Vishwavidyalaya, Hardwar (2006)
13. Berinde, V.: Iterative Approximation of Fixed Points. Editura Efemeride, Baia Mare
14. Kigami, J.: Analysis on Fractals. Cambridge University Press, Cambridge (2001)
15. Barnsley, M.: Fractals Everywhere. Academic Press, Inc., Boston, MA, pp. xii+396 (1988). ISBN 0-12-079062-9 MR0977274 (90e:58080)
16. Barnsley, M.F., Devaney, R.L., Mandelbrot, B.B., Peitgen, H.-O., Saupe, D., Voss, R.F.: The science of fractal images. With contributions by Yuval Fisher and Michael McGuire, pp. xiv+312. Springer, New York (1988). ISBN: 0-387-96608-0 MR0952853(92a:68145)
17. Peitgen, H.-O., Richter, P.H.: The Beauty of Fractals. Springer, Berlin (1986)
18. Huang, Z.: Mann and Ishikawa iterations with errors for asymptotically non expansive mappings. Comput. Math. Appl. 37(3), 1–7 (1999). MR1674407(2000a:47118)
19. Hutchinson, J.E.: Fractals and self-similarity. Indiana Univ. J. Math. **30**, 713–743 (1981)
20. Falconer, K.J.: Fractal Geometry: Mathematical Foumiations and Applications. Wiley, Chichester (1990)
21. Bamsley, M.F., Hurd, L.P.: Fractal Image Compression. AK Peters Ltd., Wellesley, MA (1993)
22. Mandelbrot, B.B., Cannon, J.W.: Reviews: the fractal geometry of nature. Am. Math. Monthly 91(9), 594–598 (1984). MR1540536
23. Gonzalez, R., Eugene, R.: Digital Image Processing, p. 466 (2008)
24. Thyagarajan, S.K.: Still Image and Video Compression with MATLAB, p. 97100 (2007)
25. Jitendra, R., Tarun Kumar, S., Ajith, A., Vaishali: Glossary of metaheuristic algorithms. Comput. Inf. Syst. Ind. Manag. Appl. 9, 181–205 (2017)
26. Manish, J., Ambuj, A.K.: Analysis of different fractal image compression techniques. In: Proceedings of SMART-2017
27. Agarwal, A.K., Sharma, T., Saxena, A., Ather, D.: Search based software engineering in requisite phase of SDLC: a survey. Tech. J. LBSIMDS, 95–101. ISSN -09752374
28. Agarwal, T., Agarwal, A.K., Singh, S.K.: Cloud computing security: issues and challenges. In: Proceedings of SMART-2014, pp.10–14

Web Services Classification Across Cloud-Based Applications

M. Swami Das, A. Govardhan and D. Vijaya Lakshmi

Abstract Cloud computing uses service-oriented architecture principles to design a web service which enables fast, high-performance software application services, and infrastructural services (for example, servers, networks, middleware, etc.). Cloud computing provides scalable and on-demand storage, middleware, and application as a service. To achieve high availability of cloud computing services such as software, platform, and infrastructural services, it must be scalable and extensible. Web services can be accessed via Internet, and its performance (response time) gets reduced as the network traffic and congestion increase. But cloud users prefer to access the cloud servers with high availability with low response time, while it chooses the best server among the many available. To improve the system performance with respect to a specific quality of service parameter. We proposed a model that classifies the cloud-based web applications into four categorical values. The web services enable to use shared resources. This paper explains how to choose quality parameters to design a web service, which employs QWS dataset with nine quality parameters and 2507 records and data mining techniques such data envelopment analysis, K-nearest neighbor, decision tree, fuzzy multi-attribute decision-making analysis, PNN, and BPNN classifier models. Experimental results concluded that the proposed method FMADM has better performance 91.78% than the existing methods. In future, we can extend this model to design a cloud service based on mixed QoS parameters.

Keywords Quality of service · PNN · Cloud computing · Web service classification · FMADM

M. Swami Das (✉)
Department of CSE, MREC, Hyderabad, TS, India
e-mail: msdas.520@gmail.com

A. Govardhan
JNTUH College of Engineering Hyderabad, Hyderabad, TS, India
e-mail: govardhan_cse@yahoo.co.in

D. Vijaya Lakshmi
Department of IT, MGIT, Hyderabad, TS, India
e-mail: vijayadoddapaneni@yahoo.com

K. Ray et al. (eds.), *Soft Computing: Theories and Applications*,
Advances in Intelligent Systems and Computing 742,
https://doi.org/10.1007/978-981-13-0589-4_23

1 Introduction

Web services are software applications using which a client (web user) can access services using networked communication from remote servers using remote procedure calls. Rapid development of web applications includes business to business, E-commerce, and online applications. Enterprise applications are used to provide Quality of Service (QoS), the quality parameter plays a significant role to choose best web service applications [1]. Web applications development uses Service-Oriented Architecture (SOA), and these can be connected with upon cloud-based services.

Cloud computing is a model that makes use of SOA principles, fast Wide Area Network (WAN), high performance, virtualization, bandwidth, software, and infrastructure facilities. Cloud computing service provides on-demand network access, computing resources (for example, servers, services, network, storage, middleware, and others) using Internet availability with minimum management effort [2]. Cloud computing exposes resources across the network in a shared pool. It allows visualizing, configuring, and managing the resources effectively.

Scalability is to extend the resources on demand, service provider to add or delete computing resources (memory, storage, CPU, and network), and pay-per-use the cloud services. Cloud server with web applications is ability to access multiple customers to allow the servers in data centers in isolated manner. The cloud computing is a solution that brings the cloud users cost-effective solution with minimum maintenance and management, service as pay and use of resources [3]. Internet of Things using Smart connected a device which uses sensors to collect the data and communicated to the cloud data server and Internet.

The Government of India is planning to establish 98 smart cities across the country. It includes infrastructural facilities provided to 24 capital cities, 18 cultural, five port cities, and three Educational and Health hubs. The city is an urban region with high advanced infrastructural facilities, real estate, communication, and market viability. Technology platforms and automated sensor networks are using data centers [4]. Worldwide Internet of Things (IoT) industry is estimated at $300 billion and number of communicated electronic devices around 27 billion users by 2020. The users will utilize electronic devices to access web services including household supplies, banking, web apps, cloud services, and other applications every day operated using sensor connected with Internet communication [5]. Billion numbers of IoT connected devices play a significant role in data collection, data sharing, web services, and cloud services. Cloud offers everything as a service business model for IoT and big data. Cloud computing implementation uses service-oriented architecture. XML, UDDI, WSDL, and some of the web technologies used in web services. Service provider (web server), service requester (client), and UDDI are most essential parts of service-oriented architecture [6, 7].

Cloud computing requires high availability, security, minimum response time with more reliable sources of information of data (cloud servers) [8]. The cloud server required to be low cost, with pay and use services with high availability, scalability

IT, and development resources. The best practice is to create realistic goal of cloud server such as

1. Keep minimum cost.
2. Easy to consume, economical IT infrastructural facilities.
3. Software as a service: pay as you go model, integration solution.
4. Integration application minimizing development, implementation, maintenance of resources.
5. Security for cloud computing to be increased.
6. Ensure high performance and high availability of the data with minimum response time.
7. Platform as service, connectivity to standard enterprise applications, legacy systems, web services, and databases with maximum connectivity points [4, 9].

QoS dataset has nine attributes with 2507 records; these can be classified into Class-A (high quality represents platinum), Class-B (gold), Class-C (Silver), and Class-D (poor quality represents bronze) [10, 11] using various classification techniques.

The remainder of the paper is organized as follows: Sect. 2 gives the overview of literature, Sect. 3 gives the classification using data mining techniques, Sect. 4 gives the results and discussions, and finally, Sect. 5 concludes the paper.

2 Literature Survey

In 1991, Tim Berners Lee Mark Weiser, British computer scientist and software engineer, created first website editor/browser. The development of Internet and web applications in 1995, Internet goes commercial Amazon, Ebay, e-Commerce Company providing business to business, consumer to consumer and consumer to sales and services using Internet web applications. The Software development organizations and consumers demand to use the quality parameters to design and develop web, cloud-based applications. The future challenges of web services are to deal with minimum response time, high availability, high throughput, and heterogeneity, more security features.

Cloud computing is on-demand computing business expansion, IT infrastructure, maintenance, management of cost-effective commercially available service model, with virtual infrastructure resources. This provides shared resource processes and data to computers and other devices on demand. Cloud computing model is shown in Table 1; it has deployment models, service attributes, and service layers such as Platform as a Service (PaaS), Infrastructure as a Service (IaaS), and Software as a Service (SaaS). IaaS is the lowest layer including operating system, services, network, storage, switch, routers, and specific workload, measured in CPU cycles and network compounds. Example firewalls, load balancer. Ubiquitous computing forces the computer to live world with people, research areas distributed computing, mobile computing, and artificial intelligence, soft computing, and IoT.

Table 1 Cloud computing models

Deployment models	Community	Hybrid
	Private	Public
Service models	IaaS	PaaS
	SaaS	
Service attributes	Resource pooling	Measured service
	On-demand service	Scalability

Cloud computing provides scalability, availability, low response time, reliability, and interoperability. Problems and challenges of cloud computing are security, loss of control (ownership), intermediately (transactions third parties on Internet), data storage, vendor relation issues, integration and service levels, high availability and fast response time, etc. Web servers can be connected to public Internet services. As traffic increases, congestion increases, congestion increases as a result in low response time. Redirection is the process, selecting the best server on user request, web server can redirect the browser to another server instance [12, 13].

Cloud computing IT-related resources such as computing systems, storage devices, network system security, cloud applications, platform as service. Services for mobiles, laptops, PC using Internet with cost-effective solutions.

In future, schools, colleges, universities, and research and development are going to access cloud resources at affordable cost. Use of IT infrastructure consume their business grow, business resources, multimedia content providers in distributed various consumers for a lower price 3G, 4G, 5G network providers with high-speed Internet facilities [14]. The cloud deployment models are classified as public, private, and hybrid clouds.

Public cloud: Unrestricted potential users designed for group enterprise pay as use manner. The disadvantages are availability, latency, security, privacy, and regularity problem.

Private cloud: The user centric to access to software service fully configured, production-ready software, self-deployment model. The goal of the private cloud is to provide scalable, available services without staff support can create images, packing services. It includes virtual services, run corporate firewalls, organization or third party manages infrastructure trusted consumer service, security, and high priority.

Hybrid cloud: It uses the combination of extended public cloud components and internal resources in a coordinated fashion. This permit choice to consulate virtualizes automate internal IT resource significant integration and coordination between internal and external environments.

Internet of Things will use Radio Frequency IDentification (RFID) for tracking the objects, with sensors, gathering the information, storing, and processing the information in various applications.

India has a huge mobile market enabling people to connect to billions of devices invoking web services in the future. For example, people use IoT with data, pro-

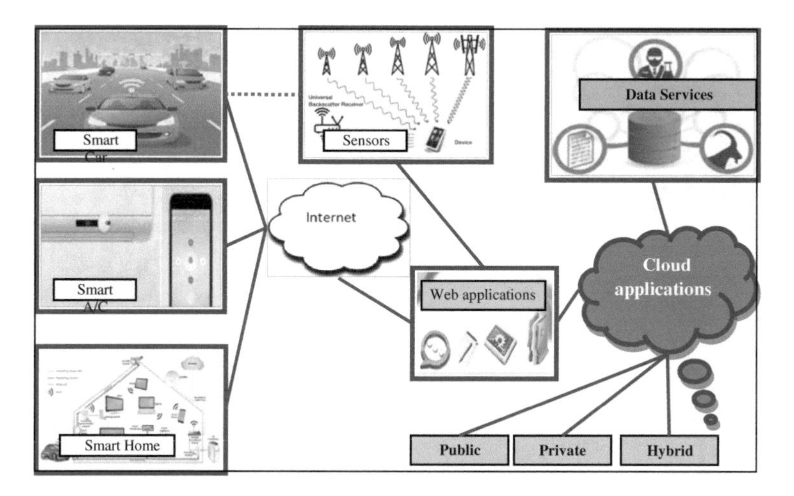

Fig. 1 IoT connected devices using sensors, transmission, process, and data services

cess, and resources in various domains like health, performance monitors, connected vehicles, smart grids, and other services. The large-scale deployment of IoT systems and IT security architecture requires new security approach and innovative solutions about information exchange over Internet spreads in various domains.

The applications of IoT include passenger security in railways, roadways, safety, and security with route optimization and critical sensing device operations. Integrating application such as traffic signals, reduced congestion, improve emergency services response time, parking, and lighting power thus saves the cost saving, city services including efficient service delivery, environment monitoring, and connecting to cars. Challenging task is security of IoT, connecting any device to any other device with security, confidentiality, integrity, and high availability with data protection to various attacks [15].

Figure 1 shows the various IoT connected devices using sensors and communication networks through sensing, transmission, processing, and data services.

Sensing is an electronic device to receive the input. Transmission is the communication channel that uses the technologies like Internet, Wi-Fi, heterogeneous IoT networking, etc. Processing information in web server or cloud computing using SOA architecture to access the functionalities are data and services. Technology supports include integrated chips, information security, processing of web servers (nodes), and information analysis.

Major tasks for IoT are summarized in the following:

1. Building frameworks: Designing frameworks for IoT, web, and cloud application.
2. Techniques: IoT standard framework, networking, services, and information security.
3. Formulate industry application standard: Implement the framework techniques with application programs, web services, and cloud applications to give solutions

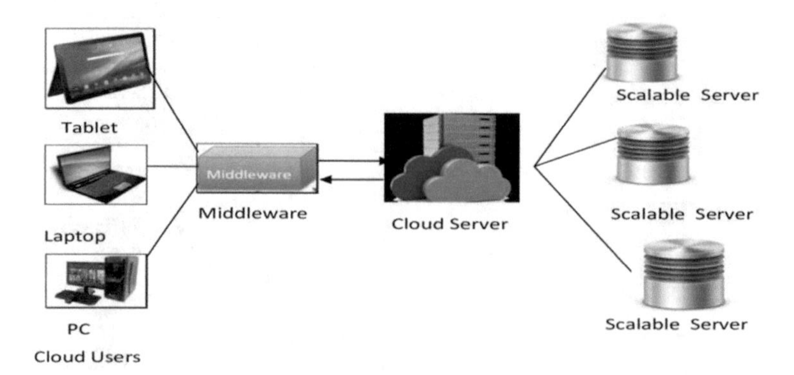

Fig. 2 Cloud computing resources access from best server

to problems. Cloud computing user resources can be accessed by selecting best server using classification techniques is shown in Fig. 2. Example of IoT applications includes smart washing machine, glucose monitoring, smart sprinkler control, smart home security, smart lighting, smart A/C, blood pressure monitor, smart cardio, smart door lock, etc.

Cloud Service Set: The set of all services (for example, cloud providers cloud computing technology Microsoft Azure and other cloud services).

Cloud Services $Cs = \{S_1, S_2,\ldots,S_i,\ldots S_n\}$ where $S_1, S_2,\ldots S_n$ are cloud services at various locations.

Cloud Location: C_{loc} is the location of the cloud in distance.

User Service Set: The set of all services (for example, language translators, image editors, etc.)

$U_s = \{U_{1s}, U_{2s},\ldots U_i \ldots U_{ns}\}$ where $U_i \in U_s$

Algorithm: 1 To Access Cloud best service among web servers

Input: User service U_i among Cloud services $Cs = \{S_1, S_2,\ldots, S_i,\ldots S_n\}$ in C_{loc} locations.

Output: User service Cs cloud resource with cloud location with best service.

1. Initialize the variables.
2. Read the Cloud user requests.
3. Choose the Cloud server with shortest distance.
4. Process Load balancing among replicated cloud servers.
5. Web server or cloud server has lookup index for frequently access items or data.
6. Access and assign the cloud server has best server (after web services classifications).

Information can be accessed from best cloud server to clients to improve performance [13]. Figure 3 describes the quality of web service parameters required to provide web service classifications to improve the web services in cloud-based

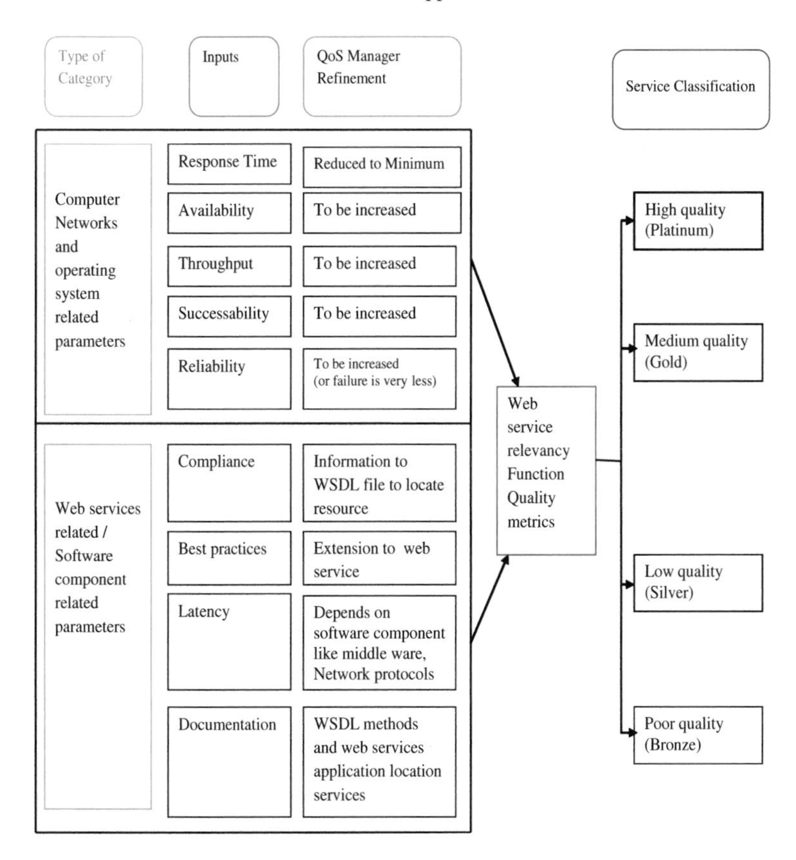

Fig. 3 QoS web services performance parameters and service classification

applications. Further, Algorithm 1 describes accessing procedures for the best cloud web service using various data mining techniques.

3 Classification Using Data Mining Techniques

Classification is the data mining technique widely used in various applications. Authors have considered QWS dataset which contains nine attributes and 2507 records. The proposed model is shown in Algorithm 2.

Algorithm: 2 Web Service Classifications
Input: QWS dataset with quality of parameters.
Output: Web service classification Class-1, Class-2, Class-3 and Class-4.

1. Initialize the variables.
2. Read the data matrix.

3. Preprocessing for given data matrix (if required, for example, dimensionality reduction or normalization, etc.).
4. Convert the given input into normalized data.
5. Read the Training dataset (Labeled dataset).
6. Read the testing dataset (Unlabeled dataset).
7. Apply the Classifier method according to the feature selection.
8. Display the Classifier label for testing dataset.

Data Preprocessing: In general, real-world dataset is incomplete, noisy, and inconsistent. By using preprocessing techniques, the data is transformed into standard; consistent is converted and complete format. The web service dataset during preprocessing WSDL documents, and UDDI services, or any services need and information into standard format, for example, the dataset to be preprocessed into dimensionality reduction.

The various Classification methods adopted in the proposed method are (1) data envelopment analysis, (2) K-nearest neighbor, (3) decision tree, (4) fuzzy multi-attribute decision-making, (5) probabilistic neural network, and (6) back propagation and neural network techniques.

3.1 Data Envelopment Analysis (DEA)

The Performance and the efficiency of multicriteria decision analysis of organization systems measured using data envelopment analysis which uses a linear, mathematical programming model. It also allows you to classify the decisions into smaller number of category. DEA uses mathematical programming to analyze the different Decision-Making Units (DMUs) independently.

DMUs help to measure the performance of organizational units by considering increase output or decrease input. The efficiency is considered until the convex hull of the DMU is reached [16, 17]. The efficiency of DEA is calculated by ratio of total outputs by inputs units represented by Eq. 1.

$$\text{Efficiency} = \text{Output/Input} \tag{1}$$

3.2 k–Nearest Neighbor Algorithm

The k-nearest neighbor method is used to classify the web service data. Figure 4 shows the selection of the dataset, feature selection using the proposed method, and measured the results into categories. The technique is used to classify the dataset into unknown instances by relationships or distance functions. Manhattan, Euclidean, and Minkowski distance functions are used to find the nearest neighbor. A point in "n" dimensional space for "k" tuples represents each row of the dataset. These tuples are

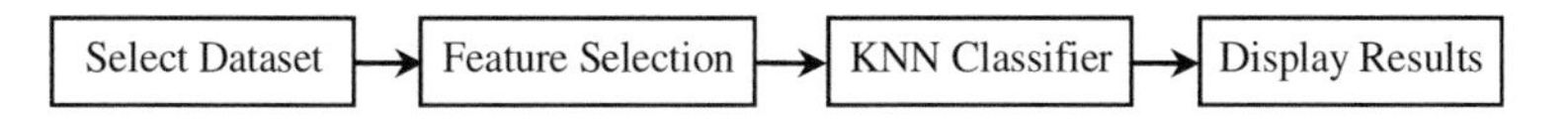

Fig. 4 KNN classifier for QWS data

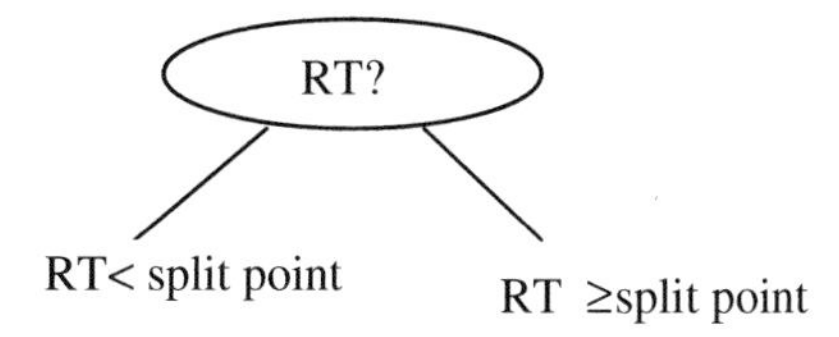

Fig. 5 Partition scenario

closed to unknown tuples. The k-nearest tuples are unknown tuples [6, 18, 19] to be determined by known tuples [11], and the results are shown in Table 4.

3.3 *Decision Tree*

The decision tree can be defined as a function that has decision rules with conditions is shown in Fig. 5. The expression can be represented as the decision tree where the leaf node represents the solutions and nonleaf node represents the conditions in the conjunction in if clause can be defined by set of rules [20, 21]. Example If<Exprssion1> and <Expression2> and <Expression3> the solution.

3.4 *Fuzzy Multi-Attribute Decision-Making (FMADM)*

The rank of the web service is found by using FMADM Analysis by considering a set of inputs related to quality. A finite set of inputs let us (X), where input X_1, X_2,...X_n are taken into consideration by MADM to produce a rank or goals. Let us assume (G), where G_i is G_1, G_2,...G_n. The proposed web services classifications have two-fold task: (i) perform classification and (ii) locate service description. In each and every decision with respect to all goals, criteria of web service case. Decision alternatives are used to rank according to aggregated values of justifications. FMADM Model has the following three stages: (i) compute the weights, (ii) find fuzzy relationships, (iii) apply data membership values for evolution criteria. Product and compensatory operators also used in Eqs. 2 and 3.

The product operator is represented as

$$\dddot{\mu}'_{Ai(x)} = \prod_{i=1}^{m} \mu i(x), \text{x} \in X \tag{2}$$

Compensatory and operator is defined as follows:

$$\dddot{\mu}'_{Ai(x)} = \left(\prod_{i=1}^{m} \mu_X\right)^{1-\gamma} \left(1 - \prod_{i=1}^{m} (1 - \mu_X)\right)^{1-\gamma} \quad \mathrm{x} \in \mathrm{X}, 0 \le \gamma \le 1 \tag{3}$$

In FMADM model, a domain expert or data-driven methodology is used to define the required membership functions of attributes with minimum, maximum, and average values of QWS data in Eqs. 4–12. The results are shown in Tables 2, 3 and 4.

3.5 *Probabilistic Neural Network (PNN)*

PNN is a classifier which maps the input model into number of classifications. PNN can be implemented in numeral algorithm with analysis operations ordered into four layers which are, (1) input, (2) pattern, (3) hidden or summation layer, and (4) output layer. The target category class label is identified in the method illustrated in [18, 21, 22].

PNN classifier uses nine attributes of QWS dataset into input layers of nine input variable dataset. Input layer, accepts the input, pattern layer stores the pattern or class label, and finally reaches to output layer predicting the class labels.

$$\text{Response time } \mu_{RT} = \begin{cases} 1 & \text{if RT} < 380 \\ \frac{4990-RT}{4990-380} & \text{if RT } \le 4990 \\ 0 & \text{if RT} > 4990 \end{cases} \tag{4}$$

$$\text{Availability } \mu_A = \begin{cases} \text{A/80 if A} < 80 \\ 1 \quad \text{if A } \ge 80 \end{cases} \tag{5}$$

$$\text{Throughput } \mu_{Th} = \begin{cases} \text{(Th - 26) /26} & \text{ifTh} < 26 \\ 1 & \text{if Th } \ge 26 \end{cases} \tag{6}$$

$$\text{Successability } \mu_S = \begin{cases} \text{(S) /80 if } \text{S} < 80 \\ 1 \quad \text{if } S \ge 80 \end{cases} \tag{7}$$

$$\text{Reliability } \mu R = \begin{cases} \text{R /80 if } \text{R} < 80 \\ 1 \quad \text{if } R \ge 80 \end{cases} \tag{8}$$

$$\text{Compliance } \mu c = \begin{cases} \text{C /30 if } \text{C} < 30 \\ 1 \quad \text{if } C \ge 30 \end{cases} \tag{9}$$

$$\text{Best practices } \mu bp = \begin{cases} \text{BP /50 if } \text{BP} < 50 \\ 1 \quad \text{if BP } \ge 50 \end{cases} \tag{10}$$

Table 2 FMADM analysis experimental results

WS	M_{RT}	M_A	M_{TH}	M_S	M_R	M_C	M_{BP}	M_L	M_D	Min	Product
S1	1	1	0.273077	1	0.9125	1	1	0.96632	0.355556	0.273077	0.085614
S2	0.977874	1	0.615385	1	0.9125	1	1	1	0.022222	0.022222	0.012203
S3	0.361952	1	0.053846	1	0.9125	1	1	1	1	0.053846	0.017784
S4	1	1	0.461538	1	0.8375	1	1	1	0.988889	0.461538	0.382244
S5	1	1	0.073077	1	0.9125	1	1	0.997963	1	0.073077	0.066547
S6	1	1	0.065385	1	0.8375	1	1	1	0.677778	0.065385	0.037115
S7	1	1	0.05	1	0.8375	1	1	1	0.044444	0.044444	0.001861
S8	1	0.95	0.107692	0.95	0.75	1	1	1	0.088889	0.088889	0.006479

μpow (w1)	μpow (w2)	μpow (w3)	μpow (w4)	μpow (w5)	μpow (w6)	μpow (w7)	μpow (w8)	μpow (w9)	Min	Product	1-Product	Ci	Class
1	1	0.8171	1	0.9818	1	1	0.9969	0.8712	0.8171	0.6968	1	0.834	2
0.9975	1	0.9272	1	0.9818	1	1	1	0.6019	0.6019	0.5466	1	0.739	3
0.8932	1	0.6347	1	0.9818	1	1	1	1	0.6347	0.5567	1	0.746	3
1	1	0.8866	1	0.9651	1	1	1	0.9985	0.8866	0.8545	1	0.924	1
1	1	0.6656	1	0.9818	1	1	0.9998	1	0.6656	0.6534	1	0.808	2
1	1	0.6542	1	0.9651	1	1	1	0.9494	0.6542	0.5995	1	0.774	3
1	1	0.6275	1	0.9651	1	1	1	0.6602	0.6275	0.3998	1	0.632	4

Table 3 FMADM analysis for QWS dataset classifications

Service classification	Compensatory operator (Ci)	Number of records classified
Class-A (High quality—Platinum)	If (Ci > 0.9) then Class = "1"	277
Class-B (Gold)	If (Ci >= 0.8 And Ci <= 0.9) then class = "2"	706
Class-C (Silver)	If (Ci >= 0.7 And Ci <= 0.8) then class = "3"	974
Class-D (Low quality—Bronze)	If (Ci < 0.7) then Class = "4"	550

Table 4 Classification accuracy

Classifier	Accuracy (%)
Fuzzy multi-attribute decision-making analysis (FMADA)	91.78
Data envelopment analysis (DEA)	90.02
K-nearest neighbor (KNN)	89.78
Decision tree (DT)	89.59
Probabilistic neural network (PNN)	91.5
Back propagation and neural network (BPNN)	90.12

$$\text{Latency}\quad \mu_L = \begin{cases} 1 & \text{if L} < 50 \\ \frac{4140-L}{4190} & \text{if } 50 \le L \le 4190 \\ 0 & \text{if L} > 4190 \end{cases} \tag{11}$$

$$\text{Documentation}\quad \mu D = \begin{cases} \text{D}/90 & \text{if} \quad \text{D} < 90 \\ 1 & \text{if D} \ge 90 \end{cases} \tag{12}$$

3.6 Back Propagation Neural Network (BPNN)

BPNN is a neural network learning technique which is widely used for classification and prediction. It accepts a set of inputs and emits results related to the different layers which are learning layer, pattern layer, and output layer.

Fig. 6 Classification of web services using KNN and other methods

4 Results and Discussions

Applications of Cloud computing are Google cloud, Microsoft cloud computing services, Amazon web services, and HP cloud services. Google cloud provides many services to cloud customers. Examples include Gmail, Google documents, Google analytics, Good Adworks, and Picasa. Gmail, Email with storage are Google docs, Services spreadsheets, documents, etc. Google analytics is used to monitor the traffic onto the web services. Picasa is a tool used for uploading images on cloud. Microsoft provides cloud services, for example, Windows Azure. Amazon Web Service (AWS) provides all business sizes, web services, storage, IT infrastructure, global network, database functions, and Amazon elastic map reduce service. HP cloud service provides services through workload, data storage, Networks, relational database, and user security. Salesforce cloud service provides solutions as cloud service and sales cloud, CRM, sales customizing tools and services [13]. Cloud computing applications that use SOA architectures to access resources in the cloud environment, to access cloud resources with a minimum response time with quality attributes. To choose the best cloud server in web services using service classification algorithms, for example, QWS Dataset [11] analyzed and the results shown in Table 4. The framework is implemented in Java/J2EE methods KNN, decision trees, FMADMA work carried out by Fuzzy relationships, memberships, product, and compensatory operators. DEA, PNN, and BPNN using tools [17, 23–25] carried out and results are depicted in Figs. 6, 7, and 8, and also in Tables 2, 3, and 4, respectively. FMADM has better accuracy of 91.78%, when compared to other contemporary methods.

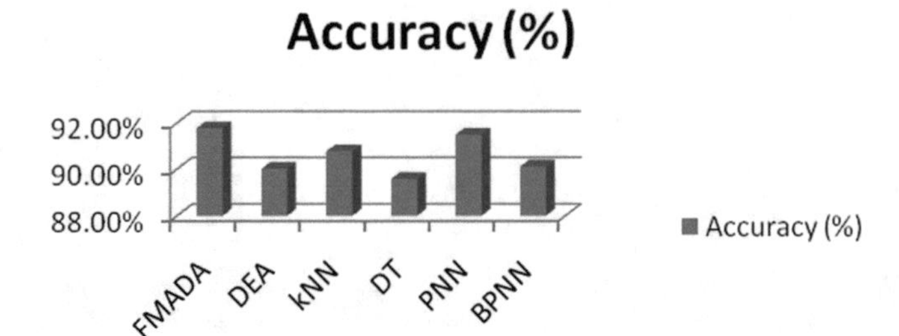

Fig. 7 Web services classification performance

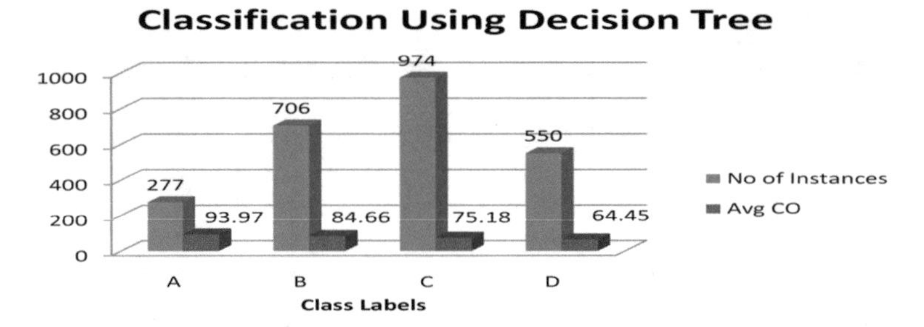

Fig. 8 Classification of web services using WEKA

5 Conclusion

Cloud computing in India has highest adaption rate for mobile users. Cloud IT-related resources computing, storage, network, security, and platform as service. The objective is to provide scalable, high availability of web service with security, high performance, and maximizes connectivity points. Cloud computing uses SOA as a service. Selection of best web service for functional and nonfunctional parameters plays a significant role to choose best service among the group of services. QoS web services are important factor to differentiate cloud and web service providers. In this paper, applications of web services over different classifications were discussed, and their performance is depicted in Table 4. We specifically applied high-quality data mining techniques using KNN, DEA, decision tree, FMADM, PNN, and BPNN among different classifier models [11]. Among different classifiers, FMADM method shows high-performance results. We found that response time, reliability, successability, availability, and throughput are the most important parameters. Different classifiers are described in Fig. 3 and are shown in Table 4. In future, we can extend this model to have the best cloud service based on mixed quality of parameter, for example, high availability and security, high response time and extensibility, etc.

Acknowledgements Author would like to personally thank Dr. Eyhab Al-Masri, Assistant Professor, University of Washington for providing the QWS dataset, and also to Dr. Ramakanta Mohanty, Professor, Department of IT, KMIT, Hyderabad for his timely suggestions to carry out this work.

References

1. Swami Das, M., Govardhan, A., Vijaya lakshmi, D.: QoS of web services architecture. In: The International Conference on Engineering & MIS 2015 (ICEMIS '15), vol. 66, pp. 1–8. ACM (2015)
2. Papazoglou, M.P.: Web Services & SOA Principles and Technology, 2nd edn. Pearson Publications (2012)
3. http://www.thehindu.com/todays-paper/tp-features/tp-opportunities/how-governments-can-benefit-from-cloud-computing/article4552817.ece (2003)
4. http://www.financialexpress.com/opinion/cloud-computing-for-firms-here-is-why-it-is-a-challenge-to-harness-its-potential/808544/ (2017)
5. https://esj.com/Articles/2009/08/18/Cloud-Best-Practices.aspx (2009)
6. Michael Raj, T.F., Siva Pragasam, P., Bala Krishnan R., Lalithambal, G., Ragasubha, S.: QoS based classification using K-Nearest Neighbor algorithm for effective web service selection. In: IEEE International Conference on Electrical, Computer and Communication Technologies (ICECCT), Coimbatore, pp. 1–4 (2015)
7. Pratapsingh, R., Pattanaik, K.K.: An approach to composite QoS parameter based web service selection. In: 4th International Conference on Ambient Systems, Networks and Technologies, pp. 470–477. Elsevier Publications (2013)
8. Swamidas, M., Govardhan, A., Vijayalakshmi, D.: QoS web service security dynamic intruder detection system for HTTP SSL services. Int. J. Comput. Sci. Inf. Secur. (IJCSIS) 14(S1), 1–5 (2016)
9. https://www.itu.int/en/ITU-T
10. http://en.wikipedia.org/wiki/Quality_of_service#Application
11. www.uoguelph.ca/~qmahmoud/qws/dataset
12. http://www.nishithdesai.com/fileadmin/user_upload/pdfs/Cloud_Computing.pdf
13. Kaur, S., Kaur, K., Singh, D.: A framework for hosting web services in cloud computing environment with high availability. In: IEEE International Conference on Engineering Education: Innovative Practices and Future Trends (AICERA), Kottayam, pp. 1–6 (2012)
14. Youssef, A.E.: Exploring cloud computing services and applications. J. Emerg. Trends Comput. Inf. Sci. **3**(6), 838–847 (2012)
15. http://cloudcomputing.syscon.com/node/1764445
16. Ramanathan, R.: An Introduction to Data Envelopment Analysis, pp. 22–44. Sage Publications, New Delhi, India (2003)
17. http://www.deafrontier.net/deasoftware.html
18. Han, J.: In: Kamber, M. (ed.) Data Mining Concepts and Techniques, 2nd edn., pp. 286–347. Elsevier Publications (2006)
19. Das, M.S., Govardhan, A., Lakshmi, D.V.: An approach for improving performance of web services and cloud based applications. In: International Conference on Engineeing & MIS (ICEMIS), Agadir, pp. 1–7 (2016)
20. http://en.wikipedia.org/wiki/Decision_tree
21. Das, M.S., Govardhan, A., Lakshmi, D.V.: A classification approach for web and cloud based applications. In: International Conference on Engineering & MIS (ICEMIS), Agadir, pp. 1–7 (2016)
22. Shreepad, S., Sawant, P., Topannavar, S.: Introduction to probabilistic neural network–used for image classifications. Int. J. Adv. Res. Comput. Sci. Softw. Eng. 279–283 (2015)
23. Neuroshell2 tool. http://www.inf.kiew.ua/gmdh-home
24. http://www.cs.waikato.ac.nz/WekaTool
25. https://www.ibm.com/software/analytics/spss/products/.../downloads.html
26. Mohanty, R., Ravi, V., Patra, M.R.: Applications of fuzzy multi attribute decision making analysis to rank web services. In: IEEE Conference CISM, pp. 398–403 (2010)
27. Mohanty, R., Ravi, V., Patra, M.R.: Web service classifications using intelligent techniques. Expert Syst. Appl. Int. J. Elsevier, 5484–5490 (2010). https://doi.org/10.1016/j.eswa.2010.02.063

28. AL-Masri, E., Mahmoud, Q.H.: Investing web services on the world wide web. In: 17th International ACM Conference on World wide web, Beijing, pp. 795–804 (2008)
29. http://www.ise.bgu.ac.il/faculty/liorr/hbchap9.pdf
30. Kusy, M., Kluska, J.: Probabilistic neural network structure reduction for medical data classification. In: Lecture Notes in Computer Science, vol. 7894, pp. 118–129 (2013)
31. http://www.financialexpress.com/industry/transforming-hr-with-cloud-computing/895835/ (2017)

Performance Evaluation and Analysis of Advanced Symmetric Key Cryptographic Algorithms for Cloud Computing Security

Abdul Raoof Wani, Q. P. Rana and Nitin Pandey

Abstract Cloud computing environment is adopted by a large number of organizations, so the rapid transition toward the cloud has fueled concerns on security perspective. Encryption algorithms play the main role in solving such kind of problems in the cloud computing environment. The problem with these encryption algorithms is that they consume a substantial amount of CPU time, memory, and other resources. This paper provides the performance comparison of seven popular symmetric algorithms: AES, DES, 3DES, Blowfish, RC4, IDEA, and TEA. This comparison is done on the basis of encryption time, decryption time, memory usage, flexibility, security, and scalability. The results achieved can be used in cloud-based application and services for the efficient working of cloud computing environment.

Keywords Cloud computing · Symmetric key cryptography · DES · 3DES Blowfish · RC4 · IDEA · TEA

1 Introduction

With rising concerns regarding the cloud computing and security of data, the prominent security algorithms especially symmetric algorithms could be widely used in cloud application services which involve encryption techniques. Cryptography is used in hiding information from intruders and storing it confidentially so that only those users are able to use it whom it is intended for and communicate this information securely. The use security algorithms minimize security concerns with the

A. R. Wani (✉) · N. Pandey
Amity University, Noida, Uttar Pradesh, India
e-mail: wanirauf@gamail.com

N. Pandey
e-mail: Npandey@gmail.com

Q. P. Rana
Jamia Hamdard University, New Delhi, India
e-mail: qprana@jamiahamdard.ac.in

K. Ray et al. (eds.), *Soft Computing: Theories and Applications*,
Advances in Intelligent Systems and Computing 742,
https://doi.org/10.1007/978-981-13-0589-4_24

help of cryptographic and authenticating techniques [1]. Cryptography is the process of crafting message securely altering the data to be sent with encrypting the plain text by taking user data and then executing the reverse process called as decryption which is returning back to original text. The cryptography can resolve the problems in cloud computing regarding network data and server security [2].

Encryption is the fundamental tool for protecting sensitive information. The goal of cryptography is keeping data secure from unauthorized users. With the swift development in science of encryption, an innovative area of cryptography can be classified as symmetric key cryptography [3]. Single key, one key also known as symmetric key cryptography uses the same key at both encryption and decryption process. Due to the use of single key for encrypting, the big quantity of data can be processed at a very fast speed [4]. There is no defined process within the cloud service providers for safeguarding and securing data from threats and attacks. The target of the cyber attackers is end user data which is being secured by the cloud using encryption techniques which are intended to make it impossible for the attacker to decrypt the ciphertext. The long length of the key makes harder to decrypt the classified text and makes them secure as compared to short keys.

2 Related Work

With the rise in the attacks, the emphasis is done by the clouds service providers at the users end to make data secure. Due to the inconsistency in the selection of encryption-decryption algorithms, low priority has been given to the cloud performance. Cloud performance and data security can be achieved by using the appropriate cryptographic algorithm at end user. For unintentional and accidental use of algorithms, it is important to do the algorithm analysis to check the competency of that particular algorithm that may result in degradation of performance in encryption or decryption process. For applications which use real-time data, an algorithm which might take long time would prove a hindrance for such applications and such algorithms end up consuming a lot of power for computing and storage to execute, thus making the algorithm unusable in that environment.

Various symmetric key algorithms have been developed in the past years and some of these algorithms work very efficiently, but still have some overheads. There has been a great revolution in field of internet security especially cryptography but this particular work deals only with symmetric key cryptographic algorithms which will be used in the field of cloud computing security. We have tried to find the results of different cryptographic algorithms in various ways and on different parameters. Different symmetric key algorithms were compared on the basis of energy consumption, and it was found that although AES is faster than other algorithms, there is 8% increase in power consumption [5, 6]. We can reduce the number of rounds to save power but it can compromise the protocol so this practice is not popular and is disliked by security experts. Performance analysis of various algorithms shows that blowfish has a better performance, and AES requires more processing power than other algorithms [7]. This work deals with the comparison of various symmetric algorithms and their advantages and it was concluded that AES is good for both high and low-level languages [8]. This paper presented the comparison between AES, DES, BLOWFISH, 3DES.RC2 and RC5. On different parameters like data types, key size battery power consumption, and encryption-decryption speed, it was concluded that blowfish has a better power consumption and speed than of the algorithms. Other papers discussed the performance DES, AES, RC2, BLOWFISH, and RC6 based on various simulation results and concluded that the algorithms should be well documented for better results [9]. This work studied the various algorithms on cloud network using single processor and concludes that RSA consumes most resources among all of them and MD5 the least.

3 Security Algorithm Overview

3.1 AES

AES symmetric block cipher Feistel structure that means it uses the same key for both encryption and decryption AES algorithm accepts a block size of 128 bits and a choice of three 128, 192, 256 key length permuted with variable 10, 12, and 14 rounds. The variable nature of Rijndael provides it with a great security, and the key size up to 256 gives it a resistance to the future attacks [10].

3.2 Blowfish

Blowfish is a Feistel structure symmetric key algorithm. It has a 64-bit block size, and the key varies from 32 to 448 bits. It uses 16 rounds and has large key dependent s box. There are four S boxes in blowfish algorithm, and the same algorithm is used in inverse for decryption [11]. Blowfish security lies in the key size providing high level of security. It is invincible against different key attacks because of many rounds which are being used by master key making such attacks infeasible.

3.3 DES

DES is a symmetric key cryptographic algorithm which is used for encrypting electronic data. DES is a block cipher which uses 64 bits of keys but only 56 bits of keys are effective and rest of the bits are used for parity. It consists of 16 round and two initial and final permutations [12]. The 56-bit key size generating 7.2 * 1016 possible keys gives strength to DES in typical threat environments.

3.4 3DES

3DES is named exactly as it is should be and as the name suggests because of three iterations of the DES encryption. The algorithm is based on a Feistel structure providing the three key options. The key length is 168 bits permuted in 16 subkeys which are 48 bits in length containing 8 s blocks. It uses the same algorithm for decryption [13]. 3DES uses a large size that is 168 bits to encrypt than DES, and the operations are performed three times in 3DES and two or three dissimilar keys offering with 112 bits of security which are used in avoiding so-called man in the middle attacks.

3.5 RC4

RC4 is a stream cipher which generates pseudorandom stream of bits. The key streams are used for encryption using bitwise exclusive OR. The decryption process of RC4 is performed in the same way. There is a permutation of 256 possible keys with two 8-bit index pointer in RC4. The permutation is done with a variable key length using key scheduling algorithm. The key size varies between 40 and 2048 bits. The strength of the RC4 is due to the location of the table which is hard to find and used only once.

3.6 IDEA

IDEA is fast and fairly secure and is strong to both differential and linear analysis [14]. In public domain, it is one of the secure block ciphers, which is based on substitution-permutation concept [14]. IDEA uses six 16-bit subkeys and half round uses four, a total of 52 of 8.5 rounds. IDEA under certain hypothesis is having a strong resistance against cryptanalysis [15]. It uses multiple group operations, which increases the strength against most of the attacks. The 128-bit key size makes it one of the strong security algorithms. There is no weakness related to linear or algebraic attacks which have been reported yet [16].

3.7 TEA

The Tiny Encryption Algorithm (TEA) is having few lines of code with simple structure and is easy to implement. The algorithm is a block cipher and consists of few lines of code. It is having a Feistel structure and operates on two 32-bit unsigned integers and 64 rounds [17]. TEA has a simple key structure, and key structure is same for each cycle. The TEA algorithm is having the same security as IDEA and also consists of 128-bit key size.

4 Performance Evaluation Matrix

The experimental design was performed on laptop with Core i7 processor on Windows 10 environment ranging from 83.3 Kb to 1.54 Mb. The evaluation parameters are as follows:

- Encryption time,
- Decryption time,
- Memory usage,
- Flexibility,
- Scalability, and
- Security.

5 Methodology

The experimental design was performed on a Core i7 machine with Windows 10 environment. The research was conducted in the suited environment according to the problem being investigated. Different file sizes were used ranging from 83.3 Kb to 15.48 Mb for the performance analysis process. The language used to check the space and complexity was java. Time and space complexity depends on lots of things like hardware, operating system, processors, etc., but we have only taken execution time into consideration. The objective we tried to achieve in terms of memory occupied by an algorithm during the course of execution was obtained by the methods like

getruntime().freememory()
getruntime().totalmemory()

with the runtime memory management options compiling heap memory, stack memory, etc. The other objective was how much time an algorithm takes right from the input of a file to the desired output.

The aim of this work was to check the performance of different symmetric cryptographic algorithms on the basis of various parameters like encryption and decryption time memory usage, flexibility, scalability, and security. All the implementations were done in the accurate manner to get the fair results. Exploratory research was used to evaluate these encryption algorithms and comparison was done on each of those algorithms.

6 Results and Analysis

We used java environment for simulation. We have taken two parameters time and memory for the simulation setup. The tables below represent the speed of AES, DES, 3DES, BLOWFISH, RC4, IDEA, and TEA algorithm to encrypt the data of same length. The results are obtained with the different file sizes by running the simulation shows the impact of changing the algorithms. There are 10 data files used to conduct this experiment and comparison of these cryptographic algorithms (Figs. 1, 2 3, 4 and 5, Tables 1, 2).

Experimental results of encryption algorithms are shown which shows all algorithms use same text files for ten experiments. By analyzing the table RC4 is taking less encryption time while as the 3DES is taking maximum encryption time. In the

```
C:\Users\Misger\Desktop\Encryption Algos\src>java EncryptionAlgos
################################################################
Encryption Algos Menu
Enter 1 for DES
Enter 2 for Tripe DES
Enter 3 for AES
Enter 4 for BLOWFISH
Enter 5 for RC4
Enter 9 to Exit
3
Time taken for encryption : 723
Text Encryted : [B@73035e27
Memory Usage : -60697448
Time taken for dycryption : 62

################################################################
```

Fig. 1 Simulation results of AES

```
C:\Users\Misger\Desktop\Encryption Algos\src>java EncryptionAlgos
################################################################
Encryption Algos Menu
Enter 1 for DES
Enter 2 for Tripe DES
Enter 3 for AES
Enter 4 for BLOWFISH
Enter 5 for RC4
Enter 9 to Exit
4
Time taken for encryption : 531
Text Encryted : [B@75412c2f
Memory Usage : -59702576
Time taken for dycryption : 234

################################################################
```

Fig. 2 Simulation results of blowfish

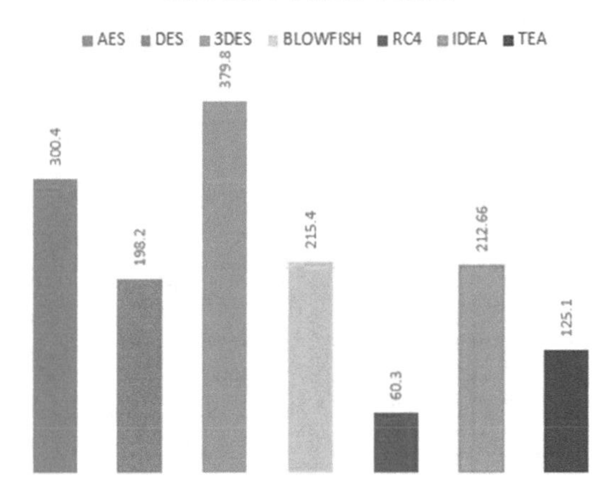

Fig. 3 Encryption time in milliseconds

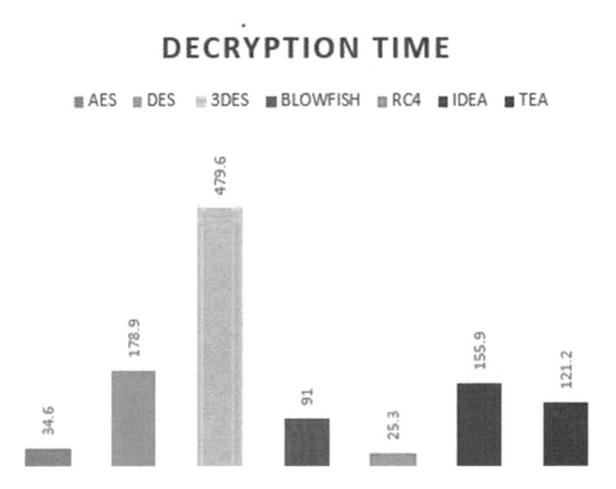

Fig. 4 Decryption time in milliseconds

second table, RC4 and AES are having very less decryption time while 3DES is having maximum of all the algorithms. Table 3 depicts the memory usage of all the algorithms in which IDEA and TEA are having very less memory usage while as RC4 is taking the maximum memory of all the algorithms.

Table 1 Encryption time (ms)

KB	AES	DES	3DES	Blowfish	RC4	IDEA	TEA
83.3	625	41	43	8	15	16	54
108	31	47	47	16	16	125	15
249	468	47	47	47	15	32	16
333	32	47	64	281	15	63	10
416	46	48	94	343	31	127	16
1370	141	110	243	78	47	78	41
2740	62	172	361	93	62	205	97
5480	78	296	749	156	63	325	170
10003	723	484	749	531	141	397	357
15483	798	690	1401	601	198	758	475
Average	300.4	198.2	397.8	215.4	60.3	212.6	125.1

Table 2 Decryption time (ms)

KB	AES	DES	3DES	Blowfish	RC4	IDEA	TEA
83.3	24	10	25	17	5	16	15
108	15	16	31	10	6	943	16
249	16	32	31	15	4	31	10
333	16	46	17	16	7	31	12
416	10	47	125	17	10	47	16
1370	31	78	187	62	12	45	40
2740	47	141	359	78	17	129	96
5480	31	225	678	156	48	218	158
10003	62	484	1346	234	55	351	304
15483	88	680	1988	305	89	597	545
Average	34.4	178.6	497.6	91	25.3	155.9	121.2

Table 3 Memory usage (KB)

KB	AES	DES	3DES	Blowfish	RC4	IDEA	TEA
83.3	11014064	10081304	992712	580824	980952	259080	2097168
108	985312	10390693	20853952	1142800	1961832	2097168	259080
249	11244008	11169376	11169376	10968120	3211080	5497720	2097168
333	2230250	1061603	4256520	10791776	4629096	2097168	259000
416	4186640	5167096	7498084	3010032	33160	3124168	209040
1370	7063848	11187024	1321440	85711216	8675472	5172418	2349040
2740	6361832	125688808	10698640	17228432	12568808	2097168	2347008
5480	34660616	9927424	18077072	16506648	11823224	20971862	8963325
10003	60697448	60721016	120146520	59702576	130702776	12031904	17514192
15483	407289063	70648440	138223590	76209224	142526000	2097168	32708440
Average	54573308.1	31604278.4	33323790.6	28185164.8	317112400	5544582.4	6880346.1

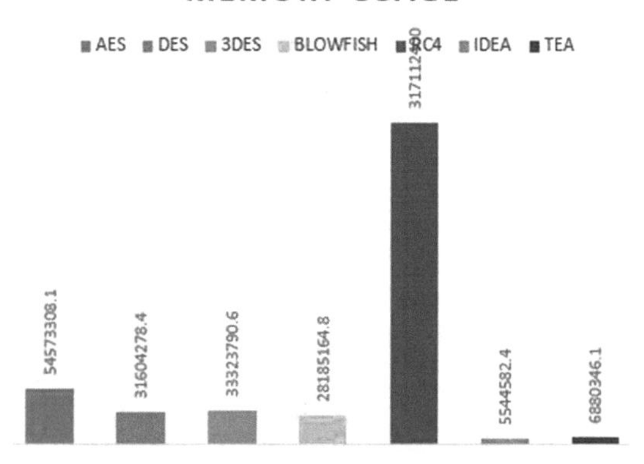

Fig. 5 Memory usage in kilobytes

7 Level of Security

Level of security of a particular algorithm depends on key size; the greater the key size, stronger the algorithm and encryption.

Encryption algorithms	Plain text/cipher (text)	Length key length (bits)	No of rounds (bits)
DES	64 bits	56	16
3DES	64 bits	168	48
AES	128 bits	128, 192, 256	10, 12, 14
Blowfish	64 bits	32–448	16
RC4	40–2048 bits	Variable	256
IDEA	64 bits	128	8.5
TEA	64 bits	128	32 cycles

8 Flexibility and Security

Algorithm	Flexibility	Modification	Comments
DES	No	None	No modifications are supported by DES
3DES	Yes	168	The DES key is extended to 168 bits
AES	Yes	128, 192, 256	The AES is expandable and support modifications
Blowfish	Yes	32–448	The structure of BLOWFISH is extendable to 448 bits
RC4	Yes	variable	Modifications supported
IDEA	No	128	No modifications supported
TEA	No	128	No modifications supported

9 Conclusion

With the cloud computing emerging new technology the security still remains one of the biggest challenges in cloud computing environment. Usage of security algorithms and guaranteeing these security algorithms are implemented properly and accurately employed in order to safeguard end user security. These encrypting algorithms play important role in communication security where encryption-decryption time and memory usage are major issues of concern. The performance evaluation of selected cryptographic algorithms like AES, DES, 3DES, RC4, BLOWFISH, IDEA, and TEA was carried out based on the encryption-decryption and memory usage, flexibility, scalability, and security. Based on the text files used and the experimental results RC4 was taking less encryption time while as 3DES has taken maximum encryption time. The decryption time of RC4 and AES is less than other algorithms. In terms of memory usage, IDEA and TEA are taking very less memory, and RC4 is taking maximum of all the algorithms. During the analysis, it was found that AES will be best among all the algorithms in terms of flexibility, security, memory performance, and usage. Although other algorithms were also effective, considering all the parameters such as security, flexibility, encryption-decryption time, scalability, and memory usage AES will be the best among all algorithms.

References

1. Meyer, C.H.: Cryptography-a state of the art review. In: CompEuro'89., VLSI and Computer Peripherals. VLSI and Microelectronic Applications in Intelligent Peripherals and their Interconnection Networks', Proceedings. IEEE (1989)
2. Krutz, R.L., Dean Vines, R.: Cloud Security: A Comprehensive Guide to Secure Cloud Computing. Wiley Publishing (2010)
3. Ruangchaijatupon, N., Krishnamurthy, P.: Encryption and Power Consumption in Wireless LANs-N. In: The Third IEEE Workshop on Wireless LANS (2001)
4. Hirani, S.: Energy Consumption of Encryption Schemes in Wireless Devices Thesis, University of Pittsburgh, April 9, 2003. Accessed 1 Oct 2008
5. Thakur, J., Kumar, N.: DES, AES and blowfish: symmetric key cryptography algorithms simulation based performance analysis. Int. J. Emerg. Technol. Adv. Eng. **1**(2), 6–12 (2011)
6. Penchalaiah, N., Seshadri, R.: Effective comparison and evaluation of DES and Rijndael Algorithm (AES). Int. J. Comput. Sci. Eng. **2**(05), 1641–1645 (2010)
7. Elminaam, D.S.A., Abdual-Kader, H.M., Hadhoud, M.M.: Evaluating the performance of symmetric encryption algorithms. IJ Netw. Secur. **10**(3), 216–222 (2010)
8. Mushtaque, M.A.: Comparative analysis on different parameters of encryption algorithms for information security. JCSE Int. J. Comput. Sci. **2**(4) (2014)
9. Ebrahim, M., Khan, S., Khalid, U.B.: Symmetric algorithm survey: a comparative analysis. arXiv:1405.0398 (2014)
10. Schneier, B.: The Blowfish Encryption Algorithm-One-Year Later, p. 137 (1995)
11. Singh, S.P., Maini, R.: Comparison of data encryption algorithms. Int. J. Comput. Sci. Commun. **2**(1), 125–127 (2011)
12. Standard, Data Encryption.: Federal information processing standards publication 46. National Bureau of Standards, US Department of Commerce (1977)

13. Stallings, W.: Cryptography and Network Security: Principles and Practices. Pearson Education India (2006)
14. Lai, X., Massey, J.L.: A proposal for a new block encryption standard. In: Workshop on the Theory and Application of Cryptographic Techniques. Springer, Heidelberg (1990)
15. Elbaz, L., Bar-El, H.: Strength Assessment of Encryption Algorithms. White paper (2000)
16. Forouzan, B.A., Mukhopadhyay, D.: Cryptography and Network Security (Sie). McGraw-Hill Education (2011)
17. Xie, H., Zhou, L., Bhuyan, L.: Architectural analysis of cryptographic applications for network processors. In: Proceedings of the IEEE First Workshop on Network Processors with HPCA-8 (2002)

A Survey of CAD Methods for Tuberculosis Detection in Chest Radiographs

Rahul Hooda, Ajay Mittal and Sanjeev Sofat

Abstract Tuberculosis is a highly infectious disease and the second largest killer worldwide. Every year, millions of new cases and deaths are reported due to tuberculosis. In the developing countries, tuberculosis suspect cases are enormous, and hence, a large number of radiologists are required to perform the mass screening. Therefore attempts have been made to design computer-aided diagnosis (CAD) systems for automatic mass screening. A CAD system generally consists of four phases, namely, preprocessing, segmentation, feature extraction, and classification. In this paper, we present a survey of the recent approaches used in different phases of chest radiographic CAD system for tuberculosis detection.

Keywords Computer-aided diagnosis · Automatic screening · Medical imaging Chest X-rays

1 Introduction

Tuberculosis (TB) is highly infectious disease and the second largest killer worldwide after HIV/AIDS. According to World Health Organization's Global TB Report 2016, there were 10.4 million new cases and 1.8 million deaths reported in 2015 due to TB. India with almost 23% of the global total has the largest number of TB cases. Tuberculosis is a contagious disease which is caused by the *bacillus Mycobacterium tuberculosis*. TB bacteria can be present in any part of the body; however since it is favorable in high oxygen condition, it mainly affects the lung region and is known as pulmonary TB (PTB). PTB is generally spread from an infected person to normal person through coughing and sneezing. There is no single sign of the presence

R. Hooda (✉) · S. Sofat
Department of Computer Science & Engineering,
PEC University of Technology, Chandigarh, India
e-mail: r89hooda@gmail.com

A. Mittal
University Institute of Engineering and Technology, Panjab University, Chandigarh, India

K. Ray et al. (eds.), *Soft Computing: Theories and Applications*,
Advances in Intelligent Systems and Computing 742,
https://doi.org/10.1007/978-981-13-0589-4_25

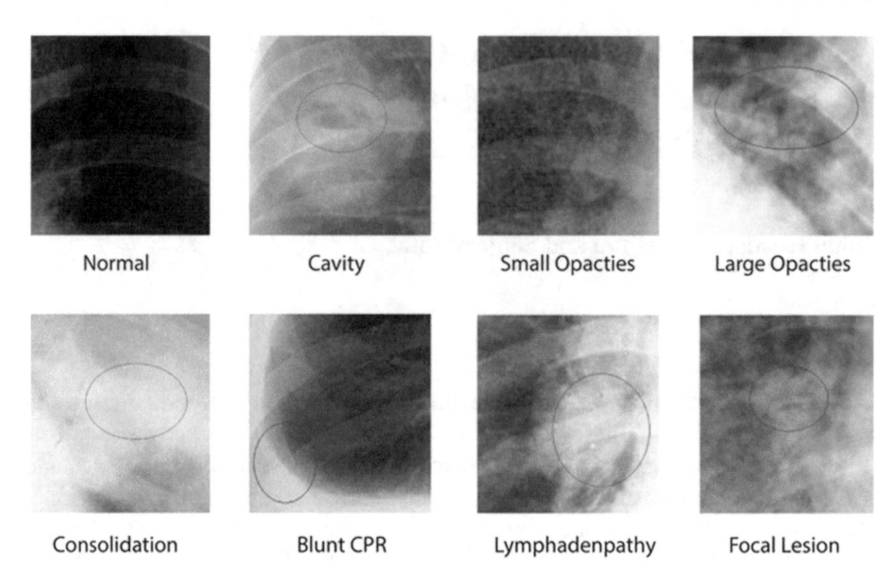

Fig. 1 Different types of TB manifestations

of TB and it can have different pathological patterns depending on many factors. Different types of TB manifestations are cavities, small opacities, large opacities, consolidation, focal lesions, and nodules which are shown in Fig. 1. TB can also affect the shape of the lung by having an abnormal shape of the costophrenic angle.

For the successful treatment and containment of TB, it should be detected as early as possible. Traditional chest X-ray (CXR) is still the most widely used radiological procedure (even after the introduction of various new imaging modalities) for screening TB due to its low radiation dose, low cost, wider acceptability, and extensive deployment. CAD systems that can analyze the digital CXRs automatically can be of great help in undeveloped countries where medical services cannot be provided actively. These systems are designed to help the physicians in the diagnosis of abnormalities based on computer algorithms. Development of a CAD system can be broken into a number of discrete steps which can be addressed independently. The work on CAD systems has been started in the 2000s. Different CAD systems have been proposed after that but till now not a complete system has been developed which is used in practice today. List of the CAD surveys which have been done till now is presented in Table 1. In this paper, some important works in the field of chest radiographic CAD systems have been discussed.

A CAD system basically consists of four phases: preprocessing, segmentation, feature extraction, and classification. Preprocessing is performed to enhance the image quality and to reduce the intensity variation among the image dataset. They are also used in CAD systems for suppression of anatomical regions to clearly visualize the abnormalities or interested region. Next phase is segmentation where different anatomical structures are segmented so that region-of-interest can be obtained. For

Table 1 List of CAD surveys done till date

Authors	Year	Study
Duncan and Ayache [1]	2000	Study on the developments in the domain of medical image analysis since 1980s
Ginneken et al. [2]	2001	Survey on CAD in chest radiographs
Mehta et al. [3]	2005	Survey on CAD in lung cancer detection
Sluimer et al. [4]	2006	Survey on computer analysis on lung CT scans
Doi [5]	2007	Survey on various CAD techniques
Li et al. [6]	2009	Survey on detection of malignant lung nodules
Ginneken et al. [7]	2009	Comparative study of anatomical structure segmentation techniques in chest radiographs
Jaeger et al. [8]	2013	Survey of methods for automatic screening of tuberculosis from chest radiographs

TB detection, lung segmentation is performed to obtain lung region and discard other anatomic structures visible on the CXR. Feature extraction is the next phase in which relevant features are extracted and then the selection algorithm is performed to obtain best performing features and form the feature vector. Extracted features are based on the local pixel level or on the region level. The last phase of building CAD system is classification which refers to categorizing objects into different classes. This step has two stages: training stage, in which classifier is built and trained with the help of labeled examples, and testing stage in which classifier is used to solve two class-classification problem which divides each pixel or region into normal and abnormal parts.

In the next three sections, work done till now in the above mentioned four steps of CAD system for tuberculosis is discussed.

2 Preprocessing

The first step of any CAD system is to apply a series of preprocessing algorithms to the input image. The main aim of applying these algorithms is to enhance the quality of the image so that blurred lung boundary is clearly visible. The quality enhancement obtained by applying preprocessing steps can vastly improve the performance of remaining steps of a CAD system. Typical preprocessing steps for CXRs are image enhancement and bone suppression. Image enhancement is done to enhance the contrast of the image near lung boundary and of the lung tissues. One of the most common techniques for image enhancement is histogram equalization, in which histogram of the image is analyzed and contrast of low-contrast regions is increased. Some studies have also applied *image sharpening*, which basically uses high-frequency details to enhance the display of digital images. In the literature, sharpening of images is performed using different methods such as selective unsharp masking [9], automatic spatial filtering [10] and adaptive contrast enhancement algorithm [11].

Preprocessing also involves suppression of different structures to clearly visualize the region-of-interest or the abnormalities present in the CXR. Structures which are considered for suppression are ribs and clavicles [12–14], bone [15–17] and foreign objects [18]. Clavicles are a major hindrance to find the top boundary of the lung region whereas rib cage produces local minima which affect the performance of lung segmentation methods. Removal of these structures helps in improving the interpretation of the CXRs by radiologists as well as by CAD systems. Images with suppressed bone structures can be generated by applying image processing techniques, image subtraction or by using dual-energy imaging. Hogeweg et al. [12, 13] proposed a method to suppress the clavicles and ribs using intensity model created by using image profile sampled perpendicular to the structure to be suppressed. Simko et al. [14] proposed a method to remove the shadow of the clavicle bones and measure the performance of nodule detection system. Schalekemp et al. [15] and Li et al. [17] concluded that the bone suppressed images improve the performance of observer to analyze the CXRs. Freedman et al. [16] measured the performance of visualization software for bone suppression and concluded that the use of software significantly improves the performance of radiologists in detecting nodules and lung cancer. Hogeweg et al. [18] proposed a k-NN classifier-based pixel classification method to remove the foreign objects like brassier clips, jewelry, pacemakers, etc., which may be present in CXRs. These foreign objects significantly reduce the performance of automated systems.

3 Lung Segmentation

Segmentation of anatomical structures is important to restrict the analysis to the region-of-interest (ROI) which subsequently helps in detecting the extent of disease. Lung field segmentation (LFS) of CXRs, which is one of the preliminary steps of pulmonary disease diagnosis, outputs a lung contour for further processing. LFS is also used in localization of pathology, quantification of tissue volumes, the study of anatomical structures, and lung disease diagnosis [8]. Accurate LFS is important because it decreases the false positives in the automatic analysis using CAD systems. Many LFS techniques have been proposed for CXRs in the last four decades but it is still a difficult task to precisely segment the lung fields due to reasons like the variation of lung shapes, variation in image quality, and ambiguity of lung boundary due to the existence of superimposed structures. LFS methods can be split up into three broad classes described below:

Rule-based segmentation methods: These methods employ predefined anatomical rules to segment the lung field. These methods employ a sequence of rules or algorithms to obtain the desired results. They include low-level processing which refers to region and/or pixel intensity-based segmentation methods, such as region growing, thresholding, morphological operations, edge detection and linking and dynamic programming [7]. Cheng and Goldberg [19] proposed one of the earliest LFS methods. The method uses horizontal and vertical profiles (HVP) of the CXRs to determine minimum rectangular frame enclosing lungs. Li et al. [20] thereafter used gradient of these profiles to obtain discrete lung edges. These edges are used to segment smooth contour using iterative boundary point adjustment. Duryea and Boone [21] also used a similar approach to segment lung fields but used images at a reduced scale. In [22, 23], thresholding-based techniques are used after HVP analysis to obtain lung boundary which is then made continuous by using a smoothing algorithm. Later, Ahmad et al. [24] proposed a method in which lung boundary is roughly segmented by applying Gaussian derivative filter and is refined using fuzzy C-means (FCM) clustering method.

Pixel-based classification methods: These methods consider LFS as a pixel-level classification problem and hence train a classifier to label each pixel as either lung or non-lung using a labeled dataset. Different statistical and texture features are used so that each pixel can be classified. The classifier is trained on the extracted features and then tested on a separate dataset. McNitt-Gray et al. [25] investigated the effect of number and quality of handcrafted features on the performance of different classifiers. It is found that best-selected feature set provides performance comparable to complete feature set. These findings stress upon the profound impact of features on the machine learning algorithm's ability to create a good classifier model. Tsujii et al. [26] used different spatial features and adaptive-sized hybrid neural network (ASHNN) to achieve lung segmentation. The classification using ASHNN is propitious as it can control the complexity and have good convergence. Ginneken and Romney [27] presented a technique in which a rule-based technique is initially used to extract information about lung edges. This information is combined along with intensity and entropy measures to form feature set subsequently used by modified k-NN classifier to obtain final lung boundary. There is also one unsupervised pixel classification approach presented by Shi et al. [28] and is based on Gaussian kernel-based FCM.

Deformable model-based methods: These methods are based on the concept of internal and external forces. The internal force from an object shape ensures that the curve is bendable and stretchable, whereas the external force from image appearance drives the curve to the desired image characteristics, such as terminations, edges, and lines. Cootes et al. [29] proposed a parametric method, namely, *active shape model* (ASM) which deforms itself based on the characteristics of the object class it represents. Ginneken et al. [30] proposed modified ASM by using optimal local features and nonlinear k-NN classifier in place of traditional normalized first-order derivative profiles and linear Mahalanobis distance. Xu et al. [31] used modified gradient vector flow (GVF)-based ASM for performing lung segmentation. Some other ASM-based modified methods are proposed in [32–34]. ASM's limitation of not considering textural appearance is removed by *active appearance models* (AAM) which takes into account both the shape and appearance of the object while constructing prior models. Ginneken [7, 27] has compared the performance of various segmentation methods including AAM. Some other parametric methods are presented in [35–37]. In the literature, some geometric model-based methods are also presented. Annangi et al.

[38] have used a level set method to perform LFS in which initial rough boundary is obtained by applying Otsu thresholding. Whereas Lee et al. [39] proposed an unsupervised LFS method based on multi-resolution fractal feature vector and modified level sets. Candemir et al. [40, 41] proposed a graph cut-based method, which is also a geometric model. In this method, initial contour is obtained by using content-based image retrieval approach and scale invariant feature transform (SIFT)-flow nonrigid registration technique, which is refined by applying graph cut method.

4 Feature Extraction and Classification

After the lung segmentation is performed, features are obtained from the selected region-of-interest. These features are then passed to feature selection algorithm and important features are selected to form the feature vector. In the literature, it is found out that only texture and geometrical features are mostly considered to detect different TB patterns. Thereafter, a classifier is used, which is trained on some images and provide its classification on the set of test images to detect whether an abnormality is present in the lung region. In this section, different studies about the TB detection will be discussed. Some papers have discussed abnormality or patterns due to TB as a whole while other techniques manage only a specific type of abnormality or pattern.

In the literature, some methods have been presented which can detect specific manifestations like cavities, focal opacities. Sarkar et al. [42] proposed a method to detect cavitation and infiltration using an iterative local contrast approach. Kosleg et al. [43] presented a method to identify miliary TB based on template matching performed in the frequency domain. Each image is correlated with 16 different template images and each template image has different threshold set. Similarly, Leibstein et al. [44] proposed a method to detect TB nodules using local binary pattern and Laplacian of Gaussian techniques. Song and Yang [45] detected focal opacities in CXRs using a region growing method based on *seed point* is used to locate focal opacity. Tan et al. [46] presented a first-order statistical approach in which texture features and decision tree classifier are used to detect TB. Xu et al. [47] proposed a two-scale cavity detection method in which cavity candidates are firstly identified by performing feature classification at the course level and then segmented using active contour-based segmentation. Finally, at a fine scale, classification is performed using support vector machine (SVM) to remove false positives using a different set of features.

Ginneken et al. [48] proposed a method for automatic detection of TB abnormalities using local texture analysis. In this method, lung fields are divided into several overlapping regions and then for each of the regions, different texture features such as moments of Gaussian derivatives and difference features are extracted. Thereafter, classification is performed using k-nearest neighbor at region level and the weighted multiplier is used for each region to obtain final abnormality score. Hogeweg et al. in [49] presented improved TB detection method in which textural, focal, and shape abnormalities are analyzed separately and thereafter combined into one TB score.

Melendez et al. [50] proposed a multiple instance learning classifier which used weakly labeled images for TB detection. In weakly labeled approach, exact location or outline of manifestation is not provided to the system to learn however image label is provided that is whether the CXR is normal or abnormal. Arzhaeva et al. [51] proposed a classification method in which distribution of texture features is measured, and then histograms which are generated are compared with the stored ones to obtain dissimilarities. These dissimilarities are classified using linear discriminant classifier.

5 Conclusion

CAD systems for automatic TB detection have shown tremendous growth over the last few years. However, much work still needs to be done as these systems for automatic mass screening [52, 53] are still in development stage and practical performance needs to be evaluated once they are employed in the developing countries. Also in these systems features are manually extracted which creates a hindrance to enhance the performance. Therefore, nowadays, researchers have started to move toward deep learning approaches which extract features automatically. Deep learning-based CAD systems are end-to-end approaches and thus lung segmentation is not necessarily required in these systems. These systems will be the future of CAD systems and improve the state-of-art performance in different medical domains.

References

1. Duncan, J.S., Ayache, N.: Medical image analysis: progress over two decades and the challenges ahead. IEEE Trans. Pattern Anal. Mach. Intell. **22**(1), 85–106 (2000)
2. Van Ginneken, B., Romeny, B.T.H., Viergever, M.A.: Computer-aided diagnosis in chest radiography: a survey. IEEE Trans. Med. Imaging **20**(12), 1228–1241 (2001)
3. Mehta, I.C., Ray, A.K., Khan, Z.J.: Lung cancer detection using computer aided diagnosis in chest radiograph: A survey and analysis. IETE Tech. Rev. **22**(5), 385–393 (2005)
4. Sluimer, I., Schilham, A., Prokop, M., van Ginneken, B.: Computer analysis of computed tomography scans of the lung: a survey. IEEE Trans. Med. Imaging **25**(4), 385–405 (2006)
5. Doi, K.: Computer-aided diagnosis in medical imaging: historical review, current status and future potential. Comput. Med. Imaging Graph. **31**(4), 198–211 (2007)
6. Li, B., Tian, L., Ou, S.: CAD for identifying malignant lung nodules in early diagnosis: a survey. Sheng wu yi xue gong cheng xue za zhi = Journal of biomedical engineering = Shengwu yixue gongchengxue zazhi, **26**(5), 1141–1145 (2009)
7. Van Ginneken, B., Stegmann, M.B., Loog, M.: Segmentation of anatomical structures in chest radiographs using supervised methods: a comparative study on a public database. Med. Image Anal. **10**(1), 19–40 (2006)
8. Jaeger, S., Karargyris, A., Candemir, S., Siegelman, J., Folio, L., Antani, S., Thoma, G.: Automatic screening for tuberculosis in chest radiographs: a survey. Quant. Imaging Med. Surg. **3**(2), 89 (2013)
9. Sezn, M.I., Teklap, A.M., Schaetzing, R.: Automatic anatomically selective image enhancement in digital chest radiography. IEEE Trans. Med. Imaging **8**(2), 154–162 (1989)

10. Tahoces, P.G., Correa, J., Souto, M., Gonzalez, C., Gomez, L., Vidal, J.J.: Enhancement of chest and breast radiographs by automatic spatial filtering. IEEE Trans. Med. Imaging **10**(3), 330–335 (1991)
11. Chang, D.C., Wu, W.R.: Image contrast enhancement based on a histogram transformation of local standard deviation. IEEE Trans. Med. Imaging **17**(4), 518–531 (1998)
12. Hogeweg, L., Sanchez, C.I., van Ginneken, B.: Suppression of translucent elongated structures: applications in chest radiography. IEEE Trans. Med. Imaging **32**(11), 2099–2113 (2013)
13. Hogeweg, L., Sánchez, C.I., de Jong, P.A., Maduskar, P., van Ginneken, B.: Clavicle segmentation in chest radiographs. Med. Image Anal. **16**(8), 1490–1502 (2012)
14. Simkó, G., Orbán, G., Máday, P., Horváth, G.: Elimination of clavicle shadows to help automatic lung nodule detection on chest radiographs. In: 4th European Conference of the International Federation for Medical and Biological Engineering, pp. 488–491. Springer, Berlin, Heidelberg (2009)
15. Schalekamp, S., van Ginneken, B., Meiss, L., Peters-Bax, L., Quekel, L.G., Snoeren, M.M.,… Schaefer-Prokop, C.M.: Bone suppressed images improve radiologists' detection performance for pulmonary nodules in chest radiographs. Eur. J. Radiol. **82**(12), 2399–2405 (2013)
16. Freedman, M.T., Lo, S.C.B., Seibel, J.C., Bromley, C.M.: Lung nodules: improved detection with software that suppresses the rib and clavicle on chest radiographs. Radiology **260**(1), 265–273 (2011)
17. Li, F., Hara, T., Shiraishi, J., Engelmann, R., MacMahon, H., Doi, K.: Improved detection of subtle lung nodules by use of chest radiographs with bone suppression imaging: receiver operating characteristic analysis with and without localization. Am. J. Roentgenol. **196**(5), W535–W541 (2011)
18. Hogeweg, L., Sánchez, C.I., Melendez, J., Maduskar, P., Story, A., Hayward, A., van Ginneken, B.: Foreign object detection and removal to improve automated analysis of chest radiographs. Med. Phys. **40**(7) (2013)
19. Cheng, D., Goldberg, M.: An algorithm for segmenting chest radiographs. In: Proceedings of the SPIE, vol. 1001, pp. 261–268, Oct, 1988
20. Li, L., Zheng, Y., Kallergi, M., Clark, R.A.: Improved method for automatic identification of lung regions on chest radiographs. Acad. Radiol. **8**(7), 629–638 (2001)
21. Duryea, J., Boone, J.M.: A fully automated algorithm for the segmentation of lung fields on digital chest radiographic images. Med. Phys. **22**(2), 183–191 (1995)
22. Pietka, E.: Lung segmentation in digital radiographs. J. Digit. Imaging **7**(2), 79–84 (1994)
23. Armato, S.G., Giger, M.L., Ashizawa, K., MacMahon, H.: Automated lung segmentation in digital lateral chest radiographs. Med. Phys. **25**(8), 1507–1520 (1998)
24. Ahmad, W.S.H.M.W., Zaki, W.M.D.W., Fauzi, M.F.A.: Lung segmentation on standard and mobile chest radiographs using oriented Gaussian derivatives filter. Biomed. Eng. Online **14**(1), 20 (2015)
25. McNitt-Gray, M.F., Huang, H.K., Sayre, J.W.: Feature selection in the pattern classification problem of digital chest radiograph segmentation. IEEE Trans. Med. Imaging **14**(3), 537–547 (1995)
26. Tsujii, O., Freedman, M.T., Mun, S.K.: Automated segmentation of anatomic regions in chest radiographs using an adaptive-sized hybrid neural network. Med. Phys. **25**(6), 998–1007 (1998)
27. Van Ginneken, B., ter Haar Romeny, B.M.: Automatic segmentation of lung fields in chest radiographs. Med. Phys. **27**(10), 2445–2455 (2000)
28. Shi, Z., Zhou, P., He, L., Nakamura, T., Yao, Q., Itoh, H.: Lung segmentation in chest radiographs by means of Gaussian kernel-based FCM with spatial constraints. In: Sixth International Conference on Fuzzy Systems and Knowledge Discovery, 2009. FSKD'09, vol. 3, pp. 428–432, Aug, 2009. IEEE
29. Cootes, T.F., Taylor, C.J., Cooper, D.H., Graham, J.: Active shape models-their training and application. Comput. Vis. Image Underst. **61**(1), 38–59 (1995)
30. Van Ginneken, B., Frangi, A.F., Staal, J.J., ter Haar Romeny, B.M., Viergever, M.A.: Active shape model segmentation with optimal features. IEEE Trans. Med. Imaging **21**(8), 924–933 (2002)

31. Xu, T., Mandal, M., Long, R., Basu, A.: Gradient vector flow based active shape model for lung field segmentation in chest radiographs. In: Annual International Conference of the IEEE Engineering in Medicine and Biology Society, 2009. EMBC 2009, pp. 3561–3564, Sept, 2009. IEEE
32. Wu, G., Zhang, X., Luo, S., Hu, Q.: Lung segmentation based on customized active shape model from digital radiography chest images. J. Med. Imaging Health Inform. **5**(2), 184–191 (2015)
33. Lee, J.S., Wu, H.H., Yuan, M.Z.: Lung segmentation for chest radiograph by using adaptive active shape models. Biomed. Eng. Appl. Basis Commun. **22**(02), 149–156 (2010)
34. Xu, T., Mandal, M., Long, R., Cheng, I., Basu, A.: An edge-region force guided active shape approach for automatic lung field detection in chest radiographs. Comput. Med. Imaging Graph. **36**(6), 452–463 (2012)
35. Dawoud, A.: Lung segmentation in chest radiographs by fusing shape information in iterative thresholding. IET Comput. Vision **5**(3), 185–190 (2011)
36. Shao, Y., Gao, Y., Guo, Y., Shi, Y., Yang, X., Shen, D.: Hierarchical lung field segmentation with joint shape and appearance sparse learning. IEEE Trans. Med. Imaging **33**(9), 1761–1780 (2014)
37. Zhang, G.D., Guo, Y.F., Gao, S., Guo, W.: Lung segmentation in feature images with gray and shape information. In: Applied Mechanics and Materials, vol. 513, pp. 3069–3072. Trans Tech Publications (2014)
38. Annangi, P., Thiruvenkadam, S., Raja, A., Xu, H., Sun, X., Mao, L.: A region based active contour method for x-ray lung segmentation using prior shape and low level features. In: 2010 IEEE International Symposium on Biomedical Imaging: From Nano to Macro, pp. 892–895, Apr, 2010. IEEE
39. Lee, W.L., Chang, K., Hsieh, K.S.: Unsupervised segmentation of lung fields in chest radiographs using multiresolution fractal feature vector and deformable models. Med. Biol. Eng. Comput. **54**(9), 1409–1422 (2016)
40. Candemir, S., Jaeger, S., Palaniappan, K., Antani, S., Thoma, G.: Graph-cut based automatic lung boundary detection in chest radiographs. In: IEEE Healthcare Technology Conference: Translational Engineering in Health & Medicine, pp. 31–34, Nov, 2012
41. Candemir, S., Jaeger, S., Palaniappan, K., Musco, J.P., Singh, R.K., Xue, Z., … McDonald, C.J.: Lung segmentation in chest radiographs using anatomical atlases with nonrigid registration. IEEE Trans. Med. Imaging **33**(2), 577–590 (2014)
42. Sarkar, S., Chaudhuri, S.: Automated detection of infiltration and cavitation in digital chest radiographs of chronic pulmonary tuberculosis. In: Proceedings of the 18th Annual International Conference of the IEEE Engineering in Medicine and Biology Society, 1996. Bridging Disciplines for Biomedicine, vol. 3, pp. 1185–1186. IEEE (1997)
43. Koeslag, A., de Jager, G.: Computer aided diagnosis of miliary tuberculosis. In: Proceedings of the Pattern Recognition Association of South Africa (2001)
44. Leibstein, J.M., Nel, A.L.: Detecting tuberculosis in chest radiographs using image processing techniques. University of Johannesburg (2006)
45. Song, Y.L., Yang, Y.: Localization algorithm and implementation for focal of pulmonary tuberculosis chest image. In: 2010 International Conference on Machine Vision and Human-Machine Interface (MVHI), pp. 361–364, Apr, 2010. IEEE
46. Tan, J.H., Acharya, U.R., Tan, C., Abraham, K.T., Lim, C.M.: Computer-assisted diagnosis of tuberculosis: a first order statistical approach to chest radiograph. J. Med. Syst. **36**(5), 2751–2759 (2012)
47. Xu, T., Cheng, I., Long, R., Mandal, M.: Novel coarse-to-fine dual scale technique for tuberculosis cavity detection in chest radiographs. EURASIP J. Image Video Process. **2013**(1), 3 (2013)
48. Van Ginneken, B., Katsuragawa, S., ter Haar Romeny, B.M., Doi, K., Viergever, M.A.: Automatic detection of abnormalities in chest radiographs using local texture analysis. IEEE Trans. Med. Imaging **21**(2), 139–149 (2002)

49. Hogeweg, L., Sánchez, C.I., Maduskar, P., Philipsen, R., Story, A., Dawson, R.: van Ginneken, B.: Automatic detection of tuberculosis in chest radiographs using a combination of textural, focal, and shape abnormality analysis. IEEE Trans. Med. Imaging **34**(12), 2429–2442 (2015)
50. Melendez, J., van Ginneken, B., Maduskar, P., Philipsen, R.H., Ayles, H., Sánchez, C.I.: On combining multiple-instance learning and active learning for computer-aided detection of tuberculosis. IEEE Trans. Med. Imaging **35**(4), 1013–1024 (2016)
51. Arzhaeva, Y., Hogeweg, L., de Jong, P. A., Viergever, M.A., van Ginneken, B.: Global and local multi-valued dissimilarity-based classification: application to computer-aided detection of tuberculosis. In: International Conference on Medical Image Computing and Computer-Assisted Intervention, pp. 724–731, Sept, 2009. Springer, Berlin, Heidelberg
52. Jaeger, S., Karargyris, A., Antani, S., Thoma, G.: Detecting tuberculosis in radiographs using combined lung masks. In: 2012 Annual International Conference of the IEEE Engineering in Medicine and Biology Society (EMBC), pp. 4978–4981, Aug, 2012. IEEE
53. Jaeger, S., Antani, S., Thoma, G.: Tuberculosis screening of chest radiographs. In: SPIE Newsroom (2011)

Static Image Shadow Detection Texture Analysis by Entropy-Based Method

Kavita, Manoj K. Sabnis and Manoj Kumar Shukla

Abstract For applications such as object identification and object tracking, the primary goal is that the system should be able to track only the object of interest. If it is unable to do, it would lead to false tracking. The images so obtained by the tracking system may have a number of objects with similar shapes and colors. These images can be further outdoor images with illumination conditions so as to cast the shadow of the object. Of these objects present, the objects that map very closely to the object of interest is its shadows. If the image along with its shadow is allowed to enter the system, it may lead to false tracking. This is so because the image acquisition system cannot differentiate between shadow and its object on its own. Therefore, a stage called as shadow detection and elimination stage had to be introduced between the image acquisitions and processing stage in the tracking system. When the shadow detection taxonomy is examined it gives four standard methods of detection based on intensity, color, geometry, and texture. Most of the images map their requirement with first three methods and a very few which does not satisfy the requirements of these three methods adopt the texture-based method of detection. Texture-based method is used mainly in medical fields and other specialized applications. This paper makes an attempt to use texture-based method for shadow detection in case of static images.

Keywords Tracking · Shadow detection · Histogram · Entropy · Texture-based method

---

Kavita
JVWU, Jharna, Jaipur, Rajasthan, India
e-mail: drkavita@jvwu.ac.in

M. K. Sabnis (✉)
CS & IT Department, JVWU, Jaipur, Rajasthan, India
e-mail: manojsab67@yahoo.co.in

M. K. Shukla
Amity School of Engineering Noida, Noida, UP, India
e-mail: manojshukla001@yahoo.co.in

K. Ray et al. (eds.), *Soft Computing: Theories and Applications*,
Advances in Intelligent Systems and Computing 742,
https://doi.org/10.1007/978-981-13-0589-4_26

1 Introduction

The introduction section specifies the flow in which the paper proceeds. After finalizing the image types, the shadow types are discussed so as to select the shadow type that has to be detected. Then, the shadow detection taxonomy is discussed so as to present the standard methods of which texture is one of them. Texture analysis at the image level is then discussed at length. Then on discussing the existing methods, the suggestions as improvement are presented in the form of the proposed algorithm. The results presented are evaluated and compared with the existing ones, which are then specified for comparison in form of graphs and tables.

2 Image Type

On consideration of the image type, dynamic images have the objects on the move. This type of images can have a number of frames which aids the analysis process by using background subtraction due to which only the object of interest can be obtained easily for further analysis [1, 2]. Similarly, there are static images, where a single image is present with a limited number of pixels in that image. These images operate with no scope of background subtraction which, in turn, limits the analysis process [2, 3].

Thus, working with static images as a challenge and hence a lot of work is done on dynamic images in the field of tracking, navigation, etc. Thus, in static images, shadow detection has been explored in this paper [4].

3 Shadow Type

The shadows comprises of being self, attached, overlapping, or cast. Out of these shadows, the attached- and the overlapping-type shadows cannot be handled by a single method. Therefore, a hierarchical approach is required for them. These types of shadow are considered to be outside the working domain for the paper under consideration. The self-shadows being a small overlap within the shadow can be ignored, thus cast shadow on the background having the shape similar to that of the object forms the main source for false tracking. Therefore, this paper focuses on the detection and elimination of these cast shadows [5–7].

Cast shadow is made up of umbra and penumbra. The umbra is the central dark uniform part and the penumbra is the varying intensity region which extends from the umbra boundary to the shadows outer edges. Thus, to compare the shadow and its object, both the umbra and the penumbra is required [8].

4 Shadow Detection Taxonomy

The taxonomy was suggested by Prati et al. which specifies various working features which further helped to define the four standard shadow detection methods [9, 10]. At the primary level, two approaches were defined, the deterministic (D) and the statistical approach (S). The statistical was further bifurcated into parametric (SP) and nonparametric (SNP). The SP model is usually not preferred because a number of parameters available in the image may not be favorable for texture analysis. On the other hand, SNP model if used, the pixels in it have to be trained. As static images are used, the number of pixels available for training is less. Therefore, more number of iterations are required which may further increase the computational load.

In case of deterministic approach, DM, i.e., the deterministic model based or DNM deterministic nonmodel based is available. In case of-DM based model, the image model created will be valid for that conditions of environment and illumination and if there is any change in any of the two parameters considered then it will make the model imbalanced.

Thus, DNM has two versions DNM1 and DNM2 of which DNM2 is selected as it gives penumbra detection. Also, it can further be used with HSV color model.

5 Texture

Texture processing is required in applications like automatic inspection where defects in images, carpets, and automatic paints can be detected. It is also used for automatic extraction of medical images for classification of normal tissues from the abnormal ones, for document imaging, character recognition, postal address recognition, and interpretation of maps [11, 12]. In such wide and useful applications, texture analysis can be used in the field of image processing. In this, it is possible to recognize the texture of images or videos. This leads the image processing field to extend its domain to a number of texture-based algorithms forming the standard texture-based methods. For this, it is required to understand the physical concept of the texture. In general, texture is considered to be a fundamental concept that attributes to almost everything in nature. From the human perception, texture analysis initially was dependent on the look and feel concept. The texture was then defined as spatially repetitive structure of surfaces formed by repeating a particular element or several elements in different relative spatial positions [11, 12]. Considering three texture samples as shown in Fig. 1 named as Texture Images represented are presented below.

ptThe image on the left in the figure has the texture domain which can be seen and understood where the variations in the parameter are not to that extent that it can be measured. Similarly, the image in the center has texture in repeated pattern and are in the measurable form. In the image to the right, the texture maps are more

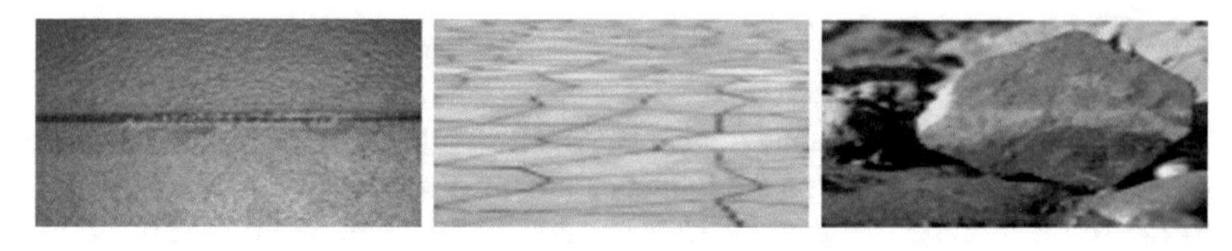

Fig. 1 Texture images

closely to the image processing domain. In such type of images, local variations can be measured in terms of colour and geometry. Texture can now be defined in a very simplified manner as the area containing variations of intensities that form repeated patterns.

6 Texture Analysis

Representing the existing images texture mathematically in a proper way for its use is called as the texture analysis. It can also be referred as a class of mathematical procedures and models that characterizes the spatial variations within an image as a means for extracting information. As the image is made of pixels, the texture in this case can be defined as consisting of mutually related pixels or group of similar pixels. This group of pixels is called as texture primitives or texture elements, i.e., texels [13, 14].

7 Image-Level Texture Analysis

In case of image analysis, assumptions are made saying that light source is uniform, then the intensity of the image will also be uniform. Due to texture analysis, it is possible to remove this assumption as regions of images do not always give the same intensities, but it depends on the variations of intensities given by image pixels at that point which is basically called as texture of that region.

Thus it can be analyzed as, image has regions, regions have textures, and these regions define local spatial organization of spatially varying spectral values that are repeated in the region. These textures contents have to be quantified for their measurement. For this, descriptors are defined which specify the smoothness, coarseness, roughness, and regularity. This cannot be measured but only felt at a qualitative level. Also, this qualitative analysis differs from person to person.

This leads to the redefining of texture in the form of features that can be obtained in a measurable form. This leads to the definition of informal qualitative structural feature like fineness, coarseness, smoothness, granularity, linearity, directionality, roughness, regularity, and randomness [12, 14].

This human-level classification could not be used as a formal definition in case of image texture analysis within the domain of image processing.

8 Model-Based Analysis

In case of image processing, for the texture analysis of the image computational model has to be developed depending on the texture features exhibited by that image. After several decades of research and development in the field of texture analysis and synthesis, a number of computational characteristics and properties have been defined for images. Of all these properties, CBIR standards have been defined that accepts six texture-based properties called as the Tamura's texture features.

These features include coarseness, contrast, degree of directionality, line likeness, regularity, roughness, and Markov random. Out of all these features applied to the field of image processing, the two main features used are coarseness and contrast. These two are represented below [12–14].

Coarseness: In an image, there are some primitive elements which form the texture. These primitive elements are differentiated by their gray levels based on these characteristics. The coarseness in an image is defined as the distance between notable spatial variations between the gray levels of the pixels. They are further classified accordingly to the size of the primitive elements. Such variations further form the texture.

Contrast: In an image, there are always some variations of the gray level on pixel-to-pixel basis. Contrast, as the name suggests is a measure of the gray level variation in an image followed by the extent to which these distributions are based toward black and white colour.

9 Texture-Based Method

The texture-based method works on the principle that the shadow does not change the texture of the background on which it falls whereas the objects do change the texture of the background.

The texture-based algorithm so formed does not work in isolation but are used as one of the stages of hierarchical method of shadow detection. Texture-based method is only used when there is no geometrical relationship existing between the object and its shadow.

Working within the domain of image processing the best condition is by histogram and entropy techniques. The combination of these two techniques is used for the proposed algorithm represented in the paper.

10 Algorithm

The existing algorithm is first presented and then improvements in the same are suggested as the proposed algorithm.

10.1 Existing Algorithm

Various existing algorithms of D. Xu, Tian, Leone and Distante Sanin et al, and G. N. Srinivasan and Shobha. G are considered as a base to put forward the proposed algorithm [10, 15, 16].

10.2 Proposed Algorithm

The various modules used in the proposed algorithm are image selection, image processing, histogram calculations, probability, sum, entropy calculations, thresholding, comparison, and output representation.

Image Selection — Texture can be used as a method where it is not possible to use colour, intensity, or geometry-based techniques. It works on the principle that less variations are shown by the background pixels with shadows as compared to non-shadow pixels. Thus, the algorithm is designed on these lines, i.e., to detect from the images those pixels with minimum intensity, variations which can be then classified as the shadow.

Image Processing — The input image is converted into gray scale and also its dimensions are calculated.

Histogram — The gray level histogram of the test image is obtained.

Probability — The probability calculation for each histogram level is done as P(i) = H(i)/(Size of image)

Sum Calculations — Four sum probability, sum values are calculated of which two are the natural and other two are for determination of the log.

Entropy Calculations — In an image, how many pixel changes take place can be found out by its entropy which defines this measure of randomness.

Thresholding — Select *m* equal to one. Find its entropy and NaN entropy and then select the largest value of *m* as the threshold level.

Comparison — For k equal to 1–255, compare entropy (k) and entropy (m) and select the largest among this as the threshold value. Now, for K equal to 1–255 and for i equal to k–256, find two minimum entropy values for each row between two adjacent columns.

Mask Formation	Input size mask with all zeros is formed and on this mask, for each row between two nearby columns, if the adjacent pixel entropy is minimum then set that pixel to one else it is maintained as zero. Now repeat this process for k equal to 1–255.
Digital Mask	In the mask, set the pixel marked as one to logic one and the remaining pixels which are zero are set to logic zero.
Shadow Detection	On the mask, logic one represents the shadow while those pixels indicated as logic zero are the objects. Now, keep the mask on the image and wherever the mask is set to one, the corresponding pixels in the image is set to zero thus removing the shadow areas. These pixels can be represented with some colour so as to show the shadow eliminated areas clearly.

11 Results

The results are represented from two angles; the quantitative and the qualitative. The quantitative is represented in the form of accuracy and the qualitative analysis in the form of metrics selected. The quantitative results are represented by three images as represented in Fig. 2 named as texture results. The images on the left indicate three input test images. The images at the center are their binary intermediate outputs. The images on the right are their respective outputs of which the right top and right center are the images and that on right bottom is the binary images, respectively.

12 Evaluation

The evaluation of the available results is done at the quantitative and qualitative levels. The quantitative evaluation is the measure of accuracy. Three accuracies are considered, the producer, user and the combined accuracy. They are specified as [10, 17–21].

Producer accuracy with two parameters, ηs and ηp are as represented by Eqs. (1) and (2), respectively, where

$$\eta\mathrm{s} = \mathrm{TP}/(\mathrm{TP} + \mathrm{FN}) \tag{1}$$

$$\eta\mathrm{n} = \mathrm{TP}/(\mathrm{FP} + \mathrm{TN}) \tag{2}$$

User accuracy also with two parameters, Ps and Pn are as represented by Eqs. (3) and (4), respectively, where

$$\mathrm{Ps} = \mathrm{TP}/(\mathrm{TP} + \mathrm{FP}) \tag{3}$$

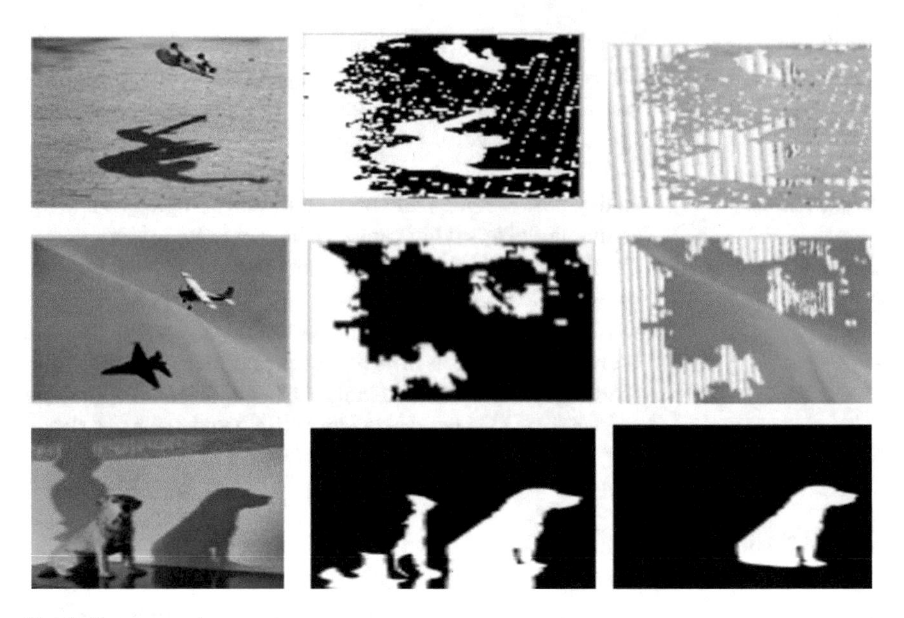

Fig. 2 Texture results

$$\text{Pn} = \text{TN}/(\text{TN} + \text{FN}) \tag{4}$$

Combined accuracy as T is represented by Eq. (5) as,

$$\text{T} = (\text{TP} + \text{TN})/(\text{TP} + \text{TN} + \text{FP} + \text{FN}) \tag{5}$$

The results of accuracy are represented in Table 1 named as quantitative evaluation on the following page as follows:

The qualitative evaluation is done with a set of metrics which are applicable to the texture-based method. The Average Measuring Value (AMV) and the Relative Measuring Value (RMV) are compared with these metrics and mapped on a scale from one to five where one is excellent, two is good, three is fair, four is poor with five very poor. The metrics used in the Table 2 named as metric evaluation are RN (Robustness to Noise), OI (Object Independence), SI (Scene Independence), CC (Computational Load), SID (Shadow Independency), PD (Penumbra Detection), ID (Illumination Independency), DT (Detection/Discrimination Tradeoff), CS (Chromatic Shadow), SC (Shadow Camouflage), and ST (Shadow Topology) respectively [9, 10, 18].

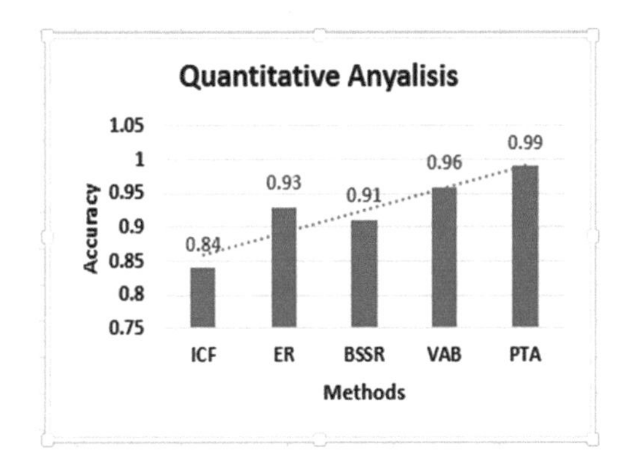

Fig. 3 Texture quantitative analysis

13 Result Analysis

The results obtained by qualitative and quantitative evaluation are now represented in graphical form for detailed analysis. In case of quantitative analysis, the producer accuracy value for existing algorithms like invariant colour feature is 0.84, edge ratio method is 0.93, for BSSR method is 0.91, and varying area-based method is 0.96.

Compared to this, the accuracy for the proposed textured-based algorithm is 0.99. This is as shown in the Graph Number 3 named as texture quantitative analysis.

The qualitative analysis shows that of all the even available metrics almost nine metrics have value two, i.e., Good. This gives almost an 80% improvement over the existing algorithm. This is as shown in the Graph number 4 called as the qualitative metrics.

Table 1 Quantitative evaluation

Sr. no	Input image	Producer accuracy		User accuracy		Combined accuracy
		Ns	Nn	Ps	Pn	T
1		0.95	0.2472	0.0369	0.98	0.9219
2		1	0.0395	0.0809	1	0.6903
3		1	0.1418	0.1940	1	0.6293

Table 2 Metric evaluation

	RN		OI		SI		CC		SID		PD		ID		DT		CS		SC		ST	
	E	P	E	P	E	P	E	P	E	P	E	P	E	P	E	P	E	P	E	P	E	P
AMV	3	2	3	2	3	2	3	4	3	2	3	2	3	2	3	2	4	4	1	1	2	2
RMV	5	5	4	4	4	4	3	3	4	4	3	3	3	3	2	2	3	3	1	1	2	2

Fig. 4 Qualitative metrics

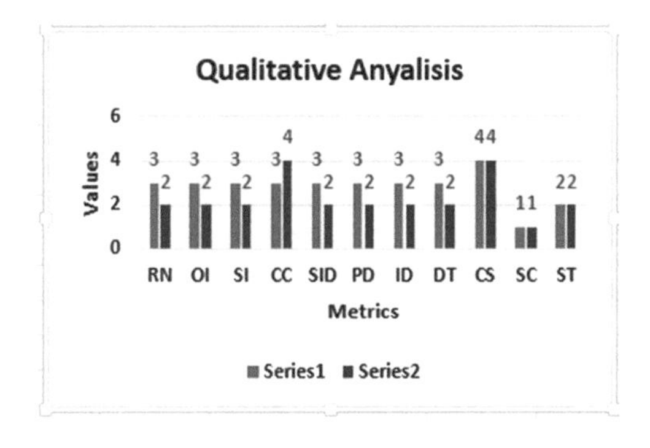

14 Conclusion

The proposed algorithm shows that there is an improvement in the results as compared to the existing algorithm as seen by the qualitative and quantitative analysis. This texture method, mainly used for CBIR is successfully used for shadow detection within the domain of image processing.

15 Future Scope

Any single method cannot satisfy all the image-type requirements. Thus, the future scope is that a system needs to have a hierarchical method having two stages of implementation namely the coarse and fine method of design.

References

1. Rosin, P.L, Ellis, T.: Image difference threshold strategies and shadow detection. Institute of remote sensing applications, Joint Research Center, Italy
2. Yao, J., Zhang, Z.: Systematic static shadow detection. In: Proceedings of the 17th International Conference on pattern Recognition, 1051–4651. IEEE, Computer Society (2004)
3. Stauder, J., Mech, R., Ostermann, R.: Detection of moving cast shadows for object segmentation. IEEE, Trans. Multimed. **1**(1), 65–76 (1999)
4. Madsen, C.B, Moeslund, B.T, Pal, A, Balasubramanian, S.: Shadow detection in dynamic scenes using dense stereo information and an outdoor illumination model. In: Computer Vision and Media Technology Lab, Aalborg university Denmark, pp. 110–125 (2009)
5. Ullah, H., Ullah, M., Uzair, M., Rehmn, F.: Comparative study: The evaluation of shadow detection methods. Int. J. Video Image Process. Netw. Secur. **2**(10):1–7
6. Salvador, E., Cavallaro, A., Ebrahimi, T.: Shadow identification and classification using invariant colour models. Signal Processing Laboratory (LTS), Swiss Federal Institute of Technology (EPFL), CH-1015 Lausanne, Switzerland
7. Jyothirmai, M.S.V., Srinivas, K., Rao, V.S.: Enhancing shadow area using RGB colour space. IOSR J. Comput. Eng. July Aug, **1**(2), 24–28 (2012). ISSN:2278-0661

8. Stauder, J., Mech, R., Ostermann, J.: Detection of moving cast shadows for object segmentation. IEEE Trans. Multimed. **1**(1), 65–76 (1999)
9. Prati, A., Mikie, I., Trivedi, M.M., Cucchiara, R.: Detecting moving shadows formulation, algorithms and evaluation. Technical Report-Draft Version, 1–39
10. Sanin, A., Sanderson, C., Lovell, B.C.: Shadow detection: a survey and comparative evaluation of recent methods. Pattern Recogn., Elsevier, **4**(45), 1684–1695 (2012). ISSN 0031–3203
11. Tuceryan, M., Jain, A.K.: Texture Analysis. In: The Handbook of Pattern Recognition and Computer Vision, 2nd edn., pp. 207–248 (1998). World scientific Publication Co
12. Srinivasan, B.G.N., Shobha, G.: Statistical texture analysis. In: Proceeding of World Academy of Science, Engineering and Technology, vol. 36, 1264–1269, Dec, 2008. ISSN 2070–3747
13. Mateka, A., Strzelecki, M.: Texture analysis method review. Technical University of Looz, Institute of Electronics, Report, Brussel (1998)
14. Lin, H.C., Chiu, C.Y., Yang, S.N.: Texture analysis and description in linguistic terms. In: ACCV2002, The 5th Asian Conference on Computer Vision, Melbourne, Australia, 23–25 Jan (2002)
15. Moving Cast Shadow Detection. In: Vision Systems Segmentation and Pattern Recognition, pp. 47–58
16. Leone, A., Distante, C.: Shadow detection for moving objects based on texture analysis. J. Pattern Recogn. (2006). ISSN 0031-3203; Pattern Recogn. **40** (2007), 1222–1233
17. Huang, J.B., Chen, C.S.: Moving cast shadow detection using physics based features, pp. 2310–2317. IEEE (2009). 978-1-4244-3991-1/09/
18. Lakshmi, S., Sankaranarayanan, V.: Cast shadow detection and removal in a real time environment, pp. 245–247. IEEE (2010). ISDN 978-1-4244-9008-0/10/
19. Withagen, P.J., Groen, F.C., Schutte, K.: Shadow detection using physical basis. Intelligent Autonomous System Technical Report, pp. 1–14 (2007)
20. Sun, B., Shutao, Li.: Moving cast shadow detection of vehicle using combined colour models. IEEE (2010)
21. Chung, K.L., Lin, R.Y., Huang, Y.H.: Efficient shadow detection of colour ariel image based on successive thresholding scheme. Trans. Geo Sci. Remote Sens., 0196–2892, **2**(42), 671–682. IEEE (2009)

A PSO Algorithm-Based Task Scheduling in Cloud Computing

Mohit Agarwal and Gur Mauj Saran Srivastava

Abstract Cloud computing is one of the most acceptable emerging technologies, which involves the allocation and de-allocation of the computing resources using the Internet as the core technology to compute the tasks or jobs submitted by the users. Task scheduling is one of the fundamental issues in cloud computing and lots of efforts have been made to solve this problem. For the success of any cloud-based computing model, efficient task scheduling mechanism is always needed which, in turn, is responsible for the allocation of tasks to the available processing machines in such a manner that no machine is over- or under-utilized while executing them. Scheduling of tasks belongs to the category of NP-Hard problem. Through this paper, we are proposing the particle swarm optimization (PSO)-based task scheduling mechanism for the efficient distribution of the task among the virtual machines (VMs) in order to keep the overall response time minimum. The proposed algorithm is compared using the CloudSim simulator with the existing greedy and genetic algorithm-based task scheduling mechanism and results clearly shows that the PSO-based task scheduling mechanism clearly outperforms the others techniques which are taken into consideration.

Keywords Cloud computing · Virtual machines · Particle swarm optimization
Task scheduling · Makespan

1 Introduction

With the advancement in the field of information technology, the need of computing model which helps the users to carry out their day-to-day computation of task also arises with the passage of time. Cloud computing model presents itself as the

---

M. Agarwal (✉) · G. M. S. Srivastava
Department of Physics & Computer Science, Dayalbagh Educational Institute, Agra, India
e-mail: rs.mohitag@gmail.com

G. M. S. Srivastava
e-mail: gurmaujsaran@gmail.com

K. Ray et al. (eds.), *Soft Computing: Theories and Applications*,
Advances in Intelligent Systems and Computing 742,
https://doi.org/10.1007/978-981-13-0589-4_27

solution to such demand and since its inception; this computing paradigm gains a lot of popularity both in academia as well as in industry [1]. The term *Cloud* may be defined as the pool of the computing resources like memory units, processing elements, networking components, etc., which are allocated and released in order to execute the tasks submitted by the users [2]. The prominent characteristics like ubiquitous, economical, scalable, on-demand access and elastic in nature are responsible for the migration of the business from the traditional model to the cloud-based model [3, 4]. Cloud computing is said to be an extension of the existing technologies like distributed computing, grid computing and utility computing [1] with the distinguishing features like virtualization, virtual machine migration and much more. Cloud computing is successful in providing the new business opportunity not only to the service providers but also to the users of such services by the means of the platform for delivering Software as a Service (SaaS), Platform as a Service (PaaS) and Infrastructure as a Service (IaaS). It allows the user to focus on their core business activity instead of investing their time and wealth in setting the IT infrastructure required to perform their business activity. Cloud computing model enables the service users to use the underlying IT infrastructure purely on the basis of plug and pay, i.e. users only need to pay for the duration of time for which they consume the services not for the entire period of time. The users also do not to need to bother about the location and number of the computing resources required for the processing of their tasks or jobs what they are submitting. Scheduling used to play a very significant role in the process of optimization as they involve the distribution of the load or tasks on the VMs in order to maximize their utilization and minimizing the overall execution time [5].

The remainder of this paper is organized as follows: Sect. 2 presents the related work in the field. In Sect. 3, particle swarm optimization is discussed while Sect. 4 helps in understanding the problem of task scheduling in cloud computing environment. Section 5 presents the experimental results and discussion followed by the conclusion in Sect. 6.

2 Related Work

Task scheduling involves the allocation of the tasks submitted by the users to the available virtual machines (VMs) in order to execute such task with the intent of keeping the overall response time as minimum as possible. The problem of task scheduling in cloud computing environment lies into the category of the NP-Hard problems [6]. Lots of sincere efforts have been made to explore the efficient solution of this problem and as the field of cloud computing is relatively new, still space is left for the new work. In this section, we are going to give the brief idea of the work which has been done. The author in [7] proposed the first-fit strategy and the same is used by the many cloud systems including Eucalyptus [8]. The strategy proposed in [7] nearly solves the problem of starvation along with the reduction in makespan for the jobs but fails to support the optimum usage of the involved resources. Mohit et al. [9] proposed the genetic algorithm-aware task scheduling in order to reduce the overall response time or makespan. The proposed strategy also results in the optimal utilization of the computing resources. In [10] the author used the Stochastic hill climbing algorithm in order to solve the underlying problem of load balancing. Hsu et al. [11] proposed the energy-efficient task scheduling mechanism for the better efficiency of data centers. The authors in [12] present the Cuckoo search-based model for the efficient task scheduling and show its supremacy over other techniques which were taken into the consideration.

3 Proposed Methodology

Particle swarm optimization (PSO) was originally proposed by Kennedy and Eberhart [9] in 1995 for the optimization. PSO is a stochastic optimization technique [4] which works on the principle of social behavior of fish schooling and birds flocking [13]. In PSO, particles used to fly in the search space and position of the particles changes due to socio-psychological tendency, i.e., emulation of an individual success by other individuals.

Let $\overrightarrow{P_i}(t)$ denote the position of ith particle, at iteration t. The position of this particle changed to $\overrightarrow{Pi}(t+1)$ by adding velocity $\overrightarrow{V_i}(t+1)$ as shown in Eq. 1.

$$\overrightarrow{P_i}(t+1) = \overrightarrow{P_i}(t) + \overrightarrow{V_i}(t+1) \tag{1}$$

Here velocity, as defined in Eq. 2 represents the socially exchanged information.

$$\overrightarrow{V_i}(t+1) = \omega * \overrightarrow{V_i}(t) + C_1 r_1 \left(\overrightarrow{pbest_i} - \overrightarrow{P_i}(t)\right) + C_2 r_2 \left(\overrightarrow{gbest_i} - \overrightarrow{P_i}(t)\right) \tag{2}$$

where

ω inertia weight;
$C_1 and C_2$ cognitive and social learning parameters respectively;
$r_1 and r_2$ random values and lies between [0, 1];
$\overrightarrow{pbest_i}$ personal best position of particle i;
$\overrightarrow{gbest_i}$ global best position.

4 The Problem

4.1 Task Scheduling

Task scheduling in cloud computing may be defined as the process of allocation or mapping of the task or jobs submitted by the users to the VMs in order to keep the

Table 1 Task scheduling pattern

Tasks	T_1	T_2	T_3	T_4	…	T_n
Virtual Machine (VM)	M_3	M_1	M_1	M_2	…	M_m

overall response time minimum. The jobs or tasks require the computing resources for their processing and sometimes it may be the case that required computing resource may not available at the particular location. In such situation, the task needs to be migrated to some resource rich location and this migration should be done in an efficient manner so that the time require for the execution remains minimum.

Let there are n tasks (T_1, T_2, T_3,..., T_n) and m virtual machines (M_1, M_2, M_3,..., M_m) for execution of such tasks with the condition that $n > m$. Such situation results into the m^n ways of allocating such tasks to the available machines. Our task is to present an efficient mechanism using *PSO* algorithm, which will allocate the given number of tasks to the machines in order to keep the response time as low as possible (Table 1).

4.2 Mathematical Formulation

Our objective of task scheduling in cloud computing is to reduce overall execution time; Execution time may be defined as the time required by a task to complete its execution over the allocated VM.

$$T_{exei} = \sum_{j=0}^{m} T_{ji} + C_{ji} \tag{3}$$

where

T_{exei} finish time for ith Virtual Machine
i ith Virtual Machine
j jth Task
T_{ji} execution time required by jth task when processed on ith Virtual Machine.
C_{ji} (TaskIsize + TaskOsize)/Bandwidth.

5 Experiments and Result Analysis

We have compared our proposed PSO-based task scheduling mechanism with established techniques like genetic algorithm as presented in [14] and greedy-based scheduling techniques using the CloudSim simulator [15] which has been designed

Table 2 Virtual machine configuration (Sample)

VMId	MIPS	Bandwidth
M_1	1000	1000
M_2	1000	1000
M_3	2000	1000
M_4	3000	1000
M_5	5000	1000

Table 3 Task configuration (Sample)

TaskId	MI	TaskISize	TaskOSize
T_1	10000	1000	900
T_2	20000	500	700
T_3	10000	1000	1050
T_4	30000	1200	1100
T_5	50000	1300	1200
…			
T_n	100000	1250	1500

Table 4 PSO parameters

Parameters	Values
C_1	1.6
C_2	1.6
r_1	[0, 1]
r_2	[0, 1]
ω	0.9

and developed by University of Melbourne, Australia. Tables 2 and 3 present the sample configuration of VMs and task, respectively, while Table 4 provides the various parameters required for the PSO to execute. The experiments involve various tasks whose size or length lies in the range of (10000–100000) million instructions (MI) and processing power of VMs lies in the (1000–5000) million instructions per seconds (MIPS).

Several experiments with different parameters have been performed to evaluate the performance of PSO algorithm over the other well-established scheduling strategies.

Figure 1 shows the comparison of the execution time obtained for the different sets of tasks. 10, 20 30, 40 and 50 tasks with different task lengths and other involved parameters are executed on the 5 VMs with different configurations as present in Table 2. Execution time using Eq. 3 obtained by the PSO-based task scheduling mechanism is minimum in comparison to others in all the cases.

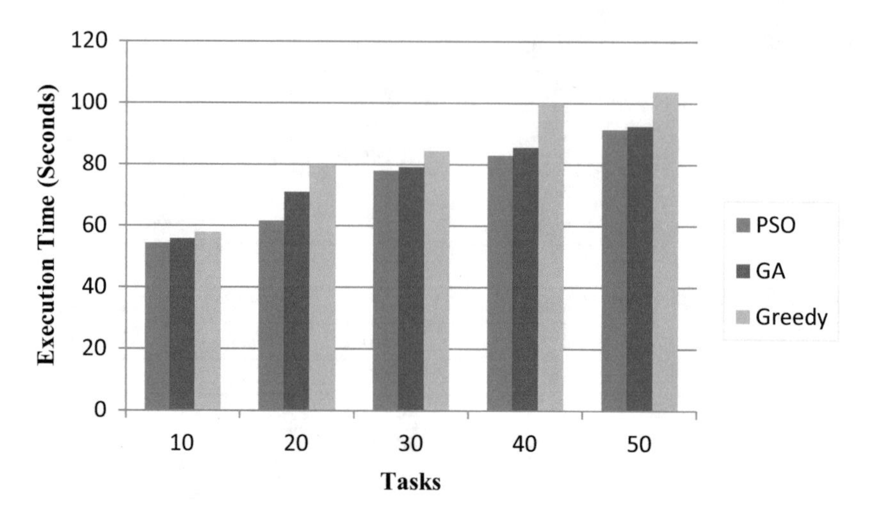

Fig. 1 Execution time for the different number of tasks executing on 5 VMs

6 Conclusion

This paper presents the PSO-based task scheduling mechanism for the cloud computing model where the number of computing resources and tasks submitted by the users changes dynamically. So in such a dynamic environment like cloud computing, it is important to utilize the underlying resources optimally otherwise this may lead to the huge economic loss to the cloud service provider. PSO algorithm proves themselves better for such kind of optimization problems and results also established this fact, as PSO-based task scheduling outperforms the other task scheduling techniques. For future studies, we will try to find the suitability of this algorithm for the quality of services (QoS) parameters like cost, energy, and bandwidth.

References

1. Sadiku, M.N., Musa, S.M., Momoh, O.D.: Cloud computing: opportunities and challenges. IEEE Potentials **33**(1), 34–36 (2014)
2. Alhamazani, K., Ranjan, R., Mitra, K., Rabhi, F., Jayaraman, P.P., Khan, S.U., Bhatnagar, V.: An overview of the commercial cloud monitoring tools: research dimensions, design issues, and state-of-the-art. Computing **97**(4), 357–377 (2015)
3. Duan, Q., Yan, Y., Vasilakos, A.V.: A survey on service-oriented network virtualization toward convergence of networking and cloud computing. IEEE Trans. Netw. Serv. Manage. **9**(4), 373–392 (2012)
4. Shi, Y., Eberhart, R.C.: Parameter selection in particle swarm optimization. In: International Conference on Evolutionary Programming, pp. 591–600. Springer, Berlin, Heidelberg (1998)
5. Zomaya, A.Y., Teh, Y.H.: Observations on using genetic algorithms for dynamic load-balancing. IEEE Trans. Parallel Distrib. Syst. **12**(9), 899–911 (2001)

6. Michael, R.G., David, S.J.: Computers and Intractability: A Guide to the Theory of NP-Completeness, pp. 90–91. WH Free. Co., San Fr (1979)
7. Brent, R.P.: Efficient implementation of the first-fit strategy for dynamic storage allocation. ACM Trans. Programm. Lang. Syst. (TOPLAS) **11**(3), 388–403 (1989)
8. Nurmi, D., Wolski, R., Grzegorczyk, C., Obertelli, G., Soman, S., Youseff, L., Zagorodnov, D.: The eucalyptus open-source cloud-computing system. In: Proceedings of the 2009 9th IEEE/ACM International Symposium on Cluster Computing and the Grid, pp. 124–131, May, 2009. IEEE Computer Society
9. Kennedy, J., Eberhart, R.: Particle swarm optimization. In: IEEE International Conference on Neural Networks, pp. 1942–1948 (1995)
10. Mondal, B., Dasgupta, K., Dutta, P.: Load balancing in cloud computing using stochastic hill climbing-a soft computing approach. Procedia Technol. **4**, 783–789 (2012)
11. Hsu, Y.C., Liu, P., Wu, J.J.: Job sequence scheduling for cloud computing. In: 2011 International Conference on Cloud and Service Computing (CSC), pp. 212–219. IEEE (2011)
12. Agarwal, M., Srivastava, G.M.S.: A cuckoo search algorithm-based task scheduling in cloud computing. In: Advances in Computer and Computational Sciences, pp. 293–299. Springer, Singapore (2018)
13. Shi, Y., Eberhart, R.C.: Empirical study of particle swarm optimization. In: Proceedings of the 1999 Congress on Evolutionary Computation, 1999. CEC 99, vol. 3, pp. 1945–1950. IEEE (1999)
14. Agarwal, M., Srivastava, G.M.S.: A genetic algorithm inspired task scheduling in cloud computing. In: The Proceedings of 2nd IEEE Conference on Computing, Communication and Automation 2016 (2016)
15. Calheiros, R.N., Ranjan, R., Beloglazov, A., De Rose, C.A., Buyya, R.: CloudSim: a toolkit for modeling and simulation of cloud computing environments and evaluation of resource provisioning algorithms. Softw. Pract. Exp. **41**(1), 23–50 (2011)

A Novel Approach for Target Coverage in Wireless Sensor Networks Based on Network Coding

Pooja Chaturvedi and A. K. Daniel

Abstract Wireless sensor network (WSN) is a type of infrastructure network, which is used to monitor and collect the characteristics of the environmental phenomenon such as military applications and environmental observations. The fundamental functionality of the nodes in a sensor network is to collect the historical data from the nearby environment and send it to the base station. Network coding is an efficient mechanism used to increase the network throughput and hence performance by mixing the packets from several sources at the intermediate nodes and hence reduces the number of transmission. The proposed protocol is an enhancement of energy-efficient coverage protocol (EECP) for target coverage by incorporating the network coding technique. The simulation results show that the proposed protocol improves the network performance in terms of number of transmissions, energy consumption and throughput by a factor of 63%, 62% and 61% respectively.

Keywords Wireless sensor networks · Target coverage · Network coding
Performance · Throughput · Network lifetime

1 Introduction

WSN is an ad hoc network consisting of a number of sensor nodes which are responsible for sensing the information from the environment in a distributed manner and transmit it to a central node known as base station [1, 2]. In the conventional networks, the traditional mechanism of data propagation is routing in which the intermediates utilize the concept of store and forward until the data reaches to the base station. There was no consideration of processing of the sensed information. But the researchers

---

P. Chaturvedi (✉) · A. K. Daniel
Department of Computer Science and Engineering, M. M. M. University of Technology, Gorakhpur, India
e-mail: chaturvedi.pooja03@gmail.com

A. K. Daniel
e-mail: danielak@rediffmail.com

K. Ray et al. (eds.), *Soft Computing: Theories and Applications*,
Advances in Intelligent Systems and Computing 742,
https://doi.org/10.1007/978-981-13-0589-4_28

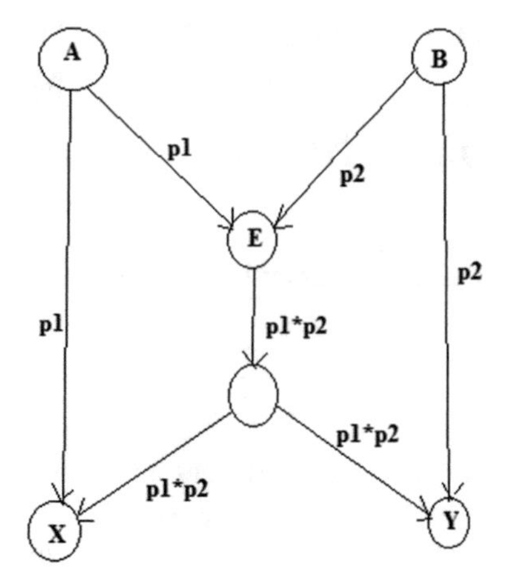

Fig. 1 Network coding application

have now started to study the impact of processing at the intermediate nodes to increase the network throughput. The processing of the information at intermediate node is known as network coding [3]. The topology of the sensor networks rises the problem of congestion and resource constraints as the traffic pattern in this is from the nodes to base station, i.e., many to one. Network coding is considered as a technique which aims to mitigate these problems by employing mathematical operations at the intermediate nodes.

The organization of the paper is as follows: concepts of network coding in Sect. 2, related work in Sect. 3, proposed protocol in Sect. 4, simulation results and analysis in Sects. 5 and 6 concludes the paper.

2 Network Coding

The network coding technique is considered as the application of the mathematical operations to mix the packets from the source nodes with the objective of minimizing the number of transmissions. The advantages of network coding technique can be analysed on a butterfly network in which multiple sources want to transmit their information to multiple destinations as shown in Fig. 1.

In the above figure, nodes A and B have packets $p1$ and $p2$ to transmit to the nodes X and Y. In the absence of network coding, each node will require two transmissions for each packet transmission one for sending the packet and other for acknowledgement which results in overall four transmissions. If network coding is applied at each node then the packets received at E are coded in one packet and transmitted to the X and Y nodes which results in overall three transmissions. Hence, the network coding technique reduces the number of transmission required for the data transmissions.

Another advantage of the network coding is ensuring reliability. The common approach used to ensure reliability is to use feedback mechanism which is not efficient as it consumes bandwidth. To ensure the reliability, the packets from the same source are encoded using random linear coding. Consider the node A wants to send $p1$ and $p3$ to node X and if the link reliability is 2/5. The packets are coded as $p1+p3$, $2p1+p3$, $p1+2p3$, $3p1+p3$, $p1+3p3$. The link reliability ensures that the two packets out of the five coded packets are received correctly at the destination X node [4, 5].

3 Related Work

Target coverage is a major problem in the field of wireless sensor networks which aims to monitor a set of predefined targets such that the network lifetime is improved by considering the scarce resources in the network. The point in the target region is said to be covered if it lies within the sensing range of at least one node. The coverage problem can be classified as *1*-coverage, *2*-coverage or k-coverage depending on the number of nodes which can monitor a point in the target region [6].

In [7], the authors have proposed a node scheduling protocol in which the number of set covers to monitor all the targets with the desired confidence level. The nodes are included in the set cover on the basis of node contribution, trust values and coverage probability. The dynamism and uncertainty of the network are considered by incorporating the trust concept. The trust values of the nodes are determined as the weighted average of the direct trust, recommendation trust and indirect trust. The direct trust is further calculated as the weighted average of the data trust, communication trust and energy trust [8–12].

The motivation of this paper is to design a scheduling protocol for target coverage such that the network lifetime is improved while monitoring all the targets for maximum duration along with the consideration of scarce resources. The network coding is a mechanism which can significantly reduce the energy consumption. So, in this paper, we aim to improve the *EECP* protocol by incorporating the network coding technique which aims to perform the processing on the collected data using some mathematical operations and enhance the network performance.

4 Proposed Protocol

For a network of s nodes and t targets, the target coverage problem aims to increase the network lifetime such that all the targets are monitored with the desired confidence level. The observation probability P_{ij}^{obs} of a target t_i by a sensor node s_j is determined as

$$p_{ij}^{obs} = \text{cov}(i, j) \times T_{ij} \tag{1}$$

where cov (i, j) represents the coverage probability of the target i with respect to the node j and T_{ij} represents the trust value of the target i with respect to node j.

Table 1 Proposed protocol

Terms s: source, d: destination, th: threshold, d(pi, pj): difference between the packets pi and pj. Output: *p'*: encoded packet
Algorithm
1. Determine the source *s* and destination *d*
2. Determine the packet p_i to transmit from *s* to *d*.
3. $d(p_i \text{-} p_j) > th$ then perform encoding
$p' = p_1 \oplus p_2$
forward the encoded packet to next node until it reaches the destination node d.

Based on the observation probability, the base station determines the schedule of the set covers which are activated periodically to monitor the targets.

In this paper, we propose a network coding protocol to reduce the number of transmissions and hence increase the network performance in terms of throughput and energy consumption. The two level coding mechanisms are adopted: one at the cluster level and at the set cover. At the cluster level, *XOR* operation is used to mix the packets at the intermediate nodes. At the set cover level, the leader node is selected in each set cover based on the distance from the base station and it performs random linear coding. The proposed protocol for network coding is shown in Table 1.

5 Simulation Result and Analysis

The performance of the proposed protocol is evaluated through several experiments in terms of number of transmissions and energy conservation. A network of 6 nodes and 3 targets is deployed randomly and uniformly in the target region of dimensions 50 * 50 as shown in Fig. 2.

The other simulation parameters considered are shown in Table 2.

Based on the observation probability, the various set covers are obtained as shown in Table 3.

The leader node is selected within each set cover which transmits the collected data to the base station directly. The farther nodes transmit the data to the leader node which performs the network coding at the received packets and transmit it to the base station. The number of transmissions required to transmit a single data packet to the base station for the various set cover is as shown in Table 4.

The number of transmission required to transmit a data packet in various set covers is shown in Fig. 3. The total number of transmissions required to transmit the 800 data packets for the existing and proposed is 600 and 225, respectively. The results show that the proposed protocol reduces the number of transmissions by 63%.

The energy consumption in various set cover for both the approaches is as shown in Table 5.

Table 2 Simulation parameters

Parameter	Values
No. of nodes	6
No. of targets	3
Area of region	50 * 50
Data rate	2 kbps
No. of communications	1000
No. of data packets	800
Packet size	50 bits
Initial energy of node	10 J
Energy consumption in transmission	0.66 J
Energy consumption in reception	0.33 J
Energy consumption in sleep state	0.003 J

Table 3 Set covers obtained

Set cover	Node set
C1	{4}
C2	{1,4}
C3	{1,6}
C4	{3,4}
C5	{1,3,6}
C6	{3,4,6}
C7	{4,5}
C8	{1,5,6}

Table 4 Number of transmissions required to transmit a single data packet

Set cover	Without network coding	With network coding
C1	1	1
C2	3	2
C3	4	2
C4	3	2
C5	11	3
C6	9	3
C7	6	2
C8	11	3

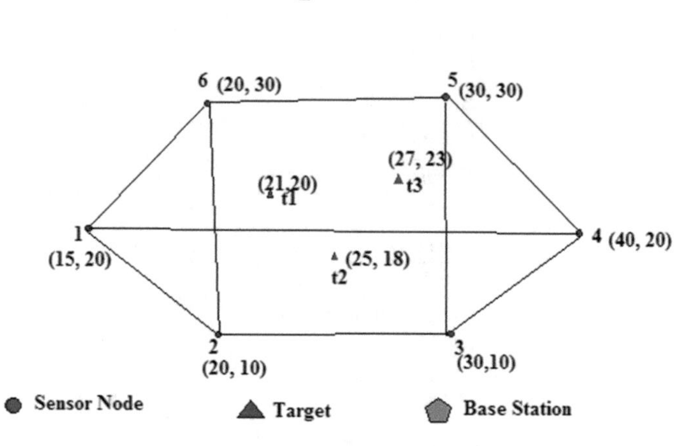

Fig. 2 Deployed network

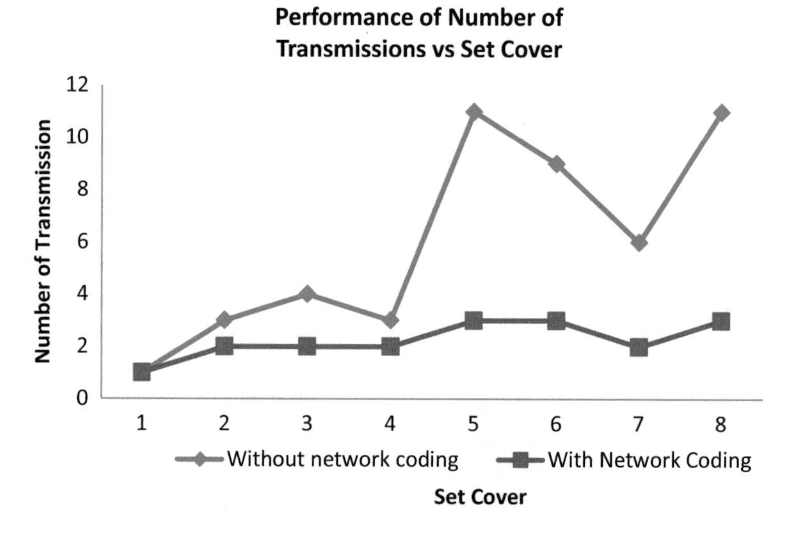

Fig. 3 Performance of number of transmissions versus set covers

The performance of the energy consumption in various set covers is shown in Fig. 4. The results show that the proposed approach shows the decrease in energy consumption as compared to the existing approach.

The performance of the proposed protocol is also evaluated in terms of packet throughput, which represents the number of packets transmitted per unit time. The results show that the packet throughput for the existing and proposed approaches is 7 and 18, respectively.

Table 5 Energy consumption

Set cover	Without coding	With coding
C1	0.6	0.6
C2	1.8	1.2
C3	2.4	1.2
C4	1.8	1.2
C5	6.6	1.8
C6	5.4	1.8
C7	3.6	1.2
C8	6.6	1.8

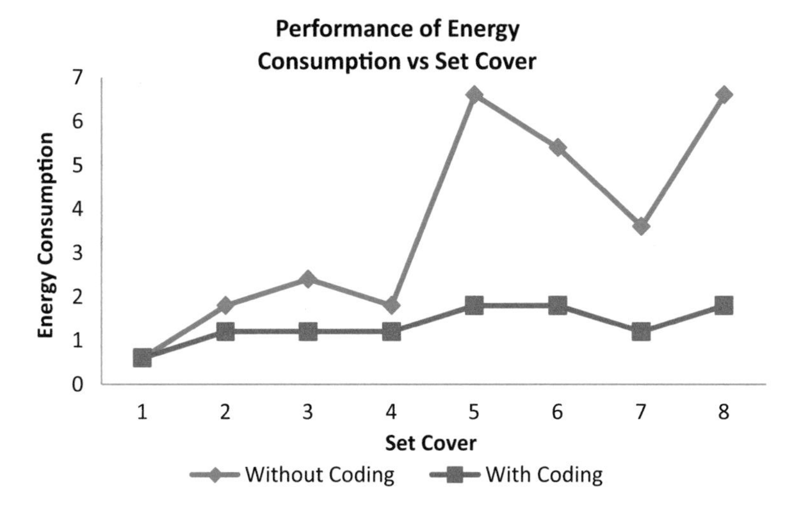

Fig. 4 Performance of energy consumption versus set cover

The overall energy consumption for both the approaches is 29 J and 11 J, respectively. The results show that the proposed protocol reduces the energy consumption by 62% than the existing approaches.

6 Conclusion

The paper proposes a network coding mechanism at the cluster level and set cover level for the node scheduling protocol for target coverage. The leader node in each set cover performs the coding of the packets at the intermediate nodes. The simulation results show that the proposed protocol improves the network performance in terms of number of transmissions, energy consumption, throughput and average energy by a factor of 63%, 61% and 62%, respectively.

References

1. Akyildiz, I.F., Su, W., Sankarasubramaniam, Y., Cayirci, E.: Wireless sensor networks: a survey. Comput. Netw. **38**(4), 393–422 (2002)
2. Yick, J., Mukherjee, B., Ghoshal, D.: Wireless sensor networks survey. Comput. Netw. **52**, 2292–2330 (2008)
3. Thai, M.T., Wang, F., Du, D.-Z.: Coverage problems in wireless sensor networks: designs and analysis. Int. J. Sens. Netw. **3**(3), 191–200 (2008)
4. Chou, P.A., Wu, Y., Jain, K.: Practical network coding. In: Proceedings of the Conference on Communication Control and Computing, Oct. 2003
5. Dong, Q., Wu, J., Hu, W., Crowcroft, J.: Practical network coding in wireless networks. In: Proceedings of the 13th Annual ACM International Conference on Mobile Computing and Networking, Montr´eal, QC, Canada, pp. 306–309, 9–14 Sept. 2007
6. Taghikhaki, Z., Meratnia, Havinga, P.J.M.: A trust-based probabilistic coverage algorithm for wireless sensor networks. In: 2013 International Workshop on Communications and Sensor Networks (ComSense-2013). vol. 21, pp. 455–464. Procedia, Computer Science (2013)
7. Chaturvedi, P., Daniel, A.K.: An energy efficient node scheduling protocol for target coverage in wireless sensor networks. In: The 5th International Conference on Communication System and Network Technologies (CSNT-2015), Apr 2015
8. Jiang, J., Han, G., Wang, F., Shu, L., Guizani, M.: An efficient distributed trust model for wireless sensor networks. In: IEEE Transactions on Parallel and Distributed Systems (2014)
9. Lim, H.S., Moon, Y.S., Bertino, E.: Provenance based Trustworthiness Assessment in Sensor Networks, DMSN'10, Singapore, 13 Sept 2010
10. Saaty, T.L.: How to make a decision. The analytical hierarchy process. Eur. J. Oper. Res. **48**, 9–26 (1990)
11. Heinzelman, W.R., Chandrakasan, A., Balakrishnan, H.: Energy-efficient communication protocol for wireless microsensor networks, HICSS'00: Proceedings of the 33rd Hawaii International Conference on System Sciences. vol. 8, p. 8020. IEEE Computer Society, Washington, DC, USA (2000)
12. Rajpurohit, J., Vaishali, T.K.S., Abraham, A.: Glossary of metaheuristic algorithms. Int. J. Comput. Inf. Syst. Ind. Manag. Appl. **9**, 181–205 (2017)

Natural Vibration of Square Plate with Circular Variation in Thickness

Amit Sharma and Pravesh Kumar

Abstract In this paper, the authors study natural vibration of non-uniform and non-homogeneous square plate on clamped boundary condition. The authors characterize circular variation in thickness in x-direction and bi-linear temperature variation along both the axes. For non-homogeneity, the author viewed linear variation in density parameter along x-direction. Rayleigh–Ritz technique has been applied to obtained frequency equation and first two modes of vibration of square plate for different values of taper constant, non-homogeneity constant and temperature gradient. All the results are presented in tabular form.

Keywords Natural vibration · Circular variation · Bi-linear temperature

1 Introduction

Plates are the common structural components, which have been widely used in various engineering fields such as marine industries, aerospace engineering, ocean engineering and optical instruments. Thus, for the predesign of the engineering structures, it is very important to have the knowledge of vibration characteristics of the plates.

Leissa [1] provided vibration of different structures (plates of different shape) on different boundary (clamped, simply supported and free) conditions in his excellent monograph. Chopra and Durvasula [2, 3] investigated free vibration of simply supported trapezoidal plates (symmetric and non-symmetric) using Galerkin method. Srinivas et al. [4] analysed vibration of simply supported homogeneous and laminated thick rectangular plates. Zhou et al. [5] discussed three-dimensional vibration analyses of rectangular thick plates on Pasternak foundation. Zhou et al. [6] used

A. Sharma (✉)
Amity University, Gurgaon, Haryana, India
e-mail: dba.amitsharma@gmail.com

P. Kumar
Rajkiya Engineering College, Bijnor, India
e-mail: praveshtomariitr@gmail.com

K. Ray et al. (eds.), *Soft Computing: Theories and Applications*,
Advances in Intelligent Systems and Computing 742,
https://doi.org/10.1007/978-981-13-0589-4_29

Chebyshev polynomial and Ritz method to study three-dimensional vibration analysis of thick rectangular plates. Khanna et al. [7] studied the effect of temperature on vibration of non-homogeneous square plate with exponentially varying thickness. Sharma et al. [8] studied the effect of circular variation in Poisson's ratio to the natural frequency of rectangular plate. Sharma et al. [9] studied the vibration on rectangle (orthotropic) plate with two-dimensional temperature and thickness variation. Sharma and Sharma [10] provided a mathematical modelling of vibration on parallelogram plate with non-homogeneity effect. Vibrational study of square plate with thermal effect and circular variation in density has been discussed by Sharma et al. [11]. An improved Fourier series method is presented by Zhang et al. [12] to analyse free vibration of the rectangular (thick laminated composite) plate with non-uniform boundary conditions. Liu et al. [13] studied vibration characteristic of rectangular thick plates on Pasternak foundation with arbitrary boundary conditions using three-dimensional elasticity theories. Sharma [14] studied the natural vibration of parallelogram plate with linear variation in thickness under temperature environment.

The present paper provides the effect of circular variation in thickness to the vibrational frequency of square plate (clamped along the four edges). The authors also study the effect of temperature (bi-linear) and density (linear) variation on frequency. First, two modes of vibration is calculated and presented in the form of tables.

2 Analysis

The differential equation for non-homogeneous tapered plate is

$$\frac{\partial^2 M_x}{\partial x^2} + 2\frac{\partial^2 M_{xy}}{\partial x \partial y} + \frac{\partial^2 M_y}{\partial y^2} = \rho g \frac{\partial^2 \phi}{\partial t^2} \tag{1}$$

where

$$\left.\begin{aligned} M_x &= -D_1\left(\frac{\partial^2 \phi}{\partial x^2} + \nu \frac{\partial^2 \phi}{\partial y^2}\right) \\ M_y &= -D_1\left(\frac{\partial^2 \phi}{\partial y^2} + \nu \frac{\partial^2 \phi}{\partial x^2}\right) \\ M_{xy} &= -D_1(1-\nu)\frac{\partial^2 \phi}{\partial x \partial y} \end{aligned}\right\} \tag{2}$$

Substitute Eq. (2) in Eq. (1), we get

$$\left[\begin{array}{l} D_1\left(\frac{\partial^4 \phi}{\partial x^4} + 2\frac{\partial^4 \phi}{\partial x^2 \partial y^2} + \frac{\partial^4 \phi}{\partial y^4}\right) + 2\frac{\partial D_1}{\partial x}\left(\frac{\partial^3 \phi}{\partial x^3} + \frac{\partial^3 \phi}{\partial x \partial y^2}\right) + 2\frac{\partial D_1}{\partial y}\left(\frac{\partial^3 \phi}{\partial y^3} + \frac{\partial^3 \phi}{\partial y \partial x^2}\right) \\ + \frac{\partial^2 D_1}{\partial x^2}\left(\frac{\partial^2 \phi}{\partial x^2} + \nu \frac{\partial^2 \phi}{\partial y^2}\right) + \frac{\partial^2 D_1}{\partial y^2}\left(\frac{\partial^2 \phi}{\partial y^2} + \nu \frac{\partial^2 \phi}{\partial x^2}\right) + 2(1-\nu)\frac{\partial^2 D_1}{\partial x \partial y}\frac{\partial^2 \phi}{\partial x \partial y} \end{array}\right] + \rho g \frac{\partial^2 \phi}{\partial t^2} = 0 \tag{3}$$

For solution of Eq. (3), we can take

$$\phi(x, y, t) = \Phi(x, y) * T(t) \tag{4}$$

Substituting Eq. (4) in Eq. (3), we get

$$T\left[\begin{array}{l} D_1\left(\frac{\partial^4\Phi}{\partial x^4} + 2\frac{\partial^4\Phi}{\partial x^2\partial y^2} + \frac{\partial^4\Phi}{\partial y^4}\right) + 2\frac{\partial D_1}{\partial x}\left(\frac{\partial^3\Phi}{\partial x^3} + \frac{\partial^3\Phi}{\partial x\partial y^2}\right) + 2\frac{\partial D_1}{\partial y}\left(\frac{\partial^3\Phi}{\partial y^3} + \frac{\partial^3\Phi}{\partial y\partial x^2}\right) \\ + \frac{\partial^2 D_1}{\partial x^2}\left(\frac{\partial^2\Phi}{\partial x^2} + \nu\frac{\partial^2\Phi}{\partial y^2}\right) + \frac{\partial^2 D_1}{\partial y^2}\left(\frac{\partial^2\Phi}{\partial y^2} + \nu\frac{\partial^2\Phi}{\partial x^2}\right) + 2(1-\nu)\frac{\partial^2 D_1}{\partial x\partial y}\frac{\partial^2\Phi}{\partial x\partial y} \end{array}\right] \tag{5}$$
$$+ \rho g\,\Phi\,\frac{\partial^2 T}{\partial t^2}$$

Now separating the variables, we get

$$\frac{\left[\begin{array}{l} D_1\left(\frac{\partial^4\Phi}{\partial x^4} + 2\frac{\partial^4\Phi}{\partial x^2\partial y^2} + \frac{\partial^4\Phi}{\partial y^4}\right) + 2\frac{\partial D_1}{\partial x}\left(\frac{\partial^3\Phi}{\partial x^3} + \frac{\partial^3\Phi}{\partial x\partial y^2}\right) + 2\frac{\partial D_1}{\partial y}\left(\frac{\partial^3\Phi}{\partial y^3} + \frac{\partial^3\Phi}{\partial y\partial x^2}\right) \\ + \frac{\partial^2 D_1}{\partial x^2}\left(\frac{\partial^2\Phi}{\partial x^2} + \nu\frac{\partial^2\Phi}{\partial y^2}\right) + \frac{\partial^2 D_1}{\partial y^2}\left(\frac{\partial^2\Phi}{\partial y^2} + \nu\frac{\partial^2\Phi}{\partial x^2}\right) + 2(1-\nu)\frac{\partial^2 D_1}{\partial x\partial y}\frac{\partial^2\Phi}{\partial x\partial y} \end{array}\right]}{\rho g\,\Phi} = -\frac{1}{T}\frac{\partial^2 T}{\partial t^2} = k^2 \tag{6}$$

where two expressions are set to constant k^2 because both the expressions are independent in (x, y) and t.

Now taking first and last expression of Eq. (6), we get

$$\left[\begin{array}{l} D_1\left(\frac{\partial^4\Phi}{\partial x^4} + 2\frac{\partial^4\Phi}{\partial x^2\partial y^2} + \frac{\partial^4\Phi}{\partial y^4}\right) + 2\frac{\partial D_1}{\partial x}\left(\frac{\partial^3\Phi}{\partial x^3} + \frac{\partial^3\Phi}{\partial x\partial y^2}\right) + 2\frac{\partial D_1}{\partial y}\left(\frac{\partial^3\Phi}{\partial y^3} + \frac{\partial^3\Phi}{\partial y\partial x^2}\right) \\ + \frac{\partial^2 D_1}{\partial x^2}\left(\frac{\partial^2\Phi}{\partial x^2} + \nu\frac{\partial^2\Phi}{\partial y^2}\right) + \frac{\partial^2 D_1}{\partial y^2}\left(\frac{\partial^2\Phi}{\partial y^2} + \nu\frac{\partial^2\Phi}{\partial x^2}\right) + 2(1-\nu)\frac{\partial^2 D_1}{\partial x\partial y}\frac{\partial^2\Phi}{\partial x\partial y} \end{array}\right] \tag{7}$$
$$- \rho k^2 g\,\Phi = 0$$

Equation (7) represents differential equation of motion for non-homogeneous isotropic plate and $D_1 = \frac{Yg^3}{12(1-\nu^2)}$ is called flexural rigidity of the plate.

3 Assumptions and Frequency Equation

We are using Rayleigh–Ritz technique (i.e. maximum strain energy V_s must be equal to kinetic energy T_s) to solve frequency equation, therefore we have

$$\delta(V_s - T_s) = 0 \tag{8}$$

where

$$T_s = \frac{1}{2}k^2 \int_0^a \int_0^a \rho g\, \Phi^2\, dydx \tag{9}$$

$$V_s = \frac{1}{2}\int_0^a \int_0^a D_1 \left\{ \left(\frac{\partial^2 \Phi}{\partial x^2}\right)^2 + \left(\frac{\partial^2 \Phi}{\partial y^2}\right)^2 + 2\nu \frac{\partial^2 \Phi}{\partial x^2}\frac{\partial^2 \Phi}{\partial y^2} + 2(1-\nu)\left(\frac{\partial^2 \Phi}{\partial x \partial y}\right)^2 \right\} dydx \tag{10}$$

Vibration of plate is a very vast area. It is not possible to study vibration at once; therefore, the present study requires some limitations in the form of assumptions.

(i) We consider the temperature variation on the plate is linear along x-axis and y-axis, therefore

$$\tau = \tau_0\left(1 - \frac{x}{a}\right)\left(1 - \frac{y}{a}\right) \tag{11}$$

where τ and τ_0 are known as temperature above the mention temperature at any point on the plate and at origin, i.e. $x = y = 0$. For engineering material, the modulus of elasticity is

$$Y = Y_0(1 - \gamma\tau) \tag{12}$$

where Y_0 is the Young's modulus at $\tau = 0$ and γ is known as slope of variation.

Substituting Eq. (11) in Eq. (12), we get

$$Y = Y_0\left\{1 - \alpha\left(1 - \frac{x}{a}\right)\left(1 - \frac{y}{a}\right)\right\} \tag{13}$$

where α, $(0 \leq \alpha < 1)$ is known as temperature gradient, which is the product of temperature at origin and slope of variation, i.e. $\alpha = \gamma\tau_0$.

(ii) The thickness of the plate is assumed to be circular in x-direction as shown in Fig. 1.

$$g = g_0\left\{1 + \beta\left(1 - \sqrt{1 - \frac{x^2}{a^2}}\right)\right\} \tag{14}$$

where β, $(0 \leq \beta \leq 1)$ is known as tapering parameter of the plate and $g = g_0$ at $x = 0$.

(iii) Since plate's material is considered to be non-homogeneous, for this the authors considered linear variation in density as

$$\rho = \rho_0\left(1 + m\frac{x}{a}\right) \tag{15}$$

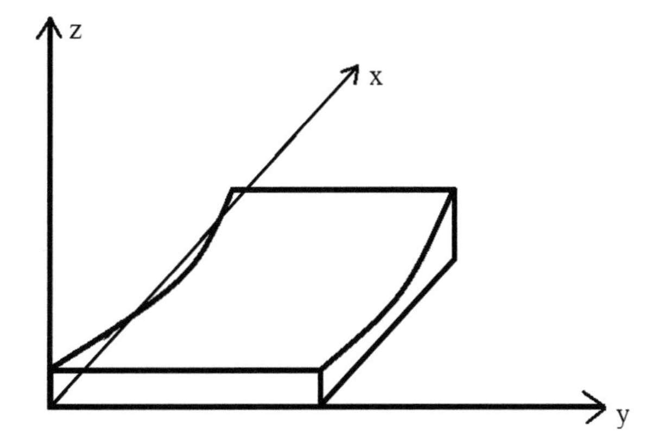

Fig. 1 Square plate with one-dimensional circular variation in x-direction

where m, $(0 \leq m \leq 1)$ is known as non-homogeneity constant.

Using Eqs. (13) and (14), the flexural rigidity of the plate becomes

$$D_1 = \frac{Y_0 g_0^3 \left[1 - \alpha\left\{1 - \frac{x}{a}\right\}\left\{1 - \frac{y}{a}\right\}\right] \left\{1 + \beta\left(1 - \sqrt{1 - \frac{x^2}{a^2}}\right)\right\}^3}{12\left(1 - \nu^2\right)} \tag{16}$$

(iv) Here, we are computing vibrational frequency on clamped plate, i.e., (C-C-C-C). Therefore, boundary conditions of the plate are

$$\Phi = \frac{\partial \Phi}{\partial x} = 0, \ \text{at}\, x = 0, \ a \text{ and } \Phi = \frac{\partial \Phi}{\partial y} = 0, \ \text{at}\, y = 0, \ a \tag{17}$$

The two-term deflection function which satisfies Eq. (17) could be

$$\Phi = \left[\left(\frac{x}{a}\right)\left(\frac{y}{a}\right)\left(1 - \frac{x}{a}\right)\left(1 - \frac{y}{a}\right)\right]^2 \left[B_1 + B_2\left(\frac{x}{a}\right)\left(\frac{y}{a}\right)\left(1 - \frac{x}{a}\right)\left(1 - \frac{y}{a}\right)\right] \tag{18}$$

where B_1 and B_2 represent arbitrary constants.

4 Solution for Frequency Equation

Now we are converting x and y into non-dimensional variable X and Y as

$$X = \frac{x}{a}, Y = \frac{y}{a} \tag{19}$$

Using Eqs. (19), (9) and (10) become

$$T_s^* = \frac{1}{2}k^2\rho_0 g_0 a^2 \int_0^1\int_0^1 (1+mX)\left\{1+\beta\left(1-\sqrt{1-X^2}\right)\right\}\Phi^2\,dY\,dX \tag{20}$$

$$V_s^* = \frac{Y_0 g_0^3}{24a^2(1-\nu^2)}\int_0^1\int_0^1 \begin{bmatrix} [1-\alpha\{1-X\}\{1-Y\}] \\ \left\{1+\beta\left(1-\sqrt{1-X^2}\right)\right\}^3 \\ \left\{\left(\frac{\partial^2\Phi}{\partial X^2}\right)^2 + \left(\frac{\partial^2\Phi}{\partial Y^2}\right)^2 + 2\nu\frac{\partial^2\Phi}{\partial X^2}\frac{\partial^2\Phi}{\partial Y^2} + 2(1-\nu)\left(\frac{\partial^2\Phi}{\partial X\partial Y}\right)^2\right\} \end{bmatrix} dY\,dX \tag{21}$$

Using Eqs. (20) and (21), (8) becomes

$$\left(V_s^* - \lambda^2 T_s^*\right) = 0 \tag{22}$$

Here $\lambda^2 = \frac{12(1-\nu^2)\rho_0 k^2 a^4}{Y_0 g_0^2}$ is known as the frequency parameter.

Equation (22) consists of two unknowns constants, i.e. B_1, B_2 because of substitution of deflection function Φ. These constants can be determined by

$$\frac{\partial\left(V_s^* - \lambda^2 T_s^*\right)}{\partial B_i} = 0,\ i = 1,\ 2 \tag{23}$$

After simplifying Eq. (23), we get homogeneous system of equation as

$$\begin{bmatrix} d_{11} & d_{12} \\ d_{21} & d_{22} \end{bmatrix}\begin{bmatrix} B_1 \\ B_2 \end{bmatrix} = \begin{bmatrix} 0 \\ 0 \end{bmatrix} \tag{24}$$

where $d_{11}, d_{12} = d_{21}$ and d_{22} involve parametric constants and frequency parameter. To get a non-trivial solution, the determinant of the coefficient matrix of Eq. (24) must be zero. So, we get equation of frequency as

$$\begin{vmatrix} d_{11} & d_{12} \\ d_{21} & d_{22} \end{vmatrix} = 0 \tag{25}$$

Equation (25) is quadratic equation from which we get two roots as λ_1 (first mode) and λ_2 (second mode).

5 Results and Discussion

The first two modes of vibration are calculated for non-homogeneous square plate for different values of plate parameters (non-homogeneity, taper constant and temperature gradient). The following values of the parameter are used for numerical calculations.

Table 1 Thickness (taper constant β) variation in plate versus vibrational frequency (λ)

β	α = m = 0.0		α = m = 0.4		α = m = 0.8	
	λ_1	λ_2	λ_1	λ_2	λ_1	λ_2
0.0	35.99	140.88	31.17	122.00	27.21	106.49
0.2	37.64	147.03	32.59	127.62	28.63	111.76
0.4	39.43	153.80	34.32	133.79	30.16	117.53
0.6	41.33	161.14	36.06	140.47	31.78	123.76
0.8	43.34	169.00	37.88	147.61	33.47	130.41
1.0	45.43	177.34	39.78	155.18	35.22	137.44

Table 2 Non-homogeneity (m) variation in plate's material versus vibrational frequency (λ)

m	α = β = 0		α = β = 0.4		α = β = 0.8	
	λ_1	λ_2	λ_1	λ_2	λ_1	λ_2
0.0	35.99	140.88	37.63	146.83	39.71	155.25
0.2	34.32	134.32	35.86	139.86	37.83	147.74
0.4	32.86	128.60	34.32	133.79	36.19	141.22
0.6	31.57	123.56	32.96	128.45	34.75	135.49
0.8	30.42	119.06	31.75	123.70	33.47	130.41
1.0	29.39	115.03	30.67	119.44	32.32	125.87

$$\rho_0 = 2.80 * 10^3\ \text{Kg/m}^3,\ \ \nu = 0.345,\ \ g_0 = 0.01\ \text{m},\ \ y_0 = 7.08 * 10^{10}\ \text{N/m}^2$$

First two modes of natural frequency for continuous increment in tapering parameter are tabulated in Table 1 for three different cases $(i)\ m = \alpha = 0, (ii)\ m = \alpha = 0.4$ and $(iii)\ m = \alpha = 0.8$. For all the three cases, the authors notice that frequency modes increase when tapering parameter increases. The rate of increment is less due to circular variation in thickness. It is also noticed that frequency modes decrease when the combined value of non-homogeneity (m) and temperature gradient (α) increases from 0 to 0.8 for all the values of thickness parameter (β).

First two modes of natural frequency for continuous increment in non-homogeneity are tabulated in Table 2 for three different cases $(iv)\ \alpha = \beta = 0$, $(v)\ \alpha = \beta = 0.4$ and $(vi)\ \alpha = \beta = 0.8$. A continuous decrement is noticed in frequency modes as non-homogeneity (m) increases from 0 to 1. The frequency of both modes increases when the combined value of thermal gradient (α) and tapering parameter (β) increases from case (iv) to case (vi) for all the values of non homogeneity (m).

Table 3 provides the natural frequency corresponding to thermal gradient for three different cases $(vii)\ m = \beta = 0, (viii)\ m = \beta = 0.4$ and $(ix)\ m = \beta = 0.8$. When the temperature gradient increases from 0 to 0.8, frequency modes decreases for all the three cases. Also, frequency mode increases when the combined value of non homogeneity constant and taper constant increases from case (vii) to case (ix).

Table 3 Temperature (α) variation on plate versus vibrational frequency (λ)

α	$m = \beta = 0$		$m = \beta = 0.4$		$m = \beta = 0.8$	
	λ_1	λ_2	λ_1	λ_2	λ_1	λ_2
0.0	35.99	140.88	35.96	140.14	36.53	141.97
0.2	35.08	137.31	35.15	137.00	35.79	139.17
0.4	34.15	133.65	34.32	133.79	35.03	136.31
0.6	33.19	129.88	33.47	130.50	34.26	133.39
0.8	32.19	126.01	32.60	127.12	33.47	130.41

6 Conclusion

Based on the results discussion, the authors conclude the following points.

(i) The present study provides the influence of structural parameter (thickness, non-homogeneity and temperature) of the plate on vibrational frequency of square plate.
(ii) Due to circular variation in thickness in present paper, variation in frequency is less when compared to parabolic and exponential variation in thickness.
(iii) When the non-homogeneity in the plate material increases, frequency modes decrease.
(iv) The frequency mode decreases when temperature on the plate increases.

References

1. Leissa, A.W.: Vibration of plate. NASA SP-160 (1969)
2. Chopra, I., Durvasula, S.: Vibration of simply supported trapezoidal plates. I. symmetric trapezoids. J. Sound Vib. **19**, 379–392 (1971)
3. Chopra, I., Durvasula, S.: Vibration of simply supported trapezoidal plates. II. un-symmetric trapezoids. J. Sound Vib. **20**, 125–134 (1971)
4. Srinivas, S., Joga Rao, C., Rao, A.K.: An exact analysis for vibration of simply-supported homogeneous and laminated thick rectangular plates. J. Sound Vib. **12**(2), 187–199 (1970)
5. Zhou, D., Cheung, Y.K., Lo, S.H., Au, F.T.K.: Three dimensional vibration analysis of rectangular thick plates on Pasternak foundation. Int. J. Numer. Methods Eng. **59**(10), 1313–1334 (2004)
6. Zhou, D., Cheung, Y.K., Au, F.T.K., Lo, S.H.: Three dimensional vibration analysis of thick rectangular plates using Chebyshev polynomial and Ritz method. Int. J. Solids Struct. **39**(26), 6339–6353 (2002)
7. Khanna, A., Deep, R., Kumar, D.: Effect of thermal gradient on vibration of non-homogeneous square plate with exponentially varying thickness. J. Solid Mech. **7**(4), 477–484 (2015)
8. Sharma, A., Raghav, A.K., Sharma, A.K., Kumar, V.: A modeling on frequency of rectangular plate. Int. J. Control Theory Appl. **9**, 277–282 (2016)
9. Sharma, A., Sharma, A.K., Raghav, A.K., Kumar, V.: Effect of vibration on orthotropic visco-elastic rectangular plate with two dimensional temperature and thickness variation. Indian J. Sci Technol. **9**(2), 7 pp. (2016)
10. Sharma, A., Sharma, A.K.: Mathematical modeling of vibration on parallelogram plate with non homogeneity effect. Roman. J. Acoust. Vib. **XIII**(1), 53–57 (2016)

11. Sharma, A., Sharma, A.K., Rahgav, A.K., Kumar, V.: Vibrational study of square plate with thermal effect and circular variation in density. Roman. J. Acoust. Vib. **XIII**(2), 146–152 (2016)
12. Zhang, H., Shi, D., Wang, Q.: An improved Fourier series solution for free vibration analysis of the moderately thick laminated composite rectangular plate with non-uniform boundary conditions. Int. J. Mech. Sci. **121**, 1–20 (2017)
13. Liu, H., Liu, F., Jing, X., Wang, Z., Xia, L.: Three dimensional vibration analysis of rectangular thick plates on Pasternak foundation with arbitrary boundary conditions. Shock Vib. **2017**, 10 pp. (2017)
14. Sharma, A.: Vibrational frequencies of parallelogram plate with circular variations in thickness. Soft Comput. Theor. Appl. Adv. Intell. Syst. Comput. **583**, 317–326 (2018)

Majority-Based Classification in Distributed Environment

Girish K. Singh, Prabhati Dubey and Rahul K. Jain

Abstract Distributed database system is used to store large datasets and dataset is partitioned and stored on different machines. Analysis of data in distributed system for decision support system becomes very challenging and is an emerging research area. For decision support system, various soft computing techniques are used. Data mining also provides a number of techniques for decision support system. Classification is a two-step data mining technique in which a model is developed using available datasets to predict the class label of new data. Support vector machine is a classification technique, which is based on the concept of support vectors. In this technique, after finding a classification model, class label of new record can be assigned. The trained classification model will give correct assignment if it is developed using entire dataset. But, in distributed environment, it is very difficult to bring all the data on a single machine and then develop a model for classification. Many researchers have proposed various methods for model built up in distributed system. This paper presents a majority-based classification after the development of SVM model on each machine.

Keywords Classification · Data mining · Distributed system · SVM

G. K. Singh · P. Dubey
Department of Computer Science and Applications,
Dr. Harisingh Gour University Sagar, Sagar, India
e-mail: gkrsingh@gmail.com

P. Dubey
e-mail: prabhatidubey15@gmail.com

R. K. Jain (✉)
B.T.I.R.T. College Sagar, RGPV University Bhopal, Bhopal, India
e-mail: rahulkumarjain16@gmail.com

K. Ray et al. (eds.), *Soft Computing: Theories and Applications*,
Advances in Intelligent Systems and Computing 742,
https://doi.org/10.1007/978-981-13-0589-4_30

1 Introduction

Advancement in digital devices and data collection techniques has increased the rate of creation of data. In other words, large volume of data can be created in a very short span of time. Maintenance of the large datasets and retrieving meaningful information from these large datasets is a major issue.

1.1 Data Mining

Data mining deals with problems like analysis of data to acquire knowledge from large datasets. Data mining refers to extracting or mining the knowledge/information from large amount of data. It is a pattern discovering process from a large dataset. Data analysis tasks such as decision-making, pattern evolution are major problems which can solve by data mining techniques. In data mining, different methods have been used such as Machine Learning, Artificial Intelligence, Statistics and other field. One of the major tasks of data mining is classification and predication.

1.2 Classification

The Classification is a data mining technique used to predict and classify the unknown data to a class. Classification is used to classify data records to get a decision, based on its learning models. A model is learned from the training dataset to predict the class of unseen data. In classification, training data tuple contains a dependent attribute and a set of predictor attributes. Predictor attributes contains some information about data. The value of dependent attribute or class is predicted by the learned model using predictor attributes values.

Many techniques have been proposed for classifications like Decision Tree [1], Artificial Neural Network [2], Support Vector Machine (SVM) [3], Rule-Based Classification [4], Genetic Algorithms [5], Rough Set Theory [6], Fuzzy Set Approach [7], k-Neighbour Classifiers [8], etc. Researchers have reported that SVM is the best classifier in many cases [9]. Selection of training data also played an imported role in accuracy of the used techniques. The best way is to use all available data as training dataset. If data is stored in distributed environment, then this is not possible to fetch all datasets to a single machine and then learned the model. So, the concept of distributed classification emerged as new challenging area to researcher.

Training a model for classification in distributed environment is a challenging task. There are many issues like collecting all data on a single machine or sampling-based training dataset. This paper presents majority-based classification of unseen data, where each machine has its own trained SVM model for classification.

1.3 Distributed Database

With the advancement in technology, the amount of data becomes very large and is not possible to store in single machine. So, researcher suggests the concept of storing of data in distributed fashion [10]. In distributed environment, a single logical database is spread physically across various computers in multiple locations and that are connected by a data communications network. It may be stored in multiple computers, located in the same physical location or may be dispersed over a network of interconnected computers. Distributed database system provides accessibility of data to a user (program) from one location to another location. A major advantage of distributed database system is that it reduces storage cost.

Rest of the paper is organized as follows: Sect. 2 presents the concepts of SVM and in Sect. 3 various approaches for classification in distributed environments have been discussed. Proposed majority-based classification has been given in Sects. 4 and 5 discusses the result and analysis of the proposed method. Finally Sect. 6 concludes the paper.

2 Support Vector Machine

The concept of support vector machine (SVM) was proposed by Vladimir Vapnik, Bernhard Boser and Isabelle Guyon in 1992 [3]. Statistical learning theory is considered as ground work of SVM which was done by Vladimir Vapnik and Alexei Chervonenkis in 1960. The current standard soft margin was proposed by Vladimir Vapnik et al in 1993 and published in 1995 [11].

SVM can be used for both classification and prediction. A Support Vector Machine (SVM) can be imagined as a surface that defines a decision boundary between various points of data, which represent examples plotted in multidimensional space according their feature (attribute) values. The goal of an SVM is to create a decision boundary, called a hyperplane, which leads to homogeneous partitions of data on either side. It categorized data tuples which related to same class through a hyperplane and separate them in different regions. A hyperplane can be found by using support vectors and margins. Support vectors are essential training tuples and define margins. Hyperplane is defined as minimum distance between support vectors. Any training tuples that fall on hyperplane are called support vector.

Let D be the dataset given as (X_1, y_1), (X_2, y_2), …, (X_d, y_d) where X_i is the set of training tuples with associated class labels, y_i. Each y_i can has one of the two values either +1 or −1, i.e., $y_i \in \{1, -1\}$. In other words, each tuple has class value either +1 or −1. There are infinite number of separating lines that can be drawn and among them a separating line will have the minimum classification error on unseen tuples. The problem of deciding which separating line is best is addressed by searching Maximum Marginal Hyperplane. Hyperplane which has maximum margin can classify tuples more accurately.

A separate hyper plane can be written as

$$W \cdot X + b = 0 \quad (1)$$

where $W = (w_1, w_2, \ldots, w_n)$ is a weight vector, where n is the number of attributes and b is a scalar.

If training tuples are 2-dimensional, then let x_1 and x_2 are the respective values of attributes A_1 and A_2 for X. If b as an additional weight is w_0, then equation of hyper plane can be written as

$$w_0 + w_1x_1 + w_2x_2 = 0 \quad (2)$$

If the point (x_1, x_2) lies above the separating hyperplane, then it satisfies

$$w_0 + w_1x_1 + w_2x_2 > 0 \quad (3)$$

and if point lies below the separating hyperplane then it satisfies

$$w_0 + w_1x_1 + w_2x_2 < 0 \quad (4)$$

The weights can be adjusted so that the hyperplane defining the sides of the margin can be written as:

$$H_1 : w_0 + w_1x_1 + w_2x_2 \geq +1 \text{ for } y_i = +1 \quad (5)$$

$$H_2 : w_0 + w_1x_1 + w_2x_2 \leq -1 \text{ for } y_i = -1 \quad (6)$$

These equations represent two hyperplanes H_1 for +1 (yes) and H_2 for -1 (no).

The problem of finding a hyperplane, which separates the dataset is now converted to find vector W and b so that

$$y_i(W \cdot X_i + b) > 0 \quad (7)$$

If the two classes are linearly separable there exists a hyperplane that satisfies Eq. (7) and in this case, it is always possible to rescale W and b so that

$$\min_{1 \leq i \leq N} y_i(W \cdot X_i + b) \geq 1 \quad (8)$$

From (8), it can be observed that the distance from the closest point to the hyperplane is 1/||W|| and so (8) can be written as

$$y_i(W \cdot X_i + b) \geq 1 \quad (9)$$

More than one hyperplane would exist for linearly separable dataset among them a best hyperplane, which has maximal distance to the closet point is called optimal separating hyperplane (OSH) [11]. Since the distance of optimal separating hyperplane to the closest point is 1/||W||, the OSH can be found by maximizing 1/||W|| or minimizing $||W||^2$ under constraint (8). To minimize $||W||^2$ Lagrange multipliers and Quadratic Programming (QP) optimization methods has been used. If λ_i, $i = 1, \ldots, N$ are the non-negative Lagrange multipliers associated with constraint (8), the optimization problem becomes one of maximizing [12]:

$$L(\lambda) = \sum \lambda_i - \frac{1}{2} \sum_{i,j} \lambda_i \lambda_j \, y_i \, y_j \, (X_i \cdot X_j) \tag{10}$$

under the constraints $\lambda_i > 0$, $i = 1, \ldots, k$

If $\lambda^a = (\lambda_1^a, \ldots \lambda_k^a)$ is an optimal solution of the maximization problem (10) then the optimal separating hyperplane can be expressed as:

$$W^a = \sum y_i \, \lambda_i^a \, X_i \tag{11}$$

The support vectors are the points for which $\lambda_i^a > 0$ when the equality in (9) holds. If the dataset is not linearly separable, a slack variable ξ_i, $i = 1, \ldots, k$ can be introduced with $\xi_i \geq 0$ [13]. Slack variables are defined to transform an inequality expression into an equality expression with an added slack variable. The slack variable is defined by setting a lower bound of zero (>0). After introducing slack variable Eq. (9) can be written as

$$y_i(w \cdot x_i + b) - 1 + \xi_i \geq 0 \tag{12}$$

The solution to find a generalized OSH, also known as soft margin hyperplane, can be obtained from the following three the conditions:

$$\min_{W,b,\xi_1,\ldots\xi_N} \left[\frac{1}{2|W|^2} + c \sum_{i=1}^{i=N} \xi_i \right] \tag{13}$$

$$y_i(W \cdot X_i + b) - 1 + \xi_i \geq 0 \tag{14}$$

$$\xi_i \geq 0 \quad i = 1, \ldots k \tag{15}$$

The first term in (12) is same as in the linearly separable case, and controls the learning capacity and the second term controls the number of misclassified points. The parameter c is chosen by the user. If the value of c is large then there is a chance of high errors.

Sometimes, it may not be possible to have a hyperplane that can be defined by linear equations on the training dataset, in that case above techniques for linearly separable data can be extended to allow nonlinear decision surfaces. Boser et al. introduced a technique which maps input data into a high-dimensional feature space through some nonlinear mapping [3]. By transforming a lower dimensional data into a higher dimensional space, it may be possible to find linear hyperplanes. Suppose there is a function $f()$ which maps dataset into a higher dimensional feature space and so X can be replace by $f(X)$ and Eq. (10) can be written as

$$L(\lambda) = \sum \lambda_i - \frac{1}{2} \sum_{i,j} \lambda_i \lambda_j y_i y_j (f(X_i) \cdot f(X_j)) \tag{16}$$

To minimize the computational complexity in feature space, the concept of the kernel function K can be used and is given by Cristianini and Shawe-Taylor [14], Cortes and Yapnik [13] such that

$$K(X_i, X_j) = f(X_i) \cdot f(X_j) \tag{17}$$

Now, the equation becomes

$$L(\lambda) = \sum \lambda_i - \frac{1}{2} \sum_{i,j} \lambda_i \lambda_j y_i y_j K(X_i, X_j) \tag{18}$$

A number of kernel functions are possible that can be used in SVM classifier. Originally, SVM was designed for two-class problems.

3 Distributed Approach of Classification Using SVM

In classification, generally a model is developed to classify data and/or to extract information for decision making (future prediction). If dataset which will be used for training is available at a single machine, then it is easy to train a model using entire dataset and this model will be efficient to classify new unseen tuples.

In distributed database system, dataset is partitioned into subsets and each subset is stored in one machine in the distributed system and classification process becomes difficult as dataset is not available on single machine. In this situation, one heuristic approach is transferring the entire datasets from different machine to single machine and classification procedure is as usually applied. This approach has two limitations: transfer of datasets to a single machine and accommodating entire dataset on a single machine. Transfer of data to a single machine where classification model will be built up has many issues like transmission channel utilizing, data noising and data loss. Moreover, process of classification will not start unless data from all machines are not received to particular machine. To solve these issues, many researchers have proposed many approaches for classification in distributed database system.

Let us formulate the problem of classification in distributed system. Let the database D has N objects and these objects are partitioned into D_1, D_2, ..., D_n subsets. Each subset is stored on a single machine for convenience without any loss of generality it can be assumed that D_1 is stored on machine M_1, D_2 is stored on machine M_2,... and D_n is stored on machine M_n. Now a classification model has to be developed based on the entire dataset D so that when an unseen tuple comes to any machine a label should be assigned using developed classification model.

In 1999, Syed et al. [15] proposed incremental learning of SVMs on distributed dataset. They perform SVM learning model on each distributed data node and send developed model's supported vectors to a central node which performs again some learning on support vectors to obtain final set of support vectors. In this approach, SVM algorithm of learning is run on each node M_j ($j = 1$ to n) and obtained support vector of each model is MSV_j ($j = 1$ to n). After that these obtained support vector (MSV_j) are sent to a central machine. Central machine applies an algorithm on all support vectors to create a union set of support vectors with corresponding weight and hence generating separating hyperplanes. Now, the question about this approach is whether the support vectors finally obtained are will be same as if the SVM is run on entire dataset, i.e. $\bigcup_{i=1}^{n} \mathrm{MSV}_i$ is same as $\mathrm{SV}(\bigcup_{i=1}^{n} D_i)$.

Caragea et al. [16] shows that if datasets are not individually true representative of data at different machines, then union of the set of support vectors obtain from each data source does not show sufficient relation and model for learning from distributed data $[\mathrm{SV}(\bigcup_{i=1}^{n} D_i) \neq \mathrm{SV}(D)]$. This approach of incremental learning of SVM works reasonably good if each data partition is a true representative of D.

In 2000, Caragea et al. [16] proposed a new SVM learning scheme on distributed database. To perform distributed classification, first build classification models by using SVM learning algorithm at each machine of distributed database and send convex hulls instances value of training tuples of each model to a central machine. A convex hull in SVM learning algorithm is defined by outer boundaries of instance tuples in a class label. So SVM learning algorithm finds two convex hulls for two respective classes in each model. On central machine, SVM learning algorithm is performed to obtain union of convex hull instances value of each machine with corresponding classes. In this approach, SVM algorithm of learning is run on each node M_j ($j = 1$ to n) and convex hulls vertices values of each model is $\mathrm{Conh}(M_j)$ ($j = 1$ to n). $\mathrm{Conh}(M_j(+))$ and $\mathrm{Conh}(M_j(-))$ are two convex hull vertices values for a machine j. On central machine, SVM algorithm is applied on all convex hull vertices set to obtain the union of positive and negative values respectively. So, this is $\mathrm{Conh}\ (\bigcup_{i=1}^{n} \mathrm{Conh}(D_i)) = \mathrm{Conh}\ (\bigcup_{i=1}^{n} (D_i))$. It has some limitations while applying, growing number of dimensions of convex vertices.

In 2005, Caragea et al. [17] presents a method based on a set of global support vector, according to their approach central machine sends a set of global support vectors instance tuples (CSV) to each machine of distributed database, initially global support vector (CSV) is empty. All datasets D_j ($j = 1$ to n) adds the support vector values to its dataset ($D_j = D_j \bigcup \mathrm{CSV}_k$) and perform SVM learning algorithm. The resultant support vector ($\mathrm{MSV}_{jk} = \mathrm{MSV}_{jk}(D_j \bigcup \mathrm{CSV}_k)$) then sends to central machine by each node. At central machine, support vector from all nodes are joined together and find a new set of central support vector CSV_{k+1} by applying SVM learning. This method is performed iteratively as far as no changes are seen in central machine's support vector (CSV). So $\mathrm{CSV}_{k+1} = \mathrm{SVM}(\mathrm{CSV}_k \bigcup \mathrm{MSV}(D_1 \bigcup \mathrm{CSV}_k) \ldots\ldots. \bigcup \mathrm{MSV}(D_k \bigcup \mathrm{CSV}_k))$. In this approach, number of iterations increased the size of central machine's support vectors value. Iterations would be performed because central machine's resultant support vector should be as SVM(D) ⊆ SVM(CSV) to SVM(D) = SVM(CSV). Frequent changes in distributed database nodes can increase the number of iterations.

In 2010, Forero et al. [18] proposed another approach. In this approach to develop support vectors in distributed system SVM learning is performed at each machine using available datasets (D_j) and determined local support vectors of each node. These locally support vector are exchanged with its neighbours nodes iteratively. So after exchanging iteratively, final support vectors values of each node will be convergent and based on consensus of its entire neighbours. This distributed approach prediction depends on the consensus decision of all neighbour nodes because each node eliminates some non-functional values of support vectors and contains classifying values of support vector which decide class labels of unseen tuples on consensus of each other.

All the techniques discussed here have their own advantages and disadvantages, like acceptable set of support vectors, utilization of network resources and accommodation of new data in training datasets. In next section, a new approach of SVM classification has been proposed which focuses on classification of new unseen data rather than development of global set of support vector machine.

4 Majority-Based Approach of Classification Using SVM

In distributed classification, first a set of support vector is obtained which is globally acceptable to all machine and then unseen data are classified using the support vector machine. Various methods as discussed in Sect. 3 focuses on first step that is development of globally acceptable set of support vector machine. The proposed method focuses on assignment of class using local support vectors of each machine. Proposed method is based on the following two principles.

i. Each machine is independent and the historical data generated is not influenced by the other machine.
ii. If data which is used as training data on a machine is not influenced by other machines then support vectors at that machine is also not influenced.

In proposed approach, a model is developed on each distributed node. These models can be separately developed without exchanging their data to each other. So this removes overhead of communication and transaction cost. Also deal with privacy and data safety like issues. The proposed classification techniques will work in the following steps.

Step 1 Each machine trained support vector machine with data available on that particular machine.
Step 2 Let ith machine receive an unseen data and this has to be labelled.
Step 3 ith machine sends unseen data to all machine in system.
Step 4 After receiving unseen data, each machine assigned the label using their learned model including ith machine.

Step 5 Each machine sends the assigned label to the unseen data to calling machine, i.e. *i*th machine.

Step 6 Once all assigned labels are received by *i*th machine from each machine, label of the unseen data is decided on the basis of majority. If there is the tie, then class label of assigned by *i*th machine will be label of unseen data.

5 Results and Analysis

To evaluate the performance of proposed method, two datasets Bank Note Authentication and Diabetic Retinopathy Debrecen dataset are used that are available in UCI repository. Bank note authentication data [12] is provided by Helene Dajrksen. Data were created from captured images of bank notes. This dataset contains 1372 number of tuples and each tuple has 5 attributes value to represent the tuple. Where 4 attributes express the value of attribute information that are the variance of Wavelet Transformed image (continuous), skewness of Wavelet Transformed image (continuous), Kurtosis of Wavelet Transformed image (continuous) and entropy of image (continuous) respectively. The remaining one attribute is representing class label of tuple in two categories 0 and 1.

Diabetic Retinopathy Debrecen Dataset [14] is provided by Dr. Balint Antal and Dr. Andras Hajdu. This dataset has features extracted from the Messidor image set to predict whether an image contains signs of diabetic retinopathy or not. This dataset has 1151 number of tuples. Each tuple has 20 attributes; these are integer and real values. First 19 attributes explain the tuples contain information and last number of attribute is representing class label of tuple which is 0 and 1. Detail of partition of both datasets is given in Table 1.

In experimental distributed system, there are 4 machines say, M_1, M_2, M_3 and M_4 with each machine has D_1, D_2, D_3 and D_4 dataset respectively. MATLAB has been used as software tool to develop SVM. Same dataset transfer on one machine and then SVM is run on a single machine to compare the proposed distributed method with single central machine classification system. Each machine partitions its own

Table 1 Description of training and testing tuples on different machines

Machine	Bank note authentication data			Diabetic Retinopathy Debrecen dataset		
	Total tuple	Training tuple	Test tuple	Total tuple	Training tuple	Test tuple
M_1	342	308	34	282	252	30
M_2	342	308	34	282	252	30
M_3	342	308	34	282	252	30
M_4	346	310	36	305	274	31
Total	1372	1234	138	1151	1030	121

		Actual value		
		T	F	Total
Prediction Outcome	T'	True Positive	False Positive	T'
	F'	False Negative	True Negative	F'
Total		T	F	

Fig. 1 Confusion matrix representation

data into two parts one for training and second for testing. On each machine 90% of total data case consider as training data while remaining 10% data are test data. The selection of tuples for training and test is done randomly. On central machine, the same sets of training and test data are used.

The proposed method of classification is compared with central classification system using confusion matrix. In Predictive Analytics, a table of confusion, also known as a confusion matrix, is a table with two rows and two columns that reports the number of True Negatives, False Positives, False Negatives and True Positives. A confusion matrix is a visualization tool typically used in supervised learning. Each row of the matrix represents the instances in a predicted class, while each column represents the instances in an actual class. One benefit of a confusion matrix is that it is easy to see if the system is confusing two classes (i.e. commonly mislabeling one as another) (Fig. 1).

First the experiment is performed for bank note authentication data. In first step each machine trains SVM with available training tuples. To classify test data T_1, machine M_1 distributed T_1 to all other machines M_2, M_3 and M_4. Each machine performed the classification according to their developed models. After classification, resultant class label of tuples of T_1 is sent back to machine M_1 by all the machines. M_1 has 4 results to determine final class labels of all tuples of T_1 on majority basis. In this dataset, obtained result is same for each tuple by all machines. So, the class label of each tuple is easily found on majority basis without any conflict. Similarly, classification is performed for test tuples T_2, T_3 and T_4. Each machine predicts the same class labels for each tuple so calling machine predicts class label on majority basis without any issue.

Now training tuples of M_1, M_2, M_3 and M_4 are transferred to a central machine and SVM model is trained. Test tuples T_1, T_2, T_3 and T_4 are tested on trained SVM model and the result is presented as confusion matrix. Table 2 represents the confusion matrix of all test tuples $T_1 \bigcup T_2 \bigcup T_3 \bigcup T_4$ of bank note authentication dataset on distributed system and centralized machine.

Similarly, experiment on Diabetic Dataset is performed and the result is given in Table 3.

The prediction accuracy of the proposed distributed classification method is almost same as the performance of centralize SVM trained model. As shown in Table 4, accuracy of prediction of both systems is same for bank note authentication dataset. Accuracy of prediction for Diabetic Retinopathy dataset by the proposed method is slightly lacking from the accuracy obtained in centrally developed SVM system.

Table 2 Confusion matrix of all test tuples of bank note authentication dataset

Distributed datasets					Centralize database			
	Actual class				Actual class			
Prediction outcome	Class	1	0	Total	Class	1	0	Total
	1	61	2	63	1	61	2	63
	0	0	75	75	0	0	75	75
	Total	61	77	138	Total	61	77	138

Table 3 Confusion matrix of all test tuples of Diabetic Debrecen dataset

Distributed datasets					Centralize database			
	Actual class				Actual class			
Prediction outcome	Class	1	0	Total	Class	1	0	Total
	1	50	6	56	1	52	3	55
	0	19	46	65	0	17	49	66
	Total	69	52	121	Total	69	52	121

Table 4 Accuracy comparison of different methods

Dataset	Proposed distributed method (%)	Centralize classification method (%)
Banknote authentication	98.55	98.55
Diabetic retinopathy	79.34	83.47

6 Conclusions

This paper presented a novel and affordable approach to solve the problem of classification in distributed system using majority-based classification. In the proposed method, SVM model is developed at all machines then a label was assigned to unseen data based on majority basis by each machine. A class label for a particular tuple is considered which is mostly predicted by all machines. Prediction accuracy of the proposed system is almost same as of centralized system. This approach reduces the communication of data between various machines. Incremental learning in distributed database system may be an interesting future work.

References

1. Quinlan, J.R.: Induction of Decision Trees, pp. 81–106. Kluwer Academic Publishers Boston (1986)
2. Rumelhart, D.E., Hinton, G.E., Williams, R.J.: Learning internal representations by error propagation. Parallel Distributed Processing. MIT Press (1986)

3. Boser, B.E., Guyon, I.M., Vapnik, V.N.: A training algorithm for optimal margin classifiers. In: Proceedings of the 5th Annual ACM Workshop on Computational Learning Theory, New York, USA, pp. 144–152 (1992)
4. Hong, J., Mozetic, I., Michalski, R.S.: Incremental learning of attribute-based descriptions from examples: the method and user's guide. In: Report ISG 85-5, UIUCDCS-F-86-949. University of Illinois at Urbana-Champaign (1986)
5. Goldberg, D.E., Holland, J.H.: Genetic Algorithms and Machine Learning, vol. 3, pp. 95–99. Kluwer Academic Publishers. Machine Learning (1988)
6. Ziarko, W., Shan, N.: Discovering attribute relationships, dependencies and rules by using rough sets. System sciences. In: Proceedings of the Twenty-Eighth International Conference (1995)
7. Zadeh, L.A.: Commonsense knowledge representation based on fuzzy logic. In: IEEE Computer. IEEE Computer Society, pp. 61–67 (1983)
8. Dasarathy, B.V.: Nearest Neighbor (NN) Norms: NN pattern classification techniques. In: IEEE Computer Society Press (1991)
9. Yu, H., Yang, J., Han, J.: Classifying large datasets using SVMs with hierarchical clusters. In: KDD '03 Proceeding of the Ninth ACM SIGKDD International Conference on Knowledge Discovery and Data Mining, pp. 306–315 (2003)
10. Stockinger, H.: Distributed database management systems and the data grid. In: MSS '01. Eighteenth IEEE Symposium (2001)
11. Scholkopf, B., Burges, C., Vapnik, V.: Extracting Support Data for a Given Task, pp. 252–257. AAAI (1995)
12. Osuna, E., Freund, R., Girosi F.: An Improved Training Algorithm for Support Vector Machines. In: Proceedings of the Seventh IEEE Conference on Neural Networks for Signal Processing, pp. 276–285 (1997)
13. Cortes, C., Yapnik, V.: Support Vector Machine, vol. 20, pp. 273–297. Kluwer Academic Publisher. Machine Learning (1995)
14. Shawe-Taylor, J., Christianini, N.: Kernel methods for pattern analysis. Cambridge University Press (2000)
15. Syed, N.A., Huan, S., Kah, L., Sung, K.: Increment learning with support vector machines. KDD Knowledge Discovering and Data Mining, New York, pp. 271–276 (1999)
16. Caragea, D., Silvescu, A., Honavar, V.: Incremental and distributed learning with support vector machines. In: AAAI, p. 1067 (2004)
17. Caragea, C., Caragea, D., Honavar, V.: Learning support vector machines from distributed data sources. In: AAAI, pp. 1602–1603 (2005)
18. Forero, P.A., Cano, A., Giannakis, G.B.: Consensus-based distributed support vector machines. JMLR 1663–1707 (2010)
19. https://archive.ics.uci.edu/ml/datasets/banknote+authentication (2017). Accessed 2017
20. https://archive.ics.uci.edu/ml/datasets/Diabetic+Retinopathy+Debrecen+Data+Set (2017). Accessed 2017

Detection of Advanced Malware by Machine Learning Techniques

Sanjay Sharma, C. Rama Krishna and Sanjay K. Sahay

Abstract In today's digital world most of the anti-malware tools are signature based, which is ineffective to detect advanced unknown malware, viz. metamorphic malware. In this paper, we study the frequency of opcode occurrence to detect unknown malware by using machine learning technique. For the purpose, we have used kaggle Microsoft malware classification challenge dataset. The top 20 features obtained from Fisher score, information gain, gain ratio, Chi-square and symmetric uncertainty feature selection methods are compared. We also studied multiple classifiers available in WEKA GUI-based machine learning tool and found that five of them (Random Forest, LMT, NBT, J48 Graft and REPTree) detect the malware with almost 100% accuracy.

Keywords Metamorphic · Anti-malware · WEKA · Machine learning

1 Introduction

A program/code which is designed to penetrate the system without user authorization and takes inadmissible action is known as malicious software or malware [1]. Malware is a term used for Trojan Horse, spyware, adware, worm, virus, ransomware, etc. As the cloud computing is attracting the user day by day, the servers are storing enormous data of the users and thereby luring the malware developers. The threats

S. Sharma (✉) · C. Rama Krishna
Department of Computer Science and Engineering, National Institute of Technical Teachers Training and Research, Chandigarh, India
e-mail: sanjay.cse@nitttrchd.ac.in

C. Rama Krishna
e-mail: rkc_97@yahoo.com

S. K. Sahay
Department of Computer Science and Information System, BITS, Goa Campus, Pilani, India
e-mail: ssahay@goa.bits-pilani.ac.in

K. Ray et al. (eds.), *Soft Computing: Theories and Applications*,
Advances in Intelligent Systems and Computing 742,
https://doi.org/10.1007/978-981-13-0589-4_31

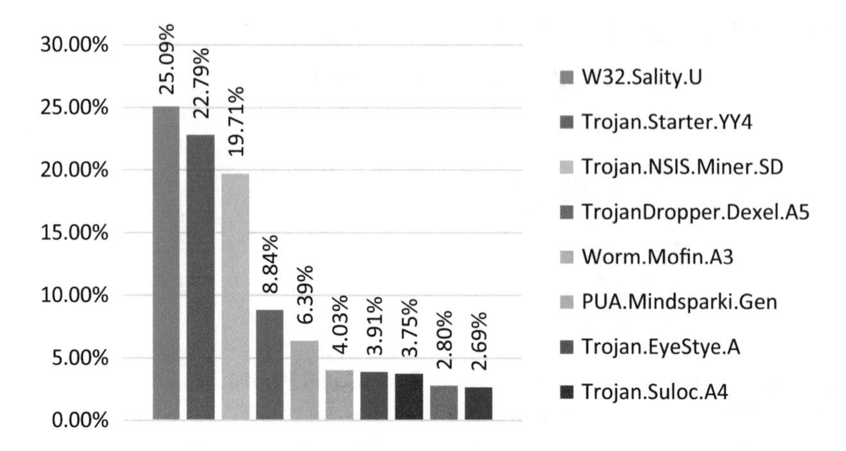

Fig. 1 Top 10 Windows malware

and attacks have also increased with the increase in data at Cloud Servers. Figure 1 shows the top 10 Windows malware reported by Quick Heal [2].

Malwares are classified into two categories—first generation malware and second generation malware. The category of malware depends on how it affects the system, functionality of the program and growing mechanism. The former deals with the concept that the structure of malware remains the same, while the later states that the keeping the action as is, the structure of malware changes, after every iteration resulting in the generation of new structure [3]. This dynamic characteristic of the malware makes it harder to detect, and quarantine. The most important techniques for malware detection are signature based, heuristic based, normalization and machine learning. In the past years, machine learning has been an admired approach for malware defenders.

In this paper, we investigate the machine learning technique for the classification of malware. In the next section, we discuss the associated work; Sect. 3 describes our approach comprehensively, Sect. 4 includes experimental outcomes and Sect. 5 contains inference of the paper.

2 Related Work

In 2001, Schultz et al. [4] introduced machine learning for detection of unknown malware based on static features, for feature extraction author used PE (Program Executables), byte n-g and Strings. In the year 2007, Danial Bilar [5] introduced opcode as a malware detector, to examine opcodes frequency distribution in malicious and non-malicious files. In year 2007, Elovici et al. [6] used Program Executable (PE) and Fisher Score (FS) method for feature selection and used Artificial Neural Network (5 g, top 300, FS), Bayesian Network (5-g, top 300, FS), Decision Tree (5-g, top 300,

FS), BN (using PE) and Decision Tree (using PE) and obtained 95.8% accuracy. In the year 2008, Moskovitch et al. [7] used filters approach for feature selection. They used Gain Ratio (GR) and Fisher Score for feature selection and Artificial Neural Networks (ANN), Decision Tree (DT), Naïve Bayes (NB), Adaboost.M1 (Boosted DT and Boosted NB) and Support Vector Machine (SVM) classifiers and got 94.9% accuracy.

In the year 2008 again, Moskovitch et al. [8] presented an approach in which they used n-g (1, 2, 3, 4, 5, 6 g) of opcodes as features and used Document Frequency (DF), GR and FS feature selection method. They used ANN, DT, Boosted DT, NB and Boosted NB classification algorithms, out of this ANN, DT, BDT outperformed, preserving the low level of false positive rate.

In 2011, Santos et al. [9] inferred that supervised learning requires labelled data, so they proposed semi-supervised learning to detect unknown malware. In 2011, Santos et al. [10] again came with the frequency of the appearance of operational codes. They used information gain method for feature selection, and different classifiers, i.e. DT, k-nearest neighbour (KNN), Bayesian Network, Support Vector Machine (SVM), among them SVM outperforms with 92.92% for one opcode sequence length and 95.90 for two opcode sequence length. In the year 2012, Shabtai et al. [11] used opcode n-g pattern feature and to identify the best feature they used Document Frequency (DF), G-mean and Fisher Score method. In their approach, they used many classifiers, in which Random Forest outperforms with 95.146% accuracy.

In 2016, Ashu et al. [12] presented a novel approach to identify unknown malware with high accuracy. They analysed the occurrence of opcodes and by grouping the executables. Authors studied 13 classifiers found in the WEKA machine learning tool, out of them a Random forest, LMT, NBT, J48, and FT were examined in depth and got more than 96.28% malware detection accuracy. In 2016, Sahay et al. [13] grouped executables on the base of malwares size by using Optimal k-means clustering algorithm, and these groups used as promising features for training (NBT, J48, LMT, FT and Random Forest) the classifiers to identify unknown malware. They found that detection of unknown malware by the proposed approach gives accuracy up to 99.11%.

Recently some authors worked on malware dataset released for kaggle dataset [14]. In the year 2016, Ahmadi et al. [15] took Microsoft malware dataset and used hex dump-based features (n-g, metadata, entropy, image representation and string length) as well as features extracted from disassembled file (metadata, symbol frequency, opcodes, register, etc.) and XGBoost classification algorithm. They reported ~ 99.8% detection accuracy. In 2017, Drew et al. [16] used The Super Threaded Reference Free Alignment-Free N sequence Decoder (STRAND) classifier to perform classification of polymorphic malware. In their approach, they presented ASM sequence model and obtained accuracy greater than 98.59% using 1tenfold cross-validation.

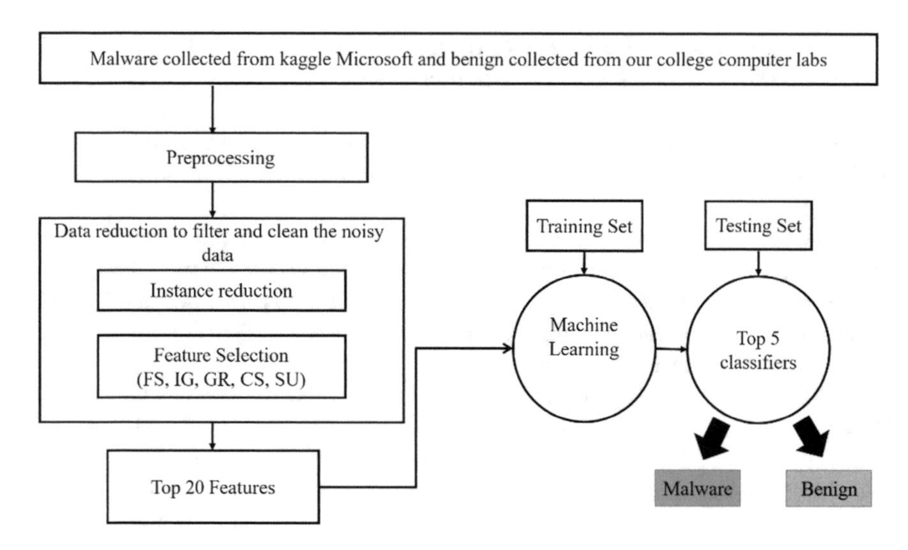

Fig. 2 Flowchart for malware detection

3 Methodology

To detect the unknown malware using machine learning technique, a flowchart of our approach is shown in Fig. 2. It includes preprocessing of dataset, promising feature selection, training of classifier and detection of advanced malware.

3.1 Building the Dataset

Microsoft released approximately half terabyte for kaggle Microsoft Malware Classification Challenge [14] containing malware (21653 assembly codes). We downloaded malware dataset from kaggle Microsoft and collected benign programs (7212 files) for the Windows platform (checked from virustotal.com) from our college's lab. In our experiment, we found that as dataset grows, there is an issue of scalability. This issue increases time complexity, storage requirement and decreases system performance. To overcome these issues, reduction of data set is necessary. Two approaches can be used for data reduction, viz. *Instance Selection (IS)* and *Feature Selection (FS).* In our approach, Instance Selection (IS) is used to reduce the number of instances (rows) in dataset by selecting most appropriate instances. On the other hand, Feature Selection is used for the selection of most relevant attributes (features) in dataset These two approaches are very effective in data reduction as they filter and clean, noisy data which results in less storage, time complexity and improve the accuracy of classifiers [17, 18].

	Opcode weight interval	no. of malwares	no. of benigns
2	1-50	39	12
3	51-100	11	3
4	101-150	11	4
5	151-200	33	18
6	201-250	43	1
7	251-300	26	1
8	351-400	18	2
9	401-450	23	1
10	451-500	11	1
11	501-550	31	33
12	551-600	129	8
13	601-650	368	5
14	651-700	356	38
15	701-750	303	24
16	751-800	111	12
17	801-850	73	7
18	851-900	193	23

Fig. 3 Opcode weight interval over period of 50

3.2 Data Preparation

From the earlier studies [12], we have found that opcodes contain a more meaningful representation of the code, so in proposed approach, we use opcodes as features. Malware dataset contains 21653 assembly codes of malware representation, a combination of 9 different families, i.e. Ramnit, Lollipop, Kelihos_ver3, Vundo, Simda, Tracur, Kelihos_ver1, Obfuscator.ACY, Gatak. Collected benign executables were disassembled using *objdump* utility available in Linux system to get the opcodes.

In the malware dataset, we have found that maximum size of assembly code is 147.0 MB, so all the benign assembly above the 147.0 MB are not considered for the analysis. From earlier studies, we found that there are 1808 unique opcodes [12] so in our approach, there are 1808 features for machine learning. Then, the frequency of each opcode in every malware and the benign file is calculated. After that in every malware and benign file total opcodes weight is calculated. Then it is noticed that there are 91.3% malware file and 66% benign file which contains opcodes weight below 40000. So to maintain the proportion of malware and benign all the files under 40000 weight is selected. After this step, 19771 and 4762 malware and benign files are left for analysis.

The next step is to remove noisy data from malware for that we have calculated the malware and benign files in the 500 intervals of opcodes weight. Those intervals in which there are no benign files, malware files are also deleted in that interval. In this way, further intervals 100, 50, 10 and 2 of opcodes weights are created as shown in Fig. 3 to remove the noise from malware. Finally, dataset contains 6010 Malware and 4573 benign files.

3.3 Feature Selection

Feature selection is an important part of machine learning. In the proposed approach, there are 1808 features among them many do not donate to the accuracy and even decrease it. In our problem, reduction of features is crucial to maintaining accuracy. Thus, we first used *Fisher Score (FS)* [19] for feature selection and later four more feature selection techniques were also studied. The five feature selection method employed in this approach which functions according to the filters approach [20]. In this method, correlation of each feature with the class (Malware or benign) is quantified, and its contribution to classification is calculated. This method is independent of any classification algorithm unlike wrapper approach and allows to compare the performance of different classifiers. In this approach, *Fisher Score (FS), Information Gain (IG), Gain Ratio (GR), Chi-Square (CS) and Uncertainty Symmetric (US)* is used. Based on these feature selection measures, we have selected top 20 features as shown in Table 1.

Table 1 Top 20 features

Rank	Information gain	Gain ratio	Symmetrical uncertainty	Fisher score	Chi-square
1	jne	jne	jne	je	jne
2	je	je	je	jne	je
3	dword	dword	dword	start	dword
4	retn	retn	retn	cmpl	retn
5	jnz	jnz	jnz	retn	jnz
6	jae	jae	jae	dword	jae
7	offset	offset	offset	test	offset
8	jz	movl	movl	cmpb	movl
9	movl	cmpl	jz	xor	jz
10	cmpl	jz	cmpl	jae	cmpl
11	int	movzwl	movzwl	movzwl	int
12	movzwl	movb	movb	ret	movzwl
13	movb	sete	sete	jbe	movb
14	sete	int3	int3	movl	sete
15	int3	testb	testb	andl	int3
16	testb	setne	cmpb	lea	testb
17	setne	cmpb	setne	cmp	cmpb
18	cmpb	andl	andl	testb	setne
19	andl	incl	incl	incl	andl
20	incl	movzbl	movzbl	setne	incl

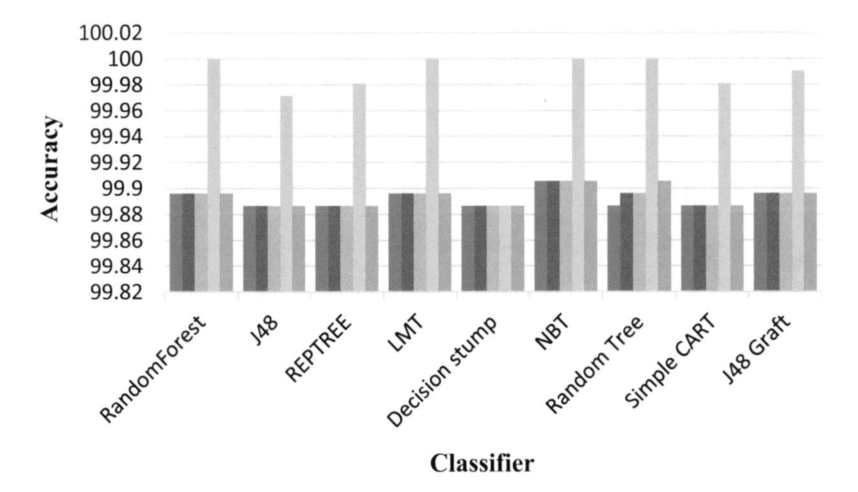

Fig. 4 Accuracy of classifiers concerning different feature selection methods

3.4 Training of the Classifiers

After the feature selection, next step is to find the best classifier for the detection of advanced malware. Next step is to compare different classifiers on FS, IG, GR, CS and US using top 20 features. We studied nine classifiers, viz. Decision Stump, Logistic Model Tree (LMT), Random Forest, J48, REPTREE, Naïve Bayes Tree (NBT), J48 Graft, Random Tree, Simple CART available in WEKA. WEKA is an open-source GUI-based machine learning tool. We run all these classifiers on each feature selection technique using tenfold cross-validation to train the classifiers. Figure 4 shows the accuracy of each classifier concerning feature selection method. From the Fig. 4, it is clear that Fisher score method is best in among all and got accuracy 100% in case of Random Forest, LMT, NBT and Random Tree. So in our proposed, Fisher Score performs better than other methods, viz. Information Gain (IG), Gain Ratio (GR), Symmetrical Uncertainty and Chi-square.

3.5 Unknown Malware Detection

In an earlier section, we have noticed that Random Forest, LMT, NBT, J48 Graft and Random Tree achieved maximum accuracy, so we selected these five classifiers for depth analysis. We have randomly selected 3005 malware and 2286 benign programs which are 50% of the overall dataset. Table 2 shows the results of top 5 classifiers.

Table 2 Performance of top 5 classifiers with Fisher score feature selection method

Classifiers	True positive (%)	False negative	False positive	True negative (%)	Accuracy (%)
Random forest	100	0	0	100	100
LMT	100	0	0	100	100
NBT	100	0	0	100	100
J48 graft	100	0	0	100	100
REPTREE	99.98	0.04%	0.05%	99.95	99.96

4 Experimental Results

As mentioned in Sect. 3, malware is already in assembly code only benign are disassembled.

Then opcodes occurrence is calculated for all malware and benign programs. In next noise from malware, data is removed by creating an interval of opcodes weight, i.e. 500, 100, 50, 10, 5 and 2 for malware and benign files. Interval in which there are no benign files, malware files are deleted. To find the dominant features or to remove irrelevant feature we used five feature selection methods and found that there are 20 features which are dominating in the classification process. Figure 4 shows that Fisher Score outperforms among five feature selection methods.

The analysis of top five classifiers, viz. Random Forest, LMT, Random Tree, J48Graft and REPTree done in WEKA to find their effectiveness, regarding True Positive Ratio (TPR), True Negative Ratio (TNR), False Positive Ratio (FPR), False Negative Ratio and Accuracy, defined by Eqs. 1, 2, 3, 4 and 5.

$$TPR = \frac{True\ Positive}{Total\ Malware} \tag{1}$$

$$TNR = \frac{True\ Negative}{Total\ Benign} \tag{2}$$

$$FPR = \frac{False\ Positive}{Total\ Benign} \tag{3}$$

$$FNR = \frac{False\ Negative}{Total\ Malware} \tag{4}$$

$$Accuracy = \frac{TP + TN}{TM + TB} \times 100 \tag{5}$$

where

True Positive: the no. of malware correctly detected.
True Negative: the no. of benign correctly detected.
False Positive: the no. of benign identified as malware.
False Negative: the no. of malware identified as benign.

Table 2 shows the result obtained by the top 5 classifiers. The study shows that the selected five classifiers accuracy is more or less same.

5 Conclusion

In this paper, we have presented an approach based on opcodes occurrence to improve malware detection accuracy of the unknown advanced malware. Code obfuscation technique is a challenge for signature-based techniques used by advanced malware to evade anti-malware tools. Proposed approach uses Fisher score method for the feature selection and five classifiers used to uncover the unknown malware. In the proposed approach Random forest, LMT, J48 Graft, and NBT detect malware with 100% accuracy which is better than the accuracy (99.8%) reported by Ahmadi et al. (2016). In future, we will implement proposed approach on different datasets and will perform in the deep analysis for the classification of advanced malicious software.

Acknowledgements Mr. Sanjay Sharma is thankful to Dr. Lini Methew, Associate Professor and Dr. Rithula Thakur Assistant Professor, Department of Electrical Engineering for providing computer lab assistance time to time.

References

1. Sharma, A., Sahay, S.K.: Evolution and detection of polymorphic and metamorphic malware: a survey. Int. J. Comput. Appl. **90**(2), 7–11 (2014)
2. Solutions, E.S., Heal, Q.: Quick Heal Quarterly Threat Report | Q1 2017. http://www.quickheal.co.in/resources/threat-reports (2017). Accessed 13 June 2017
3. Govindaraju, A.: Exhaustive Statistical Analysis for Detection of Metamorphic Malware. Master's project report, Department of Computer Science, San Jose State University (2010)
4. Schultz, M.G., Eskin, E., Stolfo, S.J.: Data Mining Methods for Detection of New Malicious Executables (2001)
5. Bilar, D.: Opcodes as predictor for malware. Int. J. Electron. Secur. Digit. Forensics **1**(2), 156–168 (2007)
6. Elovici, Y., Shabtai, A., Moskovitch, R., Tahan, G., Glezer, C.: Applying machine learning techniques for detection of malicious code in network traffic. In: Annual Conference on Artificial Intelligence, pp. 44–50. Springer, Berlin, Heidelberg (2007)
7. Moskovitch, R., Stopel, D., Feher, C., Nissim, N., Japkowicz, N., Elovici, Y.: Unknown malcode detection and the imbalance problem. J. Comput. Virol. **5**(4), 295–308 (2009)
8. Moskovitch, R., et al.: Unknown malcode detection using OPCODE representation. In: Intelligence and Security Informatics. LNCS, vol. 5376, pp. 204–215. Springer, Berlin, Heidelberg (2008)
9. Santos, I., Nieves, J., Bringas, P., G.: Semi-supervised learning for unknown malware detection. In: International Symposium on Distributed Computing and Artificial Intelligence, vol. 91, pp. 415–422. Springer, Berlin, Heidelberg (2011)
10. Santos, I., Brezo, F., Ugarte-Pedrero, X., Bringas, P.G.: Opcode sequences as representation of executables for data-mining-based unknown malware detection. Inf. Sci. **231**, 64–82 (2013)
11. Shabtai, A., Moskovitch, R., Feher, C., Dolev, S., Elovici, Y.: Detecting unknown malicious code by applying classification techniques on OpCode patterns. Secur. Inf. **1**(1), 1 (2012)

12. Sharma, A., Sahay, S.K.: An effective approach for classification of advanced malware with high accuracy. Int. J. Secur. Its Appl. **10**(4), 249–266 (2016)
13. Sahay, S.K., Sharma, A.: Grouping the executables to detect malwares with high accuracy. In: Procedia Computer Science, First International Conference on Information Security & Privacy 2015, vol. 78, pp. 667–674, June 2016
14. Kaggle: Microsoft Malware Classification Challenge (BIG 2015). Microsoft. https://www.kaggle.com/c/malware-classification (2015). Accessed 10 Dec 2016
15. Ahmadi, M., Ulyanov, D., Semenov, S., Trofimov, M., Giacinto, G.: Novel feature extraction, selection and fusion for effective malware family classification. In: ACM Conference Data Application Security Privacy, pp. 183–194 (2016)
16. Drew, J., Hahsler, M., Moore, T.: Polymorphic malware detection using sequence classification methods and ensembles. EURASIP J. Inf. Secur. **2017**(1), 2 (2017)
17. Derrac, J., García, S., Herrera, F.: A first study on the use of co evolutionary algorithms for instance and feature selection. In: International Conference on Hybrid Artificial Intelligence Systems, pp. 557–564. Springer, Berlin, Heidelberg (2009)
18. Blum, A.L., Langley, P.: Selection of relevant features and examples in machine learning. Artif. Intell. **97**(1–2), 245–271 (1997)
19. Golub, T.R., et al.: Molecular classification of cancer: class discovery and class prediction by gene expression monitoring. Science **286**(5439), 531–537 (1999)
20. Dietterich, T.G.: Machine learning in ecosystem informatics and sustainability. In: Proceedings of the Twenty-First International Joint Conference on Artificial Intelligence, pp. 8–13 (2009)

Inter- and Intra-scale Dependencies-Based CT Image Denoising in Curvelet Domain

Manoj Diwakar, Arjun Verma, Sumita Lamba and Himanshu Gupta

Abstract The most demanded tool to identify diagnosis in medical science is computed Tomography (CT). Radiation dose acts as the major factor in the degradation of CT image quality, in terms of noise. Hence in curvelet domain, an inter-scale and intra-scale thresholding-based noisy CT image quality improvement is proposed. In the proposed scheme, inter- and intra-scale dependencies are applied parallel in high-frequency coefficients. These filtered high-frequency coefficients are analyzed by obtaining correlation values. Using correlation analysis, an aggregation is performed between both filtered high-frequency coefficients. Denoised image has been retrieved using inverse curvelet transform. Comparison of the proposed method with existing methods has been performed. The result analysis of the proposed method shows that its performance is better than existing ones concerning visual quality, peak signal-to-noise ratio (PSNR), and image quality index (IQI).

Keywords Image denoising · Curvelet transform · Inter- and intra-scale dependencies · Correlation analysis

M. Diwakar (✉) · S. Lamba · H. Gupta
Department of Computer Science and Engineering, U.I.T., Uttaranchal University, Dehradun, India
e-mail: manoj.diwakar@gmail.com

S. Lamba
e-mail: sumitadatsme@gmail.com

H. Gupta
e-mail: forever.himanshugupta@gmail.com

A. Verma
Bundelkhand Institute of Engineering and Technology, Jhansi, India
e-mail: arjunverma.cse@gmail.com

K. Ray et al. (eds.), *Soft Computing: Theories and Applications*,
Advances in Intelligent Systems and Computing 742,
https://doi.org/10.1007/978-981-13-0589-4_32

1 Introduction

In Computed Tomography, the soft and hard tissues of the human body are detected by the detector when X-rays are projected in different angles. Detectors get the weak signals when higher density tissues are observed (like bones) and it changes those weak signals into electrical signals [1]. The digital/analog convertor converts these electrical signals into digital data. With the help of digital/analog convertor, digital matrix data may be altered into a tiny box which ranges from black to white gray. Finally, the CT images are rebuilt by logically computing the raw data which was attained from different angles of X-rays, using Radon Transform [2].

Among numerous existing methods [3–8] of noise suppression in CT images, decent results are gained from denoising based on wavelet transform. Borsdorf et al. [9] proposed a noise reduction method in CT images based on wavelet transform. In this method, two noisy CT images are taken as input. These input images have identical CT data but dissimilar noise. Once the wavelet coefficients are obtained, the correlational values are identified and thresholding is performed for CT image denoising. Correlation analysis is also tested on the basis of gradient approximation. The results show that nearly 50% noise reduction is possible without losing the information structure. Recently, Fathi et al. [8] proposed an efficient image denoising algorithm in wavelet packet domain. In this, optimal linear interpolation is combined with adaptive thresholding methods. Medical as well as natural noisy images were evaluated and the result analysis shows that the proposed method works well for both types of images as compared to formerly existing methods of wavelet-based denoising.

Li et al. [12] proposed a NLM (nonlocal means) algorithm for CT image denoising. The computational cost is increased for the reason that NLM filtering is modified and better outcomes are obtained from the resultant image as compared to the standard NLM methods. Wu et al. [13] proposed a parallelized NLM method for image denoising. Here, each pixel was filtered on identical basis along with diverse levels of noise reduction. This method delivers even and sharp denoised CT images but the drawback is that the operation time is very high and the minor particulars in the noisy image are removed. The process time of the proposed method is directly proportional to the image size.

Rabbani [10] proposed an algorithm for image denoising in which using the steerable pyramid along with Laplacian Probability Density Function (PDF), every sub-band coefficient is suppressed. In this framework, a maximum a posterori (MAP) estimator-based image denoising was performed which relies on the zero mean Lapalacian random variables. The result analysis shows that proposed algorithm works for denoising the image.

With different formerly existing filtering rules for medical images, it is too difficult to choose and decide the better one in terms of noise suppression and edge preservation. With this consideration, a new method has been proposed to reduce the pixel noise with structure preservation using curvelet domain in which the merits of inter- and intra-scale shrinkage rule has been inherited. The flow of the paper is: Curvelet transform overview is given in Sect. 2. The proposed work is represented in Sect. 3. Experimental evaluations are shown in Sect. 4 which includes the comparative study with former methods of denoising and at last Sect. 5 has the conclusions.

2 Curvelet Transform

Using multi-scale ridgelets transform, Curvelet provides more multidirectional, multi-resolution and co-localized the frames shape in time and frequency domain [12]. The process of curvelet transform has been performed using smooth partitioning, sub-band decomposition, ridgelet analysis and renormalization. In curvelet transform, signals are decomposed into multiple sub-bands to obtain variable positions and orientations at different levels in time domain as well as in frequency domain. By combining the properties of wavelet and ridgelet transforms, curvelet transform has been introduced. In curvelet transform, decomposition into scales is performed using wavelet transform. After decomposition, each scale is divided into several number of blocks. Large-scale wavelet components are divided into large size blocks and the wavelet components are divided into blocks of smaller size. Further, ridgelet transform is used over each block to obtain the curvelet coefficients. This process can also be considered as localized ridgelet transform. This localized process helps to extract more sharp features such as curves. For curvelets translation at each scale and angle, curvelet transform has been implemented using USFFT or wrapping method.

3 Proposed Methodology

Generally, CT images have been degraded with Gaussian noise. Hence, a CT image denoising model is proposed which inherits the properties of inter-scale and intra-scale dependencies in curvelet domain.

Let, noisy CT image $Y(i, j)$ can be represented as follows:

$$Y(i, j) = W(i, j) + \eta(i, j) \tag{1}$$

where, $W(i, j)$ is a clean image and $\eta(i, j)$ is an additive noise. In the proposed model, the curvelet transform is used to decompose the input noisy CT image. The proposed model is focused on high-frequency sub-bands. Over the high-frequency sub-bands, inter- and intra-scale shrinkage rules are applied, in parallel. For intra-scale shrinkage rule, Bayes Shrink rule is used while, Bivariate shrinkage rule has been used in inter-scale shrinkage rule. The similarity of both filtered curvelet coefficients have been obtained using correlation analysis. Through correlation analysis, it was analyzed that similarity of both filtered curvelet coefficients are different. To gain better outcomes, correlation-based aggregation method has been performed.

For both thresholding methods, the noise variance (σ_η^2) has been estimated using robust median estimator [3], as given below:

$$\sigma_\eta^2 = \left[\frac{median(|Y(i,j)|)}{0.6745}\right]^2 \tag{2}$$

where, $Y(i, j) \in HH$ sub-band.

3.1 Thresholding Rule Using Intra-scale Dependency

The thresholding rule has been performed only over the high-frequency coefficients [4]. For that, the estimated threshold value [5] can be obtained, as

$$\lambda = \frac{\sigma_\eta^2}{\sigma_W} \tag{3}$$

The variance σ_W^2 of noiseless image can be calculated [6] as

$$\sigma_W^2 = max(\sigma_Y^2 - \sigma_\eta^2, 0) \tag{4}$$

where $\sigma_Y^2 = \frac{1}{b}\sum_{i=1}^{b} Y_i^2$ and b represents block size. Thresholding can be performed by applying soft thresholding method [7] as

$$G := \begin{cases} sign(Y)(|Y| - \lambda), & |Y| > \lambda \\ 0, & Otherwise \end{cases} \tag{5}$$

3.2 Thresholding Rule Using Inter-scale Dependency

To suppress the noise from high-frequency sub-bands, a local adaptive thresholding is performed using bivariate shrinkage function. Let w_{2k} be the parent of w_{1k} (w_{2k} is the complex wavelet coefficient at the same position as the kth complex wavelet coefficient w_{1k}, but at the next coarser scale). Then

$$y_{1k} = w_{1k} + \eta_{1k} \;\; and \;\; y_{2k} = w_{2k} + \eta_{2k} \tag{6}$$

where y_{1k} and y_{2k} are noisy complex wavelet coefficients, w_{1k} and w_{2k} are noiseless complex wavelet coefficients and η_{1k} and η_{2k} are additive noise coefficients.

To denoise the coefficients of high-frequency sub-bands, bivariate shrinkage function [11] is used and can be expressed as

$$\hat{w}_{1k} = \frac{(\sqrt{y_{1k}^2 + y_{2k}^2} - \lambda_k)_+}{\sqrt{y_{1k}^2 + y_{2k}^2}} \cdot y_{1k} \tag{7}$$

where $\lambda_k = \frac{\sqrt{3}\sigma_{\eta_k}^2}{\sigma_{W_k}}$. The marginal variance of $\sigma_{W_k}{}^2$ and noise variance $\sigma_{\eta_k}{}^2$ for each kth complex wavelet coefficient can be computed using Eqs. (4) and (2), respectively. The function $(a)_+$ can be defined as

$$(a)_+ := \begin{cases} 0, & if \quad a < 0 \\ a, & if \quad a > 0 \end{cases} \tag{8}$$

3.3 Correlation-Based Aggregation

After thresholding, both filtered curvelet coefficients have been received for all high-frequency components. It is hard to analyze the best filtered curvelet coefficients. For better outcomes, a new patch-based aggregation function is introduced using correlation analysis over the both filtered curvelet coefficients. It can be expressed as

$$R(i,j) := \begin{cases} \sqrt{Avg(R_1^2(i,j), R_2^2(i,j))}, & C \leq \tau \\ Max(R_1(i,j), R_2(i,j)), & Otherwise \end{cases} \tag{9}$$

where R_1 and R_2 are the filtered curvelet coefficients of intra- and inter-scale shrinkage-based denoising. C is defined for correlation values which are obtained between R_1 and R_2. τ is a defined threshold value which is estimated by calculating average value of correlation values.

4 Results and Discussion

The experimental results has been evaluated on noisy CT images of 512 × 512 size. The CT images shown in Fig. 1a–d are taken from public access database (https://eddie.via.cornell.edu/cgibin/datac/logon.cgi). The proposed method is performed to all test images which are corrupted by additive Gaussian noise at four different noise levels (σ): 10, 20, 30, and 40. Figure 2a–d shows the dataset of noisy CT images with $(\sigma) = 20$. Over the noisy input images, curvelet transform has been performed to get high-frequency coefficients. To suppress noise in high-frequency sub-bands, both inter- and intra-scale shrinkage rules are used in parallel and results of both are extracted using patch-based aggregation using correlation analysis. In experimental results, the average defined threshold value (τ) for all CT test images is 0.89. For patch-based aggregation, the patch size 7 × 7 has been used. The results of proposed model has been compared with popular recent existing methods. The

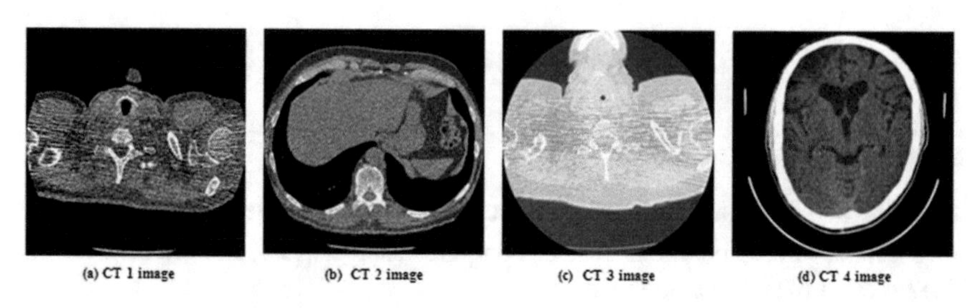

Fig. 1 Original CT image data set

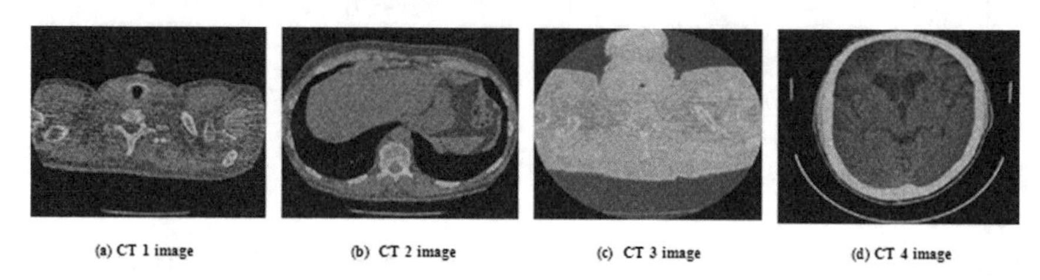

Fig. 2 Noisy CT image data set ($\sigma = 20$)

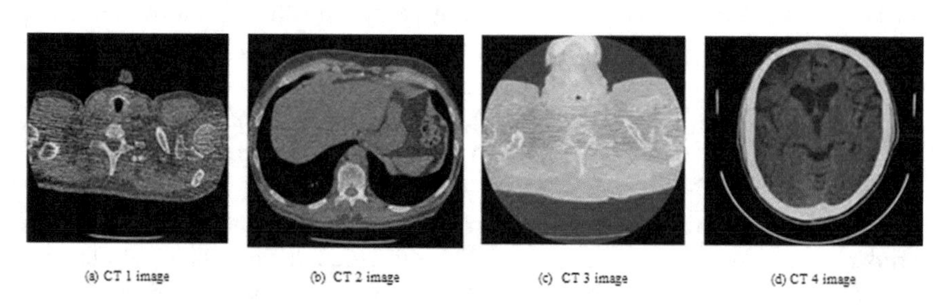

Fig. 3 Results of curvelet transform-based denoising (CBT) [12]

existing methods for comparison are Curvelet transform-based denoising (CBT) [12] and curvelet-based NLM filtering (CNLM) [13]. Figures 3a–d, 4a–d and 5a–d are showing the results of Curvelet transform-based denoising [12] and curvelet-based NLM filtering [13] and proposed method respectively. Image quality index (IQI) and PSNR are also examined on CT image dataset for proposed method as well as methods [12, 13]. Table 1 shows the PSNR (in dB) and IQI values for the proposed method and also methods of [12, 13]. With the analysis of Table 1, it can be concluded that most the times, the proposed method is superior to compared methods. The results of [12] has been analyzed and observed that as noise level increases, the edges over the homogenous are blurred. Similarly, the results of [12] has been analyzed and observed that as noise level increases, the small sharp structures are lost. However, the proposed model for higher noise gives better outcomes in comparison to [12, 13].

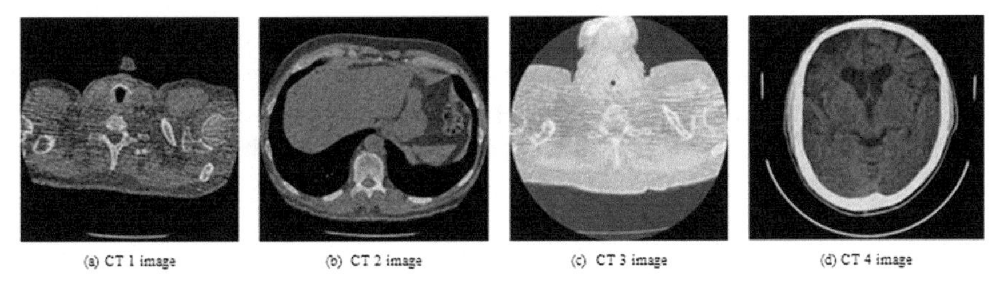

(a) CT 1 image (b) CT 2 image (c) CT 3 image (d) CT 4 image

Fig. 4 Results of curvelet-based NLM filtering (CNLM) [13]

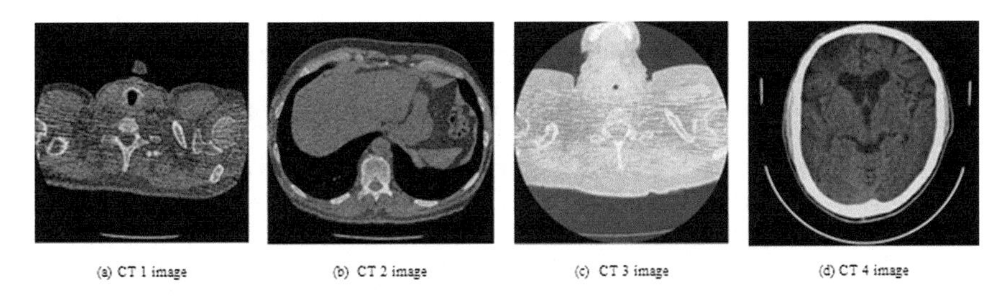

(a) CT 1 image (b) CT 2 image (c) CT 3 image (d) CT 4 image

Fig. 5 Results of proposed model

Table 1 PSNR and IQI of CT-denoised images

Image	σ	PSNR				IQI			
		10	20	30	40	10	20	30	40
CT1	CBT	32.21	28.64	26.31	24.03	0.9987	0.9232	0.8919	0.8351
	CNLM	32.86	30.05	26.63	25.80	0.9991	0.9486	0.9130	0.8771
	Proposed	**33.17**	**30.64**	**26.71**	**26.09**	**0.9992**	**0.9697**	**0.9201**	**0.8898**
CT2	CBT	31.89	29.51	26.12	23.91	0.9972	0.9239	0.8919	0.8195
	CNLM	32.61	29.81	26.35	**24.32**	0.9979	0.9371	0.9101	**0.8712**
	Proposed	**32.67**	**29.98**	**26.47**	24.29	**0.9981**	**0.9623**	**0.9314**	0.8607
CT3	CBT	33.59	29.75	27.47	26.61	0.9981	0.9412	0.9213	0.8565
	CNLM	33.86	30.13	27.51	26.34	0.9983	0.9656	0.9412	0.8882
	Proposed	**34.02**	**30.59**	**27.63**	**26.63**	**0.9986**	**0.9779**	**0.9614**	**0.9105**
CT4	CBT	35.39	32.95	30.12	26.96	0.9982	0.9713	0.9519	0.8695
	CNLM	**35.82**	33.12	30.43	27.01	**0.9991**	0.9768	0.9623	0.8981
	Proposed	35.61	**33.31**	**30.72**	**27.43**	0.9989	**0.9787**	**0.9681**	**0.9271**

5 Conclusions

This paper gives a brief literature of CT image denoising and also gives a new proposed CT image denoising model in curvelet domain. Here, two thresholding methods (Bayes and Bivariate) have been performed over the high-frequency coefficients in noisy CT images. An advantage of both methods have been inherited in the proposed model using patch-based correlation analysis. The results are excellent in terms of structure preserving and noise reduction. The quality of results may help to reduce the radiation dose for CT imaging. The values of PSNR and IQI indicated that proposed method is also giving better results as compared to existing methods which shows that clinical details in denoised CT images are more preserved using the proposed model.

References

1. Lall, M.: Cone beam computed tomography: the technique and its applications. Int. J. Dent. Clin. 5.2 (2013)
2. Chae, B.G., Sooyeul, L.: Sparse-view CT image recovery using two-step iterative shrinkage-thresholding algorithm. ETRI J. **37**(6), 1251–1258 (2015)
3. Jaiswal, A., Upadhyay, J., Somkuwar, A.: Image denoising and quality measurements by using filtering and wavelet based techniques. AEU-Int. J. Electron. Commun. **68**(8), 699–705 (2014)
4. Jain, P., Tyagi, V.: LAPB: locally adaptive patch-based wavelet domain edge-preserving image denoising. Inf. Sci. **294**, 164–181 (2015)
5. Kumar, M., Diwakar, M.: CT image denoising using locally adaptive shrinkage rule in tetrolet domain. J. King Saud Univ.-Comput. Inf. Sci. (2016)
6. Eslami, R., Radha, H.: Translation-invariant contourlet transform and its application to image denoising. IEEE Trans. Image Process. **15**(11), 3362–3374 (2006)
7. Khare, A., Tiwary, U.S., Pedrycz, W., Jeon, M.: Multilevel adaptive thresholding and shrinkage technique for denoising using Daubechies complex wavelet transform. Imaging Sci. J. **58**(6), 340–358 (2010)
8. Fathi, A., Naghsh-Nilchi, A.R.: Efficient image denoising method based on a new adaptive wavelet packet thresholding function. IEEE Trans. Image Process. **21**(9), 3981–3990 (2012)
9. Borsdorf, A., Raupach, R., Flohr, T., Hornegger, J.: Wavelet based noise reduction in CT-Images using correlation analysis. IEEE Trans. Med. Imaging **27**(12), 1685–1703 (2008)
10. Rabbani, H., Nezafat, R., Gazor, S.: Wavelet-domain medical image denoising using bivariate laplacian mixture model. IEEE Trans. Biomed. Eng. **56**(12), 2826–2837 (2009)
11. Sendur, L., Selesnick, W.I.: Bivariate shrinkage functions for wavelet-based denoising exploiting interscale dependency. IEEE Trans. Signal Process. **50**(11), 2744–2756 (2002)
12. Deng, J., Li, H., Wu, H.: A CT Image denoise method using curvelet transform. In: Communication Systems and Information Technology, pp. 681–687. Springer, Berlin, Heidelberg (2011)
13. Wu, K., Zhang, X., Ding, M.: Curvelet based nonlocal means algorithm for image denoising. AEU-Int. J. Electron. Commun. **68**(1), 37–43 (2014)

A Heuristic for the Degree-Constrained Minimum Spanning Tree Problem

Kavita Singh and Shyam Sundar

Abstract Given a connected, edge-weighted and undirected complete graph $G(V, E, w)$, and a positive integer d, the degree-constrained minimum spanning tree (dc-MST) problem aims to find a spanning tree of minimum cost in such a way that the degree of each node in T is at most d. The dc-MST problem is a $\mathcal{NP}$-Hard problem. In this paper, we propose a problem-specific heuristic ($\mathcal{H}$_DCMST) for the dc-MST problem. $\mathcal{H}$_DCMST consists of two phases, where the first phase constructs a degree-constrained spanning tree (T) and the second phase examines the edges of T for possible exchange in two stages followed one-by-one in order to further reduce the cost of T. On a number of TSP benchmark instances, the proposed $\mathcal{H}$_DCMST has been compared with the heuristic BF2 proposed by Boldon et al. (Parallel Comput. 22(3):369–382, 1996) [3] for constructing spanning trees with d = 3. Computational experiments show the effectiveness of the proposed $\mathcal{H}$_DCMST.

Keywords Degree-constrained · Spanning tree · Heuristic

1 Introduction

Given a connected, edge-weighted and undirected complete graph $G(V, E, w)$, where V is a set of vertices, E is a set of edges and w is an associated cost ($w(e)$) to each edge $e \in E$, and a positive integer d, the degree-constrained minimum spanning tree (dc-MST) problem aims to find a spanning tree of minimum cost in such a way that the degree of each node in T should not exceed d.

K. Singh · S. Sundar (✉)
Department of Computer Applications, National Institute of Technology Raipur,
Raipur 492010, CG, India
e-mail: ssundar.mca@nitrr.ac.in

K. Singh
e-mail: ksingh.phd2015.mca@nitrr.ac.in

K. Ray et al. (eds.), *Soft Computing: Theories and Applications*,
Advances in Intelligent Systems and Computing 742,
https://doi.org/10.1007/978-981-13-0589-4_33

The dc-MST problem is a $\mathcal{NP}$-Hard problem for $d \geq 2$ [7]. Due to degree-constraint, the dc-MST problem finds practical relevance in the context of backplane wiring among pins where any pin could be wrapped by at most a fixed number of wire-ends on the wiring panel, in communication networks where the maximum degree in a spanning tree is a measure of vulnerability to single-point failures [12], and VLSI routing trees [3].

The dc-MST problem is a well-studied problem. Narula and Ho [9] proposed a primal and a dual heuristic procedure and a branch and bound algorithm for the dc-MST problem. Later, two general heuristics and a branch and bound algorithm were proposed in [13]. Boldon et al. [3] proposed four heuristics based on Prim's algorithm [10]. All four heuristics consist of two phases. The first phase which is common to all four heuristics constructs a minimum spanning tree (MST); however, the second phase (also called blacklisting phase) for each heuristic differs by four blacklisting functions (BF1, BF2, BF3 and BF4). The role of blacklisting phase is to penalize the weights of those tree edges that are incident to vertices with the degree exceeding d. By doing so, such edge in tree is discouraged from appearing in the next spanning tree. This procedure continues until a feasible spanning tree is obtained; however, the authors [3] allowed their four heuristics for at most 200 iterations in order to obtain a feasible spanning tree. Among all four heuristics, the heuristic BF2 performs best.

Among metaheuristic techniques, various versions of genetic algorithms based on Prufer-encoding [14], a $|V| \times (d-1)$ array encoding [8] and edge-set encoding [11] have been proposed for the dc-MST problem. A number of ant colony optimization approaches [1, 4, 5] as well as particle swarm optimization algorithms [2, 6] have been presented for the dc-MST problem.

This paper proposes a problem-specific heuristic ($\mathcal{H}$_DCMST) for the dc-MST problem. $\mathcal{H}$_DCMST consists of two phases, where the first phase constructs a degree-constrained spanning tree (T) and the second phase examines the edges of T for possible exchange in two stages followed one-by-one in order to further reduce the cost of T. Being problem-specific heuristic, on a number of TSP benchmark instances, $\mathcal{H}$_DCMST is compared with the the heuristic BF2 proposed by Boldon et al. [3] for constructing spanning trees with $d = 3$.

The rest of this paper is organized as follows: Sect. 2 describes the proposed hybrid heuristic $\mathcal{H}$_DCMST for dc-MST problem. Computational results are reported in Sect. 3. Finally, Sect. 4 contains some concluding remarks.

2 Problem-Specific Heuristic for the dc-MST Problem

This section describes our proposed problem-specific heuristic ($\mathcal{H}$_DCMST) for the dc-MST problem. One can notice two important characteristics about the problem-structure of the dc-MST problem which are as follows: first, the cost of edges in tree (say T) should be minimal; and second, the number of edges incident to a vertex in T should be in such a way that the degree of any vertex in T should not be greater than d. Such characteristics of the dc-MST problem motivated us to design a problem-specific heuristic focusing on the d-number of minimum edge-cost incident to a node.

The proposed $\mathcal{H}$_DCMST consists of two phases in which the first phase constructs a tree (T), and the second phase examines the edge e_{ij} of T in a particular order for possible *edge-exchange* with an another edge e_{kl} whose edge-cost is lesser or equal to $w(e_{ij})$ (see description of Second Phase).

Prior to the first phase of $\mathcal{H}$_DCMST, two attributes ($W_d[v]$ and $V_d[v][d]$) associated with each vertex $v \in V$ are maintained, where $W_d[v]$ is the weight on a given vertex $v \in V$ that is computed by adding the weight of d-number of minimum edge-cost incident to v; and $V_d[v][d]$ for a given vertex $v \in V$ contains a list of those d-number of vertices in order adjacent to v having minimum edge-weight (in non-decreasing order). The description of two phases of $\mathcal{H}$_DCMST is as follows:

First Phase: Initially, a degree-constrained spanning tree (T) and the set S are empty; create a copy, say U, of V; label each vertex $v \in V$ *unmarked* ($Mark[v] \leftarrow 0$); and set the degree of each vertex $v \in V$ zero ($deg[v] \leftarrow 0$). Select a vertex, say v_1, with minimum $W_d[v_1]$ from the set U. In case of tie (i.e., more than one vertex of same minimum $W_d[v]$), a tie-breaking rule is applied. As per this rule, from the set of more than one vertex of same minimum $W_d[]$, select that vertex (say v_1) having a minimum edge-cost with other vertex in $V_d[v_1][d]$. After this, add v_1 to S and delete v_1 from U. Label v_1 *marked* ($Mark[v_1] \leftarrow 1$). Hereafter, iteratively at each step, select a vertex v_2 with minimum $W_d[v_2]$ from U (apply tie-breaking rule in case of tie which is mentioned above in this Section). Now, select a vertex $v_x \in S$ from the list of $V_d[v_2][d]$ having minimum edge-cost with v_2. Add an edge $e_{v_2 v_x}$ to T. After this, add v_2 to S and delete v_2 from U. Label v_2 *marked* ($Mark[v_2] \leftarrow 1$).
Due to addition of an edge in T, increment $deg(v_2)$ and $deg(v_x)$ in T by one. It is to be noted that if $|deg(v_2)|$ is equal to d, then in such a situation, an *UPDATE* method is called which is as follows:

Algorithm 1: The pseudocode of the First Phase of $\mathcal{H}$_DCMST

```
Input : A connected, weighted and undirected complete graph G = (V, E, w), and d
Output: A degree-constrained spanning tree T
1  T ← φ, S ← φ, U ← V, flag1, flag2 ← 0, count ← 0, Ws ← φ;
2  Compute the value of Wd[v] and Vd[v][d] associated with each vertex v ∈ V;
3  for (each vertex i in V) do
4      Mark[i] ← 0;
5      deg[i] ← 0;
6  Select a vertex v1 of minimum Wd[v1] from U; // Apply tie-breaking rule (see First Phase) in case of tie
   Mark[v1] ← 1;
7  S ← S ∪ v1;
8  U ← U \ v1;
9  while U ≠ φ do
10     Select a vertex v2 of minimum Wd[v2] from U; // Apply tie-breaking rule (see First Phase) in case of tie
11     flag1 ← 0;
12     for (each vertex vx ∈ Vd[v2][d]) do
13         if (Mark[vx] = 1) then
14             flag1 ← 1;
15             break;
16     if (flag1 = 1) then
17         Mark[v2] ← 1;
18         S ← S ∪ v2;
19         U ← U \ v2;
20         T ← T ∪ e_v2vx;
21         deg[v2] ← deg[v2] + 1;
22         deg[vx] ← deg[vx] + 1;
23         if (deg[v2] = d) then
24             UPDATE; // see the description of UPDATE method in the First Phase of H_DCMST
25         if (flag2 == 1) then
26             U ← U ∪ {Ws};
27             Ws ← φ;
28             count ← 0;
29             flag2 ← 0;
30     else
31         U ← U \ v2;
32         Ws ← Ws ∪ v2;
33         flag2 ← 1;
34         count ← count + 1;
35 if (count > 0) then
36     for ( each vertex i ∈ Ws) do
37         Find an edge (eij) of minimum cost in E, connecting i to a vertex j ∈ S;
38         T ← T ∪ eij;
```

As per this method, update the set $W_d[v]$ of each node $v \in U$ that contains v_2. While updating $W_d[v]$ of such node $v \in U$, we consider the next minimum edge-cost connecting v to another vertex (say v_y), discarding the edge-cost connecting a vertex v to vertex v_2. Where v_y can be one from either S or U. While updating $W_d[v]$ of such node $v \in U$, simultaneously also update its corresponding $V_d[v][d]$.

This whole procedure is repeated again and again until U becomes empty. At this juncture, a degree-constrained spanning tree T is constructed.

It is to be noted that two situations may occur during execution of $\mathcal{H}$_DCMST. The first situation may occur, when no edge, connecting v_2 to one of its *marked* vertex in its $V_d[v_2][d]$, exists for selection in order to construct T, then to handle this situation, such vertex v_2 is kept in a *waiting state* which is temporarily kept in a set, say W_s. After this, another vertex $\in U$ is given a chance to participate in constructing T. If this participation adds new edge to T, then all vertices in *waiting state* move to U. The second situation may occur, when U becomes empty and still there is at least one vertex, say v_w, in W_s (*waiting state*), then to handle this situation, start searching an edge in G that connects a vertex $\in W_s$ to a vertex $\in S$ in such a way that no degree-constraint is violated. Once the edge is searched, it is added to T. This procedure continues until no vertex remains in W_s.

Algorithm (1) describes the pseudocode of first phase of $\mathcal{H}$_DCMST. **Second Phase**: Once T is constructed from the first phase, the second phase follows the idea of *edge-exchange* to further reduce the cost of T. Although using *edge-exchange* is a common idea; however, the second phase examines the edges of T for possible *edge-exchange* in two stages followed one-by-one in such a way that *edge-exchange* (deletion of an edge $e_{ij} \in T$ and inclusion of edge $e_{kl} \in E$), if it exists, will result into T and the degree-constraint at vertices k or l or both in T will not be violated. In the first stage, consider each vertex i whose $|deg(i)| = d$. For each edge $e_{ij} \in T$ incident to i ($|deg(i)| = d$), search an appropriate edge $e_{kl} \in E$, whose edge-cost is less than or equal to that of e_{ij} for exchange. If the search is successful, then the edge e_{ij} is exchanged with e_{kl} in T. Once the first stage is completed, then the second stage is applied. In the second stage, consider each vertex i whose $|deg(i)| < d$. For each edge $e_{ij} \in T$ incident to i ($|deg(i)| < d$), search an appropriate edge $e_{kl} \in E$ whose edge-cost is less than that of e_{ij} for exchange. If the search is successful, then the edge e_{ij} is exchanged with e_{kl} in T. The *edge-exchange* idea is repeatedly applied until no candidate edge in T exists for possible *edge-exchange*.

2.1 Illustration of $\mathcal{H}$eu_dc-MST with an Example

To illustrate the proposed problem-specific heuristic ($\mathcal{H}$_DCMST) for the dc-MST problem, we consider two graph instances for constructing tree with $d = 3$. The first graph (G_1) instance presented by Narula and Ho [9] consists of $|V| = 9$, distance matrix (upper triangle) shown in Table 1. This instance was solved heuristically (optimal value 2256) by Narula and Ho [9] and Savelsbergh and Volgenant [13]. The second graph (G_2) instance presented by Boldon et al. [3] consists of $|V| = 6$, distance matrix (upper triangle) shown in Table 3. Boldon et al. [3] obtains the value 39 (optimal value 37) through heuristically.

For the first graph (G_1), the $\mathcal{H}$_DCMST starts with first phase. Table 2 maintains entries for various stages of first phase of $\mathcal{H}$_DCMST. The entry $W_d[v]$ of each vertex v in the first row of Table 2 contains the sum of first three edge-cost adjacent to its vertex v. Stage-1 (Fig. 1a) in Table 2 denotes that vertex 3 (✓) is selected as it has minimum $W_d[3] = 600$ entry. Hereafter, iteratively, at each step, $\mathcal{H}$_DCMST adds a new edge to T. At Stage 2, vertex 1 is selected as it has the next minimum $W_d[1]$ entry. Edge e_{13} (Fig. 1b) is added to T. Continuing this iterative process, at Stages 3, 4, 5, vertices 0, 5, 2 are selected and their associated edges e_{01} (Fig. 1c), e_{35} (Fig. 1d), e_{21} (Fig. 1e) are added to T. Since here, the degree of vertex 1 in T becomes 3 which is equal to d, therefore, Stage 6 shows an update in the set $W_d[v]$ of

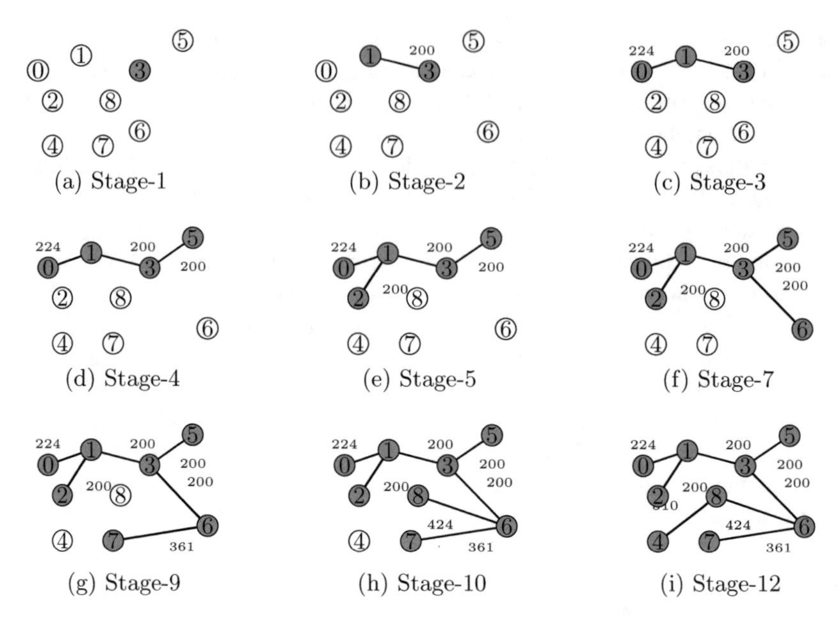

Fig. 1 Various stages of first phase of $\mathcal{H}$_DCMST for G_1

each node $v \in U$ that contains vertex 1. While updating $W_d[v]$ of such node $v \in U$, simultaneously also update its corresponding $V_d[v][d]$. Again at Stage 7, vertex 6 is added to selected and edge e_{63} (Fig. 1f) is added to T. Since here, the degree of vertex 3 in T becomes 3 which is equal to d, therefore, Stage 8 shows an update in the set $W_d[v]$ of each node $v \in U$ that contains vertex 3. While updating $W_d[v]$ of such node $v \in U$, simultaneously also update its corresponding $V_d[v][d]$. Continuing

Table 1 Distance matrix of G_1

Node	0	1	2	3	4	5	6	7	8
0	–	224	224	361	671	300	539	800	943
1	–	–	200	200	447	283	400	728	762
2	–	–	–	400	566	447	600	922	949
3	–	–	–	–	400	200	200	539	583
4	–	–	–	–	–	600	447	781	510
5	–	–	–	–	–	–	283	500	707
6	–	–	–	–	–	–	–	361	424
7	–	–	–	–	–	–	–	–	500
8	–	–	–	–	–	–	–	–	–

Table 2 Various stages of first phase of $\mathcal{H}$_DCMST for G_1

Node	0	1	2	3	4	5	6	7	8
$W_d[]$	748	624	824	600	1294	766	844	1361	1434
Stage-1	–	–	–	✓	–	–	–	–	–
Stage-2	–	✓	–	–	–	–	–	–	–
Stage-3	✓	–	–	–	–	–	–	–	–
Stage-4	–	–	–	–	–	✓	–	–	–
Stage-5	–	–	✓	–	–	–	–	–	–
Stage-6 (update)	–	–	–	–	1357	–	844	1361	1434
Stage-7	–	–	–	–	–	–	✓	–	–
Stage-8 (update)	–	–	–	–	1523	–	–	1361	1434
Stage-9	–	–	–	–	–	–	–	✓	–
Stage-10	–	–	–	–	–	–	–	–	✓
Stage-11(update)	–	–	–	–	1676	–	–	–	–
Stage-12	–	–	–	–	✓	–	–	–	–

this iterative process, at Stages 9, 10, vertices 7, 8 are selected and their associated edges e_{76} (Fig. 1g), e_{86} (Fig. 1h) are added to T. Since here, the degree of vertex 6 in T becomes 3 which is equal to d, therefore, Stage 11 shows an update in the set $W_d[v]$ of each node $v \in U$ that contains vertex 6. While updating $W_d[v]$ of such node $v \in U$, simultaneously also update its corresponding $V_d[v][d]$. At Stage 12, vertex 4 is selected and its associated edges e_{48} (Fig. 1i) are added to T. At this juncture, T with value 2319 is constructed (see Fig. 1i) (Table 3).

To further reduce the cost of T, the second phase applies the idea of *edge-exchange*, which is demonstrated in Fig. 2. As per the first stage of *edge-exchange*, it examines the edges incident to vertex (say v with $deg(v) = 3$) for possible exchange with an edge of lesser cost or same cost in E as described in Sect. 2. Edge $e_{01} \in T$ can be exchanged with $e_{02} \in E$ (see Fig. 2a–c). No further edge-exchange is possible in the first stage; however, in the second stage of *edge-exchange*, edge e_{48} can be exchanged with e_{41} (see Fig. 2c–e). Since here, no further edge-exchange is possible for further improvement in T, it stops. The final tree T with value 2256 is shown in Fig. 2e. It should be noted that the value of T obtained by our proposed heuristic is optimal which was also obtained by Savelsbergh and Volgenant [13].

In a similar way for the second graph G_2, one can construct a spanning tree T with $d = 3$ through $\mathcal{H}$_DCMST. Table 4 maintains entries for various stages of first phase of $\mathcal{H}$_DCMST. Various stages of first phase of $\mathcal{H}$_DCMST for G_1 are depicted in Fig. 3a–f. The second phase based on idea of *edge-exchange* is applied; however, it does not find even a single candidate edge for possible exchange. At this juncture, the final tree T with value 38 is same shown in Fig. 3f. It should be noted that the value of T obtained by our proposed heuristic is better than that of heuristics proposed by Boldon et al. [3].

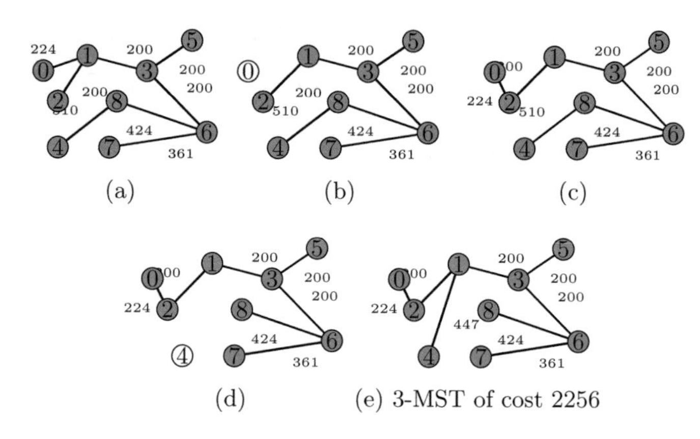

Fig. 2 *edge-exchange* of $\mathcal{H}$_DCMST for G_1

Table 3 Distance matrix of G_2

Node	0	1	2	3	4	5
0	–	9	15	17	8	19
1	–	–	12	18	6	20
2	–	–	–	13	10	21
3	–	–	–	–	7	5
4	–	–	–	–	–	22
5	–	–	–	–	–	–

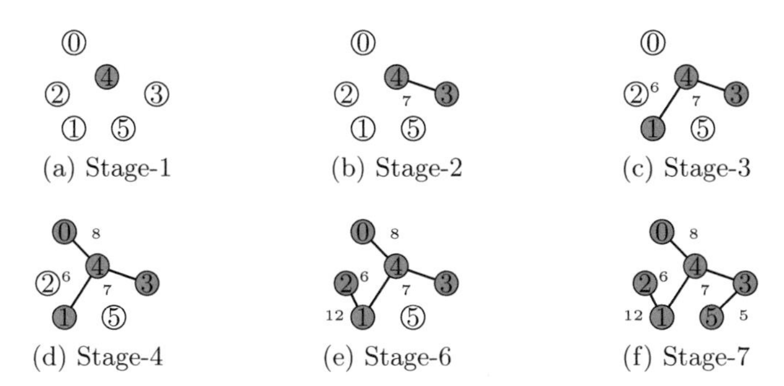

Fig. 3 Various stages of first phase of $\mathcal{H}$_DCMST for G_2

Table 4 Various stages of first phase of $\mathcal{H}$_DCMST for G_2

Node	0	1	2	3	4	5
$W_d[]$	32	27	35	25	21	44
Stage-1	–	–	–	–	✓	–
Stage-2	–	–	–	✓	–	–
Stage-3	–	✓	–	–	–	–
Stage-4	✓	–	–	–	–	–
Stage-5 (update)	–	–	40	–	–	44
Stage-6	–	–	✓	–	–	–
Stage-7	–	–	–	–	–	✓

3 Computational Results

The proposed $\mathcal{H}$_DCMST has been implemented in C. All experiments have been carried out on Linux with the configuration of 3.2 GHz $\times$ 4 Intel Core *i5* processor with 4 GB RAM. To test $\mathcal{H}$_DCMST, like [3], 34 TSPLIB datasets have been used. $\mathcal{H}$_DCMST has been compared with the heuristic BF2 [3]. The heuristic BF2, which is one of their four heuristics based on blacklisting functions, performs best. Since Boldon et al. [3] reported the results on only 10 TSP benchmark instances, therefore, to compare $\mathcal{H}$_DCMST with the heuristic BF2 on 34 TSP benchmark instances, we re-implement the heuristic BF2 (referred to as BF2). Subsequent subsections discuss about TSP benchmark instances and a comparison study of $\mathcal{H}$_DCMST and BF2 on these TSP benchmark instances.

3.1 TSP Benchmark Instances

Table 5 reports the results of $\mathcal{H}$_DCMST and BF2 on TSP benchmark instances. The range of the size of instances varies from 124 to 5934. The number of vertices in each such instance is mentioned on the name of its instance. All these instances are on points in a plane. These points are floating point numbers. We consider the distance in integral value which is computated as follows:

$$val_F = \sqrt{((x[i]-x[j]) \times (x[i]-x[j])) + ((y[i]-y[j]) \times (y[i]-y[j]))}$$
$$val_I = (int)val_F$$
$$var = |val_F - val_I|$$
$$if(var \geq 0.5)\ dis \leftarrow val_I ++$$
$$else\ dis \leftarrow val_I$$

where val_F is the distance between two Euclidean points ((x[i], y[i]), (x[j], y[j])) in float values; dis is the distance in integer values; The TSP instances can be found on http://elib.zib.de/pub/mp-testdata/tsp/tsplib/tsp/index.html. Constructing minimum spanning tree without considering degree-constraint (MST) on these instances, it was found that the maximum degree in such MST is 4 [3]. MST on each such instance gives the lower bound on the weight of d-MST for all values of d. BF2 [3] and $\mathcal{H}$_DCMST are applied to construct spanning tree with d = 3 (or 3-MST) on these instances.

3.2 Comparison of $\mathcal{H}$_DCMST with BF2 [3]

The results of experiments for 3-MST are summarized in Table 5 on 34 TSP benchmark instances. In Table 5, the first two columns (Instance and lower bound (MST)) denote the name of each instance and its lower bound value obtained by MST, respectively, the next three two columns present the results (obtained by BF2, BF2+*edge-exchange* and $\mathcal{H}$_DCMST) and total execution time (*TET*) for each instance respectively. Comparing with BF2 on 34 instances, $\mathcal{H}$_DCMST is better on all instances. Out of 34 instances, BF2 fails to converge to a feasible solution for 10 instances (in 200 iterations [3]). To have a fair comparison, we apply *edge-exchange* idea on feasible solutions of those instances obtained by BF2. Combining BF2 with *edge-exchange* will be referred to as BF2+*edge-exchange*. Experimental results in Table 5 show that still $\mathcal{H}$_DCMST is better on 23 instances, equal of 9 instances and worse on 2 instances in comparison with BF2+*edge-exchange*. One can observe in Table 5 in terms of computational time (TET) that BF2 and BF2+*edge-exchange* are faster than $\mathcal{H}$_DCMST on those instances where BF2 succeeds in finding feasible solutions; however, $\mathcal{H}$_DCMST always converges to a feasible solution and finds better solution on these TSP benchmark instances.

Table 5 Results of BF2, BF2+*edge-exchange* and $\mathcal{H}$_DCMST on TSP benchmark instances for d=3

Instances	Lower bound (MST)	BF2	TET	BF2+*edge-exchange*	TET	$\mathcal{H}$_DCMST	TET
d493.tsp	29272	–	–	–	–	**29297**	0.0
d657.tsp	42490	42582	0.0	42506	0.0	**42500**	0.0
d1291.tsp	46931	46956	0.0	**46931**	0.0	**46931**	1.0
d1655.tsp	56542	56730	0.0	**56542**	0.0	**56542**	0.0
d2103.tsp	76331	76331	0.0	**76331**	1.0	**76331**	3.0
fl1400.tsp	16831	–	–	–	–	**16831**	1.0
fl1577.tsp	19344	–	–	–	–	**19352**	1.0
lin318.tsp	37906	–	–	–	–	**37914**	0.0
p654.tsp	29456	29519	0.0	**29469**	0.0	**29469**	0.0
pcb442.tsp	46358	46673	0.0	46472	0.0	**46364**	0.0
pcb3038.tsp	127302	128449	0.0	127575	3.0	**127419**	9.0
pr124.tsp	50535	50535	0.0	**50535**	0.0	**50535**	0.0
pr136.tsp	88964	88964	0.0	**88964**	0.0	**88964**	0.0
pr144.tsp	49466	49466	0.0	**49466**	0.0	**49466**	0.0
pr264.tsp	41142	41147	0.0	41147	0.0	**41143**	0.0
pr299.tsp	42488	42887	0.0	42577	0.0	**42526**	0.0
pr439.tsp	92193	92338	0.0	**92221**	0.0	**92221**	0.0
pr2392.tsp	342269	345037	0.0	343147	1.0	**342595**	5.0
rat575.tsp	6248	6294	0.0	6259	0.0	**6250**	0.0
rat783.tsp	8125	8183	0.0	8136	0.0	**8129**	0.0
rd400.tsp	13638	13844	0.0	**13643**	0.0	**13643**	0.0
rl1304.tsp	222849	223058	0.0	**222853**	0.0	222854	1.0
rl1323.tsp	239986	240139	0.0	239999	0.0	**239986**	1.0
rl5915.tsp	521871	–	–	–	–	**522306**	91.0
rl5934.tsp	513952	516533	2.0	514460	31.0	**514100**	92.0
u574.tsp	32078	32227	0.0	32120	0.0	**32084**	0.0
u724.tsp	37959	38101	0.0	38011	0.0	**37980**	0.0
u1060.tsp	195463	–	–	–	–	**195625**	0.0
u1432.tsp	145977	–	–	–	–	**146018**	0.0
u1817.tsp	54286	–	–	–	–	**54286**	2.0
u2152.tsp	61492	–	–	–	–	**61492**	3.0
u2319.tsp	232200	–	–	–	–	**232200**	3.0
vm1084.tsp	209247	209940	0.0	209390	0.0	**209367**	0.0
vm1748.tsp	294628	297363	0.0	**295020**	1.0	295034	2.0

4 Conclusion

This paper presents a problem-specific heuristic ($\mathcal{H}$_DCMST) for the degree-constrained minimum spanning tree (dc-MST) problem. $\mathcal{H}$_DCMST constructs a degree-constrained spanning tree (T) in its first phase, and then the edges of T in a particular order are examined for possible exchange in order to further reduce the cost of T. $\mathcal{H}$_DCMST has been compared with the heuristic BF2 proposed by Boldon et al. [3] for constructing spanning trees with $d = 3$ on TSP benchmark instances. Experimental results show that $\mathcal{H}$_DCMST performs better than the heuristic BF2 in terms of finding high solution quality on these instances.

Acknowledgements This work is supported by the Science and Engineering Research Board—Department of Science & Technology, Government of India [grant no.: YSS/2015/000276].

References

1. Bau, Y., Ho, C.K., Ewe, H.T.: An ant colony optimization approach to the degree-constrained minimum spanning tree problem. In: Proceedings of Computational Intelligence and Security, International Conference, CIS 2005, Xi'an, China, 15–19 Dec 2005, Part I, pp. 657–662 (2005)
2. Binh, H.T.T., Nguyen, T.B.: New particle swarm optimization algorithm for solving degree constrained minimum spanning tree problem. In: Proceedings of PRICAI 2008: Trends in Artificial Intelligence, 10th Pacific Rim International Conference on Artificial Intelligence, Hanoi, Vietnam, 15–19 Dec 2008, pp. 1077–1085 (2008)
3. Boldon, B., Deo, N., Kumar, N.: Minimum-weight degree-constrained spanning tree problem: heuristics and implementation on an SIMD parallel machine. Parallel Comput. **22**(3), 369–382 (1996)
4. Bui, T.N., Zrncic, C.M.: An ant-based algorithm for finding degree-constrained minimum spanning tree. In: Proceedings of Genetic and Evolutionary Computation Conference, GECCO 2006, Seattle, Washington, USA, 8–12 July, 2006, pp. 11–18 (2006)
5. Doan, M.N.: An effective ant-based algorithm for the degree-constrained minimum spanning tree problem. In: Proceedings of the IEEE Congress on Evolutionary Computation, CEC 2007, 25–28 Sept 2007, Singapore, pp. 485–491 (2007)
6. Ernst, A.T.: A hybrid lagrangian particle swarm optimization algorithm for the degree-constrained minimum spanning tree problem. In: Proceedings of the IEEE Congress on Evolutionary Computation, CEC 2010, Barcelona, Spain, 18–23 July 2010, pp. 1–8 (2010)
7. Garey, M.R., Johnson, D.S.: Computers and Intractability: A Guide to the Theory of NP-Completeness. Freeman, W.H (1979)
8. Knowles, J.D., Corne, D.: A new evolutionary approach to the degree-constrained minimum spanning tree problem. IEEE Trans. Evol. Comput. **4**(2), 125–134 (2000)
9. Narula, S.C., Ho, C.A.: Degree-constrained minimum spanning tree. Comput. OR **7**(4), 239–249 (1980)
10. Prim, R.: Shortest connection networks and some generalizations. Bell Syst. Tech. J. **36**, 1389–1401 (1957)
11. Raidl, G.R., Julstrom, B.A.: A weighted coding in a genetic algorithm for the degree-constrained minimum spanning tree problem. In: Applied Computing 2000, Proceedings of the 2000 ACM Symposium on Applied Computing, Villa Olmo, Via Cantoni 1, 22100 Como, Italy, 19–21 Mar 2000, vol. 1, pp. 440–445 (2000)

12. Ravi, R., Marathe, M., Ravi, S., Rosenkrantz, D., III, H.H.: Many birds with one stone: multi-objective approximation algorithms. In: Proceedings of 25th Annual ACM STOCS, pp. 438–447 (1993)
13. Savelsbergh, M.W.P., Volgenant, T.: Edge exchanges in the degree-constrained minimum spanning tree problem. Comput. OR **12**(4), 341–348 (1985)
14. Zhou, G., Gen, M.: A note on genetic algorithms for degree-constrained spanning tree problems. Networks **30**(2), 91–95 (1997)

Resource Management to Virtual Machine Using Branch and Bound Technique in Cloud Computing Environment

Narander Kumar and Surendra Kumar

Abstract Resource allocation is a piece of resource administration process and primary goal of it is to adjust the load over Virtual Machine (VM). In this paper, the resource allocation is made on the premise of the Assignment Problem arrangement techniques like branch and bound. The branch and bound algorithmic approach has been used to find best solutions for an allocation of resources and promising the optimal solution of the optimization problem, which is figured for cloud computing. This paper likewise gives the expected outcomes, the usage of the proposed algorithm and comparison between the proposed algorithm and the previous algorithms like FCFS, Hungarian, etc.

Keywords Resource allocation · Virtual machine · Branch and bound algorithm Assignment problem · Cloud computing

1 Introduction

Cloud computing is a model of the organization whether private, public or hybrid in nature. It gives a paperless specialized means in terms of networks, servers and capacity. Distributed computing permits the cloud computing, to break down a world-wide operation into a few errands, and afterward send them to handling frameworks. The necessities of the web clients are regularly different and rely on upon the assignments. In any case, resource allocation turns out to be more unpredictable in a domain made out of heterogeneous resources and it relies upon the prerequisites of clients. Resource allocation is one of the interesting issues of cloud computing research these days [1]. Resources incorporate taking after issues provisioning, allocation, displaying, adaption, mapping, estimation out of this Resource is most influencing issue.

N. Kumar (✉) · S. Kumar
Department of Computer Science, BBA University (A Central University), Lucknow, India
e-mail: nk_iet@yahoo.co.in

S. Kumar
e-mail: kumar.surendra1989@gmail.com

K. Ray et al. (eds.), *Soft Computing: Theories and Applications*,
Advances in Intelligent Systems and Computing 742,
https://doi.org/10.1007/978-981-13-0589-4_34

Fundamentally, resource allocation implies dissemination of resource monetarily among contending gatherings of individuals or projects. Resource assignment has a critical effect in cloud computing, particularly in pay-per-utilization organizations where the quantities of resources are charged to application suppliers. For the best way to find the resource allocation problem, it can be used to solve with branch and bound algorithm for the optimal solutions of the undertakings together with their gear to every workstation means to achieve the correct arrangements ascribing to hypothetical significance of our issue [2].

2 Review of Work

Using operation research approaches reduces the unwanted issues. The main thing to focus on is maximizing the frequency signals at transmission and to reduce the spectrum congestions. The main two objectives are achieved by the branch and bound approach, first step is to maximize assignments of frequency and stations. Second, minimize the values of criterion functions, derived functions and techniques used for accurate results [3]. Branch and bound are used to reduce the overall resource contribution scheduling time. It can find an initial space for searching. Branch and bound can search parallel and improves the searching performance during the searching phases [4]. There are two tests that take place; first can check the resourcefulness and appropriateness, second make comparison between linear models to nonlinear model for the accurate results [5]. For overcoming these problems, branch and bound techniques are used to get the accurate optimal solutions. These bounding techniques are evaluated numerically [6]. Branch and bound proceeds iteration and decreases spaces. It can fully guarantee for an optimal solution and this methods applies on the electrical vehicles in grid form for the simulation learning [7]. A result-based comparison between works to get optimal solution by the branch and bound has been discussed in [8]. The casual issue is appeared to be a convex optimization which permits acquiring the lower bound [9]. In the OFDMA, the frame used high altitude platforms benefits for expending the quantity of client that get the multicast from decomposing the bound in branch and bound can calculate the complex problem using decision trees [10]. For solving this issue, a substituting enhancement-based algorithm, which applies branch and bound and simulate tempering in illuminating sub problems at every optimization step [11]. Broad simulations are led to demonstrate that the proposed calculation can fundamentally enhance the execution of vitality proficiency over existing optimized solutions [12]. To minimize the computational multifaceted nature, a near-optimal cooperative routing algorithm is given, in which it gives the solution of the issue by decoupling the energy distribution issue and the route determination issue [13]. The receptive pointer position scheduling algorithms, utilizes the plant state data to dispatch the computational resources in a way that enhances control executions [14]. To guarantee the productivity of the deep searching algorithm, particular bounding and branching methodologies utilizing a devoted vertex shading methodology and a particular vertex sorting approaches is presented [15].

Notwithstanding giving a hypothetically focalized wrapper for progressive hedging connected to stochastic mixed integer programs, computational outcomes show that for some difficult issue occurrences branch and bound can solve the enhanced solutions in the wake of investigating some nodes [16]. The Integer–Vector–Matrix constructs branch and bound algorithms in light of the graphics processing unit, addressing the inconsistency of the algorithm as far as workload, memory get to patterns and control stream [17]. Getting motivation from various advances in solving of this famous issue and build up another rough streamlining approach, which depends on the colonialist aggressive algorithm hybridized with an effective neighbourhood search [18].

3 Formulation and Implementation

Branch and bound is a technique for exploring an implicit directed graph. Using of implicit directed graphs is generally acyclic or evens a free in nature. We are achieving the optimal solution to some problems. At each node we are calculating bound on the most possible values of solutions that might lie farther on in the graph. The optimal calculated bound is also used to select which open path looks the most likely, so it can be branched first.

In the assignment problem R agents are to be assigned R task to perform. If agents is $1 \leq a \leq R$ (is assigned task b, a $1 \leq a \leq R$) then the cost of performing this particular task will be c_{ab}. Given the compute matrix of cost the problem is to assign agents to tasks so as to minimize the total cost of executing the R tasks. The Branch and Bound methods are normally based on some relaxation of the ILP (Integer Linear Programming Model) model. In the following optimization models, the variables x_{ab} are either excluded from the model or prevented by setting $c_{ab} = 4$.

$$\text{Max} \sum_{a=1}^{R} \sum_{b=1}^{R} c_{ab}\ x_{ab}\ \text{Subject to} \sum_{b=1}^{R} x_{ab}\ \text{For every a} = 1, \ldots, \text{R}. \tag{1}$$

$$\sum_{b=1}^{R} x_{ab} \quad \text{For every b} = 1, \ldots, \text{R}$$

$$X_{ab} = 0 \text{ or } 1 \quad \text{a, b} = 1, \ldots, \text{R} \tag{2}$$

For finding optimal solution, we can use branch and bound method. Here, it leads a tree of choice through which each branch moves to a one best possible way to proceed to the loss from the present nodes. We observe the branches by finding at the lower bound of every present loss then proceeds with that branch, i.e. the most minimal bound. The steps will stops when, found the possible solution and no other nodes is further traverse in the chosen tree that has bring lower bound and he does to get the optimal solution.

Table 1 The main table of resources (VM) and request time

VM_capacity	Request_time
10	100
5	45
20	40
40	70
12	144

Table 2 The table of resources (VM) and request time

V/R	R1	R2	R3	R4	R5
V1	10	4.5	4	7	14.4
V2	20	9	8	14	28.8
V3	5	2.25	2	3.5	7.2
V4	2.5	1.125	1	1.75	3.6
V5	8.33	3.75	3.33	5.85	12

Table 3 Shows the selected optimal request time at Stage II

V/R	R1	R2	R3	R4	R5
V1	10	4.5	4	7	14.4
V2	20	9	8	14	28.8
V3	5	2.25	2	3.5	7.2
V4	2.5	1.125	1	1.75	3.6
V5	8.33	3.75	3.33	5.85	12

Analysis is done using a sample o understand the working of algorithm and specifications, for example are as follows [19] (Table 1).

Following is Execution Time Matrix Ex[V, R] of all Resource Requests and VMs (Table 2).

Step 1: We start by taking a node at lower bound which is the minimum execution time. In this method, take a minimum execution time with respective column if we take a node first R1V4 that is 2.5. The same process is for second node R2V4 that is 1.125. Third node R3V4 is 1. Fourth node R4V4 is 1.75. Fifth node R5V4 is 3.6. If we add all the allocated nodes, then we get the minimum optimal solution.

$$R1V4 + R2V4 + R3V4 + R4V4 + R5V5 = 9.975.$$

This does not mean that it is a possible solution. It is just a lowest possible solution that is guaranteed; it is equal or greater than 9.975. It is also a root node. We start to make our decision tree with the lowest bound 9.975.

Step 2: we select the R1V1 node to make the best minimum allocations in virtual 113 machine and select the V1 row and R1 column then remaining nodes will be optimized 114 for minimum execution time (Table 3).

By optimizing the selected region, they get the lowest execution time for the best allocations in the region are R2V1 + R3V2 + R3V3 + R3V4 + R3V5 = 18.33.

Table 4 Shows the optimal request time at Stage III

V/R	R1	R2	R3	R4	R5
V1	10	4.5	4	7	14.4
V2	20	9	8	14	28.8
V3	5	2.25	2	3.5	7.2
V4	2.5	1.125	1	1.75	3.6
V5	8.33	3.75	3.33	5.85	12

Table 5 Shows the optimal request time at Stage IV

V/R	R1	R2	R3	R4	R5
V1	10	4.5	4	7	14.4
V2	20	9	8	14	28.8
V3	5	2.25	2	3.5	7.2
V4	2.5	1.125	1	1.75	3.6
V5	8.33	3.75	3.33	5.85	12

Table 6 Shows the lowest optimal request time at Stage V

V/R	R1	R2	R3	R4	R5
V1	10	4.5	4	7	14.4
V2	20	9	8	14	28.8
V3	5	2.25	2	3.5	7.2
V4	2.5	1.125	1	1.75	3.6
V5	8.33	3.75	3.33	5.85	12

Step 3: There we find a lowest value at R1V4 + R2V4 + R3V2 + R4V4 + R5V4 = 16.975 then we calculate this value for resources allocations in the virtual machine and process the best minimum request execution time. The same process will take place for the further optimization (Table 4).
Step 4: At the lowest point at the region R1V4 + R2V4 + R3V2 + R4V4 + R5V4 = 16.975 lowest bound nodes are 16.975. Further evaluating the lowest bound for the best possible execution to get the resource allocation in the virtual machines and get the lowest execution at the R1V4 + R2V4 + R3V4 + R4V3 + R5V4 = 11.725 optimization point (Table 5).
Step 5: At this execution point, we get the lowest bound nodes as 16.975. Further, evaluating the lowest bound for the best possible execution to get the resource allocation in the virtual machines and get the lowest execution at the R1V3 + R2V3 + R3V3 + R4V3 + R5V4 = 16.35 optimization point (Table 6).
Step 6: After the previous lowest point, we get the lowest bound nodes as 16.35. Further evaluating and reaching the lowest bound for the best possible execution to get the allocations to resources in the virtual machines and get the lowest execution at

Table 7 Shows the lowest optimal request time at Stage VI

V/R	R1	R2	R3	R4	R5
V1	10	4.5	4	7	14.4
V2	20	9	8	14	28.8
V3	5	2.25	2	3.5	7.2
V4	2.5	1.125	1	1.75	3.6
V5	8.33	3.75	3.33	5.85	12

Fig. 1 The lowest execution request time for the resources in VM

the R1V5 + R2V4 + R3V4 + R4V4 + R5V4 = 15.805 optimization point (Table 7 and Fig. 1).

Since assignment problem through Branch and Bound is used to assign the workload jobs in respect of optimized solution are one-to-one basis. The final optimized time is 15.805. It is better than the first come first serve algorithm and the Hungarian algorithm so to get the most minimum request time at a large resources allocation in the virtual machines.

4 Comparison of Optimization Techniques

Here algorithm is compared with FCFS algorithm, Hungarian algorithm, and Branch and Bound. In FCFS algorithm, first request is allocated to place resources. This algorithm is in sequential mode so served as a Queue. We ascertain execution time through FCFS algorithm can executed and give the optimized result are as below (Tables 8 and 9).

The execution time in the Hungarian algorithm is promising the minimum optimal solution in the given in the tables (Tables 10 and 11).

From the FCFS algorithm, we get the minimum request time as 34.75 s and calculated from the Hungarian algorithm the minimum execution time as 26.95 s and then we proposed an algorithm and get the more minimum execution time for the resource allocations. From the proposed formulation, we get the resource request

Table 8 The execution time with lowest point from FCFS algorithm

V/R	R1	R2	R3	R4	R5
V1	10	4.5	4	7	14.4
V2	20	9	8	14	28.8
V3	5	2.25	2	3.5	7.2
V4	2.5	1.125	1	1.75	3.6
V5	8.33	3.75	3.33	5.85	12

Table 9 Execution time from the FCFS algorithm

VM	Resources	Execution time
V1	R1	10
V2	R2	9
V3	R3	2
V4	R4	1.75
V5	R5	12
Total		34.75

Table 10 Lowest execution time point from the Hungarian algorithm

V/R	R1	R2	R3	R4	R5
V1	10	4.5	4	7	14.4
V2	20	9	8	14	28.8
V3	5	2.25	2	3.5	7.2
V4	2.5	1.125	1	1.75	3.6
V5	8.33	3.75	3.33	5.85	12

Table 11 Execution time from the FCFS algorithm

VM	Resources	Execution time
V1	R2	4.5
V2	R3	8
V3	R1	5
V4	R5	3.6
V5	R4	5.85
Total		26.95

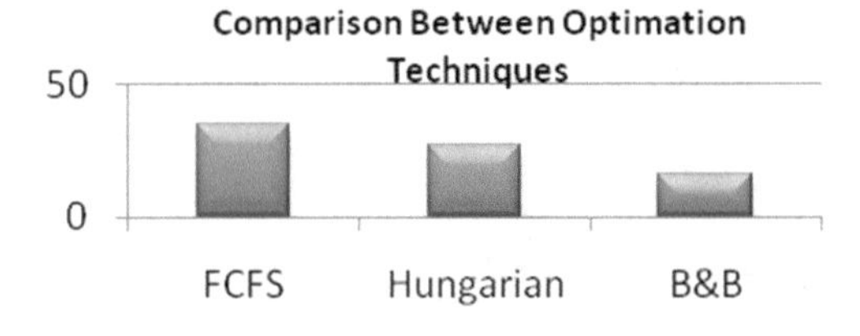

Fig. 2 A comparison between FCFS, Hungarian and Branch and Bound

time as 15.805 s for the executions. It is provided the much better and faster execution time and maximum resources allocation in cloud environment (Fig. 2) .

5 Conclusion and Future Work

Several virtual machine allocation techniques in related works have been reviewed and the various techniques to make them more efficient and optimal utilization for the resources have been found. In this paper, we present a comparison between the branch and bound techniques with the other previous techniques used in resources allocation and the optimal or minimum resource allocation times for the appropriate optimum resources utilization. Different QoS parameters are not viewed as like Migration, Cost, etc., and evaluation of this approach in Real Cloud Environment is a piece of future work.

References

1. Sharma, O., Saini, H.: State of art for energy efficient resource allocation for green cloud datacenters. Int. Sci. Press I J C T A **9**(11), 5271–5280 (2016)
2. Ogan, D., Azizoglu, M.: A branch and bound method for the line balancing problem in U-shaped assembly lines with equipment requirements. Elsevier J. Manuf. Syst. **36**, 46–54 (2015)
3. Gerard, T., Capraro, P., Bruce, B.: Frequency assignment for collocated transmitters using the branch-and-bound technique. IEEE, print ISBN: 978-1-5090-3158-0. https://doi.org/10.1109/isemc.1972.7567666
4. Chen, M., Bao, Y., Fu, X.: Efficient resource constrained scheduling using parallel two-phase branch-and-bound heuristics. IEEE Trans. Parallel Distrib. Syst. 1045-9219 (2016). https://doi.org/10.1109/tpds.2016.2621768
5. Lourencao, A.M., Baptista, E.C., Soler, E.M.: Mixed-integer nonlinear model for multiproduct inventory systems with interior point and branch-and-bound method. IEEE Latin Am. Trans. **15**(4), 1548-0992 (2017). https://doi.org/10.1109/tla.2017.7896403
6. Weeraddana, P.C., Codreanu, M., Latva-aho, M.: Weighted sum-rate maximization for a set of interfering links via branch and bound. IEEE Trans. Signal Process. **59**(8). Print ISSN: 1053-587X (2011). https://doi.org/10.1109/tsp.2011.2152397
7. Luo, R., Bourdais, R., van den Boom, T.J.J.: Integration of resource allocation coordination and branch-and-bound. IEEE (2015). ISBN: 978-1-4799-7886-1. https://doi.org/10.1109/cdc.2015.7402885
8. Nguyen, T.M., Yadav, A., Ajib, W.: Resource allocation in two-tier wireless backhaul heterogeneous networks. IEEE Trans. Wirel. Commun. (2016). Print ISSN: 1536-1276. https://doi.org/10.1109/twc.2016.2587758
9. Touzri, T., Ghorbel, M.B., Hamdaoui, B.: Efficient usage of renewable energy in communication systems using dynamic spectrum allocation and collaborative hybrid powering. IEEE Trans. Wirel. Commun. **15** (2016). Print ISSN: 1536-1276. https://doi.org/10.1109/twc.2016.2519908
10. Ibrahim, A., Alfa, A.S.: Using Lagrangian relaxation for radio resource allocation in high altitude platforms. IEEE Trans. Wirel. Commun. **14** (2015). Print ISSN: 1536-1276. https://doi.org/10.1109/twc.2015.2443095
11. Li, Z., Guo, S., Zeng, D.: Joint resource allocation for max-min throughput in multicell networks. IEEE Trans. Veh. Technol. **63** (2014). Print ISSN: 0018-9545. https://doi.org/10.1109/tvt.2014.2317235
12. Li, P., Guo, S., Cheng, Z.: Max-Min lifetime optimization for cooperative communications in multi-channel wireless networks. IEEE Trans. Parallel Distrib. Syst. (2014). ISSN: 1045-9219. https://doi.org/10.1109/tpds.2013.196

13. Mansourkiaie, F., Ahmed, M.H.: Optimal and near-optimal cooperative routing and power allocation for collision minimization in wireless sensor networks. IEEE Sensors J. **16** (2016). Print ISSN: 1530-437X. https://doi.org/10.1109/jsen.2015.2495329
14. Gaid, M.E.M.B., Cela, A.S., Hamam, Y.: Optimal real-time scheduling of control tasks with state feedback resource allocation. IEEE Trans. Control Syst. Technol. **17** (2009). Print ISSN: 1063-6536. https://doi.org/10.1109/tcst.2008.924566
15. Wu, Q., Hao, J.K..: A clique-based exact method for optimal winner determination in combinatorial auctions. Inf. Sci. **334–335**, 103–121 (2016)
16. Barnett, J., Watson, J.P., Woodruff, D.L.: BBPH: using progressive hedging within branch and bound to solve multi-stage stochastic mixed integer programs. Oper. Res. Lett. **45**, 34–39 (2017)
17. Gmys, J., Mezmaz, M., Melab, N., Tuyttens, D.: A GPU-based branch-and-bound algorithm using integer–vector–matrix data structure. Parallel Comput. **59**, 119–139 (2016)
18. Yazdani, M., Aleti, A., Khalili, S.M., Jolai, F.: Optimizing the sum of maximum earliness and tardiness of the job shop scheduling problem. Comput. Ind. Eng. **107**, 12–24 (2017)
19. Parmar, A.T., Mehta, R.: An approach for VM allocation in cloud environment. Int. J. Comput. Appl. 0975-8887, **131**(1) (2015)

Anomaly Detection Using K-means Approach and Outliers Detection Technique

A. Sarvani, B. Venugopal and Nagaraju Devarakonda

Abstract The main aim of this paper is to detect anomaly in the dataset using the technique Outlier Removal Clustering (ORC) on IRIS dataset. This ORC technique simultaneously performs both K-means clustering and outlier detection. We have also shown the working of ORC technique. The datapoints which is far away from the cluster centroid are considered as outliers. The outliers affect the overall performance and result so the focus is on to detect the outliers in the dataset. Here, we have adopted the preprocessing technique to handle the missing data and categorical variable to get the accurate output. To select the initial centroid we have used Silhouette Coefficient.

Keywords Outlier removal clustering · Silhouette coefficient · K-means clustering · Outlyingness factor · Box plot

1 Introduction

The data taken from number of sources probably of different types which are non-trivial, interesting and hidden knowledge can be effectively handled by data mining. The growth of databases with varied dimensions and complexity is exponential, due to high knowledge sharing. Hence, here we need automation of large amount of data. The user or application makes decision based on the results of analysis on data. Here, the problem lies in outlier detection and hence, we shift our focus on anomaly detection. The data points which deviates from the defined normal behaviour called outliers. Detecting the data points which are not according to the expected behaviour

A. Sarvani (✉) · N. Devarakonda
Lakireddy Balireddy College of Engineering, Mylavaram, India
e-mail: sarvani.anandarao@gmail.com

N. Devarakonda
e-mail: dnaraj_dnr@yahoo.co.in

B. Venugopal
Andhra Loyola Institute of Engineering and Technology, Vijayawada, India
e-mail: srees.boppana@gmail.com

K. Ray et al. (eds.), *Soft Computing: Theories and Applications*,
Advances in Intelligent Systems and Computing 742,
https://doi.org/10.1007/978-981-13-0589-4_35

is the main focus of the outlier detection. The outlier technique is used to detect the data which is abnormal from the rest of the data.

Outliers are often considered as a consequence of clustering algorithms in data mining. The data mining technique considers outliers as points which do not belong to any defined clusters. Inherently background noise in a cluster is considered as outliers. Outliers are points which act uniquely from the normal behaviour and are not considered as part of a cluster nor a part of the background noise. Some outlier points may be located nearer and some are located distant from the data points. The outlier which is distant from the data points will have great impact on the overall result. The outlier point which is distant from the other data points need to be detected and removed. So to get the effective clusters we need to use both the clustering and outlier technique. Outlier detection is used in number of areas like such as military surveillance, intrusion detections in cyber security, fraud detection in financial systems, etc.

2 Classification of Outliers

1. Point outliers
2. Contextual outliers
3. Collective outliers.

2.1 Point Outliers

This is the most familiar type of outlier. If a particular data point or instance is not according to the expected behaviour then that outlier is called point outlier.

2.2 Contextual Outliers

When a particular data point or instance is not according to the expected behaviour in particular context then it is called as contextual or conditional outlier. When we are defining the problem formulation and structure of data set, we should also include the meaning of a context. Here, we use contextual attributes and also behavioural attributes. This contextual attributes define the context of the data points. If we consider location of the data sets then longitude and latitude data sets are the contextual attributes. This behavioural attributes define the non-contextual of the data points. For example, temperature of a sunny day is a behavioural attribute in a particular context. We can detect the deviation from the normal behaviour using these behavioural attributes within a specific context. For effective result of contextual outlier technique we need to define both contextual and behavioural attributes.

2.3 The Collective Outliers

When a particular set of data points or instances is not according to the expected behaviour with respect to the entire data set then it is called collective outliers. Here, we do not consider the abnormal behaviour of the individual data points but only consider the abnormal behaviour of group of data points.

Here, we insert the outlier detection procedure into the K-means-based density estimation. This technique recursively removes the data points which are far away from most of common data points. In this paper, we have used the Outlier Removal Clustering (ORC) algorithm. This algorithm performs both data clustering and outlier detection parallel. To provide the accurate results here we have used the K-means clustering and outlier detection and the outlier are shown in box plot. This ORC is divided into a two-step process. In the first step, K-means clustering is applied on the data points. Then in the second step data points which are distinct from the cluster centroids are removed which gives the clusters without any anomaly.

In this paper, we have taken the iris dataset from UCI repository to perform the K-means and outlier detection and to remove the anomalies. Here, we consider the data points which are far away from the cluster centroid as the anomalies.

3 Literature Survey

In [1], this paper gives the overview of various data mining techniques. These techniques can be applied to the intrusion detection system (IDS). Performance of various data mining techniques on IDS is also shown in this paper. This paper has proven that the outlier technique has produced the best result among other techniques. This outlier technique has produced accurate result in detection of anomaly and in transmission of data. In [2] this paper, the author has taken the stock data to detect the anomaly data points. Here they have used the K-means and point outliers to detect the outlier points. In [3] they have taken the nonparametric model for estimation to develop an algorithm which adds outlier removal into clustering. The methodology used in this paper is compared with simple K-means and traditional outlier removal technique. This paper has proven that new methodology have given the better result compared to traditional methods. In [4] the weight-based approach is used and enhanced K-means clustering to detect the finest outliers also. Here we will group the similar items into a single cluster. By this technique, we can detect the outlier in short time even for large datasets. In [5] the author has given detail and clear idea of number of data mining techniques for outlier detection. In [6] the author has presented a novel technique which uses both the k, l values to detect the outlier. Here k indicates number of cluster and l indicates the number of outliers. We then present an iterative algorithm and prove that it converges to local optima. The proposed technique in this paper show that distance measure used for K-means is Bregman divergence by replacing the Euclidean distance. In [7] the focus is on detecting the denial of service attack using various data mining techniques. This had taken log file to perform the experiment and had applied pattern recognition file. Here, we use threshold value to detect the outlier. We detect that there is an attack when number of similar type of request crosses the threshold value. In [8] a technique which simultaneously performs the clustering and outlier detection is presented. Here K-means clustering is done. The results are shown by comparing with the traditional methods. In [9] the K-Nearest Neighbour (KNN) method is used to detect the anomaly from the given dataset. This technique can be used to find the malpractices and crime within limited resources. This paper concentrated on detecting the credit card fraud using KNN method. The outlier detection produces best results with large online datasets. This paper has proven that KNN method has given the accurate and efficient result.

4 Problem Statement

The anomaly in the data point affects the accuracy of the overall result. Detecting and removal of anomaly is the best solution. By using the cluster information of such anomalies, our approach detects the anomalies in the IRIS dataset.

5 Methodology

We have started our experiment by reading the iris dataset consisting of 5 columns and 150 rows. The first column has the maximum value of 7.9 and minimum value of 4.3, for the second column maximum value is 4.2 and minimum 2.0, for third column maximum value is 6.9 and minimum value is 1.0, for fourth column maximum value is 2.5 and minimum value is 0.3, the data points which crosses these ranges they are considered as outlier and this we found using the outlier removal clustering. Here, we have used distance factor from cluster centroid to detect the outliers from the data points. First, we find the vector which is having the maximum distance from the cluster centroid.

$$d_{\max} = \max\left\{\left\|x_i - c_{pi}\right\|\right\},\ \mathrm{i} = 1, \ldots, \mathrm{N}. \tag{1}$$

Then for the vector, we calculate the outlyingness factors.

$$o_i = \frac{\left\|x_i - c_{pi}\right\|}{d_{\max}} \tag{2}$$

After collecting the dataset, we have applied pre-processing techniques to handle the missing data and categorical variables. Next by using the Silhouette Coefficient, we have selected the initial centroids for clustering. Next by applying outlier

Fig. 1 Flowchart

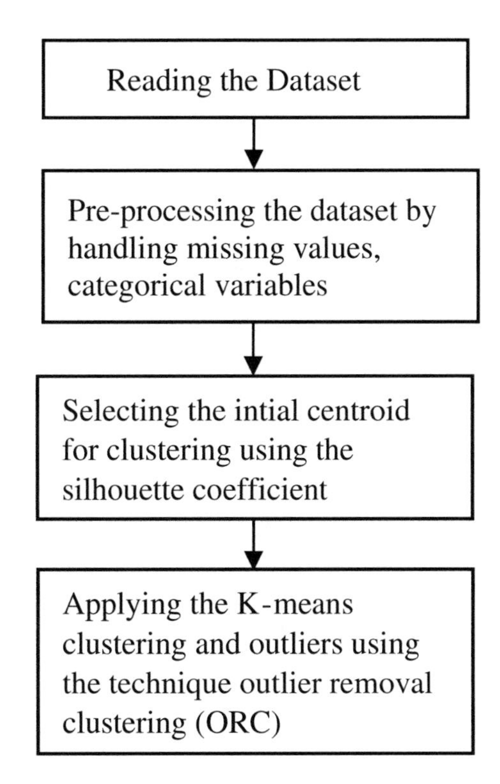

removal clustering (ORC), we had simultaneously clustered the data and can detect the anomaly in the Iris dataset (Fig. 1).

5.1 Reading the Dataset

Here we took the iris dataset that consists of 150 iterations and 5 columns. In this last column is the class name.

5.2 Pre-processing of Data

Missing value Handling—The K-means clustering cannot handle the missing values. Even if we have one single iteration with missing value it should be resolved to bring out the accurate result. If possible we can also avoid the iterations having the missing value from clustering. In some cases, it is not possible to avoid the missing value then we need to assign a value to the missing observation. Care must be taken while replacing the missing value, replaced missing value should not be misleading. For example, we cannot replace the height of the person as −1 and age of person to 99999 this leads to misleading

Categorical variables—K-means working is based on distance measure so it can handle only the numerical data. To apply K-means on our dataset first we need to change categorical variable into the numerical data. This can be done in two ways:

- The ordinal variables are replaced with their appropriate numerical values. For example small/medium/high can be replaced with 1/5/10, respectively. As we perform the normalization this will not have the great impact on the overall result.
- The cardinal variables must be replaced with binary numerical variables. The number of binary numerical variable should be equal to the number categorical classes. For example cardinal variable are cycle/bike/car/auto can be converted into four binary numerical variables.

5.3 Selecting the Initial k Value

Cluster quality using Silhouette Coefficient Silhouette coefficient comes under quality measure of clustering. Any clustering technique can adopt this Silhouette coefficient this is not only designed for K-means clustering

$$S_i = \frac{x(i) - y(i)}{\max(x(i), y(i))} \tag{3}$$

Here S_i is the Silhouette Coefficient of observation i, x(i) shows us that how well the observation 'i' is matched with that cluster behaviour and y(i) is the lowest average dissimilarity to any other cluster i, where i is not the member of that class. If silhouette coefficients are equal to + 1 then it indicates that clustering had done correctly. If silhouette coefficients equal to 0 indicates that the observation is on or very nearer on boundary present in between the two neighbouring clusters and if silhouette coefficients equal to −1 indicate that observations are assigned to the wrong clustering. Based on this silhouette coefficients, we can do the K-means clustering perfectly by avoiding continues change in the centroids. Initially we can select the correct clusters using this silhouette coefficients value.

5.4 Outlier Removal Clustering (ORC)

This produces the code book, such that it very closes to the mean vector parameters which can be calculated from our original data. This has two stages. K-means clustering is applied in the first stage and we calculate the outlyingness factor for each data point in the second stage. This outlyingness factor gives the distance of the data point from the cluster centroid. These two stages repeat a number of times until we get the desired result. Here, initially ORC detects the data point which is far away from the cluster centroid. First $d_{\max}$ and then o_i is calculated.

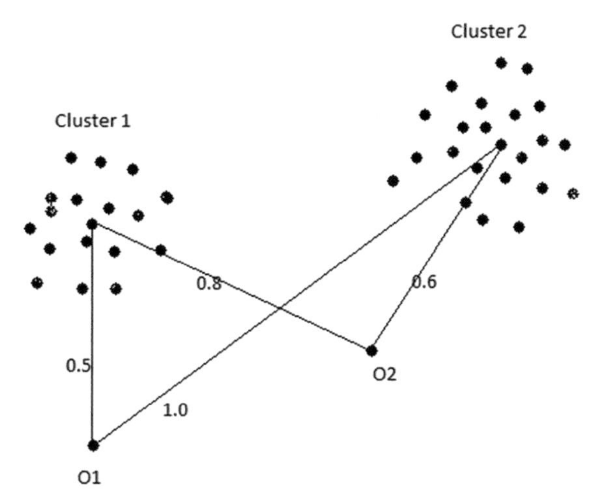

Fig. 2 Representing the outlier

$$d_{\max} = \max_{x} \left| x_x - c_p \right|, \quad i = 1, \ldots, N \tag{4}$$

x_i is one of the data points and c_p is the cluster centroid

$$o_i = \frac{\left| x_x - c_p \right|}{d_{\max}} \tag{5}$$

o_i is the outlyingness factors.

The outlyingness factor which can be calculated from the above Eq. 3 and this is normalized to scale [0, 1]. If the value is high, it impacts the overall result and this is called outlier point. An example is shown in Fig. 2.

In the above figure, there are two clusters say cluster 1 and cluster 2. In this, we have two outlier data points which does not belong to any of two clusters. Those are o1, o2 and these two data points are far from the cluster centroid.

Algorithm of ORC

C ← Select the best centroid by trying different centroids with K-means clustering
for x ← 1, …, I do
$d_{\max} = \max_x \{ \left| x_x - c_p \right| \}$
for y ← 1, …, N do
$o_i = \left| x_y - c_p \right| / d_{\max}$
If $o_i > T$ then
X = X/$\{x_x\}$
end if
end for
C ← K-means (K, C)
end for

The data point whose value of O_i >T is the outlier and removed from the dataset.

6 Experiment

This experiment is done on iris dataset to detect the outlier. This dataset consists of 5 columns in that 4 columns are sepal length in cm, sepal width in cm, petal length in cm, petal width in cm and fifth column is class name Iris Setosa or Iris Versicolour or Iris Virginica. The data point which does not come under any of three classes is considered as outlier. This technique of ORC has given the better results than the traditional approach. The outliers detected in the dataset are represented on box plot and on scatter plot. Here we have divided the data points into 3 clusters. The data point which does not belong to any of the three clusters is detected as outliers. This experiment in done on Windows 10 operating system with RAM 2 GB, hard disk of 500 GB and on the processor Intel core i4.

The dataset is divided into three clusters:

Cluster_0: Consists of 37 data points
Cluster_1: Consists of 64 data points
Cluster_2: Consists of 49 data points.

6.1 *Outliers on Box Plot Graph*

In Fig. 3 the data points 11.2 and 10.0 are outliers as they are not in the range of column 0. The data points 8.2, 7.1, 5.5 are outliers as they are not in the range of column 1. The data point 13 is outlier as it is not in the range of column 2. Using box plot, we had represented the outliers which are identified in each and every column and the result in the scatter graph is shown in Fig. 4.

6.2 *Result in the Scatter Graph*

See Fig. 4.

6.3 *Outliers in Clusters*

In Fig. 5, we had represented the outliers detected in each and every cluster along with the data points and Fig. 6 shows only the outliers detected. Here we had clustered the data points into three clusters.

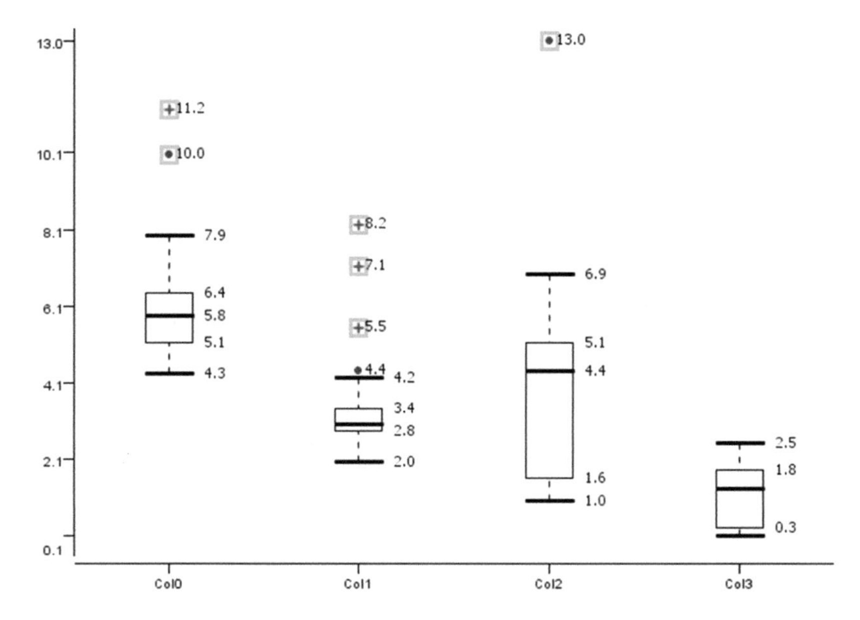

Fig. 3 Outliers represented on box plot

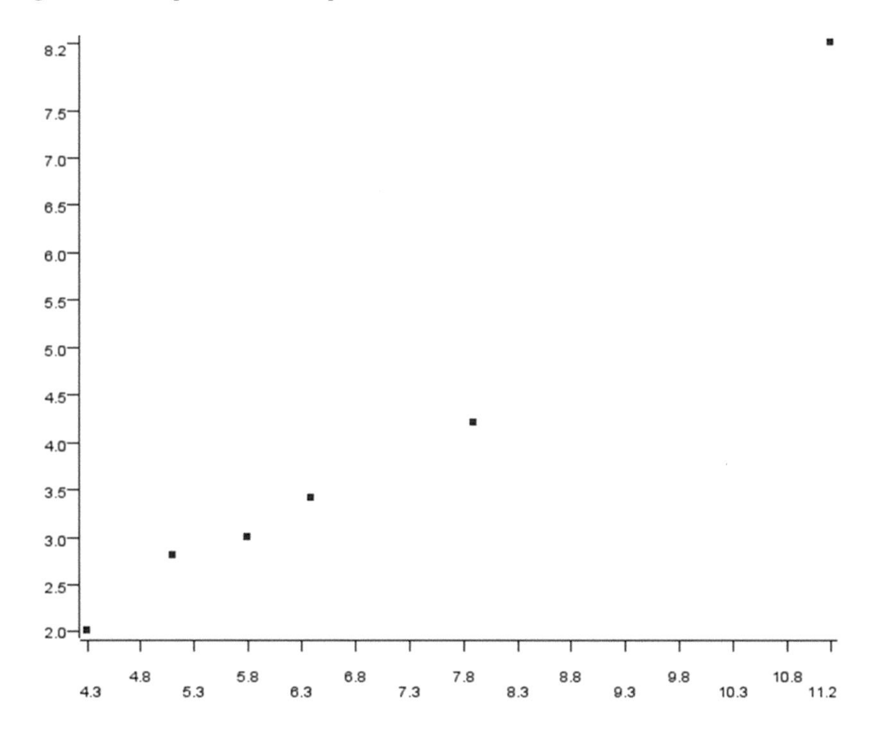

Fig. 4 Outlier represented on scatter graph

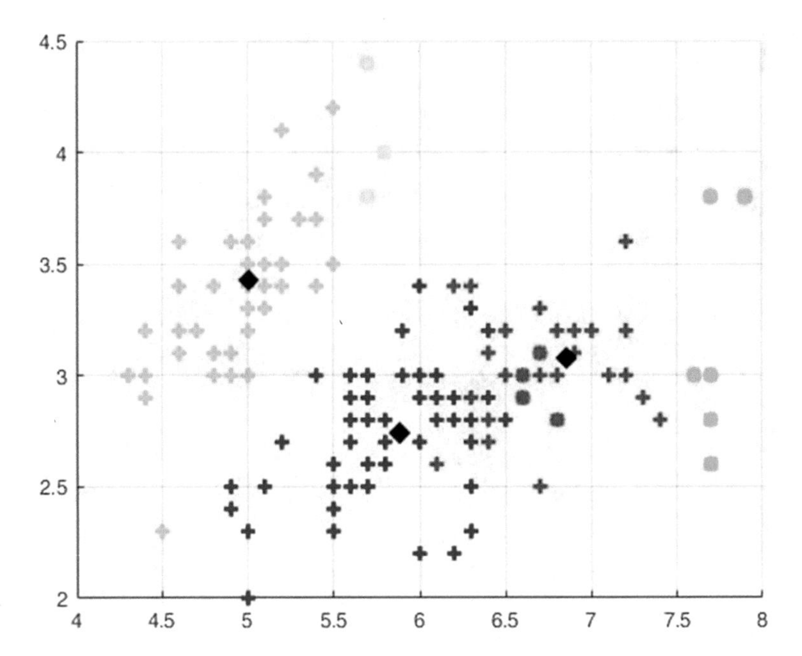

Fig. 5 Outliers among the scattered data points

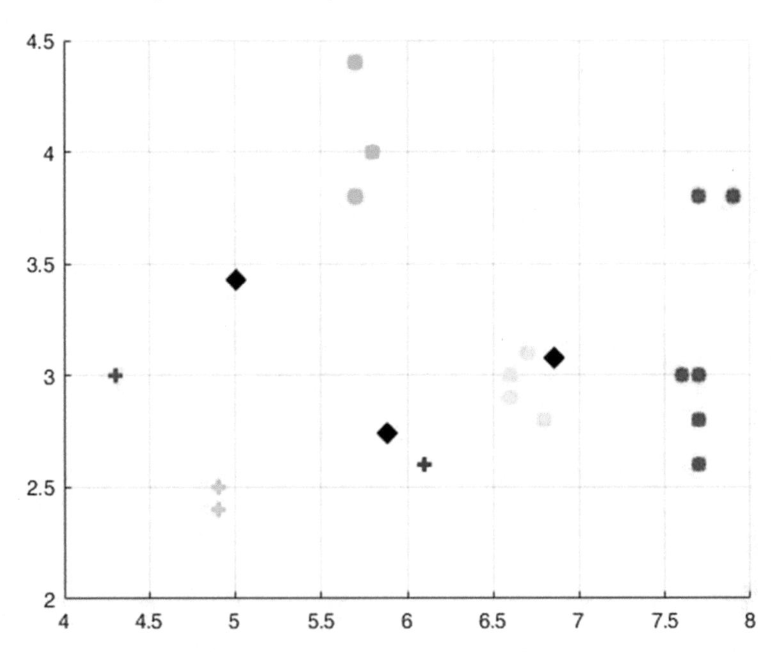

Fig. 6 Only outliers

7 Conclusion

This paper focused on detecting the anomaly using the ORC technique. To prove that our technique has given the better result and accurate result we have taken the IRIS dataset and on that we have applied the K-means and outlier detection method. Here, we have taken the data point as outlier if the distance from centroid is above the threshold value. The data points which are outliers are represented on the box plot and also on scatter plot. Here, we have also concentrated on the pre-processing of the dataset and also on the technique for selecting the best centroid. Finally, we proved that our methodology has produced the accurate and satisfactory result.

References

1. Razak, T.A.: A study on IDS for preventing denial of service attack using outliers techniques. In: Proceedings of the 2016 International Conference on Engineering and Technology (ICETECH), pp. 768–775 (2016)
2. Zhao, L., Wang, L.: Price trend prediction of stock market using outlier data mining algorithm. In: Proceedings of the Fifth International Conference on Big Data and Cloud Computing (BDCloud), pp. 93–98. IEEE (2015)
3. Marghny, M.H., Taloba, A.I.: Outlier detection using improved genetic k-means (2011)
4. Jiang, M.F., Tseng, S.S., Su, C.M.: Two-phase clustering process for outliers detection. Pattern Recogn. Lett. **22**(6), 691–700 (2001)
5. Bansal, R., Gaur, N., Singh, S.N.: Outlier detection: applications and techniques in data mining. In: Proceedings of the 6th International Conference on Cloud System and Big Data Engineering (Confluence), pp. 373–377. IEEE (2016)
6. Chawla, S., Gionis, A.: k-means: a unified approach to clustering and outlier detection. In: Proceedings of the 2013 SIAM International Conference on Data Mining, Society for Industrial and Applied Mathematics, pp. 189–197 (2013)
7. Khan, M.A., Pradhan, S.K., Fatima, H.: Applying data mining techniques in cyber crimes. In: Proceedings of the 2nd International Conference on Anti-Cyber Crimes (ICACC), pp. 213–216. IEEE (2017)
8. Hautamäki, V., Cherednichenko, S., Kärkkäinen, I., Kinnunen, T., and Fränti, P.: Improving k-means by outlier removal. In: Proceedings of the Scandinavian Conference on Image Analysis, pp. 978–987. Springer, Berlin (2005)
9. Malini, N., Pushpa, M.: Analysis on credit card fraud identification techniques based on KNN and outlier detection. In: Proceedings of the Third International Conference on Advances in Electrical, Electronics, Information, Communication and Bio-Informatics (AEEICB), pp. 255–258. IEEE (2017)

A Steady-State Genetic Algorithm for the Tree t-Spanner Problem

Shyam Sundar

Abstract A *tree t-spanner*, in a given connected graph, is a spanning tree T in which the distance between every two vertices in T is at most t times their shortest distance in the graph, where the parameter t is called stretch factor of T. This paper studies the *tree t-spanner problem* on a given connected, undirected and edge-weighted graph G that seeks to find a spanning tree in G whose stretch factor is minimum among all tree spanners in G. This problem is $\mathcal{NP}$-Hard for any fixed $t > 1$. In the domain of metaheuristic techniques, only genetic algorithm (GA) based on generational population model is proposed in the literature. This paper presents a steady-state genetic algorithm (SSGA) for this problem, which is quite different from the existing GA in the literature, not only in the population management strategy, but also in genetic operators. Genetic operators in SSGA use problem-specific knowledge for generating offspring, making SSGA highly effective in finding high-quality solutions in comparison to the existing GA. On a set of randomly generated instances, computational results justify the effectiveness of SSGA over the existing GA.

Keywords Tree spanner · Evolutionary algorithm · Genetic algorithm
Steady-state · Weighted graph · Problem-specific knowledge

1 Introduction

A *tree t-spanner*, in a given connected graph $G(V, E)$, is a spanning tree T in which the distance between every two vertices, viz. x and y in T is at most t times their shortest distance in G. The t is a parameter called stretch factor of T. The stretch of x and y in T is the ratio of distance between x and y in T to their distance in G, and the stretch factor is the maximum stretch taken over all pairs of vertices in G. The *tree t-spanner problem* is a combinatorial optimization problem that aims to find a

S. Sundar (✉)
Department of Computer Applications, National Institute of Technology Raipur,
Raipur 492010, CG, India
e-mail: ssundar.mca@nitrr.ac.in

K. Ray et al. (eds.), *Soft Computing: Theories and Applications*,
Advances in Intelligent Systems and Computing 742,
https://doi.org/10.1007/978-981-13-0589-4_36

spanning tree of G whose stretch factor t is minimum among all tree spanners of G. This problem is also called the minimum max-stretch spanning tree problem [7] for *unweighted* graphs.

This paper considers the *tree t-spanner problem* for *connected, undirected* and *weighted* graph $G(V, E, w)$, where V is a set of vertices; E is a set of edges; and for each edge, there exists an edge weight which are positive rational numbers. The objective of this problem considered in this paper is to find a spanning tree of G of minimum stretch factor among all tree spanners of G. Hereafter, the *tree t-spanner problem* for *connected, undirected* and *weighted* graph $G(V, E, w)$ will be referred to as T_[t-SP]^{w}.

1.1 Literature Survey

Peleg and Ullman [14] introduced *t-spanner* (spanning subgraph H) of a graph in which the distance between every pair of vertices is at most t times their distance in the graph. The stretch of u and v vertices in H is defined as the ratio of the distance between u and v in H to their distance in G. A solution to this problem presents distance approximation property related to pairwise vertex-to-vertex distances in the graph by its spanning subgraph, and finds practical applications in communication networks, distributed systems, motion planning, network design and parallel machine architectures [1–3, 11, 14, 15].

A spanning subgraph (say T) of a connected graph G is a *tree t-spanner* if T is both a *t-spanner* and a *tree*. A spanning tree of G is always a tree spanner, where a tree spanner is a *tree t-spanner* for some $t \geq 1$. Tree spanner also finds a significant application, i.e. a tree spanner with small stretch factor can be utilized to perform multi-source broadcast in a network [2], which can greatly simplify the message routing at the cost of only small delay in message delivery. Later, Cai and Corneil [5] studied graph theoretic, algorithmic and complexity issues about the tree spanners of *weighted* and *unweighted* connected graphs. They showed on *weighted* graphs that a *tree* 1-*spanner*, if it exists, is a minimum spanning tree that can be obtained in polynomial time; whereas for any fixed $t > 1$, they also proved that the problem of finding a *tree t-spanner* in a *weighted* graph is a $\mathcal{NP}$-Hard problem. In case of *unweighted* graphs, they showed that a *tree* 2-*spanner* can be constructed in polynomial time; whereas, for any fixed integer $t \geq 4$, the *tree t-spanner* in a *unweighted* graph is a $\mathcal{NP}$-Hard problem. One can find various related tree spanner problems in detail in [10].

In the literature, approximation algorithms on some specific types of *unweighted* graph [4, 13, 22] have been developed for the *tree t-spanner problem*; however, such specific types of graph may not be encountered in every situation. For any fixed $t > 1$, T_[t-SP]^{w} is a $\mathcal{NP}$-Hard problem for *connected, undirected* and *weighted* graph. For such $\mathcal{NP}$-Hard problem, metaheuristic techniques are the appropriate approaches that can find high-quality solutions in a reasonable time. In the literature, only genetic algorithm (GA) [12] has been proposed so far for the T_[t-SP]^{w}.

GA [12] is based on generational population model, which uses multi-point crossover and mutation genetic operators for generating offspring.

This paper presents a steady-state genetic algorithm (SSGA) for the solution of the *tree t-spanner problem* for *connected, undirected* and *weighted* graph (T_[t-SP]^{w}), which is quite different from the existing GA [12] in the literature, not only in the population management strategy, but also in genetic operators. Genetic operators in SSGA use problem-specific knowledge for generating offspring, making SSGA highly effective in finding high-quality solutions in comparison to GA [12]. On a set of randomly generated graph instances, computational results show the superiority of SSGA over GA.

The rest of this paper is organized as follows: Sect. 2 describes a a brief introduction of SSGA, whereas Sect. 3 describes SSGA for the T_[t-SP]^{w}. Computational results are reported in Sect. 4. Finally, Sect. 5 contains some concluding remarks.

2 Steady-State Genetic Algorithm (SSGA)

Genetic algorithm (GA) is an evolutionary algorithm that is based on the principles of natural evolution [9]. GA has been applied successfully to find global near optimal solutions of numerous optimization problems. This paper presents a steady-state genetic algorithm (SSGA) [6] for the T_[t-SP]^{w}. The reason behind choosing SSGA is that SSGA follows steady-state strategy for the population management which is different from generational replacement strategy for the population management. It is to be noted that generational GA for the T_[t-SP]^{w} was proposed in [12]. In SSGA, if a new child solution generated either from crossover or mutation is found to be unique from the current individuals of the population, then this child solution replaces a worst individual of the population, otherwise it is discarded. Through this approach, one can easily avoid the occurrence of duplicate solutions in the population. In addition, this approach also guarantees the survival of highly fit solutions over the whole search, helping in faster convergence towards higher solution quality in comparison to that of generational replacement [6].

3 SSGA for the T[t-SP]^{w}

The main components of SSGA for the T_[t-SP]^{w} are as follows:

3.1 Encoding

Edge-set encoding [17] is used to represent each solution (chromosome) of the T_[t-SP]^{w}, as each solution is a spanning tree. The encoding consists of a set of

$|V|$-1 edges to represent a spanning tree, where $|V|$ is the set of vertices in G. The reason behind selection of this encoding is of twofolds: first, it offers high heritability and locality; and second, problem-specific heuristics can be easily incorporated.

3.2 Initial Solution Generation

Generation of each initial solution (spanning tree T) of the population follows a procedure similar to Prim's algorithm [16]. The tree T starts with selecting a vertex randomly from V called *root* vertex r and grows until T spans all the vertices of V. Each step, instead of adding a new edge of least cost that joins a new vertex to the growing T, a new random edge that joins a new vertex to the growing T is added. When this procedure completes, the edges in T construct a spanning tree.

3.3 Fitness

Once the solution (spanning tree T) is constructed, the fitness of T is computed in terms of stretch factor t.

Since a spanning tree is a tree spanner, therefore, when the spanning tree T is constructed, the fitness of T, in terms of stretch factor t, is computed. Before computing the fitness of T, lowest common ancestors for all pairs of vertices in T with its *root* vertex r (the first selected vertex during construction of T), are preprocessed through least common ancestor algorithm [8], where lowest common ancestor for any two vertices, say x and y in T is the common ancestor of vertices x and y that is located farthest from its *root*, and is referred to as $LCA(x, y)$. In doing so, one can get lowest common ancestor query for any pair of vertices, say x and y, in T, i.e. $LCA(x, y)$ in constant-time. Distance $d_T(x, y)$ between any two vertices x and y in T can be computed as

$$d_T(x, y) = d_T(r, x) + d_T(r, y) - 2 \times (d_T(r, LCA(x, y))) \tag{1}$$

One can easily determine the maximum stretch factor t of T taken over all pairs of vertices (x and y) in G, i.e.

$$\frac{d_T(x, y)}{d_G(x, y)} \forall (x, y) \in G \tag{2}$$

Once the maximum stretch factor t of T which is also the fitness of T is determined, the uniqueness of current initial solution (T) is tested against all individuals (solutions) of the initial population generated so far. If it is found to be unique, then it is added to the initial population, otherwise it is discarded.

3.4 Selection

Binary tournament selection method is applied two times in order to select two parents for participating in the crossover. In this selection method, initially two solutions are selected randomly from the current population. Between these two selected solutions, with probability P_b, a solution with better fitness is selected as a parent, otherwise another solution with worse fitness is selected as parent. The same selection method is applied to select a solution for the mutation operator.

3.5 Crossover Operator

A problem-specific crossover operator of SSGA is designed for the T_[t-SP]^{w}. Such operator easily adapts with the tree-structure of two selected parents (p_1, p_2) for generating a feasible child solution. Simultaneously, it also allows the child solution to inherit potentially high-quality building blocks (edges or genes) of the parent solutions (p_1, p_2). Initially, the child solution, say C, is empty. Hereafter, the crossover operator proceeds as follows:

The proposed crossover starts with selecting a parent, say p_1, with better fitness from the two parents (p_1, p_2) that are selected with the help of binary tournament selection method. A path ($x \leftrightsquigarrow y$) between two vertices (say x and y), whose distance ($d_T(x, y)$) is used in computing the fitness ($\frac{d_T(x,y)}{d_G(x,y)}$) of p_1 is selected among all paths in the spanning tree T of p_1. A vertex, say r, with maximum *degree* is selected from the selected path ($x \leftrightsquigarrow y$) of the spanning tree of p_1. Set r as a $root$ vertex for the spanning tree of C.

It is possible that more than one vertex of same maximum *degree* in the selected path of the spanning tree of p_1 may exist. In case of such possibility, a tie-breaking rule is applied. As per this rule, for each such vertex, say v_m, the sum of distances of all vertices adjacent to v_m in the spanning tree of p_1 is computed. A vertex with minimum sum is selected as $root$ vertex, i.e. r. It is also possible that there may be more than one vertex with same minimum sum. In such a situation, another tie-breaking rule is applied. As per this rule, for each such vertex, say v_d, the *difference* of degree of v_d in G and degree of v_d in its spanning tree of p_1 is computed. A vertex with maximum *difference* is selected as $root$ vertex, i.e. r. All edges that connect r to its adjacent vertices in the spanning tree of p_1 are added to C.

Hereafter, if the *degree* of r in G is greater than the *degree* of r in the spanning tree of p_1, then it is checked in another selected parent, i.e. p_2 whether edge(s) that connect(s) r to its adjacent vertices in the spanning tree of p_2, and that must be different from all current edges in C exist(s). If such edge(s) exist(s), then such edge(s) is(are) added to C. Hereafter, all remaining edges of spanning trees of $p1$ and p_2 that are not in C are assigned to a set (say E_Set). Note that duplicate edges are discarded in E_Set. Vertices that are are part of current partial spanning tree (C) are assigned to a set (say S). Hereafter, at each step, the procedure, similar to Prim's algorithm [16], greedily selects an edge (e) with minimum cost from E_Set that connects a vertex (say v_p) $\in V \backslash S$ to S, and add it to the growing spanning tree C. The vertex v_p is added to S. Newly added edge e to C is now removed from E_Set. This procedure is applied repeatedly until S becomes V or $V \backslash S$ becomes empty. At this juncture, a spanning tree of C is constructed.

Note that instead of selecting an edge each time either from p_1 and p_2 in order to generate a feasible child solution C, the proposed crossover starts with selecting a vertex with maximum *degree*, as *root* vertex for C, from the spanning tree of better parent (say p_1). All good edges (good genes) of p_1 adjacent to *root* vertex become part of C. This increases the proximity of all selected edges of C, as all selected edges share *root* vertex of C. First tie-breaking rule, in case of more than one vertex with same maximum *degree* in the selected path of p_1, tries to minimize the overall fitness of C (at this juncture, C is partial). Second tie-breaking rule, in case of more than one vertex with same minimum sum, further helps in adding edges adjacent to *root* vertex from other selected parent, i.e. p_2. This further increases the proximity of all edges that are part of p_1 and p_2, and also that are adjacent to *root* vertex of C.

3.6 Mutation Operator

Mutation operator in GA is used to provide diversity in the current population. The proposed mutation operator is based on deletion and insertion of an edge. Initially, a solution from the population is selected with the help of binary tournament selection method. The selected solution is copied to an empty child solution, say C. Hereafter, a path ($x \leftrightsquigarrow y$) between two vertices (say x and y) whose distance ($d_T(x, y)$) is used in computing the fitness ($\frac{d_T(x,y)}{d_G(x,y)}$) of C is selected among all paths in the spanning tree T of C. Mutation operator first selects a random edge (say e) from the selected path ($x \leftrightsquigarrow y$) of C and then deletes this selected edge from the spanning tree of C. In doing so, C is partitioned into two components. At this juncture, C is infeasible. To make C feasible, an edge, say e', (different from e) that connects these two components of C is searched in the edge-set E of G. It is possible that there can be more than one edge in the edge-set E of G that can connect these two components of C. In such a possibility, an edge with minimum edge cost is selected for insertion in C for making C feasible. Note that *root* vertex of spanning tree of C is considered as the first end vertex of first edge of its edge-set encoding.

It should be noted that the proposed SSGA allows crossover and mutation operators in a mutual exclusive way [18–21] rather than one-by-one in each generation for the generation of a child solution. With probability P_x, SSGA proceeds crossover operator, otherwise it proceeds mutation operator. The notion behind this one is that crossover operator employing problem-specific heuristic generates a child solution inheriting potentially high-quality building blocks (edges or genes) of the parent solutions, and mutation operator based on deletion and insertion of an edge generates a child solution. If the proposed SSGA proceeds crossover and mutation operators together, then it is very likely that the resultant child solution may lose a potentially high-quality building block (edge or gene).

Once the spanning tree C is generated from either crossover or mutation operator, its fitness is computed (see Sect. 3.3).

3.7 *Replacement Policy*

Each newly generated child solution C is tested for its uniqueness against all solutions of current population. If it is unique, then it is added immediately to the current population by replacing the worst individual of the population; otherwise, it is discarded.

The search scheme of SSGA for the T_[t-SP]^w is explained in Algorithm 1.

Algorithm 1: Search scheme of SSGA for the T_[t-SP]^w

```
Initialize each solution of the population (see Section 3.2);
best ← Best solution in terms of fitness in the population;
while (Termination criteria is not met) do
    if (u01 < Px) then
        Select two solutions as parents (p1, p2) (see Section 3.4);
        Apply crossover operator on p1 and p2 to generate a child solution (C) (see Section 3.5);
    else
        Select a solution (p1) (see Section 3.4) ;
        Apply mutation operator on p1 to generate a child solution (C) (see Section 3.6);
    Compute the fitness of C (see Section 3.3);
    if (C is better than best) then
        best ← C;
    Apply replacement policy;
return best;
```

---

4 Computational Results

The proposed SSGA for the Tree_[t-SP] has been implemented in C. A set of randomly generated Euclidean graph instances (described in next subsection) has been used to test the proposed SSGA. All computational experiments have been carried out on a Linux with the configuration of Intel Core *i5* 3.20 GHz $\times$ 4 with 4 GB RAM. In all computational experiments with the proposed SSGA, a population of 400 has been used. We have set $P_b = 0.7$ and $P_x = 0.7$. All these parameter values are set empirically after a large number of trials. These parameter values provide good results though they may not be optimal for all instances. SSGA has been allowed $|V| \times 1000$ generations to execute for each graph instance.

Since the number of graph instance used in GA [12] are two only and such graphs are also not available. Therefore, a set of Euclidean graph instances are generated randomly. In addition, to test the effectiveness and robustness of the proposed SSGA against GA [12] for this problem, GA [12] has been re-implemented. Similar values of parameters in GA [12] are considered. For example, the population size is equal to the size of vertices of the graph under consideration; crossover probability = 0.9; and mutation probability = 0.2. Since the maximum number of generations in GA [12] is 300 which is considered to be less in the domain of metaheuristic techniques. To match with SSGA in terms of computational time (approximately) (see Table 1), GA is allowed 2000 generations to execute for each graph instance. Both SSGA and GA have been also executed 10 independent runs for each graph instance in order to check their robustness. In subsequent subsections, a description of generation of Euclidean graph instances that have been used for testing SSGA and GA, and a comparison study of results obtained by SSGA and GA on these generated graph instances have been reported.

4.1 Graph Instances

A set of Euclidean graph instances $G(V, E)$ with $|V| = \{50 \text{ and } 100\}$ are generated randomly in 100×100 plane. Generation of each graph instance is as follows: each point representing a vertex in G is selected randomly in 100×100 plane. The Euclidean distance (rounded to two decimal positions) between two vertices, say v_1 and v_2, presents its edge weight or edge cost. For each $|V|$, three different complete graph instances $G_i(V_i, E_i)$ with $i = \{1, 2 \text{ and } 3\}$ are generated. With the help of each generated graph $G_i(V_i, E_i)$, further three sparse graphs with different edge-density, i.e. $|E_{i1}|$, $|E_{i2}|$, $|E_{i3}|$ are generated, where $|E_{i1}| = 0.8 \times |E_i|$; $|E_{i2}| = 0.6 \times |E_i|$; and $|E_{i3}| = 0.4 \times |E_i|$. It is to noted that $0.8 \times |E_i|$ ($0.6 \times |E_i|$ and $0.4 \times |E_i|$) means 20% (40% and 60%) edges of E_i that are selected randomly are not part of E_{i1} (E_{i2} and E_{i3}). It leads to 12 graph instances for each $|V|$. So, overall total instances generated for two different $|V|$ are $2 \times 12 = 24$.

4.2 Comparison Study of Results Obtained by GA [12] and SSGA

In this subsection, a comparison study of results obtained by SSGA and GA [12] on a set of graph instances is carried out for the Tree_[t-SP]. Table 1 reports the results obtained by SSGA and GA [12] on a set of randomly generated graph instances. In Table 1, the column *Instance* presents the name of each graph instance in the form of *Y1_Y2_Y3*; columns $|V|$ and *Edge-Density* present the characteristics of graph in terms of total number of vertices in G and the edge-density respectively; and three columns for each approach, i.e. SSGA and GA report the best value (*Best*), the average solution quality (*Avg*), standard deviation (*SD*) and the average total execution time (*ATET*) obtained over 10 runs. Note that a graph instance *Y1_Y2_Y3* (say *50_0.2_1*) in column *Instance* of Table 1, $Y1$ (50) is $|V|$ of G; $Y2$ (0.2) is $X\%$

Table 1 Results of GA [12] and SSGA for T_[t-SP]^{w} on a set of graph instances

Instance	Characteristics		GA				SSGA			
	$\|V\|$	*Edge-Density (%)*	*Best*	*Avg*	*SD*	*ATET*	*Best*	*Avg*	*SD*	*ATET*
50_0.0_1	50	0	19.01	25.06	4.41	3.55	**5.90**	**6.09**	0.11	2.05
50_0.2_1	50	20	15.96	21.34	3.51	3.55	**4.98**	**5.23**	0.22	1.82
50_0.4_1	50	40	11.64	17.37	3.10	3.54	**4.89**	**5.10**	0.14	2.22
50_0.6_1	50	60	10.08	11.73	1.52	3.48	**5.55**	**6.20**	0.40	1.94
50_0.0_2	50	0	15.75	29.03	7.89	3.60	**7.39**	**8.25**	0.87	1.81
50_0.2_2	50	20	13.91	25.33	6.00	3.68	**6.05**	**6.68**	0.29	1.83
50_0.4_2	50	40	15.02	18.76	3.16	3.59	**4.88**	**5.76**	0.36	1.94
50_0.6_2	50	60	9.02	13.01	2.17	3.45	**5.38**	**5.74**	0.18	1.93
50_0.0_3	50	0	22.63	28.26	4.56	3.54	**7.03**	**7.41**	0.24	2.04
50_0.2_3	50	20	13.93	23.35	7.92	3.51	**5.16**	**5.62**	0.24	2.09
50_0.4_3	50	40	13.93	23.35	7.92	3.51	**5.16**	**5.62**	0.24	2.09
50_0.6_3	50	60	10.20	13.12	2.00	3.44	**4.61**	**5.03**	0.15	2.05
100_0.0_1	100	0	99.93	130.98	36.71	23.31	**7.26**	**9.88**	1.47	18.13
100_0.2_1	100	20	69.92	131.66	58.08	23.33	**7.84**	**8.44**	0.49	19.06
100_0.4_1	100	40	73.27	109.27	36.81	23.38	**8.37**	**9.01**	0.43	20.10
100_0.6_1	100	60	52.21	65.47	8.87	23.47	**8.93**	**9.75**	0.57	19.42

(continued)

Table 1 (continued)

Instance	Characteristics		GA				SSGA			
	|V|	*Edge-Density* (%)	*Best*	*Avg*	*SD*	*ATET*	*Best*	*Avg*	*SD*	*ATET*
100_0.0_2	100	0	93.09	165.08	51.95	24.21	**7.93**	**9.90**	1.17	19.30
100_0.2_2	100	20	71.04	141.65	51.99	24.40	**8.77**	**9.90**	0.73	19.63
100_0.4_2	100	40	64.88	107.02	44.76	23.95	**7.84**	**9.22**	0.86	20.30
100_0.6_2	100	60	59.83	74.67	7.85	24.72	**7.15**	**7.62**	0.57	22.37
100_0.0_3	100	0	124.26	141.03	9.71	25.07	**9.07**	**10.32**	1.19	20.01
100_0.2_3	100	20	112.00	137.24	19.85	25.09	**8.73**	**9.63**	0.70	22.07
100_0.4_3	100	40	80.39	102.81	30.61	24.80	**10.18**	**11.07**	0.95	20.53
100_0.6_3	100	60	60.75	75.68	11.62	24.83	**9.09**	**10.65**	0.91	19.79

(20% in its column E) random edges of total number of edges from its corresponding complete graph with $|Y1|$ vertices (i.e. *50_0.0_1*) are not considered in *50_0.2_1*; and $Y3$ (1) presents different graph instance (in terms of edge-density) with the same $|V|$, i.e. 50. Hence, three different instances with different edge-densities have been created from the same complete graph $Y3$ with vertices $|Y1|$ and $Y2 = 0.0$ by allowing changes in $Y2$, i.e. $Y2 = \{0.2, 0.4, 0.6\}$. One can observe in Table 1 that the solution quality, in terms of *Best* and *Avg*, obtained by SSGA is many times better than the solution quality, in terms of *Best* and *Avg*, obtained by GA on all graph instances. Particularly, on larger instances, SSGA performs much better than GA.

One can also observe from Table 1 that the problem-specific crossover and mutation operators of SSGA play crucial roles in finding high-quality solutions Overall, high-quality solutions evolve from the results of effective coordination among problem-specific genetic operators and other components of SSGA.

5 Conclusions

This paper presents a steady-state genetic algorithm (SSGA) for the *tree t-spanner problem* on a given connected, undirected and edge-weighted graph. Genetic operators (crossover and mutation) in SSGA use problem-specific knowledge for generating offspring, making SSGA overall more effective in finding high-quality solutions in comparison to the existing genetic algorithm (GA) [12] in the literature. Particularly, the problem-specific crossover operator is designed in such a way that it favours the offspring to inherit potentially high-quality building blocks (edges or genes) of two selected parent solutions. On a set of randomly generated Euclidean graph instances, computational results show that the solution quality obtained by SSGA is many times better than the solution quality obtained by the existing GA on each instance. Particularly, SSGA consistently performs better and better than GA as the size of instances increases.

As a future work, other metaheuristic techniques may be developed for this problem.

Acknowledgements This work is supported by the Science and Engineering Research Board—Department of Science & Technology, Government of India [grant no.: YSS/2015/000276].

References

1. Althöfer, I., Das, G., Dobkin, D., Joseph, D., Soares, J.: On sparse spanners of weighted graphs. Discrete Comput. Geom. **9**, 81–100 (1993)
2. Awerbuch, B., Baratz, A., Peleg, D.: Efficient broadcast and light-weight spanners (1992)
3. Bhatt, S., Chung, F., Leighton, F., Rosenberg, A.: Optimal simulations of tree machines. In: Proceedings of 27th IEEE Foundation of Computer Science, pp. 274–282 (1986)
4. Cai, L.: Tree spanners: spanning trees that approximate distances (1992)

5. Cai, L., Corneil, D.: Tree spanners. SIAM J. Discrete Math. **8**, 359–387 (1995)
6. Davis, L.: Handbook of Genetic Algorithms. Van Nostrand Reinhold, New York (1991)
7. Emek, Y., Peleg, D.: Approximating minimum max-stretch spanning trees on unweighted graphs. SIAM J. Comput. **38**, 1761–1781 (2008)
8. Harel, D., Tarjan, R.: Fast algorithms for finding nearest common ancestors. SIAM J. Comput. **13**, 338–355 (1984)
9. Holland, J.H.: Adaptation in Natural and Artificial Systems: An Introductory Analysis with Applications in Biology, Control, and Artificial Intelligence. Michigan Press, University, MI (1975)
10. Liebchen, C., Wünsch, G.: The zoo of tree spanner problems. Discrete Appl. Math. **156**, 569–587 (2008)
11. Liestman, A., Shermer, T.: Additive graph spanners. Networks **23**, 343–364 (1993)
12. Moharam, R., Morsy, E.: Genetic algorithms to balanced tree structures in graphs. Swarm Evol. Comput. **32**, 132–139 (2017)
13. Peleg, D., Tendler, D.: Low stretch spanning trees for planar graphs (2001)
14. Peleg, D., Ullman, J.: An optimal synchronizer for the hypercube. In: Proceedings of 6th ACM Symposium on Principles of Distributed Computing, Vancouver, pp. 77–85 (1987)
15. Peleg, D., Upfal, E.: A tradeoff between space and efficiency for routing tables. In: Proceedings of 20th ACM Symposium on Theory of Computing, Chicago, pp. 43–52 (1988)
16. Prim, R.: Shortest connection networks and some generalizations. Bell Syst. Tech. J. **36**, 1389–1401 (1957)
17. Raidl, G.R., Julstrom, B.A.: Edge sets: an effective evolutionary coding of spanning trees. IEEE Trans. Evol. Comput. **7**, 225–239 (2003)
18. Singh, A., Gupta, A.K.: Two heuristics for the one-dimensional bin-packing problem. OR Spectr. **29**, 765–781 (2007)
19. Sundar, S., Singh, A.: Metaheuristic approaches for the blockmodel problem. IEEE Syst. J. (Accepted) (2014)
20. Sundar, S.: A steady-state genetic algorithm for the dominating tree problem. In: Simulated Evolution and Learning—10th International Conference, SEAL 2014, Dunedin, New Zealand, 15–18 Dec 2014. Proceedings, pp. 48–57 (2014)
21. Sundar, S., Singh, A.: Two grouping-based metaheuristics for clique partitioning problem. Appl. Intell. **47**, 430–442 (2017)
22. Venkatesan, G., Rotics, U., Madanlal, M.S., Makowsky, J.A., Rangan, C.P.: Restrictions of minimum spanner problems. Inf. Comput. **136**, 143–164 (1997)

A Triple Band $ Shape Slotted PIFA for 2.4 GHz and 5 GHz WLAN Applications

Toolika Srivastava, Shankul Saurabh, Anupam Vyas and Rajan Mishra

Abstract A triple band $ shape slotted planar inverted-F antenna (PIFA) is proposed for ISM band 2.4/5 GHz applications. The results have been compared between the antenna with slotted ground plane and the antenna with intact ground plane. Various improvements in the simulated results, such as large bandwidth, large gain, and higher radiation efficiency, have been achieved in the antenna with slotted ground as compared to non-slotted ground PIFA antenna. The size of the antenna is $100 \times 40 \times 1.6\ \text{mm}^3$. FR4 epoxy is taken as substrate of the antenna. Three resonant frequencies are obtained at 2.4, 3.76, and 4.68 GHz. Respective bandwidths obtained at these frequencies are 81.5 MHz., 102 MHz., and 683 MHz. The results have been simulated on Ansoft HFSS 13.0. Proposed antenna applications include dual-band WLAN operation at 2.4 and 5 GHz.

Keywords Ansoft HFSS · GSM
ISM band · LTE · Microstrip patch antenna
PIFA · WLAN

T. Srivastava (✉) · A. Vyas
IET, Bundelkhand University, Jhansi 284128, Uttar Pradesh, India
e-mail: toolikaec1049@gmail.com

A. Vyas
e-mail: anupam.vyas@rediffmail.com

S. Saurabh · A. Vyas · R. Mishra
Madan Mohan Malaviya University of Technology Gorakhpur, Gorakhpur 273010, Uttar Pradesh, India
e-mail: shankulmmm2013@gmail.com

R. Mishra
e-mail: rajanmishra1231@gmail.com

K. Ray et al. (eds.), *Soft Computing: Theories and Applications*,
Advances in Intelligent Systems and Computing 742,
https://doi.org/10.1007/978-981-13-0589-4_37

1 Introduction

Microstrip patch antennas are widely used in mobile phones nowadays as they have small size and low profile; they are easy to fabricate with small price. It is fabricated by etching the antenna patch of the PCB using photolithography process and can thus also called as printed antenna. An insulating dielectric material of desired relative permittivity (ε_r) is sandwiched between two metallic plates called as radiating patch and ground plane. Microstrip patch antenna can be used at microwave frequencies. As the dimension of the MPA is related to the wavelength of resonant frequency, extremely small size antenna size is possible to fabricate that can be used in complex and compact circuit design. Also, the length of antenna depends on the relative dielectric constant. As the relative dielectric constant of the substrate increases, the length of the antenna decreases [1, 2]. Using multiple feed points to the antenna, it is possible to achieve desired polarization, i.e., horizontal, vertical, LCP, RCP, or elliptical. Fringing fields associated with the microstrip patch antenna are responsible for increasing the electrical length of the antenna, and thus the resonant length of the antenna becomes slightly shorter [3–5].

A PIFA can be low profile, coupled fed, and printed that can be operated in multiple bands [6, 7] for GSM/UMTS/LTE cellular phone applications, which can cover LTE700/2300/2500, GSM850/900/1900, and UMTS2100 bands. Various slots can be made on the radiating patch in order to increase the bandwidth, efficiency, and gain of the PIFA [8, 9], thus enabling antenna to be operated for multiple applications.

Applications including body-centric wireless communications (BCWCs) and wireless body area networks (WLAN) can be achieved using PIFA [10]. Shorting structures and a folded ground plane can be used to improve the impedance matching and decrease the size of the antenna. Radiation efficiency and impedance bandwidth can be improved by using coplanar parasitic patches in a PIFA [11]. Tuning over wide frequency range can be achieved by using a varactor diode in a simple PIFA-based tunable internal antenna for personal communication handsets [12]. An engineered magnetic superstrate based on the broadside-coupled split ring resonators (SRR) can be introduced in order to improve the gain of PIFA [13].

In this paper, a PIFA for a three-band wireless communication system is described. The coaxial feed has been provided to the copper patch through ground plane. Simulation results have been shown in Sect. 3. Considerable amount of improvement was observed in return of loss radiation efficiency and radiation efficiency for the PIFA with slotted ground as compared to non-slotted ground.

2 Structure and Design of Antenna

The physical structure and parameters of the patch of proposed $ shape slotted planar inverted-F antenna are shown in Fig. 1. The FR4 epoxy is used as substrate with relative permittivity, $\varepsilon_r = 4.4$ and tangent loss, $\tan \delta = 0.02$. The dimensions of

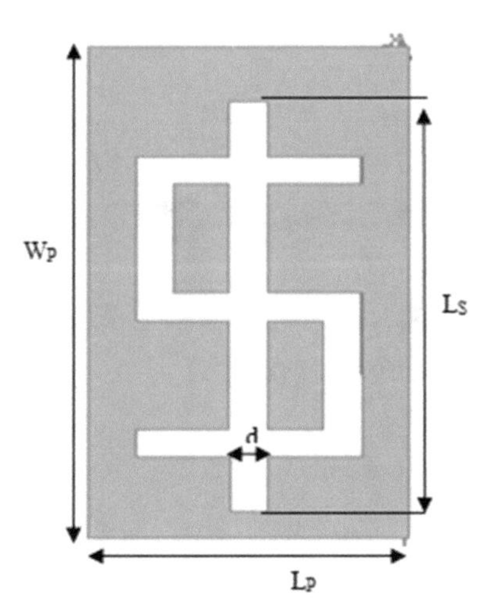

Fig. 1 Dimension of patch of the proposed antenna

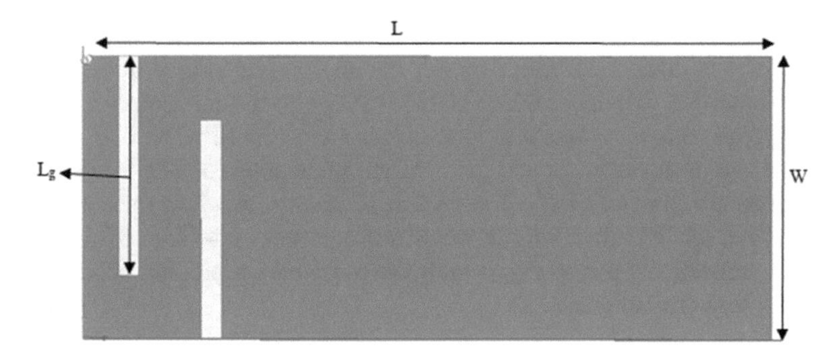

Fig. 2 Dimension of ground plane with defect

Table 1 Dimensions of proposed PIFA (All units are in mm) Ground: 100 mm × 40 mm; Substrate: FR4 epoxy (thickness = 1.6 mm)

Parameter	L	W	Lp	Ws	Ls	d	Wp	L_g	W_g
Dimension	100	40	17	12	30	2	36	31	3

the dielectric substrate are $100 \times 40 \times 1.6\ \text{mm}^3$. Two rectangular cuts of dimension $3 \times 31\ \text{mm}^2$ are made in the ground plane as demonstrated in Fig. 2. Next, Fig. 3 presents the side view of proposed antenna with shorting pin and feeding. Coaxial probe feeding technique is used which is simple and can be fed anywhere on the patch. Parameters of proposed PIFA have been listed in Table 1.

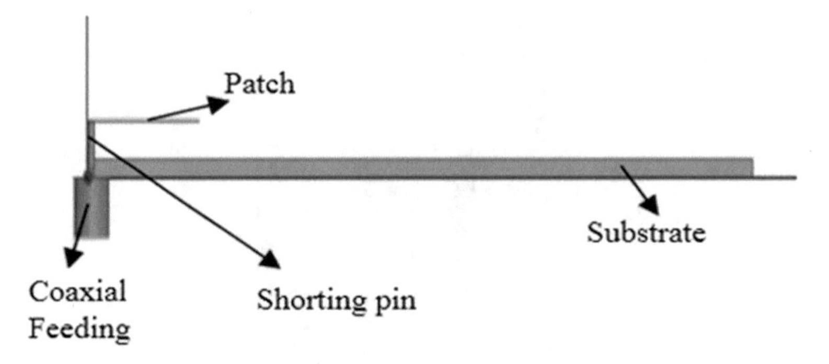

Fig. 3 Side view of the proposed antenna

3 Simulation Results

The simulated results of proposed antenna are shown below. A comparison of results between antenna without defected ground plane and with two rectangular slots in ground plane is done for return loss S_{11} parameter, radiation pattern, and gain. Figure 4 shows S_{11} parameter which shows that two bands are achieved in antenna with non-slotted ground plane and three bands are achieved when two slits of the rectangle have been cut in ground plane of the same antenna. The bands obtained in both cases are shown in Table 2. Figure 5a, b and c shows how the antenna radiates with total relative directivity at resonant frequencies 0.4 GHz, 3.76 GHz, and 4.68 GHz, respectively and in Fig. 6, gain versus frequency plot has been presented for the defected ground plane.

Fig. 4 S11 parameter of the proposed antenna

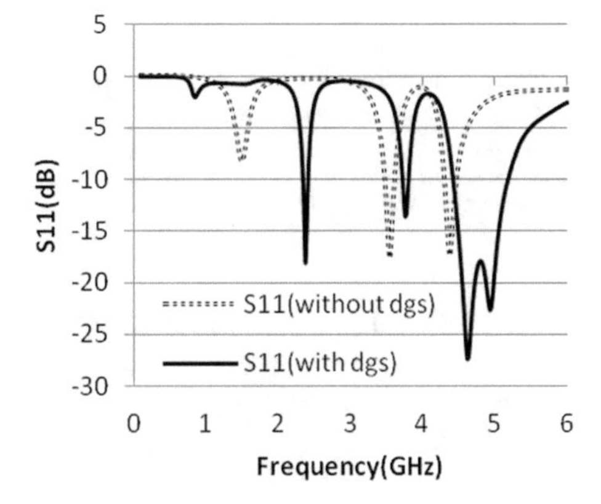

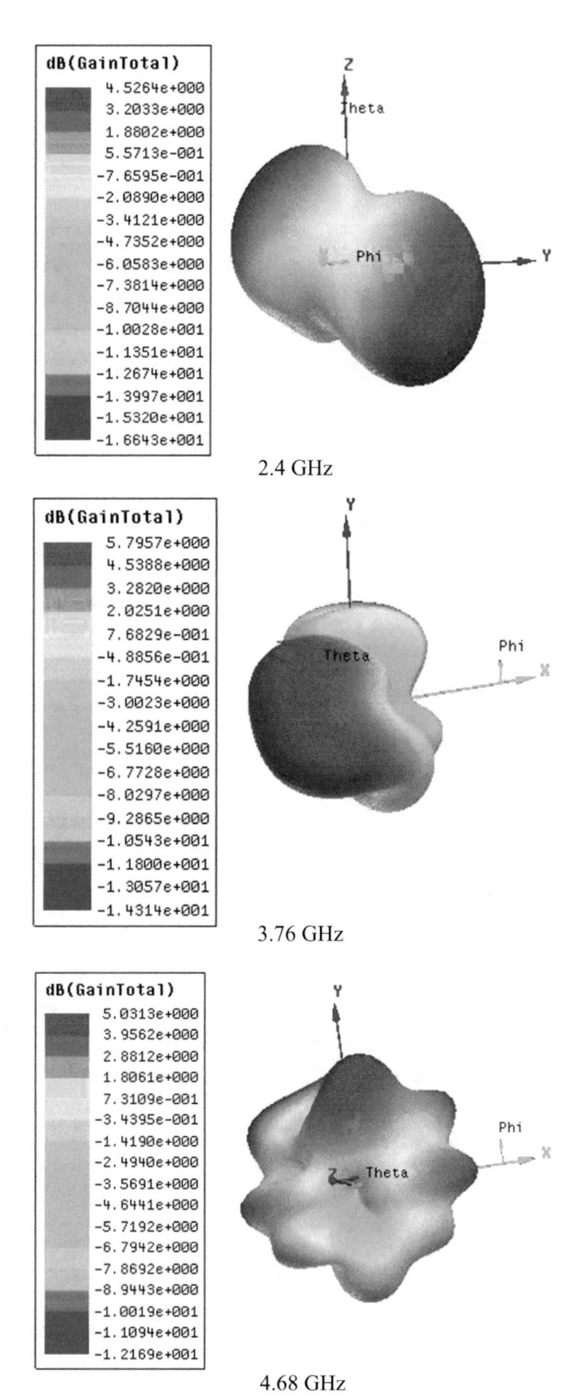

Fig. 5 3D radiation pattern of PIFA with DGS

Table 2 Bands obtained by antenna (All dimensions are in GHz)

Frequency bands	Antenna without DGS	Antenna with DGS
Band 1	3.41–3.55 (3.9%)	2.36–2.44 (3.3%)
Band 2	4.34–4.52 (3.8%)	3.73–3.83 (2.4%)
Band 3	–	4.47–5.16 (15.8%)

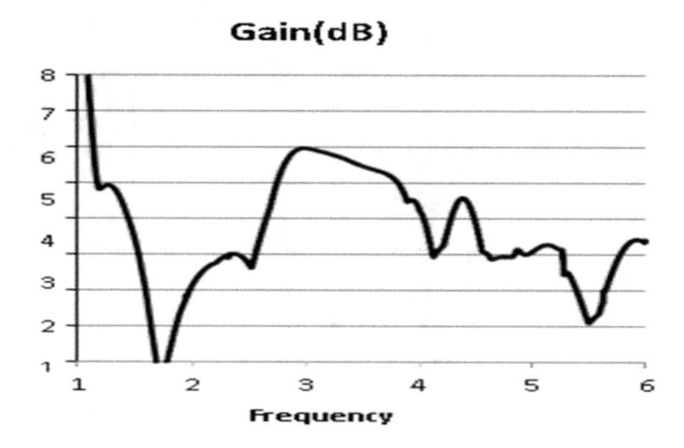

Fig. 6 Gain versus frequency plot of the proposed antenna

4 Discussion

In this paper, a PIFA is designed using defected ground structure to improve the return loss, bandwidth, gain, and directivity of antenna. A comparison is done on the basis of these parameters which are shown in results. The proposed antenna covers three bands, band 1 (2.34–2.42 GHz), band 2 (3.72–3.81 GHz), and band 3 (4.42–5.18 GHz) with percentage bandwidth of 3.3, 2.4, and 15.8%. These bands can be used for the 2.4 GHz ISM and 5 GHz WLAN band applications. Bandwidth of antenna increased from 3.9 to 15.8%. Other parameters of antenna such as directivity, gain, and radiation pattern are also enhanced by using two rectangular slots in ground plane. The relationship between co-polarization and cross-polarization has been described by the current distribution plot. The current distribution polar plot has been shown in Fig. 7a–c.

5 Conclusion

PIFA has advantages such as low profile and easy to manufacture but it has low bandwidth. A PIFA with defected ground arrangement is successfully designed for three-band application. Using defected ground structure, the bandwidth, radiation efficiency, and gain of the proposed antenna are enhanced. Good radiation character-

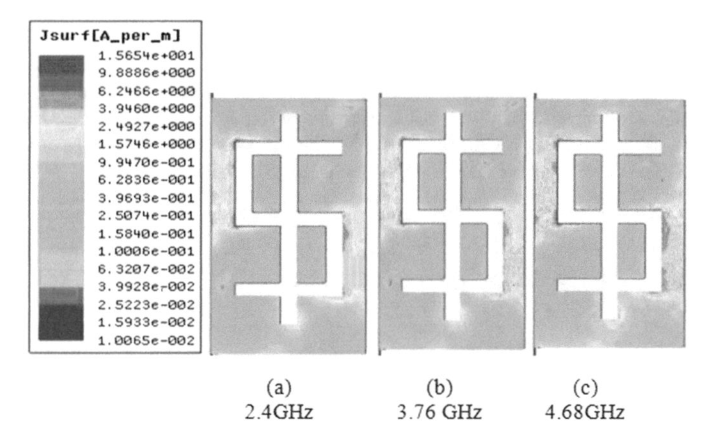

Fig. 7 Current distribution on radiating patch of PIFA with DGS

istics are achieved by using defective ground structure. Proposed antenna applications include dual-band WLAN operation at 2.4 and 5 GHz.

References

1. Belhadef, Y., Boukli Hacene, N.: Design of new multiband slotted PIFA antennas. IJCSI Int. J. Comput. Sci. **8**(4), No 1, 325–330 (2011)
2. Chattha, H.T., Huang, Y., Zhu, X., Lu, Y.: An empirical equation for predicting the resonant frequency of Planar Inverted-F Antennas. IEEE Antennas Wirel. Propag. Lett. **8**, 856–860 (2009)
3. Guha, D., Biswas, M., Antar, Y.M.M.: Microstrip patch antenna with defected ground structure for cross polarization suppression. IEEE Antennas Wirel. Propag. Lett. **4**(1), 455–458 (2005)
4. Vaughan, R.G.: Two-port higher mode circular microstrip antennas. IEEE Trans. Antennas Propag. **36**(3), 309–321 (1988)
5. Ferrero, F., Luxey, C., Staraj, R., Jaquemod, G., Yadlin, M., Fusco, V.: A novel quad-polarization agile patch antenna. IEEE Trans. Antennas Propag. **57**(5), 1563–1567 (2009)
6. Ying, L.-J., Ban, Y.-L., Chen, J.-H.: Low-profile coupled-fed printed PIFA for internal seven-band LTE/GSMIUMTS mobile phone antenna. In: Cross Strait Quad-Regional Radio Science and Wireless Technology Conference, pp. 418–421. IEEE, 26–30 July 2011
7. Luhaib, S.W., Quboa, K.M., Abaoy, B.M.: Design and simulation dual-band PIFA antenna for GSM systems. In: IEEE 9th International Multi-Conference on Systems, Signals and Devices, pp. 1–4, Mar 2012
8. Lee, Y.-H., Kwon, W.-H.: Dual-band PIFA design for mobile phones using H-type slits. IEEE Trans. Antennas Propag. Soc. Int. Symp. **3**, 3111–3114 (2004)
9. Tang, I.-T., Lin, D.-B., Chen, W.-L., Horng, J.-H., Li, C.-M.: Compact five-band meandered PIFA by using meandered slots structure. IEEE Trans. Antennas Propag. Soc. Int. Symp. 653–656 (2007)
10. Lin, Chia-Hsien, Saito, Kazuyuki, Takahashi, Masaharu, Ito, Koichi: A compact planar inverted-F antenna for 2.45 GHz on-body communications. IEEE Trans. Antennas Propag. **60**(9), 4422–4426 (2012)

11. Kärkkäinen, M.K.: Meandered multiband PIFA with coplanar parasitic patches. IEEE Microw. Wirel. Compon. Lett. **15**(10), 630–632 (2005)
12. Nguyen, V.-A., Bhatti, R.-A., Park, S.-O.: A simple PIFA-based tunable internal antenna for personal communication handsets. IEEE Antennas Wirel. Propag. Lett. **7**, 130–133 (2008)
13. Attia, H., Bait-Suwailam, M.M., Ramahi, O.M.: Enhanced gain planar inverted-F antenna with metamaterial superstrate for UMTS applications. PIERS Online **6**(6), 494–497 (2010)

A Min-transitive Fuzzy Left-Relationship on Finite Sets Based on Distance to Left

Sukhamay Kundu

Abstract We first consider the "discrete" domain of non-empty finite subsets $X = \{x_1, x_2, \ldots, x_m\}$ of $\mathcal{R} = (-\infty, +\infty)$, where each x_i has a weight or probability $P_X(x_i) > 0$. We define a min-transitive fuzzy left-relationship $\Lambda_D(X, Y)$ on this domain, which accounts for the distance to the left $L(x, y) = \max(0, y - x)$ between points $x \in X$ and $y \in Y$. Then, we extend $\Lambda_D(\cdot, \cdot)$ to finite intervals $X \subset \mathcal{R}$ with arbitrary probability density function $p_X(x)$ on them, and to fuzzy sets X with finite support, which may be a discrete set or an interval, and membership function $\mu_X(x)$.

Keywords Distance to the left · Fuzzy left-relationship · Min-transitivity
Finite set · Interval · Membership function · Probability distribution

1 Introduction

Decision-making on uncertain multidimensional data X requires a preference relationship, i.e., a fuzzy left-relationship $\mu(X, Y)$ on the underlying domain Ω that generalizes the ordinary "<"-relationship in $\mathcal{R} = (-\infty, +\infty)$. In the discrete case, $X = \{x_1, x_2, \ldots, x_m\}$ is a non-empty finite subset of $\mathcal{R} = (-\infty, +\infty)$, with either a probability distribution function $P_X(x_i)$ or a fuzzy membership function $\mu_X(x_i)$ on X. In the continuous case, X is a finite interval in $\mathcal{R}$, with either a probability density function $p_X(x)$ or a fuzzy membership function $\mu_X(x)$ on X. The usual transitivity property of the "<"-relationship in $\mathcal{R}$ is now replaced by the min-transitivity property for a fuzzy left-relationship:

$$\text{min-transitivity: } \mu(X, Z) \geq \min(\mu(X, Y), \mu(Y, Z)) \text{ for all } X, Y, \text{ and } Z.$$

S. Kundu (✉)
Computer Science and Engineering, Louisiana State University, Baton Rouge, LA 70803, USA
e-mail: kundu@csc.lsu.edu

K. Ray et al. (eds.), *Soft Computing: Theories and Applications*,
Advances in Intelligent Systems and Computing 742,
https://doi.org/10.1007/978-981-13-0589-4_38

Defining a fuzzy left-relationship which is both intuitively appealing and mathematically sound that satisfies the min-transitivity is a nontrivial task. In particular, none of the fuzzy left-relationship for intervals given in [1–3] satisfies the min-transitivity property and thus they cannot be meaningfully used in decision-making on uncertain data. A similar remark holds for the fuzzy left-relationship on fuzzy sets with finite support defined in [6]. The fuzzy left-relationship in [4] on intervals, denoted here by $\pi(\cdot, \cdot)$, on the other hand is both min-transitive and intuitively meaningful. However, $\pi(\cdot, \cdot)$ is not a perfect fit for all situations as the example shown below. The min-transitive fuzzy left-relationship $\Lambda_D(\cdot, \cdot)$ given here can properly handle situations like the one in the example. The extensions of $\Lambda_D(\cdot, \cdot)$ to finite intervals of $\mathcal{R}$ and to fuzzy subsets of $\mathcal{R}$ with finite support can solve similar problems that cannot be properly handled by $\pi(\cdot, \cdot)$ or by the fuzzy preference relationship in [6].

Example 1 Consider the test scores $B = \{50(0.2), 70(0.6), 80(0.2)\}$ for a group of boys and the test scores $G = \{60(0.4), 70(0.3), 75(0.2), 90(0.1)\}$ for a group of girls in a class test. Here, the number in the parentheses next to each test score (out of 100) gives the proportion of boys (or girls) who have that score. We want to determine whether or not the boys' performance is worse than the girls' performance, i.e., whether or not B is to the left of G and, if so, then by how much. The min-transitive fuzzy left-relationship $\Lambda_D(\cdot, \cdot)$ defined here solves this problem easily because $\Lambda_D(B, G)$ takes into account not only the amount $L(b_i, g_j) = \max(0, g_j - b_i)$ by which a boy score b_i is to-the-left of a girl score g_j but also the probabilities $P_B(b_i)$ and $P_G(g_j)$ associated with those scores. The fuzzy left-relationship $\pi(\cdot, \cdot)$ in [4] does not take into account either the distance to the left $L(b_i, g_j)$ or the arbitrary probabilities $P_B(b_i)$ and $P_G(g_j)$. Note that determining the "distance" between the weighted sets B and G, i.e., how different are the sets B and G is different from finding which group of students performed better. We address the distance problem between weighted sets elsewhere.

2 Basic Terminology

We first consider the discrete case, where $X = \{x_1, x_2, \ldots, x_m\}$ is a non-empty finite subset of $\mathcal{R}$ and P_X is a discrete probability distribution on X with each $P_X(x_i) > 0$ and $\sum_i P_X(x_i) = 1$. Note that if $P_X \neq Q_X$ are two probability distributions on X, then the pairs (X, P_X) and (X, Q_X) are considered to be different. We refer to the pair (X, P_X) simply as X when there is no confusion about P_X. This facilitates the comparison between $\Lambda_D(\cdot, \cdot)$ and $\pi(\cdot, \cdot)$. (Remark: Because the definition of the function P_X includes both its domain X and the mapping $x_i \rightarrow P_X(x_i), x_i \in X$, it would be more appropriate to use P_X as a short form for the pair (X, P_X) instead of X. Thus, $\Lambda_D(X, Y) = \Lambda_D(P_X, P_Y)$ can be regarded as a left-relationship on the probability distributions P_X and P_Y.) We write $\overline{X} = \sum P_X(x_i)x_i$ for the average or expected value $E(X)$ of X.

For $(x, y) \in \mathcal{R} \times \mathcal{R}$, we define the functions distance to the left $L(x, y) = \max(0, y - x)$ and distance to the right $R(x, y) = L(y, x) = \max(0, x - y)$. We extend these functions to a pair of (X, P_X) and (Y, P_Y) by defining

$$L(X, Y) = \sum_{i,j} P_X(x_i) P_Y(y_j) L(x_i, y_j)$$
$$= \sum_{i} P_X(x_i) \left(\sum_{x_i \le y_j} P_Y(y_j)(y_j - x_i) \right) \ge 0 \tag{1}$$

$$R(X, Y) = \sum_{i,j} P_X(x_i) P_Y(y_j) R(x_i, y_j)$$
$$= \sum_{i} P_X(x_i) \left(\sum_{x_i \ge y_j} P_Y(y_j)(x_i - y_j) \right) \ge 0. \tag{2}$$

Note that replacing "$x_i \le y_j$" by "$x_i < y_j$" in the sum for $L(X, Y)$ does not affect the value of the sum; a similar remark holds for replacing "$x_i \ge y_j$" by "$x_i > y_j$" in the sum for $R(X, Y)$. We freely make use of these different equivalent forms in various calculations in what follows. Clearly, $R(X, Y) = L(Y, X)$ and $L(X, Y) + R(X, Y) = \sum_i \sum_j P_X(x_i) P_Y(y_j)[L(x_i, y_j) + R(x_i, y_j)] = \sum_i \sum_j P_X(x_i) P_Y(y_j) |x_i - y_j|$, the average distance between the points in X and Y. In particular, $L(X, Y) + R(X, Y) = 0$ if and only if $X = Y = \{z\}$, a singleton set with $P_X(z) = P_Y(z) = 1$ and hence $P_X = P_Y$. It is clear that $L(X, Y)$ is the expected value $E(L(x, y))$ of $L(x, y)$ when we take the products $P_X(x_i) P_Y(y_j)$ as the probability distribution on the pairs $(x_i, y_j) \in X \times Y$. Likewise, $R(X, Y)$ is the expected value $E(R(x, y)) = E(L(y, x))$.

Definition 1 For non-empty finite subsets $X, Y \subset \mathcal{R} = (-\infty, +\infty)$, with probability distributions P_X and P_Y, we define the fuzzy left-relationship $\Lambda_D(X, Y) = \max(0, L(X, Y) - R(X, Y))/(L(X, Y) + R(X, Y))$ if $L(X, Y) + R(X, Y) > 0$ and $\Lambda_D(X, Y) = 0$, otherwise.

Clearly, $0 \le \Lambda_D(X, Y) \le 1$ for all X and Y. Also, if $X = \{x\}$ and $Y = \{y\}$, with $P_X(x) = P_Y(y) = 1$, then $x < y$ implies $\Lambda_D(X, Y) = 1$ and $x >= y$ implies $\Lambda_D(X, Y) = 0$. In this sense, $\Lambda_D(\cdot, \cdot)$ generalizes the "$<$"-relationship on $\mathcal{R}$. Also, note that adding an element x' to X with $P_X(x') = 0$ does not change either of $L(X, Y)$ and $R(X, Y)$ and hence $\Lambda_D(X, Y)$. A similar remark holds for adding an element y' to Y with $P_Y(y') = 0$.

Definition 2 We define a partial order "$\le$" on the non-empty finite subsets of $\mathcal{R}$ by saying $X \le Y$ if $\max\{x_i : x_i \in X\} \le \min\{y_j : y_j \in Y\}$. We say $X < Y$ if $X \le Y$ and $X \ne Y$.

Table 1 Probability distributions P_X and P_Y on $\{1, 2, 4\}$

	$x_1 = y_1 = 1$	$x_2 = y_2 = 2$	$x_3 = y_3 = 4$	Average
P_X	1/2	1/4	1/4	$\overline{X} = 2.00$
P_Y	1/4	1/4	1/2	$\overline{Y} = 2.75$

Thus, if $\max\{x_i : x_i \in X\} < \min\{y_j : y_j \in Y\}$, then $X < Y$. If $\max\{x_i : x_i \in X\} = z = \min\{y_j : y_j \in Y\}$, then $X < Y$ unless $X = \{z\} = Y$.

Lemma 1 *$L(X, Y) - R(X, Y) = \overline{Y} - \overline{X}$ and hence $\Lambda_D(X, Y) > 0$ if and only if $\overline{Y} > \overline{X}$. Also, $\Lambda_D(X, Y) = 1$ if and only if $X < Y$.*

Proof We have $L(X, Y) - R(X, Y) = E(L(x, y)) - E(L(y, x)) = E(L(x, y) - L(y, x)) = E(y - x) = \overline{Y} - \overline{X}$. Thus, $\Lambda_D(X, Y) = 1$ if and only if $L(X, Y) > 0$ and $R(X, Y) = 0$, i.e., $X < Y$. □

Example 2 Let $X = \{1, 2, 4\} = Y$ with the probability distributions P_X and P_Y as shown in Table 1. This gives $L = L(X, Y) = 1/8 + 3/4 + 2/8 = 9/8$, $R = R(X, Y) = L(Y, X) = 1/16 + 3/16 + 2/16 = 3/8$, $L - R = 3/4 = \overline{Y} - \overline{X}$, and $L + R = 3/2$. Thus, $\Lambda_D(X, Y) = (3/4)/(3/2) = 1/2$ and $\Lambda_D(Y, X) = 0$. For the sets B, G, and their associated probabilities in Example 1, we have $\Lambda_D(B, G) = 1/10.6$ and thus the boys performed slightly worse than the girls. □

3 Main Results

The following lemma shows certain monotonicity properties of $\Lambda_D(X, Y)$, which plays a critical role in the proof of min-transitivity of $\Lambda_D(\cdot, \cdot)$.

Lemma 2 *$\Lambda_D(X, Y)$ is a continuous monotonic nonincreasing function of each x_i and a continuous monotonic nondecreasing function of each y_j. Moreover, $\Lambda_D(X, Y) \to 1$ if some $x_i \to -\infty$ or some $y_j \to +\infty$ and $\Lambda_D(X, Y) \to 0$ if some $x_i \to +\infty$ or some $y_j \to -\infty$.*

Proof That $\Lambda_D(X, Y)$ is a continuous function of each x_i and y_j is obvious. We now prove the monotonicity properties of $\Lambda_D(X, Y)$. Let $p = \sum_{y_j > x_i} P_Y(y_j)$. Then, for sufficiently small $\epsilon > 0$, x_i increasing to $x_i + \epsilon$ causes $L(X, Y)$ to decrease by $p\epsilon$ and $R(X, Y)$ to increase by $(1 - p)\epsilon$. This means if $0 < \Lambda_D(X, Y) < 1$ then $L(X, Y) - R(X, Y)$ decreases by ϵ and $L(X, Y) + R(X, Y)$ decreases by $(2p - 1)\epsilon \leq \epsilon$ if $p \geq 1/2$ and $L(X, Y) + R(X, Y)$ increases if $p < 1/2$. In either case, this means $\Lambda_D(X, Y)$ decreases. It follows that if x_i is increased and $0 < \Lambda_D(X, Y) < 1$, then $\Lambda_D(X, Y)$ will decrease and a similar analysis shows that if $\Lambda_D(X, Y) = 0$ or 1, then $\Lambda_D(X, Y)$ may remain unchanged or will decrease. A similar argument also shows that if y_j is increased then $\Lambda_D(X, Y)$ will increase if $0 < LD(X, Y) < 1$ and if $\Lambda_D(X, Y) = 0$ or 1, then $\Lambda_D(X, Y)$ may remain unchanged or will increase.

To prove the last part of the lemma, we note that if we let some $x_i \to -\infty$, then both $L(X, Y) - R(X, Y)$ and $L(X, Y) + R(X, Y)$ have the dominant term $P_X(x_i)(\overline{Y} - x_i)$ and thus $\Lambda_D(X, Y) \to 1$. Likewise, if we let some $y_j \to +\infty$, then both $L(X, Y) - R(X, Y)$ and $L(X, Y) + R(X, Y)$ have the dominant term $P_Y(y_j)(y_j - \overline{X})$ and thus $\Lambda_D(X, Y) \to 1$. That $\Lambda_D(X, Y) \to 0$ for $x_i \to +\infty$ and $y_j \to -\infty$ now follows from the first part of Lemma 1. □

Theorem 1 *The fuzzy left-relationship $\Lambda_D(\cdot, \cdot)$ on non-empty finite sets $X \subset \mathcal{R}$ with probability distributions P_X satisfies the min-transitive property $\Lambda_D(X, Z) \geq$ min$(\Lambda_D(X, Y), \Lambda_D(Y, Z))$ for all X, Y, and Z.*

Proof We only need to consider the case where $\min(\Lambda_D(X, Y), \Lambda_D(Y, Z)) > 0$, i.e., both $\Lambda_D(X, Y)$ and $\Lambda_D(Y, Z) > 0$. First, we argue that we can assume without loss of generality that $\Lambda_D(X, Y) = \Lambda_D(Y, Z)$. For example, if $\Lambda_D(X, Y) < \Lambda_D(Y, Z)$, then we can keep increasing, say, $y_1 \in Y$ so that while $\Lambda_D(X, Y)$ increases we have $\Lambda_D(Y, Z)$ either remains 1 or starts decreasing. It is clear that if we increase y_1 sufficiently, then $\Lambda_D(X, Y)$ will become equal to $\Lambda_D(Y, Z)$. Likewise, if $\Lambda_D(X, Y) > \Lambda_D(Y, Z)$, then we can decrease y_1 sufficiently so that $\Lambda_D(Y, Z)$ increases and $\Lambda_D(X, Y)$ either stays 1 or decreases, but ultimately $\Lambda_D(Y, Z)$ becomes equal to $\Lambda_D(X, Y)$. In either case, for the new Y', we have $\min(\Lambda_D(X, Y), \Lambda_D(Y, Z)) \leq \min(\Lambda_D(X, Y'), \Lambda_D(Y', Z)) = \Lambda_D(X, Y') = \Lambda_D(Y', Z) > 0$ and we prove $\Lambda_D(X, Z) \geq \min(\Lambda_D(X, Y'), \Lambda_D(Y', Z))$. Henceforth, assume $Y = Y'$.

Let $L(X, Y) - R(X, Y) = a = \overline{Y} - \overline{X} > 0$, $b = L(X, Y) + R(X, Y) > 0$, $L(Y, Z) - R(Y, Z) = c = \overline{Z} - \overline{Y} > 0$, and $d = L(Y, Z) + R(Y, Z) > 0$. Then, we have $\Lambda_D(X, Y) = a/b = (a + c)/(b + d) = c/d = \Lambda_D(Y, Z)$, where $a + c = \overline{Z} - \overline{X} = L(X, Z) - L(Z, X) > 0$. The steps below show that $L(X, Z) + R(X, Z) \leq b + d$ using $|x_i - z_j| = \sum_k P_Y(y_k)|x_i - z_j| \leq \sum_k P_Y(y_k)(|x_i - y_k| + |y_k - z_j|)$. This gives $\Lambda_D(X, Z) \geq (a + c)/(b + d)$, completing the proof. □

$$
\begin{aligned}
L(X, Z) + R(X, Z) &= \sum_i \sum_j P_X(x_i) P_Z(z_j) |x_i - z_j| \\
&\leq \sum_i \sum_k P_X(x_i) P_Y(y_k) \sum_j P_Z(z_j) |x_i - y_k| + \\
&\quad \sum_j \sum_k P_Y(y_k) P_Z(z_j) \sum_i P_X(x_i) |y_k - z_j| \\
&= \sum_i \sum_k P_X(x_i) P_Y(y_k) |x_i - y_k| + \\
&\quad \sum_k \sum_j P_Y(y_k) P_Z(z_j) |y_k - z_j| \\
&= L(X, Y) + R(X, Y) + L(Y, Z) + R(Y, Z) = b + d
\end{aligned}
$$

3.1 Some Invariance Properties of $\Lambda_D(\cdot, \cdot)$

We show next that $\Lambda_D(X, Y)$ has the important properties of being both translation invariant and scale invariant. We use the following notations below. For $c \in \mathcal{R}$, let $X + c$ denote the set $\{x_i + c : x_i \in X\}$ with the probabilities $P_{X+c}(x_i + c) = P_X(x_i)$; we call $(X + c, P_{X+c})$ the translation of (X, P_X) by c. For $c \neq 0$, let cX denote the set $\{cx_i : x_i \in X\}$ with the probabilities $P_{cX}(cx_i) = P_X(x_i)$; we call (cX, P_{cX}) the scaling of (X, P_X) by c.

Lemma 3 *For all non-empty finite subsets $X, Y \subset \mathcal{R}$ with probability distributions P_X and P_Y, we have $\Lambda_D(X, Y) = \Lambda_D(X + c, Y + c)$ for all c and $\Lambda_D(X, Y) = \Lambda_D(cX, cY)$ for $c > 0$.*

Proof The lemma immediately follows from the facts that $L(X, Y) = L(X + c, Y + c)$ for all c and $cL(X, Y) = L(cX, cY)$ for $c > 0$. They imply $R(X, Y) = R(X + c, Y + c)$ for all c and $cR(X, Y) = R(cX, cY)$ for $c > 0$. (Note that for $c < 0$, $\Lambda_D(cX, cY) = \Lambda_D(cY, cX)$.) □

3.2 Monotonicity of $\Lambda_D(X, Y)$ in P_X and P_Y

For $0 < \epsilon < P_X(x_i)$ and $x_i < x_j$, let $P_X^{i,j,\epsilon}$ be the probability distribution on X obtained from P_X by shifting the probability ϵ to the right from x_i to x_j. That is, $P_X^{i,j,\epsilon}(x_k) = P_X(x_k)$ for $k \neq i, j$, $P_X^{i,j,\epsilon}(x_i) = P_X(x_i) - \epsilon$, and $P_X^{i,j,\epsilon}(x_j) = P_X(x_j) + \epsilon$. We write $X^{i,j,\epsilon}$, in short, for the pair $(X, P_X^{i,j,\epsilon})$.

Definition 3 A probability distribution P'_X is said to be a *rightshift* of P_X if it is the result of a sequence of one or more transformations of the form $P_X^{i,j,\epsilon}$.

Theorem 2 *Let P'_X be a right shift of P_X and X' denote the pair (X, P'_X) and let P'_Y be a right shift of P_Y and Y' denote the pair (Y, P'_Y). Then, $\Lambda_D(X', Y) \leq \Lambda_D(X, Y) \leq \Lambda_D(X, Y')$.*

Proof For P'_X, we only need to consider the case $P'_X = P_X^{i,j,\epsilon}$ and $\Lambda_D(X', Y) > 0$. Let $\sigma_Y(x) = \sum_{y_k > x} P_Y(y_k)(y_k - x) \geq 0$ and $\sigma'_Y(x) = \sum_{y_k \leq x} P_Y(y_k)(x - y_k) \geq 0$. Then, $L(X^{i,j,\epsilon}, Y) = L(X, Y) + \epsilon(\sigma_Y(x_j) - \sigma_Y(x_i))$ and $R(X^{i,j,\epsilon}, Y) = R(X, Y) + \epsilon(\sigma'_Y(x_j) - \sigma'_Y(x_i))$. It follows that $L(X^{i,j,\epsilon}, Y) - R(X^{i,j,\epsilon}, Y) = L(X, Y) - R(X, Y) - \epsilon(x_j - x_i) < L(X, Y) - R(X, Y)$ because $\sigma_Y(x) - \sigma'_Y(x) = \overline{Y} - x$. In particular, if $\Lambda_D(X, Y) = 0$ then $\Lambda_D(X^{i,j,\epsilon}, Y) = 0$. Now we show that if $\Lambda_D(X^{i,j,\epsilon}, Y) > 0$ then $\Lambda_D(X^{i,j,\epsilon}, Y) < \Lambda_D(X, Y)$. Let $a = \sigma_Y(x_j) - \sigma_Y(x_i) \leq 0$ and $b = \sigma'_Y(x_j) - \sigma'_Y(x_i) \geq 0$. Since $a - b = x_i - x_j < 0$, we have either $a + b \geq 0$ or $x_i - x_j \leq a + b < 0$. This means $L(X^{i,j,\epsilon}, Y) + R(X^{i,j,\epsilon}, Y) = L(X, Y) + R(X, Y) + \epsilon(a + b)$ is either $\geq L(X, Y) + R(X, Y)$ or it decreases by the amount $\epsilon|a + b| \leq \epsilon|x_i - x_j|$ = the decrease in max(0, $L(X, Y) - R(X, Y)$) as we go from P_X to $P_X^{i,j,\epsilon}$. This proves $\Lambda_D(X^{i,j,\epsilon}, Y) < \Lambda_D(X, Y)$ when $\Lambda_D(X^{i,j,\epsilon}, Y) > 0$.

For P'_Y, we likewise only need to consider the case $P'_Y = P_Y^{i,j,\epsilon}$ and an argument similar to the above shows that $\Lambda_D(X, Y) \leq \Lambda_D(X, Y^{i,j,\epsilon})$. □

4 Role of $-R(X, Y)$ in $\Lambda_D(X, Y)$

If we define $\Lambda'_D(X, Y) = L(X, Y)/(L(X, Y) + R(X, Y))$, then we no longer have the min-transitivity property for $\Lambda'_D(X, Y)$. This is seen from the sets $A = \{2, 3\}$, $B = \{-1, 5\}$, and $C = \{0, 4\}$ and the probability distribution of 1/2 for each of the two points in each of A, B, and C. We get $L(A, B) = 5/4$, $R(A, B) = 7/4$, $\Lambda'_D(A, B) = 5/12$, $L(B, C) = 6/4$, $R(B, C) = 6/4$, $\Lambda'_D(B, C) = 6/12$, $L(A, C) = 3/4$, $R(A, C) = 5/4$, and $\Lambda'_D(A, C) = 3/8 < 5/12 = \min(5/12, 6/12)$.

In this context, we briefly show that several other variations of $\Lambda_D(X, Y)$ also fail to be min-transitive. This again shows that deriving an intuitively appealing and application-wise meaningful min-transitive fuzzy relation is a nontrivial task. Let $L'(x, y) = 1$ if $x < y$ and $= 0$, otherwise. This gives $L'(X, Y) = \sum_{i,j} P_X(x_i)P_Y(y_j)L'(x_i, y_j)$, $R'(X, Y) = L'(Y, X)$, and the following "count-based" variations of $\Lambda_D(X, Y)$, with the role of distance to the left $L(x, y)$ is replaced by the count-of-left $L'(x, y)$. Clearly, each $0 \leq \Lambda_D^{(k)}(X, Y) \leq 1$.

$$\Lambda_D^{(1)}(X, Y) = L'(X, Y) \tag{3}$$

$$\Lambda_D^{(2)}(X, Y) = L'(X, Y)/(L'(X, Y) + R'(X, Y)) \tag{4}$$

$$\Lambda_D^{(3)}(X, Y) = \max\{0, L'(X, Y) - R'(X, Y)\} \tag{5}$$

$$\Lambda_D^{(4)}(X, Y) = \max\{0, L'(X, Y) - R'(X, Y)\}/(L'(X, Y) + R'(X, Y)) \tag{6}$$

To see that these are not min-transitive, let $X = \{1, 2, 4\}$, $Y = \{1, 3, 4\}$, $Z = \{2, 3\}$, each $P_X(x_i) = 1/3$, each $P_Y(y_j) = 1/3$, and each $P_Z(z_k) = 1/2$. This gives $\Lambda_D^{(1)}(X, Y) = 4/9 < 3/6 = \min\{\Lambda_D^{(1)}(X, Z), \Lambda_D^{(1)}(Z, Y)\}$, $\Lambda_D^{(2)}(X, Y) = 4/7 < 3/5 = \min\{\Lambda_D^{(2)}(X, Z), \Lambda_D^{(2)}(Z, Y)\}$, $\Lambda_D^{(3)}(X, Y) = 1/9 < 1/6 = \min\{\Lambda_D^{(3)}(X, Z), \Lambda_D^{(3)}(Z, Y)\}$, and $\Lambda_D^{(4)}(X, Y) = 1/7 < 1/5 = \min\{\Lambda_D^{(4)}(X, Z), \Lambda_D^{(4)}(Z, Y)\}$.

5 Extending $\Lambda_D(\cdot, \cdot)$ to Intervals—A Continuous Case

Given a finite interval $X = [a, b] \subset \mathcal{R}$ of length $b - a > 0$, a continuous probability density function $p_x(x) \geq 0$ on X such that $\int_X p_x(x)dx = 1$, and an integer $n \geq 1$, we can define (X_n, P_{X_n}) by taking $X_n = \{x_1, x_2, \ldots, x_n\}$, where $x_i = a + i(b - a)/n$, $1 \leq i \leq n$, and the probability distribution P_{X_n} on X_n given by $P_{X_n}(x_i) = \int_{x_{i-1}}^{x_i} p_x(x)\, dx$, where $x_0 = a$. For all sufficiently large n, (X_n, P_{X_n}) approximates (X, p_x) arbitrarily closely. In particular, given (X, p_x) and (Y, p_y) for two finite intervals $X = [a, b]$ and $Y = [c, d]$ and probability density functions $p_X(x)$ on X and $p_Y(y)$ on Y, we define

$$L(X, Y) = \lim_{n\to\infty} L(X_n, Y_n) = \int_{x=a}^{b} \int_{y=c}^{d} p_X(x) p_Y(y) L(x, y)\, dx\, dy, \quad (7)$$

$$R(X, Y) = \lim_{n\to\infty} R(X_n, Y_n) = \int_{x=a}^{b} \int_{y=c}^{d} p_X(x) p_Y(y) R(x, y)\, dx\, dy, \quad (8)$$

$$\begin{aligned} \Lambda_D(X, Y) &= \lim_{n\to\infty} \frac{\max(0, L(X_n, Y_n) - R(X_n, Y_n))}{L(X_n, Y_n) + R(X_n, Y_n)} \\ &= \frac{\max(0, L(X, Y) - R(X, Y))}{L(X, Y) + R(X, Y)} = \frac{\max(0, \overline{Y} - \overline{X})}{L(X, Y) + R(X, Y)}. \end{aligned} \quad (9)$$

This extends $\Lambda_D(\cdot, \cdot)$ in Sect. 2 for the discrete case of finite subsets $X \subset \mathcal{R}$ and probability distributions P_X on X to the continuous case of finite intervals $X \subset \mathcal{R}$ and probability density functions $p_X(x)$ on X.

Theorem 3 *The extension of $\Lambda_D(X, Y)$ to finite intervals X and Y with arbitrary probability density functions $p_X(x)$ on X and $p_Y(y)$ on Y is min-transitive.*

Proof Follows immediately from the min-transitivity of $\Lambda_D(X, Y)$ in Sect. 2 for the discrete case of finite sets $X \subset \mathcal{R}$. If for some finite intervals X, Y, and Z, we have $\Lambda_D(X, Z) < \min(\Lambda_D(X, Y), \Lambda_D(Y, Z))$ then by choosing n sufficiently large we can make (X_n, P_{X_n}), (Y_n, P_{Y_n}), and (Z_n, P_{Z_n}) approximate arbitrarily closely (X, p_X), (Y, p_Y), and (Z, p_Z), respectively. This in turn means $\Lambda_D(X_n, Y_n)$ can approximate $\Lambda_D(X, Y)$ arbitrarily closely and similarly for $\Lambda_D(Y_n, Z_n)$ and $\Lambda_D(X_n, Z_n)$. This would give in turn $\Lambda_D(X_n, Z_n) < \min(\Lambda_D(X_n, Y_n), \Lambda_D(Y_n, Z_n))$, a contradiction. This proves the theorem. □

Figure 1(i)–(vi) show the possible relative configurations of two distinct finite intervals A and B of lengths > 0. Note that for $a = 0$, Fig. 1(i) and (iii) coincide and likewise Fig. 1(ii) and (iv) coincide; similar remarks hold for $c = 0$.

We show below the computations of $L(A, B)$, $R(A, B)$, and $\Lambda_D(A, B)$ for Fig. 1(i) and (ii), assuming uniform probability density functions on each of A and B. We also assume that the leftmost point of $A \cup B$ corresponds to $x = 0$; thus, for Fig. 1(i) we have $A = [0, a + b]$ and $B = [a, a + b + c]$, and for Fig. 1(ii) we have $A = [0, a + b + c]$ and $B = [a, a + b]$. Note that for Fig. 1(i) we have $L(A, B) - R(A, B) = \overline{B} - \overline{A} = a + (b + c)/2 - (a + b)/2 = (a + c)/2$. Likewise, for Fig. 1(ii) we have $L(A, B) - R(A, B) = \overline{B} - \overline{A} = a + b/2 - (a + b + c)/2 = (a - c)/2$.

For Fig. 1(i):

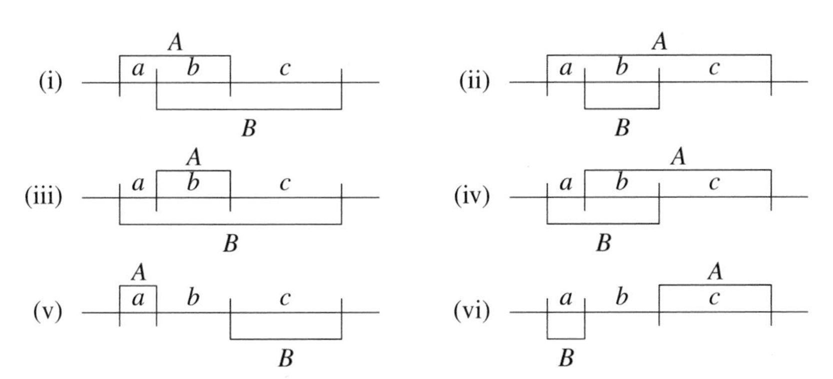

Fig. 1 Different relative configurations of two finite intervals **a** and **b** with distinct end points

$$L(A, B) = \frac{1}{(a+b)(b+c)}\left[\int_{x=0}^{a}\int_{y=a}^{a+b+c}(y-x)\,dy\,dx + \int_{x=a}^{a+b}\int_{y=x}^{a+b+c}(y-x)\,dy\,dx\right]$$

$$= \frac{1}{2(a+b)(b+c)}\left[(a+b)(b+c)(a+c) + \frac{b^3}{3}\right]$$

$$R(A, B) = \frac{1}{(a+b)(b+c)}\int_{x=a}^{a+b}\int_{y=a}^{x}(x-y)\,dy\,dx = \frac{b^3}{6(a+b)(b+c)}$$

$$\Lambda_D(A, B) = \frac{(a+b)(b+c)(a+c)}{(a+b)(b+c)(a+c)+2b^3/3} = 1 - \frac{2b^3/3}{(a+b)(b+c)(a+c)+2b^3/3}$$

For Fig. 1(ii):

$$L(A, B) = \frac{1}{(a+b+c)b}\left[\int_{x=0}^{a}\int_{y=a}^{a+b}(y-x)\,dy\,dx + \int_{x=a}^{a+b}\int_{y=x}^{a+b}(y-x)\,dy\,dx\right]$$

$$= \frac{ab+a^2+b^2/3}{2(a+b+c)}$$

$$R(A, B) = \frac{bc+c^2+b^2/3}{2(a+b+c)}, \text{ interchanging } a \text{ and } c \text{ in } L(A, B)$$

$$\Lambda_D(A, B) = \max\left(0, \frac{(a-c)(a+b+c)}{b(a+c)+a^2+c^2+2b^2/3}\right)$$

Table 2 shows the values of $\Lambda_D(A, B)$ for the cases in Fig. 1. We have also shown here the values of the min-transitive fuzzy left-relationship given in [4] and denoted here by $\pi(A, B)$, which does not make use of distance to the left $L(x, y) = \max(0, y - x)$ and is based on just the probabilities $P(x < y) = P(x \leq y)$ and $P(x > y) = P(x \geq y)$ for $x \in X$ and $y \in Y$. To be precise, we defined $\pi(X, Y) = P(x < y) - P(x > y)$. We remark that the uses of both $-P(x > y)$ in $\pi(X, Y)$ and $-R(X, Y)$ in $\Lambda_D(X, Y)$ are motivated by a construction in [7] in a different context.

Table 2 $\Lambda_D(A, B)$ and $\pi(A, B)$ for intervals A and B

Case	$\Lambda_D(A, B)$	$\pi(A, B)$
Figure 1(i):	$1 - \frac{2b^3/3}{(a+b)(b+c)(a+c)+2b^3/3}$	$1 - \frac{b^2}{(a+b)(b+c)}$
Figure 1(ii):	$\max\left(0, \frac{(a-c)(a+b+c)}{b(a+c)+a^2+c^2+2b^2/3}\right)$	$\max\left(0, \frac{a-c}{a+b+c}\right)$
Figure 1(iii):	$\max\left(0, \frac{(c-a)(a+b+c)}{b(a+c)+a^2+c^2+2b^2/3}\right)$	$\max\left(0, \frac{c-a}{a+b+c}\right)$
Figure 1(iv) and (vi):	0	0
Figure 1(v):	1	1

Theorem 4 *For finite intervals* $A, B \subset \mathcal{R}$, $\pi(A, B) < \Lambda_D(A, B)$ *except when both are 0 or 1 as in Fig. 1(iv)–(vi). Moreover,* $\pi(A, B) = \Lambda_D(A, B) = 0$ *or 1 as* $b \to \infty$ *and, for* $b \to 0$, *we have* $\pi(A, B) = \Lambda_D(A, B) = 0$ *or 1 except for Fig. 1(ii) and (iii), when* $\pi(A, B) < \Lambda_D(A, B)$.

Proof To prove that $\pi(A, B) < \Lambda_D(A, B)$ for Fig. 1(i), we show $f(b) = 3(a + b)(b + c)(a + c) - 2b(ab + bc + ac) > 0$ for all $a, b, c > 0$. But this follows from the fact that $f(0) = 3ac(a + c) > 0$ and $f'(b) = a^2 + c^2 + 2(a + c)(a + b + c) > 0$, where $f'(x) = df(x)/dx$. The inequality $\pi(A, B) < \Lambda_D(A, B)$ for Fig. 1(ii) follows from $f(b) = (a + b + c)^2 - [b(a + c) + a^2 + c^2 + 2b^2/3] = 2ac + b(a + c) + b^2/3 > 0$ for all $a, b, c > 0$. A similar argument holds for Fig. 1(iii). Finally, the results for the limiting cases $b \to \infty$ and $b \to 0$ follow immediately from Table 2. □

As one might expect from the non-min-transitivity of $\Lambda'_D(X, Y)$ in Sect. 4, the corresponding extension of $\Lambda'_D(X, Y)$ to finite intervals also fails to satisfy the min-transitivity property as shown below.

Example 3 Let $A' = [2, 3] \subset C' = [0, 4] \subset B' = [-1, 5]$, each with the uniform probability density function on it. For Fig. 1(ii), $\Lambda'_D(A, B) = (ab + a^2 + b^2/3)/(ba + bc + a^2 + c^2 + 2b^2/3)$. Also, $\Lambda'_D(A, B)$ for Fig. 1(iii) is the same as $\Lambda'_D(B, A)$ for Fig. 1(ii) and they equal $(bc + c^2 + b^2/3)/(ba + bc + a^2 + c^2 + 2b^2/3)$. Thus, we get $\Lambda'_D(A', B') = 19/56$, $\Lambda'_D(B', C') = 1/2$, and $\Lambda'_D(A', C') = 7/26 < 19/56 = \min(19/56, 1/2)$, a violation of the min-transitivity. In contrast, because $\Lambda_D(B', C') = 0$, we have $\Lambda_D(A', C') \geq \min(\Lambda_D(A', B'), \Lambda_D(B', C'))$. □

6 $\Lambda_D(X, Y)$ for Fuzzy Sets X and Y with Finite Support

For discrete fuzzy sets (X, μ_X), where $X = \{x_1, x_2, \ldots, x_m\} \subset \mathcal{R}$ is a non-empty finite set and $\mu_X(x_i) > 0$ is the membership value of x_i, let $\sum_i \mu_X(x_i) = c_X > 0$; here, c_X may not equal 1. Then, $P'_X(x_i) = \mu_X(x_i)/c_X$ is a probability distribution on X. We refer to the pair (X, P'_X) as X'. Likewise, we refer to the pair (Y, P'_Y) as Y', where Y

is a non-empty finite discrete fuzzy subset of $\mathcal{R}$ with membership values $\mu_Y(y_j) > 0$ for $y_j \in Y$, $\sum_j \mu_Y(y_j) = c_Y > 0$, and $P'_Y(y_j) = \mu_Y(y_j)/c_Y > 0$. If we take $P_X(x_i) = \mu_X(x_i)$ and $P_Y(y_j) = \mu_Y(y_j)$, which may not be probability distributions on X and Y, in the definitions for $L(X, Y)$ and $R(X, Y)$ then $L(X', Y') = L(X, Y)/(c_X c_Y)$, $R(X', Y') = R(X, Y)/(c_X c_Y)$, and $\Lambda_D(X', Y') = \Lambda_D(X, Y)$. This shows that we can use the membership values $\mu_X(x_i)$ and $\mu_Y(y_j)$ in place of the probabilities $P_X(x_i)$ and $P_Y(y_j)$ in Eqs. (1)–(2) and in Definition 1 of $\Lambda_D(X, Y)$ and this gives us a min-transitive left-relationship $\Lambda_D(X, Y)$ on discrete finite fuzzy subsets of $\mathcal{R}$.

Now consider the continuous fuzzy sets (X, μ_X), where X is a finite interval and the membership function $\mu_X(x) = 0$ for $x \notin X$. Once again, we can use $\mu_X(x)$ in place of $p_X(x)$ in the definition of $\Lambda_D(X, Y)$ for finite intervals in Sect. 5 although $\int_X p_X(x)\,dx = \int_X \mu_X(x)\,dx = c_X > 0$ may not equal 1. This gives a min-transitive fuzzy left-relationship $\Lambda_D(X, Y)$ on continuous fuzzy sets with finite support.

7 Conclusion

We have given here a new min-transitive fuzzy left-relationship $\Lambda_D(\cdot, \cdot)$ on non-empty, finite, weighted discrete subsets $X = \{x_1, x_2, \ldots, x_m\}$ of $\mathcal{R} = (-\infty, +\infty)$, where each x_i has a probability $P_X(x_i) > 0$ such that $\sum_i P_X(x_i) = 1$. We have shown that $\Lambda_D(\cdot, \cdot)$ can be generalized to the continuous case of finite intervals $X \subset \mathcal{R}$ with arbitrary probability density functions $p_X(x)$ on X. Moreover, the same formulas can be used directly to define a min-transitive left-relationship on discrete and continuous fuzzy subsets of $\mathcal{R}$ with finite support. Two distinct features of $\Lambda_D(X, Y)$ are that these are the "first" fuzzy left-relationship that (1) allow the sets (intervals) X and Y to have different probability distributions P_X and P_Y (respectively, density functions $p_X(x)$ and $p_Y(y)$), and Eq. (2) takes into account the amount $\max(0, y - x)$ by which a point $x \in X$ is to the left of a point $y \in Y$. These properties make $\Lambda_D(X, Y)$ applicable to many situations that are not handled by those in [4, 5].

References

1. Da, Q., Liu, X.: Interval number linear programming and its satisfactory solution. Syst. Eng.—Theory Pract. **19**, 3–7 (1999)
2. Hu, B.Q., Wang, S.: A novel approach in uncertain programming—part I: new arithmetic and order relation for interval numbers. J. Ind. Manag. Optim. **2**, 351–371 (2006)
3. Huynh, V.-N., Nakamori, Y., Lawry, J.: A probability-based approach to comparison of fuzzy numbers and application to target-oriented decision making. IEEE Trans. Fuzzy Syst. **16**, 371–387 (2008)
4. Kundu, S.: Min-transitivity of fuzzy leftness relationship and its application to decision making. Fuzzy Sets Syst. **86**, 357–367 (1997)
5. Kundu, S.: A new min-transitive fuzzy left-relation on intervals for a different class of problems. In: Proceedings of the SAI Intelligent Systems Conference (IntelliSys), UK, 7–8 Sept 2017

6. Nakamura, K.: Preference relations on a set of fuzzy utilities as a basis for decision making. Fuzzy Sets Syst. **20**, 147–162 (1986)
7. Orlovsky, S.A.: Decision making with a fuzzy preference relation. Fuzzy Sets Syst. **1**, 155–167 (1978)

Effective Data Clustering Algorithms

Kamalpreet Bindra, Anuranjan Mishra and Suryakant

Abstract Clustering in data mining is a supreme step toward organizing data into some meaningful patterns. It plays an extremely crucial role in the entire KDD process, and also as categorizing data is one of the most rudimentary steps in knowledge discovery. Clustering is used for creating partitions or clusters of similar objects. It is an unsupervised learning task used for exploratory data analysis to find some unrevealed patterns which are present in data but cannot be categorized clearly. Sets of data can be designated or grouped together based on some common characteristics and termed clusters, and the implementation steps involved in cluster analysis are essentially dependent upon the primary task of keeping objects within a cluster more closer than objects belonging to other groups or clusters. Depending on the data and expected cluster characteristics, there are different types of clustering algorithms. In the very recent times, many new algorithms have emerged, which aim toward bridging the different approaches toward clustering and merging different clustering algorithms given the requirement of handling sequential, high-dimensional data with multiple relationships in many applications across a broad spectrum. The paper aims to survey, study, and analyze few clustering algorithms and provides a comprehensive comparison of their efficiency on some common grounds. This study also contributes in correlating some very important characteristics of an efficient clustering algorithm.

Keywords Clustering · Proximity · Similarity · CF tree · Cluster validation

1 Introduction

Data is the goldmine in today's ever competitive world. Everyday large amount of information is encountered by organizations and people. One of the essential means of handling this data is to categorize or classify them into a set of categories

K. Bindra (✉) · A. Mishra · Suryakant
Noida International University, Plot 1, Sector-17 A, Yamuna Expressway,
Gautam Budh Nagar, Noida 203201, Uttar Pradesh, India
e-mail: kamalpreet.bindra@gmail.com

K. Ray et al. (eds.), *Soft Computing: Theories and Applications*,
Advances in Intelligent Systems and Computing 742,
https://doi.org/10.1007/978-981-13-0589-4_39

or clusters. “Basically classification systems are either supervised or unsupervised, depending on whether they assign new inputs to one of the finite number of discrete supervised classes or unsupervised categories Respectively” [1, 2]. Supervised classification is basically machine learning task of deducing a function from training dataset. Whereas unsupervised classification is the exploratory data analysis where there is no training dataset and extracting hidden patterns in dataset with no labeled Responses are achieved. The prime focus is to increase proximity of data points belonging to the same cluster and increase dissimilarity among various clusters and all this is achieved through some measure [3]. As depicted in Fig. 1. The quality of a clustering approach is also measured by its capability to discover some or all hidden patterns. Exploratory data analysis is related to a wide range of applications such as text mining, engineering, bioinformatics, machine learning, pattern recognition, mechanical engineering, voice mining, web mining, spatial data analysis, textual document collection, and image segmentation [4]. This diversity explains the importance of clustering in scientific research but this diversity can lead to contradictions due to different nomenclature and purpose. Many clustering approaches which are evolved with the intention of solving specific problems can many a times make presuppositions which are biased toward the application of interest. These favored assumptions can result in performance degradation in some problems that do not satisfy the favorable assumptions in certain premises. As an example, the famous partitioning algorithm K-means generates clusters which are hyper-spherical in shape but when the real clusters appear in varied shapes and geometric forms, K-means may no longer be effective and other clustering schemes have to be considered [5]. So the number, nature, and approach of each clustering algorithm differ on a lot of performance measures. Right from the measure of similarity or dissimilarity to feature selection or extraction to cluster validation to complexity (time/distance). Interpretation of results is equally valuable as cluster analysis is not a one-way process; in most cases, lots of trials and repetitions are required. Moreover, there are no standard or set of criteria for feature selection and clustering schemes. Some clustering techniques can also characterize each cluster in terms of a cluster prototype. The paper attempts to survey many popular clustering schemes as there exists a vast amount of work that has been done on clustering schemes and discussion of all those numerous algorithms will be out of the scope of the paper.

2 Categories of Clustering Algorithms

There exist many measures and initial conditions which are responsible for numerous categories of clustering algorithms [4, 6]. A widely accepted classification frames clustering techniques as follows:

- Partitional clustering,
- Hierarchical clustering, and
- Density-based.

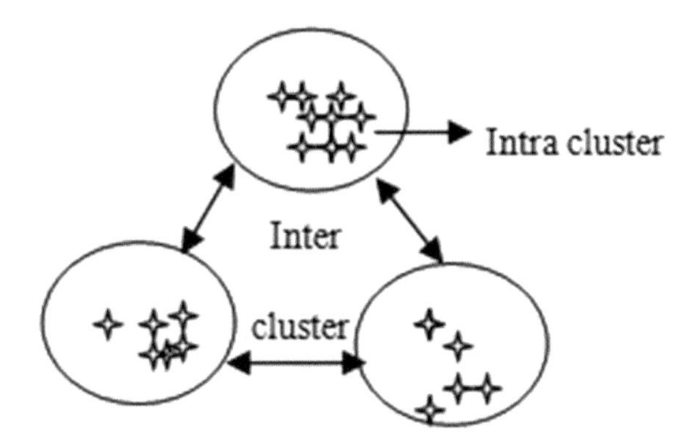

Fig. 1 Increased similarity within clusters

Table 1 Measures for finding similarity

	Popular similarity measures for data points		
	Measure	Form	Example
1	Euclidean distance	$D = \left(\sum_{i=1}^{d} \lvert x_a - x_{b_l} \rvert^{1/2}\right)^2$	k-means, PAM
2	Minkowski distance	$D = \sum_{i=1}^{d} \lvert x_a - x_b \rvert^{1/n}$	Fuzzy c-means

These classifications are based on a number of factors and few algorithms have been developed bridging the multiple approaches also. In the recent times, an extensive number of algorithms have been developed to provide solution in different fields; however, there is no universal algorithm that solves all problems. It has been very difficult to advance an integrated composition (clustering) at a technical level and profoundly diverse approaches to clustering have been seen [7]. It becomes ever crucial to discuss the various characteristics that an efficient clustering algorithm must pose in order to solve the problem at hand. Some of the characteristics are listed here.

Scalability: this is the ability of an algorithm to perform well with a large amount of data objects or say tuples, in terms of memory requirements and execution times. This feature especially distinguishes data mining algorithms from the algorithms used in machine learning [5]. Scalability remains one of the major milestones to be covered as many clustering algorithms have shown to do badly when dealing with large situations and datasets.

Similarity or dissimilarity measures: These measures are real-valued functions that quantify the similarity between two objects. Plenty of literature can be found on these measures some of the popular measures are enlisted in Table 1 [4].

Prior domain knowledge: There are many clustering algorithms which require some basic domain knowledge or user is expected to provide some input parameters, e.g., the number of clusters. However more often, the user is unable to provide such domain knowledge, and also this oversensitivity toward input parameters can degrade the performance of the algorithm as well.

3 Hierarchical Clustering

Hierarchical clustering is a paradigm of cluster analysis to generate a sequence of nested partitions (clusters) which can be visualized as a tree or so to say a hierarchy of clusters known as cluster dendrogram. Hierarchical trees can provide a view of data at different levels of abstraction [8]. This hierarchy when laid down as a tree can have the lowest level or say leaves and the highest level or the root. Each point that resides in the leaf node has its own cluster whereas the root contains all points in one cluster. The dendrogram can be cut at intermediate levels for obtaining clustering results; at one of these intermediate levels, meaningful clusters can be found. The hierarchical approach toward clustering can be divided into two classes: (a) agglomerative and (b) divisive.

Hierarchical clustering solutions have been primarily obtained using agglomerative algorithms [9, 10]. Agglomerative clustering strategies function in bottom-up manner, i.e., in this approach, merging of the most similar pair of clusters is achieved after starting with each of the K points in a different cluster. This process is repeated until all data points converge and become members of the same cluster. There exist many different variations of agglomerative algorithms but they primarily differ in how the similarity between existing clusters and merged clusters is updated. Many different agglomerative algorithms exist depending upon the distance between two clusters. Very well-known methods are as follows: (1) single linkage, (2) average linkage and (3) complete linkage technique. In single linkage, the attention is paid primarily to the area where two clusters come closest to each other, and more distant parts of the cluster and cluster's overall structure are not taken into account, whereas average linkage takes the average distance between two cluster centroids, and on the other hand, the complete linkage is more nonlocal where the entire structure of the cluster can affect merge decisions [11]. Divisive approach, on the other hand, is exactly in contrast which means it works in top-down manner. In divisive approach a cluster is recursively split after starting with all the points in the same cluster, this step is repeated until all points are in separate clusters. For a cluster with N objects, there are 2^{N-1} possible two subset divisions, which incur a high computation cost [12]. Therefore divisive clustering is considered a burden computationally Although it has its drawbacks, this approach is implemented in two popular hierarchical algorithms: DIANA and MONA [4]. Discussion of these algorithms is beyond the scope of this paper. In hierarchical algorithms, previously taken steps whether merging or splitting are irreversible even if they are erroneous. The hierarchical algorithms are relatively more prone and sensitive to noise and outliers as well as lack of robustness. To hierarchical clustering, there are many advantages and disadvantages which are reflected in many representative examples of hierarchical technique. The detailed discussion is possible only by examining popular hierarchical algorithms.

Some of the representative examples are as follows:
1. BIRCH 2. CURE 3. ROCK 4. CHAMELEON

3.1 Birch

“balanced iterative reducing and clustering using hierarchies” (birch) [13]. The Birch addresses two huge issues in clustering analysis 1. It is able to scale while dealing with large datasets and 2. Robustness against noise and outliers [13]. Birch introduces a novel data structure for achieving the above goals, (CF) tree clustering feature. CF tree can be assumed as a tuple, summarizing the information that is maintained about a cluster. Instead of directly using the original data, CF tree can compress data and develop numerous tiny points or nodes. These nodes work as tiny clusters and depict the summary of original data. This CF tree is height balanced tree with two parameters: branching factor B and threshold T. In the CF tree, every internal vertex comprises entries defined as [CF_i, $child_i$], I = 1…k, where CF_i is a summary of the cluster i and is defined as a tuple $CF_i = (N_i, LS, SS)$ where N_i = no. of data objects in the cluster, LS = the linear sum of the N data points, and SS = squared sum of the objects.

The CFs are saved as leaf nodes whereas nonleaf nodes comprise summation of all the CFs of their children. A CF tree is built dynamically and incrementally, and also it requires a single scan of the entire dataset, when an object is inserted in the closest leaf entry, and the two parameters B and T control the maximum number of children per nonleaf node and the maximum diameter of subclusters stored in the leaf node. In this manner, birch constructs a framework which can be stored in main memory, after building a CF tree the next step is to employ an agglomerative hierarchical or any of the sum of square error-based algorithm to perform clustering. Birch is capable of achieving a computational complexity of O(N). It is quite fast but again the inability to deal with nonspherical shaped clusters stands as the biggest drawback of birch.

Arbitrary shaped clusters cannot be identified by birch as it uses the principal of measuring diameter of all clusters for determining their boundary. Encountering the above drawback, *Guha, Rastogi, and Shim* developed CURE.

3.2 Cure

It was developed for identifying more complex cluster shapes [9]. It is more robust to outliers. Cure is an agglomerative method. Instead of using a single centroid it assumes many separate fixed points as clusters and a fragment of *m* is used to shrink these diverse points toward centroids. These scattered points after shrinking represent the cluster at each iteration and the pair of clusters with the closest representatives are merged together. This feature enables CURE to identify clusters correctly and makes it sensitive to outliers. This algorithm uses two enhancements (1) random sampling and 2 partitioning. These enhancements help improve cure’s scalability. CURE has a high run time complexity of $O(n^2 \log n)$ making it an expensive choice for databases.

3.3 Rock

Guha et al. suggested and proposed another agglomerative hierarchical algorithm, ROCK. This algorithm scales quite well with increase in dimensionality. With ROCK concept of "links" was introduced. "links are used for measuring proximity between a pair of data points with categorical attributes" [14]. This algorithm uses a "goodness measure" for determining how proximate or similar clusters are. A criteria function is calculated, more the value of this function the better a cluster is. This link-based approach used in rock claims to be a global approach to the clustering problem. This algorithm measures the similarity of two clusters by comparing a user-specified interconnectivity measure which is static with aggregate interconnectivity of any two clusters. ROCK is highly likely to generate ambiguous results if the choice of parameters is provided in the static model differ from the dataset being clustered. Again, clusters that are of different sizes and shapes cannot be accurately defined by this algorithm.

3.4 Chameleon

This algorithm is capable of "measuring the similarity of two clusters based on a dynamic model" [15] it improves the clustering quality by using more detailed merging criteria in comparison to CURE. This method works in two phases and measures the similarity of two clusters based on a dynamic model. During the first phase, a graph is created which contains links between each point and its N-nearest neighbor. After that during the second phase, graph is recursively split by a graph partitioning algorithm resulting in many tiny unconnected subgraphs. This algorithm iteratively combines two most similar clusters. During the second phase when each subgraph is considered as an initial subcluster, two clusters can be merged but only if the resultant cluster has similar interconnectivity and closeness to the two parent clusters prior to merging. The overall complexity depends on the amount of time required to construct an N-nearest neighbor graph combined with the time required to complete both phases. Chameleon is a dynamic merging algorithm and hence considered more functional as compared to Cure, while dealing with arbitrary shaped clusters of uneven density.

Advantages and disadvantages: As discussed in above section, hierarchical clustering can cause trouble while handling noisy high-dimensional data. In HC when merges are final, they cannot be undone at a later time preventing global optimization. After the assignment of an object or data point is done, that data point is highly unlikely to be reconsidered in future even if this assignment generates a bad clustering example. Most agglomerative algorithms are a liability in terms of storage and computational requirements. The computational complexity of most HC algorithm is at least $O(N^2)$. Not only could this HC algorithm be severely degraded when applied in high-dimensional spaces due to curse of dimensionality phenomenon. The abovementioned algorithms have each introduced their novelties toward overcoming many shortcomings of HC approach. The hierarchical approach can be a great boon when used in taxonomy tree problems, for example, when vertical relationships are present in data.

4 Partitional Clustering

Partitional clustering is highly dissimilar to hierarchical approach which yields an incremental level of clusters with iterative fusions or divisions; partitional clustering assigned a set of objects into K clusters with no hierarchical structure [5]. Research from very recent years acknowledges that partitional algorithms are a favored choice when dealing with large datasets. As these algorithms have comparatively low computational requirements [16], however, when it comes to the coherence of clustering, this approach is less effective then agglomerative approach. Partitional algorithms basically experiment with cutting data into n number of clusters so that partitioning of data optimizes a given criterion. Centroid-based techniques as used by K-MEANS and ISODATA assign some points to clusters so that the mean squared distance of points to the centroid of the chosen cluster is minimized. The sum of the squared error function is the dominant criteria function in partitional approach. It is used as a measure of variation within a cluster. One of the most popular partitioning clustering algorithms implementing SE is k-means.

4.1 K-Means

K-means is undoubtedly a very popular partitioning algorithm. It has been discovered, rediscovered, and studied by many experts from different fields, by Steinhaus (1965), Ball and Hall [17], Lloyd (proposed 1957—published 1982), and Macqueen [18]. It is distance-based and by definition, data is partitioned into predetermined groups or clusters. The distance measures used could be Euclidean or cosine. Initially, K cluster centroids [19, 20] are selected at random; k-means reassigns all the points to their closest centroids and recomputes centroids of newly created groups. This iteration continues till the squared error converges. Following steps can summarize the function of k-means.

1. Initialize a K partition based on previous information. A cluster prototype matrix $A = [a_i \ldots a_j]$ is created, where a1, a2, a3… are cluster centers. Dataset D is also initialized.
2. In the next step, assignment of each data point in the dataset (d_i) to its nearest cluster (a_i) is performed.
3. Cluster matrix can be recalculated considering the current. Updated partition or until a_i, a_j, a_k… Show no further change.

4. Repeat 2 and 3 until convergence has been reached.

K-means is probably the most wildly studied algorithm; this is the reason why there exist too many variations and improved versions of k-means; yet it is very sensitive to noise and outliers. Even if a point is at a distance from the cluster centroid, it could still be enforced to the center and can result in distorted cluster shape. K-means does not clearly define a universal method of deciding total number of partitions in the beginning; this algorithm heavily relies on the user to specify the number of clusters k. Also, k-means is not applicable to categorical data. Since k-means presumes that user will provide initial assignments, it can produce replicated results upon every iteration (the k-means ++ addresses this problem by attempting to choose better starting clusters) [4, 5].

4.2 k-Medoids

Unlike the k-means, in the k-medoids or partitioning around medoids (PAM) [21, 22] method, a cluster is represented by its medoid. This characteristic object called the medoid is the most centrally located point within the cluster [21]. Medoid shows better results against outliers as compared to centroids [4]. K-means finds the mean to define accurate center of the cluster which can result in extreme values but k-medoid calculates the cluster center using an actual point. The goal of this algorithm is to minimize the average dissimilarity of objects to their closest selected object. The following steps can sum up this algorithm:

1. **Initialize**: a random k is selected of the n data points as the medoid.
2. **Assign**: each data point should be associated with the closest medoid.
3. **Update**: for every m medoid and data point d, swapping of m and d can be done compute average dissimilarity of d to all the data points associated with m.

Steps 2 and 3 can be repeated multiple times until there is no further change left in assignments.

PAM uses a greedy search resulting in failure in finding an optimum solution. The problem of finding arbitrary shaped clusters with different densities is a problem that is faced by many agglomerative as well as partitional algorithms. The failure to generate concave shaped clusters with even distribution of data points is the biggest challenge faced by centroid-based or medoid-based approaches. The methods which are able to generate clusters of varied shapes suffer from other problems of complexity of time and space. But it is highly desired of a clustering algorithm to produce clusters which are shape-based and converge well. Figure 2 throws some light on the topic of arbitrary shaped clusters.

Fig. 2 Clusters of different shapes, sizes, and densities

4.3 Clarans

"Clustering large applications based on randomized search", this method combines the sampling techniques with PAM [23]. This method employs random searching techniques for finding clusters and no supplementary structure is used., a feature that makes this algorithm much less affected by increasing dimensionality. Many of the aforementioned techniques assume that distance function has to be an Euclidean but Clarans uses a local search technique and makes no requirement of any particular distance function. Clarans claims to identify polygon shaped objects very effectively. A method known as "IR—approximation" is used for grouping nonconvex polygon as well as convex polygon objects. It shows better performance in contrast with PAM and CURE when it comes to finding polygon shaped structures in data points and is a main memory clustering technique, while many of the other partitional algorithms are designed for out of core clustering applications.

4.4 Isodata

An interesting technique called ISODATA "Iterative self organizing data analysis technique", developed by ball and Hall [23] also evaluates k (no of clusters) and is iterative in nature. A variant of k-means algorithm, Isodata works by dynamically adjusting clusters through the process of splitting and merging depending on some thresholds defined a priori like C: no. of clusters desired, D_{min}: minimum number of data points for each cluster, V_{max}: maximum variance for splitting up clusters, and M_{min}: minimum distance measure for merging. Clusters associated with fewer than the user-specified minimum number of pixels are eliminated and lone points or pixels are either put back in the pool for reclassification or ignored as "unclassifiable". The newly calculated clusters are used as it is for next iteration. Isodata is able to handle the problem of outliers much better than the k-means through the splitting procedure and Isodata can eliminate the possibility of elongated clusters as well, not only this the algorithm does not need to know the number of clusters in advance which means less user domain knowledge is required.
Advantages and disadvantages: The most prominent advantage of partitioning-based methods is that they can be used for spherical-based clusters in small to medium sized datasets. Algorithms like k-means can tend to show high sensitivity to noise and outliers whereas other methods which show resistance against noise can prove to

Table 2 Performance evaluation of various algorithms

Algorithm	Shape	Convergence	Capability
k-means	Cannot handle arbitrary clusters	K is required in advance. Performance degrades with increased dimensionality	Simple and efficient
Birch	Cannot handle arbitrary clusters	Deals fairly with robust data. Performs strongly against noisy data	Most famous HC algorithm
Rock	Random sampling has an impact on selection of cluster shapes	Can handle large datasets	Usage of links gives better results with scattered points
Cure	Finds richer cluster shapes	Cannot scale well compared to birch. Merging phenomenon used in cure makes various mistakes when handling large datasets with comparison to chameleon	Less sensitive to outliers. A bridge between centroids-based and all points approach
Dbscan	Handles concave clusters	Targets low-dimensional spatial data	Very popular density algorithm
Denclue	Influence function can affect shapes	Faster and handles large datasets	Shows better result with outliers then dbscan
Chameleon	Fair	Scales with increase in size of data	Modify clusters dynamically. Curse of dimensionality

be computationally costly partitional algorithms suit the situations where dataset is comparatively scalable and simple. And data points are compact and well separated and easily classifiable in spherical shapes. Apart from the computational complexity, algorithms can be compared on some other common grounds, like the capability to converge to an optimum clustering solution or the potential to create clusters of different density and shapes. Although all clustering algorithms attempt to solve the same problem but there exist performance issues which can be discussed. Table 2 lists some comparisons in brief.

5 Density-Based Algorithms

The key idea of density-based clustering is such, where every N-neighborhood (for some given $N > 0$) must contain minimum number of points, or "density". The closeness of objects is not the benchmark here rather "local density" is primarily measured as depicted in Fig. 3. A cluster is viewed as a set of data points scattered in the data

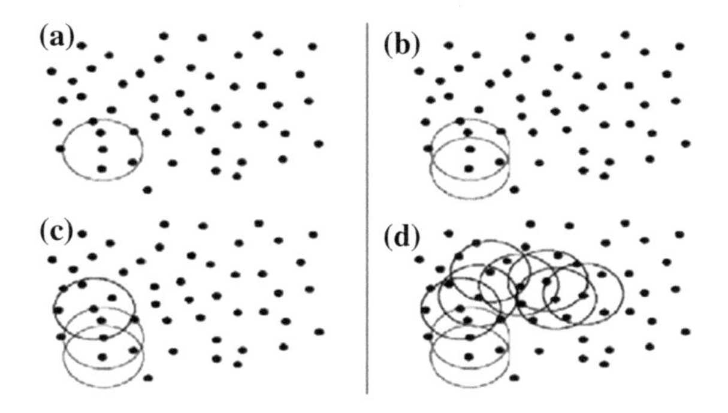

Fig. 3 Contiguous regions of low density of objects and data exist

space. In density-based clustering, contiguous region of low density of objects and data exist and the distance between them needs to be calculated. Objects which are present in low-density region are constituted of outliers or noise. These methods have better tolerance toward noise and are capable of discovering nonconvex shaped clusters. These algorithms restrict outliers in low-density areas so that irregularity in cluster shapes can be controlled. This feature makes them an excellent choice for medium sized data. Two most known representatives of density-based algorithms are density-based spatial clustering of applications with noise (DBSCAN) [24] and density-based clustering (DENCLUE) [25]. Density-based algorithms are able to scale well when administered with high-dimensional data.

5.1 Dbscan

It was proposed by Martin Ester, Hans-Peter Kriegel, Jorg Sander, and Xiawoei in 1996. Density-based spatial clustering of applications with noise requires that the density in the neighborhood of an object should be high enough if it belongs to a cluster [24]. A cluster skeleton is created with this set of core objects with overlapping neighborhood. Points inside the neighborhood of core objects represent the boundary of clusters while rest is simply noise. It requires two parameters: (1) E is the starting point and (2) Minpts is the minimum number of points required to form a dense region. The following steps can elaborate the algorithm further:

1. An unvisited random point is usually taken as the initial point.
2. A parameter $_E$ is used for determining the neighborhood (data space).
3. If there exist sufficient data points or neighborhood around the initial random point, then the algorithm can proceed and this point can be marked as visited or else the point is marked as noise.
4. If this point is considered a part of the cluster, then its $_E$ neighborhood is also the part of the cluster, and step 2 is repeated for all $_E$. This is repeated until all points in the cluster are determined.

5. A new initial point is processed and above steps are restated until all clusters and noise are discovered.

Although this algorithm shows excellent results against noise and has a complexity of $O(N^2)$, it can be a failure when tested in high-dimensional datasets and shows sensitivity to Minpts.

5.2 Denclue

Density-based clustering (DENCLUE) was developed by Hindenburg and Keim [26]. This algorithm buys heavily from the concepts of density and hill climbing [27]. In this method, there is an "influence function" which is the distance or influence between random points. Many influence functions are calculated and added up to find out the "density function" [27]. So it can be said that influence function is the influence of a data point in its neighborhood and density function is the total sum of all influences of all the data points. Clusters are determined by density attracters, local maxima of the overall density function. Denclue has a fairly good scalability, and a complexity of (O(N)) is capable of finding arbitrary shaped clusters but suffers from a sensitivity toward input parameters. Denclue suffers from curse of dimensionality phenomenon.
Advantages and disadvantages: density-based methods can very effectively discover arbitrary shaped clusters and capable of dealing with noise in data much better than hierarchical or partitional methods, and these methods do not require any predefined specification for the number of partitions or clusters but most density-based algorithms show decrease in efficiency if dimensionality of data is increased although algorithm like denclue shows some escalation while dealing with high dimensionality [4] but it is still far from completely effective.

6 Conclusion

Cluster analysis is an important paradigm in the entire process of data mining and paramount for capturing patterns in data. This paper compared and analyzed some highly popular clustering algorithms where some are capable of scaling and some of the methods work best against noise in data. Every algorithm and its underlying technique have some disadvantages and advantages and this paper has comprehensively listed them for the reader. Every paradigm is capable of handling unique requirements of user application. A lot of research and study is done in the field of data mining and subsequently cluster analysis as the future will unveil more complex database relationships and categorical data. Although there is an alarming need for some sort of benchmark for the researchers to be able to measure efficiency and validity of diverse clustering paradigms. The criteria should include data from diverse domains (text documents, images, CRM transactions, DNA sequences and dynamic data). Not just a measure for benchmarking algorithms, consistent and stable clustering is also a barrier as a clustering algorithm irrespective of its approach toward handling static or dynamic data should produce consistent results with complex datasets. Many examples of efficient clustering methods have been developed but many open problems still exist making it a playground for research from broad disciplines.

References

1. Cherkassky, V., Mulier, F.: Learning From Data: Concepts, Theory, and Methods. Wiley, New York (1998)
2. Hinneburg, A., Keim, D.A.: A general approach to clustering in large databases with noise. Knowl. Inf. Syst. **5**(4), 387–415 (2003)
3. Mann, A.K., Kaur, N.: Survey paper onclustering techniques. IJSETR: Int. J. Sci. Eng. Technol. Res. **2**(4) (2013) (ISSN: 2278-7798)
4. Everitt, B., Landau, S., Leese, M.: Cluster Analysis. Arnold, London (2001)
5. Xu, R., Wunch, D.: Survey of clustering algorithms. IEEE Trans. Neural Netw. **16**(3) (2005)
6. Berkhin, P.: Survey of clustering data mining techniques (2001). http://www.accrue.com/products/rp_cluster_review.p, http://citeseer.nj.nec.com/berkhin02survey.html.
7. Kleinberg, J.: An impossibility theorem for clustering. In: Proceedings of the 2002 Conference on Advances in Neural Information Processing Systems, vol. 15, pp. 463–470 (2002)
8. Zhao, Y., Karypis, G., Fayyad, U.: Hierarchical clustering algorithms for document datasets, min knowl disc, vol. 10, p. 141 (2005). https://doi.org/10.1007/s10618-005-0361-3
9. Guha, S., Rastogi, R., Shim, K.: CURE: an efficient clustering algorithm for large databases. In: Proceedings of the ACM SIGMOD International Conference on Management of Data, pp. 73–84 (1998)
10. Guha, S., Rastogi, R., Shim. K.: ROCK: a robust clustering algorithm for categorical attributes. In: 18th Proceedings of the 15th International Conference on Data Engineering (1999)
11. Sneath, P.: The application of computers to taxonomy. J. Gen. Microbiol. **17**, 201–226 (1957)
12. Fasulo, D.: An analysis of recent work on clustering algorithms. Department of Computer Science Engineering University of Washington, Seattle, WA, Technical Report, 01-03-02 (1999)
13. Zhang, T., Ramakrishnan, R., Livny, M.: BIRCH: an efficient data clustering method for very large databases. In: Proceedings of the 1996 ACM SIGMOD International Conference on Management of data, Montreal, Quebec, Canada, pp. 103–114, 04–06 June 1996
14. Guha, S., Rastogi, R., Shim, K.: Rock: A robust clustering algorithm for categorical attributes. Inf. Syst. **25**(5), 345–366 (2000)
15. Karypis, G., Han, E.-H., Kumar, V.: CHAMELEON: a hierarchical clustering algorithm using dynamic modeling. IEEE Comput. **32**(8), 68–75 (1999)
16. Cutting, D., Pedersen, J., Karger, D., Tukey, J.: Scatter/gather: a cluster-based approach to browsing large document collections. In Proceedings of the ACM SIGIR, Copenhagen, pp. 318–329 (1992)
17. Ball, G.H., Hall, D.J.: ISODATA–A novel method data analysis and pattern classification. Menlo park: Stanford Res. Inst, CA (1965)
18. Macqueen, J.B.: Some methods for classification and analysis of multivariate observations. In: Proceedings of 5th Berkely Symposium on Mathematical statistics and probability, **1**, 281–297 (1967)
19. He, J., Lan, M., Tan, C.-L., Sung, S.-Y., Low, H.-B.: Initialization of Cluster refinement algorithms: a review and comparative study. In: Proceeding of International Joint Conference on Neural Networks, Budapest (2004)

20. Biswas, G., Weingberg, J., Fisher, D.H.: ITERATE: a conceptual clustering algorithm for data mining. IEEE Trans. Syst. Cybern. **28C**, 219–230
21. Han, J., Kamber, M.: Data Mining Concepts and Techniques-a Reference Book, pp. 383–422
22. Pujari, A.K.: Data Mining Techniques-a Reference Book, pp. 114–147
23. He, Z., Xu, X., Deng, S.: Scalable algorithms for clustering large datasets with mixed type attributes. Int. J. Intell. Syst. **20**, 1077–1089
24. Ester, M., Kriegel, H., Sander, J., Xu, X.: A density-based algorithm for discovering clusters in large spatial databases with noise. In: Proceedings of the 2nd International Conference on Knowledge Discovery and Data Mining (KDD'96), pp. 226–231 (1996)
25. Ball, G., Hall, D.: A clustering technique for summarizing multivariatedata. Behav. Sci. **12**, 153–155 (1967)
26. Duda, R., Hart, P., Stork, D.: Pattern Classification, 2nd edn. Wiley, New York (2001)
27. Idrissi, A., Rehioui, H.: An improvement of denclue algorithm for the data clustering. In: 2015 5th International Conference Information & Communication Technology and Accessibility (ICTA). IEEE Xplore, 10 Mar 2016

Predictive Data Analytics Technique for Optimization of Medical Databases

Ritu Chauhan, Neeraj Kumar and Ruchita Rekapally

Abstract The medical databases are expanding at exponential rate; forfeit technology is required to determine hidden information and facts from such big databases. The data mining technology or Knowledge discovery from databases (KDD) tends to be the well-known technique from past decade which has provided fruitful information to discover hidden patterns and unknown knowledge from large-scale databases. Further, as medical databases are exceeding at an enormous rate data mining tends to be an effective and efficient technology to deal with inconsistent databases which include missing value, noisy attributes, and other types of attributes or factors to discover knowledgeable information for future prognosis of disease. In the current study, we have utilized predictive data analytics technique to diagnose patients suffering from liver disorder. The paper utilizes two-step clustering technology to analyze patients' disorder with different data variables to find optimal number of clusters of variant shapes and sizes. The focus of study relies on determining hidden knowledge and important factors which can benefit healthcare practitioners and scientists around the globe for prognosis and diagnosis of liver disease at an early stage.

Keywords Medical data mining · SPSS · Two-step clustering · Medical databases

1 Introduction

Medical data mining has emerged as an outstanding field in past decade to retrieve hidden knowledge and unknown information from large-scale databases to enhance medical treatment for futuristic prediction of medical diagnosis. The patterns, trends, and rules discovered from data mining techniques can be efficiently and effectively

R. Chauhan (✉) · N. Kumar · R. Rekapally
Amity Institute of Biotechnology, Amity University, Sec-125, Noida, Uttar Pradesh, India
e-mail: rituchauha@gmail.com

N. Kumar
e-mail: nkumar8@amity.edu

K. Ray et al. (eds.), *Soft Computing: Theories and Applications*,
Advances in Intelligent Systems and Computing 742,
https://doi.org/10.1007/978-981-13-0589-4_40

applied for diagnosis, prediction, and prognosis of disease [1–3]. This information will facilitate early disease detection and improve the quality of medical treatment required for patient care. Data mining techniques have been applied for a variety of domains in medical diagnosis such as early disease detection of liver cancer, prognosis evaluation at an early stage, as well as morbidity and mortality factors related to the medical condition [4–6].

The major challenge faced by researchers is real-world datasets of liver disorders, which consist of complex data form, missing value, time variant datasets, and inconsistent features which occur due to data collection, data selection, data transformation or other reasons. The prognosis and detection of liver disease using inconsistent data records can mislead to detect relevant patterns for future diagnosis of disease. Therefore, to detect early stages of liver disorder proper technological advances are required to correlate patterns which can prove beneficial for medication and treatment of disease.

The data mining application has evolved constantly for medical databases to study the effectiveness of diseases while comparing the contrasting factors such as symptoms, stage at which diagnosed, geographic factors, and course of treatment. For examples, stage at which liver cancer patient was diagnosed and treatment provided during the stay in the hospital. The factors correlated if explored can have clinical benefits for healthcare practitioners to predict future trends.

To analyze and retrieve knowledge from different forms of data different analysis techniques of data mining are utilized with statistical and graphical techniques to mine novel information from large databases. This study aims to identify patients that are suffering from liver disorder based on variable correlated features. The database of liver cirrhosis was collected from public domain repositories UCI where the data is categorized in form of instances and attributes which might have aroused due to excessive alcohol consumption mechanism. The dataset consists of 583 records which were labeled as patients suffering from liver disorder or not suffering [7]. Further, data analytical technique two-step clustering was utilized to determine the correlated factor for detection of early prognosis of disease. The scope of study is widely distributed among the sections where Sect. 2 discuss data mining technique with liver cirrhosis, Sect. 3 briefly discuss two-step clustering technique, Sect. 4 has experimental results, and last section has conclusion.

2 Extraction of Liver Cirrhosis via Data Mining Technique

Generously, data mining techniques are categorized into descriptive and predictive techniques to search for relevant patterns and derive a decision support system to discover hidden information from large clinical databases. Predictive data mining tasks are applied for prediction and classification of data, whereas descriptive data mining methods are used for discovery of individual patterns such as association rules, clustering, and other patterns which can be useful to user.

Predictive data mining technique classification is widely used to classify liver disorders. The liver disorders are usually diagnosed at later stages when the liver is partially damaged. The early diagnosis of liver disorder will reduce the mortality rate among the prevailed cases. The automatic classification technique in data mining tools can generate wide number patterns which can be used for detection of early liver disease.

Whereas, Descriptive data mining techniques such as clustering is widely used in the field of medical databases and engineering to predict future for decision-making process. Clustering can be referred as generalized tool to group the data according to similarity among the data records. It can be utilized to detect similar patterns from high dimensional databases for knowledge discovery process. For example, number of diagnosed cases for liver cancer in specific region can be diagnosed with clustering data as well as Geographic Information System to discover hidden patterns from medical as well geographic data.

The demand is to promote effective collaboration between healthcare practitioners and data miners [8, 9]. A number of scientists had worked on machine learning algorithms for decision-making in medical databases and surveyed extensive literature on medical databases for machine learning algorithms which provide an insight to improve the efficiency of medical decision-making by several decision-based systems [10–15].

But one of the major challenges is exploiting specific data mining technique to determine factors related to chronic liver diseases. Nowadays, several data mining techniques for preprocessing of data are utilized such as Feature selection, feature reduction, feature transformation, and several others but the choice of results depends on the user for future prediction [16–24]. There are several open source data mining tools such as WEKA, ORANGE, Rapid Miner, Java Data mining package, and several other which can be utilized to retrieve patterns for future prediction of liver disorders.

In this article, we have illustrated the basic cause of liver cirrhosis and treatment applied to deal with chronic disease. A lot of studies have been conducted to investigate the causes and management of liver cirrhosis. But one of the biggest challenges is to detect the cirrhosis in time and their severity as it can finally lead to hepatocellular carcinoma (HCC). Once a hepatocellular carcinoma is discovered the liver cirrhosis reached almost at incurable end stage. Usually, a Hepatocellular carcinoma stage occurs at a rate of 1–4% per year after the cirrhosis is diagnosed.

3 Materials and Methods

Data clustering is a predictive data analytics to group the data with respect to labeled class where the similar data is grouped using varied distance measures. The clustering is categorized with respect to partitioning, hierarchical, density-based, and grid-based clustering approach. Each approach utilizes different algorithms which can be synthesized depending upon type of datasets available. In the current study, we have exploited SPSS (16.0 Version for windows) to detect hidden patterns from liver cirrhosis datasets. SPSS perform varied clustering analysis techniques which include partitioning, hierarchical, and two-step clustering technique depending on type of clusters retrieved. We have utilized two-step clustering as it is able to handle large datasets and cluster the data with respect to continuous and categorical attributes. Whereas, other clustering techniques are unable to handle categorical datasets to detect clusters of varied shapes and size.

The two-step clustering procedure is widely anticipated to deal with large datasets where the algorithm determines the number of clusters itself utilizing different types of datasets. The process enables agglomerative clustering approach as one its step to determine the clusters of varied shapes and size. The two-step clustering process involves varied steps which include pre-clustering technique, this step scans primarily each data items and distinguish whether the data record can be added to formerly cluster or new cluster needs to be formed depending upon distance measure which is calculated using Euclidian distance measure and log-likelihood technique.

The Euclidian distance can be discussed as the square root of sum of squares with difference among the values or coordinates. The distance between two clusters can be measured as the Euclidian distance between the centers of clusters. Whereas the log-likelihood distance measure is applicable for continuous and categorical attributes to measure distance among them. The likelihood function is applied to measure distance among two clusters. The distance measure associated with continuous variable is liable when they have normal distribution and categorical variable should have multinomial distribution as well variables tend to be independent of each other for their values.

Further, the pre-clustering process builds a data structure tree known as Cluster Feature (CF) tree which have centers for each cluster. The CF tree consists of various nodes where each node represents the number of data entries. The leaf node tends to have a final subcluster, the tree works in recursive order from the root node where the process recursively determine the child node. After the tree reaches the leaf node the tree iteratively finds the reciprocated leaf entry in the leaf node. The CF tree updates itself if the record found was in its respective threshold value of nearest leaf value and only then it is added to CF tree, else a new leaf node is created for the value. If there exist the space in leaf node to accommodate another value then leaf is certainly divided into two values and the values are distributed among the leaves using far tests pair of seeds and the redistribution of data is pertained in respect to criterion of closeness.

The two-step clustering also deals with outliers, where outliers can be discussed as variable points that are outside the cluster and does not fit any of the cluster. The records in CF tree which are part of leaf are considered as outliers if the record has value less than the percentage of largest leaf by default the percentage value is 25%. The process invariably looks for these values and put them to another side. After the tree is created it tries to rebuild these values to fit the size of tree. If the values are unable to fit then they are considered as outliers. Further, the two-step clustering process utilizes the agglomerative clustering approach where it automatically determines the number of cluster to be formed. The agglomerative approach continuously gathers the data until single cluster formed. The approach initially starts with number of subclusters, all clusters are evaluated and compared among each other utilized minimum distance and finally merge clusters with smallest distance among each other. The process iteratively continues until all clusters are merged.

4 Experimentation

The predictive data analytics techniques were applied to determine the hidden knowledge from datasets. The UCI datasets were utilized with SPSS data mining technology for retrieval of unknown patterns. The dataset consists of patient data records which are suffering from liver disorder and not suffering. The labeled datasets have varied attributes which are discussed in Table 1.

The instances in dataset were retrieved and correlated to determine the factors which can affect the occurrence of liver disorder. The results were retrieved using two-step clustering technique to predict the accuracy of liver disorder measuring correlated patterns. The dataset was divided into two sets of labels where the selector class was determined to correlate among the varied patterns. The result shows in Fig. 1 that age at of 59 maximum patients are suffering liver disorder whereas older age group visualized lower number of cases as compared to middle age group.

The liver cirrhosis cases are much higher in cases of male patients as compared to female cases. The two-way clustering technique was able to detect 3 major clusters with respect to labeled class in SPSS where the cluster distribution is represented in Table 2. The cluster 1 represents highest number of 322 cases, cluster 3 has 166 cases, and cluster 2 has 95 records with total of 583 with 416 patients suffering from liver disorder and others 167 with no liver disease.

Table 1 Data type with definition

Attributes	Definition	Type of data
Age	Patient age at time of admission	Continuous
Gender	Patient gender	Nominal
Total bilirubin	Content of bilirubin	Continuous
DB direct bilirubin	Content of direct bilirubin	Continuous
Alkphos alkaline phosphatase	Alkaline phosphatase level	Continuous
Sgpt	Alanine aminotransferase	Continuous
Sgot	Aspartate aminotransferase	Continuous
TP total proteins	Total protein	Continuous
ALB albumin	Level of albumin	Continuous
A/G	Ratio albumin and globulin ratio	Continuous
Labeled class	Suffering from liver disease or not	Nominal

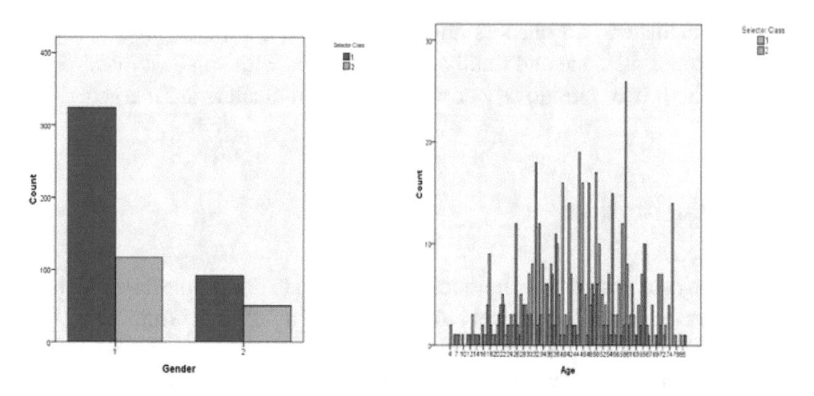

Fig. 1 Cases gender and age-wise

Table 2 Cluster distribution

		N	% of combined	% of total
Cluster	1	322	55.23156	55.23156
	2	95	16.29503	16.29503
	3	166	28.47341	28.47341
	Combined	583	100	100
Total		583		100

Table 3 Level of correlated features in each cluster

	Alk phosphatase		Sgpt		Sgot		T. Bilirubin
Cluster	Mean	Std. deviation	Mean	Std. deviation	Mean	Std. deviation	Mean
1	239.764	101.9839	51.09938	47.40496	65.81056	70.57308	1.882609
2	600.8737	444.318	263.8947	396.0475	380.5895	638.9675	11.93158
3	211.5602	93.36111	33.3253	24.77541	40.54819	36.4764	1.105422
Combined	290.5763	242.938	80.71355	182.6204	109.9108	288.9185	3.298799

In Table 3, each cluster has correlated features where cluster 2 represents the highest level of alkaline phosphatase, SGPT, SGOT and T. Bilirubin the patients in the cluster 2 with value of N 95(16%) have maximum chances of acquiring liver cirrhosis, whereas the patients relying in cluster 1 are in border line (N = 322, 55.2%) where the correlated features are above normal range of disease, however the cluster 3 is the patients not suffering (N = 166, 28.47%) from liver disorder and hence correlated features are in normal range as per specification for prognosis of disease.

Figure 2 represents the correlated feature with 95% of confidence interval with prognosis of liver cirrhosis. Each cluster with correlated features is represented where cluster 1 and cluster 2 represent are at higher risk of liver cirrhosis. The factors correlated tend to be important and if restricted to normal range can reduce prognosis

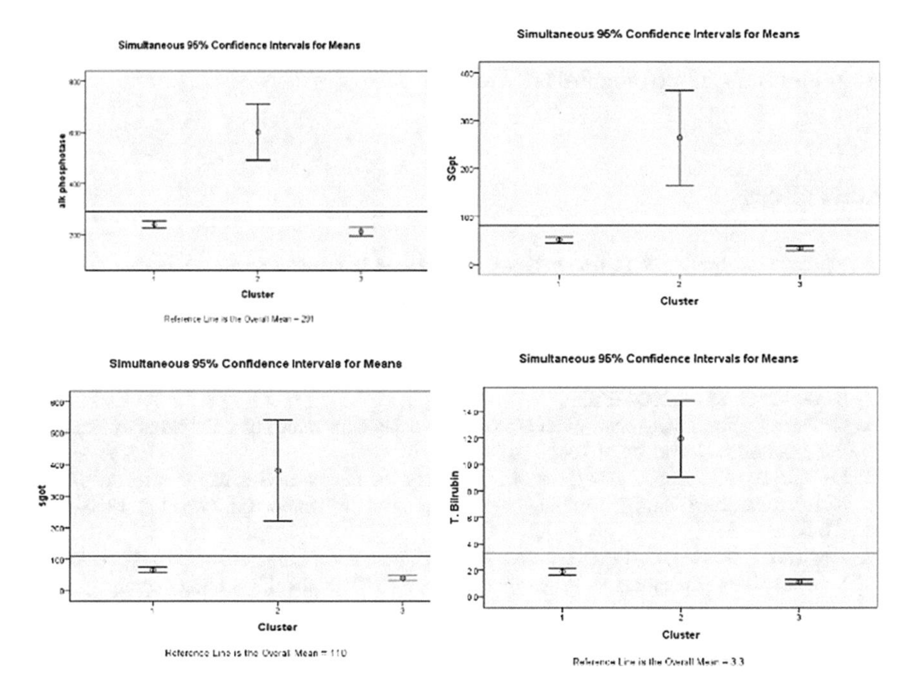

Fig. 2 Confidence interval of correlated features

of disease. The two-step clustering is an effective and efficient clustering technique to determine hidden and knowledgeable patterns from large-scale data. The results retrieved from two-step clustering can be utilized in varied application domain for future knowledge discovery of hidden patterns.

5 Conclusions

The medical data mining has constantly grown in past decade to study the exploratory behavior among the medical databases for future prognosis and detection of disease. The purpose of data mining is to discover hidden facts and knowledge from large-scale databases to facilitate researchers and scientists to identify patients risk at early stage of disease. The focus of our study is to determine effective and efficient patterns using predictive data analytics technique, i.e., two-step clustering for discovery of hidden patterns and knowledge from large-scale databases. The two-step clustering technique generated three clusters where each cluster represented correlated factors which can lead to liver disorder if analyzed at early stage and can benefit healthcare practitioners for future prognosis of disease.

References

1. Kiruba, H.R., Arasu, G.T.: An intelligent agent based framework for liver disorder diagnosis using artificial intelligence techniques. J. Theor. Appl. Inf. Technol. **69**(1) (2014)
2. Tsumoto: Problems with Mining Medical Data. 0-7695- 0792-1 I00@2000 IEEE
3. Aslandogan et. al. A.Y.: Evidence combination in medical data mining. In: Proceedings of the International Conference on Information Technology: Coding and Computing (ITCC'04) 0-7695-2108-8/04©2004 IEEE
4. Ordonez C.: Improving heart disease prediction using constrained association rules. Seminar Presentation at University of Tokyo (2004)
5. Le Duff, F., Munteanb, C., Cuggiaa, M., Mabob, P.: Predicting survival causes after out of hospital cardiac arrest using data mining method. Stud. Health Technol. Inf. **107**(Pt 2), 1256–1259 (2004)
6. Szymanski, B., Han, L., Embrechts, M, Ross, A., Zhu, K.L.: Using efficient supanova kernel for heart disease diagnosis. In: Proceedings of the ANNIE 06, Intelligent Engineering Systems Through Artificial Neural Networks, vol. 16, pp. 305–310 (2006)
7. https://archive.ics.uci.edu/ml/datasets/ILPD+(Indian+Liver+Patient+Dataset)
8. Vaziraniet H. et al.: Use of modular neural network for heart disease. In: Special Issue of IJCCT 2010 for International Conference (ACCTA-2010), Vol.1 Issue 2, 3, 4, pp. 88–93, 3–5 Aug 2010
9. Chaitrali, S., Dangare et. al.: Improved study of heart disease prediction system using data mining classification techniques. (IJCA) (0975–8887), **47**(10), 44–48 (2012)
10. Lin, R.H.: An Intelligent model for liver disease diagnosis. Artif. Intell. Med. **47**(1), 53–62 (2009)
11. Lee, H.G., Noh K.Y., Ryu K.H.: Mining biosignal data: coronary artery disease diagnosis using linear and nonlinear features of HRV. In: LNAI 4819: Emerging Technologies in Knowledge Discovery and Data Mining, May 2007, pp. 56–66
12. Guru, N., Dahiya A., Rajpal N.: Decision support system for heart disease diagnosis using neural network. Delhi Bus. Rev. **8**(1) (2007)
13. Wang, H.: Medical knowledge acquisition through data mining. In: Proceedings of 2008 IEEE International Symposium on IT in Medicine and Education 978-1-4244-2511-2/08©2008 Crown
14. Palaniappan, S., Awang, R.: Intelligent heart disease prediction system using data mining techniques. (IJCSNS) **8**(8) (2008)
15. Parthiban, L., Subramanian, R.: Intelligent heart disease prediction system using CANFIS and genetic algorithm. Int. J. Biol. Biomed. Med. Sci. **3**, 3 (2008)
16. Jin, H., Kim, S., Kim, J.: Decision factors on effective liver patient data prediction. Int. J. Bio-Sci. Bio-Technol. **6**(4), 167–168 (2014)
17. Chauhan, R., Jangade, R.: A robust model for big healthcare analytics. In: Confluence (2016)
18. Jangade, R., Chauhan, R.: Big data with integrated cloud computing for healthcare analytics. In: INDIACOM (2016)
19. Glymour, C., Madigan, D., Pregibon, D., Smyth, P.: Statistical inferenceand data mining. Commun. ACM **39**(11), 35–41 (1996)

20. Han, J.W., Yin, Y., Dong, G.: Efficient mining of partial periodic patterns in time series database. IEEE Trans. Knowl. Data Eng. (1998)
21. Kaur, H., Chauhan, R., Alam, M.A.: SPAGRID: A spatial grid framework for medical high dimensional databases. In: Proceedings of International Conference on Hybrid Artificial Intelligence Systems, HAIS 2012, springer, vol. 1, pp. 690–704 (2012)
22. Kaur, H., Chauhan, R., Aljunid, S.: Data Mining Cluster analysis on the influence of health factors in Casemix data. BMC J. Health Serv. Res. **12**(Suppl. 1), O3 (2012)
23. Chauhan, R., Kaur, H.: Predictive analytics and data mining: a framework for optimizing decisions with R tool. In: Tripathy, B.,: Acharjya, D. (eds.) Advances in Secure Computing, Internet Services, and Applications. Hershey, PA, Information Science Reference, pp. 73–88 (2014). https://doi.org/10.4018/978-1-4666-4940-8.ch004
24. Kaur, H., Chauhan, R., Alam, M.A.: Data clustering method for discovering clusters in spatial cancer databases. Int. J. Comput. Appl. **10**(6), 9–14 (2010)

Performance Analysis of a Truncated Top U-Slot Triangular Shape MSA for Broadband Applications

Anupam Vyas, P. K. Singhal, Satyendra Swarnkar and Toolika Srivastava

Abstract This attempt has been made to create a high bandwidth operational antenna by loading an U-slot on the triangular radiating patch. The novel configuration has been proposed for the broadband operation. The proposed antenna configuration has been simulated on IE3D and verified with experimental observations. The antenna characteristics like return loss, gain and efficiency and radiation pattern have been observed at various frequencies. This antenna is also providing resonance at the desirable frequency. This design is getting a high bandwidth antenna for lower microwave frequencies like 800–1300 MHz.

Keywords Bandwidth · U-slot · Broadbanding techniques

1 Introduction and Design Architecture

This is a truncated top and a triangular shape microstrip antenna, in this shape one limb is coming out from the base of it, a 4 mm by 15 mm limb is coming out, and right to it there is a slot of 4 mm by 10 mm which is etched out just to get better radiation and bandwidth. An U-slot which is one of the all-time best techniques to achieve higher bandwidth has been etched out to the upper side of the triangle, each side of the U-slot is 3 mm wide and 15 mm long, the corners of the U-slots are not a perfect

A. Vyas (✉) · T. Srivastava
Bundelkhand University Jhansi, Jhansi, UP, India
e-mail: anupam.vyas@rediffmail.com

T. Srivastava
e-mail: toolikka@gmail.com

P. K. Singhal
MITS Gwalior, Gwalior, Madhya Pradesh, India
e-mail: pks_65@yahoo.com

S. Swarnkar
SR Group of Institutions Jhansi, Jhansi, UP, India
e-mail: satya.dc07@gmail.com

K. Ray et al. (eds.), *Soft Computing: Theories and Applications*,
Advances in Intelligent Systems and Computing 742,
https://doi.org/10.1007/978-981-13-0589-4_41

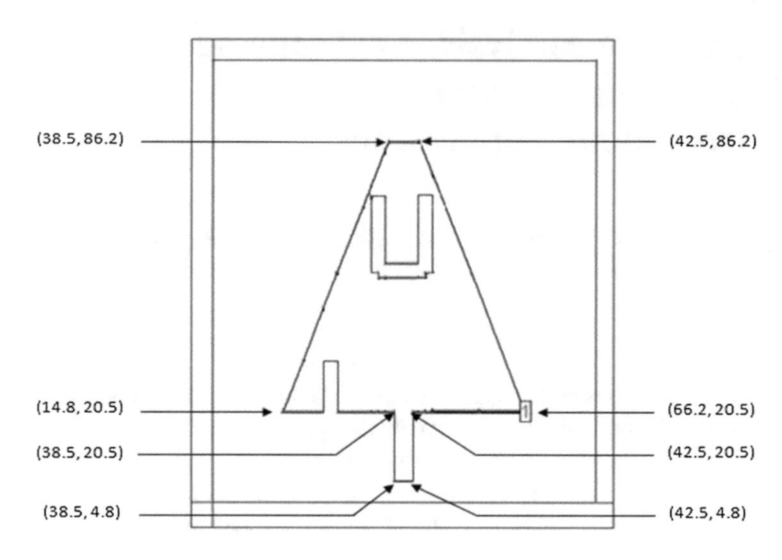

Fig. 1 Antenna design with coordinates labelling

turn they have staircase shape structure. The feed point is located at the leftmost corner of the shape, this is a single band high return loss antenna, the maximum value of return loss is −36 dB (simulated) and the frequency is 1.1 GHz. As far as the antenna architecture is concern, this configuration is novel and developed for the 1 GHz of application. Starting from point 'A' and moving in anti-clockwise direction the coordinates are (14.8, 20.5), (38.5, 20.5), (38.5, 4.8), (42.5, 4.8), (42.5, 20.5), (66.2, 20.5), (42.5, 86.2), (42.5, 96.2), (38.5, 96.2) and (38.5, 86.2).

2 Design Specifications

The parameters below show the design specifications of this novel configuration (Figs. 1, 2 and 3).

Figure 4 represents the return loss graph comparison of the software simulated results and the measured results of the novel configuration; there are two lines clearly visible on the graph: the black curve represents the return loss of the antenna which is produced by IE3D while the red curve represents the value of the return loss generated by the vector analyzer.

BW(S), i.e. Band Width (Simulated) for the second novel shape and BW(M), i.e. Band Width (Measured) can be seen across the −10 dB line on the graph. This is a single dip curve which shows that the proposed geometry is single band antenna which can be used for the applications of 1.02–1.51 GHz (Figs. 5, 6, 7, 8, 9 and 10).

Slot Dimensions	***(10 mm,3 mm), (3mm, 15 mm)*** ***(3 mm, 15 mm), (8 mm, 15 mm)*** ***(3 mm, 20 mm)***
Feed Location(X_0, Y_0)	***(66.2, 20.5)***
Broadbanding technique used	***U slotting, one narrow slot truncated top and radiating limb***

Fig. 2 Hardware design

Substrate material used	Glass epoxy
Relative dielectric constant	4.4
Thickness of the substrate	1.6 mm
Length of the patch	71.26 mm
Width of the patch	91.22 mm
Slot dimensions	(10 mm, 3 mm), (3 mm, 15 mm) (3 mm, 15 mm), (8 mm, 15 mm) (3 mm, 20 mm)
Feed location (X_0, Y_0)	(66.2, 20.5)
Broadbanding technique used	U slotting, one narrow slot truncated top and radiating limb

3 Broadbanding Techniques Used

3.1 Effects of U-Slot, Narrow Slot and Truncated Head

Out of all broadbanding techniques U-slot designing is the most effective technique to increase the bandwidth of the microstrip antenna, In this configuration also, an

Fig. 3 Hardware design with size comparison

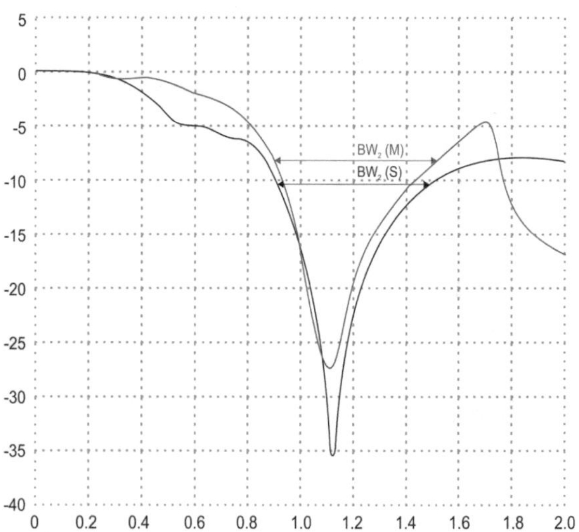

Fig. 4 Return loss graph comparison of software simulated results and measured results of microstrip antenna

U shape has been cut down in the middle, If examined closely we will find that basically U-slot is a collection of three narrow slots perpendicular consecutively with just overlapped edges and the current path disturbance produce by this U-slots enhances the performance of the antenna.

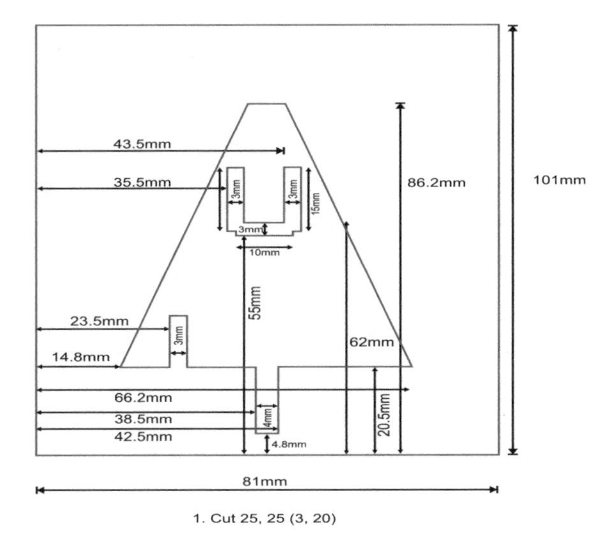

Fig. 5 Length labelled design architecture

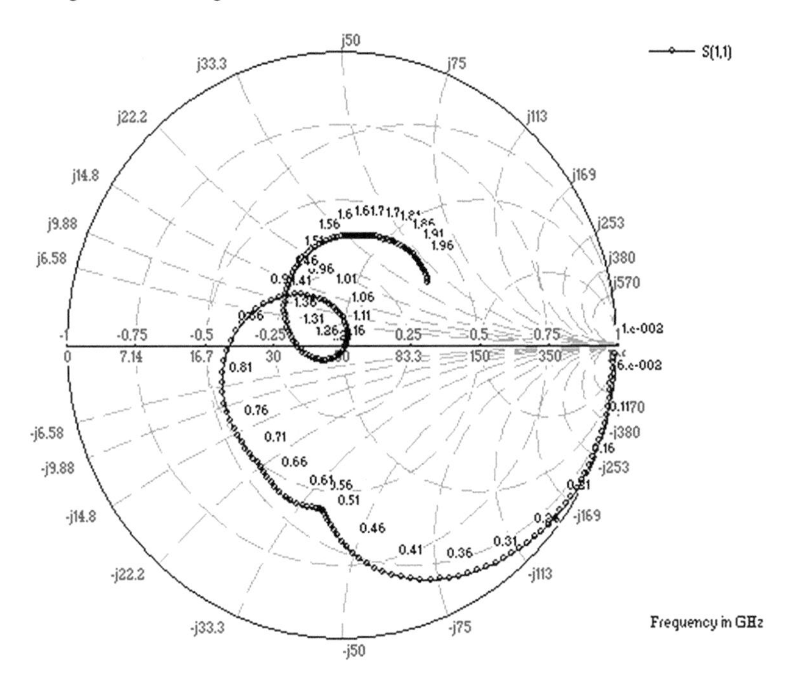

Fig. 6 Smith chart of proposed antenna

The head of the proposed geometry is also truncated with the help of a rectangular slot which also enhances the results as far as bandwidth is concern, and one more narrow slot has been dig out from the left bottom of the proposed novel geometry.

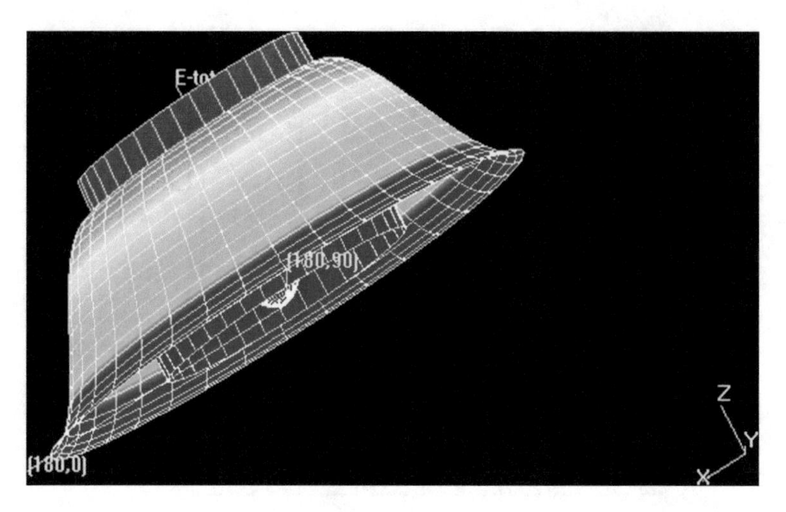

Fig. 7 Radiation pattern graph of proposed antenna

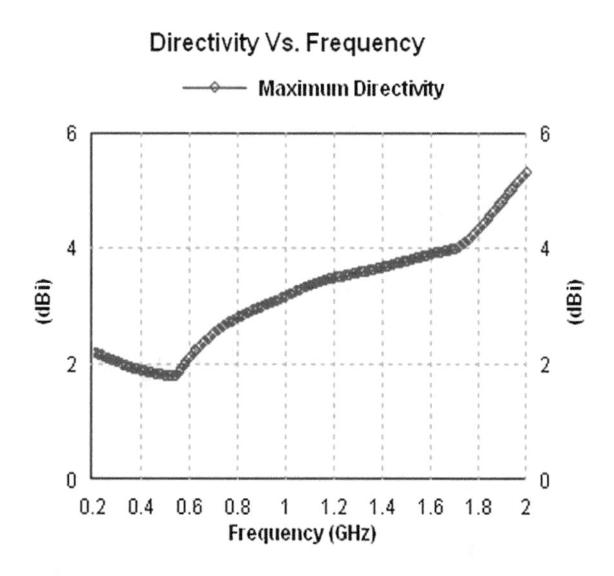

Fig. 8 Directivity versus frequency graph of proposed antenna

4 Parameters Discussion

Figure 4 shows the simulated S_{11} (dB) parameter of the novel shape, cut-off frequencies and the bandwidth are as follows:

Lower cut-off frequency (F_{ML})	0.91 GHz
Upper cut-off frequency (F_{MU})	1.51 GHz
Bandwidth of the novel shape	49.58%

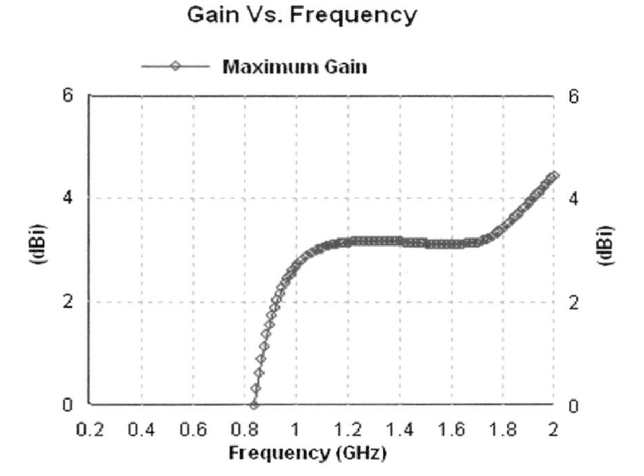

Fig. 9 Gain versus directivity graph of proposed antenna

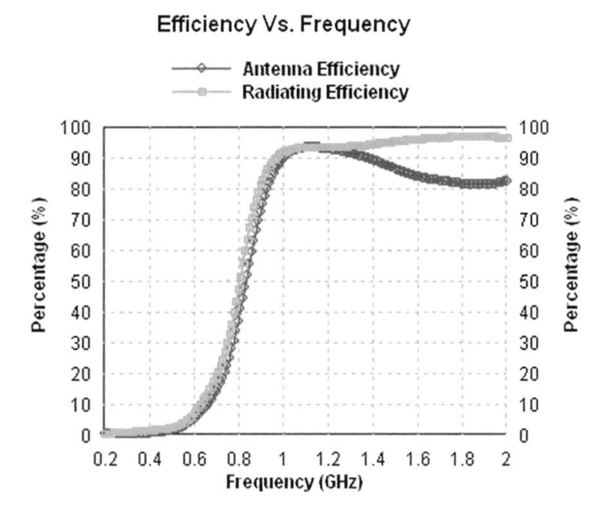

Fig. 10 Efficiency versus frequency of proposed antenna

Figure 4 shows the practically measured S_{11} (dB) parameter of the novel shape, cut-off frequencies and the bandwidth are as follows:

Lower cut-off frequency (F_{SL})	1.02 GHz
Upper cut-off frequency (F_{SU})	1.51 GHz
Bandwidth of the novel shape	34.8%

5 Conclusion

As mentioned in Sect. 4, the bandwidth found is 49.58 and 34.80% for simulated and measured results, respectively; as far as the application of this antenna is concern, this antenna is also a single band broadband antenna and can be used for broadband applications. There is a difference of 14.78% in the bandwidth of the antenna which is because of many reasons and mainly because the software simulates the antenna in an ideal environment of the transmission but the lab conditions are non-ideal and the software considers all junctions as perfect with no losses, which was not possible practically resulting in degradation.

References

1. Deschamps, G.A.: Microstrip Microwave Antennas. In: Presented at the 3rd USAF Symposium on Antennas (1953)
2. Gulton, I.L., Bassinot, G.: Flat aerial for ultra high frequencies. French Patent No. 703113 (1955)
3. Lewin, L.: Radiation from discontinuities in strip lines. Proc. IEEE **107**, 163–170 (1960)
4. Munson, R.: Single slot cavity antennas assembly. US Patent No. 3713162 (1973)
5. Munson, R.: Conformal microstrip antennas and microstrip phased arrays. IEEE Trans. Antennas Propag. **22**, 74–78 (1974)
6. Munson, R.: Conformal microstrip antennas and microstrip phased arrays. IEEE Trans. Antennas Propag. **22**, 74–78 (1974)
7. Rowe, W.S.T., Waterhouse, R.B.: Broadband microstrip patch antenna for MMICs. Electron. Lett. **36**, 597–599 (2000)
8. Hsu, W.H., Wong, K.L.: A wideband circular patch antenna. Microwave Opt. Technol. Lett. **25**, 327–328 (2000)
9. Carver, K.R., Mink, J.W.: Microstrip antenna technology. IEEE Trans. Antennas Propag. **29**(1), 2–14 (1981)
10. Richards, W.F., Lo, Y.T., Harrison, D.D.: An improved theory for microstrip antennas and applications. IEEE Trans. Antennas Propag. **AP-29**, 38–46 (1981)
11. James, J.R., Hall, P.S., Wood, C.: Microstrip Antenna Theory and Design. Peter Peregrinus, Stevenage, UK (1981)
12. Mok, W.C., Wong, S.H., Luk, K.M., Lee, K.F.: Single-layer single-patch dual-band and triple-band patch antennas. IEEE Trans. Antennas Propag. **61**, 4341–4344 (2013)
13. Sabban, A.: New broadband stacked two-layer microstrip antenna. In: Proceedings of the IEEE AP-Symposium Digest, pp. 63–66 (1983)
14. Mosig, J.R., Gardiol, F.E.: General integral equation formulation for microstrip antennas and scatters. IEEE Proc. **132**(7), pt. H, 424–432 (1985)
15. Pozar, D.M.: Microstrip antenna aperture-coupled to a microstripline. Electron. Lett. **21**, 49–50 (1985)
16. Lee, R.Q., Lee, K.F., Bobinchak, J.: Characteristics of a two-layer electromagnetically coupled rectangular patch antenna. Electron. Lett. **23**(20), 1070–1073 (1987)

17. Pinhas, S., Shtrikman, S.: Comparison between computed and measured bandwidth of quarter-wave microstrip radiators. IEEE Trans. Antennas Propag. **AP-36**(11), 1615–1616 (1988)
18. James, J.R., Hall, P.S.: Handbook of Microstrip Antennas. Stevenage. Peter Peregrinus, UK (1989)
19. Reineix, A., Jecko, B.: Analysis of microstrip patch antennas using the finite difference time domain method. IEEE Trans. Antennas Propag. **37**(11), 1361–1369 (1989)
20. Bahl, I.J., Bhartia, P.: Microstrip Antennas. Artech House, Dedham, MA (1980)

Automated Indian Vehicle Number Plate Detection

Saurabh Shah, Nikita Rathod, Paramjeet Kaur Saini, Vivek Patel, Heet Rajput and Prerak Sheth

Abstract Many countries have standardized vehicle license plates and the constraints of plates like font, size, color, spacing between characters, and number of lines are strictly maintained. Even though standards for number plate are being decided by the Indian government and the process of standardization is in process, it can be seen that every 8 out of 10 vehicles have a variation from that of the standard number plate in terms of either location of plate, fonts used in the plate or various text and designs on the vehicle as well as on plate. Because of these variations, it has been challenging to develop an automated vehicle number plate detection system which can localize number plates or fonts of number plate correctly. This paper presents novel experiments at all the important phases of number plate detection like preprocessing, number plate localization, number plate extraction, segmentation, and character recognition at last. Proposed methodology detects number plate and the characters with high accuracy of 98.75% where neural network has been used for character recognition.

Keywords Automatic number plate recognition · Optical character recognition
Neural network

S. Shah (✉) · N. Rathod · P. K. Saini · V. Patel · H. Rajput
Department of Computer Science and Engineering, Babaria Institute of Technology, BITS Edu Campus, Vadodara, Gujarat, India
e-mail: saurabhshah.ce@bitseducampus.ac.in

N. Rathod
e-mail: nikitarathod.d@gmail.com

P. K. Saini
e-mail: sainiparamjeetkaur@yahoo.com

V. Patel
e-mail: vivek.patel.95.vp@gmail.com

H. Rajput
e-mail: heetrajput1913@gmail.com

P. Sheth
Suyojan Systems, Vadodara, Gujart, India
e-mail: prerak.sheth@gmail.com

K. Ray et al. (eds.), *Soft Computing: Theories and Applications*,
Advances in Intelligent Systems and Computing 742,
https://doi.org/10.1007/978-981-13-0589-4_42

1 Introduction

With the increase of number of vehicles on the road, safety and security have been major concerns of any country. To monitor the vehicles manually based on the installed CCTV, it is most challenging and time-consuming [1, 2]. Also, because of low illumination light, vehicle speed, nonuniform number plates, and variations in fonts in Indian number plates, it is difficult to recognize the number place accurately [2–6]. The data received by digitally processing the vehicle number plate images have applications in tolling systems, vehicle tracking, speed detection, theft detection, traffic analysis, parking system, border crossings [3, 7–11].

Automatic vehicle number plate detection has been aimed to develop a system which can localize the number plate and extract the number from the Indian number plates using image processing and machine learning technique. With various preprocessing and segmentation techniques, number plate from a real-time image of vehicle is being recognized. The extracted number plate is passed to the Optical Character Recognition (OCR) system where classification and recognition of numbers and letters have been done by training the neural network. Experiments have been carried on large set of Indian vehicle number plates with verities of the size and fonts to develop a robust generalized model which can recognize almost any kind of Indian number plate.

2 Proposed Methodology

See Fig. 1.

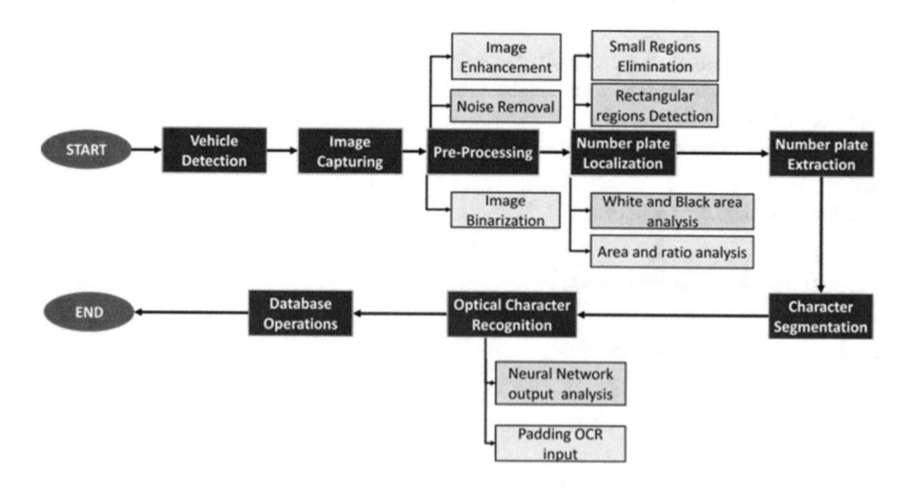

Fig. 1 Block diagram of the number plate detection and recognition figure

2.1 Preprocessing

The captured image of vehicle is given to preprocessing module. The preprocessing module is responsible for enhancing image. Our main objective is to detect number plate from the image. So, only binary form of image is required as on binarization; the number plate area will be white in color. Preprocessing includes following procedure:

RGB to grayscale conversion: The original image which is in RGB format has been converted to grayscale image as shown in Fig. 2.
Histogram equalization: Image contrast has been enhanced using histogram equalization.
Converting image to binary form: The enhanced image is then converted to binary image.
Edge detection: Edge detection has been done using Sobel edge detector as shown in Fig. 3.

Fig. 2 Grayscale image

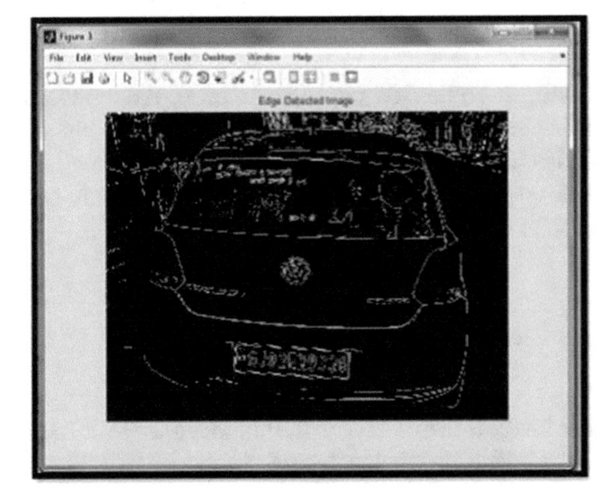

Fig. 3 Edges detected

Fig. 4 Image after applying dilation

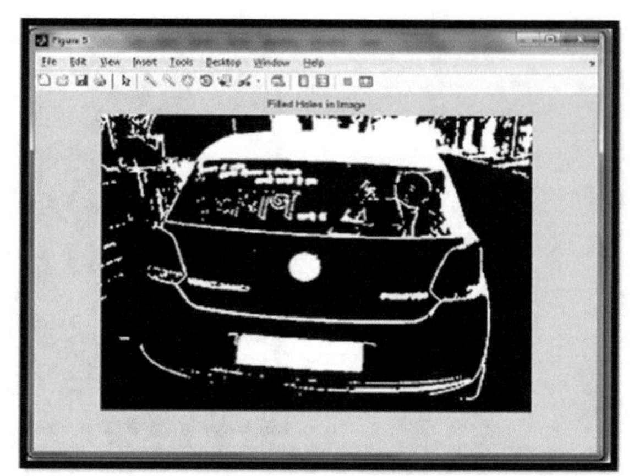

Fig. 5 Image after holes filling

Dilation: To thicken the edges to make the number plate more prominent, dilation has been performed on the image as shown in Fig. 4.
Holes filling: To remove the noise and very small areas, holes filling is performed. This step has decreased the number of regions to be analyzed for detecting number plate at the later stage as shown in Fig. 5.

2.2 *Number Plate Localization*

The image is then passed to the number plate localization module. The regions present in the image are labeled using 8 component neighborhood. This labeling information is used to create bounding box around the regions. Every bounding box is analyzed on the basis of following constraints:

Area: Area of all the Bounding boxes is being calculated and their overall average is found. The bounding boxes having area less than average are eliminated. This removes smaller noise regions.
Height to width ratio: The height to width ratio of Indian number plate is around 1:3. So, the bounding box having ratio 1:3 are considered and those having ratio different than 1:3 are eliminated.
Number of white and black pixels: The number of black and white pixels present in each bounding box is calculated and the bounding boxes having more number of black pixels compared to white pixels are eliminated.

The Bounding box satisfying these three constraints contains the location of possible number plate. The output of this module will give coordinates of the bounding box. These coordinates are passed to number plate extraction module.

2.3 *Number Plate Extraction*

The coordinates obtained from the number plate localization module are mapped with the original RGB image of the vehicle and the number plate's image is cropped as shown in Fig. 6.

Fig. 6 Extracted number plate

2.4 Segmentation

The segmentation module identifies the character regions of number plate, i.e., alphabets and digits and segments the characters. The RGB number plate image obtained from number plate extraction module has been binarized by using threshold value. Then, very small regions having area less than 50 pixels are removed. Labeling regions using 8-component neighborhood, bounding boxes are formed around regions.

The bounding boxes around characters of the number plate have almost same height with difference of few pixels as per the observation. Thus, bounding boxes having height of variation more than 10 pixels have been eliminated as shown in Figs. 7 and 8.

Fig. 7 Segmented character regions

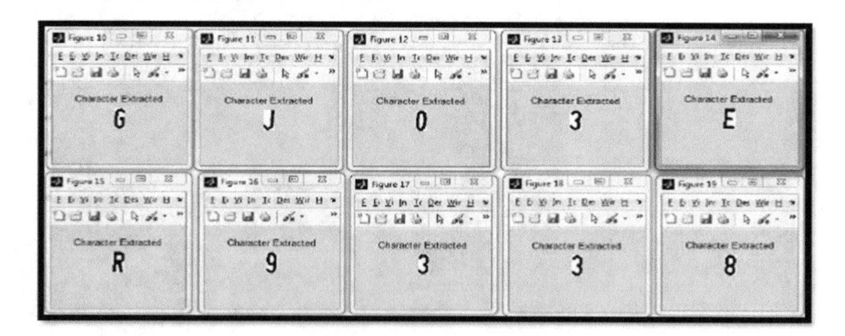

Fig. 8 Individual segmented characters

2.5 Character Recognition Using Neural Network

The segments are passed to optical character recognition module. The optical character recognition converts the segments into their correct digital form. In this module, the neural network has been used to train the character sets obtained from the standard dataset as shown in Fig. 9 and then classify the extracted characters from the number plate once the network is trained.

The neural network which has been used in our experiment is two-layer feedforward network with sigmoid function. The network is trained using dataset of different fonts by back propagation method. A variety of fonts are chosen for input dataset to increase the efficiency of optical character recognition module and to overcome the challenge of recognizing number plates having fonts different from standard number plates. The target set is a matrix of 36 classes of which 10 classes are for digits and 26 for alphabets as shown in Table 1.

The segments are converted to binary form. The segment is padded with a thin layer of white pixels. Then, they are resized according to the size of inputs given to the neural network. After resizing, the segment is converted to a row vector. The vector is then passed to the network and the output is stored in a variable. The network gives 36 values which show the probability of the segment belonging to that specific class.

Fig. 9 Dataset of characters

Table 1 Neural network structure and its parameters

Parameters	Values
Size of input data matrix	7200×400
Size of target data matrix	7200×36
Number of neurons at input layer	900
Number of neurons at hidden layer	300
Number of neurons at output layer	36

Table 2 Experimental Results

Number of car images experimented	Number of challenging images experimented	Number of correctly detected number plate	Accuracy
80	10	79	98.75%

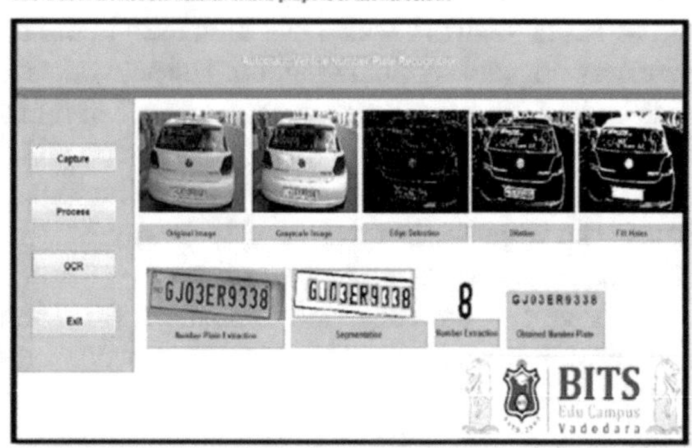

Fig. 10 Implementation of proposed methodology

Based on the maximum probability, the index is retrieved and the class is identified. Base on the class, character present in the segment is identified.

One by one, each segment is passed to the optical character recognition module and the process is repeated until all the characters are identified which results in vehicle number from the number plate. As per the experiments, the majority of the characters have been recognized correctly with the accuracy of 98.75% as shown in Table 2.

Step by Step implementation of the proposed methodology is depicted in Fig. 10.

3 Conclusion

The proposed method has been tested on varieties of the Indian vehicle number plates with different size, fonts, and color. All the experiments were performed in Matlab (version 2012a) environment and achieved high accuracy of 98.75% in recognizing the characters. It was challenging to recognize the characters from the image captured in low ambient light. Many applications could be developed based on the method proposed like tolling system, detection of vehicle thefts, and illegal number plates, creating national and generalized database for Indian vehicles.

Acknowledgements We would like to thank our mentors, Dr. Saurabh Shah (Dean, Institute of Technology, BITS Edu Campus) and Mr. Prerak Sheth (Proprietor, Suyojan Systems, Vadodara) for their constant valuable support in conceptualizing the idea and taking it further to the level of research project. Because of their constant efforts and guidance, we could achieve important milestone of completing the final year project as a product.

References

1. Sarfraz, M.Saquib, et al.: Real-time automatic license plate recognition for CCTV forensic applications. J. Real-Time Image Proc. **8**(3), 285–295 (2013)
2. Azad, R., Shayegh, H.R.: New method for optimization of license plate recognition system with use of edge detection and connected component. In: 2013 3th International eConference on Computer and Knowledge Engineering (ICCKE). IEEE (2013)
3. Chang, J.-K., Ryoo, S., Lim, H.: Real-time vehicle tracking mechanism with license plate recognition from road images. J. Supercomput. 1–12 (2013)
4. Karwal, H., and Akshay, G.: Vehicle number plate detection system for indian vehicles. In: 2015 IEEE International Conference on Computational Intelligence & Communication Technology (CICT). IEEE (2015)
5. Anagnostopoulos, Christos-Nikolaos E.: License plate recognition: a brief tutorial. IEEE Intell. Transp. Syst. Mag. **6**(1), 59–67 (2014)
6. Sulaiman, N. et al.: Development of automatic vehicle plate detection system. In: 2013 IEEE 3rd International Conference on System Engineering and Technology (ICSET). IEEE (2013)
7. Patel, C., Dipti, S., Atul, P.: Automatic number plate recognition system (anpr): a survey. Int. J. Comput. Appl. **69**(9) (2013)
8. Rasheed, S., Asad N., Omer I.: Automated Number Plate Recognition using hough lines and template matching. In: Proceedings of the World Congress on Engineering and Computer Science, Vol. 1 (2012)
9. Salgado, L. et al.: Automatic car plate detection and recognition through intelligent vision engineering. In: 1999 Proceedings of the IEEE 33rd Annual 1999 International Carnahan Conference onSecurity Technology. IEEE (1999)
10. Shukla, Dolley, Patel, Ekta: Speed determination of moving vehicles using Lucas-Kanade algorithm. Int. J. Comput. Appl. Technol. Res. **2**(1), 32–36 (2013)
11. Chhaniyara, S., et al.: Optical flow algorithm for velocity estimation of ground vehicles: a feasibility study. Int. J. Smart Sens. Intell. Syst. **1.1,** 246–268 (2008)

Named Data Network Using Trust Function for Securing Vehicular Ad Hoc Network

Vaishali Jain, Rajendra Singh Kushwah and Ranjeet Singh Tomar

Abstract Vehicular ad hoc network is based on wireless technology for providing communication among vehicles. Vehicles interact by the wireless channel which makes it vulnerable to various attacks. But security is the major concern in the Vehicular Ad hoc Network (VANET) as it is the concern of the life of an individual. In this paper, Named Data Networking (NDN) for forwarding the data packets in network is discussed. In this paper, security of the network by calculating the trust value of every node by interacting them is being done. The results show that the proposed technique has shown significant improvements in the result than previous geo-based location forwarding technique. Comparative results are shows the network improved in term of throughput, an end to end delay, good put, and PDR.

Keywords Vehicular ad hoc network (VANET) · Named data networking
Packet delivery ratio

1 Introduction

VANET is architecture which is data-centric. It has moving nodes as vehicles. Though VANET was proposed initially for only driving safety and some of the informative applications can be supported to drivers and passengers such as advertisements, traffic queries, etc. Some applications also tell about position related data which comes under geo-based routing. Although VANET has many challenges such as frequent dynamic topology and short-lived connectivity. To solve this issue NDN

V. Jain (✉) · R. S. Kushwah
Institute of Technology & Management, Gwalior, India
e-mail: vaishalijain783@gmail.com

R. S. Kushwah
e-mail: rajendrasingh.ind@rediffmail.com

R. S. Tomar
ITM University Gwalior, Gwalior, India
e-mail: er.ranjeetsingh@gmail.com

K. Ray et al. (eds.), *Soft Computing: Theories and Applications*,
Advances in Intelligent Systems and Computing 742,
https://doi.org/10.1007/978-981-13-0589-4_43

is proposed that replaces TCP/IP communication. Vehicles that used TCP/IP communicate with the centralized server through cellular networks. Many applications in VANET needed some new techniques for direct V2V communication in real-time traffic. This can be done by a new technique named NDN. Existing work shows that NDN is able to provide better support and performance improvement in data-centric applications like video conferencing; video streaming. Data dissemination is done through named data. NDN uses hourglass shape IP architecture. NDN exchanges information on the basis of interest and data packets. In the place of source and destination address packets carries data names. These data names make the communication easy. Interest and data packets go through routers; the router then forwards interest packets according to data names. The router has Pending Interest Table (PIT), Forwarding Information Base (FIB), and a content store section.

2 Named Data Networking

NDN achieves data delivery through named data instead of using host-based end to end communication. Thus it is called as named data networking. In NDN hourglass shape architecture [1] is used which is given below:

An NDN router has to maintain three important data structures—Pending Interest Table (PIT), Forwarding Interest Base (FIB), and Content Store (CS) [2].

- PIT contains the list of unsatisfied interests received by router.
- FIB is same as IP routing table that stores information about where the matching name prefixes of interests should be forwarded.
- CS stores only that data in cache which is forwarded by router.

3 Geo-Based Forwarding

This technique allows the vehicles to deliver packets in an efficient manner. NDN first disseminates interest packets all over and then data packets follow the reverse path back. Thus, it is very important to forward interest packets. By applying the combination of data naming scheme and geo-locations mechanism every vehicle became able to know the position of data source for any particular incoming interest packet, for this efficient forwarding of interest next hop should be determined. In some cases, neighbor information is not up to date and wireless communication is not reliable. That is why selected vehicles are not able to receive forwarded interest. Thus to make sure improvement in reliable data delivery, any vehicle that receives interests, should take the decision whether to forward the interest or not [3].

A multi-hop forwarding issue which is caused by frequently dynamic topology and short-lived connectivity. In wireless network, data is flooded because of the absence of Forwarding Information Base (FIB). For solving this problem, HVNDN is proposed which is a hybrid forwarding strategy for location-dependent and location-independent information. This technique shows a reduction in end to end delay [4, 5].

4 Literature Review

LFU (Least Frequently Used) and LRU (Least Recently Used) are the two caching policies used at router cache. LFU [6] keeps numbers of frequently used items at router. It catches access frequency, whereas LRU [7] keeps the details of number of recently used items. P-TAC (Push-based Traffic Aware Cache) uses links to improve cache hit rate by using those links which have the margin in transmission band. This achieves a reduction in traffic. This method has shown high performance than other methods. A CONET scheme is proposed which forwards less CDMP (Copies of Data Messages Processed) when achieving related Interest Satisfaction Rate (ISR) as the basic vehicular NDN. CONET has also reduced the overall Interest Satisfaction Delay (ISD), respectively. As traditional ad hoc networks, vehicular NDN faces many challenges too such as *Interest/Data* broadcast storm, consumer/provider movements, and so on. Previous works did not give proper dealing with data flooding. This CONET solved data flooding issue by using hop count h in interest packets and TTL into data message [8]. For maximizing information [9] in NDN hierarchical names are provided to data, thus NDN allows the networking devices to understand relations between content. For determining hot content knapsack problem is formulated in it. These are utility maximizing items. For getting better content retrieval performance, the behavior of CCN nodes for VANET is adapted. For selecting node, an algorithm is proposed which is based on minimum vertex cover set theory. This paper has also proposed CCBSR [10]. This paper proposes a novel vehicular information architecture which is based on NDN. This has improved the result in the terms content naming, addressing data aggregation, etc. [11]. This paper proposes the analysis that NDN systems detect those attacks which are vulnerable to IP-based systems. This paper detects some problems related to timestamps, large packet size processing. There is also a lack of secure key mechanism. This research work has also devised replay command attack. This work has also conducted an attack against link layer protocols. This also was unsuccessful [12]. This paper describes the initial design of V-NDN. Its implementation, challenges, and issues are discussed [13].

5 Proposed Work

The proposed work shows a secure scenario with the help of trust function. An algorithm explains the working of proposed for vehicular communication in a secure manner. Trust function is applied to the vehicular communication for providing security. If vehicles are interested to receive packet, then trust of vehicles is checked, and also how many times interest is shown by vehicles to forward list is checked. If they deny for accepting forward list then check forward list is empty or no neighbors of those vehicles are present. To find out neighbors to forward data we use how many vehicles appears in our route.

5.1 Proposed Flowchart

See Fig. 1.

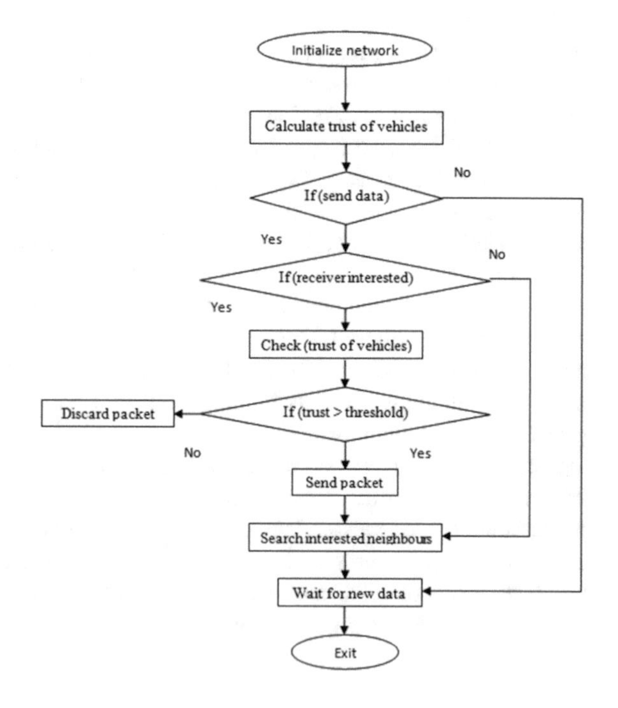

Fig. 1 Proposed flowchart

5.2 Proposed Algorithm

```
Step1:  initialize network
Step2:  calculate trust of vehicles on the basis of behavior
Step3:  if (send data) {
          If (receiver interested) {
            Check (trust of vehicles)
              If (trust > thresh hold) {
              Send packet
              Else
              Discard packet
          Else
          Search interested neighbors
        Else
        Wait for new data
Step4:  exit.
```

6 Simulation and Results

This work is done on NS-2 simulator which is a discrete open source network simulator. It is widely used by the organization and by the people who are involved in standardization of IETF protocol. It can be said as a verification model for network environment. Results show that the proposed technique, interest-based location forwarding technique, has better results than geo-based forwarding technique in terms PDR, end to end delay, throughput, and good put. Results of geo-based forwarding techniques and NDN using trust function have been compared to the given parameters in the following graph.

6.1 PDR Comparison

In the graph, the x-axis shows time and the y-axis shows PDR output. PDR varies with respect to time. At initial moments of time PDR results of both algorithms are shown, the highest peak at the initial moment of time but NDN using trust has got more than 6.000 as compared to geo-based forwarding that maximum value is less than 5.500. As the time passes, NDN using trust technique has higher value (more than 2.5) than geo-based forwarding (less than 2.5) (Fig. 2).

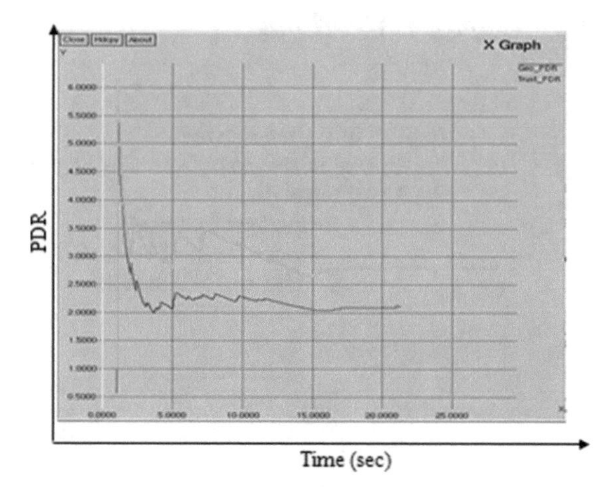

Fig. 2 PDR graph

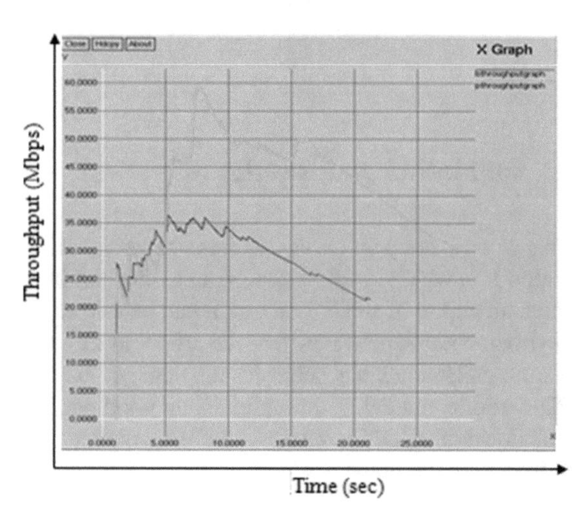

Fig. 3 Throughput graph

6.2 Throughput Comparison

The x-axis shows time and the y-axis shows throughput. The graph shows that geo-based forwarding technique has got values of throughput more than 35.000 whereas proposed technique has achieved values 60.000 which is approximately double of throughput of geo-based location forwarding. Thus, it is showing that the number of packets sent has been increased till a very high value in this proposed technique (Fig. 3).

Fig. 4 Good put graph

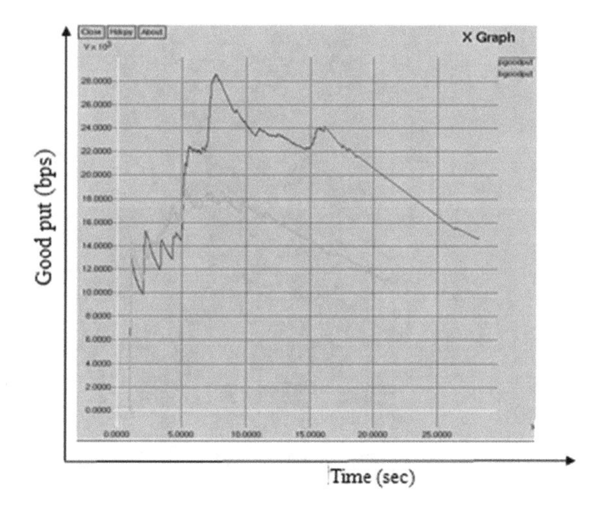

6.3 Good Put Comparison

In the given graph, good put is compared between two approaches: geo-based forwarding approach and NDN using trust. The x-axis shows time variable and the y-axis shows good put parameter (Fig. 4).

Given graph is showing that geo-based forwarding technique has been achieved till the limit of more than 18.000 value whereas the proposed technique has got the values more than 28.000. This difference is showing that proposed technique is much better in terms of good put than geo-based forwarding technique.

6.4 End to End Delay Comparison

The x-axis shows time variable and the y-axis shows end to end delay parameter. This graph is showing that geo-based forwarding technique has reached up to 2.800 which is very high delay. But when NDN using trust technique is observed than is observed that end to end delay has remained till 0.8. This decrease in delay has shown that proposed technique has been outperformed the geo-based forwarding (Fig. 5).

6.5 Results Comparison

Table 1 shows the comparative results of the geo-based forwarding strategy and named data network using trust value-based strategy with reference to the PDR, good put, end to end delay, and throughput values.

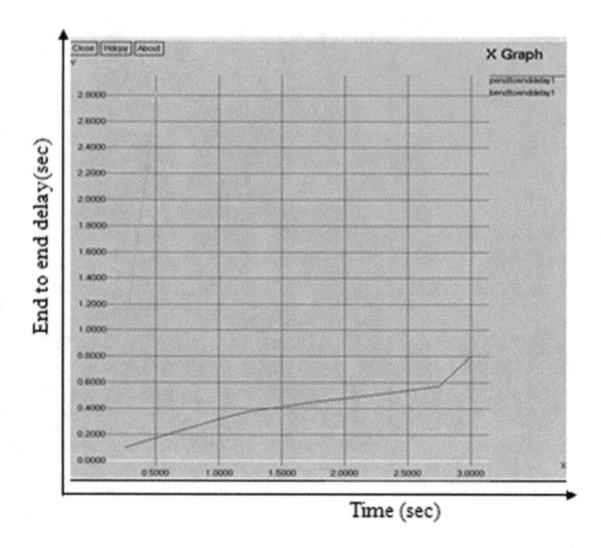

Fig. 5 End to end delay graph

Table 1 Comparison of results

Parameters	Geo-based forwarding strategy (max. value)	NDN using trust (max. value)
PDR	5.4000	6.1000
Good put	18.000	28.000
End to end delay	2.8000	0.8000
Throughput	35.000	60.000

7 Conclusion

In VANET, vehicles communicate to other vehicles by sending warning messages to indicate them about traffic congestion and roads condition. Security is the major issue to make them trusted nodes and forward data only to the trustful nodes. The work shows a secure scenario with the help of trust function. This function is applied to the vehicles for providing security. Trust function is used in named data network to calculate PDR, throughput, an end to end delay, and good put. Improvements in results are being shown in the table. But results showing higher delay in communication can be considered for future work.

References

1. Verma, G., Nandewal, A.K., and Chandrasekaran. Cluster Based Routing in NDN. In: 12th IEEE International Conference on Information Technology-New Generations, pp. 296–301. (2015)
2. Bian, C., Zhao, T., Li, X., Yan, W.: Boosting named data networking for data dissemination in urban VANET scenarios. Elsevier Veh. Commun. **2**, 195–207 (2015)
3. Micheal, M., Santha Soniya and K. Kumar. A Survey on Named Data Networking. In: 2nd IEEE Sponsored International Conference on Electronics and Communication System (ICECS), pp. 1515–1519 (2015)
4. Deng, G., Xie, X., Shi, L., Li, R.: Hybrid information forwarding in VANETs through named data networking. In: 26th IEEE International Symposium on Personal, Indoor and Mobile Radio Communications- (PIMRC): Mobile and Wireless Networks, pp. 1940–1944 (2015)
5. Li, Z., Simon, G., Gravey, A.: Caching policies for in-network caching. In: Proceedings of the 21st International Conference on Computer Communications and Networks (ICCCN), pp. 1–7 (2012)
6. Arlitt, M., Cherkasova, L., Dilley, J., Friedrich, R., Jin, T.: Evaluating content management techniques for web proxy caches. ACM SIGMETRICS Perform. Eval. Rev. **27**(4), 3–11 (2000)
7. Ahmed, S.H., Bouk, S.H., Yaqub, M.A., Kim, D., Gerla, M.: CONET: controlled data packet propagation in vehicular named data networks. In: 13th IEEE Annual Consumer Communications & Networking Conference (CCNC), pp. 620–625 (2016)
8. Dron, W., Leung, A., Uddin, M., Wang, S., Abdelzaber, T., Govindan, R.: Information-maximizing Caching in Ad Hoc networks with named data networking. In: 2nd IEEE Network Science Workshop, pp. 90–93 (2013)
9. Liu, L.C., Xie, D., Wang, S., Zhang, Z.: CCN-based cooperative caching in VANET. In: International Conference on Connected Vehicles and Expo (ICCVE), pp. 198–203 (2015)
10. Yan, Z., Zeadally, S., Park, Y.-J.: A Novel vehicular information network architecture based on named data networking (NDN). Internet Things J. **1**(6), 525–532 (2014)
11. Perez, V., Garip, M.T., Lam, S., Zhang, L.: Security evaluation of a control system using named data networking. In: 21st IEEE International Conference on Network Protocols (ICNP), pp. 1–6 (2013)
12. Grassi, G., Pesavento, D., Pau, G., Vuyyuru, R., Wakikawa, R., Zhang, L.: VANET via Named Data Networking. In: IEEE Conference on Computer Communication, (INFOCOM), pp 410–415 (2014)
13. Tiwari, P., Kushwah, R.S.: Enhancement of VANET communication range using WiMAX and Wi-Fi: a survey. Int. J. Urb Des. Ubiquitous Comput. **1**(1), 11–18 (2013)

An Intelligent Video Surveillance System for Anomaly Detection in Home Environment Using a Depth Camera

Kishanprasad Gunale and Prachi Mukherji

Abstract In recent years, the research on the anomaly detection has been rapidly increasing. The researchers were worked on different anomalies in videos. This work focuses on fall as an anomaly as it is an emerging research topic with application in elderly safety areas including home environment. The older population staying alone at home is prone to various accidental events including falls which may lead to multiple harmful consequences even death. Thus, it is imperative to develop a robust solution to avoid this problem. This can be done with the help of video surveillance along with computer vision. In this paper, a simple yet efficient technique to detect fall with the help of inexpensive depth camera was presented. Frame differencing method was applied for background subtraction. Various features including orientation angle, aspect ratio, silhouette features, and motion history image (MHI) were extracted for fall characterization. The training and testing were successfully implemented using SVM and SGD classifiers. It was observed that SGD classifier gives better fall detection accuracy than the SVM classifier in both training and testing phase for SDU fall dataset.

Keywords Computer vision · Fall detection · Feature extraction · SGD classifiers · SVM · Video surveillance

1 Introduction

The falls are significant issues of concern for elderly staying alone. It may lead to injuries, fractures, and other chronic health problem. According to the World Health Organization (WHO) reports [1], the percentage of people who are 65 and older

K. Gunale (✉)
Department of E&TC, SCOE, SPPU, Pune, India
e-mail: kgunale@rediffmail.com

P. Mukherji
Department of E&TC, Cummins College of Engineering, SPPU, Pune, India
e-mail: prachi.mukherji@cumminscollege.in

K. Ray et al. (eds.), *Soft Computing: Theories and Applications*,
Advances in Intelligent Systems and Computing 742,
https://doi.org/10.1007/978-981-13-0589-4_44

falling each year is 28–35%, and the values go from 32 to 42% for people of age above 70. Different healthcare model for aging population is needed for promptly detecting fall event and monitoring daily activities.

There are different device-based solutions for the detection of fall [2, 3], including those tools which use wearable accelerometers along with gyroscopes embedded in garments [4, 5]. The velocity of body movement or its parts is deliberated with the help of accelerometers, and negative acceleration indicates a fall event. Even though the wearable sensor-based method is not sensitive to environmental changes yet wearing it regularly and for an extended period of time causes inconvenience to once day to day activity schedule.

The smart home solutions [6, 7] were developed to overcome the problem of carrying the sensor every time with the user. In this technique, ambient devices installed including vibration sensors; sound sensors, infrared motion detectors, and pressure sensors to record daily activities at multiple positions of room. By merging information taken from all these sensors, fall can be recognized and alarmed with a high accuracy [8–11]. With the use of using a number of sensors the system cost increases. Moreover, it is cumbersome to have the presence of more than one sensor in a room. So the development of a solution for elderly person monitoring for improvement in the quality of life without significant changes in daily habits of an elderly person is necessary.

With the advancement in computer vision technology, it is now feasible to create a solution to this trouble of fall detection with the help inexpensive depth camera. Vision-dependent fall detection represents a fall action with the help of characteristic feature appearance and a classification prototype to discriminate fall from other daily activities. Despite various experiments for vision-dependent fall detection systems, it still remains a matter of concern.

In the literature, there are various systems that address the problem of fall detection using the vision-based approach. The basic concept behind every vision-based solutions includes the recognition of features like centroid, velocity, aspect ratio, motion features as well as shape features that describe the fall event.

The proposed paper uses SDU fall dataset for automatic fall detection using SVM and SGD classifiers applied to the detected features.

2 Depth-Based Human Fall Detection System

The basic flow diagram of a fall detection system is shown in Fig. 1.

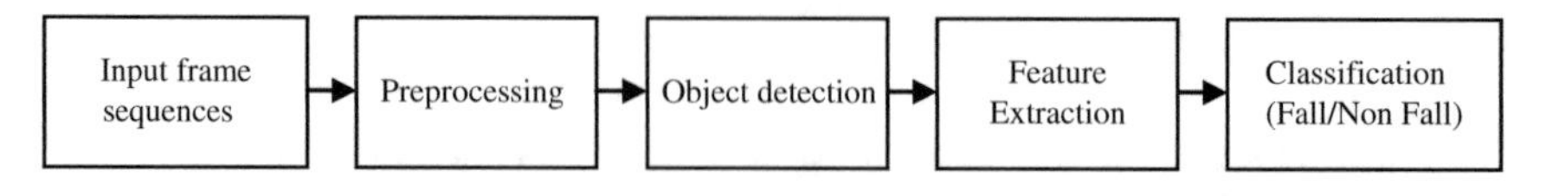

Fig. 1 Overview of the proposed algorithm

In the proposed fall detection system, the depth-based frame sequences are provided which further processed through preprocessing, segmentation, extraction of features, and classification. Each step of the proposed method is explained as below.

2.1 Preprocessing

The input frames may be affected by the noise hence the preprocessing is necessary. In the proposed approach, to remove salt and pepper noise median filter with kernel size of 5×5 is used.

2.2 Foreground Object Detection

Background subtraction is mainly used for motion segmentation as an experiment to encounter moving objects in a video by taking the distinction between the current image frame and a reference frame in a pixel by pixel manner. Otsu global thresholding method is applied to the difference image to get the binary object. The morphological operations such as erosion and dilation are applied over the binary image to get the proper shape.

2.3 Feature Extraction

The most crucial step in fall detection is detecting features that represent the fall event accurately. In the proposed system, the ellipse and rectangle are fitted over the silhouette, and the features like aspect ratio, orientation angle, silhouette below reference height, and MHI are extracted. From the observation of the feature, it is found that these are used to distinguish the fall and non-fall activities. The features are elaborate in detail below:

i. Approximating Ellipse (Orientation angle)

After the extraction of the moving human being by using background subtraction, fit an ellipse to the human by using moments to get various parameters including the orientation angle and dimension of the major and minor axis of the ellipse. When a person falls then the orientation changes notably, and deviation will become high. If the person is just walking, the deviation will reduce.

ii. Human Shape Deformation (Aspect Ratio)

In this approach, the rectangular bounding box is fitted around the human object. Thus it gives the width and height of the human object.

The feature to extract is given by Eq. (1):

$$\alpha = \frac{W}{h} \quad (1)$$

where w is the width and h is the height of the bounding box.

The aspect ratio (α) is used to discriminate standing and falling action of human being. If α is less than one, then the person is standing, but if the value of α becomes more significant than one, then it describes the fall condition.

iii. Silhouette below reference line

The silhouette position played an important role in the fall detection system. Consider the height of the complete video frame as h and consider the h/2.5 as a reference line. When the human region above the reference line becomes zero, and all the detected white pixels are below reference line, then it is considered as a fall.

iv. Motion feature (motion history image)

MHI captures the motion of a foreground human being within the given timestamp. In MHI, the original positioned motion pixels are shown in brighter than that of preceding pixel. Motion history image (MHI) as designated in [12, 13] is a materialistic method that is vigorous in depicting movements. Every pixel of the MHI is a part of the motion of duration τ as indicated in Eq. (2):

$$H_{\tau}(x; y; t) = \begin{cases} \tau & \textit{if } D(x; y; t) = 1 \\ \max(0; H(x; y; t-1) - 1) & \textit{otherwise} \end{cases} \quad (2)$$

where D(x; y; t) is a binary image sequence representing region of motion developed by image differencing method. It is considered that fall has occurred if the movement is significant and thus motion history image is used to extract the motion.

Aspect ratio, orientation angle, MHI, and silhouette below reference line are used to train SVM and SGD classifiers.

b. Classifiers

The classifiers are primarily that part of the work where actually the event is characterized by a fall or non-fall events. This system uses two classifiers SGD and SVM classifier.

i. Support Vector Machine (SVM)

Given a training dataset, the SVM, in its most basic form, finds a maximum-margin linear classification boundary. It can also be used for linearly separable data by using non-linear kernels. Kernels are functions that implicitly project the features to a high dimensional space so that they become linearly separable. The maximum-margin property ensures that it learns boundary that is most robust to outliers and does not overfit the training data. Due to these characteristics, SVMs have following advantages:

- ***Computational efficiency:*** The projection to high-dimensional space is actually implicit through the kernel and actual computations are performed in original space only. This makes SVM computation extremely efficient.
- ***Space efficiency:*** Once the classifier is learned, only a very few data points, called the support vectors, are required for all further predictions. This also means that the algorithm is robust to outliers. They can't influence the boundary a lot.
- ***Adaptability:*** Apart from the standardly known kernels such as Gaussian, polynomial and so on, one can also come up with ingenious kernels suitable for the particular problem at hand. This makes SVMs highly customizable and widely applicable.

ii. Stochastic Gradient Descent (SGD)
Stochastic Gradient Descent (SGD) is a speculative resemblance of the gradient descent optimization method for reducing a nondiscriminatory function which is drafted as a sum of differentiable functions. It is [14] a transparent yet requisite approach to the selective training of linear classifiers under convex loss functions such as (linear) Support Vector Machine and Logistic Regression.

Although SGD has been around in the machine learning community for a long duration, it has recently received noticeable attention in the context of large-scale learning. SGD classifier is used for the massive databases.

The benefits of Stochastic Gradient Descent are:

- Efficient results.
- Good performance.

3 Experiments

The proposed work is implemented using Python and OpenCV library. The efficiency results of the proposed method are calculated by considering the accuracy of training as well as the testing phase of the SGD and the SVM classifiers. The algorithm is tested on SDU fall dataset. Results are presented in qualitative and quantitative approach.

3.1 Database

The evaluation of the suggested technique was done using SDU Fall Database [15]: A Depth Video Database for Fall Detection. Two channels are recorded: RGB video (.avi) and depth video (.avi) from which Depth videos have been used in this approach. The video frame is at the size of 320×240, recorded at 30 fps in AVI format. The average video length is about 5 s.

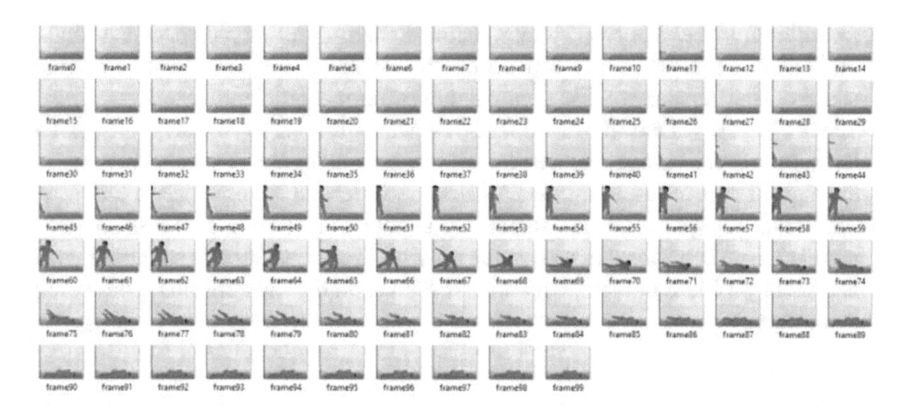

Fig. 2 Output frames extracted from one of the videos

3.2 The Results of the Proposed System

After the feature extraction, the training model is created by considering training data and their respective labels. In this approach, two machine learning algorithms are used to train the fall detection model. The testing video frames are tested with training model to classify the event into fall and non-fall.

This system observed the accuracy score of training as well as testing of both the classifier on the given depth database. The performance of the proposed system is evaluated on the basis of qualitative as well as quantitative analysis.

3.2.1 Qualitative Analysis

Step 1: Video sequence input

The video frame output for fall depth video is shown in Fig. 2.

Step 2: Foreground segmentation

Here, Fig. 3a shows original video frame, Fig. 3b depicts the result after background subtraction, and Fig. 3c shows the final outcome of Human silhouette extracted.

Step 3: Feature extraction

The four features including orientation angle, aspect ratio, silhouette below reference, and motion history image were extracted for fall characterization. For this, the fall pairs in each of the training videos frames were obtained manually.

Step 4: Fall Classification:

The classification includes two steps including Training phase and testing phase.

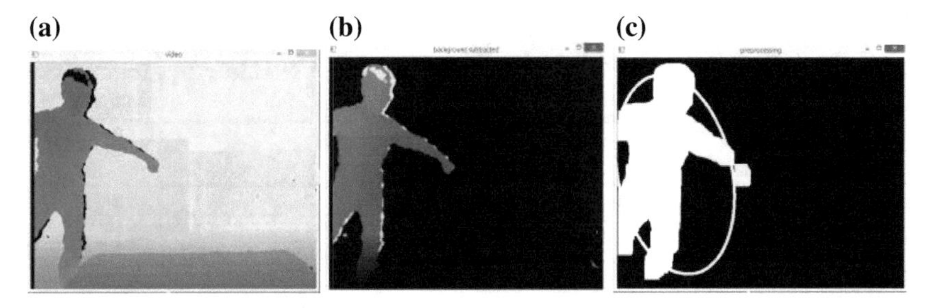

Fig. 3 The human silhouette detection: **a** input frame, **b** subtracted image, and **c** elliptical bounding box over the silhouette

1. Cross-validation Results:

The system was trained based on the given fall pairs. SVM and SGD classifiers were used to compare their training accuracies. According to the observations, SGD classifier provides better results than the SVM Classifier.

Training accuracy for SVM = 83.21%
Training accuracy for SGD = 88.59%

2. Testing Phase Results:

After training this system using 100 videos, system tested on 50 videos. The qualitative results obtained after application of the proposed approach is as shown below.

True Positive	**True Negative**	**False Negative**	**False Positive**
--FALL--		--FALL--	

3.2.2 Qualitative Analysis

The qualitative analysis of the proposed system is performed using accuracy. The accuracy is calculated as:

$$Accuracy = \frac{(TP + TN)}{(TP + TN + FP + FN)} \quad (3)$$

where TP is true positive (detecting fall event as fall), TN is true negative (detecting fall event as non-fall), FP is False positive (Detecting fall event as non-fall), and FN is false negative (detecting non-fall event as fall).

The comparison result of both the classifier training and testing accuracy (Table 1).

Table 1 Comparative results for accuracy of the classifiers

Approach	Classifier type	Training accuracy (%)	Sensitivity (%)	Specificity (%)	Testing accuracy (%)
Erdem [16]	Bays classification	–	–	–	91.89
Ma [17]		–	–	–	86.83
Aslan [18]	FV-SVM	–	–	–	88.83
Proposed	SVM	83.21	88	96	92
	SGD	88.50	94	94	94

4 Conclusion and Future Scope

In this paper algorithm for fall detection for elderly in the home, the environment has been successfully implemented using Python OpenCV. First, real-time segmentation of moving regions using background subtraction and preprocessing was performed to extract the human object. The most important step is of generating the feature that helps to characterize an event as fall or non-fall. Different features such as aspect ratio, orientation angle, MHI, and silhouette below reference line are extracted. Further, the events are classified as fall or non-fall using SVM and SGD classifier. In the training phase fall pairs were given manually, and in the testing phase, the actual fall in the video was detected. The performance of SVM and SGD classifiers to identify the fall event was compared depending upon their accuracy scores, depicting that SGD classifier gives more efficient results on a benchmark dataset.

In future, more features can be extracted as fall descriptors to classify fall very precisely. Various other more efficient classification techniques may be used for better performance of the system.

References

1. Ageing, W.H.O., Unit, L.C.: WHO global report on falls prevention in older age. World Health Org, Geneva, Switzerland (2008)
2. Mubashir, M., Shao, L., Seed, L.: A survey on fall detection: principles and approaches. Neurocomputing **100**, 144–152 (2013)
3. Skubic, M., Alexander, G., Popescu, M., Rantz, M., Keller, J.: A smart home application to eldercare: current status and lessons learned. Technol. Health Care **17**(3), 183–201 (2009)
4. Shany, T., Redmond, S., Narayanan, M., Lovell, N.: Sensors-based wearable systems for monitoring of human movement and falls. IEEE Sensors J. **12**(3), 658–670 (2012)
5. Zhao, G., Mei, Z., Liang, D., Ivanov, K., Guo, Y., Wang, Y., Wang, L.: Exploration and implementation of a pre-impact fall recognition method based on an inertial body sensor network. Sensors **12**(11), 15338–15355 (2012)
6. Demiris, G., Hensel, B.: Technologies for an aging society: a systematic review of smart home applications. Yearb. Med. Inform. **32**, 33–40 (2008)

7. Suryadevara, N., Gaddam, A., Rayudu, R., Mukhopadhyay, S.: Wireless sensors network based safe home to care for elderly people: Behaviour detection. Sensors Actuators A: Phys. **186**, 277–283 (2012)
8. Ariani, A., Redmond, S., Chang, D., Lovell, N.: Simulated unobtrusive falls detection with multiple persons. IEEE Trans. Biomed. Eng. **59**(11), 3185–3196 (2012)
9. Doukas, C., Maglogiannis, I.: Emergency fall incidents detection in assisted living environments utilizing motion, sound, and visual perceptual components. IEEE Trans. Inf. Technol. Biomed. **15**(2), 277–289 (2011)
10. Li, Y., Ho, K., Popescu, M.: A microphone array system for automatic fall detection. IEEE Trans. Biomed. Eng. **59**(5), 1291–1301 (2012)
11. Zigel, Y., Litvak, D., Gannot, I.: A method for automatic fall detection of elderly people using floor vibrations and sound-proof of concept on human mimicking doll falls. IEEE Trans. Biomed. Eng. **56**(12), 2858–2867 (2009)
12. Khan, M.J., Habib, H.A.: Video analytic for fall detection from shape features and motion gradients. In: Proceedings of the World Congress on Engineering and Computer Science 2009, vol. II WCECS 2009, 20–22 Oct 2009
13. Rougier, C., Meunier, J., St-Arnaud, A., Rousseau, J.: Fall detection from human shape and motion history using video surveillance. Dept. d'Inf. et de Rech. Operationnelle, Univ. de Montreal, Montreal, QC in Advanced Information Networking and Applications Workshops (2007)
14. http://scikitlearn.org/stable/supervised_learning.html#supervised-learning
15. http://www.sucro.org/homepage/wanghaibo/SDUFall.html
16. Akagündüz, E., Aslan, M., Şengür, A., Wang, H.: Silhouette orientation volumes for efficient fall detection in depth videos. IEEE J. Biomed. Health Inform. **21**(3), 756–763 (2017)
17. Ma, X., Wang, H., Xue, B., Zhou, M., Ji, B., Li, Y.: Depth-based human fall detection via shape features and improved extreme learning machine. IEEE J. Biomed. Health Inform. **18**(6), 1915–1922 (2014)
18. Aslan, M., Sengur, A., Xiao, Y., Wang, H., Ince, M.C., Ma, X.: Shape feature encoding via fisher vector for efficient fall detection in depth-videos. Appl. Soft Comput. **37**(C), 1023–1028 (2015)
19. Zerrouki, N., Houacine, A.: Automatic classification of human body postures based on the truncated SVD. J. Adv. Comput. Netw. **2**(1) (2014)

Vehicles Connectivity-Based Communication Systems for Road Transportation Safety

Ranjeet Singh Tomar, Mayank Satya Prakash Sharma, Sudhanshu Jha and Bharati Sharma

Abstract Wireless communication systems have now highly impact on our daily lives. The wireless communication technology has been helped billions of users in all over the world from indoor to outdoor cellular mobile networks. The generation of vehicular ad hoc network (VANET) is developing and gaining attention. There are many simulation software to allow the simulation of various media access control protocols and routing protocols. Apart from monolithic advantages in safety on the road, vehicular ad hoc network (VANET) is also proposed safety risks to the end users. Proposal of new safety concepts to retort these problems are challenging to verify due to highly dynamic real-world implementations of VANETs. To minimize problems, we have proposed VANET traffic simulation model for transportation safety. It is an event-driven simulation model, specially contrived to investigate application stage of safety entailments in vehicular communications. In VANET traffic simulation model, the message size has small and limited time value. The use of MAC protocols in VANET is to deliver the message at high reliability and low delay. In this paper, we have proposed a method in which RSU will be used for different path and measure the vehicle communication range. We have distributed RSU into clusters. In this paper, we have proposed the method which is based on safety measures and concept. We have evaluated the results of three maps on the basis of physical parameters such as current time, average travel time, average speed, average travel distance and throughput of each map. The throughput shows that in current time how much vehicle is active in every map. For safety purpose, we have presented the graph between the current times with respect to post-crash notification during simulation.

R. S. Tomar · M. S. P. Sharma (✉) · S. Jha · B. Sharma
ITM University Gwalior, Gwalior, India
e-mail: mayanksintal@gmail.com

R. S. Tomar
e-mail: er.ranjeetsingh@gmail.com

S. Jha
e-mail: er.sudhanshukumarjha@gmail.com

B. Sharma
e-mail: bharatisharma30@gmail.com

K. Ray et al. (eds.), *Soft Computing: Theories and Applications*,
Advances in Intelligent Systems and Computing 742,
https://doi.org/10.1007/978-981-13-0589-4_45

Keywords Vehicular Ad Hoc Network (VANET) · Roadside Unit (RSU)
Dedicated Short Range Communication (DSRC)

1 Introduction

VANET is a portable specially prearranged system in which the hubs are vehicles that speak with one another vehicle for trading data. VANET is a scenario of vehicles moving on the road. It is a set of ad hoc network around the permanent roadside unit (RSU) to help in network creation and communication. Vehicles are furnished with sensors and GPS framework which gather data about their position, velocity and course to be shown to all vehicles within their communication reach. The principle target is to enhance security on street and spare individuals from mishaps in brutal vehicular environment by trading safety-related messages among the vehicles. Through inter-vehicle communication (IVC), the vehicle can communicate to each other and RSU over shared wireless communication. Inter-RSU communication has a fixed channel. Currently, protocols for VANET communication are not standardized, though many protocols are proposed. IEEE 802.1.1p standard is created for utilizing of MAC protocol. These protocols are not variable, and they decrease the performance quickly in high traffic density. Vehicles fundamentally trade two sorts of messages, i.e. critical and non-critical status messages. The status message gives state data, for example, space, increasing speed and position of every vehicle [1–6]. VANET utilizes dedicated short communication range (DSRC) at 5.9 GHz for vehicle; DSRC contains seven channels with the scope of 1000 m and information rate up to 27 Mbps. DSRC contains seven 1OMHz channel and 5 MHz gatekeeper band. Out of seven channels, one is control channel for security applications and remaining six are administration channels for non-security and business applications. The location access to identify the network based on its geographic position. The protocol proposes to assign channels dynamically on the vehicle location. But IEEE 802.11 TV mode does not utilize request-to send and clear-to-send (RTS/CTS) handshaking process, which diminishes the unwavering quality extensively [7, 8].

The wireless medium efficiently does not use MAC protocol with the limiting factor. In VANETs, the movements of the vehicles are limited by traffic rules, and the directional antenna is used to help the reducing interference and collisions with ongoing transmission over neighbouring vehicular traffic but the directional antenna is not reliable for field analysis. In multichannel MAC, the channel allocation process takes place for every beacon interval for the transmission. The status messages are sent occasionally to every one of the vehicles; it can similarly be called as signal messages. The critical message gives pre-crash notice message, post-crash warning, environment and street perils. Each device chooses itself the spectrum and the best channel for each transmission. In vehicular communication, the two spectrum access at both long-term and short-term time scales along with highly dynamic channel condition decided by cognitive MAC. In vehicular system use the long-term because the long-term spectrum access CMV enhance the MAC capacity via concurrent transmission using cognitive radio [9, 10]. The throughput of the existing multichannel MAC protocol is less than CMV up to 75%. Decentralized location-based channel access protocol finds the address of the vehicle and problem of the frequency channel reuse for vehicle-to-vehicle communication in VANET. So there is a need for network connectivity-based communication model for the safety of the transportation systems.

2 Problem Description

A MAC protocol must verify the unique requirement of VANET. VANET in an environment constrained by the peculiar characteristics resulting out of high mobility of vehicles, half support of roadside infrastructure, speed in predetermined road topology, small subsets of vehicle that from independent of the ad hoc network in RSU area, and a right time condition required in the delivery of dangerous message. Vehicles move on known strait jacket roads at high speed and enter/exit RSU area in small intervals of a time. At a time number of vehicles in RSU area can change some amount of vehicles to large amount of vehicles. In RSU area the stay time is very short for vehicles. A protocol must be identified or should require some RSU support with an efficient handoff from one RSU to other and satisfy these characteristics. The communication is through transmitting over the shared wireless channel and provoke by connection delay.

3 VANET Environment in VANETSim

In VANET environment, the message size is small and limited time value. Use of the MAC protocols in VANET has the motive to deliver the message at high reliability with low delay. High mobility and fluid topology environment the connection period shared the wireless channel must be decreased. We introduce VANETSim, it provides the event-driven platform, basically designed to investigate application-level privacy and security implications in vehicular communications. In this work, we use the RSU to different path and measure the how much vehicle communicates to each other. The RSU case is part into different clusters. VANETSim aims to realistic vehicular movement on road networks and communication between routing nodes. The use of the cluster in VANETs reduces the waiting time and sending message successfully. The throughput shows that in current time how much vehicle is active in every map [9].

4 Simulation

4.1 VANETSim Overview

VANETSim is an open source, discrete event traffic and communication simulator that aims on the observe the concepts in VANET. Approximately 26,500 lines of platform-independent Java code in VANETSim. It concentrates on simulation of the application layer to achieve high performance. A GUI, the scenario creator, the simulation core and the post-processing engine these are four main components in VANETSim. The GUI provides the graphical map editor that permit the investigator to make a manipulate road maps. The map is created from imported or scratched from Open Street Map and stored in XML files, which provides interoperability with other tools. The GUI provides the simulation process in an interactive, and zoom the map that displays both roads and vehicles. In GUI, the simulation can be executed on the command line to run multiple experiments in a batch. The scenario creator offers to prepare set of experiments in which personal parameter are varied automatically. The scenario files are saved in VANETSim in XML file. The simulation core carries out the actual simulation. The simulation core is access to the map including all data, e.g. vehicles and security, privacy concept, relevant for the simulation. The communication of the vehicles simulated is known as beacon and special purpose message. We have implemented a number of privacy concepts for VANETs that consider the mix zone, silent zone and pro mix [10, 11]. The post-processing engines that created the log file during the simulation and identify and processed to create tables, and prepare the chart result to visualized and obtain the result during simulation.

4.2 Graphical User Interface (GUI)

The graphical user interface (GUI) to give facilities to configure a simulation and communicate with it while it is running. GUI provides the zoomable and scrollable visualization of the map, which shows the road area as well as vehicles and RSU (roadside unit). It indicated that the communication range of the vehicles can be activated as needed. In GUI which shows the all data as like streets name vehicles name, speed limits, current speed of vehicles. The map editor is fully integrated into the GUI. If configured the street segments by clicking on the map. We change the vehicles and other dynamic objects can be also placed by manually setting their waypoints on the map [12–14].

4.3 Performance Evaluation

A central objective guiding the development efforts related to VANETSim gives high performance on off-the-shelf hardware. In VANETSim, we provide the results so many typical simulations because it using the different-different path and different types of vehicle and events. In this simulator, we give the different path through Open Street Map (OSM) as like different cities and self-created path, Gwalior (M.P.), Basava map. All the cities or paths have own length and capacity of the vehicle. The scenario was created for general-purpose configuration: In VANETSim we set the beacon interval message is 240 ms, the Wi-Fi range of the vehicle is 100 m, and the every simulation step is 40 ms simulated time. Vehicle configuration immediately reacts and change traffic situation. We have to take two different open street maps; one is self-design open street map. This scenario is implemented on VANET simulator and also take three different paths: first one is own path, second one is Basava place and third one is Gwalior and take different measurement and prediction based on current time, active vehicle average speed, average Travel time, average travel distance and post-crash notification and take correct analysis of different roadside map with deferent parameter such as throughput, delay.

The each simulation run is covered 10,000,000 ms. We have found the current speed, simulation time, active vehicle, average travel time, average travel distance and post-crash notification.

5 Results and Discussion

We have designed the path using VANETSim simulator in Figs. 1 and 2; in simulation, we have considered the 500 vehicles and got output as like active vehicle, current time, active travel distance and post-crash notification (Figs. 3 and 4).

This map is Gwalior city map. We have taken this in open street map, and this is in open source. This map size is big, so we have taken into consideration only 1600 vehicles to run on this path and each vehicle to communicate through RSU. The RSU range is 1000 m. We have got the current time, average travel time, average travel distance, active vehicle and post-crash notification by the simulation.

In Figs. 5, 6 and 7, we have presented the relation between active vehicle, average speed of the vehicle and current time using network simulator NetSim. This graph shows the information regarding the number of vehicles with respect to current time out of which some of the vehicles are not active during communication. According to the graph, if the number of active vehicle increases, then the current time also increases logarithmically.

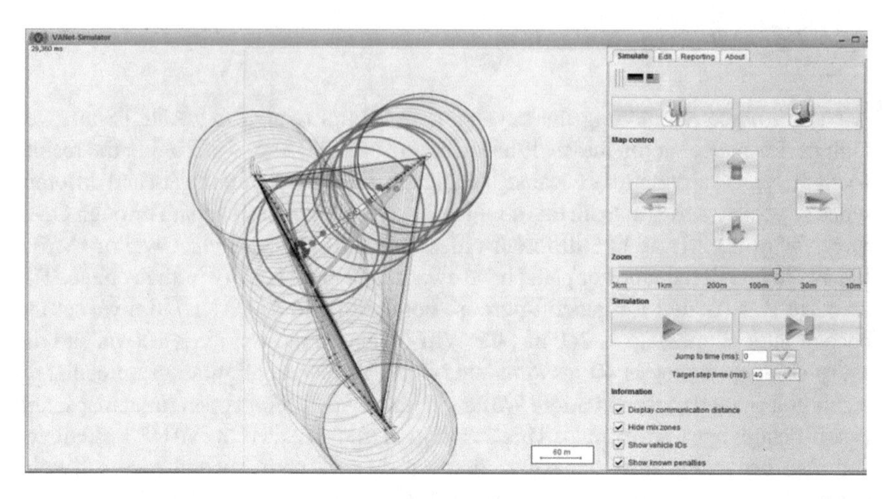

Fig. 1 Vehicles movement in VANET environment

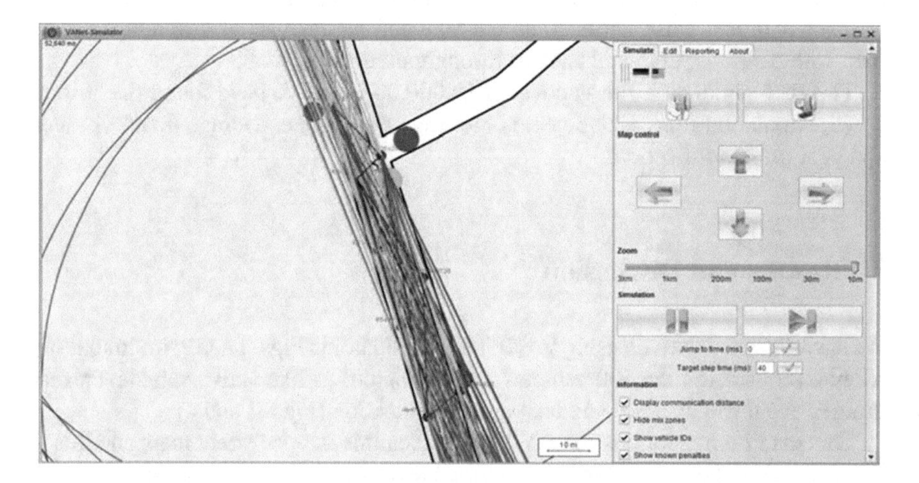

Fig. 2 Vehicles interconnection for vehicular connectivity for communication

In Fig. 5, we have presented the relation between average travel time and average travel speed using network simulator NetSim. This graph shows the active vehicle to travel in road path with average speed and average travel time. According to result, average travel time increases than average speed of vehicle decrease logarithmically. In Fig. 6, we have show the relation is average travel time and average travel distance. Some active vehicle is travelling in map to travel distance this graph is identifying the average travel distance and travel time for specific data. According to graph average travel distance initially increasing after some time constant and decrease with respect to average travel time.

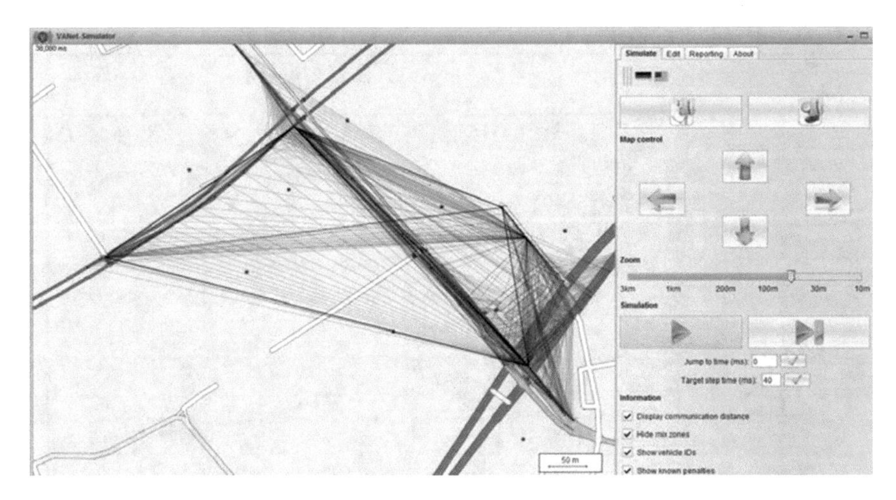

Fig. 3 Gwalior city map of vehicles movement during travelling on road

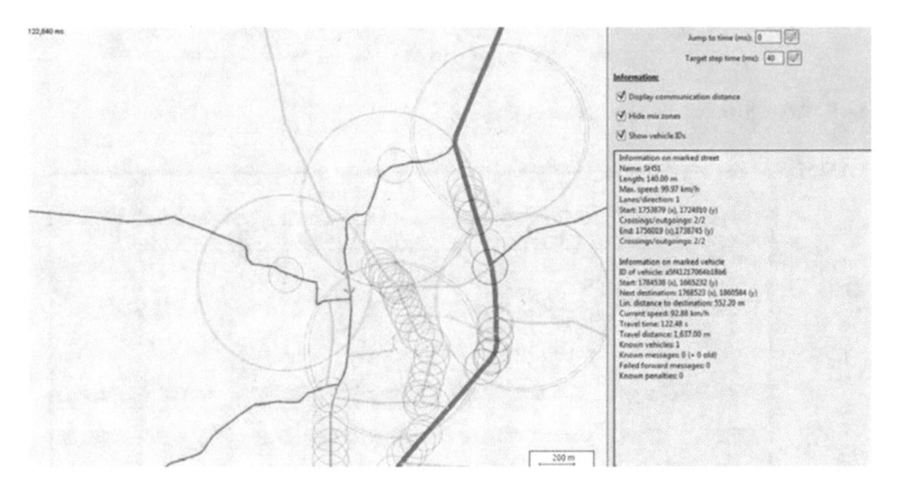

Fig. 4 Gwalior city map of vehicles connectivity during travelling on road

In Fig. 7, we have presented the relation between the post-crash notification and current time; basically, this graph is to find the active vehicles, in some vehicles are accident and communication is break. The graph shows that post-crash notification with respect to current time according to graph current time is increase than post-crash notification logarithmically increase.

In Fig. 8, we have presented the relation between the current time and the throughput during the vehicular communication.

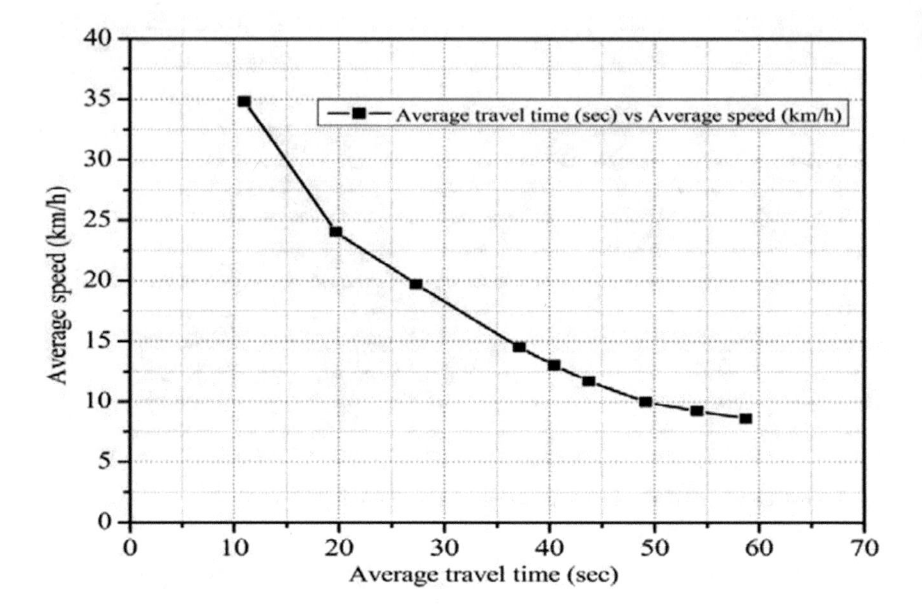

Fig. 5 Average travel time versus average speed

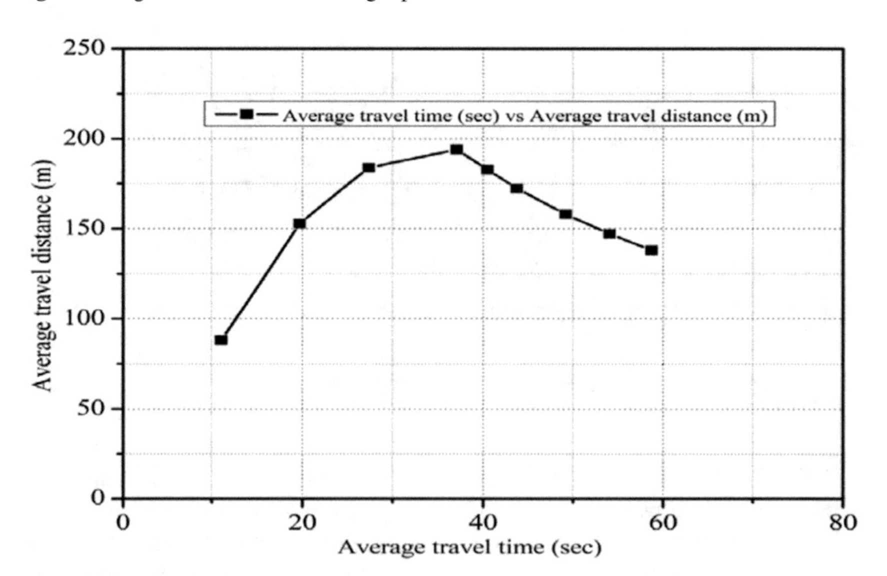

Fig. 6 Average travel time versus average travel distance

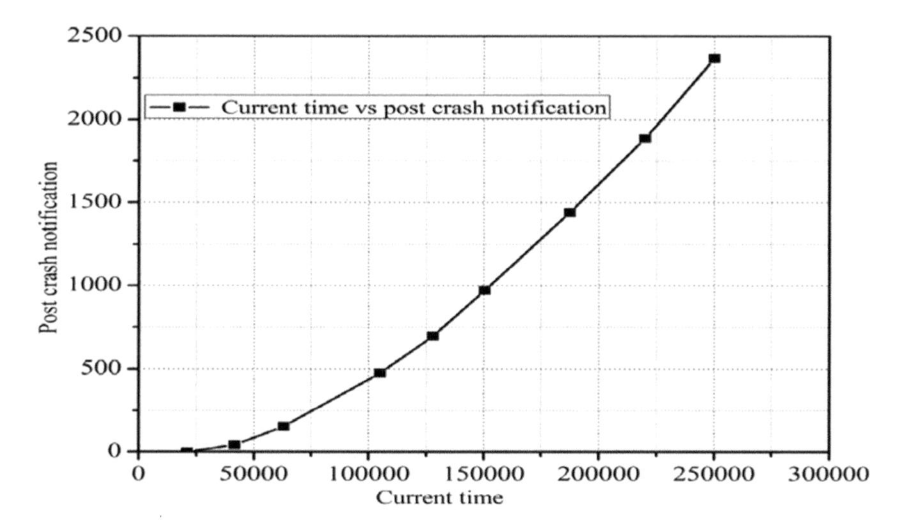

Fig. 7 Current time versus post-crash notification

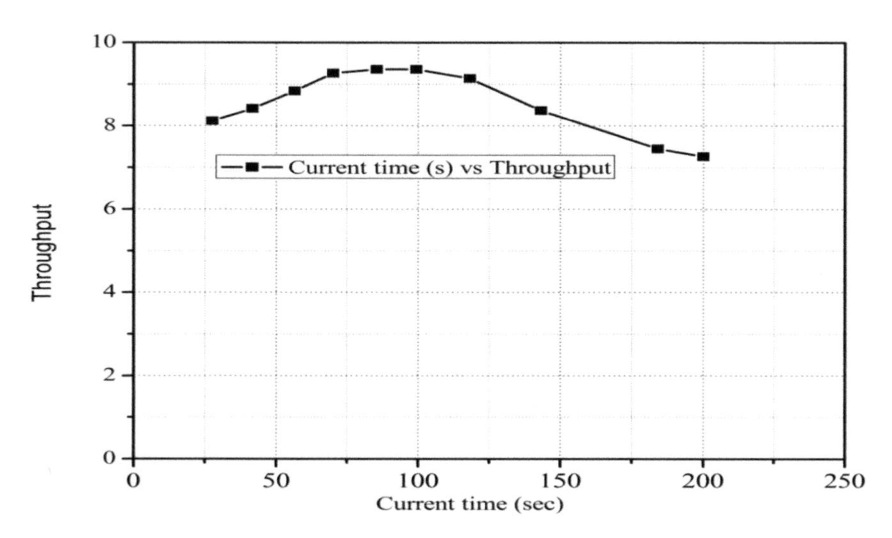

Fig. 8 Current time versus throughput

6 Conclusion

In this paper, we have presented the vehicles connectivity-based communication systems for transportation safety using VANETsim, an open source simulator specially analysis about the privacy, security and safety concerns. In this simulator, we have taken many path and street to run the vehicles and get the desired output. We have compared the various paths by this simulator and trace the difference of every result as like current speed, average speed, average travel distance, average travel time and post-crash notification. This paper has identified the result if the amount of vehicle is maximum. It is not applicable for small path and will not be provided the safety and privacy, and will not able to communicate to one vehicle to the another.

References

1. Singh, A., Kumar, M., Rishi, R. Madan, D.K.: A relative study of MANET and VANET: its applications, broadcasting approaches and challenging issues. In: International Conference on Computer Science and Information Technology. Springer, Berlin, Heidelberg, 627–632 (2011)
2. Moreno, M.T., Jiang, D., Hartenstein, H.: Broadcast reception rates and effects of priority access in 802.11-based vehicular ad-hoc networks. IEEE Trans. Veh. Technol. **61**, 1 (2012)
3. Peng, J., Cheng, L.: A distributed MAC scheme for emergency message dissemination in vehicular ad hoc networks. IEEE Trans. Veh. Technol. **56**(6), 3300–3308 (2007)
4. Yu, F., Biswas, S.: Self-configuring TDMA protocol for enhancing vehicle safety with DSRC based vehicle-to-vehicle communications. IEEE J. Sel. Areas Commun. **25**(8), 1526–1537 (2007)
5. Williams, B., Mehta, D., Camp, T., Navidi, W.: Predictive models to rebroadcast in mobile ad hoc networks. IEEE Trans. Mobile Comput. **3**(3), 295–303 (2004)
6. Zhang, H., Jiang, Z.-P.: Modeling and performance analysis of ad hoc broadcasting scheme. Perform. Eval. **63**(12), 1196–1215 (2006)
7. Fracchia, R., Meo, M.: Analysis and design of warning delivery service in inter-vehicular networks. IEEE Trans. Mobile Comput. **7**(7), 832–845 (2008)
8. Ma X., Chen, X.: Saturation performance of the IEEE 802.11 broadcast networks. IEEE Commun. Lett **11**(8), 686–688 (2007)
9. Jiang, D., Taliwal, V., Meier, A., Holfelder, W., Herrtwich, R.: Design of 5.9 GHz DSRC-based vehicular safety communication. IEEE Wirel. Commun. **13**(5), 36–43 (2006)
10. Lou, W., Wu, J.: Toward broadcast reliability in mobile ad hoc networks with double coverage. IEEE Trans. Mobile Comput. **6**(2), 148–163 (2007)
11. Bi, Y., Cai, L.X., Shen, X.: Efficient and reliable broadcast in inter vehicle communications networks: a cross layer approach. IEEE Trans. Veh. Technol. **59**(5), 2404–2417 (2010)
12. Shan, H., Zhuang, W., Wang, Z.: Distributed cooperative MAC for multi-hop wireless networks. IEEE Commun. Mag. **47**(2), 126–133 (2009)
13. Andrews, J.G., Ganti, R.K., Haenggi, M., Jindal, N., Weber, S.: A primer on spatial modeling and analysis in wireless networks. IEEE Commun. Mag. **48**(11), 156–163 (2010)
14. Palazzi, C.E., Roccetti, M., Ferretti, S.: An inter vehicular communication architecture for safety and entertainment. IEEE Trans. Intell. Transp. Syst. **11**(1), 90–99 (2010)

Some New Fixed Point Results for Cyclic Contraction for Coupled Maps on Generalized Fuzzy Metric Space

Vishal Gupta, R. K. Saini, Ashima Kanwar and Naveen Mani

Abstract The aim of this paper is to introduce the concept of cyclic $(\delta \circ \lambda)$—contraction for coupled maps on generalized fuzzy metric space. After that, some significant results of coupled maps of cyclic contraction mappings have been given on generalized fuzzy metric space. Also, we utilize the thought of continuous mappings, weakly commuting mappings, closed subset and complete subspace for demonstrating results.

Keywords Cyclic $(\delta \circ \lambda)$—contraction · Coupled maps · Cauchy sequence Generalized fuzzy metric space (FMS)

1 Introduction

Fixed point theory is a noteworthy theory in mathematics. There are valuable number of its applications in diverse branches of science and technology. One of the most significant results was given by Kirk et al. [8] in which they gave the concept of cyclic representation and cyclic contractions. It should be observed that cyclic contractions need not be continuous, which is a vital advantage of this method. Pacurar and Rus [10] presented the idea of cyclic contraction and demonstrated a result for the cyclic

---

V. Gupta (✉) · A. Kanwar
Department of Mathematics, Maharishi Markandeshwar (Deemed to be University), Mullana, Ambala, Haryana, India
e-mail: vishal.gmn@gmail.com

A. Kanwar
e-mail: kanwar.ashima87@gmail.com

R. K. Saini
Department of Mathematics, Bundelkhand University, Jhansi, UP, India
e-mail: rksaini03@yahoo.com

N. Mani
Department of Mathematics, Sandip University, Nashik, Maharashtra, India
e-mail: naveenmani81@gmail.com

K. Ray et al. (eds.), *Soft Computing: Theories and Applications*,
Advances in Intelligent Systems and Computing 742,
https://doi.org/10.1007/978-981-13-0589-4_46

contraction on complete metric space. Some fixed point results involving cyclic weaker contraction were demonstrated by Nashine and Kadelburg [9].

Definition 1 ([13]) An operator δ:[0, 1] $\rightarrow$ [0, 1] is called a comparison operator if non-decreasing δ is left continuous and $\delta(t) > t$ for all $t \in (0, 1)$ and having properties as $\delta(1) = 1$ and $\lim_{\kappa\to\infty} \delta^{\kappa}(t) = 1$ for all $t \in (0, 1)$.

Definition 2 ([10]) Let L and N be closed subsets of $M \neq \phi$. A pair of mappings $U, W{:}\diamondsuit \rightarrow \diamondsuit$ where $\diamondsuit = L \cup N$, is supposed to have a cyclic form if $U(L) \subseteq N$ and $W(N) \subseteq L$.

Karapinar et al. [6] defined the following concept of cyclic representation.

Definition 3 ([6]) Let (M, D) be a non-empty metric space. Let n be a positive integer, $B_1, B_2, \ldots, B_n$ be nonempty closed subsets of M, $\diamondsuit = \bigcup_{i=1}^{n} B_i$ and $U, W{:}\diamondsuit \rightarrow \diamondsuit$. Then cyclic representation such as $\diamondsuit$ with respect to (U, W) if

$$U(B_1) \subset W(B_2), U(B_2) \subset W(B_3), \ldots, U(B_{n-1}) \subset W(B_n), U(B_n) \subset W(B_1).$$

A fundamental point in the development of the present day thought of instability of uncertainty was the generation of a seminal paper by Zadeh [15] in which he instituted the possibility of a fuzzy set. Later on, Kramosil and Michalek [7] at first exhibited the idea of an FMS. It fills in as a guide for the development of this theory in FMS. After marginally modification in this concept, the definition of FMS is reintroduced by Georage and Veeramani [3]. In the process of generalization of FMS, Sun and Yang [11] coined the notion of generalized FMS. For more results, we refer to [1, 2, 4, 5, 12, 14].

2 Main Results

Firstly, we will define the idea of cyclic $(\delta \circ \lambda)$—contraction for coupled maps on generalized FMS. After that coupled fixed point results based on this contraction on generalized FMS will be proved.

Definition 4 Let L and N be closed subsets of $M \neq \phi$ and $\left(M, \mathsf{G}, *\right)$ be a complete generalized FMS. Let $\mathrm{K} : \lozenge \times \lozenge \rightarrow \lozenge$ be a map where $\diamondsuit = L \cup N$ which hold following conditions:

(i) a cyclic representation as $\diamondsuit = L \cup N$ *w.r.t* K,
(ii) there exists an operator λ as

$$\lambda\Big(\mathsf{G}\big(\mathrm{K}(\mu,\sigma),\mathrm{K}(\mu,\sigma),\mathrm{K}(\tau,\upsilon),s\big)\Big) \geq \delta\Big(\lambda\Big(\mathsf{G}(\mu,\mu,\tau,s)\Big)\Big);$$

$$\lambda\Big(\mathsf{G}\big(\mathrm{K}(\sigma,\mu),\mathrm{K}(\sigma,\mu),\mathrm{K}(\upsilon,\tau),s\big)\Big) \geq \delta\Big(\lambda\Big(\mathsf{G}(\sigma,\sigma,\upsilon,s)\Big)\Big),$$

for any $\mu, \tau \in L$, $\sigma, \upsilon \in N$ and δ is a comparison operator and operator $\lambda: [0,1] \to [0,1]$ is defined such that λ is non-decreasing, continuous, $\lambda(u) > 0$ for $u > 0$ and $\lambda(u) = u$ if $u = \{0, 1\}$ and $\lambda(u) \leq$ for all $u \in (0, 1)$. then K is called cyclic $(\delta \circ \lambda)$—contraction.

Theorem 1 *Let L and N be closed subsets of $M \neq \phi$ and $(M, G, *)$ be a generalized FMS which is complete and $a * b = min\{a, b\}$. Let $K: \lozenge \times \lozenge \to \lozenge$ and $U, W: \lozenge \to \lozenge$ be functions where $\lozenge = L \cup N$ which hold following conditions:*

(i) $U(\lozenge) \cap W(\lozenge) \supset K(\lozenge \times \lozenge)$,

(ii) K *is cyclic* $(\delta \circ \lambda)$*—contraction,*

(iii) *for all* $\mu, \tau \in L$ *and* $\sigma, \upsilon \in N$, *and* $s > 0$,

$$\lambda\left(G(K(\mu,\sigma), K(\mu,\sigma), K(\tau,\upsilon), s)\right) \geq \delta\left(\lambda\left(\begin{array}{l} G(U\mu, U\mu, W\tau, s) \\ * G(U\mu, U\mu, K(\mu,\sigma), s) \\ * G(U\mu, U\mu, K(\tau,\upsilon), s) \\ * G(W\tau, W\tau, K(\tau,\upsilon), s) \end{array}\right)\right).$$

(iv) *the pairs* $(K, U), (K, W)$ *are weakly commuting where* U, W *are continuous functions.*

Then there exists a fixed point of K, U *and which is unique in* $L \cap N$.

Proof Let $\mu_0 \in L$ and $\sigma_0 \in N$ be any two arbitrary elements. From condition (ii), one can get sequences $\mu_\kappa, \tau_\kappa \in L$ and $\sigma_\kappa, \upsilon_\kappa \in N$ as

$$\left.\begin{array}{l} K(\mu_\kappa, \sigma_\kappa) = U\mu_{\kappa+1} = \tau_\kappa;\ K(\sigma_\kappa, \mu_\kappa) = U\sigma_{\kappa+1} = \upsilon_\kappa, \\ K(\mu_{\kappa+1}, \sigma_{\kappa+1}) = W\mu_{\kappa+2} = \tau_{\kappa+1};\ K(\sigma_{\kappa+1}, \mu_{\kappa+1}) = W\sigma_{\kappa+2} = \upsilon_{\kappa+1} \end{array}\right\} \quad (1)$$

By taking condition (iii), we have

$$\lambda\left(G(\tau_{\kappa+2}, \tau_{\kappa+2}, \tau_{\kappa+1}, s)\right) = \lambda\left(G(K(\mu_{\kappa+2}, \sigma_{\kappa+2}), K(\mu_{\kappa+2}, \sigma_{\kappa+2}), K(\mu_{\kappa+1}, \sigma_{\kappa+1}), s)\right),$$

this implies that

$$\lambda\left(G(\tau_{\kappa+2},\tau_{\kappa+2},\tau_{\kappa+1},s)\right)\geq\delta\left(\lambda\begin{pmatrix}G(\tau_{\kappa+2},\tau_{\kappa+2},\tau_{\kappa+1},s)\\ *\,G(\tau_{\kappa+1},\tau_{\kappa+1},\tau_{\kappa},s)\end{pmatrix}\right)$$

$$\geq\delta\left(\lambda\begin{pmatrix}G(U\mu_{\kappa+2},U\mu_{\kappa+2},W\mu_{\kappa+1},s)\\ *\,G(U\mu_{\kappa+2},U\mu_{\kappa+2},K(\mu_{\kappa+2},\sigma_{\kappa+2}),s)\\ *\,G(U\mu_{\kappa+2},U\mu_{\kappa+2},K(\mu_{\kappa+1},\sigma_{\kappa+1}),s)\\ *\,G(W\mu_{\kappa+1},W\mu_{\kappa+1},K(\mu_{\kappa+1},\sigma_{\kappa+1}),s)\end{pmatrix}\right).$$

The following cases are discuses below:
If $G(\tau_{\kappa+1},\tau_{\kappa+1},\tau_{\kappa},s)>G(\tau_{\kappa+2},\tau_{\kappa+2},\tau_{\kappa+1},s)$.
Then,

$$\lambda\left(G(\tau_{\kappa+2},\tau_{\kappa+2},\tau_{\kappa+1},s)\right)\geq\delta\left(\lambda\left(G(\tau_{\kappa+2},\tau_{\kappa+2},\tau_{\kappa+1},s)\right)\right)>\lambda\left(G(\tau_{\kappa+2},\tau_{\kappa+2},\tau_{\kappa+1},s)\right).$$

This is not possible.

If $G(\tau_{\kappa+2},\tau_{\kappa+2},\tau_{\kappa+1},s)>G(\tau_{\kappa+1},\tau_{\kappa+1},\tau_{\kappa},s)$.
Then,

$$G(\tau_{\kappa+2},\tau_{\kappa+1},\tau_{\kappa+1},s)>\lambda\left(G(\tau_{\kappa+2},\tau_{\kappa+1},\tau_{\kappa+1},s)\right)\geq\delta\left(\lambda\left(G(\tau_{\kappa+1},\tau_{\kappa+1},\tau_{\kappa},s)\right)\right)$$

From this one can obtain

$$G(\tau_{\kappa+1},\tau_{\kappa+1},\tau_{\kappa},s)\geq\delta^{\kappa}\left(\lambda\left(G(\tau_{1},\tau_{1},\tau_{0},s)\right)\right).$$

Taking $\kappa\to\infty$, we have $G(\tau_{\kappa+1},\tau_{\kappa+1},\tau_{\kappa},s)\to 1$. This implies $\{\tau_\kappa\}$ and $\{\nu_\kappa\}$ are Cauchy sequence. Since completeness of L, N gives that there exist $\ell\in L, m\in N$ such that $\tau_\kappa\to\ell,\ \nu_\kappa\to m$.
From this, we conclude that

$$\{K(\mu_\kappa,\sigma_\kappa)\}\to\ell,\{K(\sigma_\kappa,\mu_\kappa)\}\to m. \tag{2}$$

From condition (ii), we have

$$G(\tau_\kappa,\tau_\kappa,K(\ell,m),s)\geq\lambda\left(G(\tau_\kappa,\tau_\kappa,K(\ell,m),s)\right)\geq\delta\left(\lambda\left(G(\mu_\kappa,\mu_\kappa,\ell,s)\right)\right).$$

As $\kappa\to\infty$, one can have

$$K(\ell, m) = \ell,\ K(m, \ell) = m. \tag{3}$$

From (i) condition, subsequences of $\{K(\mu_\kappa, \sigma_\kappa)\}$ converge to same limit such that $\{U\mu_\kappa\} \to \ell, \{W\mu_\kappa\} \to \ell$

In the same way, we have

$$\{U\sigma_\kappa\} \to m, \{W\sigma_\kappa\} \to m. \tag{4}$$

This implies

$$UK(\mu_\kappa, \sigma_\kappa) \to U\ell;\ UK(\sigma_\kappa, \mu_\kappa) \to Um \text{ and } U^2\mu_\kappa \to U\ell; U^2\sigma_\kappa \to Um$$

From the definition of weakly commuting for the pair (K, U),

$$G\big(K(U\mu_\kappa, U\sigma_\kappa), K(U\mu_\kappa, U\sigma_\kappa), U(K(\mu_\kappa, \sigma_\kappa)), s\big) \geq G\begin{pmatrix} K(U\mu_\kappa, U\sigma_\kappa), \\ K(U\mu_\kappa, U\sigma_\kappa), U(\mu_\kappa), s \end{pmatrix}.$$

By taking $\kappa \to \infty$,

$$K(U\mu_\kappa, U\sigma_\kappa) \to U\ell; K(U\sigma_\kappa, U\mu_\kappa) \to Um. \tag{5}$$

Again using condition (iii) and as $\kappa \to \infty$, one can obtain

$$U\ell = \ell; Um = m. \tag{6}$$

Now, assuming (iv) condition, we get

$$WV(\mu_\kappa, \sigma_\kappa) \to W\ell; WV(\sigma_\kappa, \mu_\kappa) \to Wm \text{ and } W^2\mu_\kappa \to W\ell; W^2\sigma_\kappa \to Wm.$$

From the definition of weakly commuting for the pair (K, W),

$$K(W\mu_\kappa, W\sigma_\kappa) \to W\ell; K(W\sigma_\kappa, W\mu_\kappa) \to Wm. \tag{7}$$

Again using condition (iii), one can obtain

$$\lambda\left(G(\mathrm{K}(\mu_\kappa,\sigma_\kappa),\mathrm{K}(\mu_\kappa,\sigma_\kappa),V(W\mu_\kappa,W\sigma_\kappa),s)\right)$$
$$\geq \delta\left(\lambda\left(\begin{array}{l} G(U\mu_\kappa,U\mu_\kappa,WW\mu_\kappa,s) \\ * G(U\mu_\kappa,U\mu_\kappa,\mathrm{K}(\mu_\kappa,\sigma_\kappa),s) \\ * G(U\mu_\kappa,U\mu_\kappa,\mathrm{K}(W\mu_\kappa,W\sigma_\kappa),s) \\ * G(WW\mu_\kappa,WW\mu_\kappa,\mathrm{K}(W\mu_\kappa,W\sigma_\kappa),s) \end{array}\right)\right)$$

As $\kappa \to \infty$, then

$$W\ell = \ell,\ \ Wm = m. \tag{8}$$

From (3), (6) and (8), we obtain

$$\mathrm{K}(\ell,m) = U\ell = W\ell = \ell;\ \mathrm{K}(m,\ell) = Um = Wm = m.$$

Let suppose $\ell \neq m$ By using (iii) condition,

$$\lambda\left(G(\ell,\ell,m,s)\right) = \left(G(\mathrm{K}(\ell,m),\mathrm{K}(\ell,m),\mathrm{K}(m,\ell),s)\right)$$
$$\geq \delta\left(\lambda\left(G(\ell,\ell,m,s)\right)\right) > \lambda\left(G(\ell,\ell,m,s)\right).$$

This is a contradiction. Therefore

$$\mathrm{K}(\ell,\ell) = U\ell = W\ell = \ell \tag{9}$$

Since $L \cap N \neq \phi$, then from (9), it follows that $\ell \in L \cap N$.

With help of condition (iii) of this theorem, one can get that there is a unique common fixed point of mappings K, U and W in $L \cap N$.

Corollary 2 *Let L and N be closed subsets of* $M \neq \phi$ *and* $(M, G, *)$ *be a generalized FMS which is complete and* $a * b = \min\{a, b\}$. *Let* $\mathrm{K} : \Diamond\times\Diamond \to \Diamond$ *and* $U : \Diamond \to \Diamond$ *be functions, where* $\Diamond = L \cup N$, *which hold following conditions:*

(i) $\mathrm{K}(\Diamond\times\Diamond) \subset U(\Diamond)$,

(ii) K *is cyclic* $(\delta \circ \lambda)$*—contraction,*

(iii) for all $\mu, \tau \in L$ *and* $\sigma, \upsilon \in N$, *and* $s > 0$,

$$\lambda\left(G(K(\mu,\sigma),K(\mu,\sigma),K(\tau,\upsilon),s)\right) \geq \delta\left(\lambda\begin{pmatrix} G(U\mu,U\mu,U\tau,s) \\ *G(U\mu,U\mu,K(\mu,\sigma),s) \\ *G(U\mu,U\mu,K(\tau,\upsilon),s) \\ *G(U\tau,U\tau,K(\tau,\upsilon),s) \end{pmatrix}\right)$$

(iv) *U is a continuous function and the pair (K,U) is weakly commuting.*

Then there exists a fixed point of V, U and which is unique in $L \cap N$.

Proof By assuming $U = W$ in Theorem 1, above result is obtained

Theorem 3 *Let $L_1, L_2, \ldots, L_n$ be closed subsets of $M \neq \phi$ and $(M, G, *)$ be a generalized FMS which is complete. Let $K : \Diamond\times\Diamond \to \Diamond$ and $U, W: \Diamond \to \Diamond$ be functions, where $\Diamond = \bigcup_{i=1}^{n} L_i$, which hold following conditions:*

(i) $U(\Diamond)\cap W(\Diamond) \supset K(\Diamond\times\Diamond)$,

(ii) *K is cyclic $(\delta\circ\lambda)$—contraction,*

(iii) *for all $\mu, \tau \in L_i$ and $\sigma, \upsilon \in L_{i+1}$ and $s > 0$,*

$$\lambda\left(G(K(\mu,\sigma),K(\mu,\sigma),K(\tau,\upsilon),s)\right) \geq \delta\left(\lambda\begin{pmatrix} G(U\mu,U\mu,U\tau,s) \\ *G(U\mu,U\mu,K(\mu,\sigma),s) \\ *G(U\mu,U\mu,K(\tau,\upsilon),s) \\ *G(U\tau,U\tau,K(\tau,\upsilon),s) \end{pmatrix}\right),$$

(iv) *the pairs $(K,U),(K,W)$ are weakly commuting, where U, W are continuous.*

Then there exists a fixed point of V, U and which is unique in $\bigcap_{i=1}^{n} L_i$.

Theorem 4 *Let L and N be closed subsets of $M \neq \phi$ and $(M, G, *)$ be a generalized FMS which is complete. Let $K: \Diamond \times \Diamond \to \Diamond$ be a function where $\Diamond = L \cup N$ which satisfies following conditions such as:*

(i) *$K : \Diamond\times\Diamond \to \Diamond$ is cyclic $(\delta\circ\lambda)$—contraction,*

(ii) *a cyclic representation as $\Diamond = L \cup N$ w.r.t K*

Then, K has a unique fixed point in $L \cap N$.

Proof Let $\mu_0 \in L$ and $\sigma_0 \in N$ be elements and let sequences $\{\mu_\kappa\}$ and $\{\sigma_\kappa\}$ be defined as

$$\mu_{\kappa+1} = K(\mu_\kappa,\sigma_\kappa), \sigma_{\kappa+1} = K(\sigma_\kappa,\mu_\kappa) \tag{10}$$

for some $\kappa \geq 0$, $\mu_\kappa \in L$ and $\sigma_\kappa \in N$.

From (i),

$$\lambda\left(G(\mu_\kappa,\mu_\kappa,\mu_{\kappa+1},s)\right)=\lambda\left(G(K(\mu_{\kappa-1},\sigma_{\kappa-1}),K(\mu_{\kappa-1},\sigma_{\kappa-1}),K(\mu_\kappa,\mu_\kappa),s)\right)$$
$$\geq\delta\left(\lambda\left(G(\mu_{\kappa-1},\mu_{\kappa-1},\mu_\kappa,s)\right)\right).$$

Using induction,

$$G(\mu_\kappa,\mu_\kappa,\mu_{\kappa+1},s)\geq\lambda\left(G(\mu_\kappa,\mu_\kappa,\mu_{\kappa+1},s)\right)\geq\delta^\kappa\left(\lambda\left(G(\mu_0,\mu_0,\mu_1,s)\right)\right).$$

For any $q > 0$,

$$G(\mu_\kappa,\mu_\kappa,\mu_{\kappa+q},s)\geq\delta^\kappa\left(\lambda\left(G\left(\mu_0,\mu_0,\mu_1,\tfrac{s}{q}\right)\right)\right)*\delta^{\kappa+1}\left(\lambda\left(G\left(\mu_0,\mu_0,\mu_1,\tfrac{s}{q}\right)\right)\right)*\ldots$$
$$*\,\delta^{\kappa+p-1}\left(\lambda\left(G\left(\mu_0,\mu_0,\mu_1,\tfrac{s}{q}\right)\right)\right).$$

By taking $\kappa\to\infty$ and using property of δ, we have $\{\mu_\kappa\}$ is a Cauchy sequence.

From (i), one can have

$$\lambda\left(G(\sigma_\kappa,\sigma_\kappa,\sigma_{\kappa+1},s)\right)=\lambda\left(G(K(\sigma_{\kappa-1},\mu_{\kappa-1}),K(\sigma_{\kappa-1},\mu_{\kappa-1}),K(\sigma_\kappa,\mu_\kappa),s)\right)$$
$$\geq\delta\left(\lambda\left(G(\sigma_{\kappa-1},\sigma_{\kappa-1},\sigma_\kappa,s)\right)\right),$$

this implies

$$G(\sigma_\kappa,\sigma_\kappa,\sigma_{\kappa+1},s)\geq\lambda\left(G(\sigma_\kappa,\sigma_\kappa,\sigma_{\kappa+1},s)\right)\geq\delta^\kappa\left(\lambda\left(G(\sigma_0,\sigma_0,\sigma_1,s)\right)\right).$$

Taking $\kappa\to\infty$ and for any $q > 0$,

$$G(\sigma_\kappa,\sigma_\kappa,\sigma_{\kappa+q},s)\geq G\left(\sigma_\kappa,\sigma_\kappa,\sigma_{\kappa+1},\tfrac{s}{q}\right)*G\left(\sigma_{\kappa+1},\sigma_{\kappa+1},\sigma_{\kappa+2},\tfrac{s}{q}\right)*\ldots*$$
$$G\left(\sigma_{\kappa+q-1},\sigma_{\kappa+q-1},\sigma_{\kappa+q},\tfrac{s}{q}\right)$$
$$\geq\delta^\kappa\left(\lambda\left(G\left(\sigma_0,\sigma_0,\sigma_1,\tfrac{s}{q}\right)\right)\right)*\delta^{\kappa+1}\left(\lambda\left(G\left(\sigma_0,\sigma_0,\sigma_1,\tfrac{s}{q}\right)\right)\right)*\ldots*$$
$$\delta^{\kappa+p-1}\left(\lambda\left(G\left(\sigma_0,\sigma_0,\sigma_1,\tfrac{s}{q}\right)\right)\right),$$

this implies that $\{\sigma_\kappa\}$ is a Cauchy sequence. The completeness of L and N implies that $\mu_\kappa \to \mu$, $\sigma_\kappa \to \sigma$ where $\mu \in L, \sigma \in N$.

We have subsequences of $\{\mu_\kappa\}$ and $\{\sigma_\kappa\}$ which converge to μ, σ, respectively. This implies $\mu, \sigma \in \chi$ where $\chi = L \cap N$. It shows that χ is closed and complete. Now, restrict the function K to χ as it can denote as $K/_\chi : \chi \times \chi \to \chi$. By using cyclic representation of K, for $\mu_\kappa \in L$, we can have $K(\mu_\kappa, \sigma_\kappa) \in N$.

Considering the definition of FMS and condition (i),

$$G(K/_\chi(\mu,\sigma), K/_\chi(\mu,\sigma), \mu, s) \geq \delta\left(ƛ\left(G\left(\mu,\mu,\mu_{\kappa+1}, s/2\right)\right)\right) * \delta\left(ƛ\left(G\left(\mu,\mu,\mu_{\kappa+1}, s/2\right)\right)\right)$$
$$\Rightarrow K/_\chi(\mu,\sigma) = \mu \in \chi.$$

By using cyclic representation of K, for $\sigma_\kappa \in N$, we can have $V(\sigma_\kappa, \mu_\kappa) \in L$. From the concept of FMS and condition (i), one can get

$$G(K/_\chi(\sigma,\mu), K/_\chi(\sigma,\mu), \sigma, s) \geq \delta\left(ƛ\left(G\left(\sigma,\sigma,\sigma_{\kappa+1}, s/2\right)\right)\right) * \delta\left(ƛ\left(G\left(\sigma,\sigma,\sigma_{\kappa+1}, s/2\right)\right)\right)$$
$$\Rightarrow V/_\chi(\sigma,\mu) = \sigma \in \chi.$$

This implies that K has coupled fixed point in $\chi = L \cap N$.

Now, we have $ƛ\left(G(\mu,\mu,\sigma,s)\right) \geq \delta\left(ƛ\left(G(\mu,\mu,\sigma,s)\right)\right) \geq ƛ\left(G(\mu,\mu,\sigma,s)\right)$.

This implies that $\mu = \sigma$. Hence, it was proved that K has common fixed point in $\chi = L \cap N$. Uniqueness of it can easily be proved from condition (iii) of this theorem. Therefore, K has unique common fixed point in $L \cap N$.

The next result can be proved with help of above theorem.

Theorem 5 *Let $L_1, L_2, \ldots, L_n$ be closed subsets of $M \neq \phi$ and $(M, G, *)$ be a generalized FMS which is complete. Let $K : \lozenge \times \lozenge \to \lozenge$ be a functions, where $\diamond = \bigcup_{i=n}^{n} L_i$, which follows following conditions:*

(i) $K : \lozenge \times \lozenge \to \lozenge$ is cyclic $(\delta \circ ƛ)$—contraction,
(ii) $\diamond = \bigcup_{i=n}^{n} L_i$ is a cyclic representation of $\diamond$ with respect to V.

Then, K has a unique coupled fixed point in $\bigcap_{i=n}^{n} L_i$.

Theorem 6 *Let L and N be closed subsets of $M \neq \phi$ and $(M, G, *)$ be a generalized FMS which is complete. Let $K : \lozenge \times \lozenge \to \lozenge$ and $U : \diamond \to \diamond$ be functions, where $\diamond = L \cup N$, which hold following conditions:*

(i) For all $s > 0$

$$ƛ\left(G(K(\mu,\sigma), K(\mu,\sigma), U\tau, s)\right) \geq \delta\left(ƛ\left(\begin{array}{c} G(\mu,\mu,\tau,s) * G(x,x,K(\mu,\sigma),s) * \\ G(\mu,\mu,U\mu,s) \end{array}\right)\right),$$

(ii) K and U are cyclic weaker $(\delta \circ ƛ)$—contractions,

(iii) $\diamond = L \cup N$ *is a cyclic representation of* $\diamond$ *w.r.t* K and U.

Then, K *and U have a unique fixed point in* $L \cap N$.

Proof Let $\mu_0 \in L$ and $\sigma_0 \in N$ be any two elements and let sequences $\{\mu_\kappa\}$ and $\{\sigma_\kappa\}$ be defined as

$$\mu_{\kappa+1} = \mathrm{K}(\mu_\kappa, \sigma_r), U(\mu_{\kappa+1}) = \mu_{\kappa+2}; \sigma_{\kappa+1} = \mathrm{K}(\sigma_\kappa, \mu_\kappa), U(\sigma_{\kappa+1}) = \sigma_{\kappa+2}$$

for some $\kappa \geq 0$, $\mu_\kappa \in L$ and $\sigma_\kappa \in N$.

From (i),

$$\lambda\left(G(\mu_{\kappa+1}, \mu_{\kappa+1}, \mu_\kappa, s)\right) = \lambda\left(G(\mathrm{K}(\mu_\kappa, \sigma_\kappa), \mathrm{K}(\mu_\kappa, \sigma_\kappa), U\mu_{\kappa-1}, s)\right) \geq \delta\left(\lambda\begin{pmatrix} G(\mu_\kappa, \mu_\kappa, \mu_{\kappa-1}, s) \\ * G(\mu_\kappa, \mu_\kappa, \mathrm{K}(\mu_\kappa, \sigma_\kappa), s) \\ * G(\mu_\kappa, \mu_\kappa, U\mu_\kappa, s) \end{pmatrix}\right),$$

As $\kappa \to \infty$, we get $\lim_{\kappa\to\infty} G(\mu_{\kappa+1}, \mu_{\kappa+1}, \mu_\kappa, s) = 1$.. This show that $\{\mu_\kappa\}$ is Cauchy sequence. Similarly, we can prove that $\{\sigma_\kappa\}$ is Cauchy sequence. Since L and N are complete subspace, so $\mu_\kappa \to \ell \in L$ and $\sigma_\kappa \to m \in N$.

Since K is a cyclic $(\delta \circ \lambda)$—contraction.

$$G(\mu_\kappa, \mu_\kappa, \mathrm{K}(\ell, m), s) \geq \lambda\left(G(\mathrm{K}(\mu_{\kappa-1}, \sigma_{\kappa-1}), \mathrm{K}(\mu_{\kappa-1}, \sigma_{\kappa-1}), \mathrm{K}(\ell, m), s)\right) \geq \delta\left(\lambda\left(G(\mu_{\kappa-1}, \mu_{\kappa-1}, \ell, s)\right)\right).$$

Considering $\kappa \to \infty$, we get $\ell = \mathrm{K}(\ell, m)$. Similarly, $m = \mathrm{K}(m, \ell)$..

Also, U is a cyclic $(\delta \circ \lambda)$ contraction which gives

$$G(\mu_\kappa, \mu_\kappa, U\ell, s) \geq \lambda\left(G(U\mu_{\kappa-1}, U\mu_{\kappa-1}, U\ell, s)\right) \geq \delta\left(\lambda\left(G(\mu_{\kappa-1}, \mu_{\kappa-1}, \ell, s)\right)\right).$$

Considering $\kappa \to \infty$, we get $\ell = U\ell$

Similarly, $m = Um$. This gives $\ell = \mathrm{K}(\ell, m) = U\ell$ and $m = \mathrm{K}(m, \ell) = Um$, which implies that K and U have common coupled fixed point.

Now, one can have

$$\lambda\left(G(\ell,\ell,m,s)\right)=\lambda\left(G\left(K(\ell,m),K(\ell,m),Um,s\right)\right)$$
$$\geq\delta\left(\lambda\begin{pmatrix}G(\ell,\ell,m,s)*\\G(\ell,\ell,K(\ell,m),s)\\*G(\ell,\ell,U\ell,s)\end{pmatrix}\right)>\lambda\left(G(\ell,\ell,m,s)\right).$$

This gives $\ell=K(\ell,\ell)=U\ell$ which show that K and U have common fixed point. Since $L\cap N\neq\phi$, then it follows that $\ell\in L\cap N$. This show that mappings K and U have common fixed point in $L\cap N$. Therefore, mappings K and U have unique common fixed point.

3 Conclusion

Fuzzy set theory and associated techniques provide an excellent tool for interfacing the real world of measurements and the conceptual world embodied by language. This theory has ample number applications in different fields. In the present study, we discussed some new coupled fixed point results for cyclic contraction on generalized fuzzy metric space.

References

1. Ali, J., Imdad, M.: An Implicit function implies several contraction conditions. Sarajevo J. Math. **17**(4), 269–285 (2008)
2. Beg, I., Gupta, V., Kanwar, A.: Fixed points on intuitionistic fuzzy metric spaces using the E.A. property. J. Nonlinear Funct. Anal. **2015**, Article ID 20 (2015)
3. George, A., Veeramani, P.: On some results in fuzzy metric spaces. Fuzzy Sets Syst. **64**(3), 395–399 (1994)
4. Gupta, V., Kanwar, A.: *V*—fuzzy metric spaces and related fixed point theorems. Fixed Point Theory Appl. **15** (2016). https://doi.org/10.1186/s13663-016-0536-1
5. Gupta, V., Kanwar, A., Gulati, N.: Common coupled fixed point result in fuzzy metric spaces using JCLR property. Smart Innovation, Systems and Technologies, vol. 43(1), pp. 201–208. Springer (2016)
6. Karapinar, E., Shobkolaei, N., Sedghi, S., Vaezpour, S.M.: A common fixed point theorem for cyclic operators on partial metric spaces. Filomat **26**(2), 407–414 (2012)
7. Kramosil, I., Michalek, J.: Fuzzy metric and statistical metric spaces. Kybernetica **11**, 336–344 (1975)
8. Kirk, W.A., Srinavasan, P.S., Veeramani, P.: Fixed points for mapping satisfying cyclical contractive conditions. Fixed Point Theory **4**(1), 79–89 (2003)
9. Nashine, H.K., Kadelburg, Z.: Fixed point theorems using cyclic weaker meir-keeler functions in partial metric spaces. Filomat **28**, 73–83 (2014)
10. Pacurar, M., Rus, I.A.: Fixed point theory for cyclic φ—contractions. Nonlinear Anal **72**(3–4), 1181–1187 (2010)

11. Sun, G., Yang, K.: Generalized fuzzy metric spaces with properties. Res. J. Appl. Sci. **2**, 673–678 (2010)
12. Sedghi, S., Altun, I.: Shobe, N: Coupled fixed point theorems for contractions in fuzzy metric spaces. Nonlinear Anal. **72**(3), 1298–1304 (2010)
13. Shen, Y.H., Qiu, D., Chen, W.: Fixed point theory for cyclic φ—contractions in fuzzy metric spaces. Iran. J. Fuzzy Syst. **10**, 25–133 (2013)
14. Shatanawi, W., Postolache, M.: Common fixed point results for mappings under non- linear contraction of cyclic form in ordered metric spaces. Fixed Point Theory Appl. **60**, 13 (2013)
15. Zadeh, L.A.: Fuzzy sets. Inform. Control **8**, 338–353 (1965)

DWSA: A Secure Data Warehouse Architecture for Encrypting Data Using AES and OTP Encryption Technique

Shikha Gupta, Satbir Jain and Mohit Agarwal

Abstract Data warehouse is the most important asset of an organization as it contains highly valuable and sensitive information that is useful in decision-making process. The data warehouse provides easy access to organizational data as it contains data from different sources. Thus, it is essential to structure security measures for the protection of data that resides in data warehouse against malicious attackers to ensure proper security and confidentiality. Security should be considered vital from initial stages of designing data warehouse and hence should be deployed. Though a lot of work is being done towards the improvement and development of data warehouse till now but very less attention is given on the implementation of the security approaches in data warehouse. This paper focuses on improving the level of security by combining One-Time Pad (OTP) encryption technique with Advanced Encryption Standard (AES) to encrypt the data before loading it into data warehouse. Finally, the proposal of the model is to incorporate OTP encryption technique in the architecture of data warehouse for enhancing its security.

Keywords Data warehouse · One-time pad (OTP) · AES · Security · Encryption Hashing · Masking function

S. Gupta (✉) · S. Jain
Department of Computer Engineering, Netaji Subhash Institute of Technology, New Delhi, India
e-mail: shikha.gpt1@gmail.com

S. Jain
e-mail: jain_satbir@yahoo.com

M. Agarwal
Department of Physics and Computer Science, Dayalbagh Educational Institute, Agra, India
e-mail: rs.mohitag@gmail.com

K. Ray et al. (eds.), *Soft Computing: Theories and Applications*,
Advances in Intelligent Systems and Computing 742,
https://doi.org/10.1007/978-981-13-0589-4_47

1 Introduction

In today's competitive environment, enterprises are continuously growing, businesses are spreading globally and organizations strive for useful information for important decision-making and improve the bottom line. In order to meet the demand of increasingly growing competitive environment, data warehouse is designed which is new paradigm, specifically intended to provide vital strategic information [1]. Data warehouse is a 'subject oriented, integrated, non—volatile and time variant collection of data in support of management's decision' [2]. Historical data is also stored in data warehouses. The data stored in warehouse is extracted from various operational systems. So, security has always been the critical issue in data warehouse for protection of essential and useful data of large organizations against unauthorized access. Various security approaches are being used in warehouse at different levels and layers. Till now, a lot of work is being done towards the improvement and development of data warehouse, but very less attention is given on the implementation of the security approaches in data warehouse.

Firstly in data warehouse, there are various sources from where data is extracted, then it undergoes in staging area for transformation where standardization, summarization, integration and aggregation on data are performed. Then this large amount of data gets loaded in data warehouse from where it is accessed by various systems like OLAP, data mining, data marts, etc. So, encryption is one such security technique, applied on data which can be employed at two places: data at rest and data travelling over a network from various sources before entering into warehouse [3]. Data over a network is more prone to attacks and encrypting data before loading it into the data warehouse is easy to implement. So to solve this, Oracle supports encryption of network traffic [3]. Enterprises are presently using standard algorithms—DES, Triple-DES and AES, and Blowfish.

One-Time Pad (OTP) [4, 5] is a symmetric algorithm in which same key is used for both sender and receiver. It is the stream cipher which encrypts one bit or byte at a time unlike block cipher AES that encrypts large chunks of data at a time. It is faster than block cipher. In this, each bit or a character of a plaintext is encrypted with random key from pad using modular addition [6]. OTP has the following requirements for perfect security:

1. Key should be random.
2. The length of input data is equal to the length of the key.
3. Key should be used only once.
4. Key should be destroyed after use so that it cannot be reused [6].

The remainder of the paper is described by following sections: Sect. 2 includes work related about various security approaches with the introduction of OTP in detail. Section 3 describes that how OTP is combined with AES and deployed in data warehouse design. Section 4 describes the proposed model and elaboration of its components. Section 5 comprises the advantages of the proposed model and its applications. Section 6 is conclusion followed by references.

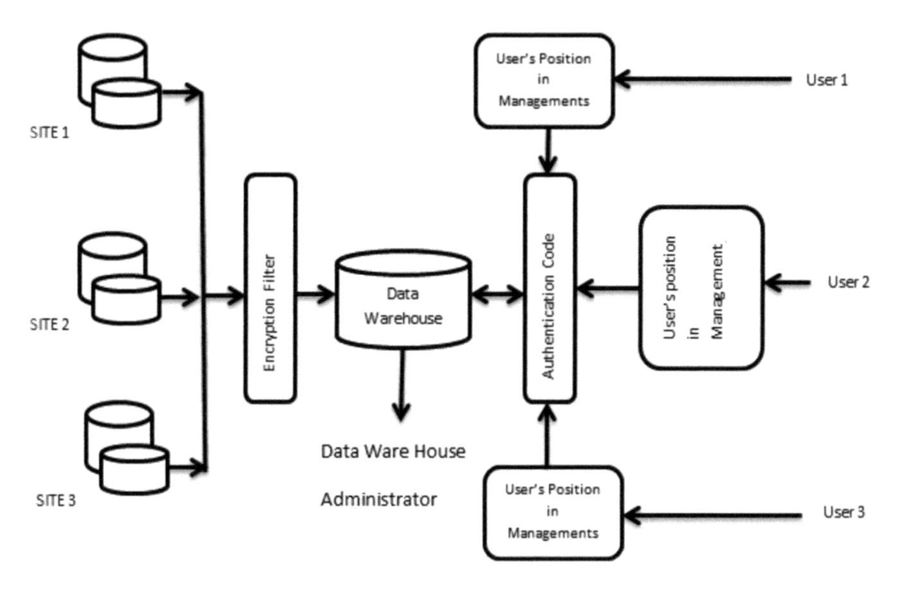

Fig. 1 Hybrid model for data warehouse security

2 Related Work

Nowadays, many security techniques are being implemented for protecting the data, i.e. sensitive in nature. It mainly focuses on multilevel security, federated databases and commercial applications [7]. Following approaches are being considered in this paper to achieve better security. Filter is the first technique used in data warehouse for security shown in Fig. 1. Basically, Encryption filter is applied on data before loading into the data warehouse. It is used for encryption of data which comes from various sources to be stored in data warehouse for decision-making [8]. It separates out the same data coming from different sources to save storage space and access time. Encryption is used to cut down any unauthorized access for maintaining data integrity and delivery performance. In case of unauthorized user, who is able to access the data but can't understand the useful information because data is present in encrypted form. Unfortunately, it has a disadvantage of response time and resource overheads. As to access the data in readable form, user has to decrypt the data first and then, they can access the data. So, it introduces the massive storage space and performance overheads [9]. It increases the cost for allocated resources for retrieving and decrypting the data [3].

In Fig. 1, another filter is introduced after data warehouse identify the level of users at different access points. It does the division of access according to user-top level, middle level and low level management for authorized access. It shows that which user can access what part of data. Presently, AES [10] algorithm is most widely used for data warehouse security. It is a fast algorithm that provides stronger encryption as compared to DES and 3DES [9]. It is a block cipher which processes the block of plaintext of 64 bits per block using key length 56 bits for encryption and produces cipher text of 64 bits per block [11]. It is easy to implement and more useful when amount of data is pre known.

OTP encrypts the input data with random key(k) using modular addition, mod26 which has tremendous properties that plays an essential part in cryptography for security. Basically, it uses the logic of simple XOR (⊕) operation. XOR (⊕), obscures the values of input data and key, hence the ciphertext gives no information about the two unknowns in the equation that the values are being added or subtracted [12]. Also, one-time pad algorithm has been designed using conventional block cipher and one-way hash algorithm [13]. This algorithm completely compensates for the inadequacies of the conventional block cipher and using the merits of one-way hash algorithm. It is safe, easy to use and extended on block cipher. This algorithm uses block cipher DES and one-way hash MD5 [14–16].

3 Motivation

OTP is the network encryption technique which can be applied on data extracted from various sources before storing the data into the data warehouse on insecure communication lines where it is prone to be attacked or unauthorized access. Now, in this paper by combining the cryptographic technique, OTP with presently using encryption technique in data warehouse that is Advanced Encryption Standard (AES) is used to encrypt the sensitive data or files stored in data warehouse for improving the security implementation [11]. It is easy to implement and use. The one-time pad is unbreakable theoretically but practically it is weak [6]. It is an encryption technique which cannot be cracked if used correctly.

The encryption function used for OTP is shown in Eq. 1

$$E_K(x) = x_1 \oplus k_1, \ \ldots \ E_K(x) = x_1 \oplus k_1 = a_1, \ \ldots \ a_n \tag{1}$$

Similarly, the decryption function is defined as shown in Eq. 2

$$D_K(a) = a_1 \oplus k_1, \ \ldots \ D_K(a) = a_1 \oplus k_1 = x_1, \ \ldots \ x_n \tag{2}$$

3.1 *Need of Combining OTP with AES*

In the AES encryption as shown in Fig. 2, encryption process has ten rounds [17]. Nine rounds repeat the following four transformations—substitution bytes, shift rows, mix columns and then add round key. In tenth round of the process, there are only three transformations used, namely, substitution bytes, shift rows and add round key [11].

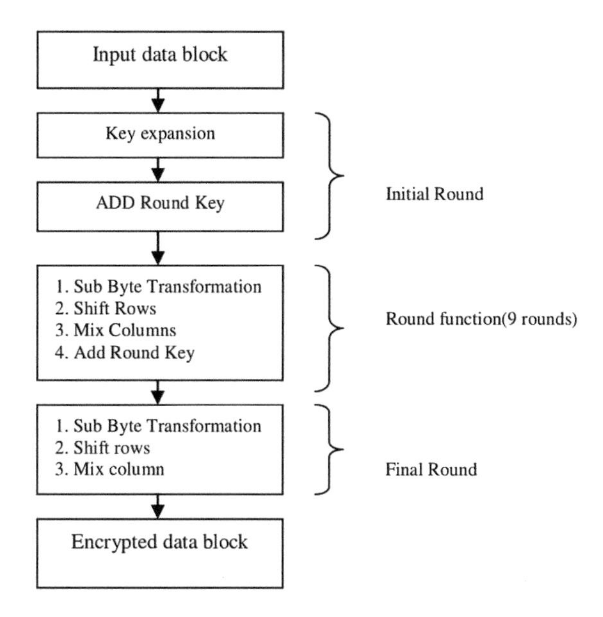

Fig. 2 AES encryption technique

- **Substitution Bytes Transformation**. It is a transformation which substitutes one state in a single cell with the corresponding cells of S-Box. The elements of S-Box are defined previously and hence, permanent. The result from S-Box is converted into decimal form.
- **Shift Rows Transformation**. In shift rows transformation, there is no shift in first row. In second row, each value shifts towards left by one place and the leftmost value moves at the right extreme position. In third row, shifts take place by two places and similarly, by three places in fourth row.
- **Mix Columns Transformation**. Mix columns transformation is performed on the basis of columns. It operates on state column by column. It multiplies one column by constant matrix c (x) to get the resultant matrix.
- **Add Round Key Transformation**. Finally, round key is added to the state using XOR operation. Input matrix and key matrix first converted into binary number and then performs XOR operation on them. The result of the XOR is then again changed into decimal form.

XOR is the logic gate arithmetic operation which is performed between the bits of plaintext and random key bits. It results true (1) if both inputs are different; else the result is false (0). It is also necessary to understand the conversion of ASCII values into binary form because in case of characters-based input, every character is first represented in its corresponding ASCII value and then converted into binary form. Then XOR is performed between binary values as shown in Table 1. One-time pad uses the mathematical operation modular arithmetic, modulo 26 on paper manually but in Boolean arithmetic, it is known as XOR operation [11]. Moreover, OTP proides better security as compared to other cryptographic algorithms because

Table 1 XOR operation on characters

Alphabets	ASCII	Binary form
Input P	80	1010000
Key H	72	1001000
Result (XOR)		0001000

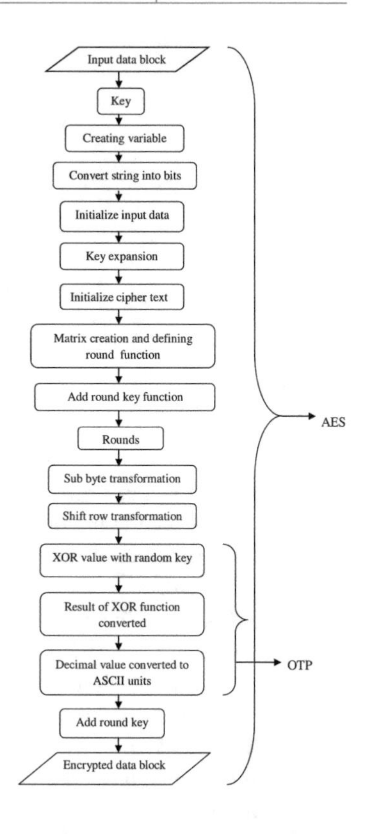

Fig. 3 Flowchart of combined approach (AES + OTP)

of random key pad. Modular arithmetic operation (XOR) used in OTP plays a very crucial role in cryptography because the resultant ciphertext reveals absolutely no information that whether the two input values either added or subtracted and keep them unknown [12]. If the result of a modulo 10 addition is 5, then there is no idea whether this is the result of 0+5, 1+4, 2+3, 3+2, 4+1, 5+0, 6+9, 7+8, 8+7 or 9+6. The resultant value 5 comes from two unknowns input of the equation which is impossible to find. Hence, OTP has an additional safety mechanism than other algorithms because of XOR modular operation. However, if the nonrecurring and random key is made along the message, it will present the effect of one-time pad which cannot be solved, even in theory [11].

So, OTP-AES flowchart is shown in Fig. 3 which is made to a data security system in accordance with the OTP process incorporated into the AES flowchart shown in Fig. 2 [11].

Basically, three steps of OTP process is substituted in AES flowchart, namely, XOR input process and key, XOR results converted to, and last the decimal value is converted into ASCII replacing the mix column transformation AES. Thus, combination of AES-OTP technique is more efficient and secure due to random key generation.

4 Proposed Approach

In above sections, various security approaches have been discussed but there are still critical areas for security that are unseen and untouched in data warehouse [18]. As, we know that data in warehouse comes from various sources and all the sensitive as well as non sensitive data are stored into the staging data area of the warehouse [15, 19]. This area is considered as a serious threat to privacy security when many internal users have different roles and responsibilities of access. As shown in Fig. 1, two filters on both sides of warehouse and encryption are used for security. Filter is used to cut redundancy. Encryption is used for representing the data in such a form so that malicious attacker cannot understand the information even if they get a chance to access. It protects the data from unauthorized access for its integrity and confidentiality. As we know, encryption causes storage space, response time and performance overheads because decryption is also required at receiver's end along with encryption and hence resources and requirements double up with considerable increase in cost. Presently, AES block cipher is used on data warehouse for encryption which has weakness that each block uses the same key. So, attackers only need to decipher one ciphertext block to decipher entire ciphertext. Due to these key reasons, a great need of appropriate model is arisen that can overcome the drawbacks of security mechanisms which are presently used in data warehouses for improvement. This model provides the access control to the right information for authorized users at right time. In Fig. 4, the following security nodes are applied, namely: OTP Encryption filter with AES, data masking function and Hash algorithm.

4.1 Encryption Filter

It is the first component used in the model before data enters into data warehouse for filtration of data and encryption. Filter is used to reduce the redundancy of data before loading into the data warehouse because data is collected from various sources. So, same data can be repeated from different sources leads to the storage space and access time overheads [8]. It analyses the data and removes repeated data. Encryption is used to protect the data on insecure communication lines. Encryption causes storage space, response time and performance overheads because decryption is also required at receiver's end along with encryption. Presently, AES block cipher is used on data warehouse for encryption which has weakness that each block uses the same

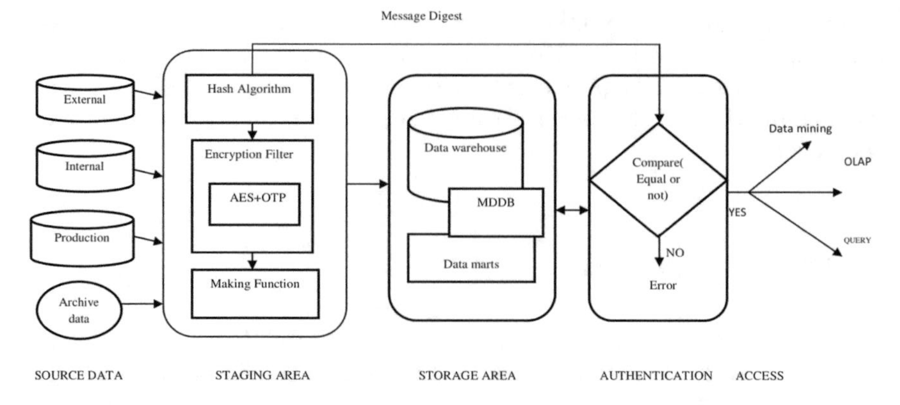

Fig. 4 Enhanced security architecture for data warehouse

key due to which it can be solved [13]. So, OTP encryption is proposed for data warehouse with AES to enhance its security because OTP uses random key generation mechanism. Hence, it is unpredictable. The resultant OTP and AES algorithm calculates encryption key for each block. So, deciphering of entire blocks is needed to decipher the entire ciphertext [13]. It is more secure algorithm.

4.2 *Masking Technique*

It is the second component of model which is applied along with encryption before data warehouse. This masking technique [20] overcomes the shortcomings of encryption using modulus operator. It balances security strength and privacy issues with performance. It provides better security and performance tradeoffs by significantly decreasing storage space and processing overheads in operations performed on data of data warehouse [9]. The formula used for masking and unmasking the data is defined in Eqs. 3 and 4 as follows [21]:

For data masking, Eq. 3 is used:

$$(B_L, D_M) = (B_L, D_M) - \big((K_{3,L} MOD\, K_2) MOD (K_2 * 2)\big) + K_2 \tag{3}$$

For unmasking the data, Eq. 4 is used:

$$(B_L, D_M) = (B_L, D_M) + \big((K_{3,L} MOD\, K_2) MOD (K_2 * 2)\big) - K_2 \tag{4}$$

- K_1, K_2 are private keys having size of 128 bytes.
- B_L defines the row and D_M defines the column in which we mask the data.
- K_3 is a public key.

4.3 Hash Algorithm

It is the last component of model which is applied before the data warehouse for authentication. It overcomes the problem of OTP encryption which does not provide message authentication and integrity [12]. It verifies the authenticity of data if the data is corrupted, either by an adversary or transmission errors. Hash Function is applied on plaintext and produces the hash output value called message digest. This message digest is encrypted with the message to produce ciphertext and then loaded into the data warehouse. The person who has the appropriate one-time pad at receiver's side is able to decrypt the message correctly and compares the corresponding hash with message digest. If hash values are same then access is allowed, else it is rejected [12]. One-way hash function should be used so that no message is disclosed by their signatures [22].

5 Advantages of Proposed Approach

The proposed model of data warehouse provides the following advantages:

1. Data security and confidentiality will be improved due to one-time pad encryption.
2. Hash algorithm will improve the authentication and integrity of message.
3. Incorporation of OTP in AES makes the unauthorized access impossible by attackers.
4. MOBAT masking technique solves the shortcomings of response time, storage space and performance overheads due to encryption.

6 Conclusion

This paper analyses different security techniques used in data warehouse. It highlights the advantages and shortcomings of existing techniques. The model mainly focuses on enhancing the security of data warehouse against unauthorized access. It keeps the characteristics of data intact in the data warehouse- integrity, confidentiality and authenticity. It brings improvement in the present scenario of data warehouse's security. Thus, it is an efficient and valid alternative for balancing the performance and security issues of data warehouse by introducing a masking function into the model. Hash algorithm provides authentication to the data warehouse. By inserting algorithm One-Time Pad (OTP) to the algorithm Advanced Encryption Standard (AES), the resulting encryption algorithm does not affect the processing time and size after encryption is the same size as the original file [11]. The model is proposed as an alternative solution to standard cryptographic algorithms such as AES, 3DES, but it does not have an intention of replacing them.

References

1. Pooniah, P.: Data Warehousing Fundamentals-A Comprehensive Guide for IT Professionals. Wiley (2001)
2. Inmon, W.H.: Building the Data Warehouse. Wiley, Chichester (1996)
3. Oracle white paper: security and the Data Warehouse, Apr 2005
4. Borowski, M., Leśniewicz, M.: Modern usage of "old" one-time pad. In: Proceedings of Communications and Information Systems Conference (MCC), Gdansk, Poland (2012)
5. Soofi, A.A., Riaz, I., Rasheed, U.: An enhanced vigenere cipher for data security. Int. J. Sci. Technol. Res. **5**(3) (2016)
6. Patil, S., Kumar, A.: Effective secure encryption scheme (One Time Pad) using complement approach. Int. J. Comput. Sci. Commun. (2014)
7. Soler, E., Trujillo, J., Fernández-Medina, E., Piattini, M.: A framework for the development of secure data warehouses based on MDA and QVT. In: Proceedings of Second International Conference on Availability, Reliability and Security, IEEE (2007)
8. Ahmad, S., Ahmad, S.: An improved security framework for data warehouse: a hybrid approach. In: Proceedings of Information Technology (ITSim), IEEE, Malaysia (2010)
9. Santos, R.,J., Bernardino, J., Vieira, M.: Balancing security and performance for enhancing data privacy in data warehouses, In: Proceedings of International Joint Conference, IEEE (2011)
10. AES: Advanced Encryption Standard, U.S. Doc/NIST, FIPS Publication 197, Nov 2002
11. Widiasari, I.R.: Combining advanced encryption standard (AES) and one time pad (OTP) encryption for data security. Int. J. Comput. Appl. **57**(20) (2012)
12. Tool: One-time Pad—Dezepagina in het Netherlands, Manuals one time pads
13. Tang, S., Liu, F.: A one-time pad encryption algorithm based on one way hash and conventional block cipher. In: Proceedings of IEEE Conference (2012)
14. Lio Zheng, X.: Research on security of MD5 algorithm application. In: Proceedings of International Conference (CST), Beijing, China (2016)
15. Ali, O., Ouda, A.: A classification module in data masking framework for business intelligence platform in healthcare. In: 7th International Proceedings of Annual Conference (IEMCON), IEEE, Vancouver, Canada (2016)
16. Rivest, R.L.: The MD5 Message Digest Algorithm, RFC 1321 (1992)
17. Sears, I.: Rijndael AES. http://www.unc.edu/marzuola/Math547_S13/Math547_S13_Projects/I_Sears_Section003_AdvancedEncryptionStandard.pdf
18. Triki, S., Ben-Abdallah, H., Harbi, N., Boussaid. O: Securing the Data Warehouse: A Semi Automatic Approach for Inference Prevention at the Design Level, Model and Data Engineering, LNCS, vol. 6918, pp. 71-84. Springer (2011)
19. Santos, R.J., Bernardino, J., Vieira, M.: A data masking technique for data warehouses. In: Proceedings of ACM Communication, IDEAS11. ACM, Portugal (2011)
20. Oracle Inc.: Data Masking Best Practices, White Paper, July (2010)
21. Santos, R.J., Bernardino, J., Vieira, M.: A survey on data security in data warehousing issues, challenges and opportunities. In: Proceedings of International Conference on Computer as a Tool (EUROCON), IEEE, Lisbon, pp. 1–4 (2012)
22. Stallings, W.: http://williamstallings.com/Extras/SecurityNotes/lectures/authent.html (2012). Accessed Dec 2012

A New Framework to Categorize Text Documents Using SMTP Measure

M. B. Revanasiddappa and B. S. Harish

Abstract This article presents a novel text categorization framework based on Support Vector Machine (SVM) and SMTP similarity mea-sure. The performance of the SVM mainly depends on the selection of kernel function and soft margin parameter C. To reduce the impact of kernel function and parameter C, in this article, a novel text categorization framework called SVM-SMTP framework is developed. In the proposed SVM-SMTP framework, we used Similarity Measure for Text Processing (SMTP) measure in place of optimal separating hyper-plane as categorization decision making function. To assess the efficacy of the SVM-SMTP framework, we carried out experiments on publically available datasets: Reuters-21578 and 20-NewsGroups. We compared the results of SVM-SMTP framework with other four similarity measures viz., Euclidean, Cosine, Correlation and Jaccard. The experimental results show that the SVM-SMTP framework outperforms the other similarity measures in terms of categorization accuracy.

Keywords Text categorization · Support vectors · Similarity measure
Similarity measure for text processing

1 Introduction

Text categorization helps in organizing and managing the electronic text documents, where electronics text documents are assigned to pre-defined categories based on similarity and content [1]. Many researchers proposed a number of statistical and computational techniques for text document categorizations viz., Bayesian [2], K-Nearest Neighbor (KNN) [3], Support Vector Machine (SVM) [4], Rule Induction

M. B. Revanasiddappa (✉) · B. S. Harish
Department of Information Science and Engineering,
Sri Jayachamarajendra College of Engineering, Mysuru, India
e-mail: revan.cr.is@gmail.com

B. S. Harish
e-mail: bsharish@sjce.ac.in

K. Ray et al. (eds.), *Soft Computing: Theories and Applications*,
Advances in Intelligent Systems and Computing 742,
https://doi.org/10.1007/978-981-13-0589-4_48

[5], Artificial Neural Networks (ANN) [6] and many more. Among these techniques, Support Vector Machine is one of the most popular Machine Learning techniques, which is reported as a best performing technique to categorize text documents [4, 7–9]. The main characteristic of SVM is Structural Risk Management (SRM) principle, which helps in discovering the hyper-plane.

In SVM, documents are categorized based on optimal separating hyper-plane [4]. The theory of hyper-plane is generalized for non-linear separable documents. The kernel functions are used to partition the non-linear separable documents correctly, which are transformed from input space to new space (higher dimensional feature space) [10, 11]. A variety of kernel functions are presented in the literature, but only certain kernel functions have succeeded in achieving better performance with SVM [12]. In order to obtain better categorization performance, it is necessary to use appropriate kernel function in SVM. The performance of the SVM mainly depends on the kernel function and soft margin parameter C. To reduce the impact of kernel function and parameter C, in this article we proposed a new text categorization framework called SVM-SMTP framework. In the proposed framework, we used similarity measure rather separating hyper-plane as categorization decision making function.

Similarity measure is used to calculate the similarity between documents, which plays a major role during categorization. Many researchers proposed different similarity (distance) measures. Some of them are Euclidean [13], Cosine [14], Jaccard [15], Correlation [16] and many more. However, we used Similarity Measure for Text Processing (SMTP) [17] measure rather than optimal separating hyper-plane to categorize text documents. SMTP is based on three properties: (i) The term occurs in both documents, (ii) The term occurs in only one document and (iii) The term does not occur in both documents. The proposed SVM-SMTP framework avoids transforming the data from input space to new space through kernel function during categorization phase. The main advantage of SVM-SMTP framework is low impact of kernel function and parameter C during categorization phase, which enhance the performance of classifier. The proposed SVM-SMTP framework comprises of two phases: Training phase and Categorization phase. During training phase, we used SVM technique to compute Support Vectors (SVs). In categorization phase, we used similarity measure in place of optimal separating hyper-plane as categorization decision making function. We used SMTP as a similarity measure to compute average distance between test documents and SVs for each class. Finally, the categorization decision is made based on the class of SVs which has minimum average distance.

The main contribution of this article is as follows:

1. Proposed a novel text categorization framework called SVM-SMTP framework
2. During categorization phase, SMTP measure is used in place of optimal separating hyper-plane to categorize text documents.

This article proceeds as follows: Sect. 2 reviews the related work. Section 3 portrays the proposed New Categorization Framework. The experimental results based on standard benchmark dataset are presented in Sect. 4. Lastly, we conclude our work with outlining future work in Sect. 5.

2 Literature Survey

The different types of categorization techniques are described in [3]. The performance of the SVM mainly depends on the kernel function and soft margin parameter C. Many researchers have proposed evolutionary algorithms to determine suitable kernel function and parameters for SVM [18, 19]. Unfortunately, these methods have failed because of its high complexity cost. Also these methods conduct iterative computation while configuring the optimal set of kernel and parameters for SVM.

Quang et al. [20] have proposed evolutionary algorithm in which SVM parameters are optimized based on genetic programming. Briggs and Oates [18] used composite kernel function in SVM for transforming data from input space to higher dimensional space. The composite kernel function provides appropriate feature space for input data. Avci [10] proposed an hybrid approach called as Genetic Algorithm-Support Vector Machine (HGASVM). This approach determines the parameter values using genetic algorithms and chooses RBF as a best kernel function for SVM. Diosan et al. [19] proposed hybrid approach based on genetic programming and SVM. The main aim of this approach is to automatically generate and adopt kernel combination. The SVM parameter optimizes by iterative process. Hence the computational complexity of hybrid approach is high. Sun [21] presents a fast and efficient method to approximate the distance between two classes. This method uses sigmoid function to select Gaussian kernel parameters. The sigmoid function tests all the combination of Gaussian kernel parameters to measure approximate distance between two classes. Sun et al. [22] presented a novel technique for tuning hyper-parameters by maximizing the distance between two classes. This method determines the optimal kernel parameters on the basis of gradient algorithm, which maximizes the distance between two classes.

All the above mentioned methods consumes more time for iterative computation, as they use evolutionary algorithms to select optimal kernel function and parameter set. In order to avoid iterative computation, Lee et al. [8] proposed a new SVM framework using Euclidean distance measure for text classification. This framework first computes the support vectors from SVM during training phase, and in classification phase, Euclidean distance measure is employed to classify text documents. Later, the average distance is computed between the new document (test document) and support vectors for each class. Lastly, test document assign to a class for which it has minimum average distance. Tsai and Chang [29] proposed Support Vector Instance Orientated Selection (SVIOS) method using support vectors. This method reduces the computation cost of SVM during training and testing phase. Harish et al. [30] proposed a new text categorization method using support vectors and Fuzzy C-Means. In this method, likelihood similarity measure is used to measure similarity between class representative and test documents during categorization phase. The cosine similarity measure is used for document clustering [31].

Lin et al. [17] proposed a novel similarity measure named as Similarity Measure for Text Processing (SMTP), which is embedded by several characteristics of existing similarity measures. This measure resolves the drawbacks of Euclidean,

Cosine, Jaccard and Correlation distance measures. SMTP successfully used in text classification and clustering techniques. Gomaa and Fahmy [23] published detailed survey work on similarity measures for text processing. Thomas and Resmipriya [24] proposed text classification method using k-means clustering technique and SMTP similarity measure. During training phase, k-means clustering is used to compute class representative (cluster centers). In classification phase, SMTP is employed to compute the similarity between cluster centers and test documents. SMTP gives more importance to absence and presence of term in a document to measures the similarity [25]. SMTP measure uses Gaussian function and standard deviation of all non-zero values. Thus, motivated from the advantages of SMTP over other similarity measure, in this article we develop a novel SVM-SMTP framework to categorize text documents.

3 Proposed Method

SVM is a well known classifier, which is widely used in text categorization. SVM categorizes the text documents by using optimal separating hyper-plane. The concept of optimal separating hyper-plane can be generalized for non-linear data by kernel function, which transforms data from input space to new space (high dimensional feature space). The proposed SVM-SMTP framework contains Training phase and Categorization phase.

3.1 Training Phase

Let us consider k number of classes C_j, $j = 1, 2, \ldots, k$, where each class contains n number of documents D_i, $i = 1, 2, \ldots, n$ with m number of terms T_l, $l = 1, 2, \ldots, m$. The text document contains set of terms (features), where all terms are not considered as important features during text categorization. Pre-processing is one of the important stage in text categorization in which stem and stop words are eliminated. Later, pre-processed text documents are represented in the form of Term Document Matrix (TDM), where each row represents document and each column represents the term. Further, TDM were fed into Support Vector Machine (SVM) to compute the Support Vectors (SVs). In SVM, as a first step, data is transformed from input space to new space and then documents are separated using hyper-plane, which is given as

$$D_i \cdot w + b = 0 \tag{1}$$

where, w is the weight, D_i is ith document and b is the bias value.

There are infinite numbers of hyper-planes present in the feature space. But SVM technique considers only one hyper-plane, which is maximum margin lay between two documents to partition the documents into two classes. The margin denotes the sum of distances of hyper-planes to the margin nearest document of each class. The hyper-plane is determined by documents, which are very nearest to each class. These documents are considered as support vectors. The numbers of SVs are less than the number of training documents. Later, these SVs considered as training set during categorization phase. The optimal separating hyper-plane is determined by the support vectors, there is a certain way to represent them for a given set of training documents. It has been described in [26] that the maximal margin can be found by minimizing $1/2\|w\|^2$, which is represented as

$$\min\{1/2\|w\|^2\} \tag{2}$$

These Support Vectors (SVs) are used in next categorization phase and the remaining documents are excluded from the training set

3.2 *Categorization Phase*

Instead of using optimal separating hyper-plane, in this phase, we introduced a new SMTP measure as categorization decision making function. Let us consider a set of test documents Q_i, where $i = 1, 2, \ldots, n$ with m number of terms T_l, $l = 1, 2, \ldots, m$. In test document, stem and stop words are eliminated during pre-processing. The SMTP computes the distance between support vectors and test documents, which is given as

$$D_{SMTP}(D_i, Q_i) = \frac{F(D_i, Q_i) + \lambda}{1 + \lambda} \tag{3}$$

where, D_i is ith Support Vector, Q_i is ith test document and λ is constant. Function $F(D_i, Q_i)$ defines

$$F(D_i, Q_i) = \frac{\sum_{l=1}^{m} N_*(D_{il}, Q_{il})}{\sum_{l=1}^{m} N_U(D_{il}, Q_{il})} \tag{4}$$

$$N_*(D_{il}, Q_{il}) = \begin{cases} 1, & if\ (D_{il} = Q_{il})\, and\, (D_{il}Q_{il} > 0) \\ 0.5\{1 + \exp\{-R\}\}, & if\ D_{il}Q_{il} > 0 \\ 0, & if\ D_{il} = 0\ and\ Q_{il} = 0 \\ -\lambda, & otherwise \end{cases}$$

$$where,\ R = \left[\frac{D_{il} - Q_{il}}{\sigma_l}\right]^2 \tag{5}$$

$$N_U(D_{il}, Q_{il}) = \begin{cases} 0,\ if\ D_{il} = 0\ and\ Q_{il} = 0 \\ 1,\ otherwise \end{cases} \tag{6}$$

where, D_{il} is the ith term of ith document (support vector) and Q_{il} is the lth term of ith test document. The SMTP considers three conditions where, first condition uses Gaussian function, in that σ_l is the standard deviation of all non-zero values for ith term in training dataset in Eq. (5). In second condition, $-\lambda$ is ignoring the magnitude of the non-zero feature value. In last condition, similarity of feature has no contribution.

The SMTP computes the distance between test document Q_i and each SVs of each class. After obtaining the distance between test document and each support vector of each class, the average distance of test document to set of SVs of each class has been computed. The average distance of each class is given by

$$D_{avg_j} = \frac{\sum_{i=1}^{n} D_{SMTP}(D_i, Q_i)}{n} \quad (7)$$

where, n is the number of support vectors of each class. Now we obtain the average distance for each class. The decision of categorization is made based on the class which has minimum distance between test document and its set of SVs. In other words, test document will be assigned with a class which has the minimum average distance between SVs and test document itself. Algorithm 1 explains individual steps involved in proposed method.

Algorithm 1: **A New Categorization Framework (SVM-SMTP)**

Input: Number of class with number of documents and number of terms (features), is constant
Output: Class Label
Step 1: Text documents are pre-processed and represented using Term Document Matrix (TDM) form
Step 2: Construct optimal separating hyper-plane using Eq. (1)
Step 3: Select nearest documents of optimal separating hyper-plane. These documents are called Support Vectors.
Step 4: Compute distance between test document and support vectors using Eq. (3)
Step 5: Compute Average distance of test document and set of support vectors of each class using Eq. (7)
Step 6: Assign test document to a class, which has lowest average distance between the test document and its set of support vectors

4 Experimental Setup

The main aim of the proposed framework is to improve the performance of categorization. The standard benchmark datasets that we are using for experiments are Reuters-21578 [27] and 20-NewsGroups [28]. The Reuters-21578 has a collection of 65 classes with 8293 documents and these documents are not equal proposition for each class. The 20-Newsgroups is a very popular standard benchmark dataset to use

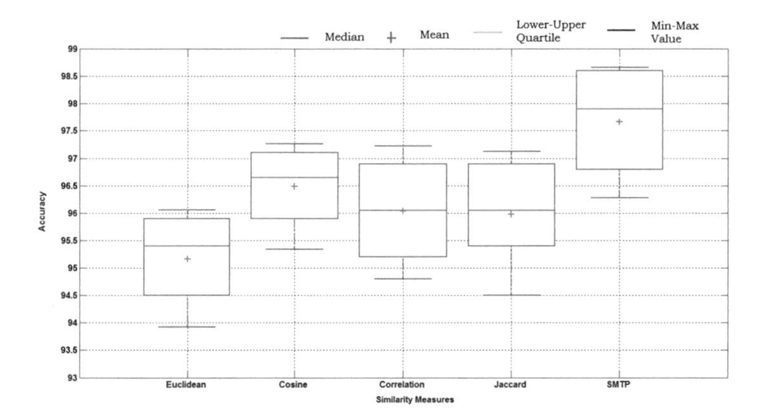

Fig. 1 Boxplots presenting the results of ten trials on Reuters-21578 dataset

in text categorization, which contains 18846 documents with pre-defined 20 classes. We set λ value as 0.5 by empirically. We used categorization accuracy to evaluate the performance of SVM-SMTP framework. We run the application for ten-fold times and selected best result among them. In order to evaluate the performance of the SVM-SMTP framework, we used four similarity measures (Euclidean, Cosine, Correlation and Jaccard) to compare with the result of proposed framework. These similarity measures are used during categorization phase to compute the average distance between test documents and support vectors of each class.

During the experimentation, 60% of each dataset is used to build the training set and 40% of remaining dataset is used for testing. We conduct experiments on all five similarity measures for both the standard datasets. All the experiments are repeated ten-fold times by selecting the input (training) documents randomly and presented the obtained results using boxplots. Since, training samples are selected randomly, the results may varies from iteration to iteration. Thus, we used boxplots to represent maximum, minimum, mean and standard deviation accuracy values of the ten trails.

Figures 1 and 2 presents the results of ten trails on Reuters-21578 and 20-NewsGroups respectively. Among five different similarity measures, SMTP (proposed framework) similarity measure outperformed during categorization phase. The proposed framework (SVM-SMTP) obtain a result of 97.66 ± 0.98 on Reuters-21578. Further, proposed framework results are also compared with other four similarity measures. The obtained results using Euclidean, Cosine, Correlation and Jaccard are 95.16 ± 0.81, 96.49 ± 0.74, 96.04 ± 0.92 and 95.98 ± 0.96 respectively. Similarly, we conducted experiments on 20-NewsGroups. The SVM-SMTP obtain a result of 87.16 ± 1.33. Further, proposed framework results are also compared with other four similarity measures. The obtained results using Euclidean, Cosine, Correlation and Jaccard are 84.50 ± 1.47, 86.63 ± 0.74, 86.44 ± 0.92, 84.72 ± 1.28 respectively. It is observed from the result Table 1, our proposed framework shows better results compared to other four similarity measures.

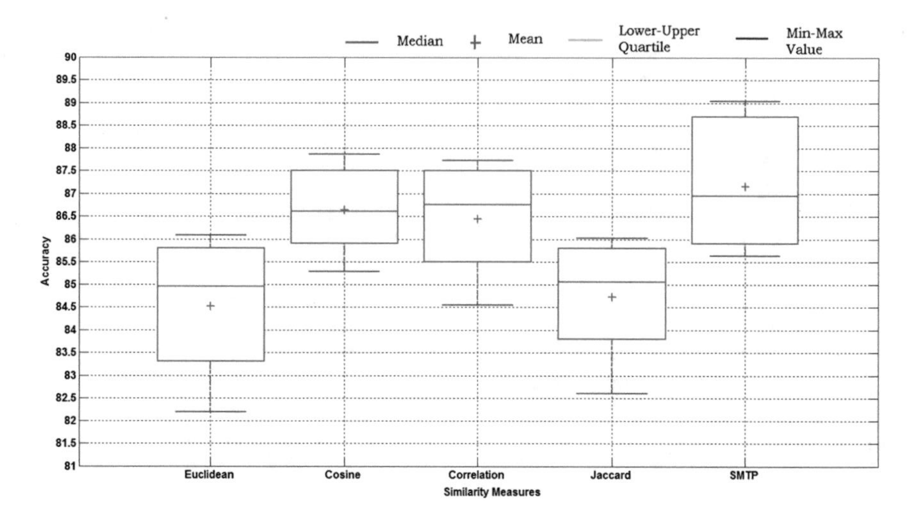

Fig. 2 Boxplots presenting the results of ten trials on 20-NewsGroups dataset

Table 1 Performance comparison of proposed framework (SVM-SMTP) with other similarity measures

Similarity Measure with SVM	Accuracy (%)					
	Reuters-21578			20-NewsGroups		
	Min	Max	Avg	Min	Max	Avg
Euclidean-SVM	93.80	96.10	95.16	81.90	86.20	84.52
Cosine-SVM	95.10	97.30	96.49	85.20	87.90	86.63
Correlation-SVM	94.60	97.30	96.04	84.20	87.70	86.44
Jaccard-SVM	94.00	97.20	95.98	82.80	86.10	84.72
SVM-SMTP	**96.20**	**98.70**	**97.66**	**85.60**	**89.20**	**87.16**

Table 1 illustrates the comparative results of the SVM-SMTP framework with other similarity measures. The minimum, maximum and average accuracies of all the ten trails are presented in Table 1. SMTP gives more importance to the presence or absence of term in the documents and also resolves limitations of other similarity measures. Euclidean fails to handle absence or presence of term in the documents. Cosine, Correlation and Jaccard does not use standard deviation to measure the similarity between documents, but SMTP considers standard deviation to determine the contribution of term value in between documents. Standard deviation is one of the best measure for variation, which computes the spread of data from average. The

SVM-SMTP achieves better results in terms of categorization accuracy over the other similarity measures on both standard benchmark datasets. The SVM-SMTP obtain an average accuracy of 97.66 and 87.16% on Reuters-21578 and 20-NewsGroups respectively.

5 Conclusion

This article, a new categorization framework called SVM-SMTP Framework is proposed to categorize text documents. The main objective of this article is introducing SMTP (Similarity Measure for Text Processing) measure in place of optimal separating hyper-plane as a categorization decision making function. Unlike, the performance of the SVM mainly depends on the selection of kernel function and soft margin parameter C. To minimize the impact of kernel function and parameter C during categorization phase. The proposed method adopts SMTP measure as a decision making function. To verify the performance of SVM-SMTP Framework, we conducted extensive experimentation on publically available datasets: Reuters-21578 and 20-NewsGroups. The experimental results reveal that the SVM-SMTP framework is superior to the other four similarity measures (Euclidean, Cosine, Correlation and Jaccard) in terms of categorization accuracy.

In future, we are intended to add feature selection technique to SVM-SMTP framework, which inturn reduces the dimensionality of feature matrix and speedup the categorization performance.

References

1. Sebastiani, F.: Machine learning in automated text categorization. ACM Comput. Surv. (CSUR). **34**(1), 1–47 (2012)
2. Domingos, P., Pazzani, M.: On the optimality of the simple Bayesian classifier under zero-one loss. Mach. Learn. **29**(2), 103–130 (1997)
3. Harish, B.S., Guru, D.S., Manjunath, S.: Representation and classification of text documents: A brief review. IJCA Special Issue RTIPPR. **2**, 110–119 (2010)
4. Joachims, T.: Text categorization with support vector machines: learning with many relevant features. In: Machine Learning: ECML-98, pp. 137–142 (1998)
5. Apt, C., Damerau, F., Weiss, S.M.: Automated learning of decision rules for text categorization. ACM Trans. Inf. Syst. (TOIS) **12**(3), 233–251 (1994)
6. Chen, C.M., Lee, H.M., Hwang, C.W.: A hierarchical neural network document classifier with linguistic feature selection. Appl. Intell. **23**, 277–294 (2005)
7. Isa, D., Lee, L.H., Kallimani, V.P., Rajkumar, R.: Text document preprocessing with the Bayes formula for classification using the support vector machine. IEEE Trans. Knowl. Data Eng. **20**(9), 1264–1272 (2008)
8. Lee, L.H., Wan, C.H., Rajkumar, R., Isa, D.: An enhanced support vector machine classification framework by using Euclidean distance function for text document categorization. Appl. Intell. **37**(1), 80–99 (2012)
9. Li, Y., Zhang, T.: Deep neural mapping support vector machines. Neural Netw. **93**, 185–194 (2017)

10. Avci, E.: Selecting of the optimal feature subset and kernel parameters in digital modulation classification by using hybrid genetic algorithm support vector machines: HGASVM. Expert Syst. Appl. **36**(2), 1391–1402 (2009)
11. Chen, Y.C., Su, C.T.: Distance-based margin support vector machine for classification. Appl. Math. Comput. **283**, 141–152 (2016)
12. Fischetti, M.: Fast training of support vector machines with gaussian kernel. Discrete Optim. **22**, 183–194 (2016)
13. Schoenharl, T.W., Madey, G.: Evaluation of measurement techniques for the validation of agent-based simulations against streaming data. In: International Conference on Computational Science, pp. 6–15. Springer, Berlin, Heidelberg (2008)
14. Al-Anzi, F.S., AbuZeina, D.: Toward an enhanced Arabic text classification using cosine similarity and Latent Semantic Indexing. J. King Saud Univ.-Comput. Inf. Sci. **29**(2), 189–195 (2017)
15. Gonzlez, C.G., Bonventi Jr, W., Rodrigues, A.V.: Density of closed balls in real-valued and autometrized boolean spaces for clustering applications. In: Brazilian Symposium on Artificial Intelligence, pp. 8–22. Springer Berlin, Heidelberg (2008)
16. Nigam, K., McCallum, A., Mitchell, T.: Semi-supervised text classification using EM. In: Semi-Supervised Learning, pp. 33–56 (2006)
17. Lin, Y.S., Jiang, J.Y., Lee, S.J.: A similarity measure for text classification and clustering. IEEE Trans. Knowl. Data Eng. **26**(7), 1575–1590 (2014)
18. Briggs, T., Oates, T.: Discovering domain-specific composite kernels. In: Proceedings of the National Conference on Artificial Intelligence, vol. 20, no. 2, pp. 732–738 (2005)
19. Diosan, L., Rogozan, A., Pecuchet, J.P.: Improving classification performance of support vector machine by genetically optimising kernel shape and hyper-parameters. Appl. Intell. **36**(2), 280–294 (2012)
20. Quang, A.T., Zhang, Q.L., Li, X.: Evolving support vector machine parameters. In: 2002 Proceedings of the International Conference on Machine Learning and Cybernetics. vol.1, pp. 548–551. IEEE (2002)
21. Sun, J.: Fast tuning of SVM kernel parameter using distance between two classes. In: 3rd International Conference on Intelligent System and Knowledge Engineering, 2008. ISKE 2008, vol. 1, pp. 108–113. IEEE (2008)
22. Sun, J., Zheng, C., Li, X., Zhou, Y.: Analysis of the distance between two classes for tuning SVM hyperparameters. IEEE Trans. Neural Netw. **21**(2), 305–318 (2010)
23. Gomaa, W.H., Fahmy, A.A.: A survey of text similarity approaches. Int. J. Comput. Appl. 68(13) (2013)
24. Thomas, A.M., Resmipriya, M.G.: An efficient text classification scheme using clustering. Proced. Technol. **24**, 1220–1225 (2016)
25. Nagwani, N.K.: A comment on a similarity measure for text classification and clustering. IEEE Trans. Knowl. Data Eng. **27**(9), 2589–2590 (2015)
26. Haykin, S., Network, N.: A comprehensive foundation. In: Neural Networks, no. 2, pp. 41 (2004)
27. Reuters-21578. http://www.daviddlewis.com/resources/testcollections/reuters21578/
28. Newsgroups. http://people.csail.mit.edu/jrennie/20Newsgroups/
29. Tsai, C.F., Chang, C.W.: SVOIS: support vector oriented instance selection for text classification. Inf. Syst. **38**(8), 1070–1083 (2013)
30. Harish, B.S., Revanasiddappa, M.B., Kumar, S.A.: A modified support vector clustering method for document categorization. In: IEEE International Conference on Knowledge Engineering and Applications (ICKEA), pp. 1–5 (2016)
31. Dhillon, I.S., Modha, D.S.: Concept decompositions for large sparse text data using clustering. Mach. Learn. **42**(1), 143–175 (2001)

A Task Scheduling Technique Based on Particle Swarm Optimization Algorithm in Cloud Environment

Bappaditya Jana, Moumita Chakraborty and Tamoghna Mandal

Abstract Cloud computing is on the edge of another revolution where resources are globally networked and can be shared user to user in easy way. Cloud computing is an emerging research domain which encompasses computing, storage, software, network, and other heterogeneous requirements on demand. Today, dynamic resource allocation and proper distribution of loads in cloud server are a challenging task. So task scheduling is an essential step to enhance the performance of cloud computing. Although lots of scheduling algorithms are used in cloud environment but still now no reasonably efficient algorithms are used. We have proposed a Modified Particle Swarm Optimisation (MPSO) technique where we have focused on two essential parameters in cloud scheduling such as average scheduling length and ratio of successful execution. According to the result analysis in simulation, Modified Particle Swarm Optimization (MPSO) technique shows better performance than Min-Min, Max-Min, and Standard PSO. Finally, critical future research directions are outlined.

Keywords Scheduling · Optimization · Virtualization · Load balancing Scalability

B. Jana (✉)
Chaibasa Engineering College (Estd. By Govt. of Jharkhand & run by Techno India under PPP), Kelende, Jharkhand, India
e-mail: bappaditya.j.in@ieee.org

M. Chakraborty
Techno India Banipur, Banipur, West Bengal, India
e-mail: sumita.03@gmail.com

T. Mandal
Indian Institute of Engineering Science and Technology
Shibpur, Shibpur, West Bengal, India
e-mail: tamoghna.iiests@gmail.com

K. Ray et al. (eds.), *Soft Computing: Theories and Applications*,
Advances in Intelligent Systems and Computing 742,
https://doi.org/10.1007/978-981-13-0589-4_49

1 Introduction

Cloud computing is a new computing technology to enhance the virtualized resources designed for end users in a dynamic environment in order to provide reliable and trusted service [1]. Cloud computing is a metered services at various qualities over virtualized networks to many end users [2]. It is a fast growing computing technology which enhances the virtualization of all IT resources as well as IT infrastructure. Cloud computing has become a fast growing research domain in last ten years and is a model for computation which provides resource in resource to cloud users, which satisfy heterogeneous demand as per requirement through Internet where Google File System (GFS), Google mail, Amazon Cloud, and Hadoop architecture are most popular cloud computing resource pool. Task scheduling problem in cloud computing is one of the most famous combinatorial optimization problem [3]. Typically, a cloud vendors wants an effective automatic load balancing technique based on distributing computing task [4] for their service delivery system. A well manage cloud has a tendency to optimize the such parameter as scalability, mobility, availability, elasticity, minimize infrastructure cost, disaster recovery, enhance throughput and storage capacity [5, 6]. Today a lot of heuristic algorithms have been used to minimize the task scheduling problem such as Genetic Algorithm (GA), Particle Swarm Optimization Algorithm (PSO), Ant Colony Optimization (ACO), Cuckoo Search (CS) algorithm, etc. (Fig. 1).

Why do customer use the cloud ?

So it is clearly observed that why do customer interested toward cloud. According to the cost reduction, speed of adoption, business process transformation cloud computing becomes an emerging IT research domain in the present scenario.

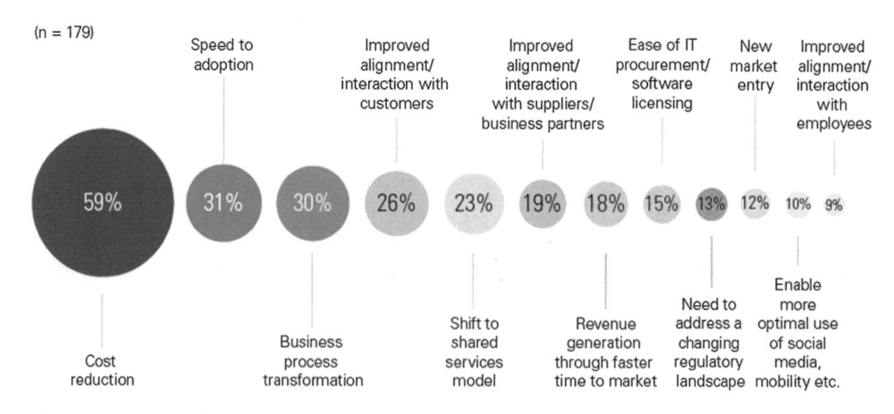

Fig. 1 KPMG International's 2012 Global Cloud Provider Survey (n = 179)

2 Cloud Service Model

Cloud service is a virtual network model, which is an on-demand service. Basic components of cloud model are user, datacenter, and cloud server. Cloud architecture includes two types of models such as cloud delivery model and cloud deployment model.

2.1 Delivery Model

Cloud delivery models are classified into four type of distinguished sub-model as follows.

2.1.1 Software as a Service (SaaS)

Cloud user executes virtually their various applications using cloud as a hosting environment. Through Internet, they can use cloud from any geographical location, for example, web browser, Google mail, Google Docs, etc. [7].

2.1.2 Platform as a Service (PaaS)

In PaaS, cloud is used as a platform for software development and supports SDLC (Software Development Life Cycle). SaaS is considered for completed or existing software whereas PaaS is used during the development of application. PaaS is applicable for both programming environment and hosting environment. Cloud is used as PaaS in Google App Engine [8, 9].

2.1.3 Infrastructure as a Service

Cloud platform is considered as a huge storage space and used for high-level computing purpose. In IaaS, cloud provider serves as a physical infrastructure as per the demand of the cloud user. The basic strategy of virtualization is to create Virtual Machines (VMs). Amazon's EC2 is one of the examples of IaaS. In Fig. 2 [10], the abovementioned services are shown.

2.1.4 Data Storage as a Service (DaaS)

DaaS is an extension of IaaS. The major purpose of DaaS is to optimize the cost of software licensee, in-house IT maintenance, dedicated server, post delivery services,

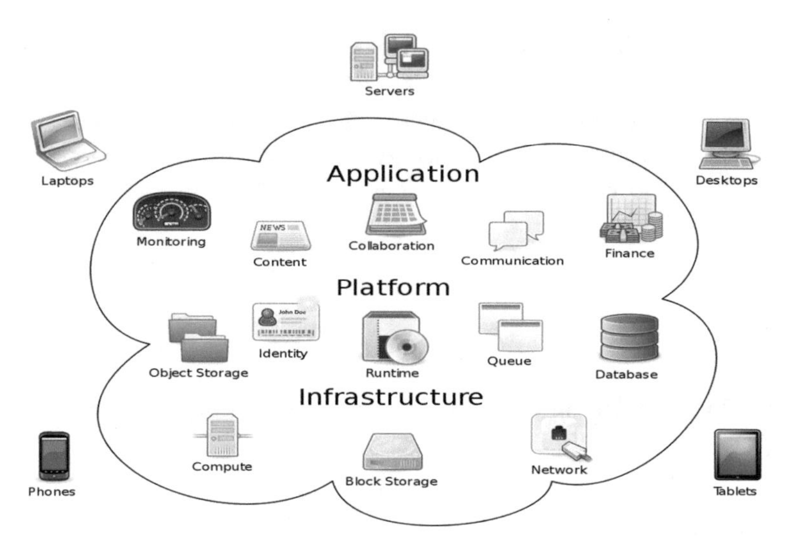

Fig. 2 Schematic diagram of delivery model

etc. DaaS provides the user the advantage of commercial RDMS. Apache, HBase, Google BigTable, and Amazon S3 are the examples of DaaS.

2.2 Deployment Model

Major cloud community has four types of deployment models as follows.

2.2.1 Public Cloud

In recent times, Public cloud model is one of the popular deployment models. Public cloud is made for basically general user. The service provider provides the accessibility of major information to public through Internet. Some of the public clouds are Google App Engine, Amazon EC2, S3, Force.com, etc.

2.2.2 Private Cloud

Private cloud is made independently for local user in a single organization to share in-house resources and it is controlled through LAN by organization. Any third party cannot access without permission. Private clouds are created in many big institute for research and teaching activities.

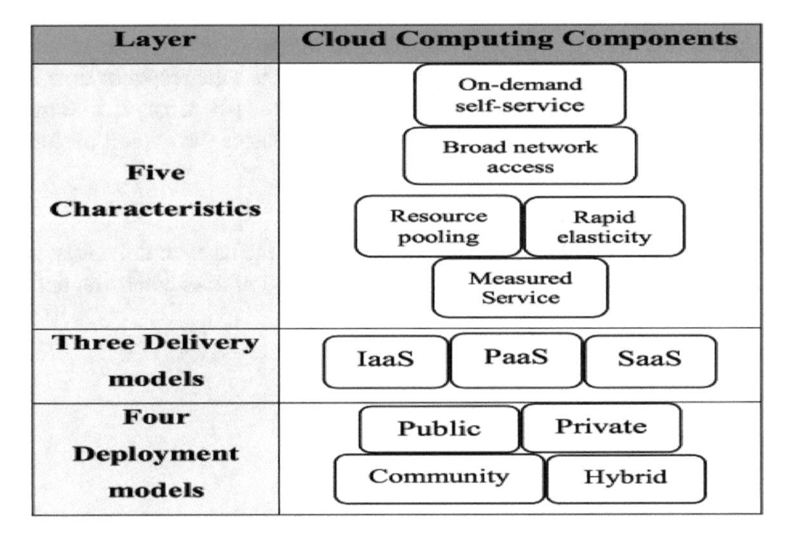

Fig. 3 Schematic diagram on cloud service architecture

2.2.3 Community Cloud

Now a day some organizations use the same cloud to share their information according to requirements with respect to some common ethics and security concern. Either a third a party or one of the organization acts as admin to control the cloud.

2.2.4 Hybrid Cloud

Hybrid cloud model consists of two or more than two clouds including public and private cloud. The major purpose of hybrid model is to optimize load balancing. Figure 3 [10] shows the overall cloud service architecture.

3 Load Balancing

To explain load balancing, let us consider all data center in cloud server behaves as a node. As cloud service demand as a service, so in most of the situation some nodes are loaded with high requirement and some nodes are loaded with comparatively too little. This situation makes queue near highly populated node, which increases the time complexity. So to minimize the time complexity, to optimize utilization of heterogeneous resources in each node, cloud service should need proper load balancing algorithm. In load balancing, our target is to create a suitable mapping between user request and resources [11, 12]. We summaries some important issues for cloud computing: 1. proper utilization of heterogeneous recourses as per demand, 2. minimize time complexity of each process, 3. decrease the response time of each process, 4. enhance satisfaction of various cloud user as per demand, 5. to maintain the equilibrium of workload of each node, and 6. enhance the overall performance of cloud the service.

Types of Load Balancing:
Load balancing in a cloud model can be classified on the basis of three way, such as according to the Present status of the cloud: Two type of load balancing according to the

3.1 Current Status of the Cloud

3.1.1 Static Load Balancing

In static load balancing algorithm, some important parameters such as processing power, storage, and requirements of data are considered. This procedure does not need real-time information as well as does not perform during execution. Some popular load balancing algorithms are used in static model.

Round-Robin Load Balancing. Shortest Job Scheduling, Max-Min Load Balancing, Min-Min Load Balancing, Combination of Opportunistic Load Balancing (OLB) and Load Balance Min-Min (LBMM) Central Load balancing Policy for Virtual Machines (CLBVM).

3.1.2 Dynamic Load Balancing

In real-time system, during execution of the process-current status of each node is changed simultaneously. So for proper scheduling, an efficient algorithm should be required. Some more frequently algorithms are used in dynamic system such as Fuzzy Active Monitoring Load Balancing (FAMLB), Active Clustering, Throttled Load Balancing, Power-Aware Load Balancing (PALB), and Biased Random Sampling Generalized Priority Algorithm [13, 14].

3.2 Initialization of Process

Load balancing according to the process initialization by sender or receiver or both that means Initialization by sender, Initialization by receiver Initialization by both sender and receiver.

3.3 Decision Strategy

Load balancing according to the decision strategy basically depends as decision taken by one node or multiple nodes or decision taken by in hierarchical order such as Centralized Load Balancing, Distributed Load Balancing and Hierarchical Load Balancing [15–17].

4 Virtual Cloud

CloudSim is a virtual framework which is used for simulation of performance of any cloud computing-based infrastructure and seamless modeling, etc. CloudSim was first developed by GRIDS laboratory in Melbourne Load balancing for virtual cloud, data center, cloud server all can be experimented by cloud successfully by different level of research [18–22]. CloudSim efficiency supports at two levels.

4.1 At Host Level

Virtual Machine (VM) policy allocation is performed at host level. It predicts the amount of processing power of each core which will be allocated in each virtual machine.

4.2 At VM Level

At VM level CloudSim performed as VM scheduling. The individual application service is assigned by VM with a fixed amount of available processing power.

5 Proposed Model

Particle Swarm Optimization technique is a meta-heuristic group-based searching algorithm which has many advantages such as strong robustness, high scalability, high flexibility and easy to realize, etc. But there are two major short comes in PSO as low convergence rate in case of large-scale optimization problem and due to strong randomness, it goes rapidly into the defects of local optima. So to overcome these short comes, we have been proposed a modified Particle Swarm Optimization (MPSO) algorithm which is the combination of PSO and Simulated Annealing Algorithm (ASA). ASA has high jumping ability to optimize partially for a better individual whereas PSO can search better Swarm with fastly. Thus, MPSO gives a comparatively high rate of convergence.

PSO Algorithm

Step-1: Initialize S_i as the swarm and initialized random position of particle within hypercube feasible space.
Step-2: Calculate performance of each particle E with respect to its current position $S_i(t)$
Step-3: According to best performance compare performance of each individual and continue it.
$E(S_i(t)) < E(P_{ibest})$:
$E(P_{ibest}) = E(Si(t))$
$E_{ibest} = S_i(t)$
Step-4: Now Compare the individual performance (E_i) with the global best particle:
If $E(S_i(t)) < E\left(P_{gbest}\right)$:
$E\left(P_{gbest}\right) = E(Si(t))$
$E_{gbest} = S_i(t)$
Step-5: Again each particle moves in new random position after changing the velocity of the particle
Step-6: Repeat Step-2, until getting the convergence.

Proposed MPSO Algorithm:

Step-1: Initialize n number of solutions randomly and initialize variables.
Step-2: Calculate the median of the current positions and select the particles
Step-3: Select the nearest node of medians and initialize that node particles as gbest(SAA)
Step-4: Start crossover with binary mutation

```
Begin:
While(False in the stop condition)
Begin
Evaluation
Velocity up gradation and mutation of position
End
```

Step-5: Select the position by local search and global search using probability.
Step-6: Two bins are partially swaped combine two least filled bins and each bin is divided into two new bins,
Step-7 Shuffle the bins in intra relation by rotating the box
Step-8: Stop the crossover with binary mutation.

Table 1 Average schedule length

Iteration number	Number of tasks	Max-Min Scheduling time	Minimum Execution time	MPSO scheduling algorithm
1	100	342	330	322
2	300	1024	1006	968
3	500	1736	1728	1632
4	700	2418	2364	2262
5	900	3096	3036	2926

Table 2 Ratio of successful execution

Iteration number	Number of tasks	Max-Min scheduling time	Minimum Execution time	MPSO scheduling algorithm
1	100	0.47	0.49	0.52
2	300	0.67	0.69	0.74
3	500	0.74	0.76	0.79
4	700	0.78	0.79	0.82
5	900	0.76	0.78	0.81

```
Step-9: Store non dominated particles for future.
Step-10: Update every position of swarm under the hyper
cubic solution space.
```

6 Results and Calculation

CloudSim is used as a VMs simulator which is designed to schedule the service requests from cloud users. Each service request is equivalent to no of task. Each VM has a particular size of memory which can execute more than millions of instructions per second. During execution memory of the virtual machine and bandwidth are dynamically changeable. Memory is varied from 256 MB to 2 GB and bandwidth 256 kbps is dynamically varied. We have considered average schedule length and ratio of successful execution for number of tasks ranging from 100 to 900 with increments of 300 in two frequently used algorithm such as Max-Min scheduling and Minimum Execution time scheduling algorithm with Particle Swarm Optimization algorithm.

From Table 1 and Fig. 4, it is observed that the average schedule length is reduced by 5.99% than Max-Min scheduling algorithm and 5.55% than Minimum Execution time for number of task 500. Similarly in Table 2 and Fig. 5, it is cleared that ratio of successful execution is increased by 6.76% Max-Min scheduling and by 3.94% than Minimum Execution time with number of tasks 500.

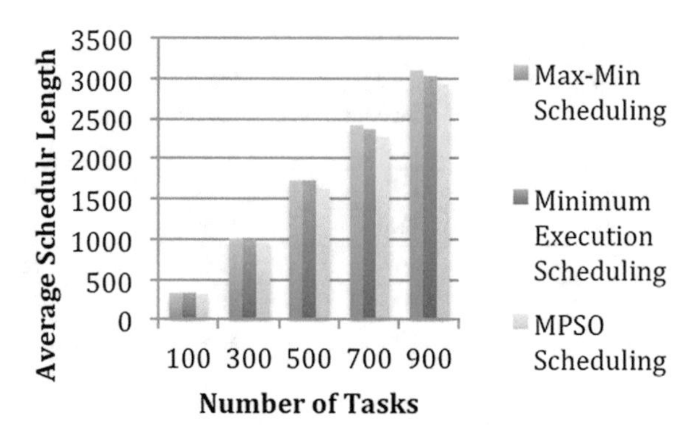

Fig. 4 Average schedule length

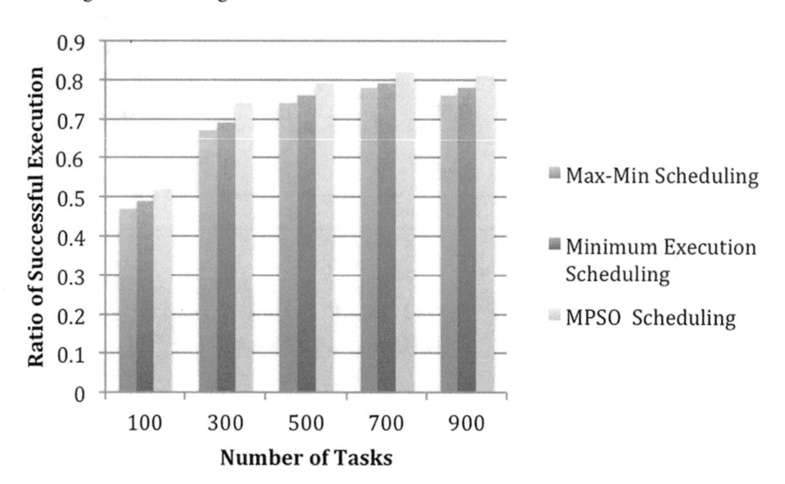

Fig. 5 Ratio of successful execution

7 Conclusions

Load balancing is the key role in enhancing the performance of the cloud computing, although lots of algorithms have been used to schedule a huge number of tasks in an efficient manner. We have mentioned two robust algorithms for cloud scheduling such as Max-Min and Minimum Execution time algorithm. In our consideration, we have tested that the result of average scheduling time and ratio of successful execution is comparatively better in Particle Swarm Optimization algorithm. So our proposed method enhances the business performance in cloud-based sector. In future, we will try to design an efficient load balancing algorithm, which is applicable for high degree of load factor as well as low degree of load factor in real-time environment.

References

1. Dillon, T., Wu, C., Chang, E.: Cloud computing: issues and challenges. In: 2010 24th IEEE International Conference on Advanced Information Networking and Applications (AINA), pp. 27–33 (2010)
2. Stephanakis, I.M., Chochliouros, I.P., Caridakis, G., Kollias, S.: A particle swarm optimization (PSO) model for scheduling nonlinear multimedia services in multicommodity fat-tree cloud networks. In: Iliadis, L., Papadopoulos, H., Jayne, C. (eds.) Engineering Applications of Neural Networks. EANN 2013. Communications in Computer and Information Science, vol. 384. Springer, Berlin, Heidelberg (2013)
3. Ramezani, F., Lu, J., Hussain, F.: Task scheduling optimization in cloud computing applying multi-objective particle swarm optimization. In: Basu, S., Pautasso, C., Zhang, L., Fu, X. (eds.) Service-Oriented Computing. ICSOC 2013. Lecture Notes in Computer Science, vol. 8274. Springer, Berlin, Heidelberg (2013)
4. Masdari, M., Salehi, F., Jalali, M., et al.: J. Netw. Syst. Manage. **25**, 122 (2017). https://doi.org/10.1007/s10922-016-9385-9
5. Beegom, A.S.A., Rajasree, M.S.: A particle swarm optimization based pareto optimal task scheduling in cloud computing. In: Tan, Y., Shi, Y., Coello, C.A.C. (eds.) Advances in Swarm Intelligence. ICSI 2014. Lecture Notes in Computer Science, vol 8795. Springer, Cham (2014)
6. Guo, G., Ting-Iei, H., Shuai, G.: Genetic Simulated Annealing Algorithm for Task Scheduling based on Cloud Computing Environment", IEEE International Conference on Intelligent Computing and Integrated Systems (ICISS), 2010, Guilin, pp. 60–63, 2010
7. Javanmardi, S., Shojafar, M., Amendola, D., Cordeschi, N., Liu, H., Abraham, A.: Hybrid Job Scheduling Algorithm for Cloud Computing Environment. In: Kömer, P., Abraham, A., Snášel, V. (eds.) Proceedings of the Fifth International Conference on Innovations in Bio-Inspired Computing and Applications IBICA 2014. Advances in Intelligent Systems and Computing, vol. 303. Springer, Cham (2014)
8. Pandey, S., Wu, L., Guru, S.M., Buyya, R.: A Particle Swarm Optimization-based Heuristic for Scheduling Workflow Applications in Cloud Computing Environments, Cloud Computing and Distributed Systems Laboratory, CSIRO Tasmanian ICT Centre, {spandey, linwu, raj}@csse.unimelb.edu.au, siddeswara.guru@csiro.au
9. Zhao, S., Lu, X., Li, X.: Quality of service-based particle swarm optimization scheduling in cloud computing. In: Wong W. (eds.) Proceedings of the 4th International Conference on Computer Engineering and Networks. Lecture Notes in Electrical Engineering, vol. 355. Springer, Cham (2015)
10. Aldossary, Sultan, Allen, William: Data Security, privacy, availability and integrity in cloud computing: issues and current solutions. (IJACSA) Int. J. Adv. Comput. Sci. Appl. **7**(4), 2016 (2016)
11. Kumar, N., Patel, P. (2016). Resource management using ANN-PSO techniques in cloud environment. In: Satapathy, S., Bhatt, Y., Joshi, A., Mishra, D. (eds.) Proceedings of the International Congress on Information and Communication Technology. Advances in Intelligent Systems and Computing, vol. 439. Springer, Singapore
12. Dillon, T., Wu, C., Chang, E.: Cloud computing: issues and challenges, In: 2010 24th IEEE International Conference on Advanced Information Networking and Applications (AINA), pp. 27–33 (2010)
13. Wu, Z., Liu, X., Ni, Z., Yuan, D., Yang, Y.: A market-oriented hierarchical scheduling strategy in cloud workflow systems. J. Supercomput. **63**(1), 256–293 (2013)
14. Yang, C.-T., Cheng, H.-Y., Huang, K.-L.: A Dynamic Resource Allocation Model for Virtual Machine Management on Cloud. In: Kim, T.-H., Adeli, H., Cho, H.-S., Gervasi, O., Yau, S.S., Kang, B.-H., Villalba, J.G. (eds.) GDC 2011. CCIST, vol. 261, pp. 581–590. Springer, Heidelberg (2011)
15. Fang, Y., Wang, F., Ge, J.: A Task Scheduling Algorithm Based on Load Balancing in Cloud Computing. In: Wang, F.L., Gong, Z., Luo, X., Lei, J. (eds.) WISM 2010. LNCS, vol. 6318, pp. 271–277. Springer, Heidelberg (2010)

16. Jin, J., Luo, J., Song, A., Dong, F., Xiong, R.: Bar: An efficient data locality driven task scheduling algorithm for cloud computing. In: Proc. of 11th IEEE/ACM International Symposium on Cluster, Cloud and Grid Computing, pp. 295–304 (2011)
17. Chi Mao, Y., Chen, X., Li, X.: Max-min task scheduling algorithm for load balancing in cloud computing. J. Springer (2014)
18. Shi, Y., Eberhart, R.C.: Parameter selection in particle swarm optimization. In: Porto, V.W., Waagen, D. (eds.) EP 1998. LNCS, vol. 1447, pp. 591–600. Springer, Heidelberg (1998)
19. Armbrust, M., Fox, A., Griffith, R., Joseph, A.D., Katz, R., Konwinski, A., et al.: A view of cloud computing. Commun. ACM **53**, 50–58 (2010)
20. Coello C. A.C., Lamont, G.B., Van Veldhuizen, D.A.: Evolutionary Algorithms for Solving Multi-Objective Problems, 2nd edn. Sprainger (2007)
21. Feng, M., Wang, X., Zhang, Y., Li, J.: Multi-objective particle swarm optimization for reseource allocation in cloud computing. In: Proceedings of the 2nd International Conference on Cloud Computing and Intelligent Systems (CCIS), vol. 3, pp. 1161–1165 (2012)
22. Bohre, A.K., Agnihotri, G., Dubey, M, Bhadoriya, J.S.: A Novel method to find optimal solution based on modified butterfly particle swarm optimization. Int. J. Soft Comput. Math. Control (IJSCMC) **3**(4) (2014)

Image Enhancement Using Fuzzy Logic Techniques

Preeti Mittal, R. K. Saini and Neeraj Kumar Jain

Abstract Image enhancement is the preprocessing task in digital image processing. It helps to improve the appearance or perception of the image so that the image can be used for analytics and human visual system. Image enhancement techniques lie in three broad categories—spatial domain, frequency domain, and fuzzy domain-based enhancement. A lot of work has been done on image enhancement. Most of the work has been done/performed on grayscale image. This paper concentrates on image enhancement using fuzzy logic approach and gives an insight into previous research work and future perspectives.

Keywords Image enhancement · Fuzzy logic techniques · Review · Future perspective · Performance measures

1 Introduction

Image enhancement is the preprocessing task of digital image processing. It is used to improve the appearance or perception of image so that enhanced image can be used for human visual system or analytics. It is required due to limiting capabilities of the hardware used for capturing the images, inadequate lighting conditions, and environmental conditions such as fog, sunlight, cloud cover, etc.

During image acquisition/digitization, images may contain vagueness or uncertainty in the form of imprecise boundaries and the intensities of colors. The quality of an image is measured objectively using mathematical functions and subjectively.

P. Mittal (✉) · R. K. Saini · N. K. Jain
The Department of Mathematical Sciences and Computer Applications,
Bundelkhand University, Jhansi, India
e-mail: preetimittal1980@yahoo.co.in

R. K. Saini
e-mail: rksaini.bu@gmail.com

N. K. Jain
e-mail: neerajjain15@gmail.com

K. Ray et al. (eds.), *Soft Computing: Theories and Applications*,
Advances in Intelligent Systems and Computing 742,
https://doi.org/10.1007/978-981-13-0589-4_50

Subjective measures are based on human perception of image and human perception is fuzzy in nature because an image is considered as good differently by different people. These lead the application of fuzzy logic in image processing because fuzzy logic deals with vagueness and uncertainty.

Image enhancement techniques are categorized into three categories—spatial domain-based enhancement, frequency domain-based enhancement, and fuzzy domain-based enhancement. In spatial domain-based enhancement, the algorithm works directly on the pixel intensity of the image. In frequency domain-based approach, the image is transformed into frequency domain such as DCT, DWT, or Fourier transformation. In fuzzy domain-based enhancement, the image is transformed into fuzzy property plane using some membership function such as Gaussian, triangular, or any other functions.

Various spatial domain enhancement techniques such as histogram equalization, histogram specification, iterative histogram modification, adaptive neighborhood equalization exist in literature. Most of the histogram enhancement techniques suffer from washed out effect and can amplify the existing noise [18]. Frequency domain techniques are good but computing a two-dimensional transform for an image (large array of intensities) is a very time-consuming task so they are not good for real-time applications [12].

The rest of the paper is organized as follows: Sect. 2 gives information about fuzzy set theory. Section 3 defines the fuzzy image. Section 4 presents the overview on architecture of fuzzy image processing. Section 5 describes performance evaluation measures used in fuzzy image processing. Section 6 discusses the previous work in detail. Future directions and conclusion are given in Sects. 7 and 8, respectively.

2 Fuzzy Set Theory

Lotfi A. Zadeh, a professor of computer science at the University of California, Berkeley, introduced the concept of fuzzy logic in 1965. Fuzzy set [34] is used as a problem-solving tool between the precision of classical mathematics and the inherent imprecision of the real world and for the problems which are having variables vague in nature. Fuzzy set is applied when the pattern indeterminacy exists because of inherent variability or vagueness rather than randomness [22].

3 Fuzzy Image Definition

An image I with the intensity level in the range [0, L − 1] can be considered as an array of fuzzy singletons and represented in fuzzy notation as

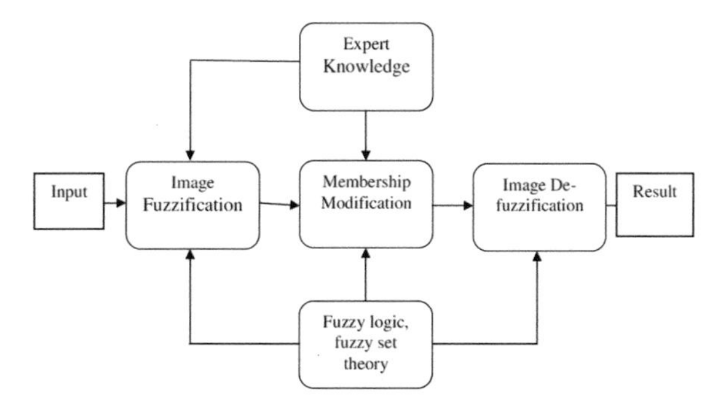

Fig. 1 Fuzzy image processing

$$I = \bigcup_{m=1}^{M} \bigcup_{n=1}^{N} \frac{\mu_{mn}}{i_{mn}} = \{\mu_{mn}/i_{mn}\} \quad (1)$$

where m = 1, 2, 3, . . ., M and n = 1, 2, 3, . . ., N.

4 Fuzzy Image Enhancement

Fuzzy image enhancement technique takes an image as input and provides an output image which is of better quality than input image. Fuzzy image processing is usually divided into three stages: image fuzzification, modification of membership values, and image defuzzification as shown in Fig. 1. Fuzzification and defuzzification have to be performed due to the absence of hardware. The power of the fuzzy image enhancement system lies in second step as the image contrast is improved by changing the membership values using transformation function.

Image fuzzification is very first task. Image is converted into fuzzy property plane using some membership function such as Gaussian, triangular, or some other functions. The image is converted from fuzzy property plane into spatial domain using inverse function of the function used for fuzzification.

5 Performance Evaluation Measures

The quality of image can be measured objectively using mathematical functions or subjectively. There are many mathematical functions or measures such as mean square error, index of fuzziness, entropy, peak signal-to-noise ratio, contrast improvement index to evaluate the quality of enhanced image.

5.1 Mean Square Error (MSE)

Mean square error (MSE) is calculated between original/acquired image and enhanced image as

$$MSE = \frac{1}{M \cdot N} \sum_{i=1}^{M} \sum_{j=1}^{N} (OI(i, j) - EI(i, j))^2 \quad (2)$$

Lower the value of MSE, lower the error, and the enhanced image is closer to the original image.

5.2 Peak Signal-to-Noise Ratio (PSNR)

PSNR is used as a quality measurement between original image and enhanced image and calculated as

$$PSNR = 10 \log_{10} \frac{L_{MAX}^2}{MSE} \quad (3)$$

L_{MAX} is the maximum intensity value (i.e., 255), and MSE is mean square error between original and enhanced image. The higher value of PSNR indicates the better quality of enhanced image.

5.3 Linear Index of Fuzziness

Linear index of fuzziness (IOF) provides the information about the fuzziness or vagueness or uncertainty present in the image by measuring the distance between fuzzy property plane and its nearest ordinary plane. Amount of fuzziness is 0, if all membership grades are 0 or 1. IOF reaches its maximum if all membership grades are equal to 0.5. An enhanced image is considered better if there is lower amount of fuzziness presented in enhanced image than original image.

$$LI = \gamma (I) = \frac{2}{M \cdot N} \sum_{i=1}^{M} \sum_{j=1}^{N} min(\mu_{ij}, 1 - \mu_{ij}) \quad (4)$$

5.4 Fuzzy Entropy

Entropy is the statistical measure of uncertainty present in the image to characterize the texture of input image and is defined in [4] as

$$E(X) = \frac{1}{M \cdot N} \sum_{i=1}^{M} \sum_{j=1}^{N} S_n \left(\mu_X \left(x_{ij} \right) \right) \tag{5}$$

where $S_n(\cdot)$ is a Shannon function.

$$S_n \left(\mu_X \left(x_{ij} \right) \right) = -\mu_X \left(x_{ij} \right) \log_2 \mu_X \left(x_{ij} \right) - \left(1 - \mu_X \left(x_{ij} \right) \right) \log_2 \left(1 - \mu_X \left(x_{ij} \right) \right) \tag{6}$$

where i = 1, 2, 3, ..., M and j = 1, 2, 3, ..., N.

5.5 *Contrast Improvement Index*

The contrast improvement index (CII) is a ratio of average values of local contrasts in original image and enhanced image [24]. Local contrast with 3 × 3 window is measured as $\frac{max - min}{max + min}$

$$CII = \frac{C_{Enhanced}}{C_{Original}} \tag{7}$$

6 Related Work

In the past few years, various researchers proposed algorithms in the field of fuzzy image enhancement. Pal et al. are said to be pioneer to use fuzzy set theory in the field of image enhancement. In 1980, Pal et al. transformed the image from spatial domain to fuzzy property domain before processing. In fuzzy property plane, an image is considered as an array of fuzzy singletons, each with a membership value denoting the degree of having some brightness level. The membership function was modified using fuzzy contrast intensification operator (INT) to enhance the contrast [22]. This concept was also applied for gray image enhancement with smoothing operation in [21]. The limitation of this INT operator is that it needs to be applied successively to the image to attain the desired enhancement.

Techniques based on modification of image histogram are the most popular among researchers due to the simplicity of image histogram. Tizhoosh and Fochem (1995) introduced a new technique based on fuzzy histogram hyperbolization (FHH). In this, a very simple function was used to fuzzify the image and FHH to change the membership values and gray levels [28]. Sheet et al. (2010) [27] and Raju and Nair (2014) [24] proposed fuzzy-based histogram equalization technique to enhance the low contrast images for better results.

The other approaches for image enhancement are by using fuzzy logic IF-THEN-ELSE rules [6, 25] and fuzzy relations [1]. Russo and Ramposi (1995) proposed an approach using fuzzy rules to sharpen details of the image and used local luminance difference as variable [25]. Choi and Krishnapuram (1995) introduced a fuzzy rule-based system to remove impulsive noise, to smooth impulsive noise, and to enhance edges [6].

In some cases, the global fuzzy image enhancement techniques fail to achieve satisfactory results. To solve this problem, Tizhoosh et al. (1997) introduced a locally adaptive version of [22, 28]. To find the minimum and maximum gray levels for the calculation of membership function, they found the minimum and maximum of subimages (30×30 windows) and interpolated these values to obtain corresponding values for each pixel [29].

To improve the limitation of INT operator, Hanmandlu et al. (1997) proposed an approach in which a new contrast intensification operator (NINT) was used for the enhancement of gray images and Gaussian type membership function with a new fuzzifier (fh) was used to model intensities into fuzzy domain. NINT does not change uniformly. It is almost linear at middle and changes marginally at extremes. Fuzzifier was obtained by maximizing the fuzzy contrast [16].

Tizhoosh et al. (1998) introduced a new algorithm for contrast enhancement of grayscale images using Sugeno's involutive fuzzy complements, called λ-complements at the place of Zadeh's complement of fuzzy sets. Membership values were modified using optimal value of λ which was optimized by maximizing index of fuzziness [30].

Cheng and Xu (2000) [4] proposed adaptive fuzzy contrast enhancement method with adaptive power variation. The image was fuzzified using S-function. The parameters a, c were measured based on the peaks in the histogram and parameter b was optimized based on the principle of maximum entropy. The global and local information about contrast were used to enhance the contrast. This method requires a lot of calculation for contrast enhancement. So to speed up the calculation, the original image was divided into subimages and the said process was applied on the subimages to get the enhanced sample mapping and the contrast of any pixel was calculated by interpolating the surrounding mapping. This technique can prevent over-enhancement. [5] used this method for mammogram contrast enhancement.

Hanmandlu et al. (2003) extended their previous work and applied NINT operator on color images by converting RGB images into HSV color model before fuzzification [15]. NINT was having a parameter t for the enhancement of color images. This parameter was calculated globally by minimizing the fuzzy entropy of the image information. These algorithms enhance the contrast of image globally/locally without considering the exposure of an image. Exposure indicates the amount of intensity exposition and characterizes the image as underexposed or overexposed image. Hanmandlu and Jha (2006) introduced a global contrast intensification operator (GINT) with three parameters: the intensification parameter (t), the fuzzifier (fh), and the crossover point (μ(c)), for enhancement of underexposed color images. The parameters were calculated globally by minimizing the fuzzy entropy of the image information with respect to the fuzzy contrast-based quality factor Qf and entropy-based quality factor Qe and the corresponding visual factors [14]. Using this method, a good improvement was seen in underexposed images.

This work was extended by Hanmandlu et al. (2009) and the image was divided into underexposed and overexposed regions based on exposure value. The parameters were calculated by minimizing the objective functions using bacterial foraging technique [17]. Hanmandlu et al. (2013) introduced a new objective measure, contrast information factor, which was optimized using particle swarm optimization to get the values of parameters for GINT for the enhancement of color images [12]. Cai and Qian (2009) proposed a technique for night color image enhancement. They enhanced the dark regions and restrained the glaring regions [2]. Verma et al. (2012) used artificial ant colony system and visual factor as objective function to optimize the parameters used for fuzzification and membership value modification [32].

Hasikin and Isa (2012) proposed a technique for enhancement of grayscale nonuniform illumination images and used flexible S-shaped function to fuzzify the intensity of image pixels and membership function was modified using power-law transformation for overexposed regions and saturation operator for underexposed regions to enhance the image. The parameters used in S-shaped functions were specified using maximized value of fuzzy entropy and index of fuzziness [18]. A. H. Mohamad (2016) introduced a new contrast intensification operator using logarithmic function with three parameters at the place of INT operator of [22] and Gaussian function to model the intensities in fuzzy property plane from spatial domain [20]. Puniani and Arora (2016) performed the fuzzy image enhancement technique for color images on L*a*b color space followed by smoothening of edges. L*a*b color space is device independent [23].

Hanmandlu et al. (2016) used Gaussian and triangular membership functions to fuzzify the S and V component of HSV color space. The S component was modified only for overexposed images. They introduced a new entropy function and visual factor as objective function to get the optimized values of parameters used in modification of membership values using particle swarm optimization method [13].

Zhang (2016) combined the fuzzy image enhancement technique proposed by [21] with the fruit fly optimization algorithm for medical images [35]. [21] set the parameters used in fuzzification manually but here, these were optimized using fruit fly optimization algorithm. Liu et al. (2017) also used [21] technique for microscopic image enhancement of Chinese herbal medicine [19].

Membership values in type-1 fuzzy sets are crisp in nature so type-1 fuzzy sets are not able to handle uncertainties where measurements are fuzzy, parameters used to model type-1 fuzzy set are fuzzy [31]. Therefore, to handle uncertainty more complex in nature, type-2 fuzzy sets are used whose membership values are fuzzy. [3, 10, 11, 31] used type-2 fuzzy sets to enhance the images and found better results than type-1 fuzzy sets. [7–9] used intuitionistic fuzzy sets for the enhancement of grayscale images and medical images.

Sharma and Verma (2017) introduced an approach for color satellite image enhancement using a modified Gaussian and sigmoid membership function to fuzzify the V (gray level intensity) component of HSV color space. A constant function was used to defuzzify the intensity values of all the regions. Contrast assessment function proposed in [33] was used to measure the performance of the algorithm. Zhou et al. (2017) proposed a hybrid method-based using fuzzy entropy and wavelet transform for the contrast enhancement of microscopic images with an adaptive morphological approach [36].

Fig. 2 **a** Original image (Peppers). **b** Enhanced image applying histogram equalization on R, G, B channels. **c** Enhanced image applying histogram equalization on V channel of HSV color space

7 Future Directions

Most of the papers used exposure (E), a crisp value, to characterize the image or images regions as overexposed ($E > 0.5$) or underexposed ($E < 0.5$). The images with exposure value 0.5 are considered as good. A fuzzy indicator should be used to distinguish the image as underexposed or overexposed.

All these techniques were designed considering one type of images. Most of the algorithms were designed considering the grayscale images and medical images. But in electronic gadgets, surveillance system, and computer vision, we deal with the color images. These algorithms cannot be generalized for all types of images, especially in the case of color images. If these algorithms are directly applied on the RGB color components, the artificial effects may rise or the color of the image may change. Image Peppers is enhanced by applying histogram equalization technique on R, G, B channels and on V channel of HSV color space after converting the RGB image into HSV color space and the natural colors of image are changed as shown in Fig. 2. A few researchers [2, 12–15, 17, 20, 23, 24, 26, 27, 32] discussed the issue of color image enhancement as per my knowledge. The work for the enhancement of images having uneven illumination is also not sorted out to the great extent.

8 Conclusion

Image enhancement is the preprocessing task in digital image processing and plays important role in human visual systems, pattern recognition, analytics, and computer vision. There are three broad categories (spatial domain, frequency domain, and fuzzy domain-based) of image enhancement. Fuzzy set is used as problem-solving tool in the systems where uncertainty and vagueness are present. Fuzzy domain-based image enhancement has three steps—fuzzification, modification of membership values, and defuzzification. The image quality is measured subjectively or objectively. Mostly, mean square error, PSNR, linear index of fuzziness, entropy, and CII are used to find the quality of an image objectively. Many researchers worked on this issue dealing with grayscale images, medical images, satellite images, and color images but a few researchers worked with the color images which are used in surveillance systems and electronic devices in our daily life. Many issues such as fuzzy indicator for the division of overexposed and underexposed region, enhancement of low contrast, and uneven illuminated images are not addressed up to a great extent. In the future work, the unaddressed issues will be solved.

References

1. Bhutani, K.R., Battou, A.: An application of fuzzy relations to image enhancement. Pattern Recogn. Lett. **16**(9), 901–909 (1995)
2. Cai, L., Qian, J.: Night color image enhancement using fuzzy set. In: 2nd International Congress on Image and Signal Processing, 2009. CISP'09, pp. 1–4. IEEE (2009)
3. Chaira, T.: Contrast enhancement of medical images using type II fuzzy set. In: 2013 National Conference on Communications (NCC), pp. 1–5. IEEE (2013)
4. Cheng, H.D., Xu, H.: A novel fuzzy logic approach to contrast enhancement. Pattern Recogn. **33**(5), 809–819 (2000)
5. Cheng, H.D., Xu, H.: A novel fuzzy logic approach to mammogram contrast enhancement. Inf. Sci. **148**(1), 167–184 (2002)
6. Choi, Y., Krishnapuram, R.: A fuzzy-rule-based image enhancement method for medical applications. In: Proceedings of the Eighth IEEE Symposium on Computer-Based Medical Systems, 1995, pp. 75–80. IEEE (1995)
7. Deng, H., Deng, W., Sun, X., Liu, M., Ye, C., Zhou, X.: Mammogram enhancement using intuitionistic fuzzy sets. IEEE Trans. Biomed. Eng. **64**(8), 1803–1814 (2017)
8. Deng, H., Sun, X., Liu, M., Ye, C., Zhou, X.: Image enhancement based on intuitionistic fuzzy sets theory. IET Image Process. **10**(10), 701–709 (2016)
9. Deng, W., Deng, H., Cheng, L.: Enhancement of brain tumor MR images based on intuitionistic fuzzy sets. In: Ninth International Symposium on Multispectral Image Processing and Pattern Recognition (MIPPR2015), pp. 98,140H–98,140H. International Society for Optics and Photonics (2015)
10. Ensafi, P., Tizhoosh, H.: Type-2 fuzzy image enhancement. In: Image Analysis and Recognition, pp. 159–166 (2005)
11. Ezhilmaran, D., Joseph, P.R.B.: Finger vein image enhancement using interval type-2 fuzzy sets. In: 2017 International Conference on I-SMAC (IoT in Social, Mobile, Analytics and Cloud) (I-SMAC), pp. 271–274. IEEE (2017)
12. Hanmadlu, M., Arora, S., Gupta, G., Singh, L.: A novel optimal fuzzy color image enhancement using particle swarm optimization. In: 2013 Sixth International Conference on Contemporary Computing (IC3), pp. 41–46. IEEE (2013)
13. Hanmandlu, M., Arora, S., Gupta, G., Singh, L.: Underexposed and overexposed colour image enhancement using information set theory. Imaging Sci. J. **64**(6), 321–333 (2016)
14. Hanmandlu, M., Jha, D.: An optimal fuzzy system for color image enhancement. IEEE Trans. Image Process. **15**(10), 2956–2966 (2006)
15. Hanmandlu, M., Jha, D., Sharma, R.: Color image enhancement by fuzzy intensification. Pattern Recogn. Lett. **24**(1), 81–87 (2003)
16. Hanmandlu, M., Tandon, S., Mir, A.: A new fuzzy logic based image enhancement. Biomed. Sci. Instrum. **33**, 590–595 (1996)
17. Hanmandlu, M., Verma, O.P., Kumar, N.K., Kulkarni, M.: A novel optimal fuzzy system for color image enhancement using bacterial foraging. IEEE Trans. Instrum. Meas. **58**(8), 2867–2879 (2009)

18. Hasikin, K., Isa, N.A.M.: Enhancement of the low contrast image using fuzzy set theory. In: 2012 UKSim 14th International Conference on Computer Modelling and Simulation (UKSim), pp. 371–376. IEEE (2012)
19. Liu, Q., Yang, X.P., Zhao, X.L., Ling, W.J., Lu, F.P., Zhao, Y.X.: Microscopic image enhancement of chinese herbal medicine based on fuzzy set. In: 2017 2nd International Conference on Image, Vision and Computing (ICIVC), pp. 299–302. IEEE (2017)
20. Mohamad, A.: A new image contrast enhancement in fuzzy property domain plane for a true color images **4**(1), 45–50 (2016)
21. Pal, S.K., King, R., et al.: Image enhancement using smoothing with fuzzy sets. IEEE Trans. Syst., Man, Cybern. **11**(7), 494–500 (1981)
22. Pal, S.K., King, R.A.: Image enhancement using fuzzy set. Electron. Lett. **16**(10), 376–378 (1980)
23. Puniani, S., Arora, S.: Improved fuzzy image enhancement using l* a* b* color space and edge preservation. In: Intelligent Systems Technologies and Applications, pp. 459–469. Springer (2016)
24. Raju, G., Nair, M.S.: A fast and efficient color image enhancement method based on fuzzy-logic and histogram. AEU-Int. J. Electron. Commun. **68**(3), 237–243 (2014)
25. Russo, F., Ramponi, G.: A fuzzy operator for the enhancement of blurred and noisy images. IEEE Trans. Image Process. **4**(8), 1169–1174 (1995)
26. Sharma, N., Verma, O.P.: A novel fuzzy based satellite image enhancement. In: Proceedings of International Conference on Computer Vision and Image Processing, pp. 421–428. Springer (2017)
27. Sheet, D., Garud, H., Suveer, A., Mahadevappa, M., Chatterjee, J.: Brightness preserving dynamic fuzzy histogram equalization. IEEE Trans. Consum. Electron. **56**(4) (2010)
28. Tizhoosh, H., Fochem, M.: Image enhancement with fuzzy histogram hyperbolization. Proc. EUFIT **95**, 1695–1698 (1995)
29. Tizhoosh, H., Krell, G., Michaelis, B.: Locally adaptive fuzzy image enhancement. Comput. Intell. Theory Appl. 272–276 (1997)
30. Tizhoosh, H., Krell, G., Michaelis, B.: Lambda-enhancement: contrast adaptation based on optimization of image fuzziness. In: The 1998 IEEE International Conference on Fuzzy Systems Proceedings, 1998. IEEE World Congress on Computational Intelligence, vol. 2, pp. 1548–1553. IEEE (1998)
31. Tizhoosh, H.R.: Adaptive λ-enhancement: type I versus type II fuzzy implementation. In: IEEE Symposium on Computational Intelligence for Image Processing, 2009. CIIP'09, pp. 1–7. IEEE (2009)
32. Verma, O.P., Kumar, P., Hanmandlu, M., Chhabra, S.: High dynamic range optimal fuzzy color image enhancement using artificial ant colony system. Appl. Soft Comput. **12**(1), 394–404 (2012)
33. Xie, Z.X., Wang, Z.F.: Color image quality assessment based on image quality parameters perceived by human vision system. In: 2010 International Conference on Multimedia Technology (ICMT), pp. 1–4. IEEE (2010)
34. Zadeh, L.A.: Fuzzy sets. Inf. Control **8**(3), 338–353 (1965)
35. Zhang, Y.: X-ray image enhancement using the fruit fly optimization algorithm. Int. J. Simul.–Syst. Sci. Technol. **17**(36) (2016)
36. Zhou, J., Li, Y., Shen, L.: Fuzzy entropy thresholding and multi-scale morphological approach for microscopic image enhancement. In: Ninth International Conference on Digital Image Processing (ICDIP 2017), vol. 10420, p. 104202K. International Society for Optics and Photonics (2017)

A Jitter-Minimized Stochastic Real-Time Packet Scheduler for Intelligent Routers

Suman Paul and Malay Kumar Pandit

Abstract In this paper, we investigate and perform detailed analysis of a stochastic real-time packet scheduler considering one of the most important QoS parameter, scheduling jitter, which generates packet loss during runtime flow of IP traffic within a router. For analysis, we take three types of IP traffic flows: VoIP, IPTV and HTTP for processing within a processor of an IP router. With a aim of keeping processor utilization to be kept at nearly 100%, during processing of internet traffic and taking practical values of jitter of a network under study, we investigate the impact of scheduling jitter (minimized and restricted to the acceptable value of VoIP traffic) on packet loss rate (PLR) during runtime. We further analyse that the PLR incurred due to scheduling jitter during processing can be further minimized using machine learning Baum–Welch algorithm and an improvement of 36.73%, 28.57% and 26.31% lower value of PLR for VoIP, IPTV and HTTP, respectively has been achieved.

Keywords Scheduling jitter · Packet loss rate · Scheduler in IP router · Network QoS · Real-time multimedia traffic

1 Introduction

1.1 Network QoS Parameters

Quality-of-Service (QoS) in a communication network plays a crucial role for guaranteeing necessary service performance and at the same time end user satisfaction. QoS is denoted as reliability and as well as availability offered by an application and by the network infrastructure that offers that service. Various network

S. Paul (✉) · M. K. Pandit
Department of ECE, Haldia Institute of Technology, Haldia, India
e-mail: paulsuman999@gmail.com

M. K. Pandit
e-mail: mkpandit.seci@gmail.com

K. Ray et al. (eds.), *Soft Computing: Theories and Applications*,
Advances in Intelligent Systems and Computing 742,
https://doi.org/10.1007/978-981-13-0589-4_51

QoS parameters are, packet loss rate (PLR), throughput, jitter (variation on packet transmit delay) [1], scheduling jitter, allocation of resources in a fair way, process utilization, availabilities of bandwidth, system latency, etc., which are involved for performance issues. In multimedia real-time IP traffic such as VoIP, IPTV etc. requires a guaranteed time-bound execution in order to meet the service level agreements, resulting in lower value of PLR. In such real-time applications, congestion in the network and as well as scheduling jitter may cause unacceptably high delay, making the processes to frequently miss deadlines. This results in increase of PLR, lower throughput, etc. In this paper, we focus on the performance analysis of a stochastic real-time packet scheduler under one of the important QoS parameter, *scheduling jitter* for three classes of multimedia IP traffic: Voice over Internet Protocol, Internet Protocol Television and Hypertext Transfer Protocol. This work is an extension, further investigation and analysis of our earlier work [2].

1.2 Timing Jitter and Its Classification

Timing jitter is treated as variation in packet transfer delay caused by router processing queuing, and the effects of contention and serialization on the path through a network. Jitter is classified into three major types, Type A, Type B and Type C, respectively. Constant value of jitter is treated as jitter of Type A. Reasons for this type of jitters are produced due to (i) sharing of load within a router, (ii) load sharing for multiple internal routes in a network offered by the service providers, (iii) load sharing for multiple access lists. Transient Jitter is treated as jitter of Type B. This type of jitters considerably causes an increase in delay for scheduling and processing of packets within the processor of a router. A single packet may cause this type of jitter. Other specific reasons of type B jitter are, congestion in the networks, change of congestion level, updates of routing tables in a router, temporarily link failure, etc. Variation of delay for a short period is treated as Type C jitter which occurs due to network congestion. Congestion in the access links and delay due to serialization enforces jitter of Type C.

1.3 Scheduling Jitter

The scheduling jitter of a processor within a router in a stochastic scheduling platform considering random arrival of Internet traffic is defined as variable deviation from ideal timing event. Scheduling jitter is defined as the delay between the time when the task shall be started, and the time when the task is being started by the processor.

2 Literature Survey

Rigorous research works are going concentrating on network packet loss behaviour and jitter in Internet traffic. The objective is to minimize the scheduling jitter in soft real-time embedded computing platforms like routers. In [3], the authors have proposed jitter and packet loss characteristics of VoIP traffic. The study is based on Markov model which states that jitter can be modelled by means of multifractal model. The authors have showed the self-similar characteristics of VoIP traffic and the model only considers VoIP traffic. The authors in [4], have demonstrated a method for reducing jitter and PLT attributes of QoS in WiLD networks. Allocation of bandwidth was implemented dynamically using weighted round robin and a fair queueing policy was accepted. However, the scheduler does not focus on processor utilization, and minimization of scheduling jitter. Palawan et al. [5] have proposed a scheduler which modifies the stream of scalable video to minimize the jitter. However, the scheduler is not utilization-driven. Recently, the authors in [6] have demonstrated a scheduling scheme which is of jitter-constrained type.

3 System Model

The proposed scheduler is a finite-state machine (FSM) based on a Hidden Markov chain decision model [7] to achieve a steady-state process utilization ratio in the order of 0.80:0.16:0.04 (*Pareto* distribution) for VoIP, IPTV and HTTP traffic, respectively. For a target steady-state distribution, we take an initial estimate of a transition probability matrix (TPM), M which is stated in (1). The elements of 'M' are calculated using the *machine learning* Metropolis–Hastings algorithm. The elements of the matrix are denoted by $p_{ij}(s)$ which state the probability of execution of class-specific processes to be their own state or to make transitions.

$$M = \begin{pmatrix} p_{11} & p_{12} & p_{13} \\ p_{21} & p_{22} & p_{23} \\ p_{31} & p_{32} & p_{33} \end{pmatrix} \tag{1}$$

In this case, we take the values of matrix elements $p_{ij}(s)$ as, $p_{11} = 0.9$, $p_{12} = 0.08$, $p_{13} = 0.02$, $p_{21} = 0.39$, $p_{22} = 0.56$, $p_{23} = 0.05$, $p_{31} = 0.42$, $p_{32} = 0.18$, $p_{33} = 0.40$.

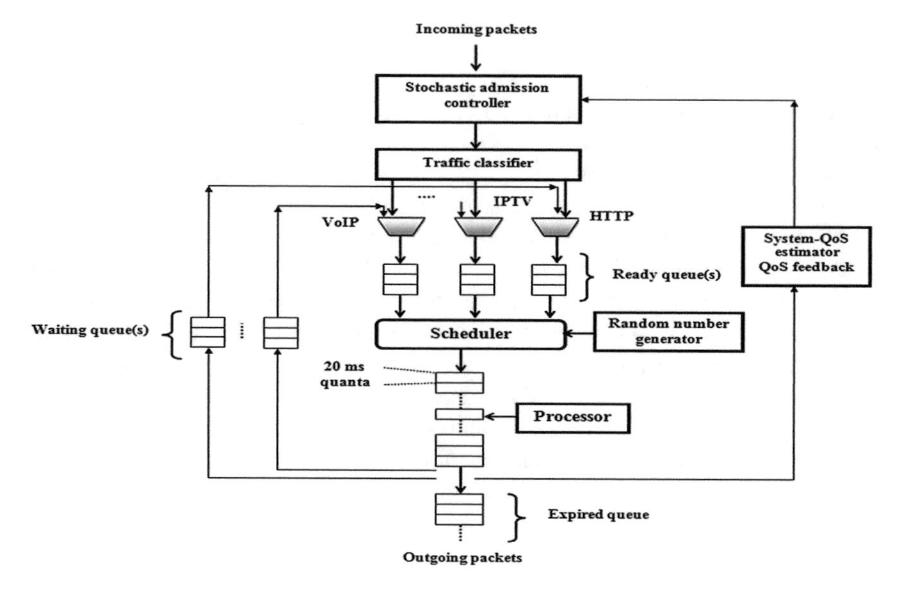

Fig. 1 Internal framework of the scheduler

Table 1 Service model parameters: deadline

Traffic class	Arrival characteristics	Deadline (ms)
VoIP	MMPP	20
IPTV	MMPP	100
HTTP-web browsing	MMPP	400

4 Proposed Jitter-Aware Stochastic Scheduling Framework

Figure 1 illustrates internal framework of the scheduler.

Three classes of IP traffic pass through the router for processing stated in Fig. 1. A traffic classifier differentiates the traffic and is sent to individual queues for each flows. The scheduler is defined as, M/BP/1/./QUEST [2]. We consider practical case, where traffic arrivals are featured as Markovian nature and are modulated by Poisson processes (MMPP). The service time distribution is of Bounded Pareto (BP). A processor executes on incoming traffic processes. Processes of traffic classes run with priorities in the order of *VoIP* > *IPTV* > *HTTP* with practical class-specific deadline. The values of deadlines are presented in Table 1.

Processes are executed within time slots. A runtime generated random number decides which class-specific process to be executed. A QoS feedback controller with help of runtime PLR and mean waiting time instructs the admission controller to restrict the traffic flows. Consideration of uniform burst time in this scheduling scheme will lead to the system utilization (U_i) as stated in Eq. (2) [8].

Table 2 Service model parameters: jitter and scheduling jitter

Traffic class	Acceptable jitter	Recommended by
VoIP	$\leq$0.5	ITU, Network service providers
IPTV	<30	ITU
HTTP	$\leq$400	ITU

$$\sum_{1}^{3} U_i = T_{BURST} \cdot \left(\frac{1}{D_{VoIP}} + \frac{1}{D_{IPTV}} + \frac{1}{D_{HTTP}} \right) \leq 1 \tag{2}$$

Here, T_{BURST} denotes the service time and D_{class} represents the deadlines. Applying the values of deadline stated in Table 1, the burst time is calculated as, T_{BURST} $\leq$16 ms. If we allow 4 ms timing jitter (scheduling jitter), T_{JITTER} then the required value of time slot, $T_{QUANTUM}$ will be, $T_{QUANTUM} = T_{BURST} + T_{JITTER} = 20$ ms.

4.1 *Methodology for Analysis of Scheduling Jitter*

Considering the above framework in Fig. 1, we further investigate how the scheduling jitter causes packet loss during runtime traffic flows. Internet traffic arrivals are stochastic in nature. The scheduling jitter is defined as the unnecessary variation of release times of stochastic processes by a packet scheduler within a router. Jitter will raise packet loss rate and lower process utilization for deadline-sensitive traffic. VoIP traffic is most sensitive to deadline compared to other traffic streams IPTV and HTTP which are described in Table 2.

Allowing an acceptable jitter as stated in Table 2, we propose and apply following jitter-aware scheduling algorithm in Sect. 4.2 to measure the scheduling jitter during execution.

4.2 *Jitter-Aware Scheduling Algorithm*

Using the TPM in (1), the jitter-aware scheduling algorithm is stated as follows.

```
1. Generate random number R, 0.01≤R≤1;
2.  Set: Time quantum T_QUANTUM: 20 ms  and T_BURST: 16 ms;
                                          where  T_QUANTUM=(T_BURST+T_JITTER)
3.  Set:  Timing jitter (scheduling jitter) : T_JITTER;      where,  0≤T_JITTER≤4 ms ;
4.  Initialize: timer, t=0
5.      for t=1, 2... (T_BURST+T_JITTER) ms,  do
6.        switch (initial_process) {
7.              CASE  initial_process:P1
8.                      if (0.01≤R≤p11) then
9.                           execute P1;
10.                     else if ((p11+0.01)≤R≤ (p11+0.08)) then
11.                          execute P2;
12.                     else execute P3;
13.                     end if;
14.             CASE  initial_process:P2
15.                     if (0.01≤ R≤p22) then
16.                          execute P2;
17.                     else if ((p22+0.01)≤R≤(p22 +0.39)) then
18.                          execute P1;
19.                     else execute P3;
20.                     end if;
21.             CASE  initial_process:P3
22.                     if (0.01≤R≤0.4) then
23.                          execute P3;
24.                     else if ((p33+0.01)≤R≤ (p33+0.42)) then
25.                          execute P1;
26.                     else execute P2;
27.                     end if;    }
28.     end for;
29.  Place Pi in expired queue;
```

5 Simulation Environment and Results

We develop simulation platform using MATLAB and DEVS suite [9], a discrete time and discrete event simulation tool. We consider an initial three state Markov model characterized by a transition probability matrix in Eq. (1). Equation (3) denotes an error probability matrix. The elements in second row describe the probability of packet loss rate (PLR). In this case, we consider the PLR is due to deadline misses and cache misses (both L1 and L2). We apply the following simulation parameters stated in Table 3:

Table 3 Simulation parameter

Parameter	Environment
Rate of arrival	100 packets/s
Burst time	16 ms
Link capacity	10 Mbps
Timing (scheduling) jitter	0.1 ms $\leq T_{JITTER} \leq$ 0.5 ms

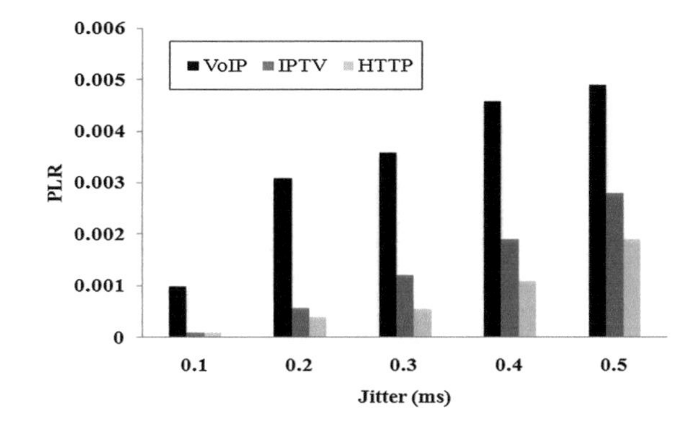

Fig. 2 PLR with increasing scheduling jitter

$$E = \begin{pmatrix} 0.98 & 0.9 & 0.8 \\ 0.02 & 0.1 & 0.2 \end{pmatrix} \tag{3}$$

As the proposed scheduler accepts most jitter-sensitive VoIP traffic, we set maximum variation of the scheduling jitter to a allowable value of 0.5 ms and observe resulting PLR shown in Fig. 2.

The simulation results show that with increasing value of scheduling jitter, the runtime VoIP traffic has maximum PLR compared to other two flows.

6 Minimization of PLR (for Minimized Scheduling Jitter) Using Machine Learning (Intelligence)

The working flow model for minimization of PLR is described in Fig. 3.

Our aim is to minimize the runtime PLR generated due to scheduling jitter. As shown in Fig. 3, the runtime PLR pattern during the execution of the scheduler is recorded. The PLR pattern is further processed by iterative machine learning Baum–Welch algorithm. For estimation, we consider three matrices—a TPM considering initial state probability of each traffic class in (1), an error probability matrix in (4). The second row of matrix $J_{(0.5)}$ indicates the recorded PLR of the system under study while running with the proposed scheduling algorithm stated in Sect. 4.2, at a maximum acceptable scheduling jitter of 0.5 ms (restriction is due to VoIP traffic). The second row in the matrix represents measured PLR for VoIP, IPTV and HTTP,

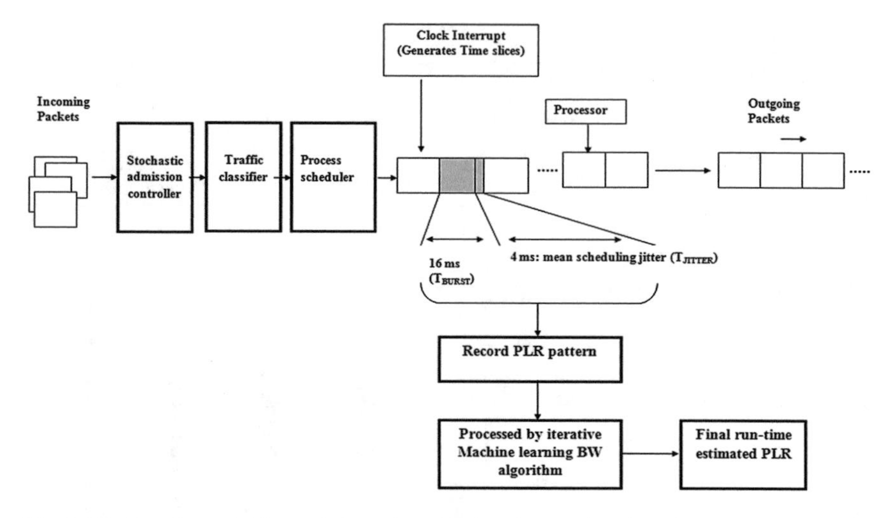

Fig. 3 Working flow model for minimization of PLR using machine learning

respectively. As the proposed scheduler reaches and maintains a guaranteed steady state of process utilization, U in the ratio of 80:16:4 for VoIP, IPTV and HTTP, respectively, therefore, we take an initial state probability vector of utilization, U. The U is stated in (5).

$$J_{(0.5)} = \begin{pmatrix} 0.9951 \ 0.9972 \ 0.9981 \\ 0.0049 \ 0.0028 \ 0.0019 \end{pmatrix} \quad (4)$$

$$U = \begin{pmatrix} 0.8 \ 0.16 \ 0.04 \end{pmatrix} \quad (5)$$

In this real-time scheduler within IP router, for learning, we consider 50 number of iterations. The newly estimated minimized PLR error probability is stated in Eq. (6). The second row denotes the newly estimated minimized PLR for VoIP, IPTV and HTTP.

$$J_{(\min)} = \begin{pmatrix} 0.9969 \ 0.9980 \ 0.9986 \\ 0.0031 \ 0.0020 \ 0.0014 \end{pmatrix} \quad (6)$$

The PLR(s) for initial and newly estimated for three types of IP traffic for a value of 0.5 ms (restricted value) scheduling jitter are shown in Fig. 4.

We observe that the PLR has been substantially minimized by 36.73%, 28.57% and 26.31% for VoIP, IPTV and HTTP, respectively.

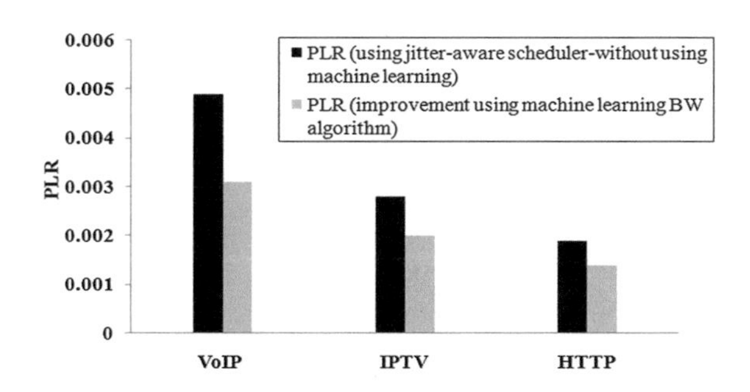

Fig. 4 Minimization of PLR using machine learning Baum–Welch algorithm

7 Conclusion and Future Work

In this paper, we have proposed a jitter-aware stochastic scheduler for internet traffic. We have performed analysis of how the scheduling jitter generates packet losses. With an aim of minimizing the PLR for a scheduling jitter, we further apply machine learning Baum–Welch algorithm and observe that an improvement of lower value of PLR of 36.73%, 28.57% and 26.31% for VoIP, IPTV and HTTP, respectively has been achieved. The proposed jitter-aware scheduler can be further implemented in a multicore processor platform of an IP router with proper load balancing among its cores.

References

1. Kim, H.-G.: Enhanced timing recovery using active jitter estimation for voice-over IP networks. Korean Soc. Internet Inf. (KSII) Trans. Internet Inf. Syst. **6**(4), 1006–1025 (2012). https://doi.org/10.3837/tiis.2012.04
2. Paul, S., Pandit, M.K.: A QoS-enhanced intelligent stochastic real-time packet scheduler for multimedia IP traffic. Multimedia Tools and Applications, pp. 1–24. Springer (2017). https://doi.org/10.1007/s11042-017-4912-6
3. Toral-Cruz, H., Pathan, A.-S.K., Pacheco, J.C.R.: Accurate modeling of VoIP traffic QoS parameters in current and future networks with multifractal and Markov models. Math. Comput. Model. **57**, 2832–2845 (2013). https://doi.org/10.1016/j.mcm.2011.12.007
4. Bhargavi, K., Bhargavi, M.: Minimizing jitter and packet loss parameters of QoS in WiLD networks using dynamic bandwidth allocation. i-manager's J. Wirel. Commun. **5**(3), 30–35 (2016). http://www.imanagerpublications.com/Article.aspx?ArticleId=10359
5. Palawan, A., Woods, J.C., Ghanbari, M.: Continuity-aware scheduling algorithm for scalable video streaming. Computers **5**(11), 1–16 (2016) (MDPI). https://doi.org/10.3390/computers5020011
6. Minaeva, A., Akesson, B., Hanzalek, Z., Dasari, D.: Time-triggered co-scheduling of computation and communication with jitter requirements. IEEE Trans. Comput. 99 (2017). https://doi.org/10.1109/tc.2017.2722443
7. Chen, C., Heath, R.B, Bovik, A.C., Veciana, G.D.: A Markov decision model for adaptive scheduling of stored scalable videos. IEEE Trans. Circ. Syst. Video Technol. **23**(6), 1081–1094 (2013). https://doi.org/10.1109/tcsvt.2013.2254896

8. Liu, C.L., Layland, J.W.: Scheduling algorithms for multiprogramming in a hard real-time environment. J. ACM **20**(1), 46–61 (1973). https://doi.org/10.1145/321738.321743
9. DEVS suite: Discrete event system simulator suite, Arizona Center of Integrative Modeling and Simulation of Arizona State University. http://acims.asu.edu/software/devs-suite

Differential Evolution with Local Search Algorithms for Data Clustering: A Comparative Study

Irita Mishra, Ishani Mishra and Jay Prakash

Abstract Clustering is an unsupervised data mining task which groups objects in the unlabeled dataset based on some proximity measure. Many nature-inspired population-based optimization algorithms have been employed to solve clustering problems. However, few of them lack in balancing exploration and exploitation in global search space in their original form. Differential Evolution (DE) is a nature-inspired population-based global search optimization method which is suitable to explore the solution in global search space. However, it lacks in exploiting the solution. To overcome this deficiency, few literatures incorporate local search algorithms in DE to achieve a good solution in the search space. In this work, we have performed a comparative study to show effectiveness of local search algorithms, such as chaotic local search, Levy flight, and Golden Section Search with DE to balance exploration and exploitation in the search space for clustering problem. We employ an internal validity measure, Sum of Squared Error (SSE), to evaluate the quality of cluster which is based on the compactness of the cluster. We select F-measure and rand index as external validity measures. Extensive results are compared based on six real datasets from UCI machine learning repository.

Keywords Differential evolution · Data clustering · Chaotic local search
Levy flight · Golden section search

I. Mishra (✉)
ABV-Indian Institute of Information Technology and Management, Gwalior, India
e-mail: iritamishra@gmail.com

I. Mishra
Indian Institute of Information Technology Allahabad, Allahabad, India
e-mail: ishanimishra.1997@gmail.com

J. Prakash
Amity University, Noida, Uttar Pradesh, India
e-mail: jayprakash.iiitm@gmail.com

K. Ray et al. (eds.), *Soft Computing: Theories and Applications*,
Advances in Intelligent Systems and Computing 742,
https://doi.org/10.1007/978-981-13-0589-4_52

1 Introduction

Clustering groups the objects of an unlabeled dataset [1] based on some similarity measure [3], which plays a significant role in applications in diverse fields. Clustering quality is often assessed by an internal validity measure, which may be based on different features of the clusters, e.g., compactness, isolation, and connectedness. Partitional clustering is one of the most commonly used approaches to solve clustering problems, which directly decomposes datasets into a number of clusters. It can be hard or fuzzy [5, 6].

In the fuzzy partitional clustering, an object may belong to every cluster with some fuzzy membership weight [5]. It is more suited to the datasets having overlapping clusters. Hard partitional clustering groups the objects of datasets into a number of nonoverlapping clusters by assigning an object to only one cluster [6]. It can be mathematically represented as

$$C_p \neq \phi \quad p = 1, \ldots, k \tag{1}$$

$$C_p \cap C_q = \phi \quad p, q = 1, \ldots, k \; and \; p \neq q \tag{2}$$

where $C = \{C_1, C_2, \ldots, C_k\}$ is a set of k number of non-empty and nonoverlapping clusters.

Many conventional/traditional algorithms for partitional hard clustering, e.g., K-means and k-Medoid, are available to solve the clustering problems as they are simple and easy to implement. However, they suffer from many drawbacks, e.g., the final solution is dependent on the initial solution, or they are easily trapped into a local optimal solution [8].

Specifically for these reasons, nature-inspired optimization methods offer more effectiveness to overcome the deficiency of the conventional clustering methods [4]. They possess several desired key features, e.g., parallel nature and self-organizing behavior [7].

They work on two key aspects, exploration and exploitation to search a good solution in the search space. Exploration is responsible for exploring solutions in the search space, whereas exploitation optimizes the solution locally so as to search a better solution near a good solution. Differential Evolution (DE), a popular global optimization method, provides a good exploration of solutions in the search space. Nevertheless, it lacks in the exploitation of solutions [1, 2, 9, 10].

To balance exploration and exploitation in global search space, few literatures incorporate local search algorithm in DE. In [10], DE is employed to diversify solutions in the search space. However, the local search algorithm, Golden Section Search (GSS), is responsible for refining the solutions in local search space. In addition, few local search algorithms Levy Flight [9] and chaotic local search [1] algorithms are also effectively employed in the literature to solve diverse problems. In this paper, we compare the effectiveness of DE with well-known local search algorithms chaotic local search (CLS), levy flight (LF), and golden section search (GSS) for data clustering. DE assists in exploring solutions in search space, and the local search algorithm exploits the solutions in the search space to obtain a refined solution. To the best of our knowledge, there has been no attempt in the literature, which reported performance study of DE with these local search algorithms for data clustering in terms of quality of solutions and convergence speed to achieve the final solution. Here, the number of clusters is known a priori in the datasets. We experiment on six real datasets from the UCI machine learning repository to judge the efficacy of these algorithms. F-measure (FM) [6] and Rand Index (RI) [8] are used to judge the accuracy of the obtained clusters. This study will be helpful to the researcher in extending DE to solve diverse real-life problems.

Rest of the paper is organized as follows. Section 2 presents a brief introduction to DE. The proposed method is detailed in Sect. 3. Section 4 presents a comparative study and discussion of the results. Finally, Sect. 5 concludes and remarks upon the possible future research directions.

2 Algorithm Background

In this section, a brief introduction of differential evolution is presented.

2.1 Differential Evolution

Storn and Price [11] propose DE which is a well-known population-based nature-inspired optimization algorithm for multi-model and continuous-valued problems. The basic DE strategy can be expressed as the notation DE/x/y/z, where x represents the target vector to be mutated (a random vector or the best vector), y is the number of difference vectors, and z indicates the crossover scheme [8, 11]. DE encompasses three operators namely mutation, crossover, and selection, which are mentioned as follows:

- Mutation: For each individual X_i of current population, a trial vector T_i is created by the mutation operator. In this sense, target vector X_{i1} is mutated with a weighted difference vector of randomly selected two individuals X_{i2} and X_{i3} such $i \neq i_1 \neq i_2 \neq i_3$ as shown in Eq. 3.

$$T_{i,d}(t) = X_{i1,d}(t) + \beta\big(X_{i2,d}(t) - X_{i3,d}(t)\big) \tag{3}$$

  where d represents the dth dimension of an individual and $\beta \in [0, 1]$ is a mutation-scale factor, which manages the amplification of difference vector; t represents the current generation number.
- Crossover: Most frequently used crossovers in DE are binomial crossover and exponential crossover. Here, offspring vector X_i^0 is generated using trial vector T_i and parent vector X_i using the binomial crossover as shown in Eq. 4.

m_{11}	m_{21}	....	m_{d1}	m_{12}	m_{22}	....	m_{d2}	.....	m_{1k}	m_{2k}	.....	m_{dk}

Fig. 1 Solution representation

$$X_i'(t) = T_{id}(t) \quad \textit{if rand } (0, 1) < CR \, or \, d = rand \; i(1, \ldots, D)$$
$$X_{id}(t) \quad \textit{otherwise} \tag{4}$$

where d represents the dth dimension of a vector and D is total number of dimensions in a vector.

- Selection: Apart from selection of individual for the mutation operation to produce the trial vector, selection operation is also performed between offspring vector and parent vector to promote fitter solution to create population for the next generation.

$$X_{id}(t+1) = X_{id}'(t) \quad \text{if} \, \mathrm{F}(X_{id}'(t)) < f(X_i) \, \text{for minimization problem}$$
$$X_{id}(t) \quad \text{otherwise} \tag{5}$$

DE attracts attention to researchers mainly because it requires very few parameters to tune and is easy to implement.

3 Designed Clustering Methods

In this section, DE with local search algorithms for data clustering problems is elaborated in detail.

3.1 Solution Representation and Initialization

A centroid-based representation to represent a candidate solution for clustering is followed as discussed in [7]. It is shown in Fig. 1. Here, every candidate solution holds k X d dimensions, where k denotes the number of clusters and d indicates the number of features in the dataset.

Here, m_{ij} represents the centroid of the j-th cluster and i indicates the i-th feature of the dataset in the solution representation. To commence the initial population for competing algorithms, the dimension of a cluster centroid of each solution is assigned by a random number between the maximum (x_{max}) and minimum (x_{min}) value of that respective feature of the dataset. In Algorithm 1, solution (i, j) represents the j-th feature of i-th cluster centroid; rand (0, 1) is a uniformly distributed random number in the range [0, 1].

Algorithm 1: Initialization of solutions
1: k number of clusters;
2: d number of dimensions;
3: for i=1:k do
4: for for j=1:d do
5: Solution (i; j) = $x_{min}(j) + rand(0; 1)$ $x_{max}(j)- x_{min}(j)$;
6: end for
7: end for

3.2 Sum of Squared Error

Internal validity measure represents the quality of a solution. As Sum of Squared Error (SSE) works well with well-separated and compact cluster [5], we select SSE as internal validity measure to evaluate quality of a solution during run. The lower value of SSE indicates the better quality of solution. SSE is expressed in Eq. 6.

$$SSE = \sum_{i=1}^{k} \sum_{\forall x0 \in ci} \|x0 - mi\|^{2}$$

Here, x0 denotes p-th object of the dataset, mi denotes centroid of the i-th cluster, and ci denotes the i-th cluster.

3.3 Procedure for Clustering Methods

To employ designed algorithms for clustering problem, initial population (of size N) is generated. Each solution in the population represents a set of cluster centroids $C = \{C_1, C_2, \ldots, Ck\}$, where k is the number of clusters. In this work, DE is selected as global optimization algorithm and DE/rand/1/bin scheme is followed among the several schemes of DE algorithm where DE stands for differential evolution; rand indicates that the target vector is selected randomly; 1 stands for the number of differential vectors selected; and bin stands for binomial crossover. Here, the employment of DE provides diversified solutions in the search space. However, DE exhibits a good exploration of solutions in the search space on the cost of exploitation. Therefore, local search algorithms with DE are incorporated. In this procedure, at the end of each iteration of DE, the obtained best solution is refined by a local search algorithm. Upon incorporating chaotic local search [1] with DE, it is named as DE-CLS.

Similarly, upon incorporating levy flight [9] with DE, it is named as DE-LF. When golden section search [10] is incorporated with DE, it is named as DE-GSS. In each designed algorithm, DE explores the solutions and local search algorithm exploits the solutions to improve solutions in the local search space (neighborhood) of an obtained solution in the search space to balance exploration and exploitation. The general procedure for these DE-based algorithms is shown.

Algorithm 2: DE for data clustering

1: Inputs:
2: Algorithmic parameters;
3: nxd data set, where n is the number of objects in the data sets and d is the dimension;
4: K is number of clusters;
5: Steps of algorithm:
6: Initialize the solutions and evaluate their fitness;
7: while (Max_nffe is not reached) do
8:     for For each solution x_i do
9:         Generate trial vector T_i by applying mutation operation (refer equation 1.3);
10:         Generate offspring x_i^0 by binomial crossover (refer equation 1.4);
11:         Evaluate fitness of offspring;
12:         Promote fitter in x_i and x_i^0 for next generation.
13:     end for
14:     Select the best solution in the population and employ local search procedure;
15: end while
16: Output: A set of K clusters of the best solution;

---

4 Experimental Results and Discussion

This section presents experimental setup, description of datasets, clustering results, and comprehensive discussion. The experiments have been performed on a system with core i5 processor and 2 GB RAM in Windows 7 environment using programs written in MATLAB R2012a.

4.1 Parameters Setting

Results of the nature-inspired algorithms are influenced by the number of control parameters. Here, we perform experiments with the set of parametric values (refer Table 1) for DE to perform a comparative study among DE-CLS, DE-LF, and DE-GSS for clustering. Parameter settings of the local search algorithms are set as suggested by the researchers in the respective research papers. Therefore, the number of solutions equals 40. The stopping criterion, i.e., maximum number of fitness function evaluation (Max nffe), is fixed to 10,000 as further processing is merely a computational overhead owing to negligible improvement.

Table 1 Control parameters

Name of parameters	Value
Population size	40
Max nffe	10000
Number of independent runs	25
Scaling factor of DE ()	0.5
Crossover rate of DE (CR)	0.9

Table 2 Datasets

Datasets	Number of clusters (k)	Number of dimensions (d)	Number of objects
Iris	3	4	150
Wine	3	13	178
Zoo	7	16	101
Dermatology	6	34	358
Yeast	10	8	1484
CMC	3	10	1473

4.2 Datasets Descriptions

The datasets are in matrix of size N with real-valued elements which are to be partitioned into k nonoverlapping clusters. Six real datasets from the UCI machine learning repository are selected. A brief summary of these datasets is presented in Table 2.

4.3 Comparison of Results

Performance of these competing algorithms is measured by the fitness function value (FFV) of solutions. The FFV of a solution is the value of SSE which indicates the quality of clusters in terms of compactness. A lower value of the FFV indicates a better solution and a lower value of the standard deviation of the FFV in different runs represents robustness of the algorithm.

Table 3 reports the quality of the solutions of the competing algorithms DE-CLS, DE-LF, and DE-GSS in terms of the best solution—the solution that achieves the lowest FFV—along with the mean and standard deviation (Std dev) of the FFVs obtained in all the independent runs. The convergence speed of these algorithms is shown pictorially in Fig. 2a–f to achieve the best-obtained solution during run.

Table 3 The best, mean, and standard deviation (Std dev) of FFV of final solutions obtained in all independent runs

Datasets		DE-CLS	DE-LF	DE-GSS
Iris	Best	79.22	79.2	79.07
	Mean	82.12	82.85	82.44
	Std dev	2.18	3.38	2.25
Wine	Best	2373962.28	2372750.86	2376829.84
	Mean	2380406.99	2383700.67	2385026.33
	Std dev	4225.23	6398.66	7398.51
Zoo	Best	191	189.48	211.91
	Mean	227.74	220	231.42
	Std dev	21.41	23.72	13.2
Dermatology	Best	22905.22	23076.32	20782.67
	Mean	23784.31	23652.6	22520.23
	Std dev	707.32	440	793.87
Yeast	Best	83.19	84.71	87.62
	Mean	90.7	88.99	92.19
	Std dev	3.64	3.41	2.79
CMC	Best	24376.11	24150.26	24515.31
	Mean	25184.22	25455.98	25347.75
	Std dev	501.02	612.7	713.28

In Table 3, the solutions obtained by the DE-LF outperform other competing algorithm on the datasets Wine, zoo, and CMC. Moreover, its convergence speed is also most promising on these datasets (refer Fig. 2a–f) with respect to other competing algorithms. DE-GSS achieves best value of FFV over other competing algorithms on the datasets, iris, and dermatology. However, DE-CLS produces better quality solutions on only one dataset, yeast with respect to other competing algorithms. In addition, DE-CLS achieves better value of mean and standard deviation (Std. dev) of FFVs in all runs on the datasets Iris, Wine, and CMC with respect to other competing algorithms.

Table 4 presents the F-measure and rand index value of final achieved solution by competing algorithms in all runs. DE-LF has the highest F-measure (FM) and rand index (RI) values over other competing algorithms for the datasets iris, wine, yeast, and CMC; there is only one exception that DE-GSS achieves the highest rand index value for yeast dataset. DE-GSS and DE-CLS report highest values of F-measure and rand index for datasets, zoo and dermatology, respectively.

In Table 3, in case of zoo dataset, though DE-CLS does not achieve better solution (lower FFV value) than the DE-LF, it achieves lower FM and RI value. It means that FFV does not exactly map FM and RI value on these datasets. It is because SSE has no absolute correlation with FM and RI when actual data distribution is not regular. In case of dermatology dataset, based on F-measure and rand index values, all adopted algorithms perform highly unsatisfactory as it is a high-dimensional dataset

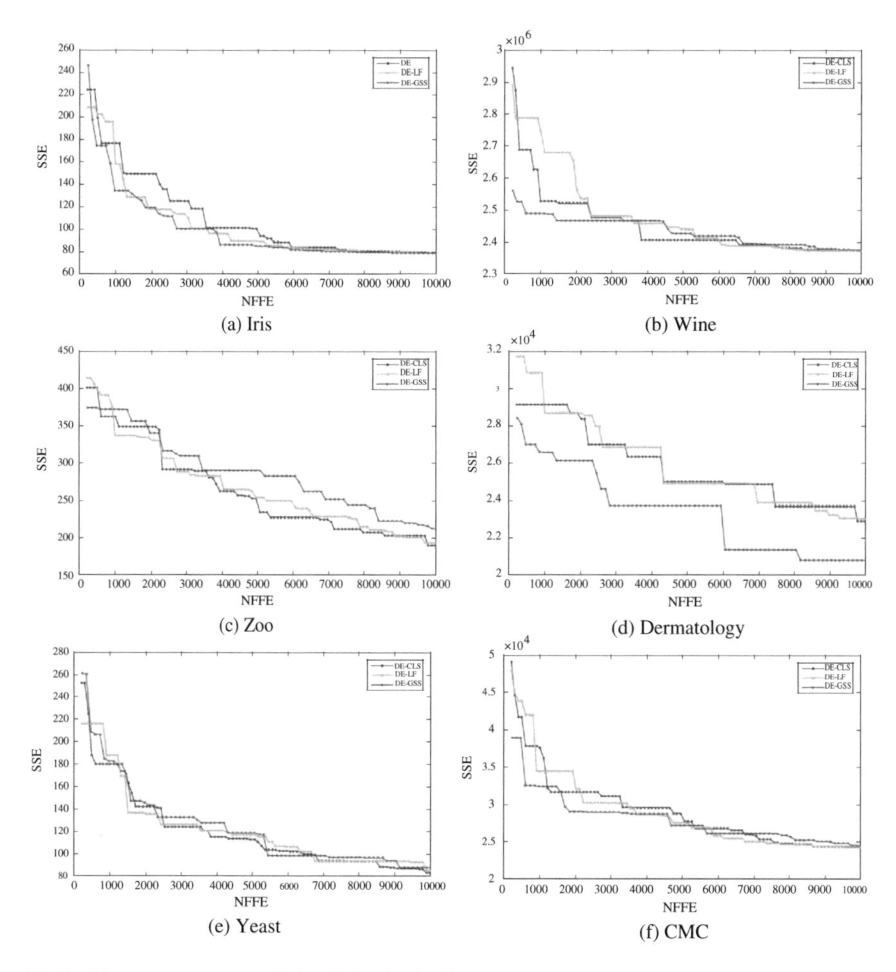

Fig. 2 Convergence speed of algorithm for the datasets

containing 34 dimensions. Overall, DE-LF achieves better value of FFV and highest values of F-measure and rand index on most of the datasets, which indicates that DE-LF balances exploration and exploitation better in global search space. Hence, it is a better alternative to solve clustering problems. However, DE-CLS is more robust as compared to other competing algorithms in terms of achieving FFV of the final solution.

Table 4 The F-measure and rand index of best solution (as per FFV)

	DE-CLS		DE-LF		DE-GSS	
Datasets	F-measure	Rand index	F-measure	Rand index	F-measure	Rand index
Iris	0.8853	0.8737	0.8918	0.8797	0.8853	0.8737
Wine	0.7148	0.7187	0.7148	0.7187	0.7148	0.7187
Zoo	0.7855	0.8865	0.7728	0.8745	0.7414	0.8566
Dermatology	0.2964	0.6919	0.2421	0.6994	0.3099	0.7012
Yeast	0.4403	0.5549	0.4668	0.5737	0.3978	0.6085
CMC	0.4002	0.5578	0.4012	0.5582	0.4002	0.5578

5 Conclusion

Evolutionary algorithms are global search optimization methods to search near-optimal solution within reasonable time. In this paper, a comparative study on DE, an evolutionary algorithm, is performed with well-known local search optimization algorithms, namely chaotic local search (CLS), levy flight (LF), and golden section search (GSS) using internal validity measure SSE for data clustering problems. Effectiveness of these algorithms is compared based on the solution quality and their convergence speed to achieve the final solution in the clustering domain over six real datasets from UCI machine learning repository. External validity measures, F-measure, and rand index are employed to evaluate the quality of final achieved solutions. Results demonstrate that DE-LF outperforms other competing algorithms in most of the datasets. However, in some cases, DE-CLS and DE-GSS provide better results than the DE-LF. Therefore, it is concluded that the DE-LF is a simple and potential algorithm to solve hard partitional clustering.

References

1. Bharti, K.K., Singh, P.K.: Chaotic gradient artificial bee colony for text clustering. Soft. Comput. **20**(3), 1113–1126 (2016)
2. Das, S., Mullick, S.S., Suganthan, P.N.: Recent advances in differential evolution—an updated survey. Swarm Evol. Comput. **27**, 1–30 (2016)
3. Hruschka, E.R., Campello, R.J.G.B., Freitas, A.A., De Carvalho, A.P.L.F.: A survey of evolutionary algorithms for clustering. IEEE Trans. Syst. Man Cybern. Part C Appl. Rev. **39**(2), 133–155 (2009)
4. Rajpurohit, J., Tarun Kumar Sharma, A.A.V.: Glossary of metaheuristic algorithms. Int. J.Comput. Inf. Syst. Ind. Manag. Appl. **9**, 181–205 (2017)
5. Jain, A.K., Murty, M.N., Flynn, P.J.: Data clustering: a review. ACM Comput. Surv. (CSUR) **31**(3), 264–323 (1999)
6. Prakash, J., Singh, P.: An effective multiobjective approach for hard partitional clustering. Memetic Comput. **7**(2), 93–104 (2015)
7. Prakash, J., Singh, P.K.: An effective hybrid method based on de, ga, and k-means for data clustering. In: Proceedings of the Second International Conference on Soft Computing for Problem Solving (SocProS 2012). Springer, pp. 1561–1572, 28–30 Dec, 2012

8. Prakash, J., Singh, P.K.: Evolutionary and swarm intelligence methods for partitional hard clustering. In: 2014 International Conference on Information Technology (ICIT), pp. 264–269. IEEE (2014)
9. Sharma, H., Jadon, S.S., Bansal, J.C., Arya, K.: Levy flight based local search in differential evolution. In: International Conference on Swarm, Evolutionary, and Memetic Computing, pp. 248–259. Springer (2013)
10. Sharma, T.K., Pant, M.: Golden search based artificial bee colony algorithm and its application to solve engineering design problems. In: 2012 Second International Conference on Advanced Computing & Communication Technologies (ACCT), pp. 156–160. IEEE (2012)
11. Storn, R., Price, K.: Differential evolution—a simple and efficient heuristic for global optimization over continuous spaces. J. Global Optim. **11**(4), 341–359 (1997)

A Review on Traffic Monitoring System Techniques

Neeraj Kumar Jain, R. K. Saini and Preeti Mittal

Abstract Increase in traffic density along with the population growth in the world has resulted in more and more congested roads, air pollution, and accidents. Growth of total number of vehicles around the world has increased exponentially during past decade. Traffic monitoring in this scenario is certainly a big challenge in many cities of India. In most of the cities, we are still dependent on human and accomplishing this herculean task manually. In controlling enormous volume of traffic with manual methods, more and more issues like availability, efficiency, and accuracy of management staff are always encountered. In this study, various traffic monitoring schemes striving to cop up and handle enormous traffic flow with optimized human intervention are studied. Applications of these schemes include identification of vehicles in traffic, sense traffic congestion on a road, measuring speed of vehicle, traffic density on intersections, the presence of VIP vehicles or ambulances, accidents on roads, path for pedestrians, and many more. Usually, intrusive and non-intrusive in situ techniques and in-vehicle technologies are used for traffic monitoring but image processing techniques win over these traditional techniques. The aim of this paper is to discuss the previous research work done to monitor the traffic using image and video processing techniques and future perspective.

Keywords Traffic monitoring · Traffic density · In situ techniques
In-vehicle technologies · Image processing

N. K. Jain (✉) · R. K. Saini · P. Mittal
Department of Mathematical Sciences and Computer Applications,
Bundelkhand University, Jhansi, India
e-mail: neerajjain15@gmail.com

R. K. Saini
e-mail: rksaini.bu@gmail.com

P. Mittal
e-mail: preetimittal1980@yahoo.co.in

K. Ray et al. (eds.), *Soft Computing: Theories and Applications*,
Advances in Intelligent Systems and Computing 742,
https://doi.org/10.1007/978-981-13-0589-4_53

Fig. 1 Year-wise vehicle registered in Delhi

Delhi Traffic Basic Facts | Vehicle Registered

Year wise Vehicle Registered in Delhi

Year	No. of Vehicle
2008	3,95,435
2009	4,14,150
2010	4,86,112
2011	5,12,988
2012	5,29,712
2013	5,31,332
2014	5,74,602
2015	4,30,603
2016	4,62,255

1 Introduction

An exponential growth of running vehicles on roads along with the population growth in the world has resulted in more and more congested roads, air pollution and accidents. Growth of total number of vehicles around the world has increased exponentially during past decade.

In Indian metro cities, traffic density on road is increasing four times faster than the population. Figure 1 shows the year-wise vehicles registration status in our national capital Delhi which depicts the enormity of vehicles and challenges in front of Delhi traffic police [1].

There are basically two types of traffic congestions—Recurring traffic congestion which appears at the same place during the same time regularly and nonrecurring traffic congestion which occurs randomly like an unplanned event [2]. This nonrecurring effect can cause a sudden traffic volume increase. Detection of nonrecurring traffic congestion is critical compared to the recurring type, because it requires real-time traffic information and evaluation thereof with appropriate traffic management decisions. Due to traffic congestion, there is obvious wastage of many resources. Traffic congestion may take place due to the following reasons [3]:

1. Too many cars for the roadway due to inadequate mass transit options or other reasons.
2. Obstacles on the road cause a blockage and merger. These can be double parking, road work, lane closure due to utility work, road narrowing down, and an accident.
3. Traffic signals out of sync many times on purpose or occasionally when the computers are malfunctioning.
4. Inadequate green/red time.
5. Too many pedestrians crossing not permitting cars to turn.
6. Too many trucks on the road due to inadequate rail freight opportunities.

7. Overdevelopment in areas where the mass transit system is already overcrowded and the road system is inadequate.
8. Delay in removal of accidental vehicles on roads.
9. Poor weather conditions.

The rest of the paper is organized as follows: Sect. 2 describes traffic monitoring systems. Section 3 discusses literature review in detail. Finally, Sect. 4 is for conclusion and future scope.

2 Traffic Monitoring System

A traffic monitoring system must tackle issues like traffic congestion, accident detection, vehicle identification/detection, automatic vehicle guidance, smart signaling, forensics, traffic density, safe pedestrian movement, emergency vehicles transit, etc.

An ideal traffic monitoring system can be designed using either of the following:

1. In situ traffic detector technologies.
2. Vehicular sensor networks (VSN) or probe vehicles (PVs), such as taxis and buses, and floating cars (FCs), such as patrol cars for surveillance.
3. Image or video processing.

2.1 In Situ Technologies

In situ traffic detector technologies are further divided into two categories—Intrusive technology and non-intrusive technology. In intrusive technologies, detectors are physically mounted at or below the road surface, installation of which causes potential disruption to traffic. These include embed magnetometers, pneumatic tube detectors, inductive detector loops, and Weigh-in-Motion (WIM) systems.

In nonintrusive technologies, detectors are mounted at or above the road surface, and their installation causes little or no disruption to traffic. Nonintrusive technologies include manual methods, video data collection, passive or active infrared detectors, microwave radar detectors, ultrasonic detectors, passive acoustic detectors, laser detectors, and aerial photography (Fig. 2).

2.2 Vehicular Sensor Networks

Vehicular sensor networks (VSNs) have been proved as a great solution for traffic monitoring. Sensing devices attached to the moving vehicles move all over the city to collect the traffic information. These moving sensing devices are attached to each other and to the traffic monitoring center as well. The collected information is passed

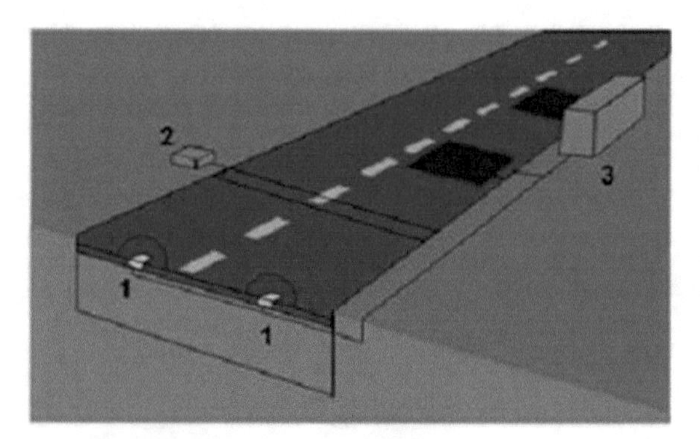

Fig. 2 Types of detectors

to city traffic monitoring center on vehicle-to-vehicle or vehicle-to-infra wireless communications basis. Traffic monitoring center takes appropriate decisions in order to ensure hassle-free traffic movement.

In-vehicle devices are generically termed as automatic vehicle location (AVL) systems. AVL devices either provide positional information whenever a suitably equipped vehicle passes a certain point in the network, or continuous information as the vehicle travels through a network. The former system typically relies on appropriate vehicles being equipped with transponders which transmit and receive information from roadside units. The latter system uses vehicles equipped with global positioning system (GPS) technology.

2.3 Image and Video Processing

Video monitoring and image processing have been widely used in traffic management for solving number of traffic issues. The traffic density estimation and vehicle classification can also be achieved using video monitoring systems. Live video feed from the cameras at traffic junctions for real-time traffic density calculation using video and image processing. It can be used efficiently in switching the traffic lights according to vehicle density on road, resulting in reducing the traffic congestion on roads, and hence lower the number of accidents. It ensures safe transit to people and reduce fuel consumption and waiting time. It will also provide significant data which will help in future road planning and analysis. In further stages, multiple traffic lights can be synchronized with each other with an aim of even less traffic congestion and free flow of traffic [4].

Some of the most demanding and widely studied applications of image or video processing are travelers' information, ramp metering, traffic monitoring, automatic vehicle guidance, accident detection, smart signaling, vehicle identification/detection, forensics, traffic density, safe pedestrian movement, etc.

Practically, it has been observed although intrusive detectors or sensors have become important in traffic applications mainly due to their rapid response, cheaper installation, operation, and maintenance, and their ability to control and monitor wide areas [5], but they have more disadvantages too like they require to dig up the road and therefore costlier to install. It also gives limited amount of information than other methods of traffic monitoring system. Radar gun is another way of getting information about the speed of vehicle but it gives less traffic information. Similarly, the pressure tubes are reliable enough but give limited information about traffic [6].

Advantages and disadvantages of these traffic monitoring techniques are tabulated as follows [7].

Technology	Strengths	Weaknesses
Inductive loop	• Flexible design to satisfy large variety of applications • Mature, well-understood technology • Large experience base Provides basic traffic parameters (e g., volume, presence, occupancy speed, headway, and gap) • Insensitive to inclement weather such as rain, fog, and snow • Provides best accuracy for count data as compared with other commonly used techniques • Common standard for obtaining accurate occupancy measurements • High-frequency excitation models provide classification data	• Installation requires pavement cut • Decreases pavement life • Installation and maintenance require lane closure • Wire loops subject to stresses of traffic and temperature • Multiple detectors usually required of monitor a location • Detection accuracy may decrease when design requires detection of a large variety of vehicle classes
Magnetometer (Two-axis fluxgate magnetometer)	• Less, susceptible than loops to stresses of traffic • Insensitive to inclement weather such as snow, rain, and fog • Some models transmit data over wireless RF link	• Installation requires pavement cut • Decreases pavement life • Installation and maintenance require lane closure • Models with small detection zones require multiple units for full lane detection • Installation requires pavement cut or boring under roadway • Cannot detect stopped vehicles unless special sensor layouts and signal processing software are used
Magnetic (induction or search coil magnetometer)	• Can be used where loops are not feasible (e.g., bridge decks) • Some models are installed under roadway without the need for pavement cuts. However, boring under roadway is required • Insensitive to inclement weather such as snow, rain, and fog • Less susceptible than loops to stresses to traffic	• Installation requires pavement cut or boring under roadway • Cannot detect stopped vehicles unless special sensor layouts and signal processing softwares are used

Technology	Strengths	Weaknesses
Microwave radar	• Typically insensitive to inclement weather at the relatively short ranges encountered in traffic management applications • Direct measurement of speed • Multiple lane operations available	• CW Doppler sensors cannot detect stopped vehicles
Active infrared (Laser radar)	• Transmits multiple beams for accurate measurement of vehicle position, speed, and class • Multiple lane operations available	• Operations may be affected by fog when visibility is less than ≈ 20 ft (6 m) or blowing snow is present • Installation and maintenance, including periodic lens cleaning, require lane closure
Passive infrared	• Multizone passive sensor's measure speed	• Passive sensor may have reduced sensitivity to vehicles in heavy rain and snow and dense fog • Some models not recommended for presence detection
Ultrasonic	• Multiple lane operations available • Capable of overheight vehicle detection • Large Japanese expedience base	• Environmental conditions such as temperature change and extreme air turbulence can affect performance. Temperature compensation is built into some models • Large pulse repetition periods may degrade occupancy measurement on freeways with vehicles traveling at moderate to high speeds
Acoustic	• Passive detection • Insensitive to precipitation • Multiple lane operations available in some models	• Cold temperatures may affect vehicle count accuracy • Specific models are not recommended with slow-moving vehicles in stop-and-go traffic
Video image processor	• Monitors multiple lanes and multiple detection zones/lanes • Easy to add and modify detection zones • Rich array of data available • Provides wide-area detection when information gathered at one camera location can be linked to another	• Installation and maintenance, including periodic lens cleaning, require lane closure when camera is mounted over roadway (lane closure may not be required when camera is mounted at side of roadway) • Performance affected by inclement weather such as fog, ram. and snow: vehicle shadows: vehicle projection into adjacent lanes: occlusion: day-to-night transition: vehicle road/contrast: and water, salt grime, icicles, and cobwebs on camera lens • Requires 30- to 50-ft (9- to 15-in) camera mounting height (in a side-mounting configuration) for optimum presence detection and speed measurement • Some models susceptible to camera motion caused by strong winds or vibration of camera mounting structure • Generally cost-effective when many detection zones within the field-of-view of the camera or specialized data are required • Reliable nighttime signal actuation requires street lighting

3 Literature Review

Kamijo et al. (2000) used spatiotemporal Markov random field (MRF), for traffic images at intersections. It models a tracking problem by determining the state of each pixel in an image and its transit, and how such states transit along both the image axes as well as the time axes. It segments and tracks occluded vehicles at a high success rate [8].

Messelodi et al. (2005) proposed a real-time vision system by analyzing monocular image sequences from pole-mounted video cameras. Their experimental results demonstrated robust, real-time vehicle detection, tracking, and classification over several hours of videos taken under different illumination conditions [9].

Lee and Baik (2006) introduced a video-based vehicle tracking system to provide information on directional traffic counts at intersections. The extracted counts were fed to estimate an origin–destination trip table which was necessary information for traffic impact study and transportation planning [10].

Aycard et al. (2011) proposed an approach for intersection safety developed in the scope of the European project INTERSAFE-2. A complete solution for the safety problem including the tasks of perception and risk assessment using on-board Lidar and stereovision sensors presented and better results were shown [11].

Wang et al. (2015) introduced an approach for real-time multi-vehicle tracking and counting using fisheye camera based on simple feature points tracking, grouping, and association. Motion similarity and neighbor-weighted grafting were used to transfer motion knowledge between long and short point trajectories [12].

Jodoin et al. (2016) proposed a tracking system to track the various road users of diverse shapes and appearances. Finite state machine handled fragmentation, splitting, and merging of the road users to correct and improve the resulting object trajectories. This tracker was tested on several urban intersection videos and it outperformed [5].

Liu et al. (2016) used a method that accurately detects vehicles using a probabilistic classification method followed by a refinement based on object segments. Both classification and segmentation methods made use of coregistered aerial RGB images and airborne LiDAR data [13].

Huang et al. (2017) introduced vehicle detection in the tunnel which is a challenging problem due to the usage of heterogeneous cameras, varied camera setup locations, low-resolution videos, poor tunnel illumination, and reflected lights on the tunnel wall. The proposed method was based on background subtraction and Deep Belief Network (DBN) with three-hidden layer architecture [6].

Tang et al. (2017) proposed a technique for vehicle detection and type recognition based on static images. This technique was highly practical and directly applicable to various operations in a traffic surveillance system. First, Haar-like features and AdaBoost algorithms were applied for feature extracting and constructing classifiers, which were used to locate the vehicle over the input image. Then, the Gabor wavelet transform and a local binary pattern operator was used to extract multiscale and multi-orientation vehicle features [14].

Ukani et al. (2017) used video surveillance systems for real-time vehicle detection and classification. Background subtraction and from each detected vehicles Scale-Invariant Feature Transform (SIFT) features were extracted. Vehicles were classified using the neural network and Support Vector Machine (SVM). SVM showed better generalization than artificial neural networks [15].

4 Conclusion and Future Scope

In this review paper, various existing traffic monitoring systems have been studied. Presently, these systems do provide a cost-effective solution, but the rate of successful operation is not good. Inductive loop detectors installed below the road fail in the case of poor road condition. IR detectors along the side of the lane for density calculation operate less efficiently where they are not applicable in the real-time process. On the other hand, several other image processing based methods are doing better and helping the present traffic control system to be more efficient. Computer vision being one of the most researched fields is for the future technologies. There is more and more scope of video and image processing techniques in traffic monitoring and analysis. Intelligent solutions to traffic issues like monitoring, automatic vehicle guidance, accident detection, smart signaling, vehicle identification/detection, forensics, traffic density, and vehicle theft can be found and used effectively using enormous capabilities of image processing and other techniques.

References

1. https://delhitrafficpolice.nic.in/about-us/statistics/
2. Nellore, K., Hancke, G.P.: A survey on urban traffic management system using wireless sensor networks. Sensors **16**(2), 157 (2016)
3. http://bklyner.com/what-really-causes-traffic-congestion-sheepshead-bay/
4. Kanungo, A., Sharma, A., Singla, C.: Smart traffic lights switching and traffic density calculation using video processing. In: 2014 Recent Advances in Engineering and Computational Sciences (RAECS), pp. 1–6. IEEE (2014)
5. Jodoin, J.P., Bilodeau, G.A., Saunier, N.: Tracking all road users at multimodal urban traffic intersections. IEEE Trans. Intell. Transp. Syst. **17**(11), 3241–3251 (2016)
6. Huang, B.J., Hsieh, J.W., Tsai, C.M.: Vehicle detection in Hsuehshan tunnel using background subtraction and deep belief network. In: Asian Conference on Intelligent Information and Database Systems, pp. 217–226. Springer (2017)
7. Leduc, G.: Road traffic data: collection methods and applications. Working papers on energy, transport and climate change, vol. 1, p. 55 (2008)
8. Kamijo, S., Matsushita, Y., Ikeuchi, K., Sakauchi, M.: Traffic monitoring and accident detection at intersections. IEEE Trans. Intell. Transp. Syst. **1**(2), 108–118 (2000)
9. Messelodi, S., Modena, C.M., Zanin, M.: A computer vision system for the detection and classification of vehicles at urban road intersections. Pattern Anal. Appl. **8**(1–2), 17–31 (2005)
10. Lee, S.M., Baik, H.: Origin-destination (od) trip table estimation using traffic movement counts from vehicle tracking system at intersection. In: IECON 2006-32nd Annual Conference on IEEE Industrial Electronics, pp. 3332–3337. IEEE (2006)

11. Aycard, O., Baig, Q., Bota, S., Nashashibi, F., Nedevschi, S., Pantilie, C., Parent, M., Resende, P., Vu, T.D.: Intersection safety using lidar and stereo vision sensors. In: 2011 IEEE Intelligent Vehicles Symposium (IV), pp. 863–869. IEEE (2011)
12. Wang, W., Gee, T., Price, J., Qi, H.: Real time multi-vehicle tracking and counting at intersections from a fisheye camera. In: 2015 IEEE Winter Conference on Applications of Computer Vision (WACV), pp. 17–24. IEEE (2015)
13. Liu, Y., Monteiro, S.T., Saber, E.: Vehicle detection from aerial color imagery and airborne lidar data. In: 2016 IEEE International Geoscience and Remote Sensing Symposium (IGARSS), pp. 1384–1387. IEEE (2016)
14. Tang, Y., Zhang, C., Gu, R., Li, P., Yang, B.: Vehicle detection and recognition for intelligent traffic surveillance system. Multimedia Tools Appl. **76**(4), 5817–5832 (2017)
15. Ukani, V., Garg, S., Patel, C., Tank, H.: Efficient vehicle detection and classification for traffic surveillance system. In: International Conference on Advances in Computing and Data Sciences, pp. 495–503. Springer (2016)

Large-Scale Compute-Intensive Constrained Optimization Problems: GPGPU-Based Approach

Sandeep U. Mane and M. R. Narsinga Rao

Abstract The large-scale compute-intensive optimization problems exist in various engineering and scientific problems and applications. Such problems contain large number of design variables which needs to optimize to obtain optimum solutions. The large-scale compute-intensive problems solved using variety of nature-inspired techniques. The DEVIIC is one of such approach found in literature, developed using VIIC technique. This paper presents design and implementation of GPGPU-based DEVIIC algorithm to address large-scale compute-intensive benchmark optimization problems, based on master–slave strategy. The proposed approach is evaluated using 12 large-scale benchmark functions found in literature. The obtained results of proposed approach compared with results found in literature, implemented existing sequential DEVIIC algorithm and proposed GPGPU-based approach. The proposed approach gives comparatively better results than results found in Sayed et al. (Inf Sci 316:457–486, 2015) [1] for functions F1, F5 to F9, and F11, and fails to obtain better results for functions F2 to F4 and F10. As the proposed approach is to develop GPGPU-based algorithm, the speedup is computed. The proposed approach significantly reduces the execution time required to obtain the best solution. The proposed approach is 23 to 35 times faster than its sequential counterpart.

Keywords Large-scale compute-intensive optimization problems · GPGPU-based evolutionary algorithm · Differential evolution algorithm · VIIC technique

---

S. U. Mane (✉) · M. R. Narsinga Rao
Department of CSE, K L University, Vaddeswaram, Guntur, AP, India
e-mail: manesandip82@gmail.com

M. R. Narsinga Rao
e-mail: ramanarasingarao@kluniversity.in

K. Ray et al. (eds.), *Soft Computing: Theories and Applications*,
Advances in Intelligent Systems and Computing 742,
https://doi.org/10.1007/978-981-13-0589-4_54

1 Introduction

Many real-life problems are computationally expensive optimization problems. For high-quality decision-making, solving computationally expensive optimization problems is valuable and challenging research area. Many computationally expensive optimization problems are large-scale optimization problems with high dimensionality. When dimensions of the problem are increasing, search space where solution is present get increases, so finding the best solution takes huge time. In many engineering applications as well as real-life problems, several objectives are to be optimized simultaneously while satisfying constraints. The lack of explicit mathematical formulas of the objectives and constraints may lead to conducting computationally expensive and time-consuming experiments. Real-world design optimization problems are typically computationally expensive, and such problems are also high-dimensional, poor models with poor accuracy, and thus degrade the optimization search [2]. Large-scale optimization problems are also treated as computationally expensive as it has huge number of variables, so finding the optimal solution for such problem takes a long time and its difficult [1, 3].

Various constrained and unconstrained benchmark problems exist for single-objective, multi-objective, many-objective, large-scale computationally expensive, and real parameter optimization, etc. These benchmark test suits are important to test the applicability and suitability of newly proposed algorithm. Special competitions, “CEC” organized to propose various optimization algorithms to solve challenging optimization problems. Li et al. have been proposed benchmark functions for CEC’13, the competition organized on large-scale global optimizations. The proposed benchmark suits are improved in functions proposed for CEC’10 competition. It contains 15 unconstrained benchmark functions. The large-scale optimization problems are computationally expensive due to factors like increase in dimensions, interaction between variables, nature of variables, etc. [4]. Sayed et al. have used CEC’13 benchmark suits and also proposed 18 test functions based on CEC’2008 test suit [1].

In last few decades, software application’s performance improved largely due to the advancement in hardware as well as advancement in programming techniques. The General-Purpose Graphic Processing Unit (GPGPU) is many-core processing chip developed, which has highly parallel many-threaded multiprocessor with huge computational power. It has twofold purpose, programmable graphics processor and scalable parallel computing platform. The GPGPU-based computational approach can be employed when computations are independent as well as problem supports data parallelism. From the literature, it is observed that use of GPGPU can improve execution speed up to 100 times or more than its sequential counterpart. The Nvidia developed a Compute-Unified Device Architecture (CUDA), to utilize the computational power of GPGPU to develop GPU-based applications. Such applications are found in different domains of engineering and science [5].

The rest of paper is organized as follows: Sect. 2 presents a review of the literature, various methods developed to solve large-scale problems, and GPGPU-based differential evolution algorithm. Section 3 describes the implementation of sequential and proposed GPGPU-based DEVIIC algorithm. The results obtained presented in Sect. 4. The conclusions of study with future work are presented in Sect. 5.

2 Literature Review

The computationally expensive optimization problems are observed in real-life, engineering, and scientific domains. This section presents brief about algorithms proposed for solving such problems.

Sayed et al. [1] developed a decomposition-based evolutionary algorithm to solve large-scale constrained optimization problems. They proposed 18 test suite problems based on CEC'10 benchmark problems, which categorized among six objective functions with three constraints. Authors mentioned that the performance of evolutionary algorithms can be improved by decreasing the number of interdependent variables between the subproblems. They achieved this by first identifying the dependent variables and then grouping them in common subproblems, and subproblems are optimized independently. The proposed approach takes 843 s when problem has 100 variables, it requires 19242.76 s when variables are 500 and when number of variables increased to 1000, it takes 71510.64 s on Intel Core i5 processor with 3.20 GHz. Yang et al. in [6] proposed an algorithm to tackle nonrigid image registration. Image registration includes the deformation process which is very complex process and computationally hard because of variance in image pixels. In this paper, they employ a block grouping as the grouping strategy to capture the interdependency between the variables. Authors have introduced a new hybrid algorithm, L-BFGS-B with Cat Swarm Optimization (CSO) for large-scale unconstrained problems. Decomposition strategy used in HLCSO is similar to Cooperative Co-evolution PSO (CCPSO2) found in the literature. In this decomposition strategy, n-dimensional search space is divided into "k" swarms and each swarm is of "s" dimensions. This strategy not works on constrained problems. The experiment performed on personal computer, which took 40.2 min for image registration process. In [7] stated that many problems in the medical science, business, and defense with location aspects can be expressed as grid-based location problem, and are computationally expensive to solve. The decomposition strategy, "relax-and-fix-based" introduced to solve large-scale problems, which significantly reduces the time required to find solution. The proposed approach takes 5351.578 s to solve 10 * 20 test grids for 13 light resources on Intel Xeon, 3.16 GHz, and 32 GB memory. Nick et al. in [8] have tackled the large-scale computationally expensive constrained optimization problem, wind power optimal capacity allocation to remote areas. The Benders decomposition method is used for the decomposition of large-scale problems, which decomposes the original problem into a master and a subproblem. The master problem is a linear problem, and subproblems are mixed integer problems. In [9], a column generation-based decomposition algorithm presented for solving large railway crew scheduling problem, which has 358 stations, 2301 trains, and 17,180 trips. The proposed approach takes 1.35 h to solve the selected problem. The extension process of CEC'2013 benchmark problems for various purposes is presented in [10]. The real-world problems have complexity in the form of size, variable interaction, and interdependency between the subcomponents. The nature of many real-world problems makes divide-and-conquer approach suitable for such problems. The large dimensionality in the problem decreases the performance of the differential evolution algorithm [11]. To achieve better performance, decomposition approach is suggested. In this paper, Hybrid Dependency Identification with Memetic Algorithm (HDIMA) model is proposed. Mallipeddi and Suganthan in [12] proposed a constraint handling technique, ensemble of constraint handling techniques (ECHT). According to author, evolutionary algorithms fail to handle constrained problems, so the additional mechanism is required for constraint handling and it is developed by authors. The new decomposition technique to handle the large-scale optimization problems has been proposed in [13]. The variable interaction is the main source of performance loss on large-scale problems. Cooperative Co-evolution (CC) is the natural solution to solve large-scale optimization problems, but due to variable interaction, performance of CC gets decreased. Authors proposed technique, Delta Grouping Method, to identify and group the interacting variables in the same subproblem. Kopanos et al. proposed a decomposition strategy for scheduling problems in multiproduct multistage batch plants. The pharmaceutical industry is selected as case study. The real-world pharmaceutical industry contains lots of stages and hundreds of batches to perform one task. The scheduling of these stages becomes very complex and computationally expensive. Authors have introduced MIP-based (Mixed integer programming) decomposition strategy to solve such large-scale scheduling problems [14].

Cao and Sun worked to speed up the optimization process of nationwide air traffic flow by using parallel framework, and Large-capacity Cell Transmission Model (CTM (L)) using a dual decomposition method. This problem has a huge number of variables and about 2326 subproblems. Authors concluded that the proposed parallel computing framework decreases the runtime of traffic flow management optimization from 2 h to 6 min [15]. Tenne in [16] presented an approach to solve expensive engineering optimization problem, the airfoil shape optimization problem.

From above literature survey, solving large-scale optimization problem is very difficult and it takes long time to obtain results. If problems have some constraints, then solving such problems become more difficult, complex, and challenging task. There are many different techniques available to solve large-scale unconstrained optimization problems but for constrained problems only a few techniques are present.

2.1 Various Techniques Used to Solve Computationally Expensive Problems

The popular techniques which are used to solve computationally expensive optimization problems and found in the literature are presented in this section. These techniques are mainly based on divide-and-conquer approach.

1. **Cooperative co-evolution with variable interaction learning** (**CCVIL**)

Cooperative co-evolutionary algorithms are an effective decomposition-based approach to solve large-scale computationally expensive optimization problems. In this, the main problem is decomposed into a number of subproblems using variable interaction; learning method directly identifies the interaction between variables. It decomposes the problem into separable and non-separable type of subproblems [17].

2. **Cooperative co-evolution with differential grouping**:

An automatic decomposition strategy called differential grouping uncover the underlying interaction structure of the decision variables and form subcomponents such that the interdependence between them is kept to a minimum [13, 18].

3. **Cooperative co-evolution algorithm with adaptive variable partitioning**:

The adaptive variable partitioning strategy is based on correlation between variables, partitions variables into subpopulations based on observed correlation [19].

4. **Cooperative co-evolution with delta grouping**:

It measures the average difference in certain variability across the entire population and uses it for identifying interacting variables. The systematic way of capturing interacting variables for more effective problem decomposition is suitable for cooperative co-evolutionary frameworks [12].

5. **Benders Decomposition technique**:

The benders decomposition is a popular approach for solving large-scale problems. It decomposes the original problem into a master and subproblems. The master problem is a linear problem. The subproblems are a mixed integer problem. The subproblems use the solution of the master problem [8].

6. **Differential evolution algorithm with variable interaction identification for constrained problems**:

The VIIC is derived from the definition of problem separability. Problem separability means that the best partitioning for a large-scale problem into subproblems minimizes the number of interdependent variables and thus increases the number of non-separable variables within each subproblem. The VIIC technique aims to find the most appropriate grouping for the variables [1].

2.2 Differential Evolution Algorithm on GPGPU from Literature

The Differential Evolution (DE) algorithm is implemented in parallel fashion using GPGPU. This section presents various problems solved using GPGPU-based DE algorithm.

In [20], parallel approach for DE algorithm with self-adapting control parameters and generalized opposition-based learning is proposed for solving high-dimensional time-consuming complex problems. The self-adapting control parameters help to avoid adjusting the control parameters manually. Experiment is carried out on system with Intel Core 2 Quad processor and NVIDIA GeForce GTX 285 GPU. The standard CEC 2008 benchmark problems with 100 to 1000 dimensions are used to measure the performance. Result shows that this approach gives good results and GPU can effectively reduce the computation time. In [21], the DE algorithm on GPU is proposed for CEC 2009 benchmark functions. Initialization of population, objective function evaluation, mutation operation and selection operation performed on device. Only three random vectors which are required for mutation operation are generated on CPU. The GTX 285 GPU and AMD dual-core processor with 2.7 GHz are used for experimentation. The result shows that proposed strategy completes the computations about 35.48 times faster than sequential approach. Kromer et al. in [22] solved linear ordering problems using DE algorithm on Tesla C2050 GPU. The proposed approach tested against well-known LOLIB dataset library. Significant speedup is achieved. Oliveira et al. in [23] stated that evolutionary algorithms require high computational resources and high processing power for some applications, due to which evolutionary algorithms do not give the satisfactory result. To overcome this problem, authors proposed parallelization of evolutionary algorithms. The proposed approach tested on ZDT test suit, and the result shows high-speed up. Qin et al. implemented an improved version of GPU-based DE algorithm. Authors mentioned that DE algorithm is promising evolutionary algorithm and highly suitable for parallelization. The proposed parallel strategy improved the device utilization and the throughput of memory access [24].

It is important to address the compute-intensive optimization problems. If the algorithm is designed in parallel, it will significantly improve the execution time. The GPUs are devices with huge computational power and its computational power can be harnessed to solve the problem in short computational time.

3 Proposed Approach

This section presents the existing sequential as well as proposed GPGPU-based parallel DEVIIC algorithm for constrained problems.

3.1 Existing Sequential DEVIIC Algorithm

The DEVIIC algorithm has been proposed by Sayed et al. in [1] by combining differential evolution algorithm with Variable Interaction Identification for Constrained (VIIC) problems technique. The VIIC technique is based on the property found in optimization problems, i.e., fully separable and partially separable. The problem separability means to determine the best partitioning scheme for a large-scale problem into subproblems is the one that minimizes the number of interdependent variables, and thus increases the number of non-separable variables within each subproblem. The VIIC technique aims to find the most appropriate grouping for the variables. When the two variables are contributing to fitness and change in one variable affects other variables, then these two variables are interdependent. The procedure of VIIC technique is available in [1]. The existing DEVIIC algorithm is implemented in a sequential fashion to evaluate the speedup factor. First, all the necessary parameters like population size, number of FEs (fitness evaluations), lower bound, upper bound and mutation factor, etc. are set.

3.2 Proposed GPU-Based DEVIIC Algorithm

The GPU-based DEVIIC algorithm proposed to solve computationally expensive problems. The master–slave approach is used to implement the proposed GPU-based parallel DEVIIC algorithm. The CPU is utilized as a master and GPU acts as a slave. The population initialization performed on a GPU to minimize the time required for initialization of large number of random numbers. Also, GPU computes the objective function value and returns the value to master. To perform mutation operation, three random vectors are generated. The weighted difference method is used in DE for mutation operation. Calculated value of constraint is validated on master. Solution which is not violating the constraints and which gives minimum function value is selected, which is global best solution for that objective function. The Kernel 1, Kernel 2, and Kernel 3 are executed on GPU, i.e., on device, acting as slave. These functions are invoked from the CPU, i.e., host, acting as master. The population is initialized on device using Kernel 1 function. It helps to save the time required to transfer initial population from CPU to GPU. The major compute-intensive part for any evolutionary algorithm is function evaluation, which is implemented on device using function Kernel 2. The Kernel 3 performs mutation operation so as to improve the performance of algorithm in terms of time.

The steps performed on device are presented below:

- Init-kernel (): It performs initialization of population and design parameters.
- Function-evolution-kernel (): The objective function evolved in this function and returns the result.
- Mutation-kernel (): It performs mutation operation on selecting three random solutions which are generated on device and return the result.

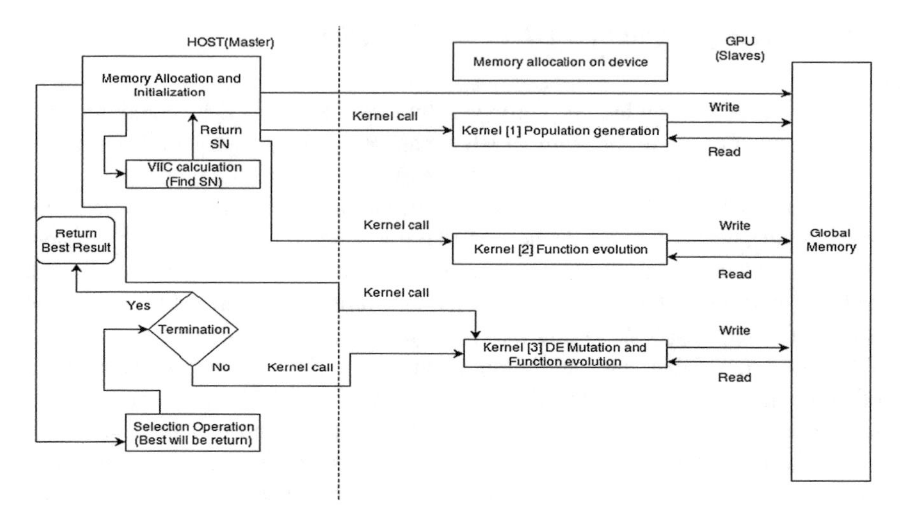

Fig. 1 Flow of GPGPU-based parallel DEVIIC algorithm

- Constraint-evolution-kernel (): The constraint also evaluated on a device and value of constraint violation with benefit points returned to the host.

Figure 1 presents the flow of GPGPU-based parallel DEVIIC algorithm.

4 Result and Discussions

The results obtained using existing DEVIIC algorithm and proposed GPGPU-based DEVIIC algorithm to solve compute-intensive large-scale benchmark optimization problems are presented in this section. The twelve out of eighteen constrained benchmark problems proposed in [1] are selected to evaluate the performance of proposed GPGPU-based DEVIIC algorithm. These benchmark problem test suits contain six objective functions and three constraints. The functions F1, F2, and F3 use objective function 1; functions F4, F5, and F6 use objective function 2; functions F7, F8, and F9 use objective function 3; and functions F10, F11, and F12 use objective function 4. The constraint 1 is for all 12 functions: the constraint 2 is for function F2, F3, F5, F6, F8, F9, F11, and F12; the constraint 3 is for function F3, F6, F9, and F12. The test functions proposed in [1] for constrained problems (scalable) are based on two functions that were used in the CEC 2008 benchmark problems, Sphere function used as separable part, and Rosenbrock's function used as non-separable part.

The experiment is performed on two different systems. The existing DEVIIC algorithm implemented on system has Intel Xeon processor, 8 GB RAM, and 2.10 GHz frequency. The GPGPU-based DEVIIC algorithm implemented using system with GeForce GTX 680 GPGPU card which has global memory of 2 GB, 8 multiprocessors, and compute capability of 3.0. The parameter's value for both the versions

Table 1 Comparison of results obtained using sequential DEVIIC and proposed GPGPU-based DEVIIC algorithm and speedup achieved

Functions	DEVIIC value [1]	Sequential DEVIIC algorithm			Proposed GPGPU-based DEVIIC			Speedup
		Best value	Mean	Time (s)	Best value	Mean	Time (s)	
F1	1.90E−98	**0.00E+00**	4.65E−01	613.40	**0.00E+00**	1.13E+01	19.66	31.19
F2	**3.00E+00**	3.47E+00	4.57E+01	871.28	3.04E+00	3.78E+01	25.45	34.23
F3	**5.00E+00**	6.17E+00	4.55E+01	952.42	5.02E+00	9.21E+01	27.19	35.02
F4	**4.14E−03**	2.84E+00	3.07E+00	426.90	2.07E+00	1.02E+01	14.43	29.58
F5	3.02E+00	**3.00E+00**	5.57E+00	455.11	**3.00E+00**	7.25E+00	16.84	27.01
F6	4.11E+00	**4.10E+00**	6.02E+00	634.82	**4.10E+00**	1.42E+01	23.90	26.56
F7	4.43E−03	**4.00E−03**	2.90E+02	423.78	**4.00E−03**	2.35E+02	14.61	29.00
F8	1.07E−05	**1.00E−05**	1.79E+01	470.31	**1.00E−05**	2.40E+01	15.32	30.68
F9	5.99E+00	**4.00E+00**	3.19E+01	687.36	**4.00E+00**	3.00E+01	21.00	32.73
F10	**9.22E−02**	1.08E+00	6.30E+01	438.62	1.40E+00	5.96E+01	15.58	28.15
F11	3.24E+00	**3.22E+00**	3.37E+01	570.10	**3.22E+00**	4.15E+01	17.88	31.87
F12	**4.30E+00**	**4.30E+00**	7.41E+00	603.01	**4.30E+00**	1.94E+00	25.63	23.52

Figures in bold indicate best value obtained

is as the population initialized randomly, population size kept as 50, the weighted difference mutation operator used with rate of 0.5, the stopping criteria set to maximum number of function evaluation as 50,000. The algorithm executed for 20 times independently; best and mean results recorded for problem size as 100 dimensions.

The Table 1 shows the comparison of results between existing results of DEVIIC taken from [1], results obtained with sequential DEVIIC, and results obtained with proposed GPGPU-based DEVIIC algorithm. From the obtained results, the sequential DEVIIC and proposed GPGPU-based DEVIIC algorithms give better results for function F1, F5, F6, F7, F8, F9, and F11 than results given in [1] for DEVIIC algorithm. The results obtained for function F12 are similar. The Table 1 also shows the time (in seconds) required to obtain the best solution for sequential DEVIIC algorithm and proposed GPGPU-based DEVIIC algorithm. The speedup obtained is also shown. The GPGPU-based DEVIIC algorithm is 23 to 35 times faster than sequential DEVIIC algorithm to solve the large-scale compute-intensive problems. The similar quality solutions are obtained by both the algorithms.

5 Conclusions and Future Work

This paper presents the design and implementation of GPGPU-based DEVIIC algorithm to solve large-scale compute-intensive benchmark optimization problems. The proposed approach uses master–slave strategy to develop GPGPU-based parallel algorithm. The performance of proposed approach is evaluated using 12 large-scale benchmark functions proposed in [1] against results given in [1], existing sequential

DEVIIC algorithm, and proposed GPGPU-based DEVIIC algorithm. The dimension of the problem is set to 100 and function evaluations used are 50,000. The proposed approach gives comparatively better results than results found in [1] for few test functions and fails to obtain better results for functions F1 to F4 and F10. As the system used for implementation of proposed approach is different than the used in [1], the time required to obtain best results is not compared with it. As the proposed approach is to develop GPGPU-based algorithm, the speedup is computed with sequential implementation of DEVIIC by us and the proposed GPGPU-based DEVIIC algorithm. The proposed GPGPU-based approach is 23 to 35 times faster than its sequential counterpart. The proposed approach significantly reduces the execution time required to obtain best solution.

As a future work, the proposed approach can be employed to solve real-time large-scale compute-intensive problems. The VIIC technique proposed in [1] can be integrated with other evolutionary or swarm algorithms and implemented on GPGPU to solve large-scale benchmark or real-time optimization problems. Also, the proposed GPGPU-based DEVIIC algorithm can be applied to solve large-scale many-objective optimization problems by performing necessary modifications.

References

1. Sayed, E., Essam, D., Sarker, R., Elsayed, S.: Decomposition-based evolutionary algorithm for large scale constrained problems. Inf. Sci. **316**, 457–486 (2015)
2. Lastra, M., Molina, D., Benitez, J.M.: A high performance memetic algorithm for extremely high-dimensional problems. Inf. Sci. **293**, 35–58 (2015)
3. Cao, Y., Sun, D.: A parallel computing framework for large-scale air traffic flow optimization. IEEE Trans. Intell. Transp. Syst. **13**(4), 1855–1864 (2012)
4. Li, X., Tang, K., Omidvar, M.N., Yang, Z., Qin, K.: Benchmark Functions for the CEC'2013. In: Proceedings of Special Session and Competition on Large-Scale Global Optimization (2013)
5. Mane, S.U., Omane, R., Pawar, A.: GPGPU based teaching learning based optimization and artificial bee colony algorithm for unconstrained optimization problems. In: IEEE International Conference on Advance Computing, pp. 1056–1061 (2015)
6. Yang, F., Ding, M., Zhang, X., Hou, W., Zhong, C.: Non-rigid multimodal medical image registration by combining L-BFGS-B with cat swarm optimization. Inf. Sci. **316**, 440–456 (2015)
7. Noor-E-Alam, M., Doucette, J.: Relax-and-fix decomposition technique for solving large scale grid-based location problems. CAIE **63**(4), 1062–1073 (2012)
8. Nick, M., Riahy, G.H., Hosseinian, S.H., Fallahi, F.: Wind power optimal capacity allocation to remote areas taking into account transmission connection requirements. Renew. Power Gen. IET **5**(5), 347–355 (2011)
9. Jütte, S., Thonemann, U.W.: Divide-and-price: a decomposition algorithm for solving large railway crew scheduling problems. EJOR **219**(2), 214–223 (2012)
10. Omidvar, M.N., Li, X., Tang, K.: Designing benchmark problems for large-scale continuous optimization. Inf. Sci. **316**, 419–436 (2015)
11. Sayed, E., Essam, D., Sarker, R.: Using hybrid dependency identification with a memetic algorithm for large scale optimization problems. Simulated Evolution and Learning, pp. 168–177 (2012)
12. Mallipeddi, R., Suganthan, P.N.: Ensemble of constraint handling techniques. IEEE Trans. Evol. Comput. **14**(4), 561–579 (2010)

13. Omidvar, M.N., Li, X., Yao, X.: Cooperative co-evolution with delta grouping for large scale non-separable function optimization. In: Proceedings of IEEE Congress on Evolutionary Computation (2010)
14. Kopanos, G., Puigjaner Corbella, L., Georgiadis, M.C.: Techniques for the efficient solution of large-scale production scheduling planning problems in the process industries. Doctoral dissertation, Universitat Politècnica de Catalunya (2011)
15. Cao, Y., Sun, D.: A parallel computing framework for large-scale air traffic flow optimization. IEEE Trans. Intell. Transp. Syst. **13**(4) (2012)
16. Tenne, Y.: A computational intelligence algorithm for expensive engineering optimization problems. Eng. Appl. Artif. Intell. **25** (2012)
17. Chen, W., Weise, T., Yang, Z., Tang, K.: Large-scale global optimization using cooperative coevolution with variable interaction learning. In: Proceedings of Parallel Problem Solving from Nature, PPSN XI, pp. 300–309 (2010)
18. Omidvar, M.N., Li, X., Mei, Y., Yao, X.: Cooperative co-evolution with differential grouping for large scale optimization. IEEE Trans. Evol. Comput. **18**(3), 378–393 (2014)
19. Ray, T., Yao, X.: A cooperative coevolutionary algorithm with correlation based adaptive variable partitioning. In: Proceedings of IEEE Congress on Evolutionary Computation, pp. 983–989 (2009)
20. Wang, H., Rahnamayan, S., Wu, Z. Parallel differential evolution with self-adapting control parameters and generalized opposition-based learning for solving high-dimensional optimization problems. J. Parallel Distrib. Comput. **73**(1) (2013)
21. De Veronese, L.P., Krohling, R.A.: Differential evolution algorithm on the GPU with C-CUDA. In: Proceedings of IEEE Congress on Evolutionary Computation (2010)
22. Kromer, P., Platoz, J., Snazel, V.: Differential evolution for the linear ordering problem implemented on CUDA. In: Proceedings of IEEE Congress on Evolutionary Computation, pp. 796–802 (2011)
23. De Oliveira, F.B., Davendra, D., Guimar.es, F. G. Multi-objective differential evolution on the GPU with C-CUDA. In: Proceedings of Soft Computing Models in Industrial and Environmental Applications, pp. 123–132 (2013)
24. Qin, A.K., Raimondo, F., Forbes, F., Ong, Y.S.: An improved CUDA based implementation of differential evolution on GPU. In: Annual Conference on Genetic and Evolutionary Computation, pp. 991–998 (2012)

A Comparative Study of Job Satisfaction Level of Software Professionals: A Case Study of Private Sector in India

Geeta Kumari, Gaurav Joshi and Ashfaue Alam

Abstract The present study was conducted to have a relative understanding of job satisfaction level of software professionals in private software industries, namely HCL Technologies Limited Noida, IBM India Pvt. Ltd., Gurgaon and Wipro Limited, Greater Noida. Job satisfaction may be referred to the attitudes and feelings of the employees about their occupation. Optimistic and constructive outlooks toward one's occupation showing job satisfaction while destructive and adverse outlooks toward the job show job dissatisfaction. Nowadays, software professionals are facing lots of challenges related to their job. Various researches indicate that job satisfaction found in software professionals has been one of the very significant factors, which is associated with the positive working behaviour toward their occupation. There has also been substantial interest in the complex association between a one's job satisfaction along with the other sides of his or her life. Conclusions of this study revealed that the most of the software professionals were satisfied with their job at HCL Technologies Limited more in terms of various job satisfaction dimensions, namely welfare facilities, appreciation and rewards, career prospect, physical working environment, communication, fringe benefits and job security, while respondents of Wipro Limited had been satisfied more in job satisfaction dimensions, namely working hours, appreciation and rewards, physical working environment and recognition and it got second place in job satisfaction, and IBM India Pvt. Ltd. has got third place in job satisfaction level at abovementioned job satisfaction dimensions namely physical working environment and career prospects.

Keywords Job satisfaction · Career prospect · Physical working environment
Welfare facilities

G. Kumari (✉) · A. Alam
Jharkhand Rai University, Ranchi, India
e-mail: geekumari@gmail.com

A. Alam
e-mail: ashfaque.alam@jru.edu.in

G. Joshi
Lal Bahadur Shastri Institute of Management, New Delhi, India
e-mail: gauravjoshi12@gmail.com

K. Ray et al. (eds.), *Soft Computing: Theories and Applications*,
Advances in Intelligent Systems and Computing 742,
https://doi.org/10.1007/978-981-13-0589-4_55

1 Introduction

Job satisfaction is a positive state of one's feelings for his or her occupation. According to Saker et al. [1], job satisfaction is a basic tool to indicate the organisational strength in terms of work excellence and it is mainly depending on the efficiency of employees. The concept of job satisfaction has various components like welfare facilities, appreciation and rewards, career prospect, physical working environment, communication, fringe benefits and job security, etc. contributing to their own impact on the basis of monetary and non-monetary scale. Locke [2] defined it as a pleasing or optimistic emotional condition, which originates from the assessment of individual's job experiences. In broad sense, job satisfaction may refer to one's optimistic sensual responses to a specific occupation. It is an emotional response to an occupation which originates from the one's comparison of achieved outcome with those which are preferred, expected or justified. Opkara [3], Logasakthi and Rajagopal [4] discovered that the workforces relish not only the pleasure from their occupation but also numerous amenities provided by the employers and also provide their extreme backing for improving the firm. The personal subdivision pays attention to all the employees in the company. The administration offers health care as well as well-being to the staffs which helps to achieve improved performance in the job as well as in working atmosphere. Srinivas [5] worked on well-being amenities and worker's pleasure level about these amenities implemented at Bosch limited, Bangalore and got that most of the amenities, for example, health care, cafeteria, working atmosphere, security conditions, etc., which is given by the enterprise satisfied most of the employees. According to Lewis [6] and Weston et al. [7], long work hours are a risk factor for a range of psychological and psychosomatic conditions, including stress, anxiety, depression and hypertension. Caruso [8] and Rogers et al. [9] revealed that long working hours have been found to reduce productivity and workplace competence, notably among medical professionals. Job satisfaction variable is power and responsibility of authority and employee empowerment, which is promoted since 1980s, is employed to create a novel form of worker participation in view of Wilkinson [10].

The past of its primary description emerges from back to 1788, when empowerment referred as the providing strength to individual to participate for administrative role. Strength ought to be boosted to the person or it ought to be witnessed in his or her administrative role. Malhotra et al. [11] recommended that empowerment is a procedure in which workforces are participating or having part in managerial decision-making. Eshun and Duah [12] told rewards as 'all forms of financial return, tangible services and benefits an employee receives as part of an employment relationship'. Establishments anticipate workforces to achieve allotted responsibilities to their satisfaction at the same time workforces also want their establishments to guarantee them of sufficient remunerations and earnings (rewards) after they obediently provide employer what is anticipated from them Becker and Billings [13], and Chen [14] suggested that the communication satisfaction comprises feel satisfied from various features of communication in an association. Duncan and Moriaty [15] revealed that power of communication is to raise employee's performance. Chen et al.

[16] suggested that investigation is deficient in finding worker happiness with communication procedure. Hence, there is the need to discover the connection between administrative communication and employee's working efficiency as communication assimilates various divisions and tasks in the association. It is the social action which connects persons to make association (Ducharme and Martin [17]). Fiedler et al. [18] done an extensive research on the difficulties connecting to job satisfaction matters directed on the staffs of global stage service providers, which established that the causes of work association communications as well as colleagues' backing have substantial optimistic association to job satisfaction. So, this investigation established that, for the service industry, colleagues' rapport may be segmented into two parts—boss–employee connection and colleagues' communications. Fletcher and Williams [19] told that workfellows' connection is the attachment, recognition and faithfulness constructed among the participants of an association that also denotes the stages of the assistants' self-confidence, faith, as well as admiration for their leaders. Pedrycz et al. [20] did research on a prototype of work satisfaction for co-operative growth procedures and it has been established that the communication as well as job sustainability are the two main reasons of work satisfaction. Davie et al. [21] did a qualitative examination of the issues affecting the job satisfaction as well as occupational growth of physiotherapists and pointed out the causes that have impact on their job satisfaction at various career phases in private practice, for increasing the backing and preservation of worker in this area. Acuña et al. [22] worked for finding on How do persona, group procedures as well as occupation characteristics are related to work satisfaction and software quality? They have established that the higher levels of satisfaction can only be accomplished when the group associates can make decision on developments and organisation of assignments. The influence of software engineers' personality traits on group environment and efficiency: A well-organised literature review is completed by Soomro et al. [23] in which they observed few factors that may be beneficial for investing the achievement as well as chances of failure of software assignments in development.

1.1 Objectives of the Study

The foremost aim of this study is to have a relative study of job satisfaction level of software professionals in private software industries, namely HCL Technologies Limited Noida, IBM India Pvt. Ltd., Gurgaon and Wipro Limited, Greater Noida.

2 Research Methodology

Research is an imaginative task. It is a methodical increase in storage of information, facts, etc. which includes an understanding of person, art, music, literature, and civilisation also the usage of this storage of wisdom for formulating novel uses.

Methodology is the regular, hypothetical investigation of the approaches employed in a field of study. It includes the hypothetical examination of the body of methods as well as ideologies connected with a division of wisdom, which incorporates notions, for example, paradigm, hypothetical prototype, stages, as well as measurable or qualitative techniques. The research methodology must be vigorous for minimising faults during information gathering as well as investigation. Owing to this, numerous methodologies, namely reviews and conferences were preferred for data gathering. The qualitative tactic includes the gathering of widespread descriptive data for gaining visions into occurrences of interest; data examination comprises the data coding and construction of a vocal synthesis. The people of the study comprise software professionals at software industries in private sectors, namely HCL Technologies Limited, IBM India Pvt. Ltd., Gurgaon, and Wipro Limited, Greater Noida. The sample population for this study was taken to be 360. Out of these 360 employees, 234 employees were male and 126 were female. The convenience sampling procedure was used for data collection. There were the pre-tested questionnaires used as instrument for data collection. The questionnaires were framed on the basis of Likert 5-point scale from strongly agree to disagree, where point-1. strongly disagree, 2. disagree, 3. neither agrees nor disagree, 4. agree and 5. strongly agree. The hypothesis testing completed with the use of one sample t-test.

3 Results and Discussion

As per the analysis of job satisfaction level, it was given on the basis of mean values of the respondents response of various job satisfaction dimensions, namely working hours, authority and responsibility, welfare facilities, appreciation and rewards, career prospects, physical working environment, communication, co-workers, fringe benefits, recognition, job condition, and job security in software industries namely HCL Technologies Limited, Noida, IBM India Pvt. Ltd., Gurgaon and Wipro Limited, Greater Noida. The results revealed from the comparison of all the job satisfaction variables among software industries have been tabulated (Table 1).

Figure 1 illustrates the mean value of the respondent's response of job satisfaction dimension for working hours in software industries, namely HCL Technologies Limited, Noida, IBM India Pvt. Ltd., Gurgaon and Wipro Limited, Greater Noida. The respondents of HCL Technologies Limited, Noida have been found with higher mean value of 3.44 as compared to IBM India Pvt. Ltd., Gurgaon Mean value of 3.15 and Wipro Limited, Greater Noida with mean value of 2.91. From Fig. 1, respondents of the HCL Technologies received more working hours as compared to other two companies. Therefore, the respondents of the Wipro Limited had been observed to be more satisfied in terms of working hours as compared with other two companies namely IBM India Pvt. Ltd., and HCL Technologies Noida.

Table 1 A comparative analysis of job satisfaction level of software professionals among HCL Technologies, IBM India Pvt. Ltd and Wipro Limited

Sl. no.	Variables Job satisfaction dimensions	HCL Technologies Limited, Noida Sample size N = 100		IBM India Pvt.Ltd., Gurgaon Sample size N = 160		Wipro Ltd. Greater, Greater Noida, Sample size N = 100	
		Mean (M)	Standard deviation (S. D.)	Mean (M)	Standard deviation (S. D.)	Mean (M)	Standard deviation (S. D.)
1	Working hours	2.90	1.521	2.91	1.485	2.90	1.521
2	Authority and responsibility	3.68	1.497	3.44	1.495	3.52	1.322
3	Welfare facilities	3.10	1.453	3.06	1.463	3.10	1.432
4	Appreciation and rewards	3.15	1.500	3.08	1.479	3.15	1.500
5	Career prospects	3.35	1.466	3.21	1.468	3.33	1.429
6	Physical working environment	3.15	1.500	3.09	1.481	3.15	1.500
7	Communication	3.22	1.433	3.13	1.435	3.21	1.351
8	Co-workers	3.30	1.528	3.18	1.503	3.30	1.528
9	Fringe benefits	3.48	1.541	3.32	1.510	3.35	1.486
10	Recognition	3.30	1.425	3.21	1.428	3.31	1.419
11	Job condition	3.30	1.425	3.27	1.400	3.26	1.375
12	Job security	3.65	1.104	3.46	1.238	3.58	1.027

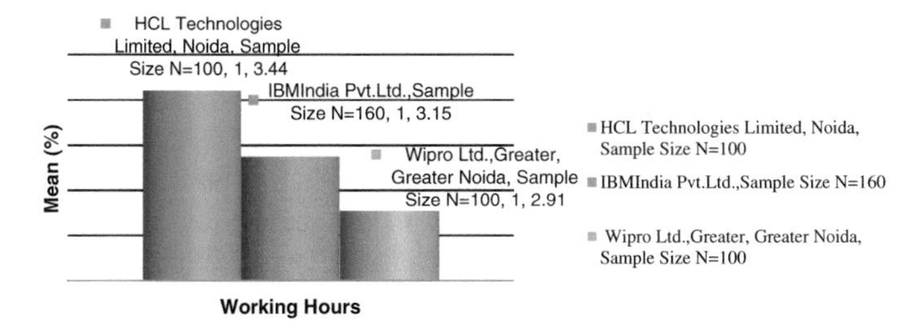

Fig. 1 Working hours

Figure 2 illustrates the mean value of the respondent's response of job satisfaction dimension for authority and responsibility in software industries, namely HCL Technologies Limited, Noida, IBM India Pvt. Ltd., Gurgaon and Wipro Limited, Greater Noida. The respondents of IBM India Pvt. Ltd, Gurgaon had been found with higher mean value of 3.68 as compared to HCL Technologies Limited, Noida Mean value of 2.9 and Wipro Limited, Greater Noida with mean value of 3.52. From Fig. 2, respondents of the IBM India Pvt. Ltd, Gurgaon got more authority and responsibility as compared to other two companies. Therefore, the respondents of the IBM India

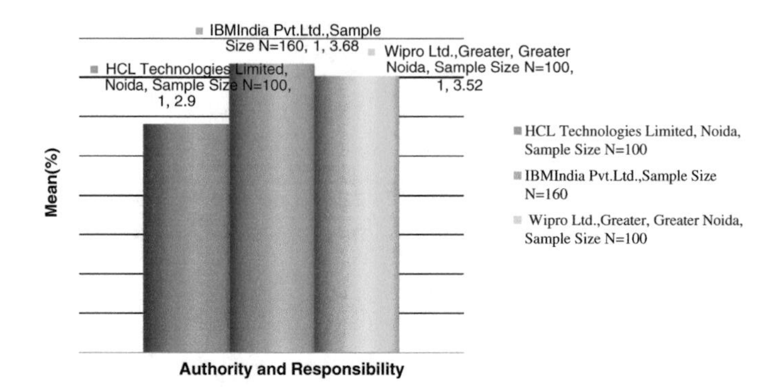

Fig. 2 Welfare facilities

Pvt. Ltd., Gurgaon had been observed to be more satisfied in terms of authority and responsibility as compared with other two companies, namely HCL Technologies Noida and Wipro Limited, Greater Noida.

Figure 3 illustrates the mean value of the respondent's response of job satisfaction dimension for welfare facilities in software industries, namely HCL Technologies Limited, Noida, IBM India Pvt. Ltd., Gurgaon and Wipro Limited, Greater Noida. In Fig. 3, the respondents of HCL Technologies Limited, Noida and Wipro Limited Greater Noida had been found with same mean value of 3.1 as compared to IBM India Pvt. Ltd., Gurgaon mean value of 3.06. Therefore, from Fig. 3, the respondents of HCL Technologies Limited, Noida and Wipro Limited Greater Noida had been found with same satisfaction level in terms of welfare facilities compared to respondents of IBM India Pvt. Ltd., Gurgaon.

Figure 4 illustrates the mean value of the respondent's response of job satisfaction dimension for appreciation and rewards in software industries namely HCL Technologies Limited, Noida, IBM India Pvt. Ltd., Gurgaon and Wipro Limited, Greater Noida. In Fig. 4, the respondents of HCL Technologies Limited, Noida and Wipro Limited Greater Noida had been observed having the same mean value of 3.15 as compared to IBM India Pvt. Ltd., Gurgaon mean value of 3.08. Therefore, from Fig. 4, the respondents of HCL Technologies Limited, Noida and Wipro Limited Greater Noida had been found same satisfaction level in terms of appreciation and rewards compared to respondents of IBM India Pvt. Ltd., Gurgaon.

Figure 5 illustrates the mean value of the respondent's response of job satisfaction dimension for career prospects in software industries, namely HCL Technologies Limited, Noida, IBM India Pvt. Ltd., Gurgaon and Wipro Limited, Greater Noida. The respondents of HCL Technologies Limited had been found with higher mean value of 3.35, while the respondents of IBM India Pvt. Ltd. had been found with less mean value of 3.21 and the respondents of Wipro Limited, Greater Noida had found mean value of 3.33. From Fig. 5, respondents of the HCL Technologies,

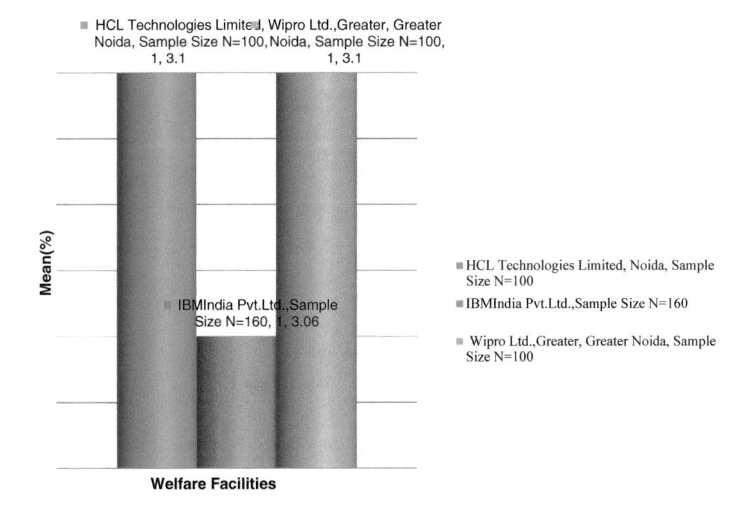

Fig. 3 Appreciation and rewards

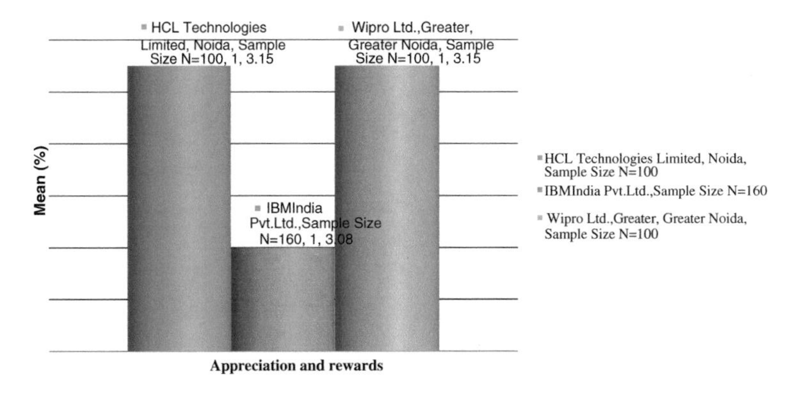

Fig. 4 Career prospects

Noida, Gurgaon got more career prospects opportunities as compared to other two companies. Therefore, the respondents of the HCL Technologies Noida had been found to be more satisfied in terms of career prospects as compared with other two companies, namely IBM India Pvt. Ltd., Gurgaon and Wipro Limited, Greater Noida.

Figure 6 illustrates the mean value of the respondent's response of job satisfaction dimension for physical working environment in software industries, namely HCL Technologies Limited, Noida, IBM India Pvt. Ltd., Gurgaon and Wipro Limited, Greater Noida. In Fig. 6, the respondents of HCL Technologies Limited, Noida and

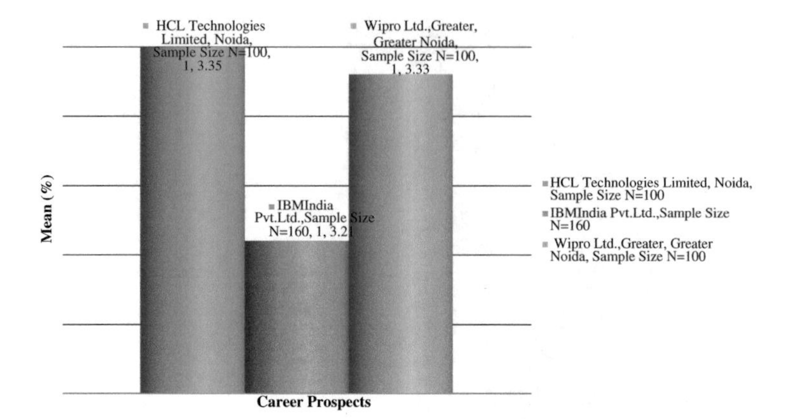

Fig. 5 Physical working environments

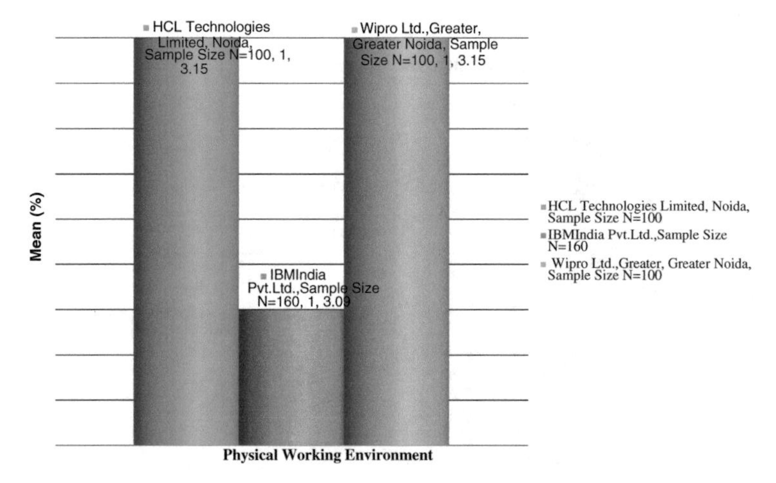

Fig. 6 Communication

Wipro Limited Greater Noida had been found with a same mean value of 3.15 as compared to IBM India Pvt. Ltd., Gurgaon mean value of 3.09. Therefore, from Fig. 6, the respondents of HCL Technologies Limited, Noida and Wipro Limited Greater Noida had been observed with same satisfaction level in terms of job satisfaction dimension physical working environment compared to respondents of IBM India Pvt. Ltd., Gurgaon.

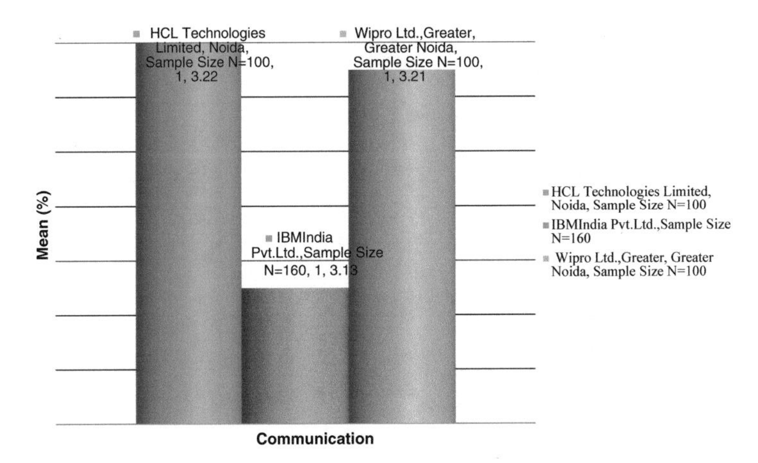

Fig. 7 Co-workers

Figure 7 illustrates the mean value of the respondent's response of job satisfaction dimension for communication in software industries, namely HCL Technologies Limited, Noida, IBM India Pvt. Ltd., Gurgaon and Wipro Limited, Greater Noida. The respondents of HCL Technologies Limited had been found with higher mean value of 3.22, while the respondents of IBM India Pvt. Ltd. had been found with less mean value of 3.13 and the respondents of Wipro Limited, Greater Noida had found mean value of 3.21. From Fig. 7, respondents of the HCL Technologies, Noida, Gurgaon received more freedom in communication with co-workers as well as management as compared to other two companies. Therefore, the respondents of the HCL Technologies Noida had been observed to be more satisfied in terms of job dimension communication as compared with other two companies namely IBM India Pvt. Ltd., Gurgaon and Wipro Limited, Greater Noida.

Figure 8 illustrates the mean value of the respondent's response of job satisfaction dimension for co-workers in software industries, namely HCL Technologies Limited, Noida, IBM India Pvt. Ltd., Gurgaon and Wipro Limited, Greater Noida. In Fig. 8, the respondents of HCL Technologies Limited, Noida and Wipro Limited Greater Noida had been found with a same mean value of 3.3 and the respondents of IBM India Pvt. Ltd., Gurgaon mean value of 3.18. From Fig. 8, the respondents of IBM India Pvt. Ltd., Gurgaon had been found with more co-worker support as compared to other two companies. Therefore, from Fig. 8, the respondents of IBM India Pvt. Ltd., Gurgaon had been found with more satisfaction level in terms of job satisfaction dimension in co-workers.

Figure 9 illustrates the mean value of the respondent's response of job satisfaction dimension for fringe benefits in software industries, namely HCL Technologies Limited, Noida, IBM India Pvt. Ltd., Gurgaon and Wipro Limited, Greater Noida. The respondents of HCL Technologies Limited, Noida had been observed with higher

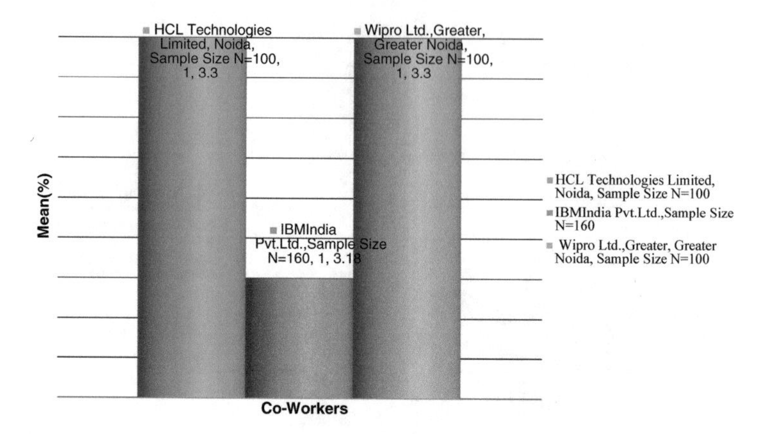

Fig. 8 Fringe benefits

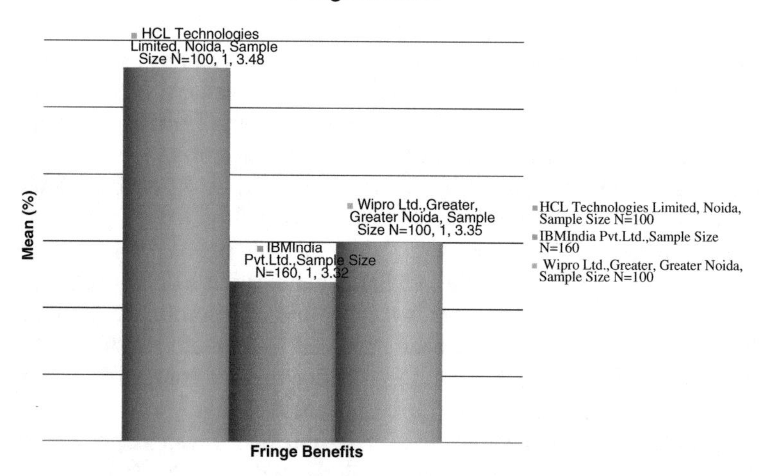

Fig. 9 Recognition

mean value of 3.48 as compared to IBM India Pvt. Ltd., Gurgaon Mean value of 3.32 and Wipro Limited, Greater Noida with mean value of 3.35. From Fig. 9, respondents of the HCL Technologies have got more satisfaction level in terms of fringe benefits as compared to other two companies.

Figure 10 illustrates the mean value of the respondent's response of job satisfaction dimension for recognition in software industries, namely HCL Technologies Limited, Noida, IBM India Pvt. Ltd., Gurgaon and Wipro Limited, Greater Noida. The respondents of Wipro Limited, Noida had been observed with higher mean value of 3.31 as compared to IBM India Pvt. Ltd., Gurgaon mean value of 3.21 and HCL

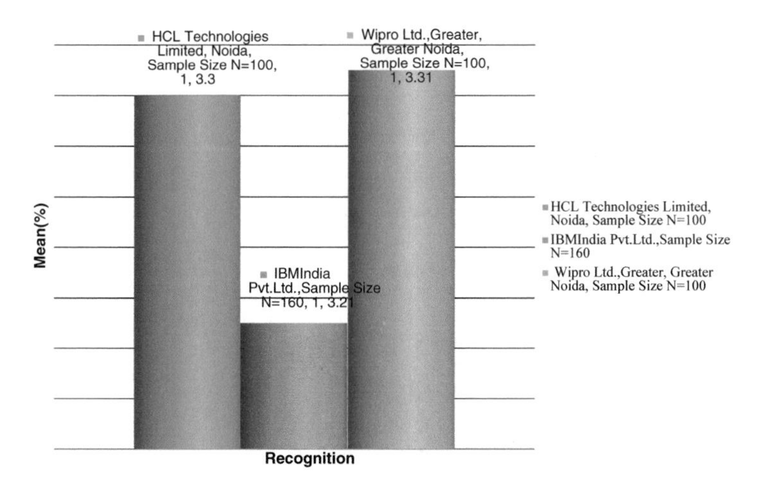

Fig. 10 Job condition

Technologies Limited with mean value of 3.3. From Fig. 10, respondents of Wipro Limited, Greater Noida are in a state of more satisfaction level in terms of recognition as compared to other two companies.

Figure 11 illustrates the mean value of the respondent's response of job satisfaction dimension for job condition in software industries, namely HCL Technologies Limited, Noida, IBM India Pvt. Ltd., Gurgaon and Wipro Limited, Greater Noida. The respondents of IBM India Pvt. Ltd, Gurgaon had been observed with higher mean value of 3.27 as compared to HCL Technologies Limited, Noida mean value of 3.3 and Wipro Limited, Greater Noida with mean value of 3.26. From Fig. 11, respondents of the IBM India Pvt. Ltd, Gurgaon have obtained more favourable job condition as compared to other two companies. Therefore, the respondents of the IBM India Pvt. Ltd., Gurgaon had been found more satisfied in terms of job condition as compared with other two companies namely HCL Technologies Noida and Wipro Limited, Greater Noida.

Figure 12 illustrates the mean value of the respondent's response of job satisfaction dimension for communication in software industries, namely HCL Technologies Limited, Noida, IBM India Pvt. Ltd., Gurgaon and Wipro Limited, Greater Noida. The respondents of HCL Technologies Limited had been observed with higher mean value of 3.65 while the respondents of IBM India Pvt. Ltd. had found less mean value of 3.46 and the respondents of Wipro Limited, Greater Noida had been found with mean value of 3.58. From Fig. 12, respondents of the HCL Technologies, Noida, Gurgaon received more job security as compared to other two companies. Therefore, the respondents of the HCL Technologies Noida had been observed to be more satisfied in terms of job dimension job security as compared with other two companies, namely IBM India Pvt. Ltd., Gurgaon and Wipro Limited, Greater Noida.

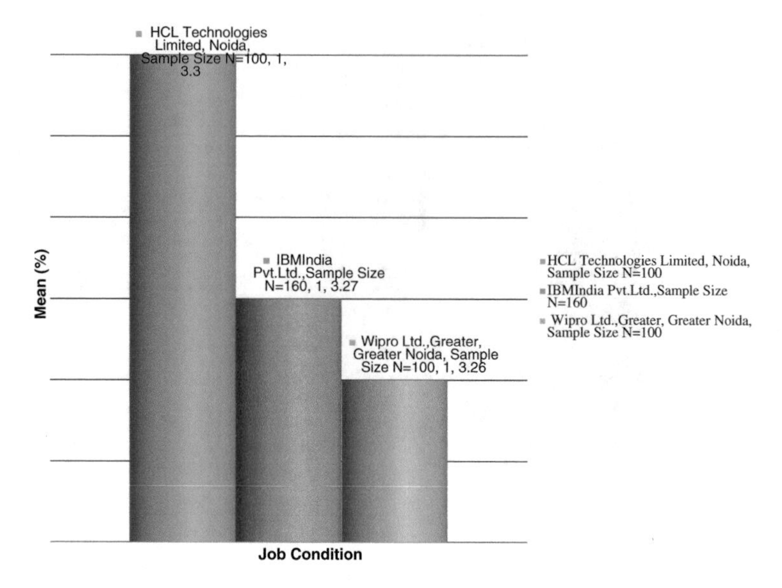

Fig. 11 Job condition

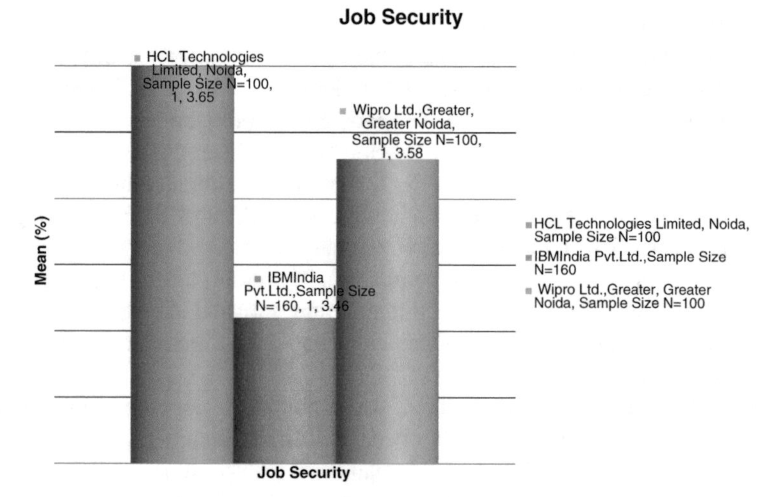

Fig. 12 Job security

4 Conclusion

The employees are the strongest assets in any organisation. If the employees get favourable working environment to do work, then they give more output to the organisation. Findings revealed that the overall job satisfaction level in HCL Technologies

Limited was more in terms of various job satisfaction dimensions, namely welfare facilities, appreciation and rewards, career prospect, physical working environment, communication, fringe benefits and job security, while respondents of Wipro Limited had been satisfied more in job satisfaction dimensions, namely working hours, appreciation and rewards, physical working environment and recognition, and it got second place in job satisfaction and IBM India Pvt. Ltd. has got third place in job satisfaction level at abovementioned job satisfaction dimensions, namely physical working environment and career prospects and the one sample test the all variables of job satisfaction, namely working hours, authority and responsibility, appreciation and rewards, career prospect, physical working environment, communication, co-workers, fringe benefits, recognition, job condition, salary and job security have significant values of (0.000). All these values are less than 0.05. So, it is concluded that all the variables of job satisfaction were found statistically significant and it is also concluded that all variables have great influence on job satisfaction level.

5 Further Scope of Research

In the further scope of research, it would be very interesting to conduct another study within the same area of research, with the incorporation of more industries, namely manufacturing industries and service sectors in general which will give more integrated results to the topic and better utility to the researchers and companies. The study will be focused on the technical and administrative supervisory level employees of certain organisations. While it could be of interest to conduct a study on more areas of work including, both the lower level and senior level employees which gives more comprehensive understanding and overview of the job stress level and job satisfaction level of the employees and even of the difference in the weak and strong organisational culture practices in different organisations.

Acknowledgements The authors acknowledge Dr. K. M. Pandey, Professor, Department of Mechanical Engineering, NIT Silchar, Assam, India for modification/correction in the manuscript.

References

1. Saker, A.H., Crossman, A., Chinmeteepituck, P.: The relationships of age and length of service with job satisfaction: an examination of hotel employees in Thailand. J. Manag. Psychol. **18**, 745–758 (2003)
2. Locke, E.A.: The nature and causes of job satisfaction. In: Dunnette, M.D. (ed.) Handbook of Industrial and Organizational Psychology, pp. 1297–1349. Rand McNally, Chicago (1976)
3. Opkara, J.O.: The impact of salary differential on managerial job satisfaction: a study of the gender gap and its implications for management education and practice in a developing economy. J. Bus. Dev. Nations 65–92 (2002)

4. Logasakthi, K., Rajagopal, K.: A study on employee health, safety and welfare measures of chemical industry in the view of Sleam region, Tamil Nadu, India. Int. J. Res. Bus. Manag. **1**(1), 1–10 (2013)
5. Shrinivas, K.T.: A study on employee's welfare facilities adopted at Bosch Limited, Bangalore. Res. J. Manag. Sci. **2**(12), 7–11 (2013)
6. Lewis, S.: The integration of paid work and the rest of life: Is post-industrial work the new leisure? Leis. Stud. **22**, 343–345 (2003)
7. Weston, R., Gray, M., Qu, L., Stanton, D.: The impact of long working hours on employed fathers and their families. Australian Institute of Family Studies Research Paper, no. 35 (2004)
8. Caruso, C.: Possible broad impacts of long work hours. Ind. Health **44**, 531–536 (2006)
9. Rogers, A., Hwang, W.T., Scott, L., Aiken, L., Dinges, D.: The working hours of hospital staff nurses and patient safety. Health Aff. **23**(2004), 202–212 (2004)
10. Wilkinson, A.: Empowerment: theory and practice. Pers. Rev. **27**(1), 40–56 (1998)
11. Malhotra, N., Budhwar, P., Prowse, P.: Linking rewards to commitment: an empirical investigation of four UK call centres. Int. J. Hum. Resour. Manag. 2095–2127 (2007). Taylor & Francis Group, Routledge
12. Eshun, C., Duah, F.K.: Rewards as a Motivation Tool for Employee Performance. Accessed 17 July 2011 from BTH2011Eshun.pdf
13. Becker, T.E., Billings, R.S.: Profiles of commitment: an empirical test. J. Organ. Behav. **14**(1993), 177–190 (1993)
14. Chen, N.: Internal employee communication and organizational effectiveness: a study of Chinese corporations in transition. J. Contemp. China **17**(54), 167–189 (2008)
15. Duncan, T., Moriaty, S.E.: A communication-based marketing model for managing relationships. J. Mark. **62** (2), 1–13 (1998). http://dx.doi.org/10.2307/1252157
16. Chen, C.R.: A Study on Fiedler's Contingency Leadership Theory, Taipei: Wu Nan Publishing Council for Economic Planning and Development 2009. 2015 Taiwan's service industry-prospects and goals. Abstracted from: http://www.find.cepd.gov.tw/tesg/reports/980708_2015. 2009 Budget for Technology Programs, Council for Economic Planning and Development, Taiwan (1989)
17. Ducharme, L.J., Martin, J.K.: Unrewarding work, co-worker support, and job satisfaction. Work Occup. **27**(2), 223–243 (2000). Fiedler, F.E., Chemers, M.M.: Improving Leadership Effectiveness (1984)
18. Fiedler, F.E., Chemers, M.M., Mahar, L.: Improving Leadership Effectiveness: The Leader Match Concept. John Wiley, New York (1977)
19. Fletcher, C., Williams, R.: Performance management, job satisfaction and organizational commitment. Brit. J. Manag. **7**, 169–179 (1996)
20. Pedrycz, W., Russo, B., Succi, G.: A model of job satisfaction for collaborative development processes. J. Syst. Softw. **84**(5), 739–752 (2011)
21. Davies, J.M., Edgar, S., Debenham, J.: A qualitative exploration of the factors influencing the job satisfaction and career development of physiotherapists in private practice. Man. Ther. **25**, 56–61 (2016)
22. Acuña, S.T., Gómez, M., Juristo, N.: How do personality, team processes and task characteristics relate to job satisfaction and software quality? Inf. Softw. Technol. **51**(3), 627–639 (2009)
23. Soomro, A.B., Salleh, N., Mendes, E., Grundy, J., Burch, G., Nordin, A.: The effect of software engineers' personality traits on team climate and performance: a systematic literature review. Inf. Softw. Technol. **73**, 52–65 (2016)

Simultaneous Placement and Sizing of DG and Capacitor to Minimize the Power Losses in Radial Distribution Network

G. Manikanta, Ashish Mani, H. P. Singh and D. K. Chaturvedi

Abstract In the present power system scenario, minimization of power losses in the distribution network is one of the interesting areas of research and modern challenges in the research community. Distribution network has large and complex structure; it produces more power losses as compared to transmission system. High power losses and poor voltage regulation are occurring at each bus when it moves away from the substation node to end node. Many methods have been implemented to minimize the power losses in the distribution system. Sitting and sizing of Distributed Generation (DG) and capacitors are new approaches used in the distribution system to minimize the power losses. However, capacity and location of DG and capacitors in the distribution system are considered independently. Improved voltage profile, increased overall energy efficiency, and reduced environmental impacts are some benefits produced by DG and capacitors. Consumer is also benefited from DG and capacitors optimization in terms of improved quality of power supply at lower cost. Sitting and sizing of DG and capacitor is a combinatorial optimization problem, and hence metaheuristics are used. An Adaptive Quantum-inspired Evolutionary Algorithm (AQiEA) approach is used for the optimization of DG and capacitors. In this paper, a new approach is considered by simultaneous placement and sizing of Distributed Generation (DG) and capacitor to minimize power losses. The effectiveness of the proposed algorithm is tested on 85-bus system. The experimental results show that AQiEA has better performance as compared with some existing algorithms.

G. Manikanta (✉) · A. Mani · H. P. Singh
Electrical & Electronics Engineering Department, A.S.E.T, Amity University, Noida, Uttar Pradesh, India
e-mail: manikanta.250@gmail.com

A. Mani
e-mail: amani@amity.edu

H. P. Singh
e-mail: hpsingh2@amity.edu

D. K. Chaturvedi
Electrical Engineering Department, F.O.E, Dayalbagh Educational Institute (Deemed University), Dayalbagh, Agra, India
e-mail: dkc.foe@gmail.com

K. Ray et al. (eds.), *Soft Computing: Theories and Applications*, Advances in Intelligent Systems and Computing 742,
https://doi.org/10.1007/978-981-13-0589-4_56

Keywords Distributed generation · Capacitor · Real power loss · AQiEA 85-bus system

1 Introduction

India is one of the leading consumers of electric power in the world after the USA, China, and Russia. In India, the discrepancy of power in power networks is mainly occurring due to losses produced by transmission and distribution systems. Distribution system losses are significantly high as compared to transmission system losses because of its high R/X ratio [1]. The role of distribution network in electric power system is to ensure that end users are provided with reliable, safe, and quality of electrical power. High power losses and poor voltage regulation are occurring at each bus when it moves away from the substation node to end node [2]. Therefore, loss reduction in distribution systems is one of the interesting area and greatest challenges for research community. Many methods have been implemented for minimizing the power losses in the distribution system: Changing transformer taps, increasing the size of conductor, load management of distribution transformers, voltage corrections, etc. The overall performance of the distribution system is improved significantly by simultaneous placement and sizing of DG and capacitors. Inappropriate placement and sizing of DG and capacitor may reduce the benefits and endangers the system operation.

Generally, DG is defined as the electric power generation source which is placed nearer to load centers. These are varied from few kW to several MWs [3]. DGs are playing an important role in minimization of power losses in distribution system. These are also termed as "clean and green" energy resources. By integrating, DG in distribution system not only reduces the power losses but also improves voltage profile and reliability of the system at technical level. As compared to conventional generators, DG uses environmental eco-friendly technologies. Fuel cells, wind turbines, reciprocating engines, micro-turbines, solar cells, and gas turbines are some of the examples of practical DG technologies. Capacitor banks are used to provide reactive power compensation in distribution system [4]. Corrections of power factor, improvement in power quality, and bus voltage regulation are some other benefits produced by capacitor placement in the distribution system.

Efficient deterministic techniques are unable to solve some complexities in engineering optimization problems. Sitting and sizing of capacitor and DG is a combinatorial optimization problem. Hence, metaheuristics are used to optimize location and capacity of DG and capacitor. Sanjib and Dipanjan [5] presented a new strategy for DG allocation under uncertainties in load condition to minimize the power losses. Fuzzy-based approach is used to find optimal location and capacity of DG. Aashish and Ganga [6] developed an efficient methodology for multiple DG installations to increase the technical and economical benefits. A novel multi-objective function is obtained for different performance evaluation indices. Karar and Naoto [7] proposed an efficient analytical method for reduction of power losses in distribution system by installing multiple DG technologies. In this case, different types of DG are considered and power factor is optimally calculated.

Sultana and Roy [8] have proposed a new metaheuristic algorithm (Teaching–Learning-Based Optimization (TLBO)) by optimal placement of capacitor to minimize energy cost and power loss. Singh and Rao [9] proposed a method to maximize the saving and minimize the real power loss by placement and sizing of capacitor in the distribution system using PSO and dynamic sensitivity analysis. Benvindo and Geraldo [10] proposed an efficient hybrid method for minimization of power losses in distribution network. Stochastic nature of DG is considered and presents a model for optimization of DG and capacitors using Tabu search and Genetic algorithm. Taghi and Amir [11] presented a multi-objective function to minimize investment cost of DG and capacitor, improve voltage and reduce power losses in distribution system by simultaneous placing and sizing of DG and capacitor. Meysam and Ahad [12] proposed an algorithm to improve voltage profile and reduce power losses by simultaneous allocation of DG and capacitor in distribution system. Karar Mahmoud [13] presented an analytical method with optimal power flow (OPF) by optimally integrating DG and capacitors into distribution system to minimize the power losses. Minnan and Jin [14] proposed two optimization models for better voltage profile and reduction in power losses by optimizing DG and capacitors. Optimal placement of DG is modelled in the first case followed by capacitor optimization model. In this paper, we have used Adaptive Quantum-inspired Evolutionary Algorithm (AQiEA) [15] with a measurement operator [16] for appropriate placement and sizing of DG and capacitor banks to minimize active power losses and maximize voltage profile in the system. AQiEA does not require fine tuning of parameters in its evolutionary operator and has performed competitively on a class of optimization problem.

2 Problem Formulation

Distribution system acts as a link between the load centers and power generating stations. Its main role is to provide reliable, safe, and quality power to the consumers at the end nodes. Due to its complex and large structure, it produces poor voltage regulation and high power losses at each bus as it moves from substation to end node. Recently, integrating Distributed Generation (DG) and capacitors into distribution system for reduction of power losses has increased considerably. The overall performance of the distribution system is improved significantly by simultaneous placement and sizing of DG and capacitors. The objective function for minimization of losses is as follows:

$$\min.\left\{\mathrm{P}_{loss}\sum_{i=1}^{m} I_i^2 R_i\right\} \tag{1}$$

where I_i and R_i are the magnitudes of currents and resistance of the ith branch. m is the total number of branches in the system.

Several constraints of the power system have been taken into consideration.

- *Constraints of Distributed Generator operation*:

$$P_{DG,i}{}^{\min} < P_{DG,i} < P_{DG,i}{}^{\max} \tag{2}$$

where $P_{DG,i}{}^{\min}$ is the lower bound and $P_{DG,i}{}^{\max}$ is the upper bound of DG output and all DG units shall function within the acceptable limit.

- *Constraints of Power injection*:

$$\sum_{i=1}^{k} P_{DG,i} \leq P_{load} + P_{losses}; k = \text{no of DG's} \tag{3}$$

The overall power produced by DG should not exceed the load and losses in the system.

- *Constraint of Power balance*:

$$\sum_{i=1}^{k} P_{DG,i} + P_{substation} + \sum_{j=1}^{k} Q_{cap,j} \leq P_{load} + P_{losses} \tag{4}$$

Total load demand and losses of the system must be equal to its generation, i.e., power generated by DG and substation power fed into the system by grid.

- *Constraints of shunt capacitor limit*:

$$Q^{c}_{max} \leq Q_{total} \tag{5}$$

where $Q^{c}_{\max}$ is the largest capacitor of allowable size and Q_{total} is the total reactive load.

3 Distribution System Load Flow

The structure of distribution network is large and complex, traditional power flow methods such as Gauss-seidel and Newton Raphson based methods, which often fail to converge in distribution networks due to its high resistance-to-reactance ratio. Distribution power flows are used in planning as well as operation stages in distribution network. It is an important tool for analyses, i.e., total system power loss and loss in each branch is also calculated. Rakesh and Nitin [17] developed a load flow

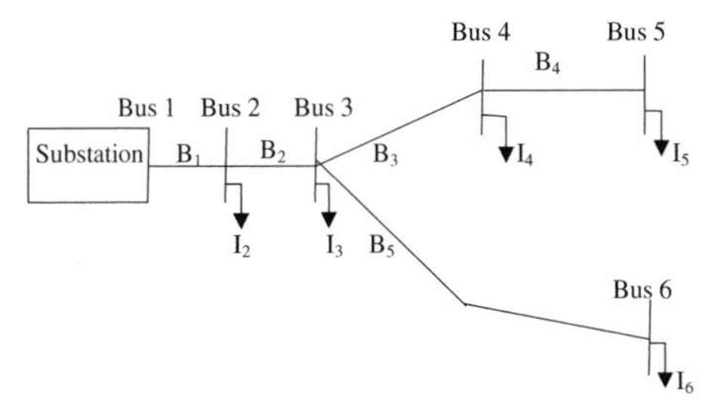

Fig. 1 Distribution system

technique for balanced radial distribution systems, which provides good convergence and computational efficiency, and load flow is independent of laterals, sub-laterals, and load models. Sanjib and Padarbinda [18] has proposed a load flow algorithm for unbalanced radial distribution networks which takes less number of iterations by modifying forward–backward load flow algorithm. Jen-Hao Teng [19] has proposed a novel technique which uses topological characteristic of distribution system and solves load flow directly. The method is known to consume less time and is more robust than existing methods and has been used in this work [19].

Two matrices are created based on graphical representation of distribution network, namely Bus Injection to Branch Current (BIBC) matrix and Branch Current to Bus Voltage (BCBV) matrix are developed (Fig. 1).

Power injections are transformed to the corresponding equivalent current injection is expressed as

$$I_j = \left(\frac{\mathrm{P_j} + j\mathrm{Q_j}}{V_j}\right)^* \tag{6}$$

where

P_j and Q_j are active and reactive loads at bus j

V_j is voltage magnitude at bus j

By applying (KCL) to above network branch, currents are formed based on equivalent current injection

$$\mathrm{B_2 = I_3 + I_4 + I_5 + I_6} \tag{7}$$

$$\mathrm{B_5 = I_6} \tag{8}$$

BIBC matrix is formed from the relation between equivalent current injection and branch currents are expressed as

$$[B] = [BIBC][I] \tag{9}$$

Similarly by applying KVL to the distribution network, the branch current and bus voltage (BCBV) are obtained as

$$\begin{aligned} V_2 &= V_1 - B_1 Z_{12} \\ V_3 &= V_2 - B_2 Z_{23} \end{aligned} \quad (10)$$

From above equations, the value of V_2 is substituted in V_3 and expressed as

$$V_3 = V_1 - B_1 Z_{12} - B_2 Z_{23} \quad (11)$$

$$[\Delta V] = [BCBV][B] \quad (12)$$

The relationship between bus current injection and bus voltage is given by

$$\begin{aligned} [\Delta V] &= [BCBV][BIBC][I] \\ &= [DLF][I] \end{aligned} \quad (13)$$

Direct Load Flow (DLF) is obtained by the multiplication of BCBV and BIBC matrices. By solving the below equations iteratively, the solution of the direct load flow is obtained.

$$I_j^m = I_j^r\left(V_j^m\right) + jI_j^i\left(V_j^m\right) = \left(\frac{P_j + jQ_j}{V_j^m}\right)^* \quad (14)$$

$$\left[\Delta V^{k+1}\right] = [DLF]\left[I^k\right] \quad (15)$$

$$\left[V^{k+1}\right] = \left[V^0\right]\left[\Delta V^{k+1}\right] \quad (16)$$

4 Proposed Algorithm

Gradient-based optimization methods are used to solve different engineering optimization problems. In some cases, these methods are failed to resolve some complexities in engineering optimization problem. Hence, metaheuristics are used to solve these complexities in engineering optimization problems. Evolutionary algorithms are inspired by nature's law of biological evolution and population-based stochastic searches. EA works on Darwinian principle, i.e., survival of fittest. The fittest one in the total population will move to the next generation. Population of solutions is evolved in EA by using operators like selection, crossover, mutation, and local heuristics to find local optima. Slow convergence, premature convergence, sensitivity to the choice of parameter, and stagnation are the major limitations in EA. Quantum-inspired Evaluation Algorithm (QEA) approach is used to overcome these limitations, by establishing a good balance between exploration and exploitation [20]. QEA is inspired by integrating some principles of Quantum mechanics into EA. It has better diversity as compared with other approaches. The smallest information element in a quantum computer is a quantum-bit (qubit) analogous to classical bits. Qubit can be represented in a quantum system with respect to the superposition of basis state. The basis states are represented in Hilbert space by a vector as $|0\rangle$ and $|1\rangle$. The qubit can be represented by vector $|\omega\rangle$ and it is defined as

$$|\omega\rangle = \Phi|1\rangle + \gamma|0\rangle \tag{17}$$

where ϕ and γ are complex numbers which specify the probability amplitudes associated with states $|1>$ and $|0>$, respectively, and should satisfy the condition

$$|\Phi|^2 + |\gamma|^2 = 1 \tag{18}$$

where $|\phi|^2$ and $|\gamma|^2$ specify the probability amplitudes of qubit to be in state "0" and "1", respectively, or superposition of both states.

The proposed Adaptive Quantum-inspired Evaluation Algorithm (AQiEA) utilizes the entanglement principle. AQiEA uses two qubits: the first qubit is used to store the design variables of solution vector and the second qubit is used to store the scaled and ranked objective function value of the solution vector. The second qubit is used as feedback in parameter/tuning-free adaptive quantum-inspired rotation crossover operator used for evolving the first qubit. In classical implementation, both qubits are entangled with each other, if any operation performed on one of these qubits will affect the other.

The ith qubit $|\omega_{1i}\rangle$ in the first set is assigned the current scaled value of the ith variable as probability amplitude ϕ_{1i}. The upper and lower limits of the variable have been used for scaling between zero and one. The amplitude γ is discarded and it is computed from Eq. (18) whenever it is needed. The qubits are stored in quantum register. Number of qubits per quantum register QR_i is equal to number of variables. The structure of QR_i is shown below:

$$\begin{aligned} QR_{1,i} &= [\Phi_{1,i,1}, \Phi_{1,i,2} \ldots \Phi_{1,i,n}] \\ &\ldots\ldots\ldots\ldots\ldots\ldots \\ QR_{1,m} &= [\Phi_{1,m,1}, \Phi_{1,m,2}, \ldots \Phi_{1,m0,n}] \end{aligned} \tag{19}$$

The second set of qubit in quantum register QR_{i+1} is used to store the scaled and ranked objective function value of corresponding solution vector in QR_i. The worst vector's second qubit set of objective function value is assigned 0, and the fittest vector's second qubit set will be given as 1. The remaining solution vector's second qubit is also ranked and given values, which are in the range of zero and one.

The classical implementations of superposition and entanglement principles are mathematically represented as follows:

$$|\omega_{2i}(t)\rangle = f_1|\omega_{1i}(t)\rangle \tag{20}$$

$$|\omega_{1i}(t+1)\rangle = f_2\big(|\omega_{2i}(t)\rangle, |\omega_{1i}(t)\rangle, |\omega_{1j}(t)\rangle\big) \tag{21}$$

where $|\omega_{1i}>$and$|\omega_{1j}>$are the first qubits associated with the ith and jth solution vectors, respectively, $|\gamma_{2i}>$is the second qubit associated with the ith solution vector, t is iteration number, and f_1 and f_2 are the functions through which both the qubits are classically entangled.

Three rotation strategies have been applied to converge the population adaptively toward global optima.

Rotation toward the Best Strategy (R-I): Except the best solution vector, it is implemented for each solution vector sequentially. All other individuals are rotated toward the best individual. All other individual solutions are improved by searching toward the best individual.

Rotation away from the Worse Strategy (R-II): For exploitation purpose, this strategy is primarily used, and best individual is rotated away from all other individuals. Around the best individual, the search takes place in all dimensions due to its rotation away from other individuals.

Rotation toward the Better Strategy (R-III): It is implemented by selecting two individuals randomly, and the inferior one is rotated toward the best individual. For exploration purpose, this strategy is primarily used.

Pseudocode of the Adaptive Quantum-inspired Evolutionary Algorithm (AQiEA) is shown below:

Pseudocode:

1. Initialize Quantum-inspired Register QR_i.
   While (! termination criterion) {
2. Compute the fitness of QR_i.
3. Quantum-inspired Register QR_{i+1} is assigned based on the fitness of QR_i.
4. Perform Adaptive Quantum Rotation based Crossover on QR_i.
5. Tournament selection criterion is applied for best individual selection.

Description:

1. The Quantum-inspired register QR_i is initialized randomly.
2. Fitness of each solution vector of Quantum-inspired register QR_i computed.
3. The second set of qubit in quantum register QR_{i+1} is used to store the scaled and ranked objective function value of corresponding solution vector in QR_i.
4. Adaptive quantum crossover is performed by using three strategies R-I, R-II, and R-III on QR_i.
5. By applying tournament selection, the best individual in population will move to the next generation.
6. Based on maximum number of iterations, termination criterion is executed.

5 Tests and Results

The effectiveness of the proposed algorithm is tested on 85-bus system. The power losses are minimized by appropriate placement and sizing of DG and capacitors. Inappropriate placement and capacity of DG and capacitors may lead the system to high power losses. Exhaustive tests have been carried out on Intel® Core TM i3

CPU @1.80 GHz processor and 4 GB of RAM capacity and implemented in Matlab 2011b. The line data and load data for 85-bus system are taken from [21]. Under normal operating conditions, the minimum and maximum voltages are 0.95 p.u and 1.05 p.u with voltage rating of $S_{base} = 100$ MVA and $V_{base} = 11$ kV, respectively. The total load of the system, i.e., active and reactive loads on the network are 2.57 MW and 2.62 MVAr. The total power losses produced by the system before placing DG and capacitor are 316.1221 kW and 198.60 kVAr with minimum voltage of 0.8713 p.u at 53rd bus. Population size for test system is 50, and convergence value is taken as 0.001. The comparison has been made on factors like reduction in power losses, percentage reduction in losses as compared to base case loss, i.e., system operating without DG, capacitors. The evaluation and optimization search are performed using adaptive quantum-inspired evolutionary algorithm.

The superiority of the proposed method is analyzed using four cases.

Case I: In this case, three DGs are optimally sized and placed for minimization of power loss through the proposed algorithm.
Case II: In this case, three capacitors are optimally sized and placed for minimization of power loss through the proposed algorithm.
Case III: In this case, simultaneous optimization of DG and capacitors are considered (one DG and two capacitors are optimally sized and placed) for minimization of power loss through the proposed algorithm.
Case IV: In this case, simultaneous optimization of DG and capacitors are considered (one capacitor and two DGs are optimally sized and placed) for the minimization of power loss through the proposed algorithm.

The proposed method reduces the power losses and improves voltage profile. In this method, all the system parameters are converted into per units for calculation purpose. The obtained results are tabulated and compared with different algorithms. The maximum allowable size of DG and capacitor are 0–1 MW and 0–1 MVAr.

Case I:

In this case, placement and sizing of DG are considered. Table 1 shows the comparative analysis of the proposed algorithm with the existing algorithms for the reduction of power losses in the system. The proposed algorithm not only minimizes the power losses but also improves the voltage profile. In this case, the proposed algorithm produces maximum power loss reduction (52%) followed by PSO (51%), SA (50%), and GA (50%) and reduces power losses from 316.1221 kW to 152.7543 kW by locating DG at 66, 26, and 49 with capacities 620 kW, 1 MW, and 455 kW. Similarly, PSO minimizes power loss to 155.93 kW by locating DG at 33, 14, and 73 with capacities 1 MW, 450 kW, and 560 kW. The voltage profile is also improved from 0.8713 p.u (without DG) to 0.9418 p.u (AQiEA). In this case, GA has minimum power loss reduction with 157.1 kW by locating DG at 30, 60, and 49 with capacities 630, 940, and 210 kW. SA has minimum increase in voltage profile, and PSO has maximum increase in voltage profile as compared with other algorithms.

Case II:

In this case, placement and sizing of capacitor are considered. Table 2 shows the comparative analysis of the proposed algorithm with the existing algorithms for

Table 1 Comparative analysis of AQiEA with other methods for Case I

	GA	SA	PSO	AQiEA
P_{loss} (kW)	157.0921	156.8557	155.9399	152.7543
Q_{loss} (kVAr)	97.5623	96.4924	95.875	94.5207
Loc	30, 60, 49	63, 26, 47	33, 14, 73	66, 26, 49
Size (MW)	0.63, 0.94, 0.21	0.83, 0.82, 0.34	1, 0.45, 0.56	0.62, 1, 0.4558
MinVV (p.u)	0.9414	0.939	0.9506	0.9481
P_{load} (MW)	0.7903	0.5803	0.5603	0.4941
Q_{load} (MVAr)	2.6222	2.6222	2.6222	2.6222
% Reduction	**50.31**	**50.38**	**50.67**	**51.68**

Table 2 Comparative analysis of AQiEA with other methods for Case II

	GA	SA	PSO	AQiEA
P_{loss} (kW)	161.0359	164.8683	157.8981	156.6324
Q_{loss} (kVAr)	98.234	101.064	96.7643	96.3713
Loc	30, 64, 27	29, 48, 11	64, 85, 32	60, 33, 85
Size (MW)	0.91, 0.69, 0.64	0.44, 0.57, 0.94	0.95, 0.137, 0.96	0.7, 0.95, 0.31
MinVV (p.u)	0.9225	0.9208	0.9186	0.9178
P_{load} (MW)	2.5703	2.5703	2.5703	2.5703
Q_{load} (MVAr)	0.3822	0.6722	0.5752	0.6222
% Reduction	**49.06**	**47.85**	**50.05**	**50.45**

reduction of power losses in the system. In this case, the proposed algorithm produces maximum power loss reduction (51%) followed by PSO (50%), SA (48%), and GA (50%) and reduces power losses from 316.1221 kW to 156.63 kW by locating capacitors at 60, 33, and 85 with capacities 700 kVAr, 950 kVAr, and 310 kVAr. Similarly, GA minimizes power loss to 161.03 kW by locating capacitors at 30, 64, and 27 with capacities 910 kVAr, 690 kVAr, and 640 kVAr. The voltage profile is also improved from 0.8713 p.u (without capacitor) to 0.9178 p.u (AQiEA). In this case, SA minimizes power loss to 164.86 kW by locating capacitors at 29, 48, and 11 with capacities 440 kVAr, 570 kVAr, and 940 kVAr. AQiEA has minimum increase in voltage profile and GA has maximum increase in voltage profile as compared with other algorithms.

Case III:

In this case, simultaneous optimization of DG and capacitors are considered (one DG and two capacitors are optimally sized and placed) for the minimization of power loss through the proposed algorithm. Table 3 shows the comparative analysis of the proposed algorithm with the existing algorithms for the reduction of power losses in the system. From Table 3, the power loss reduction is high with the proposed algorithm as compared with case I and case II. In this case, the proposed algorithm produces maximum power loss reduction (78%) followed by PSO (77%), SA (76%),

Table 3 Comparative analysis of AQiEA with other methods Case III

	GA	SA	PSO	AQiEA
P_{loss} (kW)	73.3865	75.5515	71.9186	68.1001
Q_{loss} (kVAr)	42.2717	42.4041	40.3265	38.3663
Loc	40, 66, 32	11, 28, 31	10, 29, 33	77, 32, 31
Size (MW)	0.86, 0.93, 0.94	1, 0.97, 1	1, 1, 1	1, 1, 1
MinVV (p.u)	0.9501	0.9455	0.946	0.9553
P_{load} (MW)	1.6303	1.5703	1.5703	1.5703
Q_{load} (MVAr)	0.8322	0.6522	0.6222	0.6222
% Reduction	**76.79**	**76.1**	**77.25**	**78.46**

and GA (77%) and reduces power losses from 316.1221 kW to 68.1 kW by locating two capacitors and DG at 77, 32, and 31 with capacities 1 MVAr, 1 MVAr, and 1 MW. In this case, the capacitors and DG are sized entirely. Similarly, GA minimizes power loss to 71.91 kW by locating capacitors and DG at 10, 29, and 33 with capacities 1 MVAr, 1 MVAr, and 1 MW. The voltage profile is also improved from 0.8713 p.u (without capacitor and DG) to 0.9553 p.u (AQiEA). In this case, SA minimizes power loss to 75.55 kW by locating capacitors and DG at 11, 28, and 31 with capacities 1 MVAr, 970 kVAr, and 1 MW. SA has minimum increase in voltage profile and AQiEA has maximum increase in voltage profile as compared with other algorithms.

Case IV:

In this case, simultaneous optimization of DG and capacitors are considered (two DGs and one capacitor are optimally sized and placed) for minimization of power loss through the proposed algorithm. Table 4 shows the comparative analysis of the proposed algorithm with the existing algorithms for reduction of power losses in the system. From Table 4, the power loss reduction is high with the proposed algorithm as compared with all cases. In this case, the proposed algorithm produces maximum power loss reduction (79%) followed by PSO (78%), SA (77%), and GA (77%) and reduces power losses from 316.1221 kW to 67.86 kW by locating two DGs and capacitor at 57, 32, and 32 with capacities 1 MW, 1 MW, and 1 MVAr. In this case, the DGs and capacitor are sized entirely. Similarly, PSO minimizes power loss to 68.86 kW by locating DGs and capacitor at 57, 30, and 31 with capacities 980 kW, 940 kW, and 1 MVAr. The voltage profile is also improved from 0.8713 p.u (without Capacitor and DG) to 0.9557 p.u (AQiEA). In this case, GA minimizes power loss to 71.29 kW by locating DGs and capacitor at 63, 29, and 28 with capacities 950 kW, 900 kW, and 990 kVAr. PSO has minimum increase in voltage profile and SA has maximum increase in voltage profile as compared with other algorithms.

Figures 2 and 3 show the comparative analysis for different cases with different algorithms. From Fig. 3, it is seen that Case IV has high power loss reduction as compared with other cases. Simultaneous placement of DGs and capacitor has high reduction in power losses as compared with other cases. AQiEA has maximum reduction in power losses in all cases as compared with other algorithms. Simultaneous

Table 4 Comparative analysis of AQiEA with other methods for Case IV

	GA	SA	PSO	AQiEA
P_{loss} (kW)	71.2939	72.4336	68.8647	67.8687
Q_{loss} (kVAr)	40.5286	40.731	39.1018	38.4532
Loc	63, 29, 28	32, 68, 40	57, 30, 31	57, 32, 32
Size (MW)	0.95, 0.9, 0.99	0.99, 0.9, 1	0.98, 0.94, 1	1, 1, 1
MinVV (p.u)	0.957	0.9646	0.9538	0.9557
P_{load} (MW)	0.7203	0.6803	0.6503	0.5703
Q_{load} (MVAr)	1.6322	1.6222	1.6222	1.6222
% Reduction	**77.45**	**77.09**	**78.22**	**78.53**

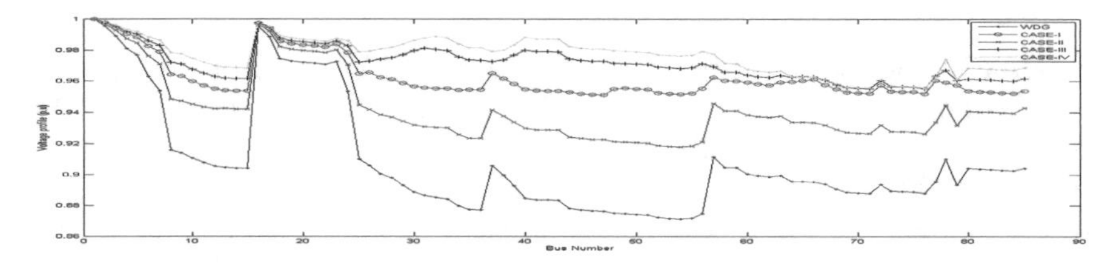

Fig. 2 Comparative analysis for voltage profile improvement

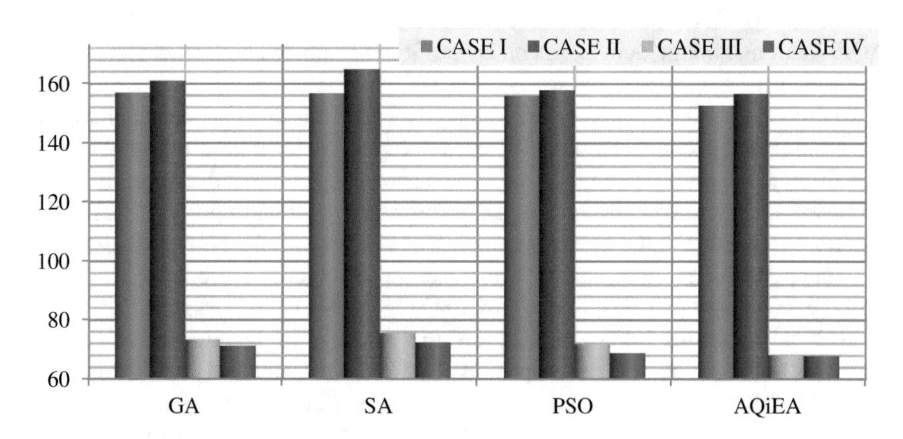

Fig. 3 Comparative analysis for power loss reduction

placement and sizing of DG and capacitor not only reduces the power losses but also improves the voltage profile.

6 Conclusion

In the present power system scenario, DG and capacitors are playing a key role in minimization of power losses in the distribution network. Placement and sizing of DG and capacitors are considered independently to reduce power losses. Simultaneous placement and sizing of DG and capacitor have maximum reduction of power losses in distribution system as compared to independent optimization. AQiEA has been used to optimize DG and capacitors. AQiEA uses two qubits instead of one qubit, which does not require any tuning of parameters and uses an adaptive quantum-based crossover operator. The effectiveness of the proposed algorithm is tested on four cases. Case IV has high reduction in power losses as compared with other cases. The experimental results show that the proposed algorithm has better performance as compared with other algorithms.

References

1. von Meier, A.: Electric Power System. Wiley, Hoboken, New Jersey (2006)
2. Yang, Y., Zhang, S., Xiao, Y.: Optimal design of distributed energy resource systems coupled with energy distribution networks. Energy **85**, 433–448 (2015)
3. Ackermann, T., Andersson, G., Sder, L.: Distributed generation: a definition. Electr. Power Syst. Res. **57**, 195–204 (2001)
4. Prakash, K., Sydulu, M.: Particle swarm optimization based capacitor placement on radial distribution systems. In: Power Engineering Society General Meeting, IEEE, Tampa, FL, pp. 1–5 (2007)
5. Ganguly, S., Samajpati, D.: Distributed generation allocation on radial distribution networks under uncertainties of load and generation using genetic algorithm. IEEE Trans. Sustain. Energy **6**, 688–697 (2015)
6. Bohre, K., Agnihotri, G., Dubey, M.: Optimal sizing and sitting of DG with load models using soft computing techniques in practical distribution system. IET Gener. Transm. Distrib. **10**, 2606–2621 (2016)
7. Mahmoud, K., Yorino, N., Ahmed, A.: Optimal distributed generation allocation in distribution systems for loss minimization. IEEE Trans. Power Syst. **31**, 960–969 (2016)
8. Sultana, S., Roy, P.K.: Optimal capacitor placement in radial distribution systems using teaching learning based optimization. Int. J. Electr. Power Energy Syst. **54**, 387–398 (2014)
9. Singh, S.P., Rao, A.R.: Optimal allocation of capacitors in distribution systems using particle swarm optimization. Int. J. Electr. Power Energy Syst. **43**, 1267–1275 (2012)
10. Pereira, B.R., da Costa, G.R.M., Contreras, J., Mantovani, J.R.S.: Optimal distributed generation and reactive power allocation in electrical distribution systems. IEEE Trans. Sustain. Energy **7**, 975–984 (2016)
11. Mahaei, S.M., Sami, T., Shilebaf, A., Jafarzadeh, J.: Simultaneous placement of distributed generations and capacitors with multi-objective function. In: 2012 Proceedings of 17th Conference on Electrical Power Distribution, Tehran, pp. 1–9 (2012)
12. Kalantari, M., Kazemi, A.: Placement of distributed generation unit and capacitor allocation in distribution systems using genetic algorithm. In: 10th International Conference on Environment and Electrical Engineering, Rome, pp. 1–5 (2011)
13. Mahmoud, K.: Optimal integration of DG and capacitors in distribution systems. In: 2016 Eighteenth International Middle East Power Systems Conference (MEPCON), Cairo, pp. 651–655 (2016)

14. Wang, M., Zhong, J.: A novel method for distributed generation and capacitor optimal placement considering voltage profiles. In: 2011 IEEE Power and Energy Society General Meeting, San Diego, CA, pp. 1–6 (2011)
15. Patvardhan Mani, C.: An improved model of ceramic grinding process and its optimization by adaptive quantum inspired evolutionary algorithm. Int. J. Simul. Syst. Sci. Technol. **11**(6), 76–85 (2012)
16. Sailesh Babu, G.S., Bhagwan Das, D., Patvardhan, C.: Real-parameter quantum evolutionary algorithm for economic load dispatch. IET Gener. Transm. Distrib. **2**(1), 22–31 (2008)
17. Ranjan, R., Das, D.: Simple and efficient computer algorithm to solve radial distribution networks. Electr. Power Compon. Syst. 95–107 (2010)
18. Samal, P., Ganguly, S.: A modified forward backward sweep load flow algorithm for unbalanced radial distribution systems. In: 2015 IEEE Power & Energy Society General Meeting, Denver, CO, pp. 1–5 (2015)
19. Teng, J.H.: A direct approach for distribution system load flow solutions. IEEE Trans. Power Deliv. **18**, 882–887 (2003)
20. Han, K.-H., Kim, J.-H.: Quantum-inspired evolutionary algorithm for a class of combinatorial optimization. IEEE Trans. Evol. Comput. **6**, 580–593 (2002)
21. Chatterjee, S., Roy, B.K.S.: An analytic method for allocation of distributed generation in radial distribution system. In: 2015 Annual IEEE India Conference (INDICON), New Delhi, pp. 1–5 (2015)

Integration of Dispatch Rules for JSSP: A Learning Approach

Rajan and Vineet Kumar

Abstract Dispatching rules have been studied from past decades, and it has concluded that these rules are composing vertebrae for various industrial scheduling applications. Generally, it is a very tedious process to develop incipient dispatching rules in a given atmosphere by indulging and implementing dissimilar models under consideration and also to appraise them through extensive research. For determining effectual dispatching rules, automatically, a pioneering approach is presented. Proposed work addresses job shop scheduling problem (JSSP) NP hard in nature with the objective of minimizing mean flow time. For achieving this, two latest dispatching rules have been introduced. These latest rules will combine the process time and total work content of a job in queue under subsequent process. Additive and alternative approaches have been taken for combining the rules. For evaluating the performance of proposed dispatching rules, a rigorous study has been carried out against S.P.T rule, WINQ rule, S.P.T and WINQ rule, and the other best existing rules in common practice.

Keywords Shortest processing time (S.P.T) · Work in next queue first (WINQ)
Longest processing time (L.P.T) · Most work remaining (M.WK.R)
Least operation numbers (L.OP.N) · Select a task (SLACK) · Earliest due date (E.D.D)

1 Introduction

For controlling, the various operations on shop floor improvements in scheduling methodologies as well as technological advances have directed to surfacing of addi-

---

Rajan (✉) · V. Kumar
Department of Mechanical Engineering, University Institute of Engineering & Technology, Maharishi Dayanand University, Rohtak, Haryana, India
e-mail: er.rajangarg@gmail.com

V. Kumar
e-mail: vineetsingla2002@gmail.com

K. Ray et al. (eds.), *Soft Computing: Theories and Applications*,
Advances in Intelligent Systems and Computing 742,
https://doi.org/10.1007/978-981-13-0589-4_57

tional effectual scheduling schemes during the recent years. In job shop environment, scheduling plays an important role in managing the shop floor system and it can have a key impact on performance of shop floor. "N" numbers of job are to be operated on "M" number of machines for a given tenure in such way that given objective can be achieved. Each job has to be processed with precise set of operation under given technical preference array called "routing". Scheduling will be termed as "static" if jobs are available at beginning of the process; otherwise, it will be termed as "dynamic" if a set of job amends transiently. If parameters are known with certainty, then the problem will be "deterministic"; otherwise, it will be "stochastic" for at least one probabilistic parameter. Release time, routing, and processing time are "stochastic" parameters and cannot be recognized, priory. Either theoretically or practically, it is not possible to evaluate optimal scheduled solution, priory, for "dynamic stochastic" scheduling problem. For processing all jobs, the prime target of planning is to trace a scheduled solution by optimizing mean flow time. Usually, following are few decisions taken with the assistance of dispatching rules:

(i) Which job has to be processed on which machine and
(ii) When will particular machine get free?

Various dispatching rules proposed by different researchers [1–3] during recent years were not found suitable for every significant criterion like mean flow time. The selection for a particular dispatching rule depends upon which the proposed crucial factor is to be modified. Recent literature survey exhibits that process time-dependent rule performs well under stiff load situation, whereas due date based rule works better under relaxed load environment [1, 4, 5]. Present work focuses on two new dispatching rules based on "additive" and "alternative" techniques designed for open job shop environment. For this, process time and work content criterion are used for subsequent processes.

2 Literature Review

Shortest processing time (S.P.T) is time-dependent rule which ignores due date information of job. S.P.T rule minimizes the mean flow time for good performance under highly loaded environment at the shop floor as in confirmation of [2, 4, 5]. Jobs are getting scheduled depending on their due date information by using "Due date" based rules. One of the best examples supporting "Due date" rule is the "Earliest due date", i.e., (E.D.D). Under light load stipulations, due date based rules furnish excellent outcomes, but under high load situations, performance of these rules depreciates [3]. Least Slack (L.S) and Critical Ratio (C.R) are good examples of combination rule as they well utilize both process time and due date information [1]. Loading of jobs is a rule which depends upon the condition of shop floor instead of nature of the job and this rule do not fall into any of previously mentioned categories. The perfect example for this is WINQ rule [2]. Many researchers have proposed to combine simple priority rules for generating more involute rules. Two customary methods are adopted for such type of cases: (i) additive method and (ii) alternative method.

(i) Additive/weighted combination method resolves the precedence by calculating equation as follows:

$$Zi = \sum_{f=1}^{g} \alpha\mathrm{f} \times (\mathrm{Qf})_{\mathrm{i}}$$

where

$(\mathrm{Qf})_{\mathrm{i}}$	Priority value of a simple priority rule "f" for job "i", f = 1, 2, …, g,
αf	Weight (or coefficient) of rule f, where αf > 0 and
Zi	Resultant priority index for job "i".

Major drawback in adopting above equation is that the priority index "Zi" is receptive to "αf". So, one has to utilize search algorithm for concluding finest values of "αf" [6, 7]. In order to obtain finest values of "αf" while using past data and search algorithms, consistency of shop floor attributes, such as "arrival pattern", "service rate", etc., is assumed.

(ii) Alternative method/hierarchical approach depends on the situational course of action. Two rules are merged together for this idiom. Thorough dialog on diverse dispatching formulae and rules can be found in [1, 2, 3]. Several dispatching rules that modernly recommended and broadly calculated are because of [8–12]. For minimizing mean flow time rule due to [10, 13] is the best rule among previously mentioned rules. In comparison, it is found that S.P.T is yet most efficient rule as far as the objective of minimizing number of tardy jobs is concerned.

3 Conventional Dispatch Rules

The following are various types of conventional dispatch rules. These rules are extensively espoused in most of the organizations.

(a) S.P.T (Shortest Processing Time)—Opt for a task having petite processing time. S.P.T exhibits good results for stiff load environment conditions.
(b) L.P.T (Longest Processing Time)—Pick an assignment having extended processing time.
(c) M.WK.R (Most Work Remaining)—Go for a job which remaining longest total processing time.
(d) L.OP.N (Least Operation Numbers)—Choose a charge having smallest operation number.
(e) SLACK—select a task which has little due date.

Proposed scheduling scheme depends on these suggested dispatch rules. Various classifications for dispatching rules are listed as follows:

(a) "Process time" dependent
(b) "Due date" dependent
(c) "Combination" dependent
(d) Neither "Process time" nor "Due date" dependent.

3.1 Assumptions Made

Some customary assumptions are to be integrated into job shop model during simulation studies [1, 2, 14]. The assumptions made are as follows:

- Only single action will be executed at a time on each machine, i.e., all jobs will be ready at beginning.
- A task once captured should be completed in all respect before taking additional one, i.e., anticipation of job will not be permitted.
- Do not perform consecutive operations on a job at same machine, i.e., one machine will process only one job at a time.
- Do not perform next operation until all pending actions will get finished on any job.
- Only machines are available resources.
- No alternate routings will be there.
- No machine breakdown will occur on shop floor.
- Jobs are autonomous as no assembly will be engaged.
- No parallel machines will occur on shop floor.

4 Expansion of Suggested Rules

Inspiration of proposed study is due to outcomes of previous imitation studies. S.P.T rule executes fine during tight load situations, whereas WINQ rule likely to offer preference to those tasks that shift on to queues with slightest backlog, before to accelerate a job [2]. S.P.T rule is proven to be pretty efficient where objective is of mean flow time. Also, it is surveyed that WINQ rule can assist in minimizing the waiting time of tasks by exploring the shop floor information of additional machines. All these surveillances have proven the leading principles for developing innovative dispatching rules.

4.1 *Additive Strategy (Rule 1)*

It is a simple rule designed on additive policies. In this, the summation of process time and total work content of jobs is accounted. Presume three jobs namely job-1, job-3, and job-4 are in queue on machine-1. Suppose job-1 and job-4 go to machine-3 and job-3 goes to machine-2 for successive action.
Consider

$TWKQ_2$ = total work content of jobs on machine-2 and
$TWKQ_3$ = total work content of jobs on machine-3

Assume:

t_{11} = process time of jobs 1 on machine-1,
t_{31} = process time of jobs 3 on machine-1 and
t_{41} = process time of jobs 4 on machine-1.

Expect machine-1 turns free and now requirement arises to select a job for next operation. The proposed rule will dispense the priority index "'i" for a specific job, viz., "Zi" as mentioned below:

For job-1: $Z_1 = t_{11} + TWKQ_3$, as job-1 is in queue on machine-3 for next process,
For job-3: $Z_3 = t_{31} + TWKQ_2$, as job-3 is in queue on machine-2 for next operation and
For job-4: $Z_4 = t_{41} + TWKQ_3$, as job-4 is in queue on machine-3 for next operation.

In accordance with proposed rule, a job with smallest value of "Zi" is selected.

4.2 *Alternative Approach (Rule 2)*

In this approach, attempt is made to utilize the process time and total work content of jobs in queue. For user-friendly interpretation, rule has been proposed with assistance of numerical explanation as presented in the previous section.
Assume:

"R_2" and "R_3" = time when machine-2 and machine-3 become free after operation.

"T" = Instant time when machine-1 becomes free.

Suggested rule will work as follows:
Step-I: "Jobs 1, 3 and 4 are in line at machine-1"
Experiment (a) verify

For job-1: $T + t_{11} \geq R_3 + TWKQ_3$ job-1 goes to machine-3 for consecutive operation,
For job-3: $T + t_{31} \geq R_2 + TWKQ_2$ job-3 joins queue at machine-2 for subsequent operation and
For job-4: $T + t_{41} \geq R_3 + TWKQ_3$ job-4 goes to machine-3 for next operation.

Verify

"A job will not wait in line for its next process after being processed on existing machine."

Make a set "ψ" with jobs which satisfy Experiment (a).

Step-II: "If 'ψ' is not an empty set."

Select a job from set "ψ" having shortest process time on machine-1;

(For such cases, S.P.T rule is operational)

Else

Select a job amongst all queued jobs on machine-1 corresponding to minimum work content, i.e., with least $\{TWKQ_2, TWKQ_3\}$.

(For such cases WINQ rule is operational).

5 Proposed Simulated Evaluation

Since proposed dispatching rules are based on additive and alternative tactics of S.P.T and WINQ, then the rule proposed by Raghu and Rajendran [10] is one of the finest rules framed till date. As far as evaluation is concerned considered, S.P.T, WINQ, and R.R rules preferably apart from two suggested rules. In order to have proper scheduling, extra rules have been inculcated for assessment, i.e., make a choice between S.P.T and WINQ rules, randomly (RAN). RAN rule should be opted to employ because such an alternative shall merge the advantages of S.P.T and WINQ rules, jointly. The measure of presentation is mean flow time.

5.1 Suggested Experimental Conditions

In an open shop environment, experiment can be performed by having at least 10 machines. In randomly created routing, each machine has same probability of being selected. Numbers of processes for every task are distributed homogeneously between 4 and 10. Following are the three process times taken:

(a) 1–50 (b) 10–50 (c) 1–100

Total work content (TWK) method [1] should be used for those experiments which are having value of allowance factor "c" = 3, 4, and 5. By using an exponential distribution, job arrivals can be created. Following can be various machine utilization levels "U_g" to be tested in all experiments, viz., 60, 70, 80, 85, 90, and 95%.

Totally, 54 simulated experimental set can be made for each dispatching rule as follows:

(i) Three types of process time distributions,
(ii) Three types of different due date settings, and
(iii) Six types of different utilization levels.

Every simulated experiment can be consisted of 20 different moves (replications), for which each move shop can be loaded continuously with job orders and that can be numbered on appearance.

6 Measurement of Best Performing Rule

For presentation of results, the mean of mean values for 20 runs could be obtained. For a given sample size, smaller number of moves and larger run length can be desirable. For this, suggested number of replications could be about 10 [15]. Proposed technique [16] can be taken as an instruction for fixing total sample size. On pursuing these instructions, 20 moves can be fixed having run length of every replication as 2000.

For further elaboration, "relative percentage increase" in mean flow time can be worked out as below:

"F_k" denotes mean flow time of job because of "k".

Where:

k = 1; specifies S.P.T rule,
k = 2; signifies WINQ rule,
k = 3; represents RAN rule (random choice between S.P.T and WINQ),
k = 4; denotes R.R rule,
k = 5; stands for Rule-1 and
k = 6; means Rule-2.

"Relative percentage increase (R.P.I)" in mean flow time regarding best performing rule can be calculated as follows:

$$\text{R.P.I} = \frac{\text{Fk} - \text{minimum}\{\text{Fk}\ ,\ 1 \leq \text{k} \leq 6\}}{\text{minimum}\{\text{Fk}\ ,\ 1 \leq \text{k} \leq 6\}} \times 100$$

Above equation is designed for obtaining different values of allowance factor as "c" = 3, 4, and 5. Mean flow time due to "k" will remain same for different values of allowance factor "c" except R.R rule. Because R.R is the only rule that uses information of due date for minimizing mean flow time.

As literature states that S.P.T and WINQ rules are competent in reducing mean flow time, proposed rule-1 exploits the integrity of both. In such situation, it is not only sought to maximize the output but also seek to load a job having less waiting time. As a result, relative performance of the rule-1 will progress if shop load will enhance.

Rule-2 will prove better than S.P.T and WINQ rules on considering process time distribution 10–50. Rule-2 will perform significantly better than WINQ rule at all levels for having process time distribution 1–50 and 1–100. Calculation of mean flow time by rule-2 will be significantly smaller than that of S.P.T rule for very high shop load level. For minimizing mean flow time, random choice among S.P.T and WINQ rules will prove much better than solitary use of S.P.T or WINQ rule at very high shop load level.

7 Conclusion

Pair of rules for dispatching in a job shop environment has been offered. Rules are based on additive and alternative approaches and use the sequence-dependent mean flow time for next operation on a job in queue. From literature, it has been found that no individual rule is efficient for minimizing all measure of performances. Proposed rule-1 is relatively important for minimizing mean flow time that integrates process time and total work content, respectively. Further, exploration can be directed toward expansion of rules which will inculcate information about process time and total work content of jobs for minimizing as many measures of performance as possible, concurrently.

References

1. Blackstone, J.H., Phillips, D.T., Hogg, G.L.: A state-of-the-art survey of dispatching rules for manufacturing job shop operations. Int. J. Prod. Res. **20**, 27–45 (1982)
2. Haupt, R.: A survey of priority rule-based scheduling. OR Spektrum **11**, 3–16 (1989)
3. Ramasesh, R.: Dynamic job shop scheduling: a survey of simulation research. OMEGA **18**, 43–57 (1990)
4. Conway, R.W.: Priority dispatching and job lateness in a job shop. J. Ind. Eng. **16**, 228–237 (1965)
5. Rochette, R., Sadowski, R.P.: A statistical comparison of the performance of simple dispatching rules for a particular set of job shops. Int. J. Prod. Res. **14**, 63–75 (1976)
6. O'Grady, P.J., Harrison, C.: A general search sequencing rule for job shop sequencing. Int. J. Prod. Res. **23**, 961–973 (1985)
7. Rajan, Kumar, V.: Hybridization of genetic algorithm & variable neighborhood search (VNGA) approach to crack the flexible job shop scheduling problem: a framework. J. Ind. Pollut. Control **33**(2), 90–99 (2017)
8. Anderson, E.J., Nyirendra, J.C.: Two new rules to minimize tardiness in a job shop. Int. J. Prod. Res. **28**, 2277–2292 (1990)
9. Baker, K.R., Kanet, J.J.: Job shop scheduling with modified due dates. J. Oper. Manag. **4**, 11–22 (1983)
10. Raghu, T.S., Rajendran, C.: An efficient dynamic dispatching rule for scheduling in a job shop. Int. J. Prod. Econ. **32**, 301–313 (1993)
11. Vepsalainen, A.P.J., Morton, T.E.: Priority rules for job shops with weighted tardiness costs. Manag. Sci. **33**, 1035–1047 (1987)
12. Rajpurohit, J., Sharma, T.K., Abraham, A., Vaishali, A.: Glossary of metaheuristic algorithms. Int. J. Comput. Inf. Syst. Ind. Manag. Appl. **9**, 181–205 (2017)
13. Rajan, Kumar, V.: Flow shop & job shop scheduling: mathematical models. Int. J. R&D Eng. Sci. Manag. **5**(7), 1–7 (2017)
14. Baker, K.R.: Introduction to sequencing and scheduling. Wiley, New York (1974)
15. Law, A.M., Kelton, W.D.: Confidence intervals for steady state simulation: I. A survey of fixed sample size procedures. Oper. Res. **32**, 1221–1239 (1984)

16. Fishman, G.S.: Estimating sample size in computing simulation experiments. Manag. Sci. **18**, 21–38 (1971)
17. Adams, J., Balas, E., Zawack, D.: The shifting bottleneck procedure for job shopscheduling. Manag. Sci. **34**, 391–401 (1988)
18. Aarts, E.H.L., Van Laarhoven, P.J.M., Lenstra, J.K., Ulder, N.L.J.: A computational study of local search algorithms for job shop scheduling. ORSA J. Comput. **6**, 118–125 (1994)
19. Balas, E., Lenstra, J.K., Vazacopoulos, A.: The one-machine problem with delayed precedence constraints and its use in job shop scheduling. Manag. Sci. **41**, 94–109 (1995)
20. Bierwirth, C.: A generalized permutation approach to job shop scheduling with genetic algorithms. OR Spektrum **17**, 87–92 (1995)
21. Blazewicz, J., Ecker, K., Schmidt, G., Weglarz, J.: Scheduling in Computer and Manufacturing Systems, 2nd edn. Springer, Berlin (1994)
22. Brucker, P.: Scheduling Algorithms. Springer, Berlin (1995)
23. Brucker, P., Jurisch, B., Sievers, B.: A branch & bound algorithm for the jobshop problem. Discret. Appl. Math. **49**, 107–127 (1994)
24. Conway, R.W., Johnson, B.M., Maxwell, W.L.: An experimental investigation of priority dispatching. J. Ind. Eng. **11**, 221–230 (1960)
25. Dell'Amico, M., Trubian, M.: Applying tabu search to the job-shop scheduling problem. Ann. Oper. Res. **41**, 231–252 (1993)
26. Dorndorf, U., Pesch, E.: Evolution based learning in a job shop scheduling environment. Comput. Oper. Res. **22**, 25–40 (1995)
27. Lawler, E.L.: Recent results in the theory of machine scheduling. In: Bachem, A., Grötschel, M., Korte, B. (eds.) Mathematical Programming: The State of the Art, pp. 202–234. Springer, Berlin (1983)
28. Lawler, E.L., Lenstra, J.K., Kan, A.H.G.R.: Recent developments in deterministic sequencing and scheduling: a survey. In: Dempster, M.A.H. (1982)
29. Lenstra, J.K., Kan, A.H.G.R. (eds.): Deterministic and Stochastic Scheduling, pp. 35–73. Reidel, Dordrecht
30. Lorenzen, T.J., Anderson, V.L.: Design of Experiments: A No-Name Approach. Marcel Dekker, New York (1993)
31. Montgomery, D.C.: Design and Analysis of Experiments. Wiley, New York (1991)
32. Kan, A.H.G.R.: Machine Scheduling Problems. Martinus Nijhoff, The Hague (1976)
33. Van Laarhoven, P.J.M., Aarts, E.H.L., Lenstra, J.K.: Job shop scheduling by simulated annealing. Oper. Res. **40**, 113–125 (1992)

White Noise Removal to Enhance Clarity of Sports Commentary

Shubhankar Sinha, Ankit Ranjan, Anuranjana and Deepti Mehrotra

Abstract The spectators of any match make large amount of noise, which is captured along with live commentary during any telecast. The removal of noise created in spectator watching any sports event will improve the audibility of the commentary. Person always faces difficulty in perceiving multiple sounds coming from different sources simultaneously, known as the Cocktail Party problem. Removing the audience noise to improve the commentary quality is the core objective of this paper. Using the signal filter functionalities provided by MATLAB and its plugins, a solution is proposed. Analysis is done by collecting the sound from different sporting events. The case studies considered in the study vary in amount of spectators and the proposed approach produces a better solution for passive sports compared to active sports.

Keywords Butterworth · Spectral subtraction · MATLAB · Audio enhancement

1 Introduction

Live sports commentary comprises various types of unwanted noise and signals. White noise is most prevalent in this type of samples [1]. The presence of sound made by spectators while viewing the match acts as a noise to live commentary. This problem is quite similar to the famous cocktail party problem where one needs to focus on the given signal sound among the music and loud noise in the surrounding.

S. Sinha · A. Ranjan (✉) · Anuranjana · D. Mehrotra
Amity University, Noida, Uttar Pradesh, India
e-mail: ankitranjan621@gmail.com

S. Sinha
e-mail: shubhankar2002@gmail.com

Anuranjana
e-mail: aranjana@amity.edu

D. Mehrotra
e-mail: mehdeepti@gmail.com

K. Ray et al. (eds.), *Soft Computing: Theories and Applications*,
Advances in Intelligent Systems and Computing 742,
https://doi.org/10.1007/978-981-13-0589-4_58

The commentary dataset of four different sports is used for noise removal and further processing.

Spectral subtraction proves useful for accomplishing the feat of removing unwanted music signals or white noise, which is prevalent in the sound samples being used. It works on the principle of spectral modeling, namely perceptual coders [2] which can reduce redundant figures occurring in the waveform of a particular sample. Multiple representations of this model result in bloated structuring [3]. To eliminate this effect, structured coding is introduced instead to create a human-like perspective on the type of sound, its origin, and unique signature. Parameters are then set according to this genre specification. To remove noise from the dataset, various filters are commonly used. Efficiency of these filters is analyzed and used in combination to reduce noise-to-signal ratio.

2 Background Work

2.1 White Noise

It is the most pervasive sort of commotion and one utilized as a part of this review. Repetitive sound is a sample that includes each recurrence inside the scope of hearing (20 Hz–20 kHz) in equal proportion [4]. The vast majority perceives this as more high-recurrence sampling than low, but this is not the case. This anomaly happens in light of the fact that each progressive octave has twice the number of frequencies as the one before this. For example, in the octave range shown below (200–400 Hz), there are 200 frequencies present [5] (Fig. 1).

2.2 Butterworth Filter

A flag handling channel designed to have the lowest possible passband filtration method within normal conventions is the main idea behind the Butterworth filter model. It was first conceived and showcased to the world in 1930 by renowned physicist Stephen Butterworth through the means of his paper: On the Theory of Filter Amplifiers. The Butterworth filter accomplishes several feats, the most noteworthy of which is the maximally level reaction to a recurrence instance in the source signal's input spectrograph, moving it to such low levels at cutoff (-3 dB) that there is no swell occurrence for the provided octave. The quality of higher frequencies dips, however, to a Q-factor value of 0.707. Moving toward the stopband from the wideband point, we notice a slight decrease in decibel level from the first stage pass to the next, decreasing at a rate of -6 dB in the first, -12 dB in the second, and -16 dB in the third octave, respectively. This results in monotonic swells in the spectrograph, in direct contrast to other similar methods [6] (Fig. 2).

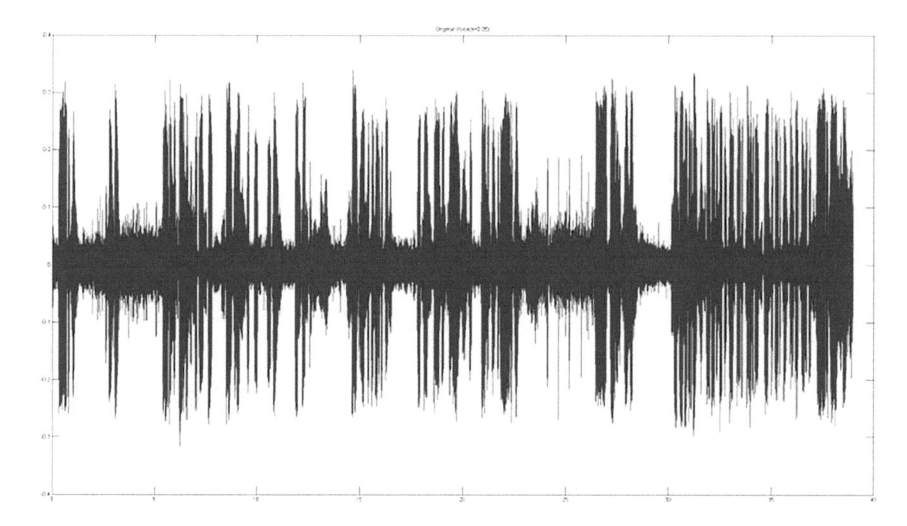

Fig. 1 Noisy speech

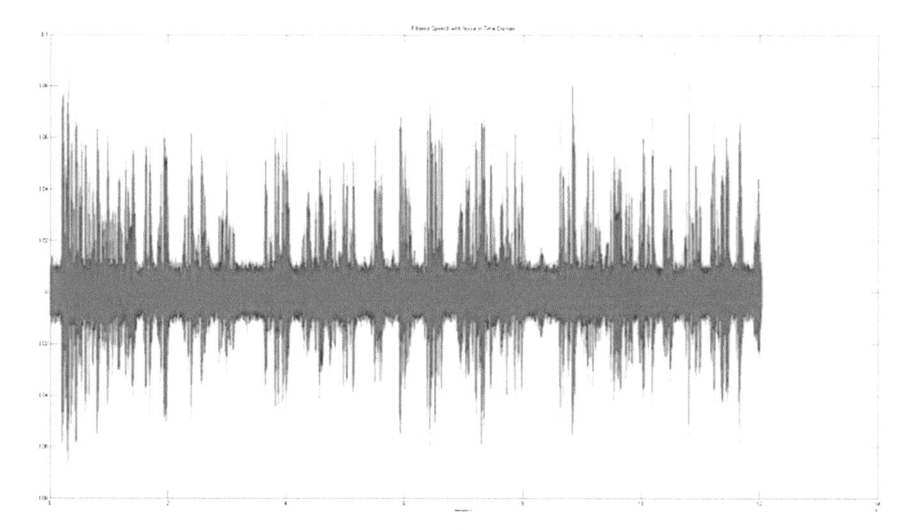

Fig. 2 Butterworth filtered speech

2.3 Spectral Subtraction

The phantom subtraction strategy is a basic and powerful technique for clamor diminishment. In this technique, a normal flag range and clamor range are evaluated in parts of recording and subtraction from each other is done so that SNR is progressed.

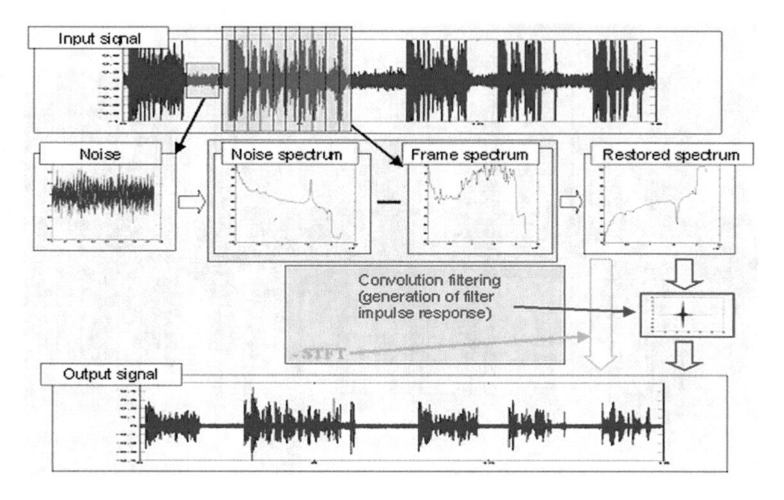

Fig. 3 Spectral subtraction method

The noisy flag y(m) is an aggregate of the desirable flag x(m) and the disorder n(m):

$$y(m) = x(m) + n(m)$$

In the recurrence space, this means as

$$Y(j\omega) = X(j\omega) + N(j\omega) \Rightarrow X(j\omega) = Y(j\omega) - N(j\omega)$$

where $Y(j\omega)$, $X(j\omega)$, and $N(j\omega)$ are Fourier changes of y(m), x(m), n(m), individually [7].

In the event that it is accepted that the flag is bent by a wideband, added substance commotion, the clamor gage is the same amid the examination and the reclamation and the stage is the same in the unique and reestablished flag, utilizing the essential ghostly subtraction, a straightforward and compelling strategy for clamor diminishment. In this strategy, commotion proportion (SNR) is moved forward. Be that as it may, a quite irritating symptom called "melodic clamor" will be incorporated into the yield [8] (Fig. 3).

3 Methodology

In this section, the workflow of noise removal process is described and shown in Fig. 4. This process has been divided into four subprocesses. The input given to the system contains the signal which is the combination of speech and white noise. The Butterworth filter and spectral subtraction algorithms are designed and implemented

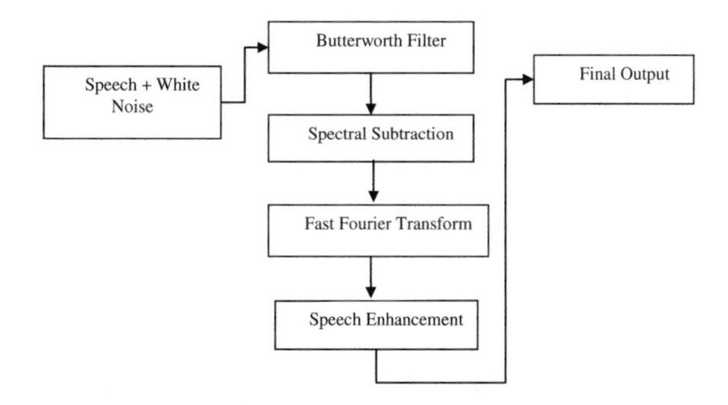

Fig. 4 Flow diagram for noise removal

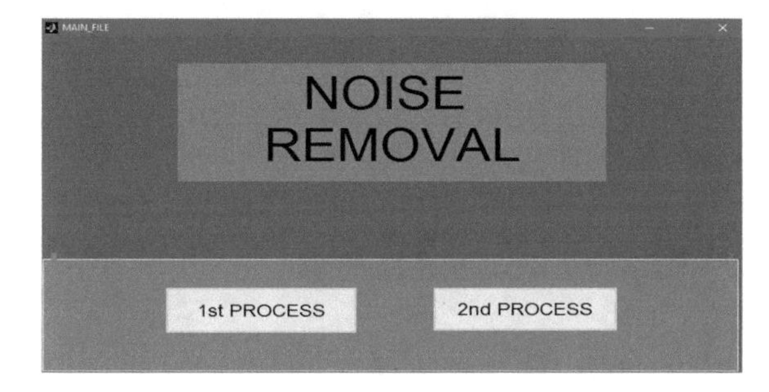

Fig. 5 Main GUI

using Filter Design and Analysis Tool (FDA Tool) in MATLAB, following which the waveform produced undergoes Fast Fourier Transform (FFA) using DSP toolbox. Then, speech enhancement is done through amplitude modulation and correction techniques. This final output is the produced after processing is complete and is saved in the output folder after a playback preview.

The DSP tools from class like time-domain analysis, fast Fourier transform, and window functions are used in speech analysis. FDA tool and spectral subtraction algorithms along with different filters are applied to original speech to reduce noise. FDA tool is a designing method for creating filters as per your specifications which can then be modified through code [9]. A popular audio enhancement algorithm along spectral subtraction is implemented in MATLAB.

Figure 5 shows the home screen which is the entry point of the noise removal; there are two buttons linked to processes defined within the main code section, with the first being an input spectrogram analysis and amplitude graph comparison along with mean square error representation; and the second being output spectrogram analysis, combined plot for noise removal after processing and a 25 s audio playback comparison between the original and processed files.

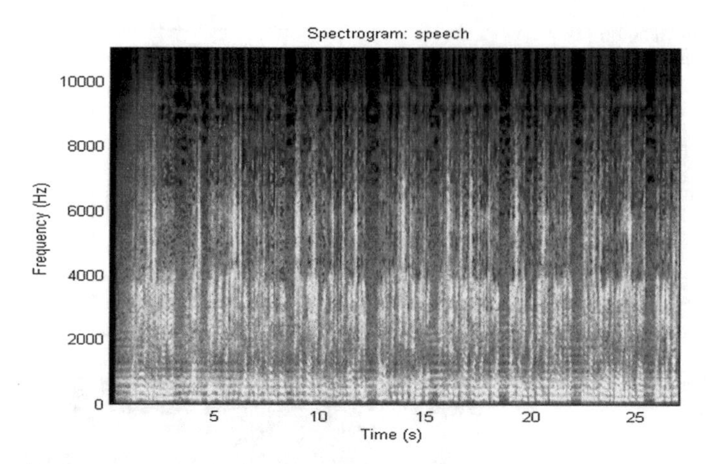

Fig. 6 Input spectrograph for noise removal

4 Experiment and Result

In this study, a self-compiled dataset incorporating 100 tracks each having a runtime of about 25 s is considered. These are stripped from various sources but have one thing in common: sports commentaries. The commentary tracks are from F1 races, IPL matches, football, and finally hockey matches of past and present timelines to conduct our research. Each of the above categories has 25 tracks each, contributing to the tally of 100 needed for this work.

The input is collected for four different types of commentary of sports which vary in number of spectators who will contribute to white noise production. In this section, the steps described in the previous section are discussed in detail for one sample of football commentary.

Step 1: Input entered in process 1 is shown in Fig. 6.

Step 2: The Butterworth filter is applied to the input signal. Input spectrograph analysis graph depicting the relation between frequency (Hz) and time (s) while scrubbing through the given sound sample and output obtained after Butterworth filter is shown in Fig. 7. The red part of first part depicts amplitude modulation and noise damping entry point for the Butterworth filter.
Amplitude graph for noise removal after processing is shown and the mean square error is given below (Fig. 8).

Step 3: To further reduce the mean square error, spectral separation followed by Fast Fourier transform is done. The output graph for the chosen sound sample is shown in Fig. 9. The objective of this step is to reduce the average MSE with the optimal value tends to 0.

Graph depicting combined plot graph drawn after the processing part is completed. The green graph represents estimated signal amplitude, while the blue is showing original amplitude values as shown in Fig. 10.

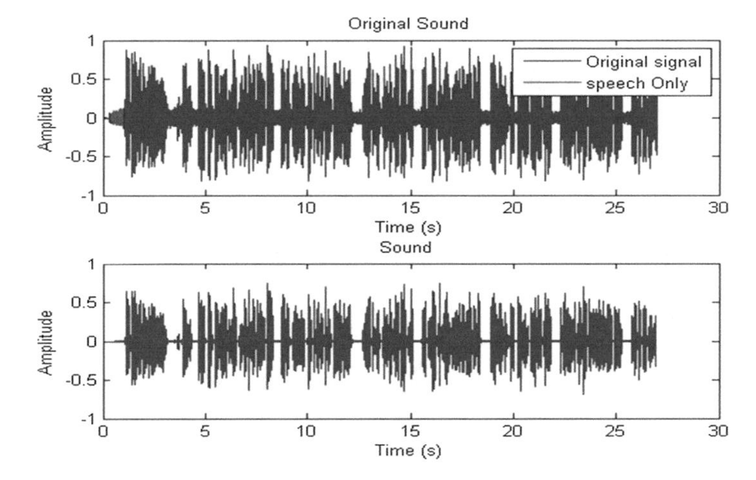

Fig. 7 Amplitude graph for noise removal

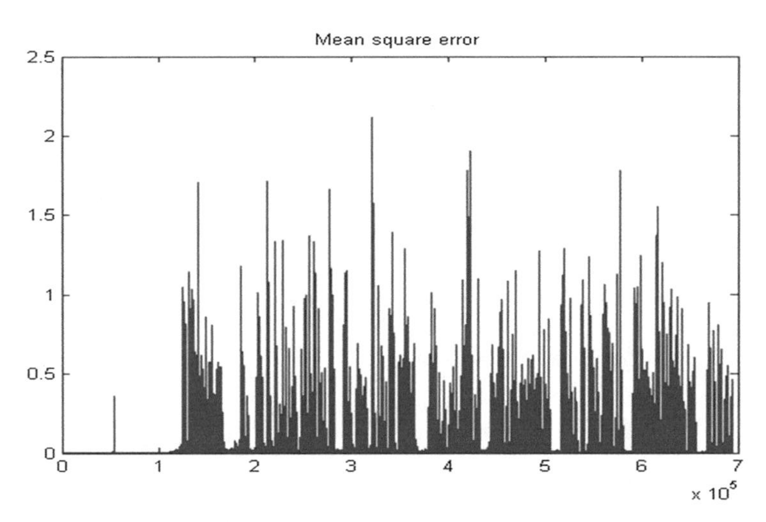

Fig. 8 Mean square error for noise removal

Estimation signal graph depicts the differences between original and processed signals while also showing the noisy signal's amplitude fluctuations (Fig. 11).

This is the output spectrograph, obtained after processing the sound sample through the Butterworth filter. The grainy portions of the graph depict noise (yellow) and speech pattern recognition (red).

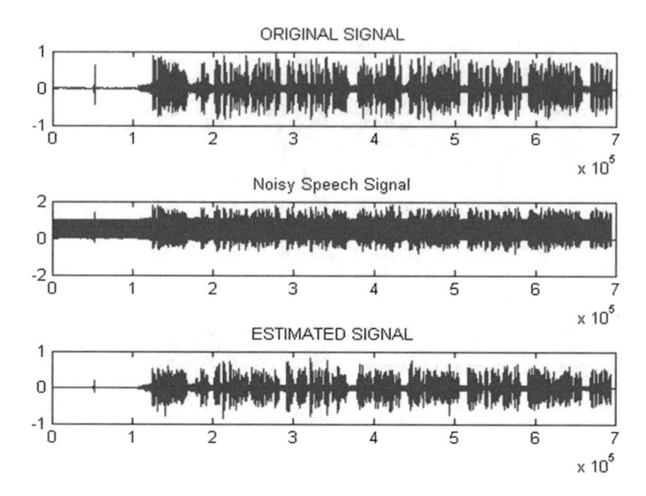

Fig. 9 Estimation signal for noise removal

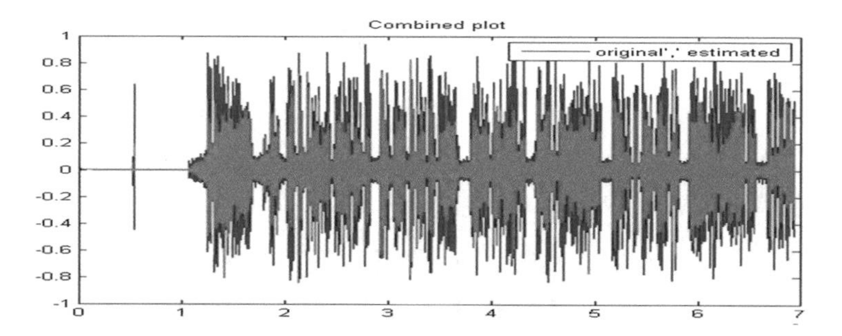

Fig. 10 Combined plot graph for noise removal

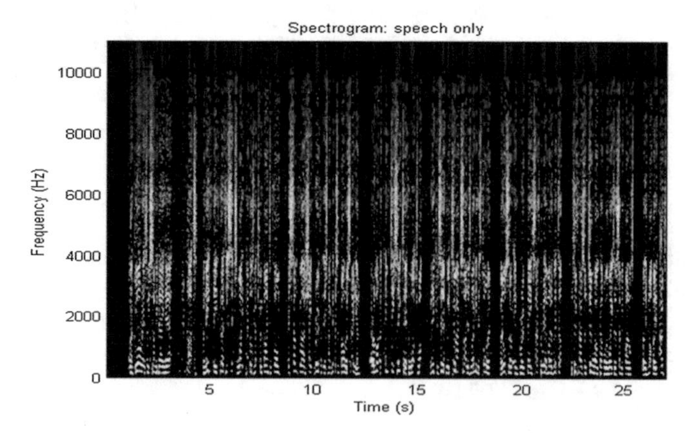

Fig. 11 Output spectrograph for noise removal

Table 1 Comparing different sports commentary on MSE

Sports	MSE
F1	0.00506
Boxing	0.01318
IPL	0.058464
Football	0.050668

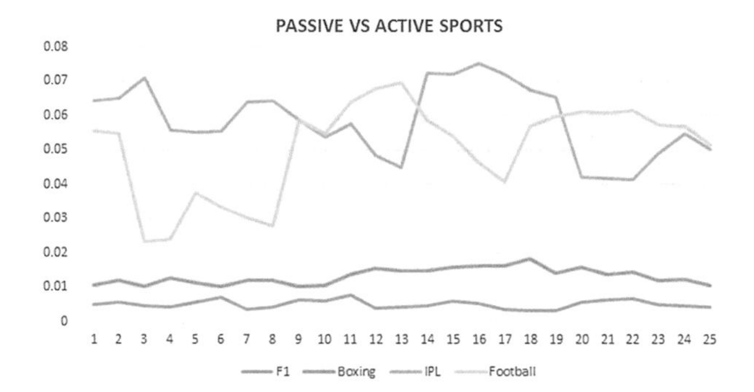

Fig. 12 Variation of MSE for active and passive sports commentary

5 Conclusion

From the results of the calculations done mean square error table shown in Table 1, it is observed that Butterworth filter is excellent when removing white noise and subtle noise types from the background. This is especially clear in the case of sports like F1 races or boxing, where the crowd really does not make as much noise as compared to more crowd-active sports such as T20 cricket or Premier League football. These vociferous crowds lead to the filter not being able to cope sufficiently and results in less than impressive numbers as shown below.

Figure 12 shows that the values of F1 and boxing are comparatively lower, and hence more fruitful for MSE calculations than IPL and football commentaries where crowd noise plays a major role. The work can be extended using post-processing filtering to further reduce the crowd noise.

References

1. Selesnick, I.W., Burrus, C.S.: Generalized digital Butterworth filter design. IEEE Trans. Signal Process. **46**(6), 1688–1694 (1998)
2. Upadhyay, N., Karmakar, A.: Speech enhancement using spectral subtraction-type algorithms: a comparison and simulation study. Procedia Comput. Sci. (2015)
3. Sim, B.L., et al.: A parametric formulation of the generalized spectral subtraction method. IEEE Trans. Speech Audio Process. **6**(4), 328–337 (1998)

4. Abd El-Fattah, M.A., Dessouky, M.I., Diab, S.M., Abd El-samie, F.E.: Speech enhancement using an adaptive Wiener filtering approach. Prog. Electromagn. Res. M. **4**, 167–184 (2008)
5. Ogata, S., Shimamura, T.: Reinforced spectral subtraction method to enhance speech signal. In: IEEE International Conference on Electrical and Electronic Technology, vol. 1, pp. 242–24 (2001)
6. Loizou, P.C.: Speech Enhancement: Theory and Practice, Ist edn. Taylor and Francis (2007)
7. Vaseghi, S.V.: Advanced Digital Signal Processing and Noise Reduction, 2nd edn. Wiley, NY, USA (2000)
8. Upadhyay, N., Karmakar, A.: The spectral subtractive-type algorithms for enhancing speech in noisy environments. In: IEEE International Conference on Recent Advances in Information Technology, ISM Dhanbad, India, pp. 841–847, 15–17 Mar 2012
9. Kamath, S., Loizou, P.: A multi-band spectral subtraction method for enhancing speech corrupted by colored noise. In: IEEE International Conference on Acoustics, Speech, and Signal Processing, Orlando, USA, vol. 4, pp. 4160–4164, May 2002

Analysis of Factors Affecting Infant Mortality Rate Using Decision Tree in R Language

Namit Jain, Parul Kalra and Deepti Mehrotra

Abstract This is a study done for the social cause that was increasing at an alarming rate and was creating a situation of panic among the people of the world, Mortality Rate. This situation was analyzed by analyzing various factors such as birth rate, literacy rate, number of health centers, etc. using the decision tree technique in R tool which illustrated trees of two different decades separately and analyzed the factors affecting the mortality rate with their contribution in driving its rate, and also the summary of decision tree will indicate its accuracy and kappa factor to judge the authenticity of the factors chosen. This will be useful to the governing bodies to get to know about the factors and work upon them for the decrement of the infant mortality rate.

Keywords Decision tree · R language · Infant mortality rate · Literacy rate
Birth rate

1 Introduction

The infant mortality rate is the number of death of the people in a particular area in a particular population out of the total population in a given time. There are various factors that drive the infant mortality rate such as birth rate, literacy rate, number of the health center, and availability of doctors in that area.

There is an emerging need to determine the affecting rate of these factors so as to counter them to reduce the infant mortality rate. Various factors that affect the

N. Jain (✉) · P. Kalra · D. Mehrotra
Amity School of Engineering and Technology, Amity University, Noida,
Uttar Pradesh, India
e-mail: namitjainrocks96@gmail.com

P. Kalra
e-mail: pkbhatia@amity.edu

D. Mehrotra
e-mail: dmehrotra@amity.edu

K. Ray et al. (eds.), *Soft Computing: Theories and Applications*,
Advances in Intelligent Systems and Computing 742,
https://doi.org/10.1007/978-981-13-0589-4_59

infant mortality rate are becoming a social problem and are affecting our lifestyle. The increasing infant mortality rate has started to become a toothache for the governments. There are various movements started worldwide to make people aware of this rising problem.

The only way to tackle this problem is to find the root causes of it which can only be found by analyzing different factors according to us that can account for the change in infant mortality rate and to know what among those factors are actually playing their role in driving the infant mortality rate.

Decision tree technique is used to analyze the factors and arrange them in the top-to-bottom approach with the factors of high influence on top and less on the bottom [1].

2 Background

Decision tree is a useful technique to determine the influence of different factors on a single entity and to judge which factors drive the variation in the entity maximum and minimum [1, 2, 5, 8]. Infant mortality rate was an uprising issue faced by the countries nowadays, especially developing countries like India, China, Brazil, etc. [3, 4, 6]. R language is a reliable tool to analyze the data as it gives the user various areas to explore which cannot be explored otherwise [5, 7, 8]. When these methodologies get combined, they perform exceptionally well and provide an efficient output.

3 Methodology

For this study, the datasets have been extracted from the repositories of Government of India available on www.data.gov.in. Infant mortality rate was analyzed using decision tree technique in R language.

Decision tree is a technique which is used to analyze a given dataset to predict the driving factors of a particular entry [2]. It is the most appropriate way to analyze a dataset as it helps us to determine the factors which are not necessary for the dataset and play an irrelevant role so that they can be removed [10]. Decision tree is arranged in a format from top to bottom with attributes with influential attributes on top and their outcomes on the bottom [6].

R language is a statistical analysis language used to analyze a dataset and predict its behavior [9]. It is basically used to find the dependencies of various attributes on a single entity and how they work accordingly to drive the behavior of the entity.

Different factors affecting the infant mortality rate are literacy rate, birth rate, population, number of doctors, and nurses in that area and number of various health centers in that particular area [5].

India/States/Union Territories	Mortality Rate	Birth Rate	Literacy Rate	Sub Centres	Primary Health Centres	Community Health Centres	doctors	Total Population
Andhra Pradesh	HIGH	21	60.47	12522	1,570	167	2137	75728
Assam	HIGH	27	54.34	379	35	31	78	26638
Bihar	HIGH	31.2	63.25	5,109	610	100	1165	82878
Chhatisgarh	HIGH	26.5	47	3909	1,643	70	2168	20796
Gujarat	MEDIUM	25	81.94	4,692	513	118	628	50597
Haryana	HIGH	26.8	64.56	172	19	5	53	21083
Jharkhand	HIGH	26.2	57.63	7,274	1,073	273	848	26909
Karnataka	MEDIUM	22.2	78.18	2435	411	86	862	52734
Kerala	LOW	17.3	81.57	2071	443	71	467	31839
Madhya Pradesh	HIGH	31	82.01	1,888	374	80	643	60385
Maharashtra	MEDIUM	20.7	69.14	3,958	330	194	2013	96752
Odisha	HIGH	23.5	67.91	3,143	1,679	254	2041	36707
Punjab	MEDIUM	21.2	76.48	5,094	909	107	949	24288
Rajasthan	HIGH	31.1	55.52	3834	1,149	270	839	56473
Tamil Nadu	MEDIUM	19.1	53.56	10,453	1,300	407	3158	62111
Uttar Pradesh	HIGH	32.1	56.64	420	72	16	67	166053
W. Bengal	MEDIUM	20.6	90.56	396	103	26	123	80221
Arunachal Pradesh	MEDIUM	22.2	66.56	366	57	9	35	1091
Delhi	LOW	18.9	63.74	397	34	21	53	13783
Goa	LOW	14.2	76.58	5,927	1,279	231	1353	1344
Himachal Pradesh	MEDIUM	21.2	70.53	2858	434	126	373	6077

Fig. 1 Dataset of different states with their corresponding metrics

Literacy rate plays a vital role in driving the infant mortality rate as the number of educated people increases, they pay attention to their loved ones and their surroundings and try to make their living healthy [4].

Birth rate is an important driving factor of infant mortality rate as the more the birth rate, the more is the population, and hence there is a high probability of getting more people dying.

Due to lack of health centers, subcenters, and staff at those centers such as doctors and nurses, the infant mortality rate rises as the quality of health providing centers decreases.

The dataset consists of data from all the states of India with their corresponding recorded infant mortality rate, literacy rate, birth rate, number of health centers, subcenters, and doctors in that state or union territory (Fig. 1).

The R language code used to analyze the dataset is described below:

```
>output.forest3<-J48(factor(ISR_Final$'Infant Mortality Rate')~ISR_Final$
'Birth Rate'+ISR_Final $ 'Literacy Rate'+ISR_Final$'SubCentres' + ISR_Final $
'Primary Health Centres' + ISR_Final $ 'Community Health Centres' + ISR_Final
$ doctors,data = ISR_Final2)
```

The dataset is analyzed using J48 decision tree technique and stored in a variable named as output.forest3.

```
> plot(output.forest3)
```

This is used to plot the output.forest3 on the decision tree and to find out the desired result.

It consists of data of the year 2001 of all the states of India and depicts the factors on which infant mortality rate depends upon (Fig. 2).

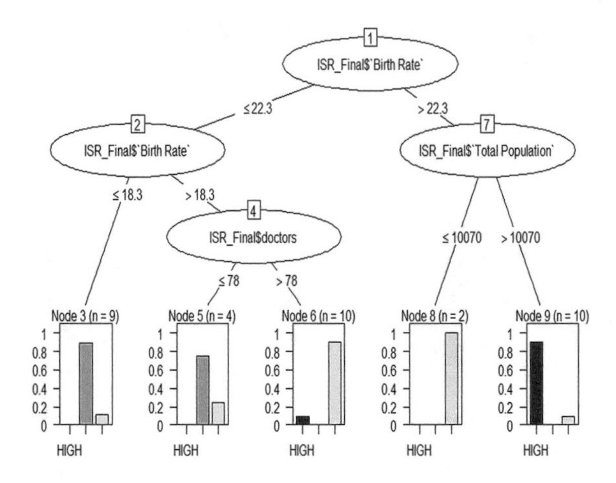

Fig. 2 Decision tree of factors affecting infant mortality rate acc. to 2001 census

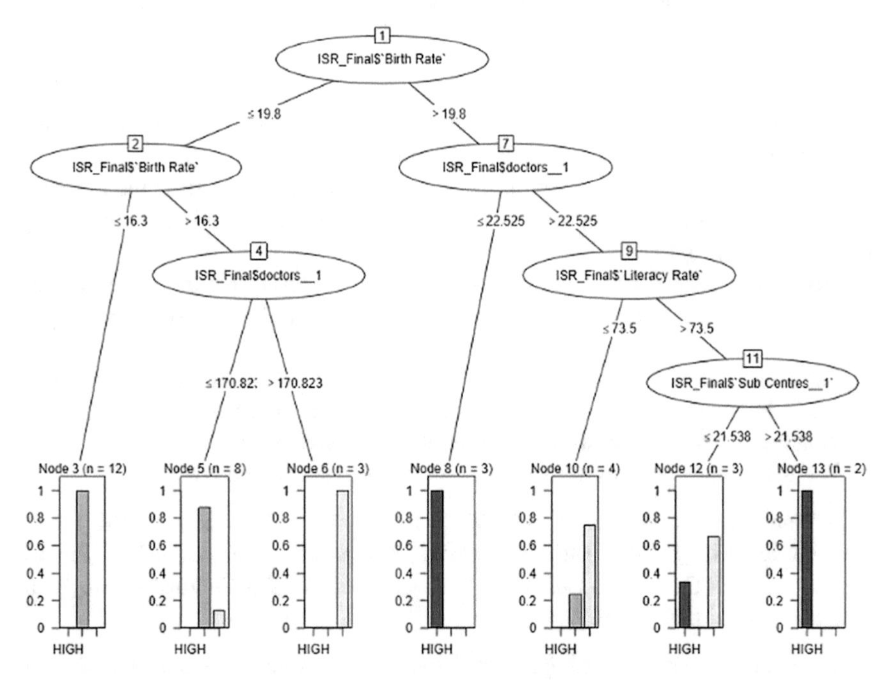

Fig. 3 Decision tree of factors affecting infant mortality rate acc. to 2011 census

It consists of data of the year 2011 of all the states of India and depicts the factors influencing infant mortality rate (Fig. 3).

The above dataset consists of a dataset of two decades so as to compare the changes that have occurred on the influencing factors and up to what extent.

```
=== Summary ===

Correctly Classified Instances          31               88.5714 %
Incorrectly Classified Instances         4               11.4286 %
Kappa statistic                          0.828
Mean absolute error                      0.131
Root mean squared error                  0.2559
Relative absolute error                 29.7671 %
Root relative squared error             54.5805 %
Total Number of Instances               35

=== Confusion Matrix ===

  a  b  c   <-- classified as
  9  0  1 |  a = HIGH
  0 11  0 |  b = LOW
  1  2 11 |  c = MEDIUM
```

Fig. 4 Summary of performance of decision tree of the year 2001

```
=== Summary ===

Correctly Classified Instances          32               91.4286 %
Incorrectly Classified Instances         3                8.5714 %
Kappa statistic                          0.8511
Mean absolute error                      0.0873
Root mean squared error                  0.2089
Relative absolute error                 22.3865 %
Root relative squared error             47.5774 %
Total Number of Instances               35

=== Confusion Matrix ===

  a  b  c   <-- classified as
  5  0  1 |  a = HIGH
  0 19  1 |  b = LOW
  0  1  8 |  c = MEDIUM
```

Fig. 5 Summary of the performance of decision tree of the year 2011

J48 decision tree is a technique used to create decision trees as it was more appropriate and gave us more accurate and reasonable results.

On analyzing the decision trees to judge their accuracy, we found out that the decision tree of the analysis of data of the year 2001 was 88.57% accurate with kappa factor nearly 0.83 and mean absolute error 0.131 which clearly indicates that the factors chosen actually are responsible for driving the infant mortality rate (Fig. 4).

On analyzing the decision tree of the analysis of data of the year 2011, there was 91.43% accurate with kappa factor nearly 0.85 and mean absolute error 0.087 which clearly indicates that the factors chosen actually are responsible for driving the infant mortality rate (Fig. 5).

4 Outcome

This study helped us to get to know the actual factors affecting the infant mortality rate and to what extent. By this, we can also categorize among the different factors, which play a more influential role.

In the experiment performed, we came to a know that how infant mortality rate is driven by the various factors.

According to 2001 census analysis, we came to know that factors such as birth rate, literacy rate, and a number of community centers affect the infant mortality rate the most, whereas all the other factors play a non-affecting role. Through analysis, we came to know that if birth rate is >22.3 and population is more than 10 million, infant mortality rate is high, whereas when it is less than 22.3 and number of community centers is >71, infant mortality rate is low but if the number of community centers is <71, infant mortality rate is generally in medium category.

According to 2011 census analysis, we came to know that factors such as birth rate, literacy rate, number of subcenters, and number of doctors affect the infant mortality rate the most, whereas all the other factors play a non-affecting role. Through analysis we came to know that if birth rate is >19.8 and number of doctors is >22.525 and literacy rate is >73.5 and number of subcenters is >21.538, then the infant mortality rate is high, whereas if birth rate is <19.8 but >16.3 and number of doctors is >170.823, the infant mortality rate is low. If birth rate is >19.8 and number of doctors is <22.525, the infant mortality rate is high. In rest of the conditions, the infant mortality rate is generally medium.

On comparison of the two decades, we can easily tell that due to reduction in birth rate, increase in literacy rate and number of health centers has led to a drastic change in infant mortality rate and finally led to decrease of the infant mortality rate. As according to 2001 census analysis, the infant mortality rate was higher for maximum part of population and the 2011 census analysis depicts the decrease of the infant mortality rate to the medium level.

This boxplot indicates the infant mortality rate w.r.t population as of census 2001 (Fig. 6), which clearly indicates that the infant mortality is generally higher and also

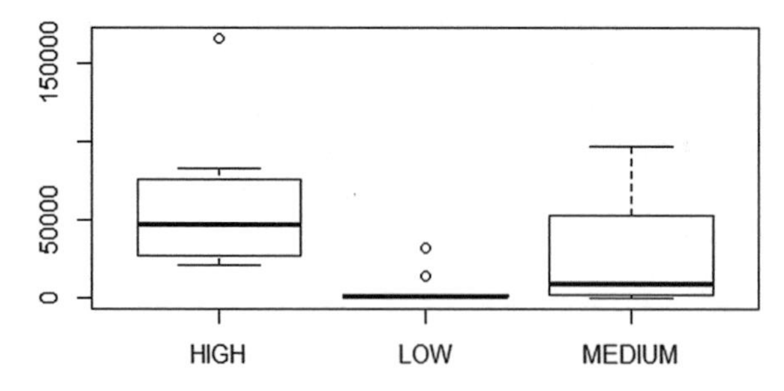

Fig. 6 Boxplot representing relation between infant mortality rate and population acc. to 2001 census

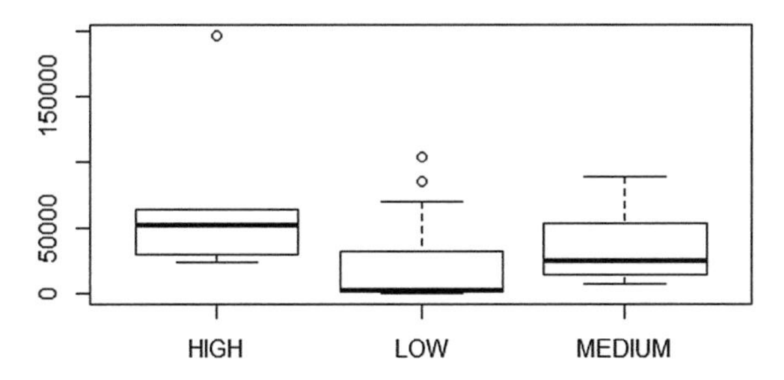

Fig. 7 Boxplot representing relation between infant mortality rate and population acc. to 2011 census

medium, whereas in Fig. 7, i.e., boxplot of infant mortality rate w.r.t population in 2011 indicates that the infant mortality is generally medium and there is an incremental change in the states with low infant mortality rate as compared to 2001. This indicates the role of population in affecting infant mortality rate.

5 Conclusion

From the following study and research, we can conclude that among the various factors affecting infant mortality rate, birth rate tops the chart, giving a clear signal that population is the major force driving the infant mortality rate. Also, lack of health centers and doctor results in higher infant mortality rate. It also made us aware of the factors which do not affect infant mortality up to an extent and can be neglected.

It helped us realize the key factors affecting the infant mortality rate and hence which are the sectors we need to work to decrease the infant mortality rate and convert our society into a healthy and fit society. We mainly need to focus on birth rate which is increasing at an alarming rate. We also need to open more and more health centers and subcenters, so as to provide a better quality of treatment to the patients which will directly affect infant mortality rate. It also indicates that if there is an increase in the number of doctors, the definitely the infant mortality rate will come down.

This study will also help NITI Aayog (Planning Commission, Government of India) to determine the field they need to work on and up to what extend while creating plans for different projects.

References

1. Izydorczyk, B., Wojciechowski, B.: Differential diagnosis of eating disorders with the use of classification trees (decision algorithm). Arch. Psychiatry Psychother. **4**, 53–62 (2016)
2. Song, Y.Y., Ying, L.U.: Decision tree methods: applications for classification and prediction. Shanghai Arch. psychiatry **27**(2), 130 (2015)

3. Wiharto, W., Kusnanto, H., Herianto, H.: Intelligence system for diagnosis level of coronary heart disease with K-star algorithm. Healthc. Inform. Res. **22**(1), 30–38 (2016)
4. Evangelista, L.S., Ghasemzadeh, H., Lee, J.A., Fallahzadeh, R., Sarrafzadeh, M., Moser, D.K.: Predicting adherence to use of remote health monitoring systems in a cohort of patients with chronic heart failure. Technol. Health Care **25**(3), 425–433 (2017)
5. Zhang, Z.: Decision tree modeling using R. Ann. Transl. Med. **4**(15) (2016)
6. Guo, L., Wang, M.C.: Data Mining Techniques for Infant Mortality at Advanced Age (2007)
7. Wang, G., Xu, Y., Duan, Q., Zhang, M., Xu, B.: Prediction model of glutamic acid production of data mining based on R language. In: Control and Decision Conference (CCDC), 2017 29th Chinese, pp. 6806–6810. IEEE (2017)
8. Datla, M.V.: Bench marking of classification algorithms: decision trees and random forests-a case study using R. In: 2015 International Conference on Trends in Automation, Communications and Computing Technology (I-TACT-15), vol. 1, pp. 1–7. IEEE (2015)
9. R Core Team (2014). R: A language and environment for statistical computing. R Foundation for Statistical Computing, Vienna, Austria. http://www.R-project.org/
10. Ghosh, S., Shukla, S., Mehrotra, D.: Application of decision tree for understanding Indian educational scenario. In: International Conference on Electrical, Electronics, and Optimization Techniques (ICEEOT), DMI College of Engineering, Chennai, pp. 4367–4372, 3–5 Mar 2016. IEEE Xplore. ISBN: 978-1-4673-9940-1

Content-Based Retrieval of Bio-images Using Pivot-Based Indexing

Meenakshi Srivastava, S. K. Singh and S. Q. Abbas

Abstract Similarity of images are measured according to their semantic contents, for instance, the searching for an image is based on its visual characteristics as well as other contents. The retrieval methods which are based only on keywords or metadata for the multimedia in fact cannot assure that the end user will get desired result. In text-based retrieval, the multimedia object's features are defined and searched in keywords. In the case of bio-mage, for example, protein structures, the information is not of textual nature and requires structural or sequential comparison of atoms present in the images, and hence presents new challenges in the designing of algorithms for handling these data. Currently available tools for searching similar protein structure face the drawback of online structure/sequence similarity check which is very time-consuming. In the present manuscript, a pivot-based efficient index algorithm has been presented which enables fast retrieval of similar protein structures. The proposed algorithm is implemented on a database created on Protein Data Bank files. The similarity on protein structures is calculated by computing pivots on structural similarity and sequential similarity. Finally, the query is performed on pivot created on combined features.

Keywords LAESA · Pivot-based indexing · Content-based retrieval
Protein structures

M. Srivastava (✉) · S. K. Singh
Amity Institute of Information Technology, Amity University, Noida,
Uttar Pradesh, India
e-mail: msrivastava@lko.amity.edu

S. K. Singh
e-mail: sksingh1@amity.edu

S. Q. Abbas
Computer Science Department, Ambalika Institute of Management and Technology,
Lucknow, Uttar Pradesh, India
e-mail: sqabbas@yahoomail.com

K. Ray et al. (eds.), *Soft Computing: Theories and Applications*,
Advances in Intelligent Systems and Computing 742,
https://doi.org/10.1007/978-981-13-0589-4_60

1 Introduction

Advances in research focus methodologies to choose the structures of bio-particles have prompted a huge increase in the sizes of the protein structure databases, for instance, Protein Data Bank (PDB) [1]. In 1992, only 1,000 structures were stored in PDB, whereas in 2002 the number of structures was over 18,000 and in 2017 there are more than 103,514 structures in the PDB [2]. The existing methods of similar protein retrieval from the structural databases are penalized due to lack of fast searching algorithms [3, 4]. Commonly available platforms are based on structural alignment, which will not be a preferred choice for searching structurally similar protein in huge size databases, because structure alignment algorithms are computationally expensive [5]. Tools and web servers such as clustal series, T coffee, BLAST (Basic Local Alignment Search Tool), FASTA, HMMER, etc. are good at the sequence alignment, whereas tools such as MAMMOTH, Dali Lite, CE (Combinatorial extension), etc. are used frequently by the scientist for structural alignment [6]. Now if a researcher is interested in comparing the structural and sequential similarity of any protein, the most common method used by them is to transfer data from one tool's site to another for later analysis. As a solution to such limitation in [7], authors have proposed a model AMIPRO for content-based retrieval of protein structures. AMIPRO allows searching visually similar protein structure, by applying intelligent vision algorithm [8]. Extended work of AMIPRO has been presented in [9], in which along with visual similarity sequential similarity is also computed on protein structure and the indexing is performed on combined features using multi-minimum product spanning tree [9, 10]. The advantage of MMPST is that structural and sequential combined queries can be performed easily. The major drawback of multi-minimum product spanning tree based index structure is that with the entry of a new element in the database the computation of complete minimum spanning tree index has to be done. In the present manuscript, **L**inear **A**pproximating and **E**liminating **S**earch **A**lgorithm (LAESA) [11, 12] is used which is a pivot-based index structure. The manuscript is organized into four sections; in Sect. 2, overview of AMIPRO is given; Sect. 3 represents the introduction of LAESA, and cluster-based implementation of LAESA. In Sect. 4, implementation and in Sect. 5 result analysis and performance of LAESA is discussed. Finally, conclusion is represented in Sect. 6.

2 Overview of AMIPRO

AMIPRO is implemented as query-by-content search system which involves interfacing among its five main modules, namely, query module, feature extraction module, the database module, the search engine module, and the visual interface which communicates with the user. A diagrammatic representation of the flow of procedures among these components is shown in Fig. 1.

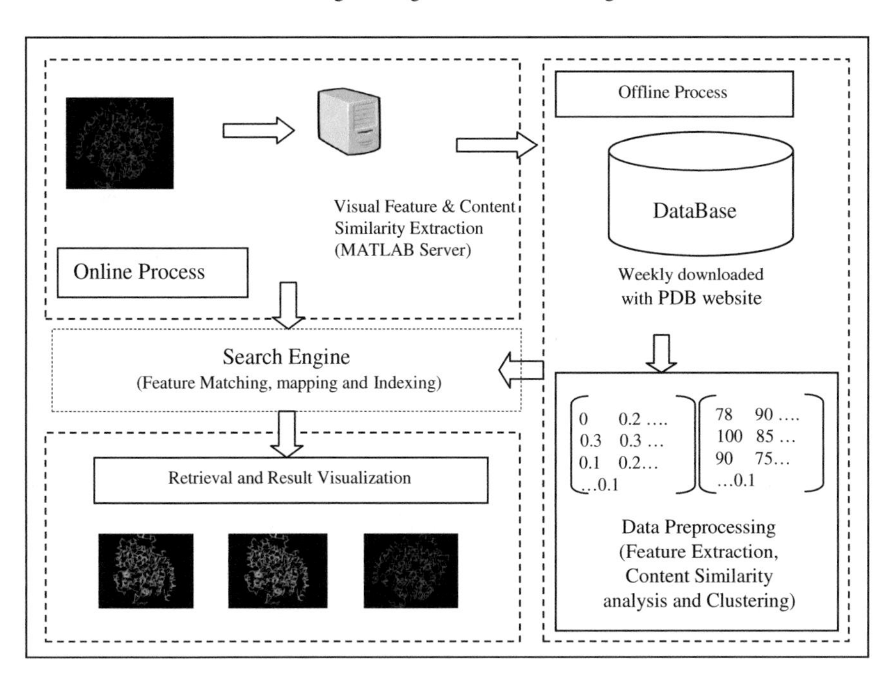

Fig. 1 The system architecture of AMIPRO [11], containing the five blocks: query image interface block, visual feature extraction block, database block, search engine block, image retrieval, and visualization block

High-order local autocorrelation features have been extracted from the protein images [6, 13]. Database module is updated weekly by RCSB PDB. 200 proteins were randomly collected and were synthesized using MATLAB Bioinformatics toolbox. Jmol [14] was used to acquire 6000 images at 128 × 128 pixels for each protein [15]. Finally, the FCM-based clustering is applied to the datasets [16]. The retrieval time is reduced by comparing the query with all the cluster centers rather than comparison with all the images in the database. Distance between the query image and the cluster center is calculated, and the clusters are ranked according to their similarity with the query image. Finally, the two clusters from the top of this ranked list are selected and the query image is directly compared with the images in these clusters. Top 3 similar images are shown in decreasing order of their similarity. Euclidian distance has been taken as the measure for calculating the similarity between the query image and center of cluster. Sequential alignment is performed using bioinformatics toolbox of MATLAB [17].

3 Introduction to Pivot-Based Indexing

Search methods can be classified into two types—pivot-based and clustering-based techniques [18]. Pivot-based search techniques choose a subset of the objects in the collection that are used as pivots. The index is built by computing the distances from each pivot to each object in the database [11, 19, 20].

3.1 Proposed Cluster Center Based LAESA Algorithm

The searching algorithm: We have classified the database of protein structure into four cluster; our proposed implementation of LAESA at first step computes the distance between the query point and the center of each cluster d(q, c), and then the clusters which have the minimum distance from the query point are selected as base property for NN search.

Improved Algorithm Basic Similarity BS-Selection: Linear AESA

Entry elements: $Se \subseteq E, n = |Se|$; {finite set of similarity values}

$BS \subseteq Se, m = |BS|$; {set of Base Similarity values}

$BS \subseteq Se,\ m = |BS|; ED€R^{nxm}$, {computed (pre) n × m array of inter prototype Euclidian distances}

$X€E$; {test sample}

Result obtained: $s * € S;\ d * €$ R; {nearest neighbor prototype and its Euclidian distance to x}

Function: ed: $E\,X\,E \rightarrow$ R; {Euclidian distance function}

Condition: Boolean; {controls the elimination of Base Similarity values}

choice: BS X (Se – BS) $\rightarrow$ Se; {selection of Base Similarity values}

Key role players/Variables: $t,\ q,\ s,\ b€Se$;

$G€R^n$; {lower bounds array, keeping track of greatest lower bound}

$dxs,\ gs,\ gq,\ gb € R$

$nc€N$; {number of Euclidian distances computed}

Begin

$$d^* := \infty; s^* := indeterminate; G := [0];$$

$$s := arbitrary\, element(BS); nc := 0;$$

while $|Se| > 0$ **do**

$\{dxs := d(x, s); S := S - \{s\}; nc := nc + 1;$ {Euclidian distance computing}

$if\ dxs < d^* then\ s^* := s; d^* := dxs; endif$ {updatings*, d^*}

q: = $indeterminate;\ gq\ := \infty\ b$: = indeterminate; gb := ∞;

For every $s € S$ **do** {eliminating and approximating loop}

If $s € BS$ **then** {updating G, if possible}

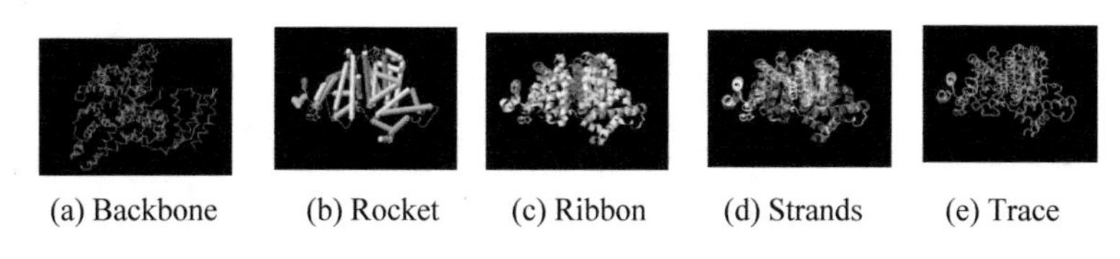

(a) Backbone (b) Rocket (c) Ribbon (d) Strands (e) Trace

Fig. 2 Various visualizations used in experiments (PDB ID-5GHK)

G[s]: = max (G[s], |E*D[s, t*] − dxs|)
End if
Gs: = G[s];
If s € BS **then**
if ($gs \geq d$ * & **Condition) then** *Se* :=*Se* − *{s}* {eliminating from BS} **else** {approximating: selecting from *B*)

  if gs < gb **then** gb:= gs; b := s **endif**
  endif
  else
  if $gs \geq d$ * **then** *S* :=*S*—*{s}* {eliminating from $S - B$}
  else (approximating: selecting from Se − BS)
    if gs < gq **then** gq := gs; q := s;
    endif

endif
endif
end of for loop.

  S1 : = choice(b, q);

end of while loop
end

4 Implementation

Prototype of the model is implemented in MATLAB [17]. Oracle Database Server version 12.1.0.2.0 and MATLAB R2015a have been used.

Protein images of four classes defined by SCOP database are Alpha (α), Beta (β), Alpha/Beta (α/β), and Alpha + Beta (α/β) [21]. Samples of protein images used in experiments are shown in Fig. 2.

Table 1 Comparison of queries on MPST and inverted index

Query	Description	Result MMPST	Result LAESA	Time in ms MMPST	Time in ms LAESA
Q3	Structural similarity = 70%	9	9	48.2	18.67
	Sequential Similarity 80%				
Q20	Structural similarity = 80%	8	8	38.3	16.23
	Sequential similarity = 90%				
Q25	Structural similarity = 90%	7	7	35.4	15.24
	Sequential similarity = 80%				

5 Results and Discussions

To check the performance metric space model on the real dataset [15] collected from RCSB PDB, a series of experiments had been carried out. Our dataset [15] is classified into four classes, so to reduce the search time, instead of searching in whole database, the distance of query image with the cluster centers is measured and the query object is searched into the clusters for which the calculated distance was measured minimum. The object which has the most extreme separation with the group focus has been picked as the possibility for base similarity. Two LAESA metric structures, one for visual comparability distance (ED) and second for content-based likeness (CD), are made. A combined index structure is additionally produced by performing element-based multiplication of ED and CD.

5.1 Query Set and Results

In set of experiments, queries like "How many PDB entries are 70% structurally similar and 90% sequentially similar?" have been executed. Performance of cluster-based implementation of LAESA is compared with MMPST index algorithm. Comparison of query time taken by cluster-based LAESA and MMPST is given in Table 1. The comparison shows clearly that retrieval time taken by LAESA index is lesser than the time taken by MMPST.

5.2 *Performance Analysis of Linear AESA*

1. LAESA [15] stores a metric of separations between database objects. Separation between all the protest is figured at the season of formation of LAESA. The structure of the LAESA framework is n × n; however, just (n − 1)/2 separations are put away, i.e., half of the lattice beneath the slanting, in light of the fact that registered separation grid fulfills the framework property and the components above and underneath the corner to corner are same.
2. For looking through a scope of question R (q, r), our usage of LAESA gets a protest, for instance, I1 (Base Similarity) which has the greatest separation from the bunch focus, the correct Euclidian separation among I1 and Q1 is figured suppose D. Presently, this separation will be utilized for pruning objects. Pruning of object I is done if the lower bound is greater than the query that is $|d(I, D) - d(q, p)| > r$.
3. The following turn is picked among all non-disposed of articles as of recently.
4. The methodology is stopped when a little game plan of non-discarded things is cleared out. Lastly, the separation of left questions is contrasted and q, and items with $d(q, D) \leq r$ are chosen.

6 Conclusion

In the present work, a novel perspective and solution to the problem of similarity mapping for bio-images (protein structures) has been presented. The visual- and content-based features of protein structures have been represented in different pivots and by creating a combined index, hence, the content-based queries can be executed in time-efficient manner. Performance of cluster-oriented LAESA and MMPST on protein image has been measured. In comparison of LAESA and MMPST, performance of LAESA was better than MMPST. Metric space-based representation of data involves pre-computation of distance between the object, so the basic drawback of existing tools which requires online similarity computation was easily removed. Our performance evaluation on both real-world and synthetic datasets with various query workloads showed significant performance gains and negligible space overhead.

References

1. Berman, H.M., Westbrook, J., Feng, Z., Gilliland, G., Bhat, T.N., Weissig, H., Shindyalov, I.N., Bourne, P.E.: The protein data bank. Nucleic Acids Res. **28**, 235–242 (2000)
2. Srivastava, M., Singh, S.K., Abbas, S.Q.: Web archiving: past present and future of evolving multimedia legacy. Int. Adv. Res. J. Sci. Eng. Technol. **3**(3)
3. Yasmin, M., Mohsin, S., Sharif, M.: Intelligent image retrieval techniques: a survey. J. Appl. Res. Technol. **14**(December 16)

4. Kim, S., Postech, Y.K.: Fast protein 3D surface search, ICUIMC (IMCOM)' 13, Jan 17–19, 2013, Kota Kinabalu, Malaysia Copyright 2013. ACM. 978-1-4503-1958-4
5. Wiwie, C., Baumbach, J., Röttger, R.: Comparing the performance of biomedical clustering methods. NCBI, Nat Methods. **12**(11), 1033–1038 (2015). https://doi.org/10.1038/nmeth.3583
6. Alberts, B., Johnson, A., Lewis, J., Raff, M., Roberts, K., Walter, P.: The Shape and Structure of Proteins (2002)
7. Srivastava, M., Singh, S.K., Abbas, S.Q., Neelabh: AMIPRO: A content-based search engine for fast and efficient retrieval of 3D protein structures. In: Somani, A., Srivastava, S., Mundra, A., Rawat, S. (eds.) Proceedings of First International Conference on Smart System, Innovations and Computing. Smart Innovation, Systems and Technologies, vol 79. Springer, Singapore (2018)
8. Nobuyuki, O.T.S.U., Kurita, T.: A new scheme for practical flexible and intelligent vision systems. In: lAPR Workshop on CV -Special Hardware and Industrial Applications, Oct 12–14, 1988, Tokyo
9. Srivastava, M., Singh, S.K., Abbas, S.Q.: Multi minimum product spanning tree based indexing approach for content based retrieval of bioimages. In: Communications in Computer and Information Science. Springer (2018)
10. Chakrabarti, K., Mehrotra, S.: The hybrid tree: an index structure for high dimensional feature spaces. In: Proceedings of the 15th International Conference on Data Engineering, 23–26 Mar 1999, Sydney, Australia, pp. 440–447. IEEE Computer Society (1999)
11. Ch´avez, E., Navarro, G., Baeza-Yates, R., Marroq, J.L.: Searching in metric spaces. ACM Comput. Surv. **33**(3), 273–321 (2001)
12. Micó, M.L. Oncina, J.: A new version of the nearest-neighbour approximating and eliminating search algorithm (AESA) with linear preprocessing time and memory requirements. Pattern Recogn. Lett. (1994). Revised 2011
13. Suzuki, M.T.: Texture. Image Classification using Extended 2D HLAC Features. KEER2014, LINKÖPING|JUNE 11–13 2014 International Conference on Kansai Engineering and Emotion Research Texture
14. Herr´aez, A.: Biomolecule in the computer: Jmol to the rescue. Biochem. Mol. Biol. Educ. **34**(4), 255–261 (2006)
15. Tetsuo Shibuya, Hisashi Kashima, Jun.: Pattern Recognition in Bio-Informatics, 7th IAPR International Conference, PRIB 2012 Proceedings
16. Nascimento, S., Mirkin, B., Moura-Pires, F.: A fuzzy clustering model of data and fuzzy c-means. In: 9th IEEE International Conference on Fuzzy Systems. FUZZ-IEEE (2000)
17. MATLAB and Statistics Toolbox Release: The Math Works Inc. Natick, Massachusetts, United States (2015)
18. Rajpurohit, J., Kumar Sharma, Tarun, Abraham, Ajith, Vaishali: Glossary of metaheuristic algorithms. Int. J. Comput. Inf. Syst. Ind. Manag. Appl. **9**, 181–205 (2017)
19. Liu, X., Ma, X., Wang, J., Wang, H., M3L.: Multi-modality mining for metric learning in person re-identification. Pattern Recogn. (2017)
20. Srivastava, M., Singh, S.K., Abbas, S.Q.: Applying metric space and pivot-based indexing on combined features of bio-images for fast execution of composite queries. Int. J. Eng. Technol. **7**(1), 110–114 (2018). https://doi.org/10.14419/ijet.v7i1.9009
21. Andreeva, A., Howorth, D., Chandonia, J.M., Brenner, S.E., Hubbard, T.J.P., Chothia, C., Murzin, A.G.: Data growth and its impact on the SCOP database: new developments. Nucleic Acids Res. **36**, D419–D425 (2007)

Road Crash Prediction Model for Medium Size Indian Cities

Siddhartha Rokade and Rakesh Kumar

Abstract Road crashes are a human tragedy, which involve immense human suffering. Road accidents have huge socioeconomic impact, especially in developing nations like India. Indian cities are expanding rapidly, causing rapid increment in the vehicle population leading to enhanced risk of fatalities. There is an urgent need to reduce number of road crashes by identifying the parameters affecting crashes in a road network. This paper describes a multiple regression model approach that can be applied to crash data to predict vehicle crashes. In this paper, crash prediction models were developed on the basis of accident data observed during a 5-year monitoring period extending between 2011 and 2015 in Bhopal which is a medium size city and capital of the state of Madhya Pradesh, India. The model developed in this paper appears to be useful for many applications such as the detection of critical factors, the estimation of accident reduction due to infrastructure and geometric improvement, and prediction of accident counts when comparing different design options.

Keywords Accident prevention · Crash prediction model · Road geometry
Multiple regression

1 Introduction

Road crashes are a human tragedy, which involve high human suffering. These road crashes impose a huge socioeconomic cost in terms of injuries. Road safety is one of the prime concerns in India. Although India has less than 1% of the world's vehicles, the country accounts for 6% of total crashes across the world and 10% of total road fatalities. Hence, the risk of fatalities also increases. The most vulnerable road users are pedestrians, cyclists, and motorcyclists [1]. There is an urgent need

S. Rokade (✉) · R. Kumar
Maulana Azad National Institute of Technology, Bhopal, India
e-mail: siddhartharokade@gmail.com

R. Kumar
e-mail: rakesh20777@gmail.com

K. Ray et al. (eds.), *Soft Computing: Theories and Applications*,
Advances in Intelligent Systems and Computing 742,
https://doi.org/10.1007/978-981-13-0589-4_61

to reduce number of road crashes by identifying the parameters affecting crashes in a road network. Dhamaniya [2] found that in mix traffic conditions, human error is the sole cause of accidents in 57% of all accidents and is a contributing factor in more than 90%. Ramadan et al. [3] studied traffic accidents at 28 hazardous locations Amman–Jordan urban roads. The following variables were found to be the most significant contributors to traffic accidents at hazardous locations: posted speed, average running speed maximum and average degree of horizontal curves, median width, number of vertical curves, type of road surface, lighting (day or night), number of vehicles per hour, number of pedestrian crossing facilities, and percentage of trucks. From the macroscopical point of view, stable and free traffic flow contributes to high level of safety, while the turbulence of traffic flow leads to more accidents [4]. Therefore, the public agencies may be interested in identifying those factors (traffic volume, road geometric, etc.) that influence accident occurrence and severity to improve roadway design and provide a safe environment [5]. For this, modeling has attracted considerable research interest in the past because of its multiple applications. Crash models are developed by statistically assessing, how variation in the numbers of crashes occurring is explained by a range of measured variables and factors, generally using advanced regression techniques. The purpose of crash modeling is to identify factors which significantly influence the number of accidents and to estimate the magnitude of the effects [6]. The data required for this tool is highly dependent on the specific crash model being developed. These models also have several applications such as estimation of the safety potential of road entities, identification and ranking of hazardous or crash-prone locations, evaluation of the effectiveness of safety improvement measures, and safety planning [7]. This paper describes a multiple regression model approach that can be applied to crash data to predict vehicle crashes. In this paper, crash prediction models were developed on the basis of accident data observed during a 5-year monitoring period extending between 2011 and 2015 in Bhopal which is a medium size city and capital of the state of Madhya Pradesh, India.

2 Data Collection and Tools

A 5-year (2011–2015) accident data was collected in the form of FIRs from police department of the Bhopal city. The data was digitized in four categories: time of accident, accident details, vehicle details, and victim details include gender and age of victims.

Selected road segments in Bhopal are located on link road 1, 2, 3, MANIT to Roshanpura Road, Bhadhbhadha Road, New Market Area, Kolar Road, and Ratibad Road. The Highway Safety Manual recommends using 30–50 sites for calibrating predictive models [8]. A total of 77 segments were selected for development of accident prediction model.

Table 1 Description of selected factors

S. no.	Factors/variable	Unit	Symbol	Description
1	Segment length	Meter	SL	Lengths of the road segment between two crossings
2	Traffic volume	PCU/hour	TV	Peak hour (decided by opinion of local residential and traffic expert) traffic volume of the segment
3	Number of lanes	Number	NOL	Number of moving traffic lanes in the segment
4	Carriageway width	Meter	CW	Width of road on which vehicles are not restricted by any physical barrier to move laterally
5	Median width	Meter	MW	Width of the median (or 0 if median not available)
6	Number of bus stops	Number	NOBS	Total number of bus stops in the segment both ways
7	Vehicular crash	Number	VA	Crash/Accident of vehicle on the selected segments

Geometric design characteristic including segment length, carriageway width, median, number of bus stop, and number of lane are prime concern to change the accident rate. Data of these geometry elements was directly collected from the field on selected road segments. The approach to geometric data collection consisted of dividing each segment as per the accident between consecutive intersections (resulting in a total of 77 segments) and gathering field information for each segment separately. Peak hour of each segment is decided on the basis of expert opinion and road user opinion. Table 1 gives a brief description of factors considered in this research. Data collected for selected factors is shown in Table 2.

3 Analysis and Results

Table 3 shows the factor analysis results for all the selected four selected areas. Table 4 shows the correlation value of independent variable with dependent variable. It shows that number of lane (0.728), carriageway width (0.640), median width (0.646), and traffic volume (0.723) has highest correlation with accidents. Table 5 shows the correlation matrix of selected independent variables.

Table 2 Road geometry and traffic volume data

S. no.	Name of road segment	SL	NOL	CW	MW	NOBS	VA	TV
1	In front of Indoor Stadium	200	2	14.7	0.42	0	19	3900
2	In front of Shoe Market	120	2	16.3	0.6	0	5	4020
3	In front of SBI TT Nagar	180	2	13.5	0.7	0	7	4100
4	Apex Bank to Shoe Market	200	2	11.5	0.6	0	18	4160
5	Rangmahal to Palash Hotel	300	2	11	0	1	8	4250
6	Palash Hotel to Ram Mandir	400	2	12.7	4.3	1	10	3750
7	Roshanpura to Rangmahal	300	4	16.2	0.2	1	6	4560
8	Rangmahal to Aapurti	200	4	13.7	0.25	1	4	4230
9	Aapurti to Union Bank	400	4	13.7	0	2	26	4160
10	Union Bank to Shiv Shakti temple	200	4	11.9	0	2	20	4112
11	Bharat Mata Square to MP Tourism	600	4	13.2	1.5	2	8	3800
12	MP Tourism to 23 Battalion	500	4	12.8	1.5	1	10	3760
13	23 Battalion to ABV Institute	500	4	13.2	1.5	2	16	3920
14	New market to Ankur School	300	4	16.2	0.55	1	16	2830
15	Ankur School to Saranngji Bungalow	400	4	17.3	0.55	0	22	2760
16	Nand Kumarji Bungalow to Sarangji Office	250	4	17.5	0.55	0	12	2700
17	Shiv Shakti to PNT Square	300	4	13.6	2	2	18	3706
18	PNT to Hajela Hospital	300	4	13.6	2	2	16	3612
19	Hajela Hospital To AK Hospital	400	4	13.2	2	1	5	3392
20	AK Hospital to Mata Mandir	300	4	13.7	2	1	3	3460
21	Mata Mandir to Dargah	400	4	13.7	3.2	1	28	3160

(continued)

Table 2 (continued)

S. no.	Name of road segment	SL	NOL	CW	MW	NOBS	VA	TV
22	Dargah to St. Mary School	300	4	13.8	3.2	0	16	3260
23	St. Mary to Water Distribution	150	4	13.6	3.2	0	20	3176
24	MANIT to Rivera Town	160	4	13.1	0.5	1	10	4130
25	Rivera Town to Mata Mandir	150	4	12.9	0.5	1	15	4235
26	Mata Mandir to Sagar Gaire	300	4	10.3	0	1	24	4160
27	Sagar Gaire to BMC Bus Stop	100	4	8.1	0	0	9	4200
28	BMC Bus Stop to Tin Shed	500	4	13.5	0.525	1	19	4360
29	Apex Bank to Roshanpura	600	4	15.7	1	2	22	4539
30	Kolar Junction to Dainik Amrat	600	4	13.5	2.1	2	5	3060
31	Dainik Amrat to National Hindi Mail	200	4	13	2	1	9	3120
32	National Hindi Mail to Union Bank	200	4	12.7	2.2	0	8	2960
33	Union Bank to Budh Vihar	100	4	12.9	2.2	1	5	2816
34	Budh Vihar to MANIT	400	4	12.6	2.1	0	8	3190
35	MANIT to Rivera Town gate	300	4	12.6	1.5	1	5	3210
36	Rivera Town gate to Top and Town	300	4	11	1.35	0	9	2920
37	Top and Town to Kamla Nagar Police Station	400	4	11.2	2.05	0	9	3096
38	Kolar police station to Nehru Nagar Square	400	4	10.2	2	1	6	3960
39	Nehru Nagar Square to Police petrol pump	500	4	11.2	2.1	1	7	3640
40	Police petrol pump to IIFM	300	4	13	1.85	2	13	3260
41	IIFM to Sher Sapata Square	800	4	11.9	1.5	2	12	3430
42	PNT Square to Harish Chowk	500	4	12.3	3.7	2	12	1930

(continued)

Table 2 (continued)

S. no.	Name of road segment	SL	NOL	CW	MW	NOBS	VA	TV
43	Harish Chowk to Ekta Market	300	4	11.8	3.3	0	16	1830
44	Ekta Market to Nehru Nagar Square	800	4	11.5	3	2	6	1890
45	Bhadhbhadha bridge to End of bridge	300	4	13.6	1.2	0	6	2750
46	Bridge to Bhadhbhadha toll naka	600	4	12.9	1.5	0	7	2650
47	Toll naka to Suraj Nagar bus stop	600	4	6.8	0	1	20	2841
48	Suraj Nagar bus stop to Bal Bharti School	600	4	6.5	0	1	14	2100
49	Bal Bharti School to Hotel Sakshi	500	4	6.7	0	1	13	2240
50	Hotel Sakshi to Sharda Vidya Mandir	800	4	6.5	0	0	13	2720
51	Sharda Vidya Mandir to Dabang Duniya	700	4	6	0	0	15	2935
52	Dabang Duniya to Bank of Baroda	600	4	6.2	0	1	13	2750
53	Bank of Baroda to Savitri petrol pump	800	4	5.8	0	1	15	2691
54	Savitri petrol pump to Bharat petrol pump	700	2	5.4	0	1	15	2530
55	Bharat petrol pump to Raj dhaba	1100	2	5.9	0	1	11	2493
56	Raj dhaba to Essar petrol pump	1000	2	6.3	0	0	12	2460
57	Essar petrol pump to Nilbad police chowki	1500	2	7	0	0	12	2560
58	Nilbad police chowki to Nanak petrol pump	1400	2	5.8	0	0	19	2690
59	Kolar Junction to Luv-Kush water supply	300	2	7	0	1	16	2616
60	Luv-Kush water supply to Sanyukta Suresh Vihar	200	4	10.5	0	1	15	2620
61	Sanyukta Suresh Vihar to Amrapali Enclave	400	4	12	0	1	13	2695

(continued)

Table 2 (continued)

S. no.	Name of road segment	SL	NOL	CW	MW	NOBS	VA	TV
62	Amrapalu Enclave to Janki Bungalow	200	4	12.7	0	1	12	2731
63	Janki bungalow to SS Tower	300	4	12.2	0.4	0	9	2749
64	SS Tower to Manoriya Hospital	400	4	12	0.4	0	8	2930
65	Manoriya Hospital to Globus city	300	4	12.3	0.4	0	14	3060
66	Globus city to Kalka Mata Mandir	200	4	11.2	0	0	17	3360
67	Police help center to Bhoj University	200	4	11.5	0	1	8	3160
68	Bhoj University to Manit Singh Maran Board	450	4	12	1.2	1	6	2916
69	Kolar police help center to Shahpura police square	200	4	12.8	0.3	1	8	2931
70	Shahpura police square to Bansal Hospital	300	4	10.9	0	2	7	3043
71	Bansal hospital to Sahayak Abhiyantriki Karyalaya	200	4	9.3	0	1	6	2950
72	Sahayak Abhiyantriki Karyalaya to Entry gate Rishabh Udhyan	100	4	9.1	0	1	5	2860
73	Entry gate Rishabh Udhyan to Shrimati Upkar Rajoriya Dwar	200	4	9.5	0	0	9	2840
74	Shrimati Upkar Rajoriya Dwar to Rajiv Gandhi Complex Manisha Market	100	4	9	0	2	4	2990
75	Kolar police help center to Parika Grahnirman Sanstha	200	4	5.7	0	1	9	1860
76	Parika Grahnirman Sanstha to Amarkunj House	200	4	6	0	1	6	1850
77	Amarkunj House to Kaliyasot Dam	300	4	5.5	0	1	12	1810

Table 3 Results of factor analysis

S. no.	Selected area	Determinant	KMO	Significance Value
Significant value		>0.001	0–1	<0.05
1	TT Nagar Police Station Area	0.084	0.461	0.001
2	Kamla Nagar Police Station Area	0.015	0.551	0.000
3	Chunna Bhatti Police Station Area	0.059	0.390	0.002
4	Ratibad Police Station Area	0.052	0.410	0.002

Table 4 Correlation value of independent (X) with dependent variable(Y)

S. no.	Independent variable	Independent variable I'D	Correlation test value
1	Segment length	X1	−0.086
2	Number of Lane	X2	−0.728**
3	Carriageway width	X3	−0.640*
4	Median width	X4	−0.646**
5	Number of bus stop	X5	−0.261*
6	Traffic volume	X6	0.723**

* indicates a moderate correlation is significant at the 0.05 level
** indicates a correlation is significant at the 0.01 level

Table 5 Correlation matrix of selected independent variables

	Number of Lane (X2)	Carriageway width (X3)	Median width (X4)	Traffic volume (X6)
Number of Lane (X2)	1			
Carriageway width (X3)	0.627**	1		
Median width (X4)	0.485**	0.454**	1	
Traffic volume (X6)	−0.527**	−0.370**	−0.435**	1

* indicates a moderate correlation is significant at the 0.05 level
** indicates a correlation is significant at the 0.01 level

4 Crash Prediction Model

Model is developed by statistically assessing how variation in the numbers of accidents occurring is explained by a range of measured variables and factors, generally using advanced regression techniques. The purpose of accident modeling is to identify factors which significantly influence the number of accidents and to estimate the magnitude of the effects [6].

To develop crash prediction model, linear regression analysis is used. If the relationship between the independent and the dependent variable is linear, then the analysis is known linear regression. If the independent variables are two or more in numbers, then the analysis is known as multiple linear regression analysis. Multiple linear regression analysis is the statistical technique most often used to derive the estimates of future accident generation, etc. [9].

From the above parameters, only three are significant for model development. The significant segments used in the analysis are carriageway width, median width, and traffic volume. Table 6 shows the result of the regression analysis. Totally, four models were used. The best model that could predict the accidents among these four models was then identified based on the criteria mentioned in above section. The model number 1 is the best model than other two models because it got the highest R, R^2 and adjusted R^2 value, low significant error and significant F-value is low.

As in this study number of accident is predicted in which human behavior is involved in many ways like perception of any situation, emotion at the time of taking decision and many more so to predict human behavior, such as psychology, typically R-squared values may be lower than 50% because humans are simply harder to predict [10]. In this study model, one is selected as best model and its R^2 value is also coming more than 50%. Model 1 with independent variables comprises the carriageway width (X3), median width (X4), and Traffic volume (X6) which are the best models that can predict accident. The accident prediction model for selected road segments using the multiple regression analysis is presented in Eq. 1 where Y = Vehicle accident.

$$\mathbf{Y} = 10.275 - 0.684 * \mathbf{X3} - 1.560 * \mathbf{X4} + 0.004 * \mathbf{X6} \tag{1}$$

5 Validation of the Model

5.1 ANOVA Test

Table 7 shows the output of the ANOVA analysis. Result of ANOVA test is f(3, 73) = 70.383 which is at the significance level 0.000 (P-value = 0.000). P-value is less than 0.005. It means that the regression equation is significant at 95% confidence limit in levels of variability within a regression model.

Table 6 Summary of outputs of regression analysis

Model no.	Predictor	R	R^2	Adjusted R^2	Standard error	Significant F-Value	P-Value of intercept and predictor			
1	X3, X4, X6	0.862[a]	0.74	0.733	3.175	0	0	0	0	0
2	X3, X4	0.754[a]	0.57	0.557	4.085	0	0	0	0	
3	X4, X6	0.812[a]	0.66	0.65	3.635	0	0	0	0	
4	X3, X6	0.827[a]	0.68	0.675	3.498	0	0	0	0	

[a] signifies the predictors (constant) X3, X4, X6

Table 7 Results of ANOVA test

Model	Sum of squares	df	Mean square	F	Sig.
Regression	2129.024	3	709.675	70.383	0.000
Residual	736.067	73	10.083		
Total	2865.091	76			

5.2 F Test

Usually, f > 4 is acceptable. The probability of the F-statistic (70.383) for the overall regression relationship is 0.000 which is less than the level of significance of 0.05. So we reject the null hypothesis at 95% confidence level that there is no relationship between the set of independent variables and the dependent variable. Significance F is less than 5%. We support the research hypothesis that there is a statistically significant relationship between the set of independent variables and the dependent variable.

6 Validation of the Model Assumptions

Most statistical tests depend upon certain assumptions for variables used in the analysis. The validation of regression assumptions is very important. The details related to the examination of assumptions of the model using regression analysis are given below.

6.1 Examining Linearity

In multiple regression analysis, the accurate relationship between variable can be estimated if the relationship between variables is linear. If the relationship between dependent and independent variables is not a linear relationship, the results of the regression analysis will underestimate the true relationship. Ratkowsky et al. [11] suggest three primary ways to detect nonlinearity. A preferable method of detection linearity is the examination of residual plots between dependent and independent. The first step of assessment of assumption begins with initial descriptive plots of dependent variable versus each of the explanatory variables, separately. The scatter plots in Fig. 1 prove the linearity assumption.

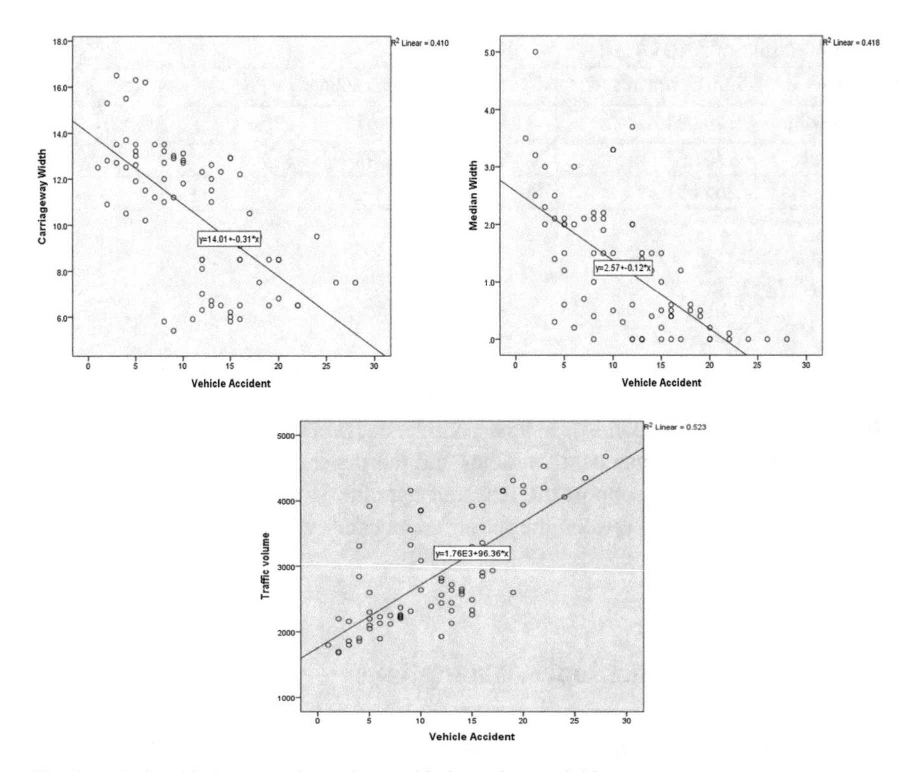

Fig. 1 Relationship between dependent and independent variables

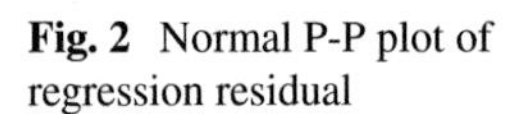

Fig. 2 Normal P-P plot of regression residual

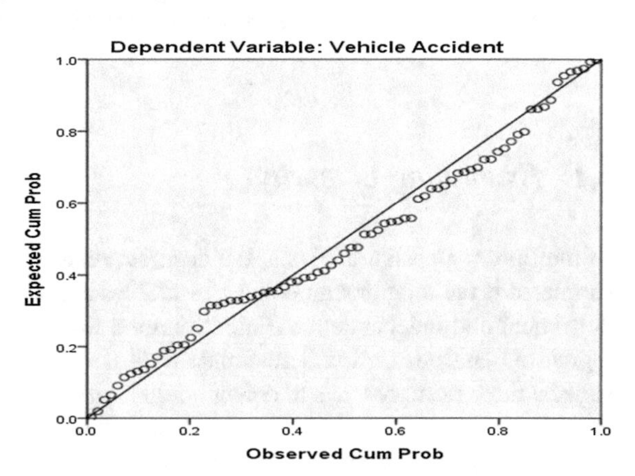

6.2 Examining Normality

Figure 2 presents normal P-P plot of regression residual. If the collected data are normally distributed, then the observed values should fall closely along the straight line. Thus, the normality assumption is proved by the P-P plot.

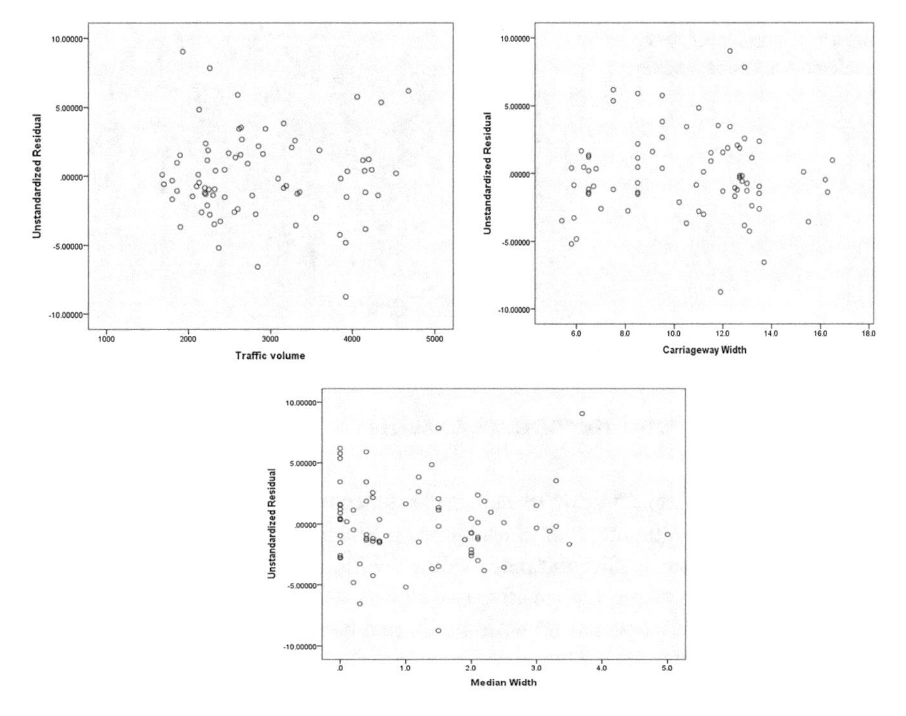

Fig. 3 Residual corresponding to the carriageway width, median width, and traffic volume

6.3 Examining Independence

The assumption of independence concerns another type of error in the residuals that is produced by estimated values of the dependent variable that are serially correlated. This is done by plotting the residuals versus each of the independent variables. In this plot, if the residuals are randomly distributed, the independence assumption is satisfied. Figure 3 presents relation between independent variables and residuals. The residuals are randomly distributed which implies that the independence assumption is satisfied.

6.4 Examining Homoscedasticity

At each level of the predictor variables, the variance of the residual terms should be constant. This just means that the residuals at each level of the predictors should have the same variance (homoscedasticity). If the residual variances are zero, it implies that the assumption of homoscedasticity is not violated. If there is a high concentration of residuals below zero or above zero, the variance is not constant and thus a systematic error exists (Fig. 4).

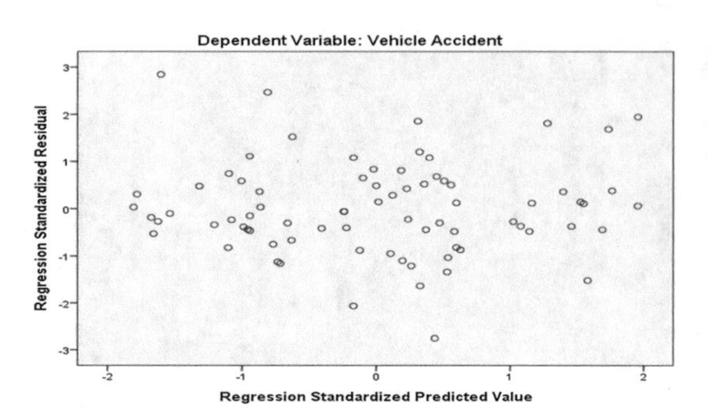

Fig. 4 Scatterplot between predicted and residual value

7 Conclusions and Recommendations

In this study, accident analysis of selected segments was carried out to find the locations with maximum number of accidents in Bhopal city. Main factors affecting vehicular crashes are traffic volume, median width, and carriageway width. The number of vehicle crashes has positive correlation with segment length and traffic volume, while it has negative correlation with number of lanes, carriageway width, median width, and number of bus stops. Result of the study shows that as the width of carriageway increases, the crash rate decreases. Also, the effect of median is negative because it separates the opposite moving traffic. Further, as width of median increases, glare due to headlight of opposite moving vehicle during night decreases. Hence, greater median width is desirable for the prevention of vehicular crashes.

The formulated regression model satisfies all the assumptions of multiple linear regression analysis indicating that all the information available is used. The crash prediction model developed can be used to predict the crashes depending upon the factors considered. Further, the model will be useful in efficient planning of new roads and modifying existing roads in order to make them safe for the users. The model developed in this paper appears to be useful for many applications such as the detection of critical factors, the estimation of accident reduction due to infrastructure and geometric improvement, and prediction of accident counts when comparing different design options.

References

1. WHO: Global status report on road safety 2015. http://www.who.int/violence_injury_prevention/road_safety_status/2015/en/
2. Dhamaniya, A.: Development of accident prediction model under mixed traffic conditions: a case study. In: Urban Public Transportation System, pp. 124–135 (2003)
3. Ramadan, T.M., Obaidat, M.T.: Traffic accidents at hazardous locations of urban roads. Jordan J. Civil Eng. **6**(4), 436–444 (2012)
4. Koornstra, M., Oppe, S.: Predictions of road safety in industrialized countries and Eastern Europe. In: International Conference Road Safety in Europe, pp. 1–22 (1992)

5. Golias, I., Karlaftis, G.: Effects of road geometry and traffic volumes on rural roadway accident rates. Accid. Anal. Prev. **34**, 357–365 (2002)
6. Elvik, R., Sorensen, S.: State-of-the-art Approaches to Road Accident Black Spot Management and Safety Analysis Of Road Networks, p. 883. Transportation institute, Oslo (2007)
7. Sawalha, Z., Sayed, T.: Statistical issues in traffic accident modeling. Can. J. Civ. Eng. **33**(9), 1115–1124 (2003)
8. Shirazi, M., Lord, D., Geedipally, S.R.: Sample-size guidelines for recalibrating crash prediction models: recommendations for the highway safety manual. Accid. Anal. Prev. **93**, 160–168 (2006)
9. Joshi, H.: Multicollinearity diagnostics in statistical modelling and remedies to deal with it using SAS [Lecture] Pune, India (2015)
10. Frost, J.: Regression analysis: how do I interprete R-squared and assess the goodness-of-fit. The Minitab Blog, 30 (2013)
11. Ratkowsky, D.A.: A statistical examination of five models for preferred orientation in carbon materials. Carbon **24**(2), 211–215 (1986)

Label Powerset Based Multi-label Classification for Mobile Applications

Preeti Gupta, Tarun K. Sharma and Deepti Mehrotra

Abstract Nowadays, we witness plethora of mobile applications running on smartphones. These mobile applications, whether native/inbuilt or web applications, face battery and processing power bottleneck. Thus, analyzing the energy consumption and RAM usage of these mobile applications become imperative, for making these applications work in longer run. The paper adopts a multi-label classification approach to study the effect of various contributory factors on energy consumption and RAM usage of mobile applications.

Keywords Multi-label classification · Mobile applications · Label powerset
Knowledge · Rule induction

1 Introduction

With the advent of smartphone technology, ever increase usage in mobile applications could be witnessed. As the smartphones are increasing their inbuilt capabilities in the form of LCD display, processing power, camera, sensors, etc., the applications for handling these capabilities have to become effective in terms of processing power and battery usage.

There are mobile applications pertaining to different areas such as entertainment, connectivity, e-commerce, navigation, etc., all varying in their functionalities and resource usage. It should be understood that all the applications are processing power and energy intensive. Regular updates of the mobile applications, though

P. Gupta (✉) · T. K. Sharma
Amity University, Jaipur, Rajasthan, India
e-mail: preeti_i@rediffmail.com

T. K. Sharma
e-mail: taruniitr1@gmail.com

D. Mehrotra
Amity University, Noida, Uttar Pradesh, India
e-mail: mehdeepti@gmail.com

K. Ray et al. (eds.), *Soft Computing: Theories and Applications*,
Advances in Intelligent Systems and Computing 742,
https://doi.org/10.1007/978-981-13-0589-4_62

much desired, induce a compromising situation between functionality and battery consumption. Hence, it can be witnessed that studying the energy usage of mobile applications has been a highly researched topic in recent times [1–4]. It is also observed that mobile applications providing similar functionalities vary significantly in energy consumption. In light of this knowledge, users can make informed decision to select a mobile application for a particular functionality.

The research work is undertaken with an objective of deducing effect of various factors on the energy usage and RAM usage of mobile applications that are categorized under different heads depending on the provided functionality, such as native, e-commerce, games and entertainment, connectivity, tools, etc. Multi-label classification (MLC) has been used for the study.

The organization of the paper is as follows. Section 2 elaborates the dataset and its preparation. Methodology adopted for rule induction is presented in Sect. 3. Section 4 discusses the relationship between the attributes, represented through IF-THEN rules. Finally, the conclusions are drawn and presented in Sect. 5.

2 Dataset Preparation

In all, 80 mobile applications were run and studied for RAM usage and energy usage under different conditions to prepare a dataset of 200 cases. PowerTutor application has been used for dataset preparation. It is an application that shows the power consumption of primary components of the system like processor, network interface, LCD display, and various applications. Applications under study have been considered under the following heads, shown in Table 1.

Using PowerTutor, the following information is collected about the mobile applications. Attributes are reflected in Table 2.

The experiment is device and network connectivity dependent. Hence, the calibrations for the device used for the study are given in Table 3. Further, the Internet service provider remains unchanged for the entire experiment.

3 Experimental Study: Multi-label Classification Using Label Powerset Approach

The study incorporates MLC represented by the function $\mathrm{f_{MLC}} : \mathrm{X} \rightarrow 2^{\mathrm{Y}}$ [5], where given $X = X_1 * X_2 * X_3 \ldots * X_d$, a "$d$"-dimensional input space of numerical or categorical features and $Y = \{\gamma_1, \gamma_2, \gamma_3, \gamma_4 \ldots \gamma_q\}$, an output space of "$q$" labels. In turn, it means that it is the problem of defining a model, mapping inputs x to binary vectors y by assigning binary value *(0/1)* to each element (label) in y.

In order to deal with MLC, there are primarily three approaches:

Table 1 Categorization of the mobile applications

Category	Example	Remarks
Native/Inbuilt	Calculator, calendar, clock, weather, gallery, recorder, compass	It comprises native applications that are built-in for smartphone. These do not usually require Internet
Tools	Polaris Office, WPS Office, CamScanner	Contains the applications that ease office automation
Games and entertainment	Mini-militia, Subway Surfers, Candy crush, Saavn, Wynk Music, Gaana, Youtube	Applications which act as a source of entertainment in the form of multimedia or games are considered
Connectivity	Whatsapp, Facebook, Snapchat, Hangout, SHAREit	Applications for communication of some form
Education	TOI, BBC News, NDTV, TED, Oxford English Dictionary	Applications related to news and teaching–learning
e-Commerce	Amazon, Flipkart, Foodpanda, Jabong, Ola, Uber	Service providers for utility items, food, garments, travel

Table 2 Attributes of the study

Attribute name	Description
Independent attributes	
Energy_LCD	Energy usage over last minute in LCD (J) calibrated over high/low
Energy_CPU	Energy usage over last minute in CPU (J) calibrated over high/low
Energy_3G	Energy usage over last minute in 3G (J) calibrated over high/low
MobileData	Whether mobile data is On/Off
WiFi	Whether connection to WiFi is On/Off
Syn	Whether synchronization is On/Off
NoUp	Whether notification updates is On/Off
MuUs	Whether multimedia usage is high/low
Dependent attributes	
Category1	Depicting low RAM usage
Category2	Depicting high RAM usage
Category3	Depicting low energy usage (Calculated over energy usage by LCD, CPU, and 3G)
Category4	Depicting high energy usage Calculated over energy usage by LCD, CPU, and 3G

Table 3 Calibrations of the device used for experimentation

Calibrations	Remarks
Battery voltage	3.67 V
Nominal voltage	3.7 V
Battery type	Nonremovable Li-Ion 4000 mAh
Camera	REAR-13 megapixel
	FRONT-5 megapixel
RAM available	3.00 GB
Mobile data rate	5.3 Mbps
WiFi data rate	510 Kbps

1. Problem Transformation method—The method deals with a multi-label problem by transforming it to a single-label problem. It can be achieved through three schemes: (i) Binary Relevance (BR) [6], (ii) Classifiers Chains (CC), and (iii) Label Powerset (LP).
2. Algorithm adaptation method—The method directly adapts algorithms to perform MLC, instead of transforming the problem into different subsets of problems (e.g., MLkNN), which is the multi-label version of kNN.
3. Ensemble approaches—Here, ensemble of base multi-label classifiers is used.

The work uses problem transformation method, more specifically LP for implementing MLC. The method of LP gives a unique class to every possible label combination that is present in the training sets. The method starts with the identification of distinct label set that appears in the training set and treating them as a new class. Hence, the multi-label training dataset is thus converted to multi-class training dataset [7]. Formally stating, consider δ_y: $2^y \rightarrow N$ as some mapping function that maps the power set of y to N (natural numbers).

Consider the training dataset as D in LP, then Eq. 1 depicts the method of considering all diverse label set in training dataset as new class:

$$D_y = \{(x_a, \delta_y(Y_a)) | 1 \leq a \leq m\} \tag{1}$$

Moreover, the set of new classes covered by D_y corresponding to Eq. 2:

$$\alpha(D_y) = \{\delta_y(Y_a) | 1 \leq a \leq m\} \tag{2}$$

The open-source machine learning framework MEKA has been employed to do MLC. Through MEKA, MLC can be achieved by using an open-source implementation of JAVA based methods. It is closely based upon WEKA framework. The study used MEKA 1.9.0 version of the tool for carrying out multi-label classification on a system with Intel(R) core (TM) i3-4005U CPU @ 1.70 GHz, 1700 MHz, 2 Core(s), 4 Logical Processor(s), and 4.00 GB installed physical memory (RAM). 80% split percentage has been used in the study, and hence the training set comprises 160

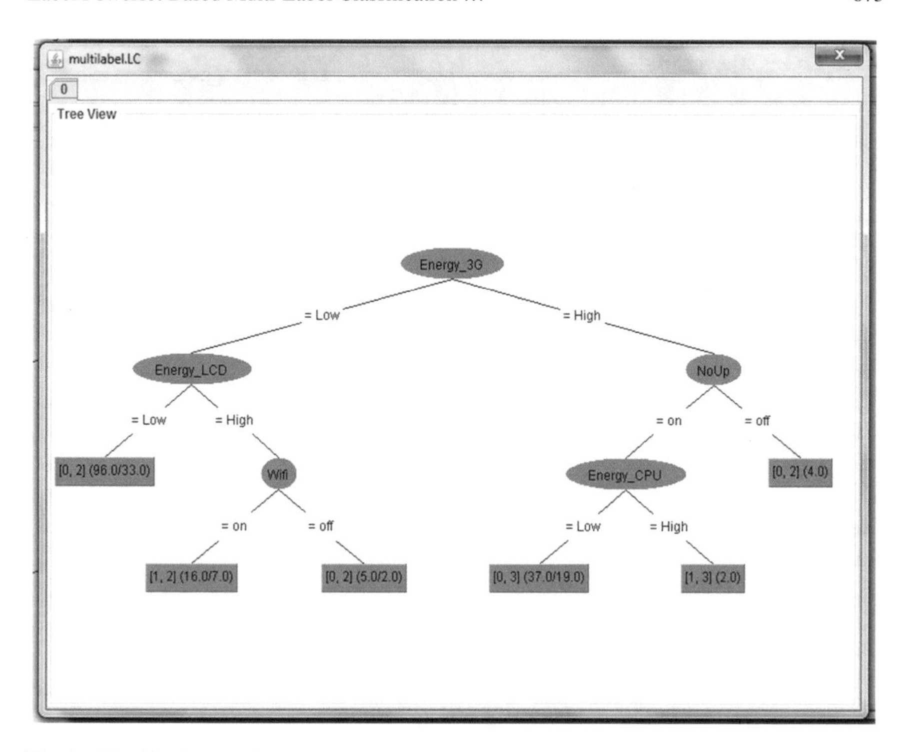

Fig. 1 Classification model

records. The following tree structure bringing out the relationship between attributes is generated and is shown in Fig. 1.

The leaf nodes denote the classes in the figure and numbers in square brackets denote the following classes:

0 Category 1—Low RAM usage,
1 Category 2—High RAM usage,
2 Category 3—Low energy usage, and
3 Category 4—High energy usage.

4 Result and Discussion

IF-THEN rules can be induced from the generated classification model. The rules that are generated are listed below:
Rule 1:
IF *Energy_3G is LOW* AND *Energy_LCD is LOW* THEN *RAM Usage is LOW* AND *Energy Usage is LOW*
Rule 2:

Table 4 Confusion matrix

		Actual class	
		Class	Not class
Predicted class	Class	*tp*	*fp*
	Not class	*fn*	*tn*

IF *Energy_3G is LOW* AND *Energy_LCD is HIGH* AND *Wifi is ON* THEN *RAM Usage is HIGH* AND *Energy Usage is LOW*
Rule 3:
IF *Energy_3G is LOW* AND *Energy_LCD is HIGH* AND *Wifi is OFF* THEN *RAM Usage is LOW* AND *Energy Usage is LOW*
Rule 4:
IF *Energy_3G is HIGH* AND *NoUp is OFF* THEN *RAM Usage is LOW* AND *Energy Usage is LOW*
Rule 5:
IF *Energy_3G is HIGH* AND *NoUp is ON* AND *Energy_CPU is LOW* THEN *RAM Usage is LOW* AND *Energy Usage is HIGH*
Rule 6:
IF *Energy_3G is HIGH* AND *NoUp is ON* AND *Energy_CPU is HIGH* THEN *RAM Usage is HIGH* AND *Energy Usage is HIGH*

In order to establish the productivity of rule, they are to be analyzed. It is said that, if a rule is of the form: IF *Ant* THEN *Const*, where *Ant* is the antecedent (a conjunction of conditions) and *Const* is the consequent (predicted class/s), the predictive performance of the rule may be summarized in the form of confusion matrix, as shown in Table 4.

The entries in the quadrants of the confusion matrix can then be interpreted as

tp = *True Positives* = *Number of tuples (instances) in test set satisfying Ant and Const*
fp = *False Positives* = *Number of tuples (instances) in test set satisfying Ant but not Const*
fn = *False Negatives* = *Number of tuples (instances) in test set not satisfying Ant but satisfying Const*
tn = *True Negatives* = *Number of tuples (instances) in test set not satisfying Ant nor Const* [8].

Further, various metrics can be established on the basis of the values of *tp, tn, fp, and fn*, as shown in Table 5 and the values for the metrics in Table 6.

It can be seen that Rule 6 exhibits highest accuracy. But accuracy measure suffers lacunas and hence cannot be called as good measure of classification. It is subjected to accuracy paradox. Hence, the rules are further evaluated over the metrics of recall and precision. Precision and recall exhibit inverse relationship; for increasing one, other has to be reduced. Giving more significance to recall or precision is situation specific. Precision is more important than recall when one would like to have less *fp* in trade-off to have more *fn*. In certain situations, getting *fp* is very costly, as compared

Table 5 Rule evaluation metrics

Metrics	Interpretation	Formula
Accuracy	Test set tuple percentage that is correctly classified	$\frac{tp + tn}{tp + tn + fp + fn}$
Recall	Measure mentions the proportion of cases in the dataset depicting the predicted class "Const" that is actually represented by the rule	$\frac{tp}{tp + fn}$
Specificity	The true negative recognition rate is represented through specificity	$\frac{tn}{tn + fp}$
Precision	Tuple percentage that the classifier labeled as positive are truly positive	$\frac{tp}{tp + fp}$

Table 6 Rule analysis

Rule	TP	FP	FN	TN	Accuracy	Recall	Precision	Specificity
1	63	33	14	50	0.7063	0.8182	0.6563	0.6024
2	9	7	40	104	0.7063	0.1837	0.5625	0.9369
3	3	2	74	81	0.5250	0.0390	0.6000	0.9759
4	4	0	73	83	0.5438	0.0519	1.0000	1.0000
5	18	19	3	120	0.8625	0.8571	0.4865	0.8633
6	2	0	11	147	0.9313	0.1538	1.0000	1.0000

Table 7 F-measure analysis

Rule no.	F-Measure (2/((1/Precision) + (1/Recall))
1	0.728324
2	0.276923
3	0.073171
4	0.098765
5	0.62069
6	0.266667

to *fn*. Further, precision and recall are not considered in isolation, as they have an inverse effect on each other, often combined into a single measure, F-measure which is a weighted harmonic mean of precision and recall. The rules are analyzed over F-measure as depicted in Table 7.

Based on the value of F-measure, calculated for individual rules as shown in Table 7, Rule 1 outshines over the others. Hence, the fact that can be established through Rule 1 is that in order to keep the RAM usage and energy consumption low, it is advisable to minimize energy usage in 3G and energy usage in LCD.

5 Conclusion

The work brings forth the relationship existing between the various attributes by inducing rules through MLC. Through multi-label classification, the effect of the independent attributes on two dependent attributes could be analyzed. Facts could be established to increase the longevity and efficiency of mobile applications through the identification of the most relevant rule. This leads to knowledge building as it could be put to best use for supplementing the process of decision-making.

References

1. Wilke, C., Richly, S., Gotz, S., Piechnick, C., Aßmann, U.: Energy consumption and efficiency in mobile applications: a user feedback study. In: IEEE International Conference on Green Computing and Communications (GreenCom), 2013 IEEE and Internet of Things (iThings/CPSCom) and IEEE Cyber, Physical and Social Computing, pp. 134–141 (2013)
2. Pang, C., Hindle, A., Adams, B., Hassan, A.E.: What do programmers know about software energy consumption? IEEE Softw. **33**(3), 83–89 (2016)
3. Dutta, K., Vandermeer, D.: Caching to reduce mobile app energy consumption. ACM Trans. Web (TWEB) **12**(1), 5 (2017)
4. Flinn, J., Satyanarayanan, M.: Powerscope: a tool for profiling the energy usage of mobile applications. In: Second IEEE Workshop on Mobile Computing Systems and Applications, 1999. Proceedings. MCSA'99, pp. 2–10 (1999)
5. Gibaja, E., Ventura, S.: Multi-label learning: a review of the state of the art and ongoing research. Wiley Interdiscip. Rev. Data Mining Knowl. Discov. **4**(6), 411–444 (2014)
6. Luaces, O., Díez, J., Barranquero, J., del Coz, J.J., Bahamonde, A.: Binary relevance efficacy for multilabel classification. Progress Artif. Intell. **1**(4), 303–313 (2012)
7. Zhang, M.L., Zhou, Z.H.: A review on multi-label learning algorithms. IEEE Trans. Knowl. Data Eng. **26**(8), 1819–1837 (2014)
8. Freitas, A.A.: A survey of evolutionary algorithms for data mining and knowledge discovery. In: Advances in Evolutionary Computing. Springer, Berlin, Heidelberg, pp. 819–845 (2003)

Artificial Bee Colony Application in Cost Optimization of Project Schedules in Construction

Tarun K. Sharma, Jitendra Rajpurohit, Varun Sharma and Divya Prakash

Abstract Artificial bee colony (ABC) simulates the intelligent foraging behavior of honey bees. ABC consists of three types of bees: employed, onlooker, and scout. Employed bees perform exploration and onlooker bees perform exploitation, whereas scout bees are responsible for randomly searching the food source in the feasible region. Being simple and having fewer control parameters, ABC has been widely used to solve complex multifaceted optimization problems. This study presents an application of ABC in optimizing the cost of project schedules in construction. As we know that project schedules consist of number of activities (predecessor and successor), variable cost is involved in accomplishing these activities. Therefore, scheduling these activities in terms of optimizing resources or cost-effective scheduling becomes a tedious task. The computational result demonstrates the efficacy of ABC.

Keywords Optimization · Evolutionary algorithms · Artificial bee colony ABC · Project scheduling

1 Introduction

Optimization problem exists in every domain may be Science, Engineering, or Management. These optimization problems are generally solved by traditional or nontraditional methods, based on the complexity of the problem. Traditional methods such

---

T. K. Sharma (✉) · J. Rajpurohit · V. Sharma
ASET, Amity University, Jaipur, Rajasthan, India
e-mail: taruniitr1@gmail.com

J. Rajpurohit
e-mail: jiten_rajpurohit@yahoo.com

V. Sharma
e-mail: varunsight@gmail.com

D. Prakash
ASAS, Amity University, Jaipur, Rajasthan, India
e-mail: divya1sharma1@gmail.com

K. Ray et al. (eds.), *Soft Computing: Theories and Applications*,
Advances in Intelligent Systems and Computing 742,
https://doi.org/10.1007/978-981-13-0589-4_63

as gradient search methods require a problem to be continuous and a differentiable, whereas nontraditional method hardly requires any domain knowledge of the problem. These methods simulate the behavior of natural species such as flock of birds, school of fishes, ants, bees, etc. and inspired by the Darwin theory of "survival of the fittest" [1]. The brief overview of nontraditional methods can be referred from [2]. Among these nontraditional methods, artificial bee colony (ABC) is recently introduced by Karaboga [3]. ABC, due to fewer number of control parameters, has been widely applied to solve many applications [4–15].

In this study, ABC has been applied to optimize the cost of project schedules in construction. Any project needs to accomplish within the given time frame. A project basically consists of several activities. Each activity is associated with certain amount of direct cost and indirect cost. If the project get finishes within the scheduled time frame, a bonus is awarded; otherwise, a penalty will be imposed. In this study, we have considered a study of a construction project that consists of seven activities. The details of the study are presented in the five sections.

The paper is structured as follows: Sect. 2 presents a brief about ABC; Sect. 3 describes the project scheduling model in construction. Parameter tuning is discussed in Sect. 4. Section 5 presents a numerical example of the case study and result analysis. Section 6 concludes the paper with future scope.

2 Artificial Bee Colony

Artificial bee colony simulates the foraging process of natural honey bees [3]. The bee colony in ABC has been divided into three groups which are named as scout, employed, and onlooker bees. Scout bees initiate searching of food sources randomly; once the food sources are found by scout bees, they become employed bees. The employed bees exploit the food sources as well as share the information about food sources (quality and quantity) to the onlooker bees (bees resting in the hive and waiting for the information from employed bees) by performing a specific dance termed as "*waggle dance*". The ABC algorithm is presented below:

2.1 Initialization of Random Food Sources

The food source (FS) or population of solutions is randomly generated in the search space using Eq. (1):

$$x_{ij} = max_j + rand(0, 1) \times \left(max_j - min_j\right) \tag{1}$$

where i represents the FS and j denotes the jth dimension. *max* and *min* are the upper and lower bounds of the search domain.

2.2 Employed Bee Process

The search equation is involved in this phase and also performs the global search by introducing new food sources $V_i = (v_{i1}, v_{i2}, \ldots v_{iD})$ corresponding to $X_i = (x_{i1}, x_{i2}, \ldots x_{iD})$ and is discussed below:

$$v_{ij} = x_{ij} + rand(-1, -1) \times \left(x_{ij} - x_{kj}\right) \tag{2}$$

where k is the randomly chosen index and different from i. Greedy selection mechanism is performed to select the population to store in a trail vector. The probability of food sources based on the objective function values (f_i) is evaluated using Eq. (3):

$$p_i = \begin{cases} \frac{1}{(1+f_i)} & if\ f_i \geq 0 \\ 1 + abs(f_i) & if\ f_i < 0 \end{cases} \tag{3}$$

2.3 Onlooker Bee Process

Onlooker bee performs local search around the food sources shared by employed bee. Onlooker bee selects the food source based on Eq. (3). It uses Eq. (2) to search for the new food sources.

2.4 Scout Bee Process

If the food source does not improve in the fixed number of trials, then employed bees turn into scout bees and randomly forage for the new food sources.

The general algorithmic structure of the ABC optimization approach is given as follows:

Initialization of the Food Sources

Evaluation of the Food Sources

Repeat

Produce new Food Sources for the employed bees

Apply the *greedy* selection process

Calculate the probability values for Onlookers

Produce the new Food Sources for the onlookers

Apply the *greedy* selection process

Send randomly scout bees

Memorize the best solution achieved so far.

Until termination criteria are met.

3 Project Scheduling Model in Construction and an Example

The project scheduling model in construction presents NLP model that consists of several activities (predecessor and successor) and variable cost is involved in accomplishing these activities. Therefore, scheduling these activities in terms of optimizing resources or cost-effective scheduling becomes a tedious task.

The objective function for optimizing the cost of project schedules is discussed in Eq. (4) which comprises generalized precedence, activity duration, and duration of project constraints [16].

$$C_{Total} = \sum_{i \in PA} C_i(D_i) + C_I(D_P) + P(D_L) - B(D_E) \tag{4}$$

where C_{Total} is the total cost involved in the project; set PA presents the set of activities i involved in the project; $C_i(D_i)$, $C_I(D_P)$, $P(D_L)$, and $B(D_E)$ represents direct, indirect, penalty, and bonus duration functions. D_i, D_P, D_L, and D_E represent the activities duration, project duration in actual, late time in completing the project, and the time project finishes early, respectively. The activities in a project are mutually linked by precedence relationship (generalized). The precedence relationship of activities is illustrated in Fig. 1.

All the activities are connected to its succeeding activities (represented by j), $j \in J$. All the activities in the project fulfill at least one of the following precedence relationship constraints:

$$\text{F}_\text{S}\ (\text{Finish} \rightarrow \text{Start}){:}\ S_i + D_i + L_{i,j} \le S_j \tag{5}$$

$$\text{S}_\text{S}\ (\text{Start} \rightarrow \text{Start}){:}\ S_i + L_{i,j} \le S_j \tag{6}$$

Fig. 1 Precedence relationship of activities in the project

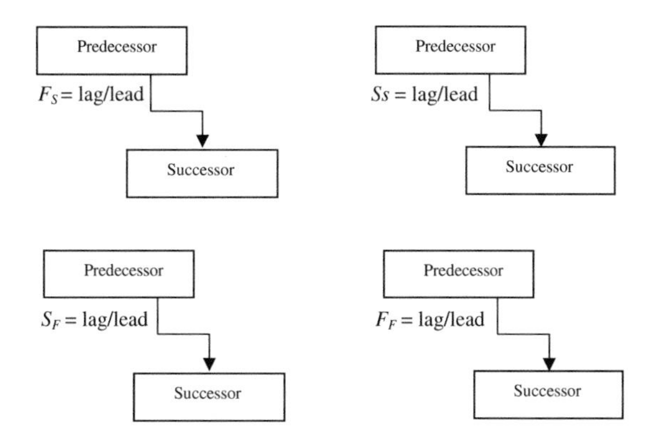

$$\mathrm{S_F}\ (\text{Start} \rightarrow \text{Finish}){:}\ S_i + L_{i,j} \leq S_j + D_j \tag{7}$$

$$\mathrm{F_F}\ (\text{Finish} \rightarrow \text{Finish}){:}\ S_i + D_i + L_{i,j} \leq S_j + D_j \tag{8}$$

where S_i, D_i, $L_{i,j}$, and S_j is the starting time of activity i, duration of the activity lead time between the activity (i and j) and starting time of succeeding activity, respectively. The D_P (actual project duration) is evaluated within the model in the following terms:

$$D_P = S_{iw} + D_{iw} - S_{ia} \tag{9}$$

where S_{iw} represents the starting time and D_{iw} represents the last activity of the project iw. The first project activity is denoted by S_{ia}. This is quite obvious that all the activities must take place within the start and finish time of the project; hence, a constraint is set to limit the completion time of all the project activities:

$$S_i + D_i - S_{ia} \leq D_P \tag{10}$$

The relationship among D_P, D_L, D_E, and D_T is presented as

$$D_P - D_L + D_E = D_T \tag{12}$$

At max, only one variable from D_L and D_E can be considered as zero (0) in any project scheduling and therefore it puts extra restraint, i.e.,

$$D_L D_E = 0 \tag{13}$$

Table 1 Generalized precedence relationship and lag times in activities

Project activity ID and description	Succeeding activity ID	Precedence relationship	Lag time (in days)
1. Underground service	2	S2S	2
2. Concrete work	3	F2S	3
3. Outer walls	4	F2S	0
4. Construction of roof	5	F2S	0
	6	F2S	0
5. Finishing of floors	7	F2S	0
6. Ceiling	7	F2F	6
7. Finishing of work	–	–	–

S2S Start to Start; *F2S* Finish to Start; *F2F* Finish to Finish

4 Parameter Tuning

Parameter tuning: In order to perform unbiased result comparison, the parameter is tuned accordingly. ABC has fewer control parameters that require to be tuned. Colony size of bees in ABC is fixed to 100 (50 each in O_b and E_b). *Limit* is tuned to 200. All the algorithms are executed in Dev C++ with the following machine configuration: *Processor*: Intel(R) Core (TM) i3-5005U CPU @ 2.00 GHz having 4 GB RAM. An inbuilt *rand*() function in C++ is used to initialize the random numbers. 25 runs are performed in each case. The number of function evaluations (NFE) has been observed at the time of termination criterion reached.

5 Numerical Example in Construction and Result Analysis

The below numerical example has been considered from [17]. In this example, there are seven activities involved in a construction project. Table 1 presents precedence relationship and the time lags between succeeding activities of the project, whereas Table 2 presents direct cost duration function along with normal/crash points.

The project needs to be finished in 47 days. Following are some assumptions:

- If project finishes before time, then a bonus of $300 will be awarded.
- If project gets delays, then a penalty of $400 per day is imposed.
- The daily indirect cost is about $200.

Now the objective is to identify the optimized project cost with activity start time and duration. The optimum solution with the help of ABC is presented in Table 3 along with optimal activity start time and duration. The total project cost found to be $28750 with a penalty cost of $400. The table also presents the activities start time and duration. There was no bonus.

Table 2 Duration options along with the functions of direct cost

Project activity ID and description	Duration in days		Direct Cost in $		Direct cost duration function
	Crash	Normal	Crash	Normal	
1. Underground service	3	6	4500	1500	$250D_1^2 - 3250D_1 + 12000$
2. Concrete work	10	12	7000	5000	$-10969.6299\ln(D_2) + 32258.5063$
3. Outer walls	8	12	3600	2000	$11664\exp(-0.1469D_3)$
4. Construction of roof	6	8	3100	2000	$-1000D_4 + 6400$
5. Finishing of floors	3	4	3000	2000	$-550D_5 + 6000$
6. Ceiling	4	6	4000	2500	$-750D_6 + 7000$
7. Finishing of work	10	14	2800	1000	$75D_7^2 - 2250D_7 + 17800$
Project	42	55	28000	16000	

Table 3 Optimal results

Project activity ID and description	Start time in days	Duration in days	Direct cost in $
1. Underground service	1	6	1500
2. Concrete work	3	12	5000
3. Outer walls	18	8	3600
4. Construction of roof	26	7	2550
5. Finishing of floors	33	4	2000
6. Ceiling	33	6	2500
7. Finishing of work	37	12	1600
Indirect cost ($)			9600
Penalty ($)			400
Bonus ($)			0
Total project cost ($)			28750

6 Conclusion

The project scheduling model in construction presents NLP model that consists of several activities (predecessor and successor) and variable cost is involved in accomplishing these activities. Therefore, scheduling these activities in terms of optimizing resources or cost-effective scheduling becomes a tedious task. In this study of a problem of cost optimization in construction, project is taken and solved. The obtained results show that artificial bee colony (ABC) is capable of identifying optimal activity start times and durations in order to optimize the overall cost of the project with optimal computational cost.

In future, we will try to implement ABC with multi-objective problems.

References

1. Pant, M., Ray, K., Sharma, T.K., Rawat, S., Bandyopadhyay, A.: Soft computing: theories and applications. In: Advances in Intelligent Systems and Computing, Springer. Singapore (2018). https://doi.org/10.1007/978-981-10-5687-1
2. Rajpurohit, J., Sharma, T.K., Abraham, A., Vaishali: Glossary of metaheuristic algorithms. Int. J. Comput. Inf. Syst. Ind. Manag. Appl. **9**, 181–205 (2017)
3. Karaboga, D.: An idea based on bee swarm for numerical optimization. Techmical Report No. TR-06). Erciyes University Engineering Faculty, Computer Engineering Department
4. Sharma, T.K., Pant, M.: Enhancing the food locations in an Artificial Bee Colony algorithm. Soft. Comput. **17**(10), 1939–1965 (2013)
5. Sharma, T.K., Pant, M.: Enhancing the food locations in an Artificial Bee Colony algorithm. In: IEEE Symposium on Swarm Intelligence (SIS), pp. 1–5. Paris, France (2011)
6. Sharma, T.K., Pant, M.: Improved search mechanism in ABC and its application in engineering. J. Eng. Sci. Technol. **10**(1), 111–133 (2015)
7. Sharma, T.K., Pant, M.: Shuffled Artificial Bee Colony algorithm. Soft. Comput. **21**(20), 6085–6104 (2017)
8. Sharma, T.K., Pant, M.: Distribution in the placement of food in Artificial Bee Colony based on changing factor. Int. J. Syst. Assur. Eng. Manag. **8**(1), 159–172 (2017)
9. Sharma, T.K., Gupta, P.: Opposition learning based phases in Artificial Bee Colony. Int. J. Syst. Assur. Eng. Manag. (2016). https://doi.org/10.1007/s13198-016-0545-9
10. Sharma, T.K., Pant, M.: Redundancy level optimization in modular software system models using ABC. Int. J. Intell. Syst. Appl. **6**(4), 40–48 (2014)
11. Sharma, T.K., Pant, M., Neri, F.: Changing factor based food sources in Artificial Bee Colony. In: IEEE Symposium on Swarm Intelligence (SIS) (pp. 1–7). Orlando, Florida, USA (2014)
12. Mao, M., Duan, Q.: Modified Artificial Bee Colony algorithm with self-adaptive extended memory. Cybern. Syst. **47**(7), 585–601 (2016)
13. Li, J., Pan, Q.P., Duan, P.: An Improved Artificial Bee Colony algorithm for solving hybrid flexible flowshop with dynamic operation skipping. IEEE Trans. Cybern. **46**(6), 1311–1324 (2016)
14. Karaboga, D., Gorkemli, B., Ozturk, C., Karaboga, N.: A comprehensive survey: Artificial Bee Colony (ABC) algorithm and applications. Artif. Intell. Rev. **42**(1), 21–57 (2014)
15. Anuar, S., Ali, S., Sallehuddin, R.: A modified scout bee for Artificial Bee Colony algorithm and its performance on optimization problems. J. King Saud Univ.—Comput. Inf. Sci. **28**(4), 395–406 (2017)
16. Klanšek, U., Pšunder, M.: Cost optimization of time schedules for project management. Econ. Res.-Ekonomska Istraživanja **23**(4), 22–36 (2010). https://doi.org/10.1080/1331677X.2010.11517431
17. Yang, I.T.: Chance-constrained time-cost trade-off analysis considering funding variability. J. Constr. Eng. Manag. **131**(9), 1002–1012 (2005)

Aligning Misuse Case Oriented Quality Requirements Metrics with Machine Learning Approach

Ajeet Singh Poonia, C. Banerjee, Arpita Banerjee and S. K. Sharma

Abstract In the recent years, techniques and approaches associated with machine learning are being proposed to improvise the aspect of security of software product. These techniques and approaches of machine learning are proposed to cater to various phases of software development process for the implementation of security. This paper investigates the various approaches and techniques of machine learning used for security purpose. The research paper further explores how the alignment of misuse care oriented quality requirements framework metrics can be done with the eligible technique/approach of machine learning for specification and implementation of security requirements during the requirements engineering phase of software development process and also highlights the outcome of this alignment. The paper also presents some areas where further research work could be carried out to strengthen the security aspect of the software during its development process.

Keywords Security · Machine learning · Artificial neural network · Security requirements engineering · Software development process

---

A. S. Poonia (✉)
Department of CSE, Government College of Engineering & Technology, Bikaner, India
e-mail: pooniaji@gmail.com

C. Banerjee
Department of CS, Pacific Academy of Higher Education & Research University, Udaipur, India
e-mail: chitreshh@yahoo.com

A. Banerjee
Department of CS, St. Xavier's College, Jaipur, India
e-mail: arpitaa.banerji@gmail.com

S. K. Sharma
Department of CSE, Modern Institute of Technology & Research Centre, Alwar, India
e-mail: sharmasatyendra_03@rediffmail.com

K. Ray et al. (eds.), *Soft Computing: Theories and Applications*,
Advances in Intelligent Systems and Computing 742,
https://doi.org/10.1007/978-981-13-0589-4_64

1 Introduction

Numerous techniques and approaches are proposed by the researchers and academicians associated with machine learning to address and improvise the aspect of security of a software product [1]. From the study of various research papers, it became clear that these proposed techniques and approaches are applicable during the various stages of software development process starting from the requirements engineering phase till the implementation phase [2–5].

It has been argued again that the aspect of security be dealt from the very beginning of software development process, i.e., the requirements engineering phase, so that a comprehensive implementation of security can take place and a more secure software can be developed [6, 7]. Also, the security should cover various aspects embedded with certain rules to provide a comprehensive view [8, 9]. Similarly, from the studies, it has been found that the machine learning approach for implementation of security during the requirements engineering phase of software development process may be the most effective, efficient, and comprehensive way toward the development of secured software [10–12].

Vulnerability identification and analysis method during the requirements engineering phase of software development process are crucial for suggesting appropriate mitigation mechanism to minimize the impact of software vulnerability exploitation [13–15]. Machine learning techniques may be utilized to identify/discover vulnerabilities and also analyze them during the requirements engineering phase [16–18]. This research paper explores the possibility of collaboration of machine learning technique with proposed misuse oriented quality requirements framework metrics to provide an improvised framework so that using the indicators and estimators derived from the collaboration the software development team may specify more comprehensive and decisive security requirements during the initial stages of software development process. It will also aid in creating security awareness among the various stakeholders involved in the development process [19].

The rest of the paper is organized as follows: Sect. 2 discusses the machine learning techniques for security implementation in requirements engineering, Sect. 3 highlights the alignment of MCOQR framework metrics with machine learning techniques, Sect. 4 deliberates upon the validation and results, and Sect. 5 provides the conclusion and future research work.

2 Machine Learning Techniques for Security Implementation in Requirements Engineering

As the system software is becoming more complex, hence, the models based on machine learning have shown practical significance in addressing the issue of security especially those encompassing failure, error, and defect predictions. Due to the multi-magnitude, the future software development process is becoming more and

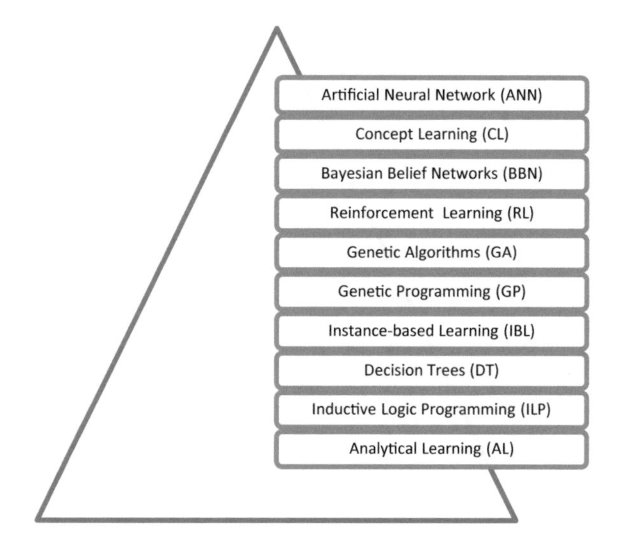

Fig. 1 Depicting various machine learning techniques

more complex [1]. And there is a need for a robust model to predict the various vulnerabilities that could exist in the software during its implementation so that a proper countermeasure could be thought off to minimize the damage.

The problem domain of such software is not very well-defined, human knowledge regarding the proper identification of problem and its requirements is limited, and there is a need that the software system when put in practical use should be ready for dynamic adaption as the condition keeps on changing [1, 17, 18]. Hence, the field of machine learning models which specially caters to the security aspect during the development of software is becoming more and more demanding [1, 16] (Fig. 1).

There are a number of machine learning techniques that could be used for the implementation of security during the requirements engineering phase of software development process. Many researchers have proposed several such learning techniques for fault prediction, defect prediction, and error prediction which could identify various vulnerabilities which could be present in the system [1, 20].

3 Alignment of MCOQR Framework Metrics with Soft Computing Technique

The proposed misuse case oriented quality requirements framework metrics [21] identifies vulnerabilities and associated misuse cases using CVSS, CVE, and CWE standards during the requirements elicitation phase of software development process. The outcome of the proposed framework metrics is in the form of six security indicators and estimators which provide guidance and help to the security team in eliciting the security requirements so that proper countermeasure can be planned

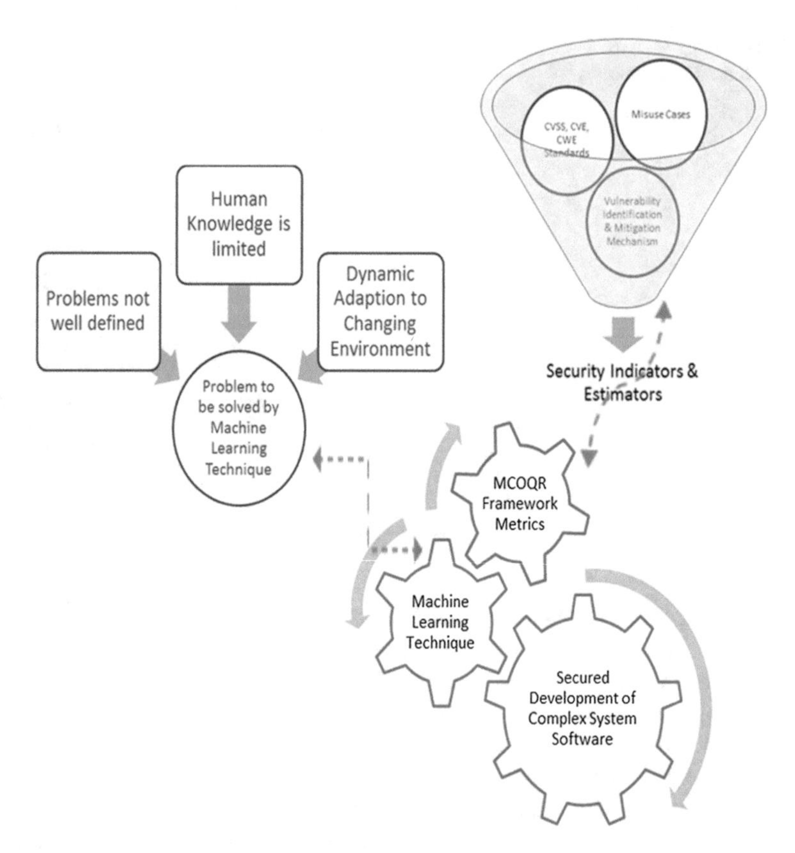

Fig. 2 Proposed framework for alignment of MCOQR framework metrics and machine learning

to either eliminate or minimize the harm that could be caused due to the breach of security during the implementation and use of software (Fig. 2).

Machine learning technique could be aligned with the proposed MCOQR framework metrics so that the security indicators and estimators could cater to that software for which the problem could not be properly defined, the human knowledge is limited, and there is a need for dynamic adaption due to changing conditions during the operation of the software. The result would be a secured development of complex system software.

4 Validation and Results

The proposed framework was applied to a real-life project from industry (on the request of the company, identity is concealed). Artificial neural network which is a machine learning technique was applied for the identification of possible vulnerabilities which could be present in the system. After the vulnerabilities were identified, the associated misuse cases were identified using CVSS metrics, CVE and CWE standards, and MCOQR metrics; framework metrics was applied to identify six different security indicators like dominant vulnerability type, level of countermeasure, influence of threat, potential source of exploit, and collateral damage estimated in terms of security threat and potential economic loss. The study shows that the level of risk involved in the development and use of software is minimized to a considerable level. Due to the page limit constraint, we are not providing the details of validation results in this paper; we will discuss in our next paper.

5 Conclusion and Future Work

The security indicators and estimators thus obtained using the alignment of MCOQR framework metrics and machine learning techniques assisted the security engineering team to specify appropriate security requirements. This framework provided the security team with all the essential indicators and estimators to plan in advance the various factors like cost, security countermeasure, schedule, etc. that is needed to develop a complex and secured software which may withstand most of security breaches that may arise due to the implementation and use of software by an organization.

Future work may include validating the proposed work on a large sample of data. Another future work may include implementation of other machine learning techniques to estimate the optimal solution according to the class and category of software under development process.

References

1. Rawat, M.S., Dubey, S.K.: Software defect prediction models for quality improvement: a literature study. IJCSI Int. J. Comput. Sci. Iss. **9**(5), 288–296 (2012)
2. Batcheller, A., Fowler, S.C., Cunningham, R., Doyle, D., Jaeger, T., Lindqvist, U.: Building on the success of building security. IEEE Secur. Priv. **15**(4), 85–87 (2017)
3. Vijayakumar, K., Arun, C.: Continuous security assessment of cloud based applications using distributed hashing algorithm in SDLC. Cluster Comput., 1–12 (2017)
4. Sharma, B., Duer, K.A., Goldberg, R.M., Teilhet, S.D., Turnham, J.C., Wang, S., Xiao, H.: U.S. Patent No. 9,544,327. Washington, DC: U.S. Patent and Trademark Office (2017)
5. Chandra, K., Kapoor, G., Kohli, R., Gupta, A.: Improving software quality using machine learning. In: 2016 International Conference on Innovation and Challenges in Cyber Security (ICICCS-INBUSH), pp. 115–118. IEEE (2016)
6. Banerjee, C., Banerjee, A., Sharma, S.K.: Use Case and Misuse Case in Eliciting Security Requirements: MCOQR Metrics Framework Perspective
7. Figl, K., Recker, J., Hidalga, A.N., Hardisty, A., Jones, A.: Security is nowadays an indispensable requirement in software systems. Traditional software engineering processes focus primarily on business requirements, leaving security as an afterthought to be addressed via

generic "patched-on" defensive mechanisms. This approach is insufficient, and software systems need to have security functionality engineered within in a similar fashion as ordinary business.... Requir. Eng. **21**(1), 107–129 (2016)
8. Banerjee, C., Pandey, S.K.: Software security rules. SDLC Perspective (2009)
9. Myyry, L., Siponen, M., Pahnila, S., Vartiainen, T., Vance, A.: What levels of moral reasoning and values explain adherence to information security rules? An empirical study. Eur. J. Inf. Syst. **18**(2), 126–139 (2009)
10. Dick, J., Hull, E., Jackson, K.: Requirements Engineering. Springer (2017)
11. Ismael, O.A., Song, D., Ha, P.T., Gilbert, P.J., Xue, H.: U.S. Patent No. 9,594,905. Washington, DC: U.S. Patent and Trademark Office (2017)
12. Witten, I.H., Frank, E., Hall, M. A., Pal, C.J.: Data Mining: Practical Machine Learning Tools and Techniques. Morgan Kaufmann (2016)
13. Poonia, A.S., Banerjee, C., Banerjee, A., Sharma, S.K.: Vulnerability identification and misuse case classification framework. In: Soft Computing: Theories and Applications, pp. 659–666. Springer, Singapore (2018)
14. Banerjee, C., Banerjee, A., Poonia, A.S., Sharma, S.K.: Proposed algorithm for identification of vulnerabilities and associated misuse cases using CVSS, CVE standards during security requirements elicitation phase. In: Soft Computing: Theories and Applications, pp. 651–658. Springer, Singapore (2018)
15. Riaz, M., Elder, S., Williams, L.: Systematically developing prevention, detection, and response patterns for security requirements. In: IEEE International Requirements Engineering Conference Workshops (REW), pp. 62–67, Sept., 2016. IEEE
16. Bozorgi, M., Saul, L.K., Savage, S., Voelker, G.M.: Beyond heuristics: learning to classify vulnerabilities and predict exploits. In: Proceedings of the 16th ACM SIGKDD International Conference on Knowledge Discovery and Data Mining, pp. 105–114. ACM (2010)
17. Alves, H., Fonseca, B., Antunes, N.: Experimenting machine learning techniques to predict vulnerabilities. In: 2016 Seventh Latin-American Symposium on Dependable Computing (LADC), pp. 151–156. IEEE (2016)
18. Webster, A.: A Comparison of Transfer Learning Algorithms for Defect and Vulnerability Detection (2017)
19. Banerjee, C., Pandey, S.K.: Research on software security awareness: problems and prospects. ACM SIGSOFT Softw. Eng. Notes **35**(5), 1–5 (2010)
20. Han, Z., Li, X., Xing, Z., Liu, H., Feng, Z.: Learning to predict severity of software vulnerability using only vulnerability description. In: 2017 IEEE International Conference on Software Maintenance and Evolution (ICSME), pp. 125–136. IEEE (2017)
21. Banerjee, C., Banerjee, A., Pandey, S.K.: MCOQR (Misuse Case-Oriented Quality Requirements) Metrics Framework. In Problem Solving and Uncertainty Modeling through Optimization and Soft Computing Applications, pp. 184–209. IGI Global (2016)

Improvising the Results of Misuse Case-Oriented Quality Requirements (MCOQR) Framework Metrics: Secondary Objective Perspective

C. Banerjee, Arpita Banerjee, Ajeet Singh Poonia and S. K. Sharma

Abstract A common notion exists that the security aspect of a software product always deals with the qualitative attributes which are quantified using some method or technique. This is done to measure the various subclassified levels of varied security facets of a software product. In doing so, most of the time due to unawareness, the specifications, protocols, and standards which are used during the quantification purpose of security are either not focused upon or updated accordingly. This leaves a void and thus promotes unnecessary gap analysis during the further course of action. This research paper shows how the proposed Misuse Case-Oriented Quality Requirements (MCOQR) framework metrics may be used to provide a comprehensive model which may encourage the software development team for the coordinated approach toward securing software application in the development process. The paper discusses the implementation mechanism of the comprehensive model. The work proposed is an extension of (MCOQR) framework metrics, and highlights one of the secondary objectives that may be achieved for a comprehensive view. The paper also highlights the areas where further research work can be carried out to further strengthen the entire system.

Keywords Security · Software security · Software product · Software development team · MCOQR metrics · MCOQR framework metrics

---

C. Banerjee (✉)
Department of CS, Pacific Academy of Higher Education & Research University, Udaipur, India
e-mail: chitreshh@yahoo.com

A. Banerjee
Department of CS, St. Xavier's College, Jaipur, India
e-mail: arpitaa.banerji@gmail.com

A. S. Poonia
Department of CSE, Government College of Engineering & Technoogy, Bikaner, India
e-mail: pooniaji@gmail.com

S. K. Sharma
Department of CSE, Modern Institute of Technology & Research Centre, Alwar, India
e-mail: sharmasatyendra_03@rediffmail.com

K. Ray et al. (eds.), *Soft Computing: Theories and Applications*,
Advances in Intelligent Systems and Computing 742,
https://doi.org/10.1007/978-981-13-0589-4_65

1 Introduction

Various tools and techniques have been proposed by the researchers and academicians for measuring the security aspect of the software during its development process [1]. Many rules have been proposed for its comprehensive implementation [2]. The aspect of security is non-functional in nature and provides an intangible value which needed to be quantified by some mechanism so that proper control measures could be adopted in case of any breach to the system [3]. The tools and techniques available for measuring the security can be worked upon during different phases of software development process [4].

Implementation and quantification of security are neglected from the very beginning of the software development process due to the misconception by the management that it may add unnecessary cost, time, and efforts to the software project schedule [4, 5]. While identifying vulnerability and its severity during the security requirements engineering, standards and protocols like CVSS, CVE, CWE, etc. are not given due to weightage which results in improper identification of vulnerabilities and calculation of its severity [6, 7].

Studies have shown that if these standards and protocols are used extensively and religiously, the process of vulnerability identification [8] and severity calculation becomes very comprehensive and effective and also prompts awareness about the importance of security among the software development team [9, 10]. Many researchers have practically shown the importance of using these protocols and standards to implement security during the requirements engineering phase of software development lifecycle [11, 12].

The research paper discusses one such proposed research work carried out where the researcher proposed a Misuse Case-Oriented Quality Requirements (MCOQR) framework metrics [13]. The rest of the paper is organized as Sect. 2 discusses the MCOQR framework metrics in brief, Sect. 3 highlights the various secondary objectives that would be met by the software development team if the proposed work is implemented from the beginning of the software development process, Sect. 4 shows the implementation mechanism for the secondary objectives highlighted, Sect. 5 explores the validation and results, and Sect. 5 provides the conclusion and future research work.

2 MCOQR Framework Metrics—A Brief

The proposed work MCOQR framework metrics was influenced by the research work carried out earlier by the researchers [14–18] and provides a more comprehensive platform for security implementation during the early stages of software development process. As shown in Fig. 1, MCOQR framework metrics uses CVSS scoring, CVE, CWE repository to identify the prospective vulnerability, and its classification and associated misuse cases to produce count, scoring, and ranking of vulnerabilities.

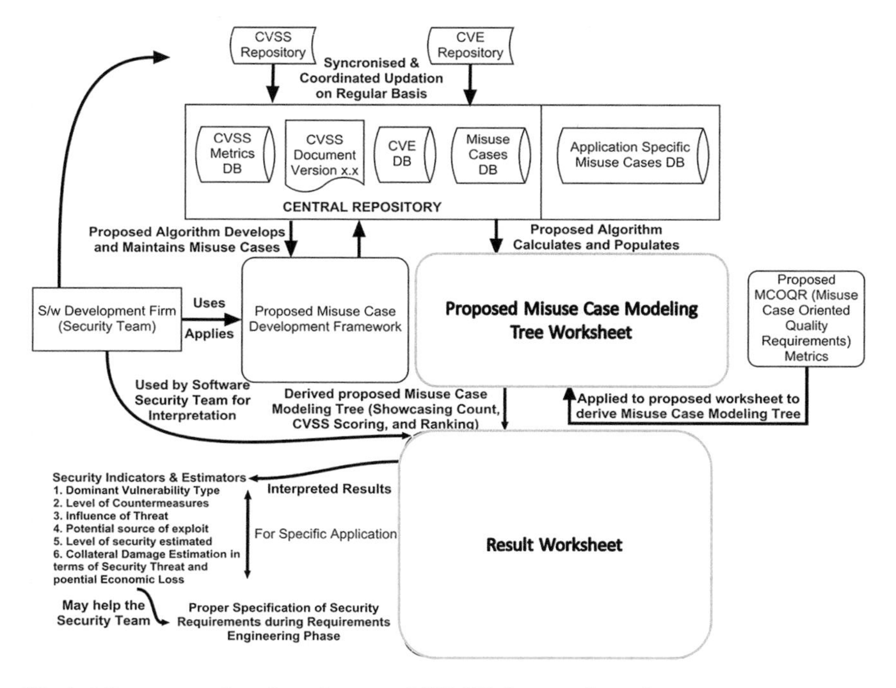

Fig. 1 Misuse case-oriented requirements (MCOQR) framework metrics

The entire process is initiated during the requirements elicitation phase of requirements engineering phase of software development process. As a primary objective, it reveals the dominant Vulnerability_Type, level of countermeasures adopted, influence of threat, potential source of exploit, level of security requirements estimated, and probable economic damage for a specific application.

3 MCOQR Framework Metrics—Secondary Objective Perspective

The proposed MCOQR framework metrics, apart from the primary objective, is also designed to meet some secondary objectives. The proposed framework metrics when practically implemented enforces a sense of awareness among the software development team for the use of protocols and standards like CVSS, CVE, CWE, etc. for the very beginning of the software development process.

MCOQR framework metrics implementation achieves the following secondary objectives:

- Induces use of security-related database which is well synchronized with official repository of CVSS, CVE, and, CWE industry accepted standards.

- Creates awareness and belongingness among the various stakeholders of the software development firm.
- Proves that investment in security during the early stages of software development is better than investing in securing the software after it is implemented onsite.

4 Implementation Mechanism

MCOQR framework metrics advocates for a central as well as application-specific repository for keeping the classified vulnerability and its associated information and in this process identifies and maintains the following databases:

- CVSS Metrics Database,
- CVE Database,
- Vulnerability Database,
- Misuse Case Database, and
- Application-Specific Misuse Case Database.

Since the entire team has individual share of involvement related to various facets of security during the development of the proposed software, hence, they are kept aware of the standard and protocols. Since the MCOQR framework metrics repository needs updates from the CVSS, CVE, and CWE official repository; hence, the staff involved in the development process is kept updated regarding the latest security updates.

5 Conclusion and Future Work

If the security is neglected during the early stages of software development process due to the misconception of the management team regarding increased cost, budget, schedule, etc. of the project, then it results in numerous security breaches after the software is implemented and in use. Also, the cost incurred in securing the software during implementation and use is massive as compared to cost saved initially due to non-implementation of security during the development of software. It also results in damaging the reputation of the organization using the software and thus the trust level of the client is decreased.

All the facts stated above indicate that MCOQR framework metrics is efficient and effective enough to create awareness about the use of security standards and protocols during the software development process. Future work may include validating the proposed work on a large sample of data. Another future work may include implementing the proposed work on various categories of software/applications.

References

1. Nikhat, P., Kumar, N.S., Khan, M.H.: Model to quantify integrity at requirement phase. Indian J. Sci. Technol. **9**(29) (2016)
2. Banerjee, C., Pandey, S.K.: Software Security Rules. SDLC Perspective (2009)
3. Banerjee, A., Banerjee, C., Pandey, S.K., Poonia, A.S: Development of iMACOQR metrics framework for quantification of software security. In: Proceedings of Fifth International Conference on Soft Computing for Problem Solving, pp. 711–719. Springer, Singapore (2016)
4. Sharma, A., Misra, P.K.: Aspects of enhancing security in software development life cycle. Adv. Comput. Sci. Technol. **10**(2), 203–210 (2017)
5. Khaim, R., Naz, S., Abbas, F., Iqbal, N., Hamayun, M., Pakistan, R.: A review of security integration technique in agile software development. Int. J. Softw. Eng. Appl. **7**(3) (2016)
6. Banerjee, C., Banerjee, A., Sharma, S.K.: Estimating influence of threat using Misuse Case Oriented Quality Requirements (MCOQR) metrics: Security requirements engineering perspective. Int. J. Hybrid Intell. Syst. **14**(1–2), 1–11 (2017)
7. Mouratidis, H., Fish, A.: Decision-making in security requirements engineering with constrained goal models. In: Computer Security: ESORICS 2017 International Workshops, CyberICPS 2017 and SECPRE 2017, Oslo, Norway, 14–15 Sept 2017, Revised Selected Papers, vol. 10683, p. 262. Springer (2018)
8. Poonia, A.S., Banerjee, C., Banerjee, A., Sharma, S.K.: Vulnerability identification and misuse case classification framework. In: Soft Computing: Theories and Applications, pp. 659–666. Springer, Singapore (2018)
9. Ando, E., Kayashima, M., Komoda, N.: A Proposal of security requirements definition methodology in connected car systems by CVSS V3. In: 2016 5th IIAI International Congress on Advanced Applied Informatics (IIAI-AAI), pp. 894–899. IEEE (2016)
10. Banerjee, C., Banerjee, A., Poonia, A. S., Sharma, S. K.: Proposed algorithm for identification of vulnerabilities and associated misuse cases using CVSS, CVE standards during security requirements elicitation phase. In: Soft Computing: Theories and Applications, pp. 651–658. Springer, Singapore (2018)
11. Salini, P., Kanmani, S.: Effectiveness and performance analysis of model-oriented security requirements engineering to elicit security requirements: a systematic solution for developing secure software systems. Int. J. Inf. Secur. **15**(3), 319–334 (2016)
12. Femmer, H.: Requirements engineering artifact quality: definition and control. Doctoral dissertation, Technische Universität München (2017)
13. Banerjee, C., Banerjee, A., Pandey, S.K.: MCOQR (misuse case-oriented quality requirements) metrics framework. In: Problem Solving and Uncertainty Modeling through Optimization and Soft Computing Applications, pp. 184–209. IGI Global (2016)
14. Faily, S., Fléchais, I.: Finding and resolving security misusability with misusability cases. Requirements Eng. **21**(2), 209–223 (2016)
15. Siddiqui, S.T.: Significance of Security Metrics in Secure Software Development. Significance **12**(6) (2017)
16. Banerjee, C., Banerjee, A., Murarka, P.D.: Measuring software security using MACOQR (misuse and abuse case oriented quality requirement) metrics: defensive perspective. Int. J. Comput. Appl. **93**(18) (2014)
17. Heitzenrater, C., Simpson, A.: A case for the economics of secure software development. In: Proceedings of the 2016 New Security Paradigms Workshop, pp. 92–105. ACM (2016)
18. Banerjee, C., Banerjee, A., Murarka, P.D.: Measuring software security using MACOQR (misuse and abuse case oriented quality requirement) metrics: defensive perspective. Int. J. Comput, Appl. **93**(18) (2014)

A Review: Importance of Various Modeling Techniques in Agriculture/Crop Production

Jyoti Sihag and Divya Prakash

Abstract Agriculture and its product are important to the sustainable of environment and world population because they provide and fulfill the essential requirement of human such as food and medicine and also provide the raw materials to many industries: jute fabric, cotton, etc. A large part of world's land with wide variation in climatic conditions is not fit for agriculture because of various natural and human causes. To fulfill the demands of rapidly growing population, there is need of healthy products along with better production techniques. Agricultural production is affected by environmental factor such as weather, temperature, etc. Weather affected the crop growth, production, yield, etc. Researchers are trying to develop sustainable system which can be dealt with many challenges of agriculture. At this time, crop modeling technique is very useful in agriculture because these techniques estimate the effect of factors on system, the interaction of the soil–plant–atmosphere system and researcher can also find that which factor is most useful in system. Modeling is a better way of transferring the research knowledge of a system to farmer/ users. The modeling can also be used to evaluate the economic impacts on agricultural land and crop. Modeling techniques can improve the quality of production, yield of crop, and minimize the impact of pest on food product.

Keywords Crop modeling · Soil–plant–atmosphere system

1 Introduction

The model is an excellent way to study about complex system, transforming important information in knowledge, and transferring this knowledge to others [1]. Models can be classified into mechanistic, static, dynamic, stochastic, deterministic, simulation, and explanatory models [2]. A crop model can be explained by a quantitative scheme

---

J. Sihag · D. Prakash (✉)
Department of Chemistry, Amity School of Applied Sciences, Amity University, Jaipur, Rajasthan, India
e-mail: dprakash@jpr.amity.edu

K. Ray et al. (eds.), *Soft Computing: Theories and Applications*,
Advances in Intelligent Systems and Computing 742,
https://doi.org/10.1007/978-981-13-0589-4_66

for predicting the growth, development, and yield of a crop, given a set of genetic features and relevant environmental variables [3]. Crop models are very important for predicting the behavior of crops and also widely used in agro-environment works such as the soil–plant growth, yield, crop monitoring, etc. [4, 5]. They also predict the processes like weather, hydrology, nutrient cycling and movement, tillage, soil erosion, and soil temperature [6–8], and quantitative information on production can only be obtained [9–11]. The models presented interaction between the plant and environment [12]. CSM played an important role in agronomy because CSM is computerized presentations of yield, production, and development. They are simulated through mathematical equations [13] and helping farmers in agronomic management [14]. Statistical models are using to crop yield prediction [15, 16]. Remote sensing and vegetation indices are used to determine crop nutrient requirements [17–19]. Remote sensing detects crop diseases [20]. Remote sensing assesses the crop yield [21–27]. The purpose of this study is to analyze the effects of crop models on different crops in different environmental conditions.

2 Modeling Techniques in Agriculture

A simulation model predicts P and K uptakes of field-grown soybean cultivars under field conditions. In this experiment, two types of soil are used: Raub silt loam and Chalmers silt loam; and the Cushman mathematical model was used to predict the nutrient uptake. The nutrients uptake by plants (K uptake in both types of soils and P uptake in Chalmers soil) was exactly predicted by the mathematical model [28].

From emergence to maturity of soybean crop, a model was developed to analyze the water, carbon, and nitrogen. Daily carbon and nitrogen accumulation were calculated as a linear function of intercepted radiation and vegetative biomass. The experimental data showed that nitrogen fixation rates were more sensitive to soil dehydration [29].

A dynamic model was elevated and validated to evaluate heat, mass transport process, energy balance, transpiration, photosynthesis, CO_2 exchange, and radiation transfer in greenhouse cucumber row crop. Each of the individual sub-models was parameterized from experimental data for a dense row cucumber crop. Validations were conducted for solar radiation, transpiration, leaf temperature, air temperature, and relative humidity.

This model helps researcher to estimate the combined influence of various factors on greenhouse environment in a microscale and can provide a wide information on climatic variables to meet specific requirements of environmental control and pest/disease management [30].

During juvenile phase and curd induction phase, a cauliflower model was described. In this study, the base temperature was estimated to be 0 °C for the curd induction phase and the optimum temperature in the PAGV data was estimated at 12.8 °C for CVS Delira and Elgon. Above and below the optimum temperature, a linear relationship between temperature and developmental rate was examined for curd induction phase. The model can be used to compare with other findings and can correct the data of plant size at transplanting [31].

A mechanistic simulation model was described for Sorghum to predict climatic risk to production under water-limited environments over a broad range of subtropical environments, using the "top-down" approach. According to data in the simulation, 64% of the variation in grain yield and 94% of variation in the biomass are tested. In these simulations, yield, biomass, and leaf area index were predicted. By curve number technique, runoff was estimated [32].

The EPOVIR model was coupled to crop growth sub-model and soil–water balance sub-model to predict the yield and virus infection by PVY and PLRV in seed potato. The EPOVIR model was integrated into the TuberPro system which predicted tuber yields by sizes. TuberPro was used for the estimation of haulm destruction dates in the seed potato production. This model can be useful to the farmers and researcher because of the evaluation of virus control [33].

A dynamic simulation model of peanut was developed to estimate the climatic risk to the production of production potential for irrigated and dryland conditions. According to data, 89% of the variation is in the pod yield. The pod yield, biomass accumulation, crop leaf area, phenology, and soil water balance were simulated in this model [34].

A dynamic simulation model (TOMSIM) was confirmed for a number of fruits per truss, planting date, truss removal, plant density, temperature, and single/double-shoot plants. The validity was assessed by linear regression of simulated cumulative fraction of dry matter partitioned to the fruits against measured fraction. TOMSIM model, due to flexible and mechanistic approach, was exhibiting good agreement between simulation and measurement for a range of conditions [35].

Neto et al. advised to use the modeling techniques in agriculture to estimate interactions of the soil–plant–atmosphere system. The models are classified into mathematical model, empirical model, and mechanistic model. They also estimated the scientific importance, management applications, and practical applications like management of cropping system, formation of stocks, making of agricultural policies, etc. of mathematical models in agriculture sciences [36].

A dynamic simulation model TOMSIM examined the growth of tomato crop and also validated for four glasshouse experiments with fruit pruning treatments and plant density. According to sensitivity analysis, temperature, plant density, sink strength of a vegetative unit, and number of fruits per truss were less effective, whereas the developmental stage of a vegetative unit at leaf pruning, SLA, CO_2 concentration, and global radiation were more effective on the growth rate of crop. After planting, the large overestimation of crop growth rate was 52% on average in the 1 month. The dry biomass was overestimated by 0–31% due to the inaccurate simulation of LAI, resulting partly from inaccurate SLA [37].

Two approaches for modelling the growth and development of cassava, Manihot esculenta Crantz, were discussed. Two models referred to the composite model which based on Chanter's (1976) equation and the other based on the spill-over hypothesis were developed and comprised. Both models were only differing to hypothesis accounting for storage root growth. In both models, the growing demands of the stem, fibrous roots, and storage roots were related to leaf demand rates. The correlation coefficients of yield prediction for the models were r = 0.898, P = 0.0385 (1), and r = 0.954, P = 0.0117 (2) analyzed [38].

Process-oriented crop growth models were used to estimate spatial and temporal yield variability over different environmental conditions. The research has tried to develop ways for spatial interactions in the CROPGRO-Soybean and CERES-Maize models. These models can provide useful information of economic return of prescriptions, cause spatial yield variability, and also estimate the value of weather information on management prescriptions [39].

Application of the CERES-Wheat Model for Within-Season Prediction of Winter Wheat Yield in the United Kingdom was estimated. In this experiment, three seasons and four locations were included for model calibration. For each site, standard meteorological data were obtained. The CERES-Wheat model was capable to predict the upper and lower limit of observed yield according to the Spearman's rank correlation for all locations and year [40].

The transpiration rate of cucumber crop was investigated at ontogeny under low and high radiation condition. The coefficient of Penman–Monteith equation was calibrated under Mediterranean climate, used to estimate the transpiration rate per soil surface area of greenhouse cucumber crop and that was used to determine the water needs of this crop and establish efficient water use. Results show that the diurnal canopy transpiration rate was 4 times higher at high radiation as compared to low radiation. The leaf transpiration rate showed that VPD was higher in the afternoon. According to Ts, in the spring cycle, as G increased throughout the cycle, the VPD dropped from 3.5 to 2.5 kPa.

The use of the simplified Penman–Monteith equation for irrigation management under high radiation conditions, with VPD values inside the greenhouse surpassing 3 kPa, could ensure that the supply of water matches the requirements of the crop and helps avoid situations of water stress, and under low radiation the evaluation of nocturnal transpiration of VPD would prevent water deficits [41].

Regression model of dry matter was investigated for solar greenhouse cucumber. In this study, environmental data such as light intensity, temperature, and day length were accumulated. Cucumber growth data (dry matter weight of leaf, stem, fruit, and petiole) were studied to develop the model. The time state variable was expressed as a logistic function about ETA and ELIA. According to result, RMSE < 6 and the R^2 value was 0.99 for model [42].

CERES-Wheat and CropSyst models were examined for water–nitrogen interactions in wheat crop. In this study, silty clay loam soil of New Delhi and semiarid, subtropical climate were used. For the evaluation of model, coefficient of determination, RMSE, MAE, and Wilmot's index of agreement were estimated. In the RMSE, IoA data indicate that CropSyst model was more suitable to forecast growth, biomass, and yield of wheat under different N and irrigation application situations as compared to CERES-Wheat [43].

Two nonlinear regression models, Gordion F1 and Afrodit F1, were developed to analyzing cucumber yield. Both models individually predict the yield of cucumber. The light intensity, temperature, and SPAD values were used in this experiment. All data were examined by R-program. The yield was well correlated with high R^2 values. The R^2 values for Gordon and Kurtar were 0.8141 and 0.9369. The environmental factors and computer systems as well as growth factors should be included in the models to get exact result [44].

Soil parameters of a crop model were estimated and observations on crops improved the prediction of agro-environmental variables. The value of soil parameters was determined from several synthetic and actual observation sets earned in different situations like winter wheat and sugar beet crops grown in different weathers, soils, and cropping conditions. These estimates are reused in the model to predict agro-environmental variables. The result showed that high wheat yield is obtained [45].

The modal of the plant growth was estimated and evaluated to yield prediction. The classical methods were implemented in the PYGMALION platform to discrete dynamic models. For sugar beet growth, new LNAS model was presented and served to explain the importance of different methods which were used for the evaluation, parameterization, and yield prediction in PYGMALION [46].

The EFuNN model predicted the yield of tomato in greenhouse. Experimentation results show that EFuNN performed better than other techniques like ANN. According to results, the EFuNN model predicted weekly fluctuations of the yield with an average accuracy of 90%. The multiple EFUNNs can be helped in crop management. Important disadvantage of EFuNN is the determination of the network parameters like number and type of MF for each input variable, sensitivity threshold, error threshold, and the learning rates [47].

A dynamic model of tomato growth was investigated to estimate the crop mineral requirement under greenhouse conditions. In this model, several regression models include in order to simulate their dynamic behavior and which were generated from measurements of the concentrations of nutrient in the various organs of the tomato plant. The results showed that the growth model adequately simulates leaf and fruit weight (EF, Index >0.95). As for harvested fruits and harvested leaves, the simulation was EF and Index <0.90 as well as in case of Ca and Mg, simulations showed indices <0.90. Simulation of minerals was suitable for N, P, K, and S as both: EF and Index >0.95 [48].

3 Conclusion

Model is a more dominated system for agriculture because it provides valuable agronomic information's such as crop physiological processes, timing of sowing, irrigation, amount of fertilizer, and their impacts on soil–crop–environment. The crop modeling is also helping to compete against many insects, diseases of crops, or estimate both potential benefits and any unwanted impacts of management actions.

They are well supported by forecasting farmer decision-making, industry planning, operations management, and consequences of management decisions on environmental issues. Until now, various types of models have developed by agronomy researchers for the evaluation of the agricultural area such as PAPRAN, GOSSYM, GLYCIM, NTRM, CROPGRO, GPFARM, APSIM, and CROPGRO [49–56] models. The modeling techniques are well-determined potential benefits, and advised for any unwanted impacts of soil management, salinity management, and fertilizer management.

Also with this, the crop model is helping in the upliftment of agriculture because it can estimate the plant–soil–atmosphere interaction and can also predict the biomass, growth, yield, weather information, nutrient potential, etc. The models can assist growers in the control of crop supplies and help in the decision-making of crop management. Modeling and simulating techniques can not only help in minimizing the uses of pesticides and fertilizer in agriculture but also can help in reducing its negative effect on environment and improving crop quality. Crop model can bring revolution in the area of agriculture. Modeler can check and develop large number of management strategies such as salinity management, soil management, crop management, nutritional management, etc. in comparatively less cost and time. Crop model can play a key role in developing the quality of products along with sustainable development.

References

1. Neto, D.D., Teruel, D.A., Reichardt, K., Nielsen, D.R., Frizzone, J.A., Bacchi, O.O.S.: Principles of crop modelling and simulation: ii the implications of the objective in model development. Scientia Agricola **55**, 51–57 (1982)
2. Murthy, V.R.K.: Basic Principles of Agricultural Meteorology. Book Syndicate Publishers, Koti, Hyderabad (2002)
3. Monteith, J.L.: The quest for balance in crop modeling. Agron. J. **88**, 695–697 (1996)
4. Gabrielle, B., Roche, R., Angas, P., Martinez, C.C., Cosentino, L., Mantineo, M., Langensiepen, M., Henault, C., Laville, P., Nicoullaud, B., Gosse, G.: A priori parameterisation of the CERES soil-crop models and tests against several European data sets. Agronomie **22**, 119–132 (2002)
5. Houles, V., Mary, B., Guerif, M., Makowski, D., Justes, E.: Evaluation of the crop model STICS to recommend nitrogen fertilization rates according to agroenvironmental criteria. Agronomie **24**, 339–349 (2004)
6. Jones, C.A., Kinirty, J.R., Dyke, P.T.: CERES-Maize: a simulation model of maize growth and development. College Station, A & M University Press, Texas, USA (1986)
7. Houghton, D.D.: Handbook of Applied Meteorology, p. 1461. Wiley, New York (1985)
8. Dent, D., Veitch, S.: Natural resources information for policy development: a national salinity map. Aust. Collab. Land Eval. Program Newslett. **9**, 19–23 (2000)
9. MacDonald, R.B., Hall, F.G.: Global crop forecasting. Am. Assoc. Adv. Sci. **208**, 670–679 (1980)
10. Matis, J.H., Saito, T., Grant, W.E., Lwig, W.C., Ritchie, J.T.: A Markov chain approach to crop yield forecasting. Agric. Syst. **18**, 171–187 (1985)

11. Bouman, B.A.M., Van Diepen, C.A., Vossen, P., Van Derwal, T.: Simulation and system analysis tools for crop yield forecasting. In: Application of Systems Approaches at the Farm And Regional Levels, vol. 1, pp. 325–340. Kluwer Academic Publishers, Dordrecht (1995)
12. Jame, Y.W., Cutforth, H.W.: Crop growth models for decision support systems. Can. J. Plant Sci. **76**, 9–19 (1996)
13. Hoogenboom, G., White, J.W., Messina, C.D.: From genome to crop: integration through simulation modeling. Field Crops Res. **90**, 145–163 (2004)
14. Shin, D.W., Baigorria, G.A., Lim, Y.K., Cocke, LaRow, T.E., O'Brien, J.J., Jones, J.W.: Assessing crop yield simulations with various seasonal climate data. Science and Technology Infusion Climate Bulletin. NOAA's National Weather Service. In: 7th NOAA Annual Climate Prediction Application Science Workshop, Norman (2009)
15. NASS: The Yield Forecasting Program of NASS by the Statistical Methods Branch. Estimates Division, National Agricultural Statistics Service, U.S. Department of Agriculture, Washington, D.C., NASS Staff Report No. SMB 06-01 (2006)
16. Lobell, D.B., Burke, M.B.: On the use of statistical models to predict crop yield responses to climate change. Agric. For. Meteorol. **150**, 1443–1452 (2010)
17. Blackmer, T.M., Schepers, J.S.: Aerial photography to detect nitrogen stress in corn. J. Plant Physiol. **148**, 440–444 (1996)
18. Blackmer, T.M., Schepers, J.S., Varvel, G.E., Mayer, G.E.: Analysis of aerial photography for nitrogen stress within corn fields. Agron. J. **88**, 729–733 (1996)
19. Blackmer, T.M., Schepers, J.S., Varvel, G.E., Walter-Shea, E.A.: Nitrogen deficiency detection using reflected shortwave radiation from irrigated corn canopies. Agron. J. **88**, 1–5 (1996)
20. Hatfield, J.L., Pinter Jr., P.J.: Remote sensing for crop protection. Crop Protection **12**, 403–413 (1993)
21. Tucker, C.J.: Red and photographic infrared linear combinations for monitoring vegetation. Remote Sens. Environ. **8**, 127–150 (1979)
22. Tucker, C.J., Elgin Jr., J.H., Mcmurtrey, J.E.: Temporal spectral measurements of corn and soybean crops. Photogr. Eng. Remote Sens. **45**, 643–653 (1979)
23. Idso, S.B., Jackson, R.D., Reginato, R.J.: Remote-sensing for agricultural water management and crop yield prediction. Agric. Water Manag. **1**(4), 299–310 (1977)
24. Idso, S.B., Jackson, R.D., Reginato, R.J.: Remote-sensing of crop yields. Science **196**, 19–25 (1977)
25. Doraiswamy, P.C., Moulin, S., Cook, P.W., Stem, A.: Crop yield assessment from remote sensing. Photogramm.Eng.Remote Sens. **69**, 665–674 (2003)
26. Hatfield, J.L.: Remote-sensing estimators of potential and actual crop yield. Remote Sens. Environ. **13**, 301–311 (1983)
27. Pinter Jr., P.J., Fry, K.E., Guinn, G., Mauney, J.R.: Infrared thermometry—a remote-sensing technique for predicting yield in water-stressed cotton. Agric. Water Manag. **6**, 385–395 (1983)
28. Silberbush, M., Barber, S.A.: Phosphorus and potassium uptake of field-grown soybean cultivars predicted by a simulation model. Soil Sci. Soc. Am. J. **48**, 592–596 (1983)
29. Sinclair, T.R.: Water and nitrogen limitations in soybean grain production I. Model Dev. Field Crops Res. **15**, 125–141 (1986)
30. Yang, X., Short, T.H., Fox, R.D., Bauerle, W.L.: Dynamic modeling of the microclimate of a greenhouse cucumber row-crop: theoretical model. Trans. ASAE **33**, 1701–1709 (1990)
31. Grevsen, K., Olesen, J.E.: Modelling cauliflower development from transplanting to curd initiation. J.Hortic.Sci. **69**, 755–766 (1994)
32. Hammer, G.L., Muchow, R.C.: Assessing climatic risks to sorghum production in water limited subtropical environment. I. Development and testing of a simulation model. Field Crops Res. **36**, 221–234 (1994)

33. Nemecek, T., Derron, J.O., Fischlin, A., Roth, O.: Use of a crop growth model coupled to an epidemic model to forecast yield and virus infection in seed potatoes. In: 2nd International Potato Modelling Conference, Wageningen. Paper for session vii: Application of Models in Crop Production (1994)
34. Hammer, G.L., Sinclair, L.R., Boote, K.J., Wright, G.C., Meinke, H., Bell, M.J.: A peanut simulation model: I. Model Dev. Testing. Agron. J. **87**, 1085–1093 (1995)
35. Heuvelink, E.: Dry matter partitioning in tomato: validation of a dynamic simulation model. Ann. Bot. **77**, 71–80 (1996)
36. Neto, D.D., Teruel, D.A., Reichardt, K., Nielsen, D.R., Frizzone, J.A., Bacchi, O.O.S.: Principles of crop modeling and simulation: I. uses of mathematical models in agricultural science. Scientia Agricola **55**, 46–50 (1998)
37. Heuvelink, E.: Evaluation of a dynamic simulation model for tomato crop growth and development. Ann. Bot. **83**, 413–422 (1999)
38. Gray, V.M.: A Comparison of two approaches for modelling cassava (Manihot esculenta crantz.) crop growth. Ann. Bot. **85**, 77–90 (2000)
39. Batchelor, W.D., Basso, B., Paz, J.O.: Examples of strategies to analyze spatial and temporal yield variability using crop models. Eur. J. Agron. **18**, 141–158 (2002)
40. Bannayan, M., Crout, N.M.J., Hoogenboom, G.: Application of the CERES-wheat model for within-season prediction of winter wheat yield in the United Kingdom. Agron. J. **95**, 114–125 (2003)
41. Medrano, E., Lorenzo, P., Guerrero, C.S., Montero, J.I.: Evaluation and modelling of greenhouse cucumber crop transpiration under high and low radiation conditions. Sci. Hortic. **105**, 163–175 (2005)
42. Song, W., Qiao, X.: A regression model of dry matter accumulation for solar greenhouse cucumber. In: Li, D. (ed.) Computer and computing technologies in agriculture, vol. I; CCTA 2007, The International Federation for Information Processing, vol. 259. Springer, Baston, MA (2007)
43. Singh, A.K., Tripathy, R., Chopra, U.K.: Evaluation of CERES-wheat and cropsyst models for water-nitrogen interactions in wheat crop. Agric. Water Manag. **95**, 776–786 (2008)
44. Kurtar, E.S., Odabas, S.M.: Modelling of the yield of cucumber (Cucumis sativus L.) using light intensity, temperature and spad value. Adv. Food Sci. **32**, 170–173 (2010)
45. Varella, H., Guerif, M., Buis, S., Beaudoin, N.: Soil properties estimation by inversion of a crop model and observations on crops improves the prediction of agro-environmental variables. Eur. J. Agron. **33**, 139–147 (2010)
46. Cournede, P.H., Chen, Y., Wu, Q., Baey, C.: Development and evaluation of plant growth models: methodology and implementation in the PYGMALION Platform. Math. Model. Nat. Phenom. **8**, 112–130 (2012)
47. Qaddoum, K., Hines, E.L., Iliescu, D.D.: Yield Prediction for Tomato Greenhouse Using EFuNN. In: ISRN Artificial Intelligence, vol. 2013. Hindawi Publishing Corporation (2013)
48. Maldonado, A.J., Mendozal, A.B., Romenus, K.D.E., Diaz, A.B.M.: Dynamic modeling of mineral contents in greenhouse tomato crop. Agric. Sci. **5**, 114–123 (2014)
49. Seligman, N.G., Keulen, H.V.: PAPRAN: a simulation model of annual pasture production limited by rainfall and nitrogen. in simulation of nitrogen behavior of soil plant system. In: Frissel, M.J., Van Veen, J.A. (eds.) Proceedings of the Workshop. Centre for Agricultural Publishing and Documentation, Wageningen, Netherlands, pp. 192–221 (1980)
50. Baker, D.N., Lambert, J.R., McKinion, J.M.: GOSSYM: a simulator of cotton crop growth and yield. South Carolina. Agric. Exp. Sta, Tech. Bull. (USA) 1089 (1983)
51. Acock, B., Reddy, V.R., Whisler, F.D., Baker, D.N.: The Soybean Crop Simulator GLYCIM: Model Documentation. USDA, Washington, D.C (1985)
52. Shaffer, M.J., Pierce, F.J.: NTRM, a soil-crop simulation model for nitrogen, tillage, and crop–residue management. US Department of Agriculture, Conservation Research Report no. 34-1 (1987)

53. Hoogenboom, G., Jones, J.W., Boote, K.J.: Modeling growth, development and yield of grain legumes using SOYGRO, PNUTGRO and BEANGRO: a review. Trans. ASAE **35**, 2043–2056 (1992)
54. Ascough, J.C., Honson, J.D., Shaffer, M.J., Buchleiter, G.W., Bartling, P.N.S., Vandenberg, B., Edmunds, D.A., Wiles, L.J., McMaster, G.S., Ahujal, L.R.: The GPFARM decision support system for the whole form/ranch management. In: Proceedings of the Workshop on Computer Applications in Water Management, Great Plains Agriculture Council Publications no. 154 and Colorado Water Resources Research Institute Information Series no. 79, Fort Collins, CO, pp 53–56 (1995)
55. McCown, R.L., Hammer, G.L., Hargreaves, J.N.G., Holzworth, D.P., Freebairn, D.M.: ASIM: a novel software system for model development, model testing, and simulation in agricultural systems research. Agri. Syst. **50**, 255–271 (1996)
56. Boote, K.J., Jones, J.W., Hoogenboom, G., Pickering, N.B.: The CROPGRO model for grain legumes. In: Tsuji G.Y., Hoogenboom G., Thornton P.K. (eds.) Understanding Options for Agricultural Production. Systems Approaches for Sustainable Agricultural Development, vol. 7. Springer, Dordrecht (1998)

Software-Defined Networking—Imposed Security Measures Over Vulnerable Threats and Attacks

Umesh Kumar, Swapnesh Taterh and Nithesh Murugan Kaliyamurthy

Abstract Software-defined networking (SDN) is a new attempt in addressing the existing challenges in the legacy network architecture, SDN is renowned due to its basic approach in managing the networks and its capability of programmability. In software-defined networks' implementation, priority remains on high security. The advantage of SDN itself opens a wide ground in posing new security threats and challenges. Focusing on the security of the software-defined networks is a prime factor as it reflects on the growth of SDN technology implementation. This paper focused and designed with an introduction on software-defined networking, its architecture, the available security solutions for the network on the various existing security solutions available for software-defined networking, and the real challenge in securing the SDN networks providing the researchers a paved platform to work on further securing the networks, and finally it concludes with the requirements of security factors and schemes in SDN.

Keywords Software-defined networking · Transmission control protocol/Internet protocol (TCP/IP) · Open flow (OF) · Open network foundation · Application programming interface (API) · Spoofing

---

U. Kumar (✉)
Department of IT, Government Women Engineering College, Ajmer, India
e-mail: umesh@gweca.ac.in

S. Taterh · N. Murugan Kaliyamurthy
AIIT, Amity University, Jaipur, Rajasthan, India
e-mail: staterh@jpr.amity.edu

N. Murugan Kaliyamurthy
e-mail: k.nitheesh.murugan@gmail.com

K. Ray et al. (eds.), *Soft Computing: Theories and Applications*,
Advances in Intelligent Systems and Computing 742,
https://doi.org/10.1007/978-981-13-0589-4_67

1 Introduction

On talking about the next generation in networking, the prime focus should be on software-defined networking. Decoupling the data plane from the control plane of the network enables the SDN design to be capable enough to provide a manageable, reliable, and flexible network [1].

This design makes the physical devices such as routers and switches to act only as a forwarding agent of the network traffic which in turn makes them vendor-independent, cost-effective, and moreover flexible and creative network design. The physical devices, being only the forwarding agents, the control facility of the data is moved to a centralized controller enabling a wider and broader view of the network. In simple, the SDN architecture provides a keen-sighted view over the network facilitating a flexible, reliable, and a better-managed network architecture in terms of the flow of data.

Focusing on the various advantages of SDN, the loopholes and the drawbacks of SDN also need primary attention. Fixing the drawbacks and increasing the advantages of SDN only will increase the implementation factor of SDN networks. With all the exciting features and flexibility that SDN provides to a network, the security factor of the network requires more functionality and high concern of improvability [1].

On various platforms where the SDN implementation factors are discussed, the experts recommend to address the security issues around SDN. The software-defined networking architecture poses various internal and external threats and vulnerabilities because of its centralized controller design, which questions the integrity and security of the software-defined networking [1]. Because the controller has the entire network on it, the controller itself can be easily used for additional attacks and collapsing the network.

One of the weak points in software-defined networking is that the advantages itself pose a variable threat to the network [2]. As discussed in the above paragraph about its centralized controller design, it is another prime functionality—the programmability also poses a strong threat to the network, due to malicious code exploits and attacks.

Additionally, the denial-of-service attacks and side-channel attacks are easily targeted in the southbound interface of the SDN network. Outraging the current scenario, cyberattacks are piece of cakes for the intruders or attackers in software-defined networks which will cause more damage to the network, on comparing with the legacy networks. Concentrating on improvising the security features in software-defined networking, each layer of SDN needs to be focused on achieving better-secured environment instead of implementing security overall to a network, because each layer has different implications and different requirements.

Main focus of security in SDN is the controller which requires dynamic, robust, and strong security policies. SDN is already a flexible, reliable, and managed network architecture, and it further needs a scalable and secured environment, for which this paper paves a pathway for the researchers to achieve the loopholes in software-defined networking.

This paper further discusses the security solutions which are available with software-defined networking and further threats and attacks possibility on SDN networks. This study concludes with a note on required steps to be taken in achieving a secured SDN network.

2 SDN Architecture Design

This part of the paper discusses in detail about the design and the security concerns of SDN architecture. The basic cling between the existing network design and the SDN is its separation of control and data plane and the facility of programmability in the control plane [1].

Figure 1 gives a clear insight on the architecture of software-defined networking where the third-party applications sit on the SDN controllers in the control plane which is decoupled from the data plane enabling improvised network operations. With reference to Fig. 1, the software-defined networking is a three-layered architecture. The data plane is the physical networking devices, controlled by the control plane through various protocols. The management plane is the software platform which is helpful to control the entire network flow. Software-defined networking focuses on dynamic flow control, network-wide visibility with centralized control, network programmability, and simplified data plane [3]. A network application program (management plane) will control the data planes in software-defined networking [4]. The physical networking devices in SDN will act as only a forwarding device and the "where and how" part will be done by the control plane [5].

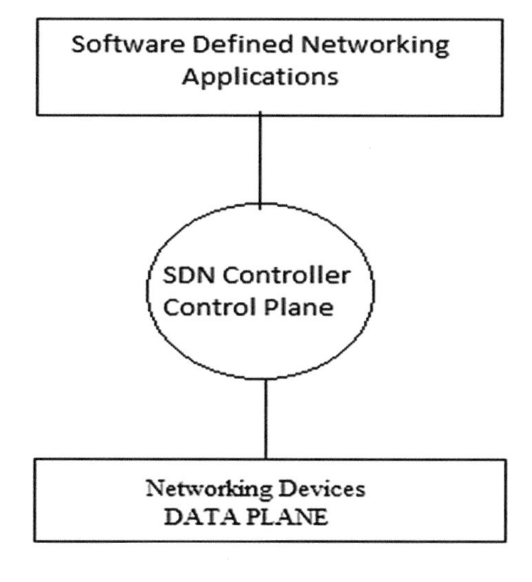

Fig. 1 SDN architecture [6]

The application layer in the SDN is responsible to provide services and run applications related to security, i.e., Intrusion detection, prevention systems, load balancing, deep packet inspections, security monitoring tools, access controls, etc.

3 Security Tools for Various SDN Layers

The SDN architecture, classified as northbound, southbound, control, data, and application layers, has various tools available to secure the network. This part of the paper discusses the existing tools available in securing SDNs. Fresco, Fortnox, Avant-Guard, and OpenWatch [7] are few tools which focus on the security design and analysis of the SDN network. They work on various layers in SDN. Fresco is an OpenFlow security-specific application which focuses on secure design development [6]. It is developed for NOX controller, in the year 2013, and works on application layer, control layer, northbound and southbound interfaces [7]. Fresco focuses on capturing the application programming interface (API) scripts to combat security threats by developing security monitoring features in an SDN architecture [1]. FortNox is also similar to Fresco, working on northbound and southbound and the control layer. It is not focusing on the application layer [6]. FortNox focuses on interpreting the security rule conflicts arising during security authorizations in a security enforcement kernel [1].

Avant-Guard focuses on the security analysis of the SDN network, working on the data layer, control layer, and northbound and southbound interfaces. OpenWatch is also similar tool like Avant-Guard focusing on the security analysis of the SDN network, however, working on the application layer also along with data, control, northbound, and southbound interfaces. These security analysis tools are proposed with a systematic approach to collect more information about the data plane. The collected information are used to gradually reduce the interactions between the data and the control plane keeping in mind the denial-of-service attacks.

Tools like Verificare and SDN debugger work on the security audit of the SDN architecture. Verificare works on the data, control, and northbound and southbound interfaces, whereas the SDN debugger works on the application layer and southbound interface.

To enforce the security policy, tools like VeriFlow and Flyover are implemented which work on the various layers of SDN architecture. Enforcement of security policy plays a vital role in SDN environments. Flyover [8] application was proposed to address the security enforcement policy by checking the flow policies in contradiction with the network policy deployed. FleXam [9] is another tool works with OpenFlow, which is designed to provide access to the controller of the OpenFlow to receive packet level information. The FleXam also seizes the low setup time and also minimizes the load of the control plane. The above is the list of few tools which are working on the security aspects of SDN.

4 Threats and Attacks in SDN Architecture

Discussing more on the existing tools and security measures in an SDN architecture gives a vast idea of the current threats and attacks which are possible in the network. The security aspects in the SDN environment are prone to be the key factor of research in today's development because of the reason that SDN environment did not consider security as a prime factor during its initial design [1].

Based on the available applications or tools in SDN, a clear factor reveals that contributions in secure design of the SDN are limited. More research works focus on the security enforcement policy rather than the security enhancement and security analysis. Threats and attacks in an SDN architecture need to be focused individually proportionate to different layers or interfaces in the SDN architecture rather than developing a common solution for SDN. There are various different possible threats, attacks, and exploits targeting each layer of the network. The various security mechanisms like access control, intrusion detection and prevention, encryption, authentication, and authorization needs to be focused based on the security requirements in each interface. Major part of SDN networks are designed based on the OpenFlow standards; our further research is focused on securing and defending OpenFlow networks.

Security threats in SDN networks are grouped under the following categories.

- Alteration,
- Modification,
- Authorization, and
- Impersonation.

Altering the operating system, the data configuration files, user data are classified under alteration. Modifying the software framework, creating a software and hardware failure, and extracting the data configuration files are classified under modification. Unauthorized access to the network is classified under authorization and impersonating as an authorized SDN controller in the network is classified under impersonation [1].

The operating system, when altered, shall cause a high impact of damage to the components like controller and forwarding nodes in SDN networks. This type of threats can be handled by ensuring high system integrity protection. In order to achieve a high system integration, trusted computing is required. This type of threat can always attack all the layers of SDN network. Like the alteration of operating system, the alteration of data configuration files is also a high impact threat factor which can cause damage to the data that are essential in performing effective SDN operations. This type of threat can always attack the control layer, the data layer, and the control–data interface. Alteration of user data is another alteration threat which troubles the user data like their profiles. Mostly, this type of threat affects the data layer. Both the data configuration and user data threats can be handled by ensuring data integrity in the SDN network.

The modification type of threat, software framework alteration, damages the middleware and the components of the software framework. Software framework alteration also attacks all the layers of the SDN network. Like the alteration of operating system threat, this threat can also be mitigated using high system integration. The software and hardware failure is one of the common threats highlighting the general software and hardware resource failure. Improving the robustness of the software and high configured hardware can be a way of mitigating the threats. The software failure can target all the layers, whereas the hardware failure affects the control and data layers of SDN architecture. One of the challenges in securing the network is based on data security. The configuration data extraction threat is a threat where the attackers gather information from various sources and various methods. The data collected is accumulated to perform subsequent attacks. In order to mitigate this threat, data integrity and confidentiality needs to be improvised. This threat targets the control and data layer including the control—data interface.

Unauthorized access to authorization is a type of threat where possibility of security breach happens. High level of security policy and security administration is required to mitigate this type of threat. This type of authorization threat can target all the layers and almost all the functionalities in the network architecture.

Impersonation of the SDN controller is one of the toughest threats in SDN which compromises the entire SDN architecture. SDN controller, being the heart of the SDN network if masqueraded, leads to a damage of entire network without the knowledge of the network itself. Techniques such as digital signatures and public key encryptions can mitigate this type of threats.

5 Securing the SDN Architecture

In the perception of securing the SDN architecture, few changes in the architecture shall be suggested based on the above classifications. Policy reinforcement can be done in the first position to secure the network. Having high enforced policies in the network shall mitigate attacks related to policy enforcement in the southbound, data, and control layers. Availability-related attacks require more attention as they are the basic steps taken in tightening the security in a network. Availability of devices and other resources in a network can cause damages in both the extremes. Availability and overload of the network. The application, northbound, and the control layer requires high level of security in the focus of availability of resources. Authorization-related attacks generally damage the southbound, data, and control layers, whereas the authentication-related attacks damage the application, northbound, and the control layer. Securing the network from both these attacks is highly efficient by providing digital certificates and using public key encryption of data. Data alteration and controller impersonation are few security breaches which attack the southbound, data, and control layers. Unauthorized rule insertions and side-channel attacks are also few attacks which cause damage to the application, northbound, and the control layer.

6 Conclusion

The emerging trend in the networking domain—Software-defined networking is highlighted for adoption from the legacy network architecture. On this point, the security factor is focused as a prime factor in reducing the implementation of SDN taking over from the legacy networks. Concentrating on the existing security aspects of the SDN architecture, the requirements of new mitigation schemes are widely discussed. Factors such as identity management and threat isolation deployment are required to mitigate the threats and attacks.

References

1. Securing Software Defined Networks: Taxonomy, Requirements and Open Issues. IEEE Commun. Mag. (2015)
2. Diego Kreutz, F.M.V.R.P.E.V.C.E.R.S.A.S.U.: Software-defined networking: a comprehensive survey. In: Proceedings of the IEEE, vol. 103, no. 1, pp. 14–76 (2015)
3. Porras, P., et al.: A security enforcement kernel for OpenFlow networks. In: Proceeding 1st Wksp. Hot Topics in Software Defined Networks, pp. 121–126. ACM (2012)
4. Seungwon Shin, L.X.S.H.G.G.: Enhancing network security through software defined networking (SDN). In: 25th International Conference on Computer Communication and Networks (ICCCN) (2016)
5. Kannan Govindarajan, K.C.M.H.O.: A literature review on software-defined networking (SDN) research topics, challenges and solutions. In: Fifth International Conference on Advanced Computing (ICoAC) (2013)
6. Murugan Kaliyamurthy, N., Taterh, S.: Understanding Software Defined Networking—A Study on the Existing Software Defined Networking Technologies And Its Security Impact. Yet to Publish (2017)
7. Seugwon Shin, P.P.V.Y.M.F.G.G.M.T.: FRESCO: modular composable security services for software-defined networks. In: ISOC Network and Distributed System Security Symposium (2013)
8. Son, S., et al.: Model checking invariant security properties in OpenFlow. In: Proceedings of IEEE ICC, 2013, pp. 1974–1979 (2013)
9. Shirali-Shahreza, S., Ganjali, Y.: FleXam: flexible sampling extension for monitoring and security applications in Openflow. In: Proceedings of 2nd ACM SIGCOMM Workshop. Hot Topics in Software Defined Networking, pp. 167–168 (2013)

Working of an Ontology Established Search Method in Social Networks

Umesh Kumar, Uma Sharma and Swapnesh Taterh

Abstract Recently, online social networking is a specific kind of promising group and has drawn masses. These networks have delivered distinct provisions for clients and permit them to fabricate a description and share their awareness and understanding. Exploring users' profiles and their statistics is the most frequent of works in social networks. We introduce a way of searching method which is depending on ontology in our paper. Here, ontology is constructed for each profile in social network and then the same is implemented for client's query. Then, these ontologies are matched using ontology matching techniques, and the similarities between them are evaluated. Lastly, according to similarities, results are developed and represented that are related to user's query. The experiments show that the presented method is ample and gives satisfactory results.

Keywords Clustering · Ontology · Social networks · Ontology matching

1 Introduction

Social networks are websites of the latest kind that presently have popularity as the most renowned websites. New applications are built as the technology advances. A key role in online communications is played by social networks. Moreover, the sharing and administration of the data along with information and knowledge are influenced by them. A systematic and effective searching methodology in social

---

U. Kumar (✉)
Department of IT, Government Women Engineering College, Ajmer, India
e-mail: umesh@gweca.ac.in

U. Sharma
Department of CSE, Banasthali Vidhyapeeth, Niwai, Tonk, Rajasthan, India
e-mail: Uma.banasthali16@gmail.com

S. Taterh
AIIT, Amity University, Jaipur, Rajasthan, India
e-mail: staterh@jpr.amity.edu

K. Ray et al. (eds.), *Soft Computing: Theories and Applications*,
Advances in Intelligent Systems and Computing 742,
https://doi.org/10.1007/978-981-13-0589-4_68

networks is critical. Hence, major search engines like Google and Yahoo are deploying services that leverage social networks [1].

Multiple services like chatting, messaging, emails, multimedia data transfer, etc. are rendered for clients by these networks so that they are able to communicate among themselves. Searching is the most carried out operation for which it is necessary to supply suitable results to the client. A large number of technologies are in this field but we cannot be certain which one supply the links corresponding to the client's inquiry precisely with less irrelevant results. So, in this paper, we try to give out a procedure which attempts to give output to users' query that is more precise and admissible.

The rest of our paper is ordered as follows: the introduction of social networks and a general summary of a number of search methods are laid out in Sect. 2. In Sect. 3, the proposed technique is narrated and in Sect. 4, the experiments and their outcomes will be examined. Finally, the conclusion of paper is in Sect. 5

2 Related Work

Internet has remarkably expanded in fast few years and has led to increase in user interactions. So, through the Internet, single users and various organizations can interchange their information and have access to massive amounts of data on the Internet effortlessly. Hence, one of the most significant uses of the global network is search querying on the Internet.

By querying on the Internet, a user can find a considerable portion of what they are searching for. The user generally states what they would be searching for in the format of a query request. A set of data in the form of text, images, and other items are provided to them depending on the query. To provide this data, it is important to carry out operations about which a large number of studies have been done and a number of methods and technologies have been presented.

Stemming: In this mechanism, initially, a catalogue of keywords contained in the user's query and the content of web pages or other such kinds of documents are first drawn out and depending on their structural and morphological characteristics, these words are compared with each other. In morphology, regarded as a field of linguistics, the internal structure of words is examined. The different kinds of a word may not influence the search output based on morphology. For example, the meaning of the word will not be influenced in case the word is used in the singular or plural form, and both results will be given out as search output applicable to the given word [2]. Hence, stemming is typically used in the information retrieval systems field and refines coverage of matching, but it also creates some noise. For example, as stated in [3], if we use the stemmer, then 'marine vegetation' and 'marinated vegetables' both stem to 'marin veget', which is not desirable.

Methods that are based on WorldNet: These methods (e.g. [4, 5]) are used for finding the nearness between two concepts of WorldNet. WorldNet is a collection of data made of English words. In this database, every word (like noun, verb and adverb), including another type of words, forms a group and explains a particular idea, till they are synonymous with each other. For each idea, there can be different groups that each one has a particular correlation with that idea. Those includes synonymous with that idea, groups that are presented in more generic sense are called as hypernyms and the groups which is expressed in more accurate way are said as hyponym [6].

WorldNet does not give us any information about several compound words although it has a reasonable comprehensive word set.

Translation models: Statistics and probability are the basis of these models. To find out candidate document that is more probable to be a transformation (or translation) of the search query is the primary goal of this model [7]. Although these models have been earlier used for various functions such as retrieval of documents [7], answering various questions [8] but on texts of short length, such as search queries, these models are less productive.

Query expansion: Here, in this technique, the query is converted from its preliminary structure into a more substantial structure which can include more ideas by attaching a set of terms which are related to the primary search query. In this way, a more appropriate content will be drawn out. These contents are drawn out from various web pages, documents and datasets that are extracted in reply to the preliminary query [9]. Therefore, the method that carries out the primary search is also crucial as it is likely that drawn out terms may be incorrect. Unwanted variations in the meaning of query from primary form may occur by adding words.

In this paper, the point in question is how to find required users' material on a social networking site. These networks are a specific kind of implicit community and for these kinds of networks there are different kinds of definitions for social networks. Some of these are as follows:

Online social network—'OSNs create online groups between people which have same appeal, action, framework, and/or society. Majority of OSNs are depending on web and give facility for users to upload profiles in the form of wording, images, and videos and communicates to each other in innumerable aspect' [10].

Social networking website—'Social networking websites are based on services of web that permit humans to make a universal or semi-universal profile limited to surrounded system and express a list of different users with whom they distribute a connection, and see and negotiate their list of links and those created by other users within the system' [11].

Online social networks, which are not like the general websites that are coordinated around their content, are organized keeping their users at the centre, i.e. social networks constitute of user profiles and interaction between them.

There are many kinds of information on social networks that generally coordinate in the form of user's account. Fast and trustworthy access to this information is one of the primary concerns of searching.

3 Proposed Method

In the present section, as ontology plays a primary role in searching, we introduce a technique for query searching where the notion of ontology and its process of matching are used. So, short definitions of the two ideas are laid:

Ontology: In common terms, ontology points to the field of philosophy which deals with the prevailing nature or being as such. Ontology is obtained by combination of words ontos that sense 'to be' and logos that sense 'word' [12]. There are various different meanings for ontology. Here, one of the most famous definitions about the otology is given by Gruber [13]: 'ontology is the traditional and obvious statement of a distributed perception'.

Matching of ontology: as presented in [14], 'Matching of Ontology is that procedure in which matching among the two basic ontologies are found. Specification of similarities among the two given ontologies is the outcome of a matching process. Various ontologies are input to the operation and specification of the equivalence between the ontologies is the output'.

Based on the presented ideas, the procedure for searching is planned. This process has certain gaits which are displayed in Fig. 1. First gait is to create ontology of profiles of users of the specified social networking site. Then, in the next step, the users' ontology of search query is created. Then, the process of matching among the ontologies obtained from the previous two steps is carried out. At last, the results are designated to each outcome and they are then arranged accordingly.

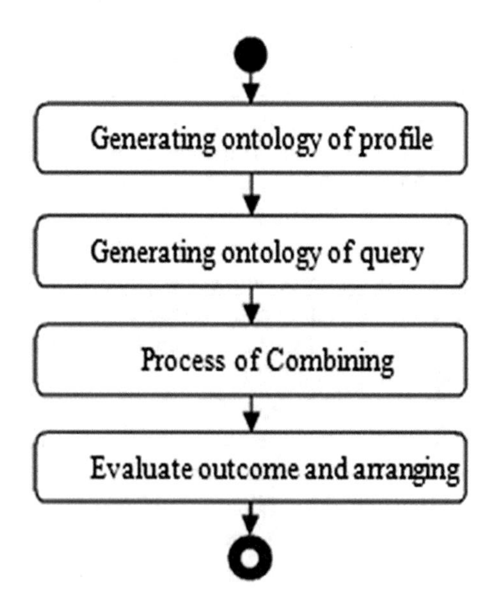

Fig. 1 Process of proposed search method

Fig. 2 Music player's ontology

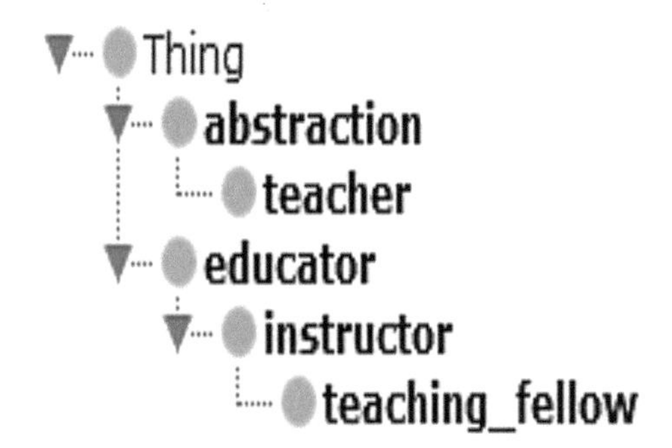

Fig. 3 Teacher's ontology

A. Creating ontology of the members' profiles

Profile of every user on a given social networking site comprises many different fields such as personal information of the user (like his name and photograph), his various interests, his job profile, educational background of the user and his contacts and links. In this paper, the job information of members of a social networking site is used to illustrate the working and efficiency of the method that is proposed here. Therefore, at first, we create ontology for each of the job titles that any of the users has. Every job title is stored as award list. Then, set of related words (semantically) are drawn out. These words are drawn out from the WordNet are categorized into three different categories: synonyms, hypernyms and hyponyms. Creation of the ontology starts after extracting the words. These words form ontology classes which are similar with each other on basis of parent-class to child-class connection. In Figs. 2 and 3, two different ontologies are displayed that are acquired by the technique given here.

As presented in Figs. 2 and 3, both ontologies have common node which is called 'Thing'. This becomes the root for these ontologies as there is use of OWL language and by default, the ontologies that are being created by use of the OWL language have this common node as the root node [15].

The number of children of thing and the total number of synonyms for a word are same. Every other word can have certain distinct meaning and for every distinct meaning, there exists a synonym set and for every synonym set there exists a hypernym set and generally a hyponym set. Various terms are there in each set. Thus, a term from each set will be chosen at random. As some of the sets contain terms that are common, it has to be constantly monitored for the chosen terms that should not be repeated as these terms are used as names for category of ontology and must not have the repeated name. The terms are then stored in a 3 * n size matrix, where n represents the total number of set of synonyms. In the matrix's second row, each cell has a word drawn out from one of the n available sets of synonyms and for each given word, one word from its hypernyms and also one word from its hyponyms are drawn out and assigned to lower cell and upper cell, correspondingly. Null value is assigned to the lower cell, if a word does not have a hyponym.

The process of construction starts after distillation of the matrix terms. If ontology is studied as a graph, the matrix's first row will be treated as children of level one or first level, second row will be considered as intermediate nodes, and the third row will be treated as leaves, i.e., branch of the ontology graph is constructed of columns of the matrix and accordingly each given cell is a subclass or derived class of its base class cell.

B. Creation of query's ontology

Query is broken down into its tokens by a tokenizer, after it is collected in the form of a string. Among these tokens, pre-positions are taken out and only keywords are kept. If needed, the main form of keywords is drawn out. Then, creation of ontology is done for each keyword. At last, these ontologies become input for the process of ontology matching.

C. Ontology matching and measurement of similarity

The input of this step is the result outputted in the first and the second steps. In this section, each ontology for profiles is matched with created ontology or ontologies for query. To calculate the similarity among given ontologies, the similarity among their entities has to be obtained first. There are many methods for this function, such as

- Methods based on string,
- Methods based on language,
- Methods based on constraints,
- Resources related to linguistic,
- Methods based on graph and
- Methods based on statistics and analysis of data [16].

From the given methods, we use methods based on strings and resources related to linguistic. For the computation of similarity measures, each class of ontology is drawn out. After that, one or more of the abovementioned techniques are applied to them. Thus, the matching between different categories are computed and then it is recorded in a matrix called as similarity matrix.

Methods based on strings: First, the edit distance among the categories of query's ontology and category of profile's ontology (in manner of two by two) are computed by the edit distance method and the outcomes are saved in the similarity matrix. Edit distance is the required number of performance that is required to change one string or one word into another string. We use the formula to compute the final edit distance which is given below, as scoring should fall in range of 0 and 1:

Levenstein = 1.0-[op/Max (Source Length, Target Length)]

Here, Op is the required number of performance required to convert a given string into another. To compute the edit distance, source and target are two strings as input. Length calculates the overall number of characters that are in the string of input.

Resources related to linguistic: To implement this type methodology, the classes of ontology of query and classes of ontology of profile classes (in manner of two by two) are collated with each other, and then the nearness measures are calculated as per the criteria given below:

1. The value of similarity is set to 1, in case both class names are identical.
2. The value of similarity is set to 0.8, in case ontology of profile class belongs to synsets of class of ontology of query.
3. The value of nearness is set to 0.5, in case ontology of profile category belongs to hyponyms of ontology of query category.
4. The value of nearness is set to 0.3, in case profile's ontology of profile category belongs to hypernyms of ontology of query category.

A nearness matrix is generated after applying these techniques on ontology of other profile. Therefore, for each ontology of profile, number of nearness matrix created is two.

Providing scores to every user: This step computes the average of the values *Sim* of each similarity matrix after taking the similarity matrixes as. Thus, two values are obtained for each user. As the result, after use of the second method is more important than that obtained in first method; for each of them, we give a weight. At last, for each member, a nearness value is computed using the given formula:

$$\mathrm{Sim}_{\mathrm{user(i)}} = \frac{(\mathrm{w}_1 \times \mathrm{a}) + (\mathrm{w}_2 \times \mathrm{b})}{\mathrm{W}_1 + \mathrm{W}_{2,}}, \mathrm{w}_1 > \mathrm{w}_2, 1 \le \mathrm{i} \le \mathrm{k}$$

$$a = \frac{\sum_{i=0}^{n} \sum_{j=0}^{m} Matrix\_Lingua[i][j]}{n \times m}$$

$$b = \frac{\sum_{i=0}^{n} \sum_{j=0}^{m} Matrix\_ED[i][j]}{n \times m}$$

where K represents the total number of members, Matrix_ED is the similarity matrix that we get in first technique and Matrix_Lingua is the similarity matrix of second technique. The *Sim* values are sorted in descending order after processing the above steps.

4 Experiment

It appears that the proposed methodology can give us a more satisfactory and dependable response. Thus, the results we get after performing the experiments are presented here to evaluate the discussed approach. In this experiment, some of the Google+ users' job title is being taken into account. Similar to other social networking sites, Google permits members to create account and also distribute their knowledge. Every social network generally has a specific structure for their accounts so they have various topics, titles and types of information. Figure 3 displays the Google outline structure. As given in Fig. 4, outline of Google is made up of around nine segments and each segment has a subpart.

In the first demonstration, the input query of the proposed method gets a single word. This method is used only for information confined to the part 'Work', so just this section is depicted in detail. 'Job title' is a field in 'Employment', which presents the user's title and its contents. As per the above sections, for every title of job, ontology is generated and after that the same process is carried out for the query of user.

At last, the procedure of joining and then scoring are performed. Figure 5 displays the outcome of the first demonstration where the user is searching for 'Teacher'.

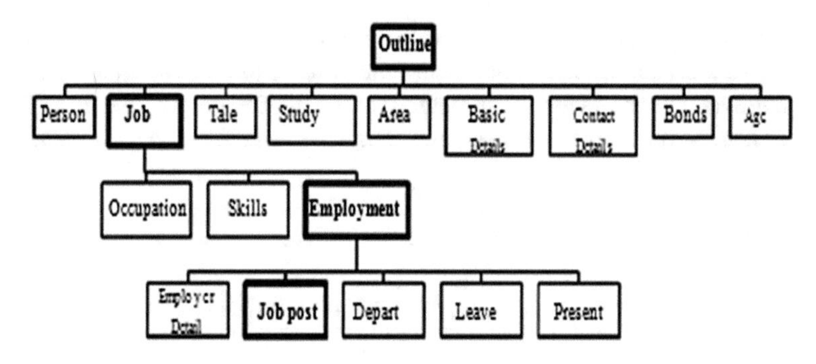

Fig. 4 Google outline formation

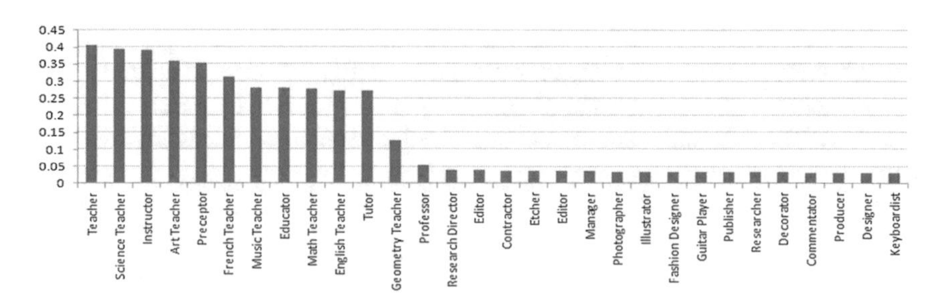

Fig. 5 Outcome of the first demonstration in which the user is searching for 'Teacher'

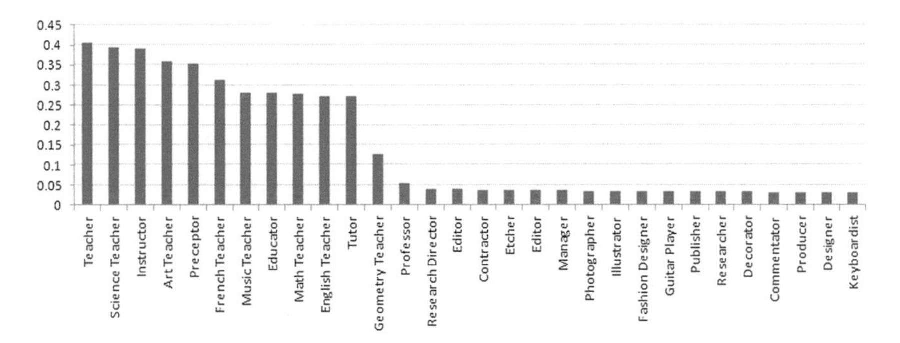

Fig. 6 Outcomes 'Expert Programmer' searches

In the second demonstration, the input for the technique proposed is a query having more than single term. If the total number of keywords drawn out from the search query is more than a single word or the top 30 outcomes are displayed in this figure, the outcome presents that the method which depends on ontology is efficient and presents outcomes in a significant way.

Number of ontologies of query is more than one, every word is matched separately with every single ontology of the profiles. Thus, every user is appointed various scores of which an average is calculated to form the user's final score. At last, the final scores are arranged in a top to bottom order. Figure 6 displays the outcome of the second demonstration where the user is searching for 'Expert Programmer'. The top 30 outcomes are displayed in this figure.

As displayed in these figures, users are arranged by the scores gained, i.e. they are arranged as per the nearness of their job name with the specific job name. The outcomes display that the method used can effectively find the suitable results and it is capable of supplying knowledge which is suitable for the search query of user.

5 Conclusion

Observing and discovering suitable knowledge and giving this knowledge to users is a critical issue in field of information retrieval. The efficient and economical search in social networking website is critical for the users. In our paper, we presented a technique which depends on ontology search in social networks. So, very first, the introduction of these networks was given along with an analysis of existing techniques for similar text. Then, the method given was presented along with the outcomes of the demonstration, which were carried out using this technique. The outcomes displayed that the method given performs efficiently and is capable of supplying trustworthy outcomes.

References

1. Burleigh, S., Cerf, V., Durst, R., Fall, K., Hooke, A., Scott, K., Weiss, H.: The interplanetary internet: a communications infrastructure for Mars exploration. In: 53rd International Astronautically Congress, The World Space Congress, Oct 2002
2. Lammerman, L., et al.: Earth science vision: platform technology challenges. In: Proceedings of International Geosciences and Remote Sensing Symposium (IGARSS 2001), Sydney, Australia (2001)
3. The sensor web project, NASA Jet Proplusion Laboratory. http://sensorwebs.jpl.nasa.gov
4. CENS: Seismic monitoring and structural response. http://www.cens.ucla.edu/Research/Applications/Seismicmonitor.htm
5. Szewczyk, R., Osterweil, E., Polastre, J., Hamilton, M., Mainwaring, A., Estrin, D.: Habitat monitoring with sensor networks. Commun. ACM **47**(6), 34–36 (2004)
6. Dobra, A., Garofalakis, M., Gehrke, J., Rastogi, R.: Processing complex aggregate queries over data streams. In: Proceedings of SIGMOD International Conference on Data Management, pp. 61–72 (2002)
7. Madden, S., Szewczyk, R., Franklin, M., Culler, D.: Supporting aggregate queries over Ad-Hoc WSNs. In: Proceedings of Workshop on Mobile Computing Systems and Applications, pp. 49–58 (2002)
8. Woo, A., Madden, S., Govindan, R.: Networking support for query processing in sensor networks. Commun. ACM **47**, 47–52 (2004)
9. Chaudhari, Q.M., Serpedin, E., Qarage, K.: On minimum variance unbiased estimation of clock offset in as two—Wa message exchange mechanism. IEEE Trans. Inf. Theory **56**(6) (2010)
10. Schmid, T., Charbiwala, Z., Shea, R., Srivastava, M.B.: Temperature compensated time synchronization. IEEE Embed. Syst. Lett. **1**(2) (2009)
11. Chen, J., Yu, Q., Zhang, Y., Chen, H.-H., Sun, Y.: Feedback-based clock synchronization in wireless sensor networks: a control theoretic approach. IEEE Trans. Veh. Technol. **59**(6) (2010)
12. Kim, H., Kim, D., Yoo, S.-E.: Cluster-based hierarchical time synchronization for multi-hop wireless sensor networks. In: 20th International Conference on Advanced Information Networking and Applications, 2006. AINA 2006, Apr 2006, vol. 2, no. 5, pp. 18–20 (2006)
13. Shao-long, D., Tao, X.: Cluster-based power efficient time synchronization in WSN. In: Proceedings of IEEE International Conference on Electro/Information Technology, pp. 147–151 (2006)
14. Ganeriwal, S., Kumar, R., Srivastava, M.B.: Timing-sync protocol for sensor networks. In: Proceeding of First International Conference on Embedded Networked Sensor Systems (2003)
15. Kim, B.-K., Hong, S.-H., Hur, K., Eom, D.-S.: Energy-efficient and rapid time synchronization for WSN. IEEE Trans. Consum. Electron. **56**(4) (2010)
16. Chauhan, S., Kumar Awasthi, L.: Adaptive time synchronization for homogeneous WSNs. Int. J. Radio Freq. Identif. Wirel. Sens. Netw. **1**(1) (2011)

Author Index

A
Abbas, S. Q, 647
Abdul Raoof Wani, 261
Abdul Wahid, 75
Abhishek Singh Rathore, 85
Ajay Mittal, 273
Ajeet Singh Poonia, 687, 693
Ambuj Kumar Agarwal, 235
Amit Sharma, 311
Anil Kumar Solanki, 205
Anita Singh, 53
Ankit Ranjan, 629
Anupam Vyas, 399, 443
Anurag Bhardwaj, 217
Anuranjana, 629
Anuranjan Mishra, 419
Archana Srivastava, 223
Arjun Verma, 343
Arpita Banerjee, 687, 693
Ashfaue Alam, 591
Ashima Kanwar, 493
Ashish Mani, 605
Avinash Sharma, 1

B
Banerjee, C, 687, 693
Bappaditya Jana, 525
Bharati Sharma, 483
Bhumika Gupta, 235

C
Chaturvedi, D. K, 605

D
Dambarudhar Seth, 159
Daniel, A. K, 303
Deepti Mehrotra, 13, 75, 629, 639, 671
Devesh Pratap Singh, 217
Divya Prakash, 679, 699

G
Gaurav Joshi, 591
Geeta Kumari, 591
Girish K. Singh, 321
Govardhan, A, 245
Gupta, S. K, 133, 147
Gur Mauj Saran Srivastava, 295

H
Harish, B. S, 515
Heet Rajput, 453
Himanshu Gupta, 343

I
Indu Maurya, 133, 147
Irita Mishra, 557
Ishani Mishra, 557

J
Jamvant Singh Kumare, 181
Jay Prakash, 557
Jitendra Rajpurohit, 679
Jyoti Sihag, 699

K
Kamalpreet Bindra, 419
Kanad Ray, 159
Kavita, 283
Kavita Singh, 351
Kishanprasad Gunale, 473

K. Ray et al. (eds.), *Soft Computing: Theories and Applications*,
Advances in Intelligent Systems and Computing 742,
https://doi.org/10.1007/978-981-13-0589-4

M
Mahira Kirmani, 63
Maiya Din, 193
Malay Kumar Pandit, 547
Manikanta, G, 605
Manish Joshi, 235
Manoj Diwakar, 343
Manoj K. Sabnis, 283
Manoj Kumar Shukla, 283
Mayank Satya Prakash Sharma, 483
Mayur Rahul, 33
Meenakshi Srivastava, 647
Mohit Agarwal, 295, 505
Mohsin Mohd, 63
Moumita Chakraborty, 525
Mudasir Mohd, 63
Muttoo, S. K, 193

N
Nagaraju Devarakonda, 375
Namit Jain, 639
Narander Kumar, 365
Narendra Kohli, 33
Narsinga Rao, M. R, 579
Naveen Mani, 493
Neeraj Kumar, 433
Neeraj Kumar Jain, 537, 569
Nida Manzoor Hakak, 63
Nikita Rathod, 453
Nithesh Murugan Kaliyamurthy, 709
Nitin Pandey, 261

P
Paramjeet Kaur Saini, 453
Parul Kalra, 75, 639
Pooja Chaturvedi, 303
Prabhati Dubey, 321
Prachi Mukherji, 473
Pravesh Kumar, 311
Preeti Gupta, 671
Preeti Mittal, 537, 569
Prerak Sheth, 453
Priyanka Gupta, 181
Pushpa Mamoria, 33

R
Raghu Nath Verma, 205
Raghvendra Singh, 159
Rahul Hooda, 273
Rahul K. Jain, 321
Rajan, 619
Rajan Mishra, 399
Rajeev Kumar Singh, 171, 181
Rajendra Singh Kushwah, 463
Rajesh Kumar, 53
Rajneesh Kumar Pandey, 123
Rajni Sehgal, 13
Rakesh Kumar, 655
Rama Krishna, C, 333
Ramchandra Yadav, 205
Ram Krishna Jha, 123
Rana, Q. P, 261
Ranjeet Singh Tomar, 463, 483
Rashi Agrawal, 33
Renuka Nagpal, 13
Revanasiddappa, M. B, 515
Ritu Chauhan, 433
Ruchita Rekapally, 433

S
Saibal K. Pal, 193
Saini, R. K, 493, 537, 569
Sandeep U. Mane, 579
Sandip Kumar Goyal, 1
Sanjay K. Sahay, 333
Sanjay Sharma, 333
Sanjeev Sofat, 273
Santosh Kumar, 123
Sanyog Rawat, 159
Sarvani, A, 375
Satbir Jain, 505
Satyendra Nath, 23
Satyendra Swarnkar, 443
Saurabh Shah, 453
Shankul Saurabh, 399
Sharma, S. K, 687, 693
Shikha Gupta, 505
Shraddha Khandelwal, 85
Shubhankar Sinha, 629
Shyam Sundar, 351, 387
Singhal, P. K, 443
Singh, H. P, 605
Singh, J. P, 45
Singh, S. K, 223, 647
Siddharth Arjaria, 85
Siddhartha Rokade, 655
Suchitra Agrawal, 171
Sudhanshu Jha, 483
Sukhamay Kundu, 407
Suman Paul, 547
Sumita Lamba, 343
Sunita, 45
Surbhi Thorat, 85
Surendra Kumar, 365
Suryakant, 419
Swami Das, M, 245
Swapnesh Taterh, 709, 717
Syed Qamar Abbas, 223

T
Tamoghna Mandal, 525
Tarun K. Sharma, 679
Tarun Kumar Sharma, 671
Toolika Srivastava, 399, 443
Tripathi, R. P, 53

U
Uday Pratap Singh, 171, 181
Uma Sharma, 717
Umesh Kumar, 709, 717

V
Vaibhav Gupta, 45
Vaishali Jain, 463
Varun Sharma, 679
Venugopal, B, 375
Vibha Yadav, 23
Vijaya Lakshmi, D, 245
Vineet Kumar, 619
Virendra Singh Kushwah, 1
Vishal Gupta, 493
Vivek Patel, 453
Vratika Kulkarni, 85

Printed in the United States
By Bookmasters